T0320980

GEOCHEMISTRY OF THE EARTH'S SURFACE

PROCEEDINGS OF THE 5TH INTERNATIONAL SYMPOSIUM ON GEOCHEMISTRY OF THE EARTH'S SURFACE/REYKJAVIK/ICELAND/16–20 AUGUST 1999

Geochemistry of the Earth's Surface

Edited by
Halldór Ármannsson
Orkustofnun, Reykjavík, Iceland

A.A.BALKEMA/ROTTERDAM/BROOKFIELD/1999

Cover photo: Oddur Sigurdsson

The texts of the various papers in this volume were set individually by typists under the supervision of each of the authors concerned.

Published by
A.A. Balkema, P.O. Box 1675, 3000 BR Rotterdam, Netherlands
Fax: +31.10.413.5947; E-mail: balkema@balkema.nl; Internet site: www.balkema.nl
A.A. Balkema Publishers, Old Post Road, Brookfield, VT 05036-9704, USA
Fax: 802.276.3837; E-mail: info@ashgate.com

ISBN 90 5809 073 6
© 1999 A.A. Balkema, Rotterdam
Printed in the Netherlands

Geochemistry of the Earth's Surface, Ármannsson (ed.)© 1999 Balkema, Rotterdam, ISBN 90 5809 073 6

Table of contents

2 Chemical weathering and climate, river catchment studies

3 Environmental geochemistry of the terrestrial environment and its effect on health

6 *Mineralogy, microbes and chemistry of weathering*

8 Geochemistry of crustal fluids and of catastrophic events

9 *Supplement*

Preface

This book contains short papers written by the participants in the Fifth International Symposium on the Geochemistry of the Earth's Surface (GES-5). The symposium was convened in Reykjavík, Iceland 16-20 August, 1999 under the auspices of the International Association of Geochemistry and Cosmochemistry (IAGC).

The GES organization was started by Yves Tardy, Daniel Nahon and Rafael Rodriguez-Clemente as the Geochemistry of Weathering and Diagenesis of Sediments (GWDS), a working group under the aegis of IAGC. The first meeting was in Granada, Spain 1986 with Rafael Rodriguez-Clemente as Secretary-General. The second meeting was at Aix-en-Provence, France 1990 with Daniel Nahon as Secretary-General. The third meeting was at Pennsylvania State University, USA 1993 with Lee Kump as Secretary-General. The fourth meeting, under the new name of GES (Geochemistry of the Earth's Surface) was at Ilkley, Yorkshire, UK 1996 with Robert Raiswell as Secretary-General. Presidents of the organization have been Yves Tardy (Granada and Aix), Robert Berner (Penn. State and Ilkley), and Blair Jones (Reykjavík and probably Honolulu).

The focus of these past meetings has been on the surface of the earth rather than geothermal and deep crustal geochemical processes, with the main contribution to the technical sessions by posters.

The papers in this book describe geochemical research, natural, experimental and theoretical, from a wide range of surface environments. The themes of the Reykjavík conference were: 1. Geochemical record of terrestrial environmental change and global geochemical cycles; 2. Chemical weathering and climate, river catchment studies; 3. Environmental geochemistry of the terrestrial environment and its effect on health; 4. Organic geochemistry; 5. Marine and sedimentary geochemistry; 6. Mineralogy, microbes and chemistry of weathering; 7. Geochemical thermodynamics and kinetics; 8. Geochemistry of crustal fluids and of catastrophic events. Each theme was introduced by invited speakers followed by poster sessions.

All the manuscripts were reviewed to ensure clarity and uniformity of presentation. The reviewers were: Halldór Ármannsson, Jón Örn Bjarnason, Hjalti Franzson, Gudmundur Ómar Fridleifsson, Hrefna Kristmannsdóttir and Magnús Ólafsson from Orkustofnun, Stefán Arnórsson, Sigurdur Reynir Gíslason and Árný Erla Sveinbjörnsdóttir from the University of Iceland, Ólafur Arnalds, Hlynur Óskarsson, Fridrik Pálmason and Halldór Sverrisson from the Agricultural Research Institute, Stefán Einarsson, Lúdvík Gústafsson and Gunnar Steinn Jónsson from the Environmental and Food Agency of Iceland, Jón Ólafsson from the Marine Research Institute and Níels Óskarsson from the Nordic Volcanological Institute. They are thanked for their effort which has greatly improved the quality of this publication.

Halldór Ármannsson
Editor

Sigurdur Reynir Gíslason
Secretary-General

Geochemistry of the Earth's Surface, Ármannsson (ed.)© 1999 Balkema, Rotterdam, ISBN 90 5809 073 6

Organization

Conference Organizing Committee

Secretary General
Sigurdur Reynir Gíslason, Science Institute, University of Iceland, Reykjavík, Iceland

Secretary
Gerlinde Xander, Science Institute, University of Iceland, Reykjavík, Iceland

Members
Árný Erla Sveinbjörnsdóttir, Science Institute, University of Iceland, Reykjavík, Iceland
Halldór Ármannsson, National Energy Authority, Reykjavík, Iceland
Hlynur Óskarsson, Agricultural Research Institute, Reykjavík, Iceland
K.Vala Ragnarsdóttir, University of Bristol, UK
Oddur Sigurdsson, National Energy Authority, Reykjavík, Iceland
Stefán Arnórsson, Science Institute, University of Iceland, Reykjavík, Iceland

International Association of Geochemistry and Cosmochemistry (IAGC)
Working Group on the Geochemistry of the Earth's Surface

President
B.Jones

Past Presidents
R.A.Berner
Y.Tardy

Members
R.Rodriguez-Clemente
D.Nahon
L.Kump
R.Raiswell
S.R.Gíslason

1 Geochemical record of terrestrial environmental change, and global geochemical cycles

Geochemistry of the Earth's Surface, Ármannsson (ed.) © 1999 Balkema, Rotterdam, ISBN 90 5809 073 6

Invited lecture: Plants, weathering and the evolution of atmospheric CO_2

R.A. Berner
Department of Geology and Geophysics, Yale University, New Haven, Conn., USA

ABSTRACT: Theoretical calculations have shown that the rise of large upland vascular plants during the Devonian might have had a major effect on atmospheric CO_2 at that time because of their role in accelerating the weathering of Ca and Mg silicates. The question is how big an effect. We have studied the quantitative role of plants in accelerating weathering in western Iceland and in experimental plots at Hubbard Brook, New Hampshire. Results, when combined with those of similar studies elsewhere, indicate that the weathering of Ca and Mg silicates is accelerated by higher plants by factors of from 3 to 10. This indicates that CO_2 should have dropped severalfold during the Devonian which should have led to greenhouse cooling and ultimately to the inception of the major glaciations that followed during the Carboniferous and Permian.

1 INTRODUCTION

On a multimillion year time scale the major process atmospheric CO_2 is exchange between the atmosphere and carbon stored in rocks.
This long-term, or geochemical, carbon cycle involves the loss of CO_2 from the atmosphere via the burial of organic matter in sediments, and the conversion of atmospheric CO_2 to dissolved HCO_3^- as a result of the weathering of Ca-Mg silicate minerals, followed by the delivery of this HCO_3^- to the oceans where it is removed as Ca and Mg carbonates. Release of CO_2 to the atmosphere takes place via the oxidative weathering of old organic matter and by the thermal breakdown of buried carbonates and organic matter, via diagenesis, metamorphism and volcanism, resulting in degassing to the earth surface.

A model (Berner 1991,1994) of the long term carbon cycle, named GEOCARB, has been constructed that attempts to quantify the various processes that affect weathering and degassing over Phanerozoic time. Factors affecting weathering include: (1) the evolution of the sun as it affects global warming; (2) the uplift of mountains as they affect relief, climate and the exposure of primary silicates to weathering; (3) the rise and evolution of vascular land plants as they affect weathering and organic carbon burial; 4) changes in continental size and position as they affect temperature and river runoff from the continents; and (5) variations in atmospheric CO_2 as they affect plant growth and global temperature and river runoff (via the atmospheric greenhouse effect) and thereby serve as a negative feedback mechanism for stabilizing both CO_2 and climate.

In performing the GEOCARB modeling it was found that the rise of large vascular land plants in upland areas during the Devonian should have had a major effect on the level of atmospheric CO_2 at that time. Large drops in CO_2 have been calculated, the magnitude of which depends strongly on the quantitative role of plants in accelerating silicate weathering. Thus, there is a great neeed to know more concerning the effect of vascular plants on the rate of weathering, not just during the geologic past, but also at present. It is the purpose of the preasent paper to summarize the work performed to date on this subject.

2 STUDIES OF THE EFFECT OF PLANTS ON WEATHERING RATE

Many previous studies have shown that weathering qualitatively should be accelerated by plants because: 1. Rootlets (+ symbiotic microflora) with high surface area secrete organic acids/ chelates which attack minerals in order to gain nutrients; 2. Organic litter decomposes to H_2CO_3 and organic acids providing additional acid for weathering; 3. On a regional scale plants recirculate water via transpiration followed by rainfall and thereby increase water/mineral contact time. There is geater rainfall in forested regions than there would be in the absence of the trees; 4. Plants anchor clay-rich soil against erosion allowing retention of water and continued weathering of primary minerals between rainfall events.

In order to evaluate quantitatively how plants affect the rate of rock weathering, it is necessary to conduct field studies that hold all other factors affecting weathering, such as bedrock lithology, climate, slope, aspect, and soil permeability, constant. This means that it is very difficult to perform an ideal experiment. However, some headway has been made by studying small geographic areas where lithology, climate., etc are reasonably constant. This includes studies of the chemistry of waters draining forested and non-forested portions of the southern Swiss Alps (Drever and Zobrist, 1992), a small high elevation area of the Colorado Rocky Mountains (Arthur and Fahey 1993), the Skorradalur area of western Iceland (K. Moulton and J. West-unpublished data). and small experimental plots at the Hubbard Brook Experimental Forest Station, New Hampshire (Bormann et al, 1998;Berner et al. 1998, Berner and Rao 1997). Results of these studies are summarized in Table 1.

Drever and Zobrist (1992) found that the riverine flux of dissolved species from a small granitic terrain in the southern Swiss Alps is a strong function of elevation. Correlated with elevation was a change from deciduous forest at the lowest elevation (200 m) through coniferous forests at higher elevations to alpine meadow and bare rocks

(with and without lichens) above tree line (2000 m). The stream flux of Ca^{++} plus Mg^{++} at the highest elevation was about twenty-five times lower than that at 200m in the deciduous forest. If one subtracts the effect of temperature due to the differences in elevation, which is about a factor of three, then the residual effect is a factor of about eight between weathering by deciduous trees and that by pure rock-water interaction or by mosses and lichens.

We have studied the quantitative role of plants in weathering in a small area of western Iceland (Skorradalur) by measuring the chemistry of of rocks, soils, and waters draining adjacent areas of basaltic rocks that are either barren (with sporadic mosses and lichens) or populated by birch and conifer trees. The study area was chosen to minimize differences in all factors affecting weathering rate (lithology, microclimate, slope) except for differences in vegetation. Iceland also offers the opportunity to study weathering in the relative absence of anthropogenic acid rain. Mass balance calculations, including corrections for rain and aerosol input and biogenic cycling, indicate that pyroxene weathering (Mg release) is enhanced by a factor of about 5 and plagioclase weathering (Ca release) by a factor of about 3 under the trees relative to the barren areas.

A collaborative study of the effect of plants on weathering has been underway for the past sixteen years at the Hubbard Brook Experimental Forest Station in New Hampshire, USA (Bormann et al. 1987, 1998, Berner and Rao 1997, Berner et al. 1998). Three 60 m^2 adjacent plots, called sandboxes, were set up in 1983 with one sandbox planted with

Table 1. Ratio of weathering fluxes, (concentrations only for Swiss Alps) in drainage waters including storage in growing trees and soil (Iceland, Hubbard Brook). "trees" = forested land; "bare" = adjacent land minimally vegetated with bryophytes, lichens, etc

Area	Ions	Ratio of Ions (trees/bare)
Southern Swiss Alps* Drever and Zobrist (1992)	$Ca^{++} + Mg^{++}$	8
Western Iceland (Moulton, West & Berner -unpublished)	Ca^{++}	3
" " "	Mg^{++}	5
Colorado Rocky Mountains (Arthur and Fahey, 1993)	Ca, Mg, Na, + K	4
Hubbard Brook, New Hampshire (Bormann et al, 1998)	Ca^{++}	10

* Data corrected for temperature difference between high and low elevations

pine seedlings, one with grasses, and one left barren, except for the sporadic occurrence of mosses and lichens. Each sandbox was filled with with an impermeable plastic base so as to force the waters draining through the sandboxes to flow to underground exit pipes where they could be sampled. sand from a nearby river and lined at the bottom In 1988 and 1989 trees were sampled for cation anlysis as were various solid weathering products (clays, iron oxides, and organic matter). In addition water chemical analyses were conducted on drainage water and rainwater over the 1983-1989 period. Results for the the period 1983-1989 indicate that, the combination of uptake by vegetation, loss in drainage and addition to solid weathering products, caused an increase in the weathering release rate of Ca by a factor of about ten for the tree-covered plot relative to the barren plot.

Additional work (Berner and Rao 1997) on weekly drainage samples from 1993 to date has found that during the spring snowmelt period, when water flow is similar beneath all plots, the flux of dissolved HCO_3^- is 2-6 times higher (depending on year) under the trees than from under the barren sandbox. The effect of the trees on HCO_3^- release can only be studied during the late winter and snowmelt period because the snowcover on the sandboxes protects them from acid rain. During May of 1998 the trees, which had attained an average height of about 5 meters, were cut down, leaving the stumps and roots in place, and alteration of drainage water chemistry has already been detected.

From the results summarized in Table 1 it can be seen that the available data indicate an accelerating effect of trees on the rate of Ca and Mg silicate weathering of a factor of roughly three to ten times. Based on this, a value of seven, chosen for the plant-acceleration factor in GEOCARB modeling, results in drops in CO_2 during the Devonian that are in agreement with independent estimates of paleolevels of CO_2 (Yapp and Poths 1996, Mora et al. 1996). This adds some credence to both the modeling and the field studies of present day weathering.

REFERENCES

Arthur, M.A. & T. J. Fahey 1993. Controls on soil solution chemistry in a subalpine forest in north-central Colorado. *Soil Sci. Soc. Am. Proc.* 57: 1123-1130.

Berner, R. A. 1991. A model for atmospheric CO_2 over Phanerozoic time. *Am. J. Sci.* 291: 339-376.

Berner, R.A. 1994. GEOCARB II: A revised model of atmospheric CO_2 over Phanerozoic time. *Am. J. Sci.* 294: 56-91.

Berner, R.A. & J-L Rao 1997. Alkalinity buildup during silicate weathering under a snow cover. *Aquatic Geochem.* 2: 301-312.

Berner, R.A., J-L Rao., S. Chang, R. O'Brien. & C.K. Keller 1998. Seasonal variability of adsorption and exchange equilibria in soil waters. *Aquatic Geochem.* 4: 273-290.

Bormann, F.H., W.B Bowden, R.S.Pierce, S.P. Hamburg, G.K., Voigt, R.C. Ingersoll & G.E. Likens 1987. The Hubbard Brook sandbox experiment. In W. R. Jordan III, M.E. Gilpin & J.D. Aber (eds) *Restoration Ecology*: 251-256. Cambridge: Cambridge Univ Press.

Bormann, B.T., D Wang, F.H., Bormann, G.,Benoit, R. April & M.C. Snyder 1998. Rapid plant induced weathering in an aggrading experimental ecosystem. *Biogeochem.* 43: 129-155.

Drever, J.I. & J. Zobrist 1992. Chemical weathering of silicate rocks as a function of elevation in the southern Swiss Alps. *Geochim. Cosmochim. Acta* 56: 3209-3216.

Mora, C.J., S.G. Driese & L.A. Colarusso 1996. Middle to late Paleozoic atmospheric CO_2 levels from soil carbonate and organic matter. *Science* 271: 1105-1107.

Yapp, C.J. & P. Hoths 1996. Carbon isotopes in continen tal weathering environments and variations in ancient atmospheric CO_2 pressure. *Earth. Planet. Sci. Lett.* 137: 71-82.

Geochemistry of the Earth's Surface, Ármannsson (ed.)© 1999 Balkema, Rotterdam, ISBN 90 5809 073 6

Invited lecture: Atmospheric CO$_2$, terrestrial ecology, and mammalian evolution

Thure E.Cerling – *Department of Geology and Geophysics, University of Utah, Salt Lake City, Utah, USA*

James R.Ehleringer – *Department of Biology, University of Utah, Salt Lake City, Utah, USA*

John Harris – *George G.Page Museum, Los Angeles, Calif., USA*

Bruce MacFadden – *Florida Museum of Natural History, Gainesville, Fla., USA*

ABSTRACT: The distribution of C$_3$ versus C$_4$ plants in the modern world is related to temperature. However, because of the sensitivity of C$_3$ to the CO$_2$/O$_2$ ratio of the atmosphere it is expected that global changes in the distribution of C$_3$ versus C$_4$ plants should accompany atmospheric CO$_2$ changes in geological time. The period from 6 to 8 million years ago was a time of C$_4$ expansion in many parts of the globe.

1 INTRODUCTION

There is strong ecological, geographical, and palaeoecological evidence that abundance and distribution of plants using C$_4$ photosynthesis are related to climate, specifically to atmospheric carbon dioxide concentration and temperature. C$_4$ plants are mainly grasses, especially the tropical and subtropical grasses (Sage et al. 1999).

2 RESULTS AND DISCUSSION

The C$_3$ photosynthetic pathway evolved under very high atmospheric CO$_2$ concentrations whereas the C$_4$ photosynthetic pathway is a much more recent development, representing adaptation to relatively low atmospheric CO$_2$ concentrations. A model of quantum yield of plants using the C$_3$ versus C$_4$ photosynthetic pathways shows that the crossover temperatures for C$_3$ plants compared to C$_4$ monocots is about 22°C for the modern condition (Cerling et al. 1997). This results from photorespiration in C$_3$ plants at low CO$_2$/O$_2$ ratios. The model further shows that the crossover is strongly dependent on both temperature and atmospheric CO$_2$ (Figure 1). C$_4$ dicots are more rare than C$_4$ monocots and have significantly lower crossover temperatures; they are favored only showing the crossover under extremely low CO$_2$ conditions, such as are found under full Glacial conditions (Ehleringer et al. 1997). The "C$_4$-world", where C$_4$ plants make up a significant fraction of tropical to temperate ecosystems, began at the end of the Miocene due to low atmospheric CO$_2$ concentrations; since then C$_3$ plants have been starved for CO$_2$. C$_4$ grasses are the dominant grass from 0 to 45 degrees latitude, and thus modern savannas are important ecosystems that characterize the "C$_4$-world" of the Plio-Pleistocene. Mammalian evolution during the Plio-Pleistocene was directly related to the low CO$_2$ concentrations of the atmosphere, because competition between C$_4$ and C$_3$ plants (and even between different C$_3$ plants) resulted from CO$_2$ starvation of C$_3$ plants due to photorespiration. Some mammals, such as equids, were already grazing on C$_3$ grasses and immediately used this new resource when it became more readily available at the end of the Miocene. Figure 2 shows that equids had already apparently adapted to a grazing diet in North America by about 15 million years ago as indicated by their development of hypsodont (high crowned) teeth, but changed to a C$_4$ diet abruptly between 6 and 7 million years ago. The change to high crowned teeth in equids was during a period of diversification within the equids,

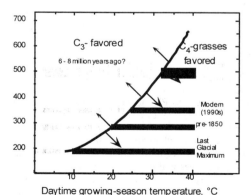

Figure 1. Model of C$_3$ versus C$_4$ photosynthesis based on relative quantum yields.

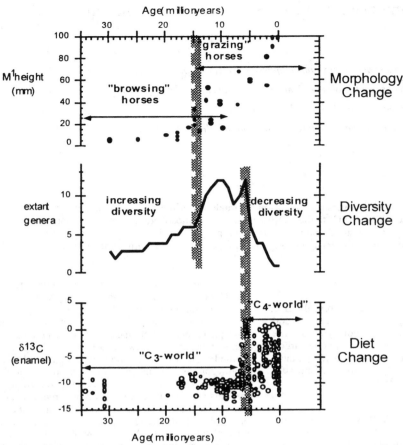

Age(million years)

M¹ height (mm)

"grazing" horses

"browsing" horses

Morphology Change

extant genera

increasing diversity

decreasing diversity

Diversity Change

$\delta^{13}C$ (enamel)

"C₄-world"

"C₃-world"

Diet Change

Age(million years)

Figure 2. Index of hypsodonty for North American equids, the number of extant genera, and $\delta^{17}C$ of equid tooth enamel from North America in the Neogene.

whereas the time of C_4 expansion was accompanied by a rapid decline in equid diversity (Figure 2).

The change to a C_4 diets in mammals was abrupt: Figure 3 shows that 10 million years ago Asian, African, and the North American equids had a C_3 diet, but changed to a C_4 diet between 6 and 8 million years ago. Entire mammalian communities on all continents (except Antarctica) underwent a great change from the "C_3-world" at the end of the Miocene to the Plio-Pleistocene "C_4-world", even in regions where C_4 plants have never been abundant (e.g., Europe).

Many vegetation changes in the Pliocene and Pleistocene have been explained as being the result of changes in seasonality or changes in aridity (e.g., Janis 1993). However, water loss from plants occurs through the stomata which is where CO_2 is taken up by plants through diffusion. Thus, reduction in photosynthesis and ultimately growth rate, could be due to lower atmospheric CO_2 concentrations instead of aridity. For this reason, changes in CO_2 concentration also should be considered in climate reconstruction because the direct impact of CO_2 on plant productivity is independent of water and temperature constraints.

3 CONCLUSIONS

Global changes in terrestrial ecology are expected to result from changes in the CO_2/O_2 ratio. At CO_2 concentrations less than about 500 ppmV C_3 plants increasingly become less efficient due to photorespiration. Therefore changes in the fraction of C_3 versus C_4 plants may be related to changes in the atmospheric CO_2 concentration. The late Miocene to early Pliocene was a period of global vegetation change as well as a period of faunal

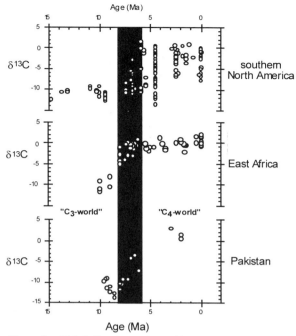

Figure 3. Global diet change in equids shown for North America, Africa, and Asia based on the $\delta^{13}C$ values of equid tooth enamel

turnover in the mammals. The last 6 to 8 million years can be characterized as a "C_4-world" which differed greatly in terrestrial ecology than the previous "C_3-world".

REFERENCES.

Cerling, T. E., J. M. Harris, B. J. MacFadden, M. G. Leakey, J. Quade, V. Eisenmann & J. R. Ehleringer. 1997. Global vegetation change through the Miocene/ Pliocene boundary. *Nature* 389: 153-158.

Ehleringer, J.R., T.E. Cerling & B. Helliker. 1997. C_4 photosynthesis, atmospheric CO_2, and climate. *Oecologia* 112: 285-299.

Janis, C. M. 1993. Tertiary mammal evolution in the context of changing climates, vegetation, and tectonic events. *Annu. Rev. Ecol. Syst.* 24: 467-500.

Sage, R. F., M. Li & R. K. Monson 1999. Taxonomic distribution of C4 photosynthesis. In Sage, R. F. & R.K. Monson (eds), *C4 Plant Biology*: . 551-584. San Diego: Academic Press.

Geochemistry of the Earth's Surface, Ármannsson (ed.)© 1999 Balkema, Rotterdam, ISBN 90 5809 073 6

Invited lecture: Carbon cycle in the past 300 years and future projections

F.T. Mackenzie & L.M.B. Ver
Department of Oceanography, SOEST, University of Hawaii, Honolulu, Hawaii, USA

A. Lerman
Department of Geological Sciences, Northwestern University, Evanston, Ill., USA

ABSTRACT: Four major perturbations owing to human activities on land, and global temperature change have affected the global carbon cycle since the year 1700. The process-driven model of the coupled biogeochemical cycles of C-N-P-S (*TOTEM*) analyzes the cycles of these elements as forced by global emissions from fossil-fuel burning, land-use change, agricultural fertilization of croplands, organic sewage discharges, and a slight temperature rise. The model results for atmospheric CO_2 change in the past 300 years agree well with the observed increase. The model estimates the increases in the delivery of land-derived carbon and its transformations in the coastal ocean, changes in the trophic status (net heterotrophy or autotrophy) of the coastal ocean, and consequences of possible changes in the oceanic thermohaline criculation. The model also estimates the atmospheric concentration and storage and transport of carbon for future decades, to the middle of the 21st century, based on different scenarios of environmental perturbations.

1 GLOBAL CARBON CYCLE

Issues of the behavior of the global carbon (C) cycle during recent Earth history and the partitioning of atmospheric CO_2 among its various sinks, as well as projections for the near future, are of fundamental concern today to scientists and policy makers. Human activities in the past 300 years have become an increasingly important geological factor with respect to the biogeochemical cycling behavior of carbon and the life-essential elements nitrogen (N), phophorus (P), and sulfur (S) associated with it. In particular, four major environmental perturbations due to human activities have come to play particularly prominent roles in the carbon cycle and its coupling to the other bioessential elements. These include: (1) C, N, and S emissions to the atmosphere from fossil fuel combustion; (2) changes in land-use activities, which include deforestation, reforestation, logging, and shifting cultivation; these changes have resulted in gaseous C emissions to the atmosphere, and increases in the dissolved and particulate loads of world rivers, including organic matter, N, and P; (3) application of nitrogen- and phosphorus-bearing fertilizers to croplands and their subsequent mobilization to the atmosphere (N) or to aquatic systems (N and P); and (4) discharges to aquatic systems of sewage containing reactive C, N, and P. In addition, the global mean surface temperature of the planet has increased during the past 300 years by approxi-

mately 1°C. This increase, at least in part, could be a result of human activities that lead to the emissions of greenhouse gases to the atmosphere and their accumulation in this reservoir.

We address some of the global issues involving the behavior of the biogeochemical cycle of C during the recent geological past and into the near future, using the Terrestrial-Ocean-Atmosphere-Ecosystem Model (*TOTEM*), a unique process-based model of the global coupled biogeochemical cycles of carbon, nitrogen, phosphorus, and sulfur (C-N-P-S). The mathematical structure of the model takes into account the time-dependent external forcings, and combinations of linear and nonlinear transport and reaction kinetics, as in the following generalized mass-balance equation:

$$dM_i/dt = F_{i,\text{ext}}(t) \pm F_{i,\text{int}}(t) - k_i(M_i,t)\,M_i, \quad (1)$$

where M_i is the mass of an element (C, N, P, or S) in a reservoir, t is time, F_i are the external input and internal fluxes that may vary with time, and k_i are rate parameters for the physical and/or chemical removal from the reservoir, reflecting the complexity of the different processes. In general, k_i depends on the reservoir mass, time, and other environmental variables and external forcings. The coupling between the C-N-P-S cycles is based on the Redfield ratios in each of the biologically-mediated transfer processes.

We begin analysis of the behavior of the carbon cycle under anthropogenic perturbations in the year

1700, with the Earth-surface system in an assumed quasi-steady state. This initial state is defined by the reservoir mass sizes and transport flux values for each element, and other defined constants. A major difference between *TOTEM* and other terrestrial carbon models and GCMs is the treatment of the observed data for atmospheric CO_2 concentration. Whereas this dataset is used as a prescribed input or forcing function in most models of global environmental change (e.g., Bruno & Joos 1997, Cao & Woodward 1998, Sarmiento et al. 1998), the time course of change of atmospheric CO_2 concentration is *not* treated as an input function in *TOTEM*. Rather, the 300-year dataset for the atmospheric C reservoir, as well as that for the other elemental reservoirs in the modeled Earth system, are outputs of the model and are a result of iterative model calculations. Numerical results using *TOTEM* for atmospheric CO_2 concentrations are then validated against the observed dataset.

2 CHANGES IN THE ATMOSPHERE AND COASTAL OCEAN

The assumption of quasi-steady-state conditions for the pre-industrial Earth allows us to establish a base line for the analysis of the effects of human perturbations and environmental stresses on the Earth system. This assumption is appropriate because the pre-industrial period, generally defined as the time prior to the year 1800, was in a quasi-steady state under a small but sustained global human disturbance of the environment. For example, the estimated rates of release of CO_2 and CH_4 to the atmosphere associated with the pre-industrial human activities of wood-fuel combustion (mostly for food preparation, heating, construction, biomass burning for metal and clay pottery production) and forest clearing for agricultural land use were almost constant over a period of 500 years prior to 1700. In the year 1200, the land-use derived CO_2 emissions, estimated to be about 0.4 Gt C/year, increased only by 0.06 %/year to the year 1700 (Kammen & Marino 1993). Fossil fuel emissions were negligible during this period (Keeling 1973), the Earth system was nearing a period of recovery from the lower global temperature of the Little Ice Age, and atmospheric CO_2 levels were relatively low (Barnola et al. 1995, Etheridge et al. 1996). Indeed, measurements from ice cores show that the concentration of atmospheric CO_2 over a period of ~1000 years prior to 1800, and including the pre-industrial era, varied by no more than 10 ppmv, and imply that the C cycle was approximately in a steady state (Raynaud & Barnola 1985, Siegenthaler et al. 1988). In our model, after the initial steady state, the system is perturbed by the four forcings listed at the beginning, that are repre-

sentative of the major pathways by which humans have perturbed the C, N, P, and S cycles and by variations in the climatic variable of temperature (Smil 1985, Charlson et al. 1992, Caraco, 1995, Galloway et al. 1995, Houghton et al. 1996, Marland et al. 1998). The four perturbations are treated as input functions in *TOTEM*.

Despite a simplistic approach to the global coupled cycles in terms of model structure and its globally aggregated reservoirs and average global fluxes, we are able to describe fully the historical behavior of the major biological elements C, N, P, and S that is consistent with biogeochemical processes. Land-use acitvities, releasing CO_2 to the atmosphere, also release N and P into soils and shallow groundwaters. Thus remineralized N and P increase photosynthetic uptake by the terrestrial phytomass, in agreement with the measured CO_2 and O_2/N_2 atmospheric data, forest inventories, and global carbon budgets (Keeling et al. 1996a, b, Houghton et al. 1998, Joos & Bruno 1998). Our analysis of the land-use perturbation also provides a mechanistic solution that explains the concurrent release of CO_2 to the atmosphere and accumulation of C in terrestrial ecosystems (Houghton et al. 1998). Interestingly, although the effect of land-use activities on nutrient remineralization has been mentioned in the literature (Melillo et al. 1993, 1996), it has yet to be incorporated in many terrestrial-ecological models.

The results of our numerical simulation of atmospheric CO_2 change during the past 300 years compare very well with observational data and with results from other models addressing the partitioning of anthropogenic CO_2 (Figure 1). The model-computed values for atmospheric CO_2 over the 300-year time course are confirmed by one of the major constraints on the presently accepted CO_2 budget — the combined atmospheric CO_2 data obtained from ice cores (Neftel et al. 1985, Friedli et al. 1986, Barnola et al. 1995, Etheridge et al. 1996) and from Mauna Loa (Keeling & Whorf 1998). In addition, there is very close agreement between the *TOTEM*-calculated fluxes for 1985 with the values reported in the literature as average rates for the period 1980 to 1989. For example, the calculated rate of anthropogenic CO_2 accumulation in the atmosphere (3.4 Gt C/year) is consistent with direct measurements of atmospheric CO_2 accumulation (average of 3.3 ± 0.2 Gt C/year; Sarmiento et al. 1992, Houghton et al. 1996). In addition, the calculated rate of oceanic accumulation for the same year (1.9 Gt C/year) is the same as that predicted from ocean-atmosphere models (1.9 Gt C/year for 1980 - 1989; Siegenthaler & Sarmiento 1993) and within the range of estimates of the oceanic sink as inferred from analysis of changes in oceanic and atmospheric $\delta^{13}C$ (average of 2.1 ± 0.9 Gt C/year over the period 1970 to 1990; Quay et al. 1992).

Keeling et al. (1996a) recently estimated the enhanced uptake of CO_2 for the period 1991 to 1994 by the Northern Hemisphere forests (2.0 ± 0.9 Gt C/year) and by the global oceans (1.7 ± 0.9 Gt C/year) from changes in atmospheric CO_2 concentration and atmospheric O_2/N_2 ratio. The rates calculated in the *TOTEM* simulation for 1992 fall within the range of these observed values: global terrestrial biotic uptake of 1.5 Gt C/year and global oceanic uptake of 2.0 Gt C/year. Finally, our calculated results for the partitioning of anthropogenic CO_2 over the time course from 1850 to present compare very well with results from GCM models (Sarmiento et al. 1992, Bruno & Joos 1997) and from models of the global carbon cycle (e.g., Hudson et al. 1994). This agreement with other workers' observational and modeling results lends some further credence to their conclusions and to some extent validates our Earth system model for the domain of land-ocean-atmosphere.

Over the past 300 years, the largest perturbation of the Earth's surface carbon system was the cumulative emission of about 440 Gt C to the atmosphere: 249 Gt C or 56% owing to fossil fuel burning and cement production, and 193 Gt C or 44% from land-use activities. Minor perturbations in the same period included the enhanced cumulative transport to the coastal zone of about 6 Gt organic C and 9 Gt inorganic C owing to land-use activities, and the cumulative loading of about 9 Gt C from municipal sewage and wastewater. The cumulative enhanced fluxes of C resulting from these perturbations included the uptake of about 130 Gt C by the global oceans and the fertilization of the terrestrial biotic reservoir of about 128 Gt C owing to rising atmospheric CO_2, remobilized N and P nutrients from land-use and application of agricultural fertilizers, and warming temperatures. The enhanced biotic uptake, however, was greatly exceeded by the loss of mass from land-use activities (193 Gt C emitted to the atmosphere and 6 Gt C transported to the coastal margin), resulting in a net loss of about 70 Gt C from the terrestrial organic C reservoirs. The atmosphere accumulated about 195 Gt of anthropogenic C, the oceans about 130 Gt C, and the coastal organic and inorganic sediments about 12 Gt C.

For the ocean realm, we confirmed the hypothesis that enhanced organic and carbonate carbon accumulation in coastal marine sediments has been a small sink for anthropogenic CO_2 during recent historical time. Prior to extensive human activities on land, the global coastal margin was likely in a near-steady state of net heterotrophy (i.e., organic matter consumption exceeding production) with an associated net evasion of CO_2 to the atmosphere of about 0.08 Gt C/year (Smith & Hollibaugh 1993). The accumulation of calcium carbonate in shallow-water, coastal margin environments was also a source of

CO_2 release to the atmosphere at that time (Wollast & Mackenzie 1989). Thus the pre-industrial net flux of CO_2 between the coastal oceans and the atmosphere owing to organic and inorganic processes was a net evasion to the atmosphere of about 0.2 Gt C/year. In addition, our modeling results show that through the initial period of simulation to about the year 1900, human perturbations on land and annual temperature variations had only a very slight effect on the trophic state of the coastal zone where total remineralization of organic C, formed *in situ* and imported from land, was greater that its fixation or formation by photosynthesis.

Over the period of 300 years of human perturbation on land, the coastal ocean maintained its state of heterotrophy and its pre-industrial role as a net source of CO_2 to the atmosphere. The export of terrestrial organic matter and its consumption in the coastal zone increasingly exceeded *in situ* production. The fate of this organic matter was remineralization to the dissolved inorganic carbon pool (DIC), burial in the coastal sediments, and to a lesser extent, export to the continental slope. Although the riverine input of inorganic nutrients also increased significantly during the same period of time, the incremental new production was not sufficient to reverse coastal zone metabolism from net heterotrophy to net autotrophy (when organic matter production exceeds consumption). Increased heterotrophy and carbonate precipitation in the coastal zone constitute enhanced sources of DIC. As CO_2 is a component of DIC, its increase in coastal ocean waters tended to oppose the pressure across the air-sea interface from rising atmospheric CO_2 concentration, thereby modulating the response of the coastal ocean to the atmospheric perturbation and modifying the magnitude of the CO_2 flux across the air-sea interface. Thus the net effect of sustained heterotrophy and increased carbonate precipitation is a reduction in the sink strength of the coastal ocean for anthropogenic CO_2. This conclusion, derived from model analysis, implies that were it not for the accumulation of anthropogenic CO_2 in the atmosphere, the global oceans would be sources of CO_2 to the atmosphere, reflecting the imbalance favoring gross respiration over gross photosynthesis and calcium carbonate accumulation.

The major changes in the carbon fluxes within the land-atmosphere-ocean system, as computed from *TOTEM* for the conditions of the four forcings and temperature change are shown in Table 1. In the 1700s and early 1800s, the terrestrial reservoirs were a net source of anthropogenic CO_2 to the atmosphere from the clearing of grasslands and extensive deforestation of Northern Hemisphere (Houghton 1983). Most of this C accumulated in the atmosphere and oceans. The transfer of perturbation effects to the coastal zone via the rivers was minimal. The period

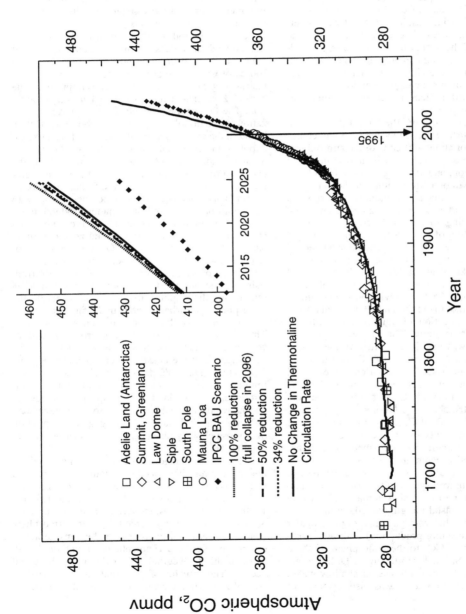

Figure 1. Rise of the atmospheric concentration of CO_2, computed for the period since 1700 to 1995, and projected for an additional 30 years to 2025. For 1700–1995, the computed curve agrees well with the reported observational data. Shown the *TOTEM*-calculated CO_2 rise (solid line) and the IPCC "Business as Usual" (BAU) projection (filled diamonds; Houghton et al. 1996). INSERT: CO_2 rise for three scenarios of reduction in the oceanic thermohaline circulation by the year 2095 by 100%, 50%, and 34%, discussed in the text.

14

Table 1. Historical summary of carbon sources, sinks, and enhanced fluxes owing to human-derived perturbations and temperature change calculated from *TOTEM*. In column 2 are the initial values prior to the start of the perturbations. In columns 3–5 are the computed magnitudes of change from initial conditions.

Carbon fluxes in background (pre-1700) and perturbed (1700–1995) states	Steady State pre-1700	Incremental change since 1700, Gt C/year		
		Pre-Industrial 1860	Industrial 1950	Recent 1995
MAJOR SOURCES OF ANTHROPOGENIC C:				
(1) Fossil fuel burning	0	0.09	1.64	6.24
(2) Land-use CO_2 emissions	0	0.58	0.79	1.76
(3) Sewage discharges to natural waters	0	0.02	0.05	0.1
MAJOR SINKS OF ANTHROPOGENIC C:				
(4) Atmospheric accumulation [1]	0	0.32	0.88	4.52
(5) Oceanic accumulation [2]	0	0.26	0.83	2.36
(6) Organic C in coastal sediments	0	0.02	0.05	0.13
(7) Net terrestrial gain (+) or loss (−) [3]	0	−0.46	−0.05	−0.56
FLUXES PERTURBED BY HUMAN ACTIVITIES:				
(8) Terrestrial CO_2 uptake	0.60	0.13 (22%)	0.76 (127%)	1.31 (218%)
(9) Oceanic CO_2 uptake (+) or loss (−)	−0.5	0.24 (48%)	0.79 (158%)	2.16 (432%)
(10) Riverine DOC and POC	0.41	0.01 (2%)	0.02 (5%)	0.12 (29%)
(11) Riverine DIC and PIC	0.56	0.02 (4%)	0.03 (5%)	0.16 (29%)
(12) Carbonate accumulation	0.25	0.01 (4%)	0.01 (4%)	0.06 (24%)
LONG-TERM, GEOLOGIC FLUXES:				
(13) Organic C burial in sediments	0.13			
(14) Carbonate C burial in sediments	0.4			
(15) Weathering	0.13			
(16) Volcanism and dust	0.10			
(17) Hydrothermal venting	0.06			
(18) Uplift of old organic C	0.37			

[1] For values in columns 3-5, Atmospheric accumulation = (1) + (2) − (8) − (9).

[2] For values in columns 3-5, Oceanic accumulation = (3) + (9) + (10) + (11) − (6) –- (12).

[3] For values in columns 3-5, Net terrestrial = (8) − (2) − (10).

of the Industrial Revolution through early post-World War II (1860-1950) was the time of a growing economy based on fossil fuels (Marland et al. 1998). The period after World War II (1950-1995) is characterized by the extensive use of synthetic agricultural fertilizers and phosphate-based detergents (Esser & Kohlmaier 1991, Smil 1991). The accelerating rates of anthropogenic perturbations on land have increased riverine transport of particulate and dissolved organic carbon to the global coastal zone, which by the year 1995 resulted in the doubling of the rate of organic C accumulation in coastal marine sediments relative to the geologic long-term rate prior to the year 1700.

3   PROJECTIONS FOR THE FUTURE

We also considered how changes in the thermohaline circulation of the ocean might affect the ocean's role in the sequestration of anthropogenic CO_2. A measure of the circulation is the volume of flow (in Sverdrups, 1 Sv = 10^6 m^3/s) in the meridional direction in the North Atlantic Ocean of the surface waters that flow from the south to the north and that have sunk because of water cooling in the high northern latitudes. The present-day thermohaline circulation is equivalent to about 18 Sv. The intensity of the thermohaline circulation, affected by the global temperature distribution, was estimated by Manabe & Stouffer (1994) for the cases of a doubled (2×) and quadrupled (4×) atmospheric CO_2 concentration. In their 4×-model scenario, by the end of 200 years, the thermohaline circulation in the North Atlantic Ocean would decrease in intensity by about 72%, equivalent to 5 Sv of flow. The ultimate cause of these changes, as projected by Manabe & Stouffer (1994), is the warming of the surface ocean brought about by rising CO_2 concentrations and an enhanced greenhouse effect. Consistent with their projections, we assume three values of reduction in thermohaline circulation at the end of a 100-year period beginning in 1995: a nearly full collapse (100% reduction), and 50% and 34% reductions by the year 2095 (Figure 1).

The computed results indicate that the greater the reduction in the rate of the thermohaline circulation, the greater the rate of increase of atmospheric CO_2 for the period 1995 to 2025. The reason for the greater concentration of CO_2 in the atmosphere is due to a decrease in the strength of the high-latitude oceanic sink for CO_2, brought about by a decrease in the production and sinking rate of deep water as a consequence of the weaker meridional circulation. The change over these three decades is not substantial, but it represents about 5 ppmv difference between the cases of an unchanged thermohaline circulation and its weakening on a century-time scale (Figure 1). In addition, if such a change in thermohaline circulation were to occur and persist, the global coastal ocean would become a stronger sink of anthropogenic CO_2 because of the rising CO_2 concentration in the atmosphere and a decreased supply of inorganic carbon-rich waters to the coastal zone by upwelling from depth.

Finally, we analyzed the effects of the procedures for decreasing greenhouse-gas emissions established by the Kyoto Protocol to the United Nations Framework Convention on Climate Change, at the meeting held in Kyoto, Japan, in 1997 (UNFCCC 1997). Results of our *TOTEM* experiment show that by the year 2010 the global mean atmospheric CO_2 concentration will have risen by about 20 ppmv, despite a 15% reduction in fossil fuel emissions relative to that of the year 2000. In comparison, a "business as usual" scenario (BAU) of the Intergovernmental Panel on Climate Change foresees an increase in atmospheric CO_2 by about 25 ppmv, with a 20% increase in emissions relative to those of the year 2000. The sluggish response by the atmospheric CO_2

reservoir to the emission-reduction protocol may be attributed to the projected continued rise in land-use CO_2 emissions and the 10-year mean residence time of CO_2 in the atmosphere. The effect of the reduction on the atmospheric CO_2 concentration becomes progressively more evident towards the mid-21st century. By 2050, fossil fuel emissions under the Kyoto Protocol scenario would be only 40% of the BAU emissions scenario, and the global mean atmospheric CO_2 concentration is projected to be 90 ppmv less than the BAU projections: about 460 ppmv under the Kyoto Protocol and 550 ppmv under the BAU scenario. It may be anticipated that this reduction in the rate of accumulation of atmospheric CO_2 would be maintained unless other feedback mechanisms act non-linearly and either increase or decrease the drawdown of CO_2: for example, such processes as enhanced terrestrial photosynthesis, enhanced terrestrial respiration, or reduction in the intensity of the oceanic thermohaline circulation. It appears, therefore, that the Kyoto protocol, setting emission limits for greenhouse gases, may be a mechanism for stemming the increase in atmospheric CO_2 but only if the physical and biogeochemical mechanisms of redistribution of CO_2 among the atmosphere, land, and ocean do not significantly change from the present.

REFERENCES

Barnola, J.-M., M. Anklin, J. Procheron, D. Reynaud, J. Schwander & Stauffer, B. 1995. CO_2 evolution during the last millennium as recorded by Antarctic and Greenland ice. *Tellus* 47B:264-272.

Bruno, M. & F. Joos 1997. Terrestrial carbon storage during the past 200 years: a Monte Carlo analysis of CO_2 data from ice core and atmospheric measurements. *Global Biogeochemical Cycles* 11:111-124.

Cao, M. & F.I. Woodward 1998. Dynamic responses of terrestrial ecosystem carbon cycling to global climate change. *Nature* 393:249-252.

Caraco, N.F. 1995. Influence of human populations on P transfers to aquatic systems: A regional scale study using large rivers. In H. Tiessen. (ed), *Phosphorus in the Global Environment: Transfers, Cycles and Management.* SCOPE 54: . 235-244. Chichester, England: John Wiley & Sons, Ltd.

Charlson, R.J., T.L. Anderson & R.E. McDuff 1992. The sulfur cycle. In S. S Butcher., R. J Charlson., G. H Orians. & G. V. Wolfe (eds), *Global Biogeochemical Cycles:* 285-300. London: Academic Press.

Esser, G. & G.H. Kohlmaier 1991. Modeling terrestrial sources of nitrogen, phosphorus, sulphur and organic carbon to rivers. In E.T. Degens S., Kempe & J.E.Richey (eds), *Biogeochemistry of Major World Rivers.* SCOPE 42: 297-322. Chichester, England: John Wiley & Sons.

Etheridge, D..M., L.P. Steele, R.L. Langenfelds, R.J. Francey, J.-M., Barnola & V.I. Morgan1996. Natural and anthropogenic changes in atmospheric CO_2 over the last 1000 years from air in Antarctic ice and firn. *J. Geophys. Res.* 101:4115-4128.

Friedli, H., H. Lotscher H., Oeschger, U. Siegenthaler & B.

Stauffer 1986. Ice core record of the $^{13}C/^{12}C$ ratio of atmospheric carbon dioxide in the past two centuries. *Nature* 324:237-238.

Galloway, J.N., W.H. Schlesinger, H. Levy II, A. Michaels & J.L. Schnoor 1995. Nitrogen fixation:, Anthropogenic enhancement-environmental response. *Global Biogeochemical Cycles* 9:235-252.

Houghton, R.A. 1983. Changes in the carbon content of terrestrial biota and soils between 1860-1980: A net release of CO_2 to the atmosphere. *Ecol. Monogr.* 53:235-262.

Houghton, R.A., E.A. Davidson, & G.M. Woodwell 1998. Missing sinks, feedbacks, and understanding the role of terrestrial ecosystems in the global carbon balance. *Global Biogeochemical Cycles* 12:25-34.

Houghton, J., L.G.M. Filho, B.A. Callander, N. Harris, A. Kattenberg, & K. Maskell 1996. *Climate Change 1995: The Science of Climate Change.* Cambridge: Cambridge University Press.

Hudson, R.J.M., S.A. Gherini, & R.A.Goldstein 1994. Modeling the global carbon cycle: Nitrogen fertilization of the terrestrial biosphere and the "missing" CO_2 sink. *Global Biogeochemical Cycles* 8:307-333.

Joos, F. & M. Bruno 1998. Long-term variability of the terrestrial and oceanic carbon sinks and the budgets of the carbon isotopes ^{13}C and ^{14}C. *Global Biogeochemical Cycles* 12:277-295.

Kammen, D. & B. Marino 1993. On the origin and magnitude of pre-industrial anthropogenic CO_2 and CH_4 emissions. *Chemosphere* 26:69-86.

Keeling, C.D. 1973. Industrial production of carbon dioxide from fossil fuels and limestone. *Tellus* 25:174-198.

Keeling, C.D. & T.P. Whorf 1998. Atmospheric CO_2 records from sites in the SIO air sampling network. Mauna Loa Observatory, Hawaii, 1958-1997, Trends: A compendium of data on global change. *http://cdiac.esd.ornl.gov/ftp/maunaloa-co2/maunaloa.co2.* Oak Ridge, TN: Carbon Dioxide Information Analysis Center, Oak Ridge National Laboratory.

Keeling, C.D., J.F.S. Chin & T.P. Whorf 1996a. Increased activity of northern vegetation inferred from atmospheric CO_2 measurements. *Nature* 382:146-149.

Keeling, R.F., S.C. Piper & M. Heimann 1996b. Global and hemispheric CO_2 sinks deduced from changes in atmospheric O_2 concentration. *Nature* 381:218-221.

Manabe, S. & R.J. Stouffer 1994. Multiple-century response of a coupled ocean-atmosphere model to increase of atmospheric carbon dioxide. *J. Climate* 7:5-23.

Marland, G., T.A. Boden, R.J. Andres, A.L. Brenkert & C.A. Johnston 1998. Global, regional, and national CO_2 emissions, NDP-030/R8, Trends: A compendium of data on global change. *http://cdiac.esd.ornl.gov/trends/emis/tre_glob.htm.* Oak Ridge, TN: Carbon Dioxide Information Analysis Center, Oak Ridge National Laboratory.

Melillo, J.M., A.D. McGuire, D. W. Kicklighter, B. Moore III, C. J. Vorosmarty & A.L. Schloss 1993. Global climate change and terrestrial net primary production. *Nature* 363: 234-240.

Melillo, J.M., I.C. Prentice, G.D. Farquhar, E.D. Schulze & O.E. Sala 1996. Terrestrial biotic responses to environmental change and feedbacks to climate. In J. Houghton, L.G.M. Filho., B.A. Callander, N. Harris, A. Kattenberg & K. Maskell (eds), *Climate Change 1995: The Science of Climate Change*: 445-481. Cambridge: Cambridge University Press.

Neftel, A., E. Moor, H. Oeschger, K.K. Turekian, & R.E. Dodge 1985. Evidence from polar ice cores for the increase in atmospheric CO_2 in the past two centuries. *Nature* 315:45-47.

Quay, P.D., B. Tilbrook & Wong C.S. 1992. Oceanic uptake of fossil fuel CO_2: Carbon-13 evidence. *Science* 256:74-79.

Raynaud, D. & J.-M. Barnola 1985. An Antarctic ice core reveals atmospheric CO_2 variations over the past few centuries. *Nature* 315:309-311.

Sarmiento, J.L., T.M.C. Hughes, R.J. Stouffer & S. Manabe 1998. Simulated response of the ocean carbon cycle to anthropogenic climate warming. *Nature* 393:245-249.

Sarmiento, J.L., J.C. Orr & U. Siegenthaler 1992. A perturbation simulation of CO_2 uptake in an ocean general circulation model. *J. Geophys. Res.* 97:3621-3645.

Siegenthaler, U. & J.L. Sarmiento1993. Atmospheric carbon dioxide and the ocean. *Nature* 365:119-125.

Siegenthaler, U., H. Friedli, H. Loetscher, E. Moor, A. Neftel, H. Oeschger & B. Stauffer 1988. Stable-isotope ratios and concentration of CO_2 in air from polar ice cores. *Annals Glaciol.* 10:1-6.

Smil, V. 1985. *Carbon, Nitrogen and Sulfur: Human Interference in Grand Biospheric Cycles.* New York: Plenum Press.

Smith, S.V. & J.T Hollibaugh. 1993. Coastal metabolism and the oceanic organic carbon balance. *Rev. Geophys.* 31:75-89.

UNFCCC 1997. Kyoto Protocol to the UN Framework Convention on Climate Change. http://www.unfccc .de/resource/docs/cop3/protocol.pdf.

Wollast, R. & F.T. Mackenzie 1989. Global biogeochemical cycles and climate. In A Berger, S. Schneider & J.-C. Duplessy (eds), *Climate and Geo-Sciences*: 453-473. Dordrecht, The Netherlands: Kluwer.

Geochemistry of the Earth's Surface, Ármannsson (ed.)© 1999 Balkema, Rotterdam, ISBN 90 5809 073 6

Invited lecture: Greenland ice cores and paleoenvironment

Á. E. Sveinbjörnsdóttir
Science Institute, University of Iceland, Reykjavík, Iceland

S. J. Johnsen
Science Institute, University of Iceland, Reykjavík & Niels Bohr Institute for Astronomy, Physics and Geophysics, Copenhagen, Denmark

ABSTRACT: This paper gives examples of past environmental changes that can be observed in the Greenland deep ice cores. The stable isotopic ratio of oxygen and deuterium is the most important proxy for paleotemperatures, whereas gases and various chemical tracers reflect changes in the atmospheric chemistry as well as processes on land, in the sea and in the atmosphere.

1 INTRODUCTION

Core drillings in polar ice sheets provide a complete collection of deep frozen precipitation samples from the past and therefore offer great potential for studies of past climate and other past environmental changes. Each annual layer of snow deposited on the surface of a cold ice cap carries with it, as it sinks slowly towards the bottom, a fingerprint of the impurities present in the atmosphere. Even the atmospheric gases are trapped in small bubbles when the snow is transformed into glacier ice. The resolution of these records is very high and allows singular events in the Earth's history, like volcanic eruptions and sharp climatic shifts, to be studied in great detail.

Refined techniques of ice core analysis allow several parameters to be obtained from the cores. The isotopic composition of the ice together with borehole thermometry provide information about past temperatures, the dust informs on storminess and source efficiency, the air pubbles on the greenhouse gases in the atmosphere, the acidity on volcanic eruption in both hemispheres and chemical tracers on various processes on land, in the sea and in the atmosphere, at the time of formation of the ice. Together these parameters give a detailed picture of the past environment and climate.

The geographic locations of the Greenland deep core sites discussed in the text are shown in Figure 1.

2 THE ISOTOPIC RECORD

2.1 *Conversion of $\delta^{18}O$ to temperature*

The isotopic ratios ($^{18}O/^{16}O$, D/H) of ice from deep cores are used as the main reference parameters in all ice core studies. They reflect the temperature at which the cloud vapour was transformed into ice crystals or snow (Dansgaard 1964), and are therefore the most important proxy for reconstructing paleotemperatures, assuming unchanged temperature and humidity at the original moisture source areas (Jouzel et al. 1998). According to Johnsen et al (1989) the present mean annual $\delta^{18}O$ (the per mil deviation of the $^{18}O/^{16}O$ ratio in a sample from the $^{18}O/^{16}O$ value in standard mean ocean water (SMOW) of the snow on the Greenland ice sheet is closely related to the mean annual surface temperature, T, in degrees Celcius, by the formula:

$$\delta = 0.67T - 13.7 \text{ ‰} \qquad (1)$$

From deep ice-core and borhole data millennial or longer-term temperature changes have been calibrated and the best estimates for the $\delta - T$ relationship is 2.0°C/‰ at -35‰ and 3.10°C/‰ at -42 ‰ (Johnsen et al. 1998).

2.2 *The GRIP $\delta^{18}O$ profile*

Over 70,000 samples have been analysed for $\delta^{18}O$ from the 3029 m long Greenland Ice Core Project (GRIP) ice core, located on top of the Greenland Ice Sheet (Summit) (Figure 1). The continuous δ record along the upper 2900 m is plotted on Figure 2 on a calculated timescale. The Holocene record in Figure 2 shows a remarkably stable δ signal apart from the δ minimum at 8210 years BP. However, when elevation changes due to varying accumulation rates and shifts of the ice margins are taken into account a cooling trend in Greenland since 8 kyr BP of some 3°C is suggested (Cuffey & Clow 1998, Johnsen et al. 1995). The older part of the GRIP-core is dominated by large and abrupt δ shifts, suggesting climatic instability in Greenland.

The warm glacial interstadials, the so-called Dansgaard-Oeschger cycles are also observed in ice cores from northwest, south and east Greenland (Dansgaard et al. 1982, 1971, Johnsen et al. 1992, 1972) as shown in Figure 3, as well as in the nearby GISP2 ice core (Grootes et al. 1993) and in North Atlantic sediment cores (Bond et al. 1993, Fronval et al. 1995, McManus et al. 1994). In Figure 3 some of the warm interstadials (IS), as defined from the GRIP δ-record (Dansgaard et al. 1993) are shown by numbers close to the δ-profiles to highlight the correlation between the cores. The Figure demonstrates that the Camp Century record shows the highest and the Renland record the lowest degree of glacial instability, suggesting different source regions for the precipitation falling at these two sites.

3 THE CHEMICAL RECORD

3.1 The methane record

According to Chappellaz et al. (1998) variations of the atmospheric CH_4 concentration are global, as the atmospheric residence time of CH_4 is ten times longer than the interhemispheric exchange time. Accordingly, methane data from Greenland and Antarctic ice cores have revealed the important climatic coupling between the North and South hemispheres (Blunier et al. 1998). A good correlation is observed between high values of methane concentration and mild Dansgaard-Oeschger events (Chappellaz et al. 1993). As the main source region for methane during the last glaciation has most probably been the wetlands in tropical and subtropical latitudes, this correlation demonstrates the global character of the fast climatic oscillations observed by the $\delta^{18}O$ record in Greenland.

3.2 The calcium record

Calcium is another parameter that correlates well with the observed $\delta^{18}O$ oscillations, in such a way that low Ca^{+2} values coincide with mild climate (Fuhrer et al 1993) (Figure 2). Calcium is mainly derived from terrestrial dusts and temporal variations are connected with large-scale changes in the transport and loading of terrestrial dusts.

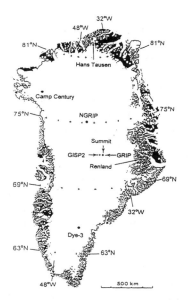

Figure 1: Deep drilling sites in Greenland, mentioned in text.

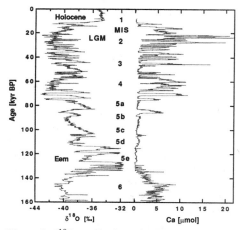

Figure 2: $\delta^{18}O$ profile along the GRIP core together with results of Ca^{+2} analyses. Low Ca^{+2} values coincides with mild climate (Fuhrer et al. 1993).

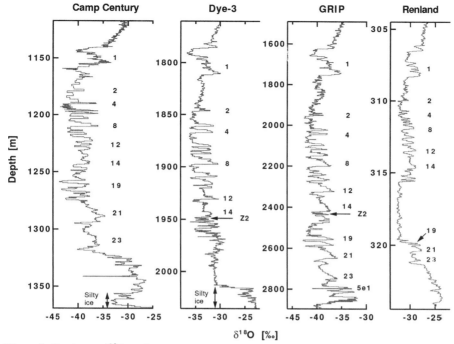

Figure 3: Continuous δ^{18}O profiles along sections of four Greenland ice cores. For locations see Figure 1.

3.3 The CO_2 record

Ice core studies show that the atmospheric CO_2 concentration has increased steadily from about 280 ppmv in the year 1750 to the present value of about 355 ppmv. Also evident from the studies is the naturally caused increase of the atmospheric CO_2 concentration from about 200 ppmv to 280 ppmv during the large climatic change from the last glacial to the post glacial epoch (Raynaud et al. 1993). The best results for the reconstruction of the atmospheric CO_2 concentration at present come from ice cores with a very low concentration of carbonates and H_2O_2, like the Antarctic ice. New results indicate that CO_2 can be produced both by chemical reactions in the ice and by oxidation of organic material (Delmas 1993). The Greenland ice cores are not well suited for CO_2 studies, because of large impurities.

3.4 Tracers of biological activity

Ammonium, organic acids and nitrate have like methane, a continental biogenic source. As ammonium and organic acids have a relatively short atmospheric lifetime, the source of these compounds measured along Greenland ice cores has to be the northern part of the North American continent. Fuhrer and Legrand (1998) have demonstrated how the records of these elements can be used to reconstruct paleo-ecosystems.

REFERENCES

Blunier, T., J. Chappellaz, J. Schwander, A. Dallenbach, B. Stauffer, T. Stocker, D. Raynaud, J. Jouzel, H.B. Clausen, C.U. Hammer & S.J. Johnsen 1998. Asynchrony of Antarctica and Greenland climate during the last glacial. *Nature*: 394:739-743.

Bond, G., W. Broecker, S.J. Johnsen, J. McManus, L. Labeyrie,J. Jouzel & G. Bonani 1993. Correlation between climate records from North Atlantic sediments and Greenland ice. *Nature* 365:143-147.

Chappellaz, J., T. Blunier, D. Raynaud, J.M. Barnola, J. Schwander & B. Stauffer 1993. Synchronous changes in atmospheric CH_4 and Grennland climate between 40 and 8 kyr BP. *Nature* 366:443-445.

Chappellaz, J., E. Brook, T. Blunier & B. Malaiz 1998. CH_4 and δ^{18}O of the O_2 records from Antarctic and Greenland ice: A clue for stratigraphic disturbance in the bottom part of the Greenland Ice Core Project and Greenland Ice Sheet Project 2 ice cores. *J. Geophys. Res.* 102 (C12):26,547-26,557.

Cuffey, K.M. & G.D. Clow 1998. Temperature, accumulation, and ice sheet elevation in central Greenland through the last deglaciation deglacial transition. *J. Geophys. Res.* 102 (C12):26,383-26,396.

Dansgaard, W. 1964. Stable isotopes in precipitation. *Tellus* 16:436-468.

Dansgaard, W., S.J. Johnsen, H.B. Clausen, D. Dhal-Jensen, N.S. Gundestrup, C.U. Hammer, C.S. Hvidberg, J.P. Steffensen, A.E. Sveinbjörnsdottir, J. Jouzel, G. Bond 1993. Evidence for general instability of past climate from a 250-kyr ice-core record. *Nature* 364:218-220.

Dansgaard, W., H.B. Clausen, N. Gundestrup, C.U. Hammer, S.J. Johnsen, P.M. Kristinsdottir & N. Reeh 1982. A New Greenland deep ice core. *Science* 218:1273-1277.

Dansgaard, W., S.J. Johnsen, H.B. Clausen & C.C. Langway Jr. 1971. Climatic record revealed by the Camp Century ice core. In K.K. Turekian (ed.), *The Late Cenozoic Glacial Ages: 37-56*. Yale Univ. Press, New Haven. Conn.

Delmas, R.A. 1993. A natural artefact in Greenland ice-core CO_2 measurements. *Tellus* 45B:391-396.

Fronval, T., E. Jansen, J. Bloemendal, S.J. Johnsen 1995. Oceanic evidence for coherent fluctuations in Fennoscandian and Laurentide ice sheets on millennium timescales. *Nature* 374:443-446.

Fuhrer, K. & M. Legrand 1998. Continental biogenic species in the Greenland Ice Core Project ice core: Tracking back the biomass history of the North Atlantic continent. *J. Geophys. Res.* 102 (C12):26,735-26,745.

Fuhrer, K., A. Neftel, M. Anklin & M.V. Maggi 1993. Continuous measurements of hydrogen peroxide, formaldehyde, calcium and ammonium concentrations along the new GRIP ice core Summit, Central Greenland. *Atmos. Environ.* Part A, 27:1873-1880.

Grootes, P.M., M. Stuiver, J.W.C. White, S.J. Johnsen & J. Jouzel 1993. Comparison of oxygen isotope records from the GISP" and GRIP Greenland ice cores. *Nature* 366:552-554.

Johnsen, S.J., D. Dahl-Jensen, W. Dansgaard & N. Gundestrup 1995. Greenland palaeotemperatures derived from GRIP bore hole temperature and ice core isotope profiles. *Tellus* Ser. B 47:624-629.

Johnsen, S.J., W. Dansgaard, H.B. Clausen & C.C. Langway 1972. Oxygen isotope profiles through the Antarctic and Greenland ice sheets. *Nature* 235:429-434.

Johnsen, S.J., W. Dansgaard & J. White 1989. The origin of Arctic presciption under present and glacial conditions. *Tellus* Ser. B 41:452-468.

Johnsen, S.J., H.B. Clausen, W. Dangaard, N.S. Gundestrup, C.U. Hammer, U. Andersen, K.K. Andersen, C.S. Hvidberg, D. Dahl-Jensen, J.P. Steffensen, H. Shoji, A.E. Sveinbjörnsdottir, J. White, & D. Fisher 1998. The $\delta^{18}O$ record along the Greenland Ice Core Project deep ice core and the problem of possible Eemian climatic instability. *J. Geophys. Res* 102:26,397-26,410.

Johnsen, S.J., H.B. Clausen, W. Dansgaard, N.S. Gundestrup, M. Hanson, P. Jonson, J.P. Steffensen & A.E. Sveinbjörnsdottir 1992. A "deep" ice core from East Greenland: 22pp.

Jouzel, J., B. Alley, K.M. Cuffey, W. Dansgaard, P. Crootes, G. Hoffmann, S.J. Johnsen, R.D. Koster, D. Peel, C.A. Shuman, M. Stievenard, M. Stuiver & J. White. 1998. On the validity of the temperature reconstruction from water isotopes in ice cores. *J. Geophys. Res* 102:26,471-26,487.

McManus, J.F., G.C. Bond, W.S. Broecker, S.J. Johnsen, L.Labeyrie & S. Higgins. 1994. High-resolution climate records from the North Atlantic during the last interglacial. *Nature* 371:326-329.

Raynaud, D., J. Jouzel, J.M. Barnola, J. Chappellaz, R.J. Delmas & C. Lorius 1993.The ice record of greenhouse gases. *Science* 259:926-934.

Geochemistry of the Earth's Surface, Ármannsson (ed.)© 1999 Balkema, Rotterdam, ISBN 90 5809 073 6

Provenance changes across the Pleistocene – Holocene boundary in the south-eastern Po Plain

A. Amorosi, M.C. Centineo, E. Dinelli & F. Lucchini
Dipartimento di Scienze della Terra e Geologico-Ambientali, Università di Bologna, Italy
F. Tateo
Istituto di Ricerca sulle Argille, CNR, Tito Scalo (PZ), Italy

ABSTRACT: The geochemical and mineralogical characterization of fine-grained deposits provides additional information on the late Quaternary depositional history of the south-eastern Po Plain. A homogeneous composition, reflecting an apenninic provenance, characterizes the investigated succession in the western part of the study area. To the east, an important compositional change across the Pleistocene-Holocene boundary marks the transition from lowstand alluvial plain deposits of apenninic provenance to transgressive and highstand deltaic and coastal sediments, showing a major supply from a mixed, alpine-apenninic (Po river) source.

1 INTRODUCTION

Provenance studies of sedimentary rocks are traditionally based on petrographic analyses of coarse-grained (gravel and sand) deposits, and previous work has paid scarce attention to the information that can derive from the detailed study of fine-grained sediments.

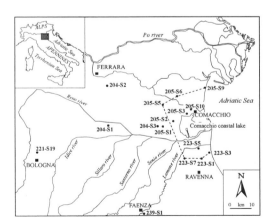

Figure 1. Study area, with indication of the sampling localities. The hatched line refers to the block diagram of Figure 2.

Severe limitations to provenance studies are often constituted by i) difficulties in defining the detailed facies architecture of the study succession, and ii) lack of a reliable chronologic framework. In order to overcome these problems, the present study was based on a well-known case history, which relies upon the late Quaternary (post-35 ka BP) deposits of the SE Po Plain, in an area comprised between Bologna and the present Po delta (Fig. 1).

The major goal of this paper is to show how the accurate geochemical and mineralogical characterization of silts and clays from different (continental, paralic, and shallow-marine) depositional environments can provide a powerful tool for provenance research, resulting in the reconstruction of the depositional history of the study area.

2 GEOLOGICAL SETTING

The drainage model of the Po Plain is the combination of a longitudinal trunk river (the Po River) with transverse tributaries from both chains. The Po Plain is the surface expression of a syntectonic sedimentary wedge bounded by the Apenninic chain to the S and by the Alps to the N.

The Po Basin fill, of Pliocene-Quaternary age, can be split into a number of depositional sequences. The late Quaternary (post-Eemian) 4th-order depositional sequence consists entirely of continental deposits in the Bologna and Faenza areas

(Fig. 1). In the eastern sector, between Ferrara and Ravenna, the continental deposits are interlayered with coastal deposits, which show a varied facies architecture (Fig. 2). Particularly, alluvial plain deposits of late Pleistocene age accumulated during the last glacial period, when the shoreline was approximately 250 km south of its present position.

A hiatal surface encompassing the Pleistocene - Holocene boundary separates these lowstand deposits from the overlying transgressive deposits, consisting of back-barrier muds and barrier sands, separated by a ravinement surface. These deposits record the landward migration of a transgressive barrier island system. At peak transgression the shoreline was located about 20 km landward of its present position. The subsequent highstand deposits are characterized by a progradational stacking pattern of prodelta, delta front, and delta plain deposits, documenting the formation of the early, wave-dominated Po delta. Eventually, renewed sedimentation in an alluvial plain took place south of Reno river, after the northward shifting of the Po river and the abandonment of the previously active delta lobe.

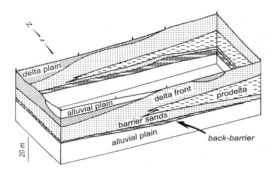

Figure 2. Schematic block diagram, showing facies architecture in the south-eastern Po Plain (see Fig. 1 for location). Horizontal distances are not to scale. Data from Amorosi et al. (1999) and Centineo (in prep.).

3 METHODS

The database includes 16 continuously-cored boreholes, performed by the Geological Survey of the Regione Emilia-Romagna. Core depths vary between 40 and 80 m.

Sampling was focused on the fine-grained portions of cores, belonging to four major facies associations (Fig. 2): i) alluvial plain, ii) delta plain, iii) back-barrier, and iv) prodelta deposits. No samples were collected within the transgressive barrier and delta front sands.

Mineralogical analyses were carried out on bulk samples using X-ray diffraction techniques.

Chemical analyses were performed by X-ray fluorescence spectrometry on pressed powder pellets.

4 RESULTS

The sediments consist of mixtures, in variable proportions, of carbonates (calcite and dolomite) and silicates (quartz, plagioclase and sheet-silicates).

Although preliminar, some remarkable variations in mineralogical composition are recorded between the late Pleistocene and Holocene deposits in the Ravenna and Ferrara areas. The major differences include higher amounts of dolomite, serpentine, and other minerals, such as amphiboles, in the Holocene back-barrier, prodelta, and delta plain clays, with respect to the underlying Pleistocene alluvial silts and clays.

Geochemical analyses show negligible compositional variations across the Pleistocene - Holocene boundary in the Bologna and Faenza areas (Fig. 3). By contrast, a larger scattering is observed in the boreholes from the Comacchio and Ravenna areas. Among the various elements, Cr and Ni are invariably enriched within the Holocene deposits, and vary in a consistent manner throughout the study succession.

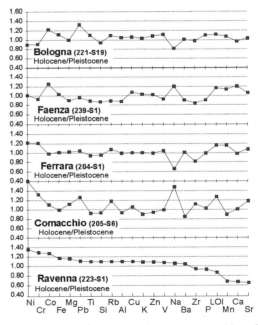

Figure 3. Comparison between the mean composition of Holocene and Pleistocene samples from selected boreholes.

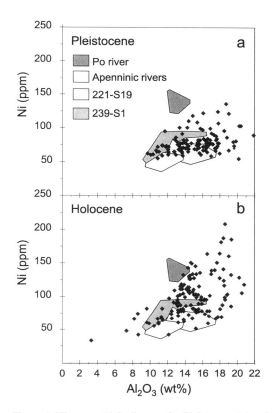

Figure 4. Ni *versus* Al$_2$O$_3$ diagram for Pleistocene (a) and Holocene (b) samples. Data from cores 221-S19 and 239-S1 are reported as reference values on the basis of their homogeneous composition (cf. Fig. 3). Fields relative to present apenninic rivers (Dinelli and Lucchini, in press) and the Po river (unpublished data) are also included.

In a Ni *versus* Al$_2$O$_3$ diagram, most Pleistocene samples (Fig. 4a) display a homogeneous distribution, and the Ni/Al$_2$O$_3$ ratios are similar to those observed in the present Apenninic rivers, between the Reno and Lamone (Dinelli and Lucchini, in press). These values are consistent with the data of samples from the Bologna (221-S19) and Faenza (239-S1) boreholes, corresponding to floodplain deposits of the Reno and Lamone alluvial plains, respectively. The similar characteristics of all Pleistocene deposits indicate that during the last glacial maximum the Apennines acted as the major sediment source for the entire study area. The slight difference in the Ni/Al$_2$O$_3$ ratios between the Bologna and Faenza cores might allow the distinction of two different sediment sources within the apenninic domain. Plotting of some of the Pleistocene samples from the Ferrara area close to an area representative of the present Po river sediments (Fig. 4a) might reflect a local, mixed apenninic-alpine (Po river) provenance.

A different picture derives from the Holocene silts and clays (Fig.4b). Highly scattered Ni/Al$_2$O$_3$ ratios and a general shifting toward higher values, similar to those observed in the present Po river, characterize the coastal deposits between Ferrara and Ravenna (cores 204-S1, 204-S2, 205-S5, 205-S6), and point to an important contribution from ophiolites and ultramafic rocks of the Po catchment basin during the transgressive and highstand phases. In contrast, some samples still display compositional features comparable to those of the Faenza and Bologna cores, thus suggesting an ongoing supply from apenninic sources in the southern part of the study area.

The depth profiles of Cr/Al$_2$O$_3$ (Fig. 5) provide further details on the evolution of sediment provenance: a sharp increase in Cr occurs at the boundary between the Pleistocene and the Holocene deposits in the Comacchio, Ravenna and partially Ferrara areas, whereas similar values are recorded across this surface in the Bologna and Faenza areas. The high Cr/Al$_2$O$_3$ ratio observed locally below the Pleistocene-Holocene boundary (Ferrara area - core 204-S2 in Fig. 5) is consistent with an interpretation of these late Pleistocene deposits as fed by an ancient Po river (see geographic position in Fig. 1). The local return to low Cr/Al$_2$O$_3$ values in the Ravenna and Ferrara areas reflects the re-establishment of an alluvial plain drained by rivers of apenninic provenance. Particularly, the comparison of Cr/Al$_2$O$_3$ ratios enables the distinction of the Reno and Lamone provinces also within the Holocene deposits (compare 223-S7 with 239-S1, and 204-S1 with 221-S19 in Fig. 5).

5 CONCLUSIONS

The detailed sedimentological, geochemical and mineralogical analysis of fine-grained late Quaternary deposits of the south-eastern Po Plain reveals the depositional history of the study area in terms of sediment provenance.

A homogeneous composition indicating an apenninic provenance is recorded in the Bologna and Faenza alluvial plain deposits related to the last 35ka. To the east (Ravenna and Comacchio), a relative increase in the amount of dolomite and serpentine across the Pleistocene-Holocene boundary suggests a change from lowstand alluvial plain sediments of apenninic provenance to a mixed contribution from both alpine and apenninic sources during the subsequent transgressive and highstand phases.

Nickel and chromium appear to constitute two

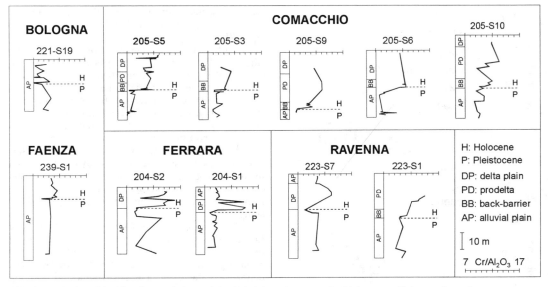

Fig. 5. Vertical profiles showing variations of the Cr/Al_2O_3 ratio across the Pleistocene-Holocene boundary.

important provenance indicators across the Pleistocene - Holocene boundary. In particular, the relatively high concentrations of Ni and Cr and the mineralogical characteristics indicate an important contribution from ophiolites and ultramafic rocks during the Holocene. This suggests an increasing importance of the Po river catchment area as major source of sediment during the development of the Po delta.

REFERENCES

Amorosi, A., M.L. Colalongo, G. Pasini & D. Preti 1999. Sedimentary response to Late Quaternary sea-level changes in the Romagna coastal plain (northern Italy). *Sedimentology,* 46, 99-121.

Dinelli, E. & F. Lucchini in press. Element dispersion patterns in the Reno river valley (northern Italy) evaluated by means of stream sediment geochemistry. *Mineralogica et Petrographica Acta,* 41.

Geochemistry of the Earth's Surface, Ármannsson (ed.)© 1999 Balkema, Rotterdam, ISBN 90 5809 073 6

Petrography, geochemistry and provenance of the Kizildere sandstones, eastern Turkey

M. Arslan

Geology Department, Karadeniz Technical University, Trabzon, Turkey

ABSTRACT: The Kızıldere formation is a volcano-sedimentary formation implying a significant volcanic activity at the time of sedimentation. Sandstones in this formation are immature, quartz-intermediate, feldspar-lithic greywackes deposited close to a source area with little physical and chemical alteration. The compositional variation is scattered and considered as a mixture of clay, quartz feldspar and carbonate. K, Fe, Ti, Rb, Zr and probably Ba, Ni and Zn were added in sheet silicates (mica and clay). Al was added in clay as well as feldspar. Rb and Ba were substantially added in clay, mica and feldspar. Y was added in apatite, zircon and partly in clay and mica. The rocks reveal low to moderately weathered sources. Trace element patterns indicate a source composed largely of active continental margin environment. The rocks were derived from acid to intermediate source rocks with some input from basic rocks. The REE patterns indicate that the source rocks are young.

1 INTRODUCTION AND GEOLOGICAL SETTING

The study area located at the north of Lake Van (Eastern Turkey) is mainly covered by collision related Neogene-Quaternary calcalkaline to alkaline volcanic rocks, associated with marine and continental sediments of the same age (Figure 1). The stratigraphy in the area starts with the Upper Miocene aged Kızıldere formation consisting of limestone, sandstone, marl, lava flows, tuff and tuffite and then follows the products of a subaerial volcanism (Arslan 1994). According to field observations and geological cross-sections, the thickness of the formation is about 450-500 m. The lower level of the formation contains lavas and sediments whereas at the upper levels, there are found mainly sandstone and clayey limestone together with volcanic material.

The aim of this study is to determine geochemical characteristics, to find out geotectonic setting, and to characterise the provenance of the Kızıldere sandstones from which the sediments were derived.

2 PETROGRAPHY

The sandstones are feldspar and lithic greywackes. The main components are the matrix (20-65%), plagioclase (10-25%), quartz (3-25%), carbonates (5-30%), lithic fragments (5-15%), opaques (magnetite and hematite, pyrite) (2-15%). Epidote, K-feldspar, chlorite, clay, apatite, zircon, detrital muscovite and biotite are also present. The rocks are mineralogically immature, suggesting deposition close to a source area with little reworking from the source area of limited physical and chemical alteration.

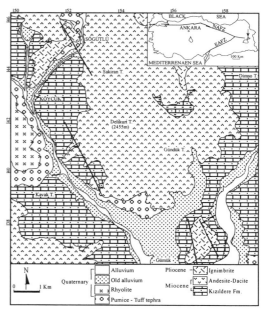

Figure 1. Location and geological map of the study area, north of Lake Van, Eastern Turkey.

In some samples, there is an abundance of feldspar (30%) and lithic particles in a matrix consisting of a fine-grained intergrowth of sericite and chlorite together with some silt-sized quartz, and feldspar. Some samples have volcanic quartz, volcanic rock fragments, zoned and/or, broken crystals of plagioclase. Some of the samples are calc-arenaceous containing up to 50% carbonate grains (mainly calcite and rare dolomite). They are developed presumably in carbonate producing areas where there is a large influx of terrigenous clastics. In the rock, calcite forms angular discrete crystals or patchy areas of pore filling, and dolomite is in discrete grains or associated with calcite.

Quartz occurs as subangular to very angular fragments. It is generally in monocrystalline grains sometimes with strongly undulose extinction, and rarely as polycrystalline aggregates. Quartz derived from volcanic rocks is typically monocrystalline with uniform extinction.

Feldspar is generally twinned plagioclase, but in some samples there is minor K-feldspar, displaying alteration into sericite and clay (kaolinite±illite), sometimes with patchy calcite. Some grains are completely replaced by calcite, perhaps these were a more calcic plagioclase and the released calcium now appears in the rock as replacement patches of calcite.

Lithic fragments are of volcanic, metasedimentary and sedimentary types. Volcanic rock fragments are altered (chloritized) and/or, devitrified glassy tuffaceous and lava fragments. Tuff fragments are yellowish brown. Lava fragments are subangular, consisting of microlites and brown glass. The metamorphic fragments are commonly quartzite. In some samples, metapelite fragments, containing biotite+chlorite+quartz and fine opaques, are also present in minor amounts. The sedimentary fragments are siltstone, mudstone and carbonate.

The matrix is a microcrystalline intergrowth of fine-grained quartz and feldspar with small amounts of chlorite, muscovite+sericite and clay, and is, in places, replaced by patchy carbonate.

3 GEOCHEMISTRY

The rocks are SiO_2 poor sandstones with Na_2O/K_2O ratios about 0.5-1.4 and relatively low MgO, Ni and Cr that underline the low proportion of material derived from mafic sources (Table 1).

Large ion lithophile elements (LILE) exhibit generally low abundances relative to acid clast-dominated sandstones (Floyd et al. 1990) and with K/Rb ratios (238-435) higher than typical upper crust values (Taylor & McLennan 1985). The high loss on ignition (LOI), CaO and Sr contents of some samples reflects the presence of carbonate clasts. The sandstone samples studied may be classified as quartz-intermediate sediments (average 68-74%

SiO_2, $K_2O/Na_2O<1$), which are indicative of Andean-type (active continental) margins.

Table 1. Some compositional parameters of the Kızıldere sandstones.

Element	MA-56C	MA-56D	MA-56E	MA-102	MA-104	MA-208
SiO_2 (wt.%)	64.4	59.1	64.2	49.6	58.3	53.3
TiO_2 (wt.%)	0.26	0.35	0.47	0.86	0.67	0.73
CaO (wt.%)	8.0	10.8	8.5	0.3	4.1	7.9
$Fe_2O_3^*$+MgO	5.10	3.68	1.72	4.38	9.32	9.76
Al_2O_3/SiO_2	0.17	0.19	0.20	0.32	0.30	0.30
K_2O/Na_2O	0.31	0.57	1.35	0.76	0.05	0.53
$Al_2O_3/(CaO+Na_2O)$	0.86	0.78	1.46	8.24	2.00	1.47
LOI (wt.%)	2.4	10.1	7.7	20.7	5.1	8.9
Sr (ppm)	134	142	255	373	339	206

3.1 Major and trace elements

The sandstones have a SiO_2 range of 58-71 wt.%, high $Fe_2O_3^*$ and MgO contents of between 1 and 3 wt.%. TiO_2 decreases as SiO_2 increases because Ti is mostly in clay minerals which decline with increasing quartz and feldspar content. SiO_2 shows positive correlation with K_2O, and suggests that the amount of K-feldspar increases as quartz increases. Similarly, SiO_2 shows a negative correlation with MgO, $Fe_2O_3^*$ and $Fe_2O_3^*$+MgO, indicating that quartz increased as clay minerals and mica declined. Al_2O_3 and SiO_2 correlation is not clear, suggesting that probably clay minerals and feldspars declined as SiO_2 increased. However, Na_2O and CaO show no clear correlation with SiO_2. Rb and Ba exhibit positive correlation with SiO_2 due to increasing amount of K-feldspar. In addition, Ba and Rb show positive correlation with K_2O and SiO_2 due to increasing amount of K-feldspar. Generally, SiO_2 versus trace element variations are scattered. The rocks can be considered as a mixture of clay, quartz, feldspar and carbonate with an end member Niggli c (27-45), mg (0-0.08), and k (0.15-0.45).

Niggli al-alk was used as an indicator of sheet mineral influence in the rocks because albite and K-feldspar each has al-alk=0, and the lack of a positive correlation of al-alk with Niggli c suggests that calcic plagioclase is not a major influence on al-alk values. The samples, except one, are high in CaO (7-12 wt.%) reflecting the presence of carbonate in the samples. By using al-alk, clay and mica influence on the composition can be obtained (Senior & Leake 1978). Therefore, K_2O, $Fe_2O_3^*$, TiO_2, Rb, Zr and probably Ba, Ni and Zn were added in sheet silicate minerals such as mica and clay. Al_2O_3 shows a positive correlation with TiO_2 and al-alk, suggesting addition of Al_2O_3 mainly in clay minerals.

CaO correlates with Sr and there is a negative correlation of c and al-alk, indicating that Ca and Sr were not originally added in the sheet silicates but probably with carbonate and/or feldspar. K_2O shows a positive correlation with Rb and Ba which suggests that both Rb and Ba were substantially added to the sediments in clay, mica and feldspar. Y shows little clear correlation with Zr, P_2O_5 and TiO_2 and al-alk which indicates that Y was probably added in apatite, zircon and partly in clay minerals and mica, i.e. a mixed source. Na_2O has a negative correlation with al-alk suggesting the presence of sodic plagioclase in the sandstones.

The chemical composition of sandstones is influenced by grain-size, degree of source weathering and diagenesis (Sawyer 1986, Wronkiewicz & Condie 1987). As the rocks studied have relatively uniform grain-size, the grain-size is not considered a major variable controlling changes in the chemical composition. Differential weathering at the source tends to mobilise and change the relative abundance of LIL elements (Nesbitt et al. 1980). A measure of the degree of weathering in the source is provided by the chemical index of alteration (CIA; Nesbitt & Young 1982). An average of CIA in the samples is 42 in the range of 30-54 respectively. The average value is lower than average upper continental crust (50) and feldspar (50), suggesting low to moderately weathered sources for the rocks. However, one sample is comparable with illite and montmorillonite (CIA=75-85), indicating that weathering effects have not proceeded to the stage where alkali and alkaline earth elements are substantially removed from clay minerals.

It has been suggested that the full range of elemental composition for greywackes in different tectonic environments can be more adequately compared by the use of upper continental crust normalised multi-element patterns (Floyd et al. 1991). The patterns exhibited for the rocks indicate a source composed largely of continental arc+active continental margin tectonic environments.

3.2 Tectonic Setting

The bulk chemistry of clastic sediments is mainly a reflection of the average composition of the source area, although other factors are also important and may variably influence element distribution (e.g. Roser & Korsch 1986) The Kızıldere Formation sandstones gave no clear tectonic setting, probably because of the Ca-rich nature of the present samples. On the K_2O/Na_2O vs. SiO_2 plot, the rocks fall within the active continental margin field (Figure 2), suggesting that the presence of arc-derived material may be discounted. The location of data points is primarily controlled by the nature of volcanism, the extend of plutonism and related erosional levels.

3.3 Chemical discrimination of provenance

Trace elements including the REE, Zr, Ti, Y, Hf and U are useful in determining the source of sediments as they have intermediate ionic potential and low ocean residence times (Henderson 1984, Taylor & McLennan 1985). The relatively high K/Rb ratios of the Kızıldere sandstones are indicative of derivation from acid and intermediate source rocks with some input from basic sources (Floyd et al. 1990).

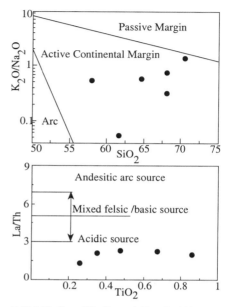

Figure 2. K_2O/Na_2O vs. SiO_2 (Roser & Korsch 1986) and TiO_2 vs. La/Th plots (Taylor & McLennan 1985).

The low levels of Cr and Ni found in the rocks again indicate little basic input into the system. The Ni/Co ratio is generally lower than 1 but one sample has 68.75, either due to weathering in the source rock that selectively enriched Ni or to a Ni-rich source rock. In addition, Th/U ratio in the rocks is lower than 4 (upper continental crust Th/U=3.8; Taylor & McLennan 1985), suggesting that they are derived from acid and intermediate source.

3.4. Rare earth elements

Rare earth elements in sediments are generally considered to reflect the nature of the source area undergoing denudation (e.g. Sawyer 1986). Two samples of the Kızıldere sandstones analysed to determine their REE content are characterised by moderate LREE enrichment (La_N/Lu_N=8-11) and less fractionated HREE (Gd_N/Yb_N=0.80-0.85). Both samples show slight positive Eu anomaly $(Eu/Eu^*)_N$=1.35-1.53, possibly as a result of plagioclase enrichment during sedimentary sorting. Such phenomenon has only been recognized in volcanogenic sediments where weathering effects are minor and plagioclase an abundant constituent.

Generally, low REE abundance and particularly $(Gd/Yb)_N < 1$ may have resulted from modest amounts of zircon. Compared to typical, well-mixed, fine-grained materials derived from the continental crust (such as NASC), the samples show a similar degree of LREE enrichment but a lower degree of HREE enrichment (Figure 3).

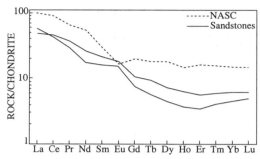

Figure 3. Chondrite-normalised (Boynton 1984) REE patterns of the Kızıldere sandstones, compared with NASC (Henderson 1984).

It is a common assertion that sedimentary rocks deposited at active continental margins will have REE characteristics similar to undifferentiated volcanic rocks of the arc with lower REE abundances, La/Sm and La/Yb ratios and without negative Eu anomalies. However, sediments deposited at active continental margins generally show REE patterns intermediate between a typical andesite pattern and PAAS. The REE patterns of the sandstones which have less fractionated LREE and depleted HREE than NASC probably represents the source rocks of young components derived from continental arc material. HREE depletion is probably the signature of either garnet, or a reflection of the mantle source REE pattern fractionation (Taylor & McLennan 1985).

4 CONCLUSIONS

Kızıldere sandstones are immature, feldspar-lithic greywackes deposited close to a source area with little physical and chemical alteration. Geochemically, they are quartz-intermediate with low proportion of material derived from mafic sources. The compositional variation is quite scattered and considered as a mixture of clay, quartz feldspar and carbonate; calcic plagioclase is not major influence on composition. K, Fe, Ti, Rb, Zr and probably Ba, Ni and Zn were added in sheet silicates (mica and clay). Sandstones studied reveal low to moderately weathered sources. Trace element patterns indicate a source composed largely of continental arc+active continental margin environments. High K/Rb and Th/U ratios, and REE

patterns indicate that they were derived from acid to intermediate source with some input from basic source rocks of young components.

REFERENCES

Arslan, M. 1994. *Mineralogy, geochemistry, petrology and petrogenesis of the Meydan-Zilan (Erciş-Van, Turkey) area volcanic rocks:* PhD Thesis 559pp. UK: Glasgow Univ.

Boynton, W.W. 1984. Cosmochemistry of the rare earth elements. In P. Henderson (ed.), *Rare Earth Geochemistry:* 63-107 Amsterdam : Elseiver.

Floyd, P.A., B.E. Leveridge, W. Franke, R. Shail & W. Dorr 1990. Provenance and depositional environment of Rhenohercynian synorogenic greywackes from the Giessen nappe, Germany. *Geol. Runds.* 79: 611-626.

Floyd, P.A., R. Shail, B.E. Leveridge & W. Franke 1991. Geochemistry and provenance of Rhenohercynian synorogenic sandstones: Implications for tectonic environment discrimination: In A.C. Morton, S. Todd & P.D.W. Haughton (eds.), *Geol. Soc. Spec. Publ.* 57: 173-188.

Henderson, P. 1984. General geochemical properties and abundances of the rare earth elements: In P. Henderson (ed.), *Rare Earth Element Geochemistry:* 1-32p. Amsterdam: Elseiver.

Nesbitt, H.W., G. Markovics & R.C. Price 1980. Chemical processes affecting alkalis and alkaline earth during continental weathering. *Geochim. Cosmochim. Acta* 44: 1659-1666.

Nesbitt, H.W. & G.M. Young 1982. Early Proterozoic climates and plate motions inferred from major element chemistry of lutites. *Nature* 299:715-717.

Roser, B.P. & R.J. Korsch 1986. Provenance signatures of sandstone-mudstone suites determined using discrimination function analysis of major element data. *Chem. Geol.* 67: 119-139.

Sawyer, E.W. 1986. The influence of source rock types, chemical weathering, and sorting on the geochemistry of clastic sediments from the Quetico metasedimentary belt, Superior Province, Canada. *Chem. Geol.* 55: 77-95.

Senior, A. & B.E. Leake 1978. Regional metasomatism and the geochemistry of the Daldradian metasediments of Connemara, western Ireland. *J. Petrol.* 19: 585-625.

Taylor, Y. & S.M. McLennan 1985. *The continental crust: Its composition and evolution:* 312 pp. Oxford: Blackwell.

Wronkiewicz, D.J. & K.C. Condie 1987. Geochemistry of Archean shales from the Witwatersrand Supergroup, south Africa: source area, weathering and provenance. *Geochim. Cosmochim. Acta* 51: 2401-2416.

Geochemistry of the Earth's Surface, Ármannsson (ed.)© 1999 Balkema, Rotterdam, ISBN 90 5809 073 6

Geochemistry of clay-rich beds, Lagonegro basin Italy – A key for basin evolution

P. Di Leo
Istituto di Ricerca sulle Argille, CNR, Tito Scalo (PZ), Italy

E. Dinelli
Dipartimento di Scienze della Terra Geologico-Ambientali, Università di Bologna, Italy

G. Mongelli & M. Schiattarella
Centro di Geodinamica, Università della Basilicata, Potenza, Italy

ABSTRACT: Composition of clay-rich beds from the Jurassic Scisti silicei Fm, Lagonegro basin (southern Apennines), have been investigated. The $Al/(Al+Fe+Mn)$ ratio and the La-Th-Sc diagram indicate that hydrothermal and mafic contributions are negligible. Four sediment types are recognized, characterized by: A- high K_2O and illite; B- high TiO_2 and HFS elements and more complex mineral assemblage (illite, kaolinite, smectite, hematite); C- abundant quartz, low K_2O, TiO_2, HFS elements; D- composite group including carbonate-rich layers and cherts. Diagenetic modifications, evaluated using a K_2O vs TiO_2 diagram, should be minimal. Most of the samples have a K_2O/TiO_2 ratio (5.8) similar to that of the upper continental crust (6.2). An exception is group B, showing lower K_2O/TiO_2 ratio (3.7) reflecting a warmer and humid climate favouring leaching of "mobile" elements and formation of kaolinite and hematite. The chemical data suggest a predominantly felsic source area for the clay-rich layers of the Scisti silicei Fm.

1. INTRODUCTION

Pelagic sediments are the results of various processes including detrital supply from continental areas, pyroclastic deposition, hydrothermal activity, authigenic precipitation and biological activity. Any of these signal may however be modified by diagenetic processes. In this study we used the chemical and mineralogical composition of clay-rich beds from the Jurassic Scisti silicei Fm. (Lagonegro basin, southern Italy) to get constraints for the basin evolution.

2 GEOLOGICAL OUTLINES AND SAMPLING

The Lagonegro units form a large part of the southern Apennines orogenic wedge (Figure 1), an Adriatic-verging fold-and-thrust belt mainly derived from the deformation of the African-Apulian passive margin (western Adria, D'Argenio et al. 1980). The palaeomargin included the Lagonegro basin, generated by continental rifting since middle Triassic times (Scandone 1975, Wood 1981).

The age of the Lagonegro pre-orogenic successions ranges from lower-middle Triassic to Oligo-Miocene boundary (Scandone 1975, Miconnet 1988, Pescatore et al. 1997). The most ancient rocks, lower-middle Triassic, are made up of shallow-water siliciclastic sediments, biogenic limestones and, toward the top, siliciclastic deposits. The overlying

pelagic succession is characterized by a predominant carbonate sedimentation up to the late Triassic, later replaced by Jurassic siliceous sedimentation (Scisti silicei Fm). During Cretaceous times a turbiditic sedimentation took place.

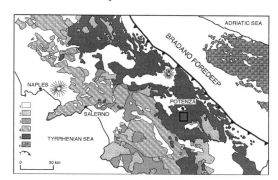

Figure 1. Geological sketch map of southern Apennines and location of the study area. Legend: 1) Plio-Quaternary clastics and Quaternary volcanics; 2) Miocene syntectonic deposits; 3) Cretaceous to Oligocene ophiolite-bearing internal Ligurian units; 4) Apenninic platform Meso-Cenozoic shallow-water carbonates; 5) Lagonegro basin Lower-Middle Triassic to Upper Miocene shallow-water and deep-sea successions; 6) Apulian platform platform Meso-Cenozoic shallow-water carbonates; 7) thrust front of the chain.

The Scisti silicei Fm is constituted of a lower marly and/or shaly member and of an upper member which is formed by thin siliceous beds (mainly radiolarites and cherts). The base of the formation shows ages varying from upper Triassic to Lias (Miconnet 1988). The Jurassic succession analysed in this work refers to the Scisti silicei outcropping in the Lucanian Apennine (Figure 1). sampling was focused on clay-rich beds from both the basal member and the upper portion. Mineralogical analyses on the bulk sample were performed by X-ray diffraction using a Cu tube, spinner and secondary monocromator. Chemical analyses were performed by X-ray fluorescence spectrometry; analytical precision better than 10 %.

3 RESULTS AND DISCUSSION

3.1 Origin of the clay-rich sediments

The Boström (1973) diagram (Figure 2) allows discrimination between various possible sources. Our samples plot close to the terrigeous end-member, similarly to what has been obtained by Amodeo et al. (1996) on the same lithologies.

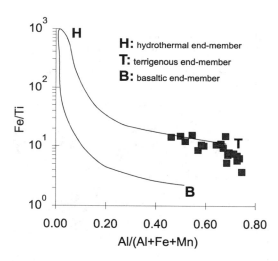

Figure 2. Boström (1973) diagram to discriminate among vaorius components of pelagic sediments.

3.2 Chemical and mineralogical discrimination within the clay-rich beds of the Scisti silicei Fm.

Q-mode cluster analysis (Figure 3) allows the distinction of four groups:
group A: this subset is characterized by enrichments in K, Rb, Zr and Nb relative to the average deep-sea

clay (DSC). The silica contents are very similar to that of the DSC whereas the other elements, and expecially the transition ones, are generally depleted (Figure 4A). The clay minerals are abundant (64-81wt %) and are dominated by illite with a low cristallinity index as confirmed by the K_2O values, which are the highest within the data-set. The calcite contents are negligible with the exceptions of the sample SS5;

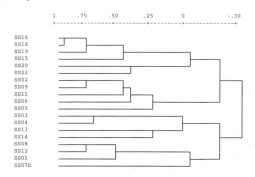

Figure 3. Q-mode dendrogram.

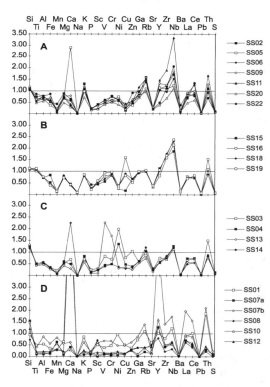

Figure 4. Spider diagrams of elements normalized to the average deep-sea clay composition (in Chester 1993).

group B: in this subset only Ti depletion relative to the DSC is not observed. Zr and Nb are enriched. Si and Rb contents are similar to those of the DSC whereas K is depleted (Figure 4B). The clay minerals assemblage is formed by illite, smectite with a high cristallinity index, kaolinite and chlorite. The abundance of the clay minerals association ranges between 70-80 % in wt. The feldspars abundance (4-6 % in wt) is not negligible, and in some sample occur traces of hematite ;
group C: in this subset Zr is depleted and Nb enrichments are minor. Vanadium, Cr and Ni are erratically distributed, with high enrichments in some samples (Figure 4C). The clay minerals assemblage is dominated by illite with a low cristallinity index. Quartz contents are relatively high (23-34 wt %);
group D: includes samples with mixed composition (Figure 4D). Some of them are carbonate-rich, whereas others are cherts. Sample SS7b, although included in this group, is mostly composed of clay minerals and is characterized by strong enrichments in high field strength elements.

3.3 Diagenetic vs provenance

To test the influence of diagenetic modification on the detrital characteristics, we used a binary diagram relating a diagenetic sensitive K_2O to the "immobile" TiO_2 (Figure 5).

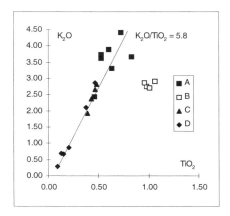

Figure 5. K_2O vs. TiO_2 diagram relative to the Scisti silicei Fm.

The majority of the samples are characterized by a $K_2O/TiO_2 = 5.8$ which is consistent with the upper continental crust composition ($K_2O/TiO_2 = 6.2$, Wedephol 1995). Thus the effects of diagenetic remobilization should not be relevant. The group B samples fall aside of the trend outlined. The lower K_2O/TiO_2 ratio (3.7) and their complex clay mineral

assemblage, characterized by the occurrence of kaolinite and hematite, point to more weathered sources. A climatic variation, towards warmer and humid conditions should be responsible for these changes (Chamley 1989).

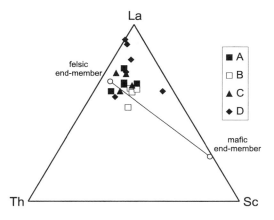

Figure 6. La-Th-Sc diagram

Assuming that in the studied succession other sediment sources than detrital are negligible and that diagenetic modifications are limited we try to define roughly the chemical characteristics of the source area using a ternary La-Th-Sc diagram (Figure 6). Although scattered the data suggest a predominantly felsic source areas for the clay-rich layers of the Scisti silicei Fm, also according to the Boström diagram (Figure 2), which excluded contributions from the basaltic end-member.

REFERENCES

Amodeo, F., Ph. Thelin, P.O. Baumgartner & H.R. Pfeifer 1996. Le argille detritiche della Formazione degli Scisti silicei (Triassico superiore-Giurassico), Bacino Lagonegrese, Italia meridionale. Origine, diagenesi ed implicazioni paleoambientali. *Atti Riunione Gruppo Sedimentologia CNR, Catania, 10-14/10/1995*: 47-50.
Boström, K. 1973. The origin and fate of ferromanganoan active ridge sediments. *Stokholm Contrib. Geol.* 27: 147-243.
Chamley, H. 1989. *Clay sedimentology.* Berlin, Springer-Verlag.
Chester, R. 1993. *Marine geochemistry.* London: Chapman & Hall.
D'Argenio, B., F. Horvath & J.E.T. Channell 1980. Palaeotectonic evolution of Adria, The African promontory. *XXVI Int. Geol. Congr., Mem. B.R.G.M.* 115: 331-351.
Miconnet, P. 1988. Evolution mesozoique du secteur de Lagonegro. *Mem. Soc. Geol. It.* 41: 321-330.
Pescatore T., P. Renda, M. Schiattarella & M. Tramutoli 1997. Stratigraphic and structural relationships between Meso-Cenozoic Lagonegro basin and coeval carbonate platforms

in southern Apennines, Italy. *8th Workshop of the ILP Task Force "Origin of Sedimentary Basins", Abstracts Volume, Palermo, June*: 7-13.

Scandone, P. 1975. The preorogenic history of the Lagonegro basin (southern Apennines). In C. Squyres (ed), *Geology of Italy, The Earth Sciences Society of the Libyan Arab Republic*: 305-315.

Wedephol, K.H. 1995. The composition of the continental crust. *Geochim. Cosmochim. Acta* 59: 1217-1232.

Wood, A.W. 1981. Extensional tectonics and the birth of the Lagonegro basin. *N. Jb. Geol. Paläont. Abh.* 161: 93-131.

Geochemistry of the Earth's Surface, Ármannsson (ed.)© 1999 Balkema, Rotterdam, ISBN 90 5809 073 6

Groundwater nitrate as a palaeoenvironmental indicator

W. M. Edmunds
British Geological Survey, Wallingford, UK

ABSTRACT: High concentrations of nitrate are preserved in unconfined groundwaters across N. Africa, both in modern and palaeowaters up to 35 kyr in age. This widespread distribution in space and time supports models for a constant vegetation distribution across much of N Africa in the Holocene and the Late Pleistocene which included N-fixing species.

1.INTRODUCTION

Large reserves of fresh groundwater underlie the present day Sahara and Sahel region, contained in large sedimentary basins comprising sandy consolidated or unconsolidated sediments ranging from Cretaceous to Quaternary in age. Except towards the Mediterranean these aquifers have never contained marine formation water and are essentially carbonate-free. The salinity of the groundwater is therefore is derived primarily from marine aerosols which have undergone evaporative concentration to variable extents. Groundwater may serve therefore as an potential archive of past environments in arid regions. The Saharan groundwaters, remain aerobic especially under phreatic conditions and, in the presence of oxygen, any nitrate remains stable, reflecting the initial conditions at the point and time of recharge.

Under strongly aerobic conditions, such as found in arid and semi-arid continental areas much of the organic carbon has reacted and material for palaeo-environmental reconstruction has largely been destroyed. In this paper those indicators that may be preserved in contemporaneously recharged groundwaters are considered, specifically the concentrations of nitrate. Results from five regions of northern Africa, where reliable radiocarbon data exist, are discussed. These data are then compared with modern environments where nitrate signals are being transferred to the shallow water table.

2. INDICATORS OF PAST ENVIRONMENTS CONTAINED IN GROUNDWATERS

There is now reliable evidence, especially from palaeohydrological indicators in lake sediments, of the main climate changes that occurred during the late Pleistocene and Holocene (Gasse et al. 1990). The late Pleistocene over the Sahara was predominantly wetter than today and was followed by an extremely dry period corresponding to the last glacial maximum (LGM) in Europe. A number of distinct wet episodes in the mid-Holocene, which correspond most probably to the northward movement of the monsoon rains, are then recorded in the sedimentary record of the Sahara. In the groundwater also a direct chronology of these wet periods is provided by the radiocarbon record which provides a low resolution archive, precise dating not always proving possible due to the dilution of the input signal by inorganic carbon (Fontes et al. 1993). Stable isotopes ($\delta^{18}O$ and $\delta^{2}H$) and noble gas ratios in groundwater also provide strong support for cooler conditions in the Sahara from groundwaters recharged prior to the LGM.

Inert chemical tracers may also provide information on input conditions. The most important of these are Cl, Br and NO_3; most other chemical constituents may be regarded as reactive tracers reflecting water-rock interaction along the flow path. At a first approximation the groundwater Cl concentration is proportional to the recharge to groundwater and supporting information on Br is usually indicative of the Cl source (whether marine or evaporite derived for example). It is also probable that the Br/Cl ratio may be related to the biomass productivity with slightly higher Br being produced in wetter years (Edmunds et al. 1992). Any organic compounds produced in the soil are likely to be oxidised and semi-arid regions are unlikely to preserve environmental information. It is possible that in non-carbonate systems the groundwater $*^{13}C$ could preserve oscillations in any changes between C_4 and C_3 plants, where these exist, although some modification of the isotopic signal due to dilution and water rock reactions would mask this signal in most groundwaters. Nitrate may be one of the best indicators of former vegetation conditions (nitrate-fixing or not) and plant communities since NO_3 may be regarded as an inert tracer under aerobic conditions found widely beneath arid regions. Cyanobacteria are also suggested as an origin of nitrate groundwaters in Australia (Barnes et al. 1992). Atmospheric baseline nitrogen concentrations are low (< 0.1 mg l^{-1} NO_3-N) and once the signal is produced in the soil the NO_3 will remain conservative and its geochemical behaviour and movement will be similar to Cl. The evolution may be followed using the NO_3/Cl ratio in shallow groundwaters.

3. METHODS

The work described here brings together studies conducted over a number of years as a part of regional investigations in Sudan, Mali, Niger Libya and Algeria including some so far unpublished data for NO_3 and ^{14}C. Nitrate has been determined by automated colorimetry, first by reduction of NO_3 to NO_2 and then measurement of a red-purple azo dye complex at 520 nm. Chloride was similarly determined using mercuric thiocyanate. The full details of field investigations are given in the regional publications cited below.

4. NITRATE IN GROUNDWATER FROM THE SAHARA AND SAHEL

4.1 Libya

High nitrate groundwaters were recorded during the exploration for groundwater in the Sirte and Kufra basins in eastern Libya (Wright & Edmunds 1969), occurring widely in aerobic waters contained mainly in the phreatic aquifers. These waters have well-constrained radiocarbon ages and all are old waters recharged during the late Pleistocene or Holocene. Nitrate concentrations in anaerobic (confined) groundwaters are below detection and it is assumed that any initial has been reduced to N_2. The nitrate concentrations in many groundwaters exceed recognised potable limits with maxima above 40 mg l^{-1} NO_3-N (Figure 1).

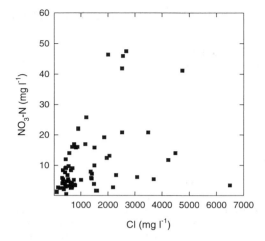

Figure 1. Nitrate distribution w.r.t. Cl in all (aerobic) groundwaters from the Sirte and Kufra basins Libya. Data from Wright & Edmunds (1969).

4.2 Sudan

An investigation was made of the Butana region of Sudan, lying between the Nile and Atbara rivers where groundwaters are found in some of the southernmost outcrops of the Nubian sandstone (Darling et al. 1987). This regional phreatic aquifer contains fresh water which has in general a Holocene age (10,300 to 5580 yr BP). Present day recharge is negligible except via wadis. The groundwaters are aerobic and contain nitrate concentrations up to 14 mg l^{-1}.

4.3 Mali

Groundwaters mainly found in the Continental Intercalaire of the Azaouad depression have been investigated geochemically and isotopically (Fontes et al. 1991). Recharge mainly of Holocene age has been recognised in the phreatic aquifer which may at least be partly derived from contemporaneous Niger river floods. The groundwaters are mainly anaerobic and contain nitrate concentrations up to 7.3 mg l^{-1} NO_3-N.

The confined groundwaters contain a significant excess of dissolved nitrogen with respect to air saturation which is considered the product of denitrification:

$$4NO_3^- + 5CH_2O + 4H^+ = 2N_2 + 5CO_2 + 7H_2O \quad (1)$$

The N_2/Ar ratios have been corrected for excess air using the noble gas ratios and then used to calculate the amount of NO_3 which was converted to N_2 gas. An equivalent of up to 10.2 mg l^{-1} NO_3-N has been reduced in this way.

4.4 *Niger*

The hydrogeochemical evolution of groundwaters contained in the multilayer aquifer of the Agadez and Dabla sandstones in NW Niger has been reported by Andrews et al. (1994). These groundwaters are mainly confined and are anaerobic with nitrate absent. The few aerobic waters contain nitrate up to 23 mg l^{-1} NO_3-N and are of Holocene age. It is calculated, as in Mali, using the corrected N_2/Ar ratios, that the anaerobic groundwaters, mainly of Holocene age, contain equivalent NO_3-N concentrations up to 6.8 mg l^{-1}.

4.5 *Algeria*

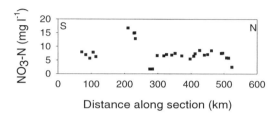

Figure 2. Profile of NO_3-N concentrations along south to north cross section through Complexe Terminal aquifer, Algeria.

The hydrogeochemistry of a south to north groundwater profile some 500 km long in Mio-Pliocene sands (Complexe Terminal), from Hassi Messaoud to the discharge area of the Chott Melrhir

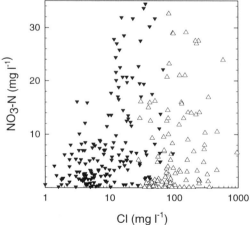

Figure 3. Nitrate distribution w.r.t. Cl in modern shallow groundwaters from N Nigeria (▼) and Senegal (Δ)

has been studied (Figure 2). This is in one of the most arid regions of the Sahara and aerobic conditions prevail in the phreatic aquifer. Nitrate

concentrations typically between 5 and 8 mg l^{-1} NO_3-N are preserved.

5. MODERN ENVIRONMENTS

In present day temperate regions baseline nitrate concentrations are rarely above 2 mg l^{-1} NO_3-N

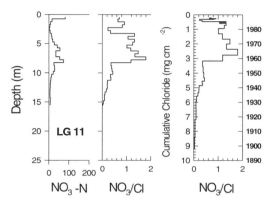

Figure 4. Unsaturated zone nitrate profile (NO_3/Cl) from NW Senegal (Edmunds and Gaye 1997).

reflecting the biological demand as well as relatively high rates of groundwater recharge. Concentrations higher than this, often exceeding the international potable limit of 11.3 mg l^{-1} NO_3-N, are the result of pollution by chemical fertilisers. The situation in semi-arid areas is illustrated (Figure 3) using data from Senegal, where high nitrate has been reported (Edmunds and Gaye 1997), and from N Nigeria - two areas representative of the Sahel. In both areas the salinity is low with groundwater recharge of 14mm and 44 mm yr^{-1} respectively. Both areas have undergone some clearance of the natural vegetation (low scrub with acacia in Senegal and grassland in Nigeria. In Senegal the average nitrate in the shallow wells is 11.1, in Nigeria 7.7 mg l^{-1} NO_3-N. These high values have arisen naturally and are considered due to the presence of leguminous vegetation, especially *Acacia* spp. and *Prosopis* as well as other undefined species.

The transfer of nitrate from the unsaturated zone to the water table may be followed in the interstitial water (Figure 4). The nitrate concentrations from one of several profiles from Senegal are here expressed as NO_3/Cl ratios to indicate the increase of nitrate above evaporative enrichment. The profile has been calibrated using the cumulative Cl and by

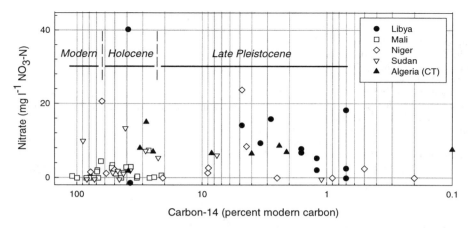

Figure 5. Nitrate in Sahara and Sahel groundwaters from the late Pleistocene to modern times

assuming that the atmospheric flux of Cl has been constant (Edmunds and Gaye 1997); this record thus represents about 100 years. The background concentrations in the first half of this century have a mean concentration of 6.5 mg l^{-1} NO$_3$-N but the strong increase in nitrate and NO$_3$/Cl is considered the result of clearing and planting with rainfed crops including groundnuts (*Arachis Hypogaea*) in the immediate post-war era.

CONCLUSIONS

Nitrate is widely preserved under water-table, phreatic conditions in arid and semi-arid climates of the Sahara and Sahel. The NO$_3$ remains inert in the presence of oxygen where it may be used to investigate the palaeoenvironmental conditions up to the limits of radiocarbon age determination (ca 35 k yr). The available data from five countries is summarised in Figure 5. In Libya, Niger , Sudan and Algeria many of the dated groundwaters contain nitrate close to or above the 11.3 mg l^{-1} limit. The frequency of high nitrate groundwaters appears to be equivalent through the Late Pleistocene and Holocene. Moreover these concentrations are close to the baseline values observed for the region in modern groundwaters, although the cultivation of leguminous crops leads to still higher NO$_3$ concentrations. The results are compatible with studies of North African palaeo-vegetation (Ritchie and Haynes 1987; Lezine 1989) which indicate no major changes in plant communities, rather a shift northwards of the Sahelian vegetation zones some 500 km during the Holocene. Thus it seems that the high nitrate concentrations are an intrinsic property

of these groundwaters supporting models of vegetation change and providing a new baseline with implications for water quality standards.

REFERENCES

Andrews J N, Fontes J-Ch, Aranyossy J-F, Dodo A, Edmunds W M, Joseph A & Travi Y. (1994). The evolution of alkaline groundwaters in the Continental Intercalaire aquifer of the Irhazer Plain, Niger. *Wat. Resour. Res.* 30: 45-61.

Barnes C.J., Jacobson, G. and Smith, G.D. 1992. The origin of high nitrate groundwaters in the Australian arid-zone. *J.Hydrol.* 137: 181-197.

Darling, W G, Edmunds, W M, Kinniburgh, D G & Kotoub, S. 1987. Sources of recharge to the Basal Nubian Sandstone Aquifer, Butana Region, Sudan. In: *Isotope Techniques in Water Resources Development* p.205-224. Vienna: IAEA.

Edmunds, W M & Wright, E P. 1979. Groundwater recharge and palaeoclimate in the Sirte and Kufra basins, Libya. *J. Hydrol.* 40, 215-241

Edmunds, W M, Gaye, C B & Fontes, J-Ch. 1992. A record of climatic and environmental change contained in the interstitial waters from the unsaturated zone of northern Senegal. In: *Internat. Symp. Isotope Techniques in Water Resources Development* p.533-549. Vienna: IAEA.

Edmunds, W.M. & Gaye, C.B. 1997. High nitrate baseline concentrations in groundwaters from the Sahel. *J. Environ. Quality* 26: 1231-1239.

Fontes, J-Ch, Andrews, J N, Edmunds, W M, Guerre, A & Travi, Y. 1991. Palaeorecharge by the Niger River (Mali) deduced from groundwater geochemistry. *Water Resour. Res.* 27: 199-214.

Gasse, F., Tehet, R., Durand, A., Gibert, E. & Fontes, J.CH. 1990. The arid-humid transition in the Sahara and the Sahel during the last deglaciation. *Nature* 346: 141-146.

Lezine, A-M. 1989. Late Quaternary vegetation and climate of the Sahel. *Quatern. Res.*, 32, 317-334.

Wright, E P & Edmunds, W M 1969. Hydrogeological Studies in Central Cyrenaica. *Unpub. Rep. to Libyan Govt.* British Geological Survey.

Geochemistry of the Earth's Surface, Ármannsson (ed.) © 1999 Balkema, Rotterdam, ISBN 90 5809 073 6

^{210}Pb dating of West Siberian geochemical background changes over the XX century

V. M. Gavshin, M. S. Melgunov, F. V. Sukhorukov, B. L. Shcherbov, V. D. Strakhovenko, I. N. Malikova, S. I. Kovalev, V. A. Bobrov & V. V. Budashkina
Analytical Center of the United Institute of Geology, Geophysics and Mineralogy, Siberian Branch of the Russian Academy of Sciences, Novosibirsk, Russia

ABSTRACT: The distribution of ^{210}Pb, ^{137}Cs, and heavy metals in sections of lake sediments, pine litter, and peat have been studied in typical West Siberian landscapes. According to the constant-rate-of-supply (CRS) model, 100-year intervals were separated in the sediments; within these intervals, concentrations of heavy metals (Cd, Pb, Zn, Cu, Mn, Cr, Ni, and Co) increased to a certain extent from the bottom upwards. It was demonstrated that partition of the chemical elements during atmospheric transfer of soil erosion products is more pronounced with decrease in ash content: Cd, Pb, and Zn concentrations increased drastically, whereas Cr, Ni, and Co concentrations remained close to the corresponding clarke values for clays. Peat was found to be the most adequate object for assessing anthropogenic component of the geochemical background.

1 INTRODUCTION

Here, we will employ the distributions of several heavy metals (Cd, Pb, Zn, Cu, Cr, Ni, Mn and Co) and technogenic radionuclides (^{137}Cs and ^{210}Pb) in soils, lake sediments, forest pine litter, and peats to trace the changes that have occurred in the geochemical background. The investigation was carried out close to the center of Asia (Figure 1).

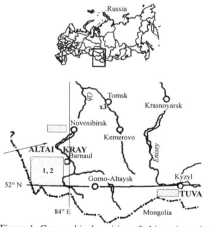

Figure 1. Geographical position of objects investigated. An arrow – the main track of the dust-storms; Rectangles - areas investigated; 1 - lake Krivoe, 2 – pine litter, 3 – peatbog.

2 METHODS

Sediments from over 25 lakes, 22 pine litters, and about 300 soil sections in plane steppe and forest-steppe landscapes and a peat bog in a taiga landscape were covered with detailed sampling. Radiocesium was determined by γ-spectrometry according to the 661.1 keV (^{137}Cs), 604, and 795.8 keV (^{134}Cs) characteristic lines at a laboratory that previously participated successfully in certification of IAEA samples (IAEA-373 and IAEA-375). ^{210}Pb was determined using HP Ge planar detector protected with lead and tungsten in the X-ray range according to 46.5 keV characteristic line. Cd, Pb, Zn, Cu, Mn, Cr, Ni and Co were determined with an atomic absorption spectrophotometer with previous testing of the technique on a wide range of reference samples of rocks, soils and sediments. We took the mean contents of chemical elements (clarke values) for clays according to Turekian & Wedepohl (1961) as upper limit of the "geochemical background", since the notion itself is too complex while clay material dominates in the mineral part of all the formations studied.

3 RESULTS

Natural soil radioactivity depends on the substrate. For example, the gamma activity of the soils formed on silty clay amounts at a height of 1 m to 115±2

Bq/kg; of the soils formed on granites, 154±5 Bq/kg; and of the soil formed on quartz sandstone, 84±2 Bq/kg. The technogenic radioactivity in the samples amounted to approximately the same values. In 1992, the global ^{137}Cs background in Altai kray was estimated as 60 mCi/km^2, being close to the corresponding data for the USA (Gavshin et al. 1993). The Semipalatinsk Nuclear Test Site is located 200 km to the southwest of this region. The radionuclides drifted to the Altai territory by the northeastern winds dominating there. As a result, certain sites of local fallout can now be found, where the radiocesium activity reaches 150—200 mCi/km^2. The radiocesium inventory in the undisturbed soils in one such site, with an area of 30×35 km^2, amounted to 90±2 mCi/ km^2; in the cultivated soils, 59 mCi/ km^2, or 65% of the native soil inventory, the distribution over the section being even (this ratio is 50% in the soils of Southern England) (He et al. 1996). On average, the 20-cm soil layer contains 95% of the entire inventory. ^{134}Cs is a specific product of the Chernobyl accident, as it is produced from the natural isotope ^{133}Cs through neutron irradiation. The ^{137}Cs/^{134}Cs ratio in the Chernobyl fallouts had by 1992 increased from the initial value of 2.0 to 13.4. The data listed in Table 1 suggest that during that time, plants accumulated radiocesium Chernobyl origin only, while its fraction in soil did not exceed 21%.

Table 1. The ^{137}Cs/^{134}Cs ratio in soils and plants

Sample	^{137}Cs, Bq/kg	^{134}Cs, Bq/kg	$\frac{^{137}Cs}{^{134}Cs}$	Date
Soil (turf)	115±5	1.9±0.3	60.5	11.06.92
Moss	110±7	8.7±2.1	12.6	27.12.91
Moss ash	615±8	39±2	15.8	Autumn-92
Ledum palustre	129±4	7.5+1.1	17.2	04.11.91
Bergeria classifolia	16±1	1.4+0.7	11.0	04.11.91
Pine needles	292±6	4.1=1.2	74.0	15.08.92
Needles ash	546±7	5+1	109	Autumn-92

^{90}Sr penetrates deeper into the soil (and lake sediments) than radiocesium; the ratio ^{137}Cs/^{90}Sr in both the soils and lake sediments decreases with depth from 1—6 to the analytical zero. The heavy metal content in the forest-steppe region is considerably lower than the clarke values for clays (Table 2).

Table 2. Heavy metal cocentration, ppm

	Cd	Pb	Zn	Cu	Mn	Cr	Ni	Co
1	0.18	19	78	25	773	92	36	9
2	0.30	20	95	45	850	90	68	19

(1) Soils of Altai (Polyakov 1996); (2) clarke values for clays.

In Kemerovo region, adjacent to Altai, considerable Pb, Cd, and Zn contamination of soils was recorded within a radius of 9 km around a zinc-smelting plant (Il'in 1990).

Lake Krivoe (52^{0}51'28''N and 81^{0}01'00''E) was chosen from all the lakes studied as most representative. Quartz sandstone forms the substrate of this small hydrologically isolated lake, while the soils of the catchment are formed on loess-like silty clays. The radiocesium inventory of the sediments amounts to 121 mCi/km^2, exceeding twofold the global background; the integral fallout of the atmospheric ^{210}Pb, 415 mCi/km^2. Both values indicate an essential contribution from the delayed input of the radionuclides, which entered the lake with wind erosion products of the catchment soils (Figure 2).

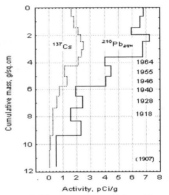

Figure 2. The chronology of lake sediments

The increase in heavy metal concentrations from the bottom of the lake sediment section upwards is not mainly connected with their falling out from the upper atmosphere, but rather with their high content in soil erosion products compared to that in the alluvial quartz sandstone, underlying the lake sediments. Therefore, it is difficult to separate the technogenic component from natural background in this case. The approximate estimations based on the CRS model (Appleby et al. 1979) demonstrate that the most drastic increase in the heavy metal and As contents is connected with reclamation of native

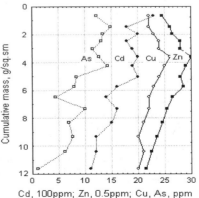

Figure 3. Heavy metals distribution in lake sediments

uncultivated lands, since 1955 (Figure 3). The turning of the virgin soils increased dramatically their wind erosion and caused black storms (Gribanov 1954), which substantially supplied the material forming lake sediments (Gavshin et al., 1999).The maximal ^{210}Pb inventory, in accordance with total Pb, in the coniferous litter is sometimes found not in the upper layer but considerably deeper. This fact together with the absence of ^{210}Pb in the living needles indicates that it is washed away with rainfalls and shifted downwards in the permeable litter section (Figure 4). In addition, the litter does not house the entire inventory, as a part of it passes into the sandstone substrate; this may indicate a rather young age for the litter (Figure 4). A 16-cm section under a 250-year-old pine tree near the village of Volchikha was examined. The radiocesium fallout density amounted to 148 mCi/km^2; that of radiolead, 333 mCi/km^2. These values are close to the values obtained for the sediments of Lake Krivoe and also confirm the contribution of soil erosion products to the inorganic component of coniferous litter. Similar to Lake Krivoe, Volchikha village is located in the path of the most frequent black storms. First, the heavy metal contents in the litter increase from the top downwards with an increase in ash content; then, with the appearance of quartz sandstone admixture, they decrease slightly.

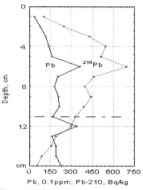
Figure 4. Distribution of Pb in peat

Peat was studied in Kirsanovskoe bog, located 30 km to the southwest of a the city of Tomsk in a plane taiga landscape far enough from any industrial objects. This bog is of a convex shape and is fed only by atmospheric falls, thus representing an ideal object for assessing the global component of the atmospheric input of radionuclides and heavy metals. The activities of ^{137}Cs and ^{210}Pb decrease gradually in the peat section down to 22 cm with a decrease in the ash content (Table 3)

On the whole, the same trend is evident in the heavy metal distribution (Table 4)

The integral fallout over the 123-year period is estimated as 135 mCi/km^2 for ^{210}Pb and 45 mCi/km^2 for ^{137}Cs. The former value is close to the corresponding value for the region of the Great Lakes of North America—147 mCi/km^2 (Kadlec & Robbins 1984); the latter, calculated for 1992, amounts to 53 mCi/km^2 and corresponds to the global background. The data listed in Tables 3 and 4 can be used to assess the contents of heavy metals in ash, fallout densities for each element (μg/cm^2), and annual flux (μg/dm^2/yr) (Table 5).

Table 3. ^{210}Pb chronology of the peat column

No*	Thickness cm	Density mg/cm^3	Ash %	^{210}Pb pCi/g	Years
1	7	27	3.6	15.2	1991-1998
2	5	63	16.3	21.7	1978-1991
3	1.5	54	12.1	20.2	1956-1978
4	1.5	54	3.0	8.67	1939-1956
5	1.5	54	2.3	5.46	1920-1939
6	2.5	54	1.8	3.73	1894-1920
7	3	51	1.3	1.46	1875-1894
8	3	51	1.2	0.95	
9	3	51	1.1	0.09	
10	3	51	1.4	0.52	
11	10	47	1.0		
12	10	47	1.2	0.11	

*Sample numbers in tables 3—6 are the same

Table 4. Heavy metals in samples of the peat section

No	Cd	Pb	Zn	Cu	Mn	Cr	Ni	Co
				ppm				
1	0.22	5.7	50	5.7	140	3.1	2.3	0.38
2	0.23	27.8	45	7.2	87	12.4	5.8	1.54
3	0.26	23.0	37	5.2	38	10.0	8.6	2.83
4	0.36	8.9	29	3.3	15	3.7	3.1	1.20
5	0.32	6.6	32	3.4	17	3.8	2.6	1.20
6	0.26	4.3	20	3.3	9.1	2.5	1.8	0.83
7	0.17	3.0	26	3.1	7.7	2.5	1.5	0.67

Table 5. Estimation of mean annual fluxes of elements

No	Cd	Pb	Zn	Cu	Mn	Cr	Ni	Co
				μg/dm^2/yr				
1	0.6	15	135	15	380	8.4	6.2	1.0
2	0.5	65	105	17	100	29	13.5	3.6
3	0.1	8.7	13.9	1.9	14	3.8	3.2	1.1
4	0.17	4.2	13.6	1.6	7	1.7	1.4	0.6
5	0.14	2.8	13.9	1.4	7	1.6	1.1	0.5
6	0.13	2.2	10.6	1.7	5		1.3	o.9	0.4
7	0.14	2.4	20.9	2.5	6	2.0	1.2	0.5

The calculated concentrations of heavy metals in ash are relatively close to the clarke values only for Cr, Ni, and Co but shifted drastically for Cd, Zn and Pb (by one order of magnitude). Normalization to Sc (Shotyk et al. 1996) was used to estimate the degree of peat enrichment. We have not analytical values for Sc but we found a sufficiently stable distribution of Co over the section to use this element for normalization. The enrichment factors for heavy metals with regard to cobalt divided by the analogous clarke values calculated demonstrate to what extent

41

the peat is enriched with one or another element relative to a comparatively stable cobalt content (Table 6).

As is evident from Table 6, a high enrichment is typical of cadmium, lead, zinc, and a rather moderate of copper.

Table 6. The enrichment factors of elements in the samples of peat section

No	Cd	Pb	Zn	Cu	Mn	Cr	Ni	Co
1	362	14.3	26.2	6.3	8.2	1.7	1.7	1
2	94	17.1	5.8	2.0	1.2	1.7	1.05	1
3	56	7.7	2.6	0.8	0.3	0.7	0.83	1
4	187	7.0	4.8	1.2	0.3	0.7	0.72	1
5	169	5.2	5.3	1.2	0.3	0.7	0.61	1
6	194	5.0	4.9	1.7	0.2	0.6	0.61	1
7	156	4.3	7.8	1.9	0.2	0.8	0.61	1

4 DISCUSSION

The substance involved in the atmospheric flux from wind erosion of soils is partitioned according to sizes of organic and mineral particles: the particles with a radius of about 26 µm prevail in air close to the Earth's surface; these with a radius of 11 µm, at a considerable height. The average radius of loess particles in the vicinity of the source is 15 µm; but less than 8 µm, 100 km and more away from the source. Saharan dust is enriched with iron and manganese compared to its source. According to Junge (1963), "the processes of physical and chemical separation act selectively in forming fine dust, therefore, its composition differs from the average composition of the soil whereof it originated". It is apparent that different elements, contained in the mass of particles or sorbed to their surface, are not always partitioned with equal efficiency. The enrichment with cadmium, zinc, and lead was most pronounced in the peat bog distant from the source of the substance, where most tiny particles sedimented from the air, whereas chromium, nickel, and cobalt contents remained virtually the same. Note that the first three chemical elements, together with mercury and arsenic, differ from others in high volatility. Their concentrations have been ever increasing in the Baltic Sea sediments over the last 60—70 years, while the distribution of nickel and cobalt remained virtually the same (Pempkowiak 1991); a similar situation was described for the sediments of a Norwegian fjord (Skei & Paus 1979). Supposing that peat reflects most precisely the dynamics of anthropogenic substance release into the atmosphere, it is most difficult to assess this fraction at the sites where the dominating substance fraction escaped the atmospheric partitioning. As is evident from Table 7, the less inorganic material ("ash") is contained in the organic—mineral mixture, the more pronounced is the selective enrichment of the substance with cadmium, lead, and zinc over the 100-years period.

Thus, if we accept the annual fluxes of chemical elements calculated for the peat bog (Table 5) as a basis, the contribution of cadmium to the upper 5-cm soil layer of the steppe and forest-steppe landscapes over 100 years might amount to 18%, lead, to 9%, zinc, to 7%, and copper, to 3%. This is in accordance with the order of anthropogenic enrichment of sediments given by Suess and Erlenkeuser: Cd>Pb>Zn>Cu (Nriagu, 1980).

Table 7. The enrichment factors in total inventory over 100 yrs

	Cd	Pb	Zn	Cu	Mn	Cr	Ni	Co
Peat	13.2	11	6.5	1.8	1.2	1.2	0.9	1
Litter	3.8	6.8	4.0	1.5	4.0	1.3	0.9	1
Lake*	1.0	1.9	1.0	0.9	1.0	1.0	0.8	1

* Lake sediments

ACKNOWLEDGEMENT

This work was supported by Russian Foundation of Fundamental Researches Grant 97-05-65235.

REFERENCES

Appleby, P.G., F.Oldfield, R.Thomson & P.Huttunen 1979. [210]Pb dating of annually laminated lake sediments from Finland. *Nature* 280: 53—55.

Gavshin, V.M., B.L.Shcherbov, M.S.Mel'gunov, V.D Strakhovenko, V.A.Bobrov & V.M.Tsibul'chik 1999. [137]Cs and ^{210}Pb in lake sediments of Steppe Altai as indicators of the dynamics of anthropogenic changes in geochemical background over XX century. *Russian Geology & Gephysics* (in press).

Gavshin, V.M., F.V.Sukhorukov, I.N.Malikova, S.I.Kovalev & B.L. Shcherbov 1993. Distribution of radionuclides over the territory of Altai kray. In: *Nuclear Tests, Environment, and the Health of Altai Population*, 1(1): 34—72. Barnaul (In Russian).

Gribanov, L.N. 1954. To the cognition of the nature of black storms in Kulunda Steppe of Altai kray, *Pochvovedenie* 6: 35—45. (In Russian).

He, Q., D.E.Walling & P.N.Owens 1996. Interpreting the ^{137}Cs profiles observed in several small lakes and reservoirs in Southern England. *Chemical Geology* 129: 115—131.

Il'in, V.B. 1990. On contamination of soils and agricultural plants with heavy metals by enterprises of nonferrous metallurgy. *Agrokhimiya* 3: 92—99. (In Russian).

Junge, C.E. 1965. *Air Chemistry and Radioactivity*. Moscow: Mir (Russian translation of original edition of 1963).

Kadlec, R.H. & J.A.Robbins 1984. Sedimentation and sediment accretion in Michigan coastal wetlands (U.S.A.). *Chemical Geology* 44: 119—150.

Nriagu J.O., 1980. *Cadmium in the environment, part1, Ecological cycling*. John Wiley & Sons, Inc.

Pempkowiak, J. 1991. Enrichment factors of heavy metals in the Southern Baltic surface sediments dated with ^{210}Pb and [137]Cs. *Environment International* 17: 421—428.

Polyakov, G.V. (Ed.). 1996. *Ecogeochemistry of Western Siberia*. Novosibirsk, SB RAS. (In Russian).

Shotyk, W., A.K.Cheburkin,., P.G.Appleby, A.Fankhauser & J.D. Kramers 1996. Two thousand years of atmospheric arsenic, antimony, and lead deposition recorded in an ombrotrophic peat bog profile, Jura Mountains, Switzerland. *Earth and Planetary Science Letters* 145: E1—E7.

Skei, J. & P.E.Paus 1979. Surface metal enrichment and partitioning of metals in a dated sediment core from a Norvegian fjord,. *Geoch. et Cosm. Acta* 43: 239—246.

Turekian, K.K. & K.H.Wedepohl 1961. Distribution of the elements in some major units in the Earth's crust. *Geol. Soc. Amer. Bull.* 72,(2): 175-191.

Geochemistry of the Earth's Surface, Ármannsson (ed.) © 1999 Balkema, Rotterdam, ISBN 90 5809 073 6

Paleogene ages of a lateritic pisolitic crust determined by Mn-oxides dating

O. Hénocque & F. Mouélé
Université d'Aix-Marseille III, CEREGE, Aix-en-Provence, France

G. Ruffet
UMR Géosciences Azur, UNSA, Nice, France

F. Colin & B. Boulangé
I.R.D., CEREGE, Aix-en-Provence, France

ABSTRACT: Laterites cover the tropical cratonic belt. They are characterized by frequent metal accumulations. In West Africa, the lateritic indurate crusts that can be observed on top surfaces of the relieves, were described as evidence of successive weathering and erosion cycles. Direct dating of supergene K-Mn oxides (cryptomelane) allow us to constrain their development ages. These minerals are supposed to trace warm and humid climates.

We performed $^{40}Ar/^{39}Ar$ dating on cryptomelanes extracted from the Tambao pisolitic crust. The Paleogene ages obtained proved that this surface has been formed mainly during two periods.

The comparison between our results and those of previous studies, performed in Brazil, underlines the occurrence of a Paleo-Eocene global weathering period. Further more, East Atlantic stratigraphic records demonstrate that Sahelian laterites ages are concordant with West African basins depositional history. We can then assert that laterites describe climatic and environmental changes.

1 INTRODUCTION

Chemical and mechanical weathering-derived processes are among the most efficient factors in modifying the Earth surface. Over geological times, weathering gradients are dependent on temperature, rainfall and tectonic stability. As a consequence, the surficial domains of the present day tropical and sub-tropical cratons have been transformed since the early Tertiary into thick supergene lateritic sequences often associated with metal accumulation. Such massive indurated crusts, which are frequently resistant to erosion (Nahon 1986, Beauvais & Colin 1993), could constitute a long-time record of past climate evolution.

Direct dating of weathering events was difficult until the development of microbeam methods. One of the most promising approaches is $^{40}Ar/^{39}Ar$ dating of isotopically closed supergene K-bearing minerals from the hollandite group (α-MnO_2). Minerals from this group have a tunnel structure (Figure 1), constituted of double chains of MnO_6 octahedra combined at corners and denoted T(2,2).

Initiated in the 60's, K-Ar dating of potassium-rich manganese oxides proved the capacity for cryptomelane to retain radiogenic argon even if dating pure mineral fractions was difficult (Chukhrov et al. 1966, Yashvili & Gukasyan 1974).

The recent development of the $^{40}Ar/^{39}Ar$ laser probe method partly resolved such problems because of the very small amount of material necessary for analysis (Vasconcelos et al. 1992, 1994, Ruffet et al. 1996). This new approach allows a fine scale sampling, making it possible to date distinct generations of K-Mn oxides (Hénocque et al. 1998).

Figure 1. Hollandite group minerals tunnel structure. The general formula is $A_x B_8 O_{16}$, nH_2O. B sites = Mn^{3+}, Mn^{4+}, Fe^{3+}, Al^{3+} ; A sites = K^+, Ba^{2+}, Na^+,...

From thermodynamical studies, it appears that cryptomelane formation needs large amount of potassium leached in weathering solutions which can result of direct transformation of illites into kaolinite. Cryptomelane is then supposed to reflect intense weathering. Because cryptomelane is an ubiquitous mineral of tropical lateritic manganese weathering profiles, we propose to expand the scope of previous studies (Vasconcelos et al. 1992, Ruffet et al. 1996) to enhance our understanding of the genesis of Mn-lateritic systems.

This work presents the analytical performances of the $^{40}Ar/^{39}Ar$ argon-ion laser probe method in "dating" samples collected in the pisolitic crust of Tambao deposit (Burkina Faso), one of the cryptomelane-richest lateritic deposit known in the world (Grandin 1976).

2 SITE DESCRIPTION

2.1 Geographical and geological setting.

The Tambao manganese deposit (Northern Burkina Faso, West Africa) is located 14°N and 0°W in Sahelian-Saharan zone. It consists of a 3 km long complex of two 80 and 50 meters hills that constitute the main relief of the area. They are structured by four manganese oxide benches that alternate within weathered Birimian metavolcanic and sedimentary series.

From landscape observations, four successive surfaces can be observed in the direct vicinity of the deposit. The highest one (between heights 340 and 325 m) consists of relics of a Mn-pisolitic crust and constitutes the oldest lateritic surface observed in the area. Pisolites result from the geochemical transformation of the underlying massive oxide layers. Such pisolitic formations were supposed to be contemporaneous with the Cretaceous Eocene Bauxites (Boulangé 1984). Three successive sandy-clayey ferruginous lateritic systems have developed laterally, from the middle of the largest hill slope to the main drainage axis in the surrounding plain. Through these formations, manganese is widely scattered and generally occurs as detrital pebbles. The corresponding surfaces have been described as erosional fan ferruginized during wet climate periods (Grandin 1976).

2.2 Weathering sequence.

About 75% of the oxides derive from a metamorphic manganese carbonate protore mainly composed of rhodochrosite ($MnCO_3$), associated to primary oxides as hausmannite (Mn_3O_4) and few manganese silicates as rhodonite ($Mn[SiO_3]$) and tephroite ($Mn_2[SiO_4]$). The rest of the oxide derive from tuff and Mn-garnets rich quartzite. The protore can be observed at about 70 to 80 meters beneath the summit of the largest hill. At that depth, the carbonated protore is directly replaced by manganese oxides and hydroxides such as manganite (MnOOH) and pyrolusite (MnO_2). Because of low amounts of potassium in the lower 20 meters of the profile, cryptomelane ($K_{1-2}Mn_8O_{16},nH_2O$) is very rare in this part of the profile and mostly occurs in small veins. The whole rock chemistry (Table 1) was determined by classical ICP-OES analyses. Manganese enrichment is very low and ranged between 1.2 and 1.4. Cryptomelane is the main oxide of the upper part of the profile. This mineral replaces the previous manganite-pyrolusite matrices. It retains a large proportion of potassium probably brought by weathering solutions. These solutions percolate from surrounding granites and tuffs and from the top pisolitic surface where K and Mn are leached.

Table 1. Mean chemical compositions for major elements and barium of the carbonated protore (#1), the pyrolusite rich ore (#2), the cryptomelane rich ore (#3) and the Mn-rich pisolitic crust (#4).

	#1	#2	#3	#4
SiO₂	5.74	2.73	2.37	16.05
Al₂O₃	1.76	1.12	2.26	20.00
Fe₂O₃	0.71	0.85	1.43	0.97
MnO	54.56	74.73	72.67	42.10
MgO	4.27	0.84	0.26	0.46
CaO	4.86	0.51	0.41	0.19
Na₂O	0.08	0.10	0.22	0.10
K₂O	0.05	0.15	4.16	3.52
TiO₂	0.11	0.09	0.21	0.36
P₂O₅	0.28	0.31	0.37	0.24
H₂O	-	0.25	0.16	-
LOI	28.09	12.72	12.40	15.15
Total	100.51	94.26	96.79	99.14
Ba	750.58	1299	1163	2740

2.3 The pisolitic crust evolution.

Pisolites are formed by geochemical evolution of the upper cryptomelane rich ore in a matrix that conserves in the first stage the massive ore characteristics. In this crust successive enrichments can be observed. First of all, absolute aluminum enrichments occurs (Figure 2) developing two facies there after called :
- "Manganiferous pisolites" (Table 1, #4) characterized by a cryptomelane-kaolinite matrix,
- "Aluminous pisolites", in which the matrix is also enriched in gibbsite.

The primary manganiferous and aluminous matrices and pisolites are after then affected by iron enrichment associated to Mn and Al leaching. During all this evolution potassium is well correlated to manganese, suggesting that it only occurs under cryptomelane form (Figure 3). This is enforced by the fact that cryptomelane is the only manganese mineral detected by X-ray diffraction.

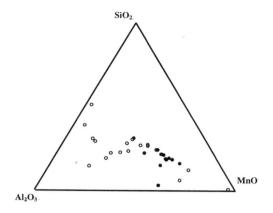

Figure 2. Ternary plot showing evolution from manganiferous pisolitic crust to kaolinite and gibbsite rich aluminous pisolitic crust (black spots : pisolites and white spots : matrices).

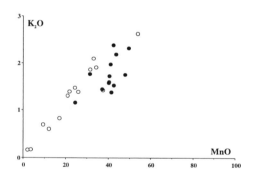

Figure 3. $K_2O\%$ vs $MnO\%$ of the whole pisolitic crust (black spots : pisolites and white spots : matrices).

3 RESULTS

3.1 $^{40}Ar/^{39}Ar$ analyses.

Dated cryptomelane samples cover the whole pisolitic surface and the interface with the underlying massive oxide ore. Most of the age spectra are disturbed. Nevertheless, plateau ages and pseudo-plateau ages were obtained for few samples. The results are related to (1) the manganiferous matrices, and (2) the pisolites.

The manganiferous matrices samples are characterized by spectra showing rather flat segments, in the medium to high temperature steps, well grouped around 46 Ma.

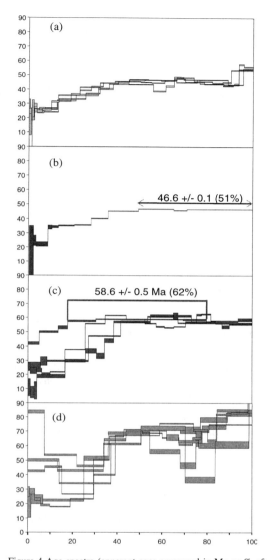

Figure 4 Age spectra (apparent ages expressed in Ma vs % of the total ^{39}Ar released) displayed by (a) samples of manganese matrices, (b) an external rim of pisolite, (c) core of pisolites within Mn-Al pisolitic crust and (d) within Al-Fe pisolitic crust.

These high temperature pseudo-plateau ages are in good agreement with the pseudo-plateau age obtained displayed by an external rim of pisolite (Figure 4 b). These age spectra show also, in the low temperature steps, concordant apparent ages around 25 Ma (Figure 4 a).

In the pisolite core samples, the age spectra are much more disturbed. Nevertheless, two groups can be defined. The first one is characterized by age spectra that allow to define concordant high temperature pseudo-plateau ages around 58.5 Ma. This age seems to be validated by a pseudo-plateau age obtained on a manganiferous glaebular nodule (Figure 4 c). The second group is characterized by strongly disturbed age spectra which do not allow to define any significant age cluster (Figure 4 d).

To summarize, two significant age clusters (58.5 and 46.6 Ma) are observed in the Tambao manganese deposit pisolitic crust. These ages have also been observed on samples from the underlying massive ore and at the interface between pisolites and cryptomelane ore. The 46.6 Ma age is concordant with ages obtained in a previous study of whole ore sample from the summit of the cryptomelane ore (Hénocque et al. 1998). On the other hand, an ore sample located at the transition between ore and pisolites provided an age spectra with a pseudo-plateau age at 59.3 ± 0.3 Ma.

3.2 Comparison with previous studies and stratigraphic records.

The Paleogene ages of the Tambao manganese deposit pisolitic crust are in good concordance with some ages obtained for Brazilian Mn-deposits by Vasconcelos et al. (1994) and Ruffet et al. (1996). This suggests that during this period, tropical weathering, on each side of the Atlantic Ocean, was governed by global climatic conditions. This hypothesis is strongly enforced by the occurrence in the Senegal and East Atlantic basins, for periods concerned by our results, of chemical deposits wich characterize the warm Paleogene climate (Chamley et al. 1988).

4 CONCLUSIONS

1. The pisolites formations are frequently observed all over the surface of most Fe, Al and Mn crust through African laterites. This study presents the first dating of African pisolites.
2. The Tambao manganese pisolitic crust may have formed through two distinct periods of the Paleogene (58.5 Ma and 46.6 Ma).
3. Our results show evidence of a climatic change characterized by a weathering interruption after 46.6 Ma.
4. The occurrence of gibbsite in such systems is linked to the degradation of the bauxites. In this way, the ages at 46.6 and 58.5 Ma constitute the lower limit for bauxite formation in the Sahelian zone.

REFERENCES

Beauvais, A. & C. Colin. 1993 Formation and transformation processes of iron duricrust systems in tropical environment. *Chem. Geol.* 106 : 77-101

Boulangé, B. 1984 Les formations bauxitiques latéritiques de Côte d'Ivoire. *Mem. ORSTOM* 175.

Chamley ,H., P. Debrabant & R. Flicoteaux 1988 Comparative evolution of the Senegal and eastern central Atlantic basins, from mineralogical and geochemical investigation. *Sedimentology* 35 : 85-103

Chukhrov, F.V., L.L. Shanin & L.P. Yermilov 1966 Feasibility of absolute-age determination for potassium-carrying manganese minerals. *Int. Geol. Rev.* 8 : 278-280

Grandin, G. 1976 Aplanissements cuirassés et enrichissements des gisements de manganèse dans quelques régions d'Afrique de l'Ouest. *Mem. ORSTOM* 82.

Hénocque, O., G. Ruffet, F. Colin & G. Féraud 1998 [40]Ar/[39]Ar dating of west African lateritic cryptomelanes. *Geochim. et Cosmochim. Acta* 62 : 2739-2756.

Nahon, D. 1986 Evolution of iron crust in tropical landscape. In S.H. Colemans & D.P. Diether (eds), *Rates of Chemical Weathering of Rocks and Minerals* : 169-191. London: Academic Press.

Ruffet, G., C Innocent., A. Michard, G. Féraud, A. Beauvais, D. Nahon & B. Hamelin 1996 A geochronological [40]Ar/[39]Ar and 87Rb/87Sr study of K-Mn oxides from the weathering sequence of Azul, Brazil. *Geochim. et Cosmochim. Acta* 60 : 2219-2232.

Vasconcelos P.M., T.A. Becker, P.R. Renne & G.H. Brimhall 1992 Age and duration of weathering by K-Ar and [40]Ar/[39]Ar analysis of supergene K-Mn oxides. *Science* 258: 451-455

Vasconcelos P.M., P.R. Renne, G.H. Brimhall & T.A. Becker 1994 Direct dating of weathering phenomena by [40]Ar/[39]Ar and K-Ar analysis of supergene K-Mn oxides. *Geochim. et Cosmochim. A   '8: 1635-1665.

Yashvili L.P. & R.K. Gukasyan 1974. Use of cryptomelane for K-Ar dating of manganese ore of the Sevkar-Sargyukh deposits, Armenia. *Dokl. Earth Sci. Sect.* 212 : 49-51

Geochemistry of the Earth's Surface, Ármannsson (ed.) © 1999 Balkema, Rotterdam, ISBN 90 5809 073 6

Interactions between the nitrogen, oxygen, and carbon cycles in coastal sediments

S. B. Joye & R. J. Wisniewski
University of Georgia, Athens, Ga., USA

S. An
Marine Science Institute, Port Aransas, Tex., USA

ABSTRACT: The sedimentary cycles of carbon, nitrogen and oxygen exhibit dynamic biogeochemical interactions. In Galveston Bay, Texas, we examined rates of benthic metabolism, denitrification, and nitrification using benthic chamber incubations, laboratory experiments, and modeling in order to identify controls on N cycling and explore interactions between N, O and C cycles. High rates of coupled nitrification-denitrification were observed during daytime chamber incubations; nighttime rates were significantly lower. Results from a 3-box model illustrated the importance of benthic photosynthesis (O_2 source for nitrification) in stimulating coupled nitrification-denitrification. Nitrification also depends on NH_4^+ availability, which is largely a function of the exchangeable NH_4^+ pool. Despite the shallow, turbid nature of Galveston Bay, benthic primary production seems to play a pivotal role in mediating N cycling and represents an important link between the N, O and C cyles.

1 INTRODUCTION

1.1 The importance of nitrogen in aquatic systems

Nitrogen, N, is an essential component of organic biomass. N cycling between ecosystem compartments and organic and inorganic forms is mediated by a suite of microbial processes, including N uptake and assimilation, N_2 fixation, ammonification, nitrification, dissimilatory nitrate reduction to ammonium, and denitrification (Joye & Hollibaugh 1995). In shallow coastal environments, a large fraction of N cycling occurs in sediments. Coupling distinct sediment processes exerts a strong influence on system-level ecological and geochemical parameters. Of the various N cycle processes, nitrification and denitrification are of particular interest since together they serve to transform N from a bioavailable form (NH_4^+) to a non-bioavailable form (N_2), resulting in net N loss from the system. Elevated denitrification rates can limit primary ecosystem production by driving primary producers to N limitation over seasonal [or annual] time scales.

1.2 Interactions between N-O-C cycles

By necessity, the N cycle is closely coupled to the cycles of oxygen, O, and carbon, C (Figure 1). Nitrification, the transformation of NH_3 to NO_2 and NO_3 requires O_2. This process is mediated by chemoauto-

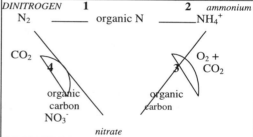

Figure 1. Linkages between important processes in the N cycle and the C and O cycles [1-N_2 fixation; 2-ammonification; 3-nitrification; and 4-denitrification]. Nitrification (3) requires O_2, which can result in coupling to oxygenic photosynthesis. In contrast, denitrification (4) is inhibited by O_2 and requires a labile organic carbon source. NO_3 derived from nitrification supports most of the denitrification in shallow sediments.

trophic bacteria (nitrifiers) that couple NH_3 oxidation to CO_2 reduction. Denitrification, the reduction of NO_3 to N_2 or N_2O, is mediated by heterotrophic bacteria that couple NO_3 reduction to organic carbon oxidation. Denitrifying bacteria are facultative anaerobes, meaning that they only denitrify when O_2 concentrations drop below a critical level (~10 µM; Tiedje et al. 1989). The fact that rates of N cycling processes depend on O_2, organic C, and CO_2, creates dynamic and important interactions between the N, O, and C cycles. However, quantifying these interactions can be difficult, particularly when benthic photosynthesis is a central process of interest.

2 METHODS AND MODELING

We used a combination of field measurements, controlled laboratory experiments, and modeling to assess interactions between N, O and C cycles in the surficial (top 10 cm) sediments of Galveston Bay, Texas. These measurements were carried out between June 1996 and October 1998 in order to assess seasonal and interannual variability.

2.1 In Situ Benthic Chamber Incubations

To determine rates of benthic metabolism and N cycling processes, gas tight benthic chambers (900 cm^2; 9 L volume; n = 2 or 4 per site) were placed onto the sediment surface by SCUBA divers. At 0 and 20 hr. time points, samples (n=3) for dissolved N_2 and O_2 quantification were collected from each chamber into gas tight 10 cc glasspak® syringes. Water samples (~200 ml) for dissolved inorganic N and C analyses were also collected. Gas samples were stored at 4°C prior to analysis. Nutrient samples were frozen prior to analysis using standard autoanalyzer techniques. Changes in dissolved O_2 and N_2 concentrations were quantified using thermal conductivity gas chromatography (An & Joye 1997, Joye & An 1998). Dissolved inorganic carbon concentration was determined using a coulometer (Joye & An 1998). Benthic fluxes were calculated as the change in concentration over the incubation period (mmol m^{-2} d^{-1}).

2.2 Laboratory Experiments

Potential denitrification (pDNF) rates were determined in anaerobic sediment slurries amended with NO_3 and organic C (usually glucose; An & Joye 1997). Potential nitrification (pNTR) rates were determined in O_2-saturated sediment slurries amended with NH_4^+ (Joye & Hollibaugh 1995). A variety of environmental factors, including temperature and O_2, sulfide, exchangeable NH_4^+, and CH_4 concentration influence rates of pDNF and/or pNTR. Here, we only discuss the influence of NO_3 on pDNF and the influence of dissolved and exchangeable NH_4 concentration on pNTR. The exchangeable NH_4 pool is by definition sediment bound, as are many nitrifying bacteria. This pool represents a localized substrate pool for attached bacteria and could influence rates of nitrification.

2.3 Modeling

A 3-box model was used to assess the influence of benthic photosynthesis on coupled nitrification-denitrification. The model describes temporal evolution of dissolved O_2 and inorganic N species (N_2, NH_4, NO_3) concentrations in benthic chambers. The three boxes represented are the overlying water trapped in the chamber (Box 1), oxic-suboxic sediment (Box 2), and anoxic sediment (Box 3). The model allowed simulation of a variety of conditions, including "no benthic photosynthesis", "constant O_2", and "full model". The "full model" represents the conditions observed in the field. Flux measurements obtained during winter (Jan.) and summer (Aug.) were compared with model results and processes responsible for the concentration changes were calculated for each box.

Full diagenetic models describing organic matter diagenesis are available, however, these models do not contain a means by which to include abrupt (hour time scale) perturbations, such as the initiation of photosynthesis, during the model run (Boudreau, 1996, Middelburg et al. 1996). Thus, our 3-box model allowed us to compare rates of benthic processes with and without benthic primary production.

The major processes influencing N_2, NH_4, NO_3, and O_2 concentrations in the boxes include diffusion, organic matter degradation, photosynthesis, nitrification, and denitrification. Diffusive exchange between boxes 1 and 2 and boxes 2 and 3 was permitted. Aerobic processes were permitted in box 2. Anaerobic processes were permitted in box 3. Since bacterial abundance in the overlying water (10^6 cells ml^{-1}) is much less than bacterial abundance in sediments (10^9 cells ml^{-1}), rates of aerobic respiration, nitrification and denitrification in the overlying water were assumed to be negligible. Apparent diffusion coefficients of gases and ions were obtained from Boudreau (1997) and were corrected for tortuosity, temperature, and salinity. The porosity was assumed to be constant (= 0.8) in boxes 2 and 3. Rates of individual processes were described according to the equations presented in Boudreau (1997), with modifications specific for this system. Model parameters were varied to reproduce the observed concentration changes in Box 1. The sensitivity of variability in each parameter for the overall result was evaluated by halving or doubling the parameter (An 1999).

Concentration changes with time in each box were described with differential equations and the equations were solved using the ordinary differential solver (ode45) in Matlab® (An 1999, An & Joye, in prep.).

3 RESULTS AND DISCUSSION

Galveston Bay is a large (1500 km^2), shallow (avg. depth ~ 2 m) estuary that exhibits high rates of water column primary production and supports large recreational and commercial fisheries. Tight coupling and rapid recycling of N between benthic and pelagic reservoirs would be expected in such a shallow estuary. However, the majority of the benthic N flux in Galveston Bay sediments is comprised of N$_2$ rather than dissolved inorganic nitrogen (DIN=NO$_3$+NH$_4$). Efficient coupling between benthic nitrification and denitrification is suggested by the low DIN and high N$_2$ fluxes (Table 1). The sediment oxygen demand is surprisingly low compared to the dissolved inorganic carbon (DIC) flux, suggesting that benthic photosynthesis is important (An 1999, An & Joye, in prep.). Rates of coupled nitrification-denitrification are highest during the daytime period, when photosynthesis is active (Table 2). This efficient coupling results in high rates of N removal relative to N loading. More than 50% of N imported to the system via rivers is lost as N$_2$ via coupled nitrification-denitrification (Joye & An 1998, An 1999).

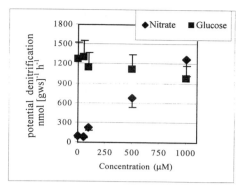

Figure 2. Potential denitrification (pDNF) rates in anaerobic slurries amended with either variable NO$_3$ + 2 mM glucose (diamonds) or 2 mM NO$_3$ + variable glucose (squares).

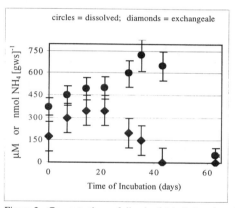

Figure 3. Concentrations of dissolved (circles) and exchangeable NH$_4$ (diamonds) in nitrifying sediment slurries. Slurries were kept anaerobic until day 21 and then O$_2$ was introduced. Nitrification was measurable 7 days later and the exchangeable NH$_4$ pool was rapidly depleted. Consumption of the dissolved NH$_4$ pool occurred after the exchangeable pool was exhausted.

Table 1. Average (1996-1998) DIN and N$_2$ fluxes, sediment oxygen demand (SOD), and DIC flux in Galveston Bay sediments.

Station[1]	N$_2$	DIN	SOD	DIC
St. 1 (n=3)	2.7[2]	1.5	-4.6	32.8
St. 3 (n=6)	1.7	-0.5	0.3	17
St. 4 (n=10)	1.8	0.2	-8.5	21.4
St. 5 (n=2)	2.8	2.0	-11.1	32.8
TC (n=10)	1.1	-0.1	-5.5	9.4
EB (n=4)	1.4	1.3	-8.5	25.4
AVERAGE[3]	**1.9 (0.7)**	**0.5 (0.9)**	**-6.3 (4)**	**23 (9)**

[1] Stations 1, 2, 4 & 5 are along the Trinity River salinity gradient (1=fresh, 5=brackish); Texas City (TC) is in the central Bay; East Bay (EB) station is the Eastern arm of the estuary. The value in parenthesis represents the number of time points used to obtain the average fluxes for each station.
[2] Fluxes in mmol m^{-2} d^{-1}; negative values denoting uptake and positive values denote release from sediments.
[3] Average (±standard deviation) of all data

Rates of pDNF in laboratory slurry incubations in proportion to available NO$_3$ concentration. Glucose did not influence pDNF rates (Figure 2). Rates of pNTR varied slightly with dissolved NH$_4$ concentration, but rates were more sensitive to exchangeable NH$_4$ concentrations (Figure 3). Nitrification rates are dependent upon the O$_2$ supply. Attempts to quantify nitrification using metabolic inhibitors (methyl fluoride and acetylene) in short-term (10-24 hr.) laboratory sediment core incubations proved unsuccessful (Joye, unpub. data). These cores were incubated in the dark and O$_2$ concentrations were rapidly exhausted, thereby limiting nitrification. The ex-

changeable NH$_4$ pool is large in Bay sediments (>500 nmol [gws]$^{-1}$ in freshwater sediments), suggesting that O$_2$ ultimately limits nitrification, and by default, denitrification, in this system. A close association between photosynthesis and nitrification is probably required to support the rates of coupled nitrification-denitrification we measured in this system. Results from the 3-box model corroborate that hypothesis (Table 2).

Gross rates of benthic processes obtained using the 3-box model agree with published values for a variety of shallow coastal sediments (see An 1999 for discussion). Summer aerobic respiration rates obtained from the CANDI diagenetic model (Boudreau 1997) were slightly larger than those obtained

Table 2. Model estimates of gross light and dark rates[1] of respiration (RSP), nitrification (NTR) and denitrification (DNF) during summer and winter at the Texas City Site.

	Light			Dark		
	RSP	NTR	DNF	RSP	NTR	DNF
Full Model[2]						
summer	70.6	12.8	11.6	5.5	0.5	1.5
winter	14.7	3.8	5.9	1.1	0.1	0.6
Constant O$_2$						
summer	68.3	12.7	12.2	16.1	1.5	1.9
winter	13.9	3.5	4.9	8.5	2.1	3.5
NO Psyn						
summer	11.0	1.4	1.8	4.8	0.5	0.6
winter	0.1	0.0	0.0	1.7	0.3	0.7

[1] Values in mmol m^{-1} d^{-1} and represent the integrated activity in boxes 2 & 3 of the model.
[2] Full model mimics conditions in the field; Constant O$_2$ maintains constant a O$_2$ concentration in Box 1; No Psyn-does not permit benthic photosynthesis to occur.

using the 3-box model but the anaerobic respiration and denitrification rates were similar (An 1999).
The 3-Box model results illustrate the dependence of nitrification and denitrification on benthic photosynthesis in this system. In both the 'full model' and 'constant O$_2$' simulations, dramatic day-night differences in rates of respiration, nitrification and denitrification are apparent and rates are not significantly different as long as O$_2$ is being produced (during the day). Notably, in the 'constant O$_2$' simulation, rates of all processes are substantially higher at night, relative to the full model where benthic photosynthesis occurs only during the day. Without benthic photosynthesis, rates of respiration, nitrification and denitrification are substantially lower during both winter and summer.

Oxygen consumption by nitrification was 26% (winter) and 24% (summer) of the total sediment O$_2$ demand and was similar to O$_2$ consumption values estimated for other coastal environments (Henriksen & Kemp, 1988). Benthic photosynthesis stimulated rates of respiration in addition to rates of N cycling.

4 CONCLUSION

The data presented here underscore the need to perform *in situ* chamber studies over day-night cycles to assess interactions between benthic photosynthesis and N cycling in coastal ecosystems. Previous studies using laboratory incubations rather than in situ methods obtained denitrification rates that were much lower than those reported here (Zimmerman & Benner 1994). Our results have important implications for understanding and predicting patterns of N cycling in coastal ecosystems. Without benthic pho-

tosynthesis, coupled nitrification-denitrification rates are substantially lower. The water column of coastal ecosystems is frequently turbid and benthic primary production in subtidal regions is often considered negligible. Our field measurements were made on calm days when the possibility of light penetration to the bottom was probably maximal and thus the rates and interactions we report may not occur during all of the year. However, simple models incorporating tidal stage, secchi depth, and incident irradiance could be used to assess seasonal and interannual variabilility of light penetration. Coupling such models to models describing benthic production and N cycling could provide improved annual estimates of benthic metabolism considering the interactions we have described.

REFERENCES

An, S. 1999. *Nitrogen cycling and benthic metabolism in Galveston Bay.* Ph. D. Dissertation., College Station, Texas, USA: Texas A & M University.
An, S. & S.B. Joye 1997. An improved gas chromatographic method for measuring nitrogen, oxygen, argon and methane in gas or liquid samples. *Mar. Chem.* 59 (1,2): 63-70.
Boudreau, B.P. 1996. A method-of-lines code for carbon and nutrient diagenesis. *Comp. and Geosci.* 22, 479-496.
Boudreau, B.P. 1997. *Diagenetic Models and their Implementation.* Springer-Verlag.
Henriksen, K. & W.M. Kemp 1988. Nitrification in estuarine and coastal marine sediments. In T.H. Blackburn & J. Sørensen (eds.), *Nitrogen cycling in coastal marine environments*, 207-249. New York: John Wiley & Sons.
Joye, S.B. & S. An 1997. *Denitrification in Galveston Bay.* Technical Report. Texas Water Development Board.
Joye, S.B. & J.T. Hollibaugh 1995. Sulfide inhibition of nitrification influences nitrogen regeneration in sediments. *Science* 270: 623-625.
Middelburg, J.J., K. Soetaert, & P.M.J. Herman 1996. Evaluation of the nitrogen isotope-pairing method for measuring benthic denitrification: A simulation analysis. *Limnol. Oceanogr.* 41: 1839-1844.
Tiedje, J.M., S. Simkins & P.M. Groffman 1989. Perspectives on measurement of denitrification in the field including recommended protocols for acetylene--based methods. *Plant Soil* 115: 261-284.
Zimmerman, A.R. & R. Benner 1994. Denitrification, nutrient regeneration and carbon mineralization in the sediments of Galveston Bay, Texas, USA. *Mar. Ecol. Prog. Ser.* 114: 275-288.

Geochemistry of the Earth's Surface, Ármannsson (ed.)© 1999 Balkema, Rotterdam, ISBN 90 5809 073 6

Chemical weathering of loess deposits and paleoclimate change

Cong-Qiang Liu, Jing Zhang & Chunlai Li
State Key Laboratory of Environmental Geochemistry, Institute of Geochemistry, Chinese Academy of Sciences, Guiyang, People's Republic of China

ABSTRACT: The chemical and Sr isotopic compositions of the acid-leachates and bulk samples of the loess deposits in central China are variable. In both bulk sample and acid-leachate, Th/U, Rb/Sr and La/Yb ratios are higher for paleosol than for loess. These element ratios for the acid-leachate of paleosol, however, are more changeable than those for loess. The $^{87}Sr/^{86}Sr$ ratios of the acid leachates range from 0.7104 to 0.7120, and clearly higher in the leachate from paleosol than in those from loess. In the loess-paleosol section, the Sr isotopic composition varies with about 100 kyr periodicity, corresponding to changes in the global glacial cycle and comparable with the variation of Sr isotopic composition of paleoseawater. These results indicate that the Sr isotopic and element ratios in loess and weathering profile of the earth's surface can be indicators of crustal weathering and paleoclimate change.

1 INTRODUCTION

Large areas of China are covered with loess-paleosol deposits. These Quaternary deposits have long been the subject of considerable attention by Quaternary scientists and geologists, since they can reveal evidence on past environments as regards climate. The studies of the loess-paleosol deposits indicate that the records of paleoclimatic change are comparable to those of deep-sea sediments. In this study, carbonate and trace element contents and Sr isotopic composition in acid (HCl) leachate and bulk samples of the Heimugou loess section at Luochuan (Liu et al. 1996) were analyzed. Chen et al. (1996) have also studied the chemical weathering of loess-paleosol deposits and its Sr isotopic composition, suggesting that the Sr isotopic composition of carbonate in loess reflects the variation of weathering intensity. Our results indicate that variations in trace element and Sr isotopic compositions of loess-paleosol deposits correspond to changes in the global glacial cycle. Thus we have obtained paleoclimate records at intervals of 100 kyrs. The results of the study suggest that the Sr isotopic and trace element compositions in loess-paleosol deposits are useful indicators of chemical weathering and paleoclimatic change.

2 SAMPLING AND CHEMICAL ANALYSIS

The samples were collected from the Heimugou loess section, Luochuan, Shaanxi Province. The whole Malan and Lishi loess sections were involved in our sampling. Each of the loess and paleosol layers was sampled in the manner of spot-to-spot sampling. Each of the samples was about 10 cm in thickness and approximately 150 g in weight.

The samples were ground as evenly as possible and 5 g of every each sample was taken for chemical analysis. In order to remove water-soluble salts from the loess samples, all samples were ultrasonically washed twice with ultra-pure water. A sufficient amount of 0.8 M HCl was added in the samples washed with ultra-pure water after they were baked to dryness, followed by 4-hour ultrasonic vibration dissolution. The final step was to filter the solution with 0.2μ filter paper. Although each sample was dissolved and leached with distilled HCl at the same conditions, experimental studies have also been carried out on the experimental results which would be influenced when different proportions of HCl and sample are used. The experimental results showed that provided the amount of HCl is sufficient to dissolve carbonates in the samples, the relative amount of HCl and sample would have no influence on the quantitative dissolution of the carbonates. This result indicates that variations in carbonate contents and their Sr isotopic composition in different samples are independent of experimental conditions.

Ca and several trace element contents in the

HCl eluate and the bulk sample were determined by ICP-AES and ICP-MS, respectively. Then, the $CaCO_3$ content of the sample was calculated in terms of the Ca content. Sr was then separated from other components by means of cation exchange and the Sr isotopic composition was analyzed with a VG-354 mass spectrometer. During the analysis the Sr standard NBS987 was analyzed, yielding a $^{87}Sr/^{86}Sr$ ratio of 0.71025 ± 2 (n=12).

3 RESULTS AND DISCUSSION

3.1 *Variation in chemical composition*

The percent content of $CaCO_3$ in the loess section varies within the range 1%-18%. Generally the $CaCO_3$ content of loess is higher than that of paleosol. This result is in agreement with those obtained by other researchers using different analytical techniques. The $CaCO_3$ content in a loess section tends to increase from the bottom upwards, indicating that during the accumulation and formation of loess the climate changed toward a dry/cold one, thus creating a climatic environment favorable to the preservation of carbonates. Whether there is any variation in $CaCO_3$ content among various loess layers or paleosol layers should not be attributed to the source of parental materials (Lu 1981), but would be determined mainly by the climatic environment at the time the loess or paleosol was formed. It is evident that there is almost no difference in the Nd isotopic composition of loess from one layer to another, but this does not mean that the material sources are obviously different (Liu et al. 1994).

Based on bulk-sample analysis, the most remarkable differences between loess and paleosol are the Th/U, Rb/Sr ratio and the Ce anomaly variations, which are generally larger in paleosol than in loess. This indicates that U relative to Th, and Sr relative to Rb, are easily removed from the weathering profile. The negative Ce anomalies in paleosol, although not large, are observable.

The paleosol leachates are largely depleted in Rb and also in Sr, and there is an apparent depletion in Ce, as compared to loess leachates (Figure 1). The Th/U, and rare earth element composition is more variable in paleosol leachates than in loess leachates. These differences also suggest preferential removal of Sr, U and tetra-valent Ce during weathering. The paleosol leachates are more abundant in middle rare earth elements than the loess leachates.

There is also fractionation between light REE's and heavy REE's during weathering. The La/Yb ratios of the bulk paleosol samples are relatively high, while those of the leachates of paleosol are relatively low and variable. This means that the heavy rare earth elements are more mobile than the light rare earth

elements, probably because the heavy REE's can form more stable carbonate complexes in solution and are easily released from the weathering profile.

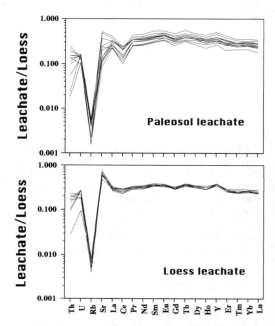

Figure 1. Trace element compositional variation in acid leachates from loess and paleosol. The data for a loess sample are used as normalizing values.

3.2 *Variation in Sr isotopic composition*

In the loess-paleosol section there is a remarkable trend showing a lower $^{87}Sr/^{86}Sr$ ratio in loess than in paleosol. In loess the ratio is as low as 0.7104 while in paleosol it as high as 0.7120, with the absolute value varying by 0.0016. The difference in $^{87}Sr/^{86}Sr$ ratio between loess and paleosol provides direct evidence suggesting that the secondary carbonate in the paleosol contains relatively more Sr derived from the weathering of aluminum silicate minerals. In the loess- paleosol section the $^{87}Sr/^{86}Sr$ ratio varies to different extents during different periods. At 500 kyr. B. P. and 900 kyrs ago the $^{87}Sr/^{86}Sr$ ratio varied over a small range, while at 500-900 kyr. B. P. the ratio varied over a large range. This phenomenon revealed that during different periods of time there existed differences in climatic fluctuations and weathering intensity.

In Figure 2, although the sampling density is too small to conduct a meaningful frequency analysis, it can be seen clearly that the $^{87}Sr/^{86}Sr$ ratio in the loess-paleosol section for the past 500

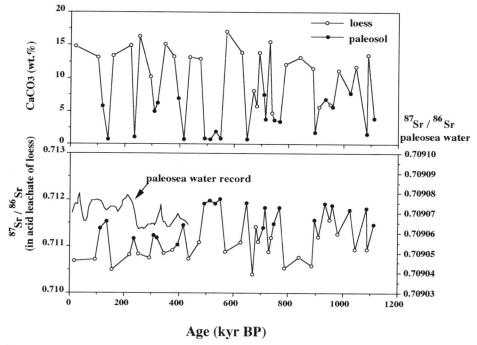

Figure 2. The distribution curves of percent contents of CaCO3 and Sr isotopies in the acid leachates of the loess-paleosol deposits. The geochronological sequence of the samples is established on the basis of the model age 5 (Liu 1985) of loess and the accumulating rates (10.78g/cm^2.10^3 a. for Lishi loess; 11.20g/cm^2.10^3 a. for Lishi paleosol) of Lishi loess and paleosol (Liu 1985).The recorded curve of paleosea water is a smooth curve established on the mean value of five data points. The data are cited from Clemens et al. (1993).

kyrs varies cyclically approximately at intervals of 1000 kyrs. As reflected in the climate curve established on the basis of oxygen isotope data from pelagic drilling cores, a cycle of 100 kyrs during the Pleistocene is just the cycle which is commonly accepted. Careful observations show that the variation of $^{87}Sr/^{86}Sr$ ratio in the loess-paleosol section at 500 kyr. B. P. is characterized by the following aspects. Firstly, there existed wider peaks on the variation curve of $^{87}Sr/^{86}Sr$ at about 500 kyr. B. P. and the peak values are also higher than those during the past 500 kyrs. A possible explanation is that there had appeared a prolonged glacial period on the continent at that time, and as a result, the intensity of weathering of the continental crust was enhanced. Secondly, at 700 kyr. B. P. the variation frequency of the $^{87}Sr/^{86}Sr$ curve was intensified, or was not clear and this may be related to a series of events which took place at about 700 kyr. B. P. That time scale is just located at the B/M (Brunhes-Matuyama) boundary, where there are present abundant microtektites (Li et al. 1993). Geomagnetic reversal and ultraterrestrial-body impact on the Earth resulted in abnormal changes of the climate. If sampling were more abundant, more extensive low might have appeared in the $^{87}Sr/^{86}Sr$ variation curve at 800-900 kyr. B. P., and this could be related to the fact that the continent had been subjected to a long-term dry/cold climatic environment at that time. Meanwhile, a low appeared in the $^{87}Sr/^{86}Sr$ variation curve at 800-900 kyr. B. P. for the last 2.5 Ma for seawater. The occurrence of this low can be explained by the obvious decrease in the river Sr flux and $^{87}Sr/^{86}Sr$ ratios of the Sr derived from continental crust weathering. Finally, the variation frequency of $^{87}Sr/^{86}Sr$ spectra 900 kyrs. ago was enhanced and the fluctuation extent became small. This feature is consistent with the characteristics of the oxygen isotope variation curve for pelagic drilling cores. The above comparison of $^{87}Sr/^{86}Sr$ variations in the loess-paleosol section with the paleoclimate records in pelagic sediments disclosed that variations in Sr isotopic composition of the secondary carbonate in the loess-paleosol section reflect the records of global climate change.

3.3 Linkage to Sr isotopic composition of paleoseawater

The $^{87}Sr/^{86}Sr$ ratio in seawater has tended to increase gradually in the last 40 Ma, indicating that the rate of global chemical weathering is increased (Capo & DePaolo 1990). Studies in recent years have revealed that in the past 450 kyrs, the $^{87}Sr/^{86}Sr$ ratio in seawater is of high-frequency variation at intervals of 100kyrs (Dia et al. 1992). What is more interesting is that the $^{87}Sr/^{86}Sr$ variation curve for secondary carbonate in the loess-paleosol section are comparable to the records of Sr isotopic compositional variation in seawater (Figure. 2). Such a consistency provides direct evidence suggesting that the variation of Sr isotopic composition in seawater is controlled by the cyclic variation of the rate of chemical weathering of the continental crust on a glacial-interglacial time scale.

The cyclic variation of $^{87}Sr/^{86}Sr$ ratio in seawater with time depends largely on the flux of Sr derived from continental crust weathering and then transported into the sea through rivers, and its isotopic composition. But till now it has not been completely understood how the flux of Sr in the rivers and its isotopic composition exert influence respectively or simultaneously on the Sr isotopic composition of seawater. As viewed from the variation characteristics of Sr isotopic composition in the loess-paleosol section, the chemical weathering rates of continental materials had once fluctuated, but did not show any tendency of progressive increase on a long-term basis. It is implied that in the past 40 Ma the increase of $^{87}Sr/^{86}Sr$ ratio in seawater is attributed chiefly to the progressive increase of Sr flux rather than that of the $^{87}Sr/^{86}Sr$ ratio in river water. However, there is no universally accepted explanation of the factors (the flux of Sr and the Sr isotopic composition) that lead to the cyclical variation of $^{87}Sr/^{86}Sr$ ratio in seawater at intervals of 100 kyrs. Loessic materials were derived from the weathering of rocks in the vast continental crust and their geochemical composition can typically represent that of the upper crustal materials. The variation of $^{87}Sr/^{86}Sr$ ratio in the loess section can well reflect the isotopic compositional variation of Sr, which was derived from terrestrial materials of different ages, in seawater. Therefore, high- resolution time-scale studies of the records of paleoclimatic fluctuations in loess-paleosol deposits will help us to gain a better understanding of the variation of Sr isotopic composition in seawater in the geological past.

4 CONCLUSIONS

1. Major and trace elements, as well as Sr isotopic compositions are significantly variable in acid leachates and bulk samples of loess-paleosol deposits,

and are related to weathering intensity and hence the paleoclimate change.

2. Sr isotopic ratios in acid-leachates are higher for paleosol than loess in the loess-paleosol profile, varying with about 100 kyrs periodicity, and thus corresponds to changes of the global glacial cycle and is comparable with the variation of Sr isotopic composition of paleoseawater.

AKNOWLEDGEMENT

This research was jointly granted by the National Outstanding Young Scientists Foundation, the State Climbing Program (95-P-39).

REFERENCES

Capo, R. C. & D. J. DePaolo 1990. Seawater strontium isotopic variations from 2.5 million years ago to the present. *Science* 249:51-55.

Chen, J., H. Wang & H. Lu 1996. Behaviours of REE and other trace elements during pedological weathering— evidence from chemical leaching of loess and paleosol from the Luochuan section in central China. *Acta Geologica Sinica* 70(1):61-72(in Chinese).

Clemens, S. C., J. W. Farrell & L. P. Groment 1993. Synchronous changes in seawater strontium isotope composition and global climate. *Nature* 363:607-610.

Dia, A. N., A. S. Cohen & R. K. O'Nions 1992. Seawater Sr isotope variation over the past 300 kyr and influence of global climate cycles. *Nature* 356:786-788.

Li, C., W. Lin & Z. Ouyang 1993. Microtektites and glassy mirospherules in loess: Their discoveries and implications. *Science in China, series B* 36:1141-1152.

Liu, C.-Q., C.-L. Li, Z.-Y. Ouyang & J. Zhang 1996. Linkage between climate change and carbonate strontium isotopic composition of Loess-paleosol deposits in China. *Abstr., 30th Int. Geol. Congr., Beijing, China* 1: 78.

Liu, C.-Q., A. Masuda & A. Okada 1994. Isotope geochemistry of Quaternary deposits from the arid lands in northern China. *Earth Planet Sci Lett* 127:25~38.

Liu, T. 1985. *Loess and The Environment*. Beijing: China Ocean Press.

Lu, Y C. 1981. Pleistocene climate cycles and variation of CaCO3 contents in a loess profile. *Scientia Geologica* 2: 122~131(in Chinese).

Geochemistry of the Earth's Surface, Ármannsson (ed.) © 1999 Balkema, Rotterdam, ISBN 90 5809 073 6

Quantitative effects of plants on weathering: Forest sinks for carbon

K. L. Moulton & R. A. Berner

Department of Geology and Geophysics, Yale University, New Haven, Conn., USA

ABSTRACT: The weathering of calcium and magnesium silicates on the continents has exerted a major control on atmospheric CO_2 over geologic time, and vascular plants may have played an important role in this process. Plants may also play a key role in the short term removal of C from the atmosphere to rivers and oceans. We have examined the role of plants in weathering in western Iceland by measuring the chemistry of waters draining adjacent areas of basaltic rocks that are either bare (with a partial cover of mosses and lichens) or forested. These areas have minimal differences in microclimate, slope, and lithology, and avoid hydrothermal waters and anthropogenic acid rain. Results include cation uptake by trees and indicate that the rate of weathering release of Ca and Mg to streams is 4 times higher in forested areas than bare areas, suggesting a major role for vascular plants in accelerating weathering and lowering atmospheric CO_2.

1 INTRODUCTION

The precipitous rise of atmospheric CO_2 during the past century as a result of burning of fossil fuels has led to an increased interest in the processes by which CO_2 is added to and removed from the atmosphere on both the short- and long-term time scales. One such process is silicate weathering. The weathering of silicate rocks on the continents results in the uptake of CO_2, thus controling its atmospheric concentration over millions of years. Plants are thought to affect this process by accelerating silicate weathering; hence, the evolution of land plants over geologic time may have influenced past levels of CO_2 in the atmosphere (Berner 1998).

Rooted vascular plants are expected to accelerate rock weathering for several reasons. Plant rootlets and associated symbionts secrete organic acids and chelates that attack primary minerals during the uptake of nutrient cations (e. g. Griffiths et al., 1994), and decay of organic matter produces organic acids and carbonic acid which provide additional acid for the attack on minerals. Plants also increase the recirculation of water by evapo-transpiration, which results in greater rainfall than would occur otherwise (Shukla and Mintz 1982) increasing soil solution flushing rates and replacing saturated solution with fresh solution whereby weathering may occur with renewed vigor. Plants and associated microbes anchor clay-rich soil against erosion, and hydrophilic organic matter and fine-grained material allow retention of water and continued weathering of

primary minerals between rainfall events (Drever 1994). Plant roots also fracture the substrate, exposing fresh surfaces to weathering and effecting changes in soil porosity, allowing for more effective water flow around primary minerals. In addition, plants affect the temperature of the soil beneath the vegetation cover; decay of vegetation may produce heat, while the forest canopy shades the soil and provides a cooling effect. Finally, plants are not 100 % efficient at recycling nutrients, which may leave the watershed in runoff or ground water (Knoll and James 1987).

The flux of solutes leaving the watershed in runoff and ground water is a sensitive indicator of the extent of silicate rock weathering (Drever 1997, Garrels 1967, Gíslason et al. 1996), and HCO_3^- may be used as a direct measure of atmospheric CO_2 uptake during silicate weathering in the absence of carbonate rocks. In this paper we quantify the effect of plants on silicate weathering and its implications for forest sinks of atmospheric C.

2 FIELD AREA AND METHODS

The field area at Skorradalur contains small (0.2 to 6 km^2) drainage basins of varying vegetational and soil cover which are all underlain by basalt. The areas with trees (conifers to the north of the lake and native birch to the north and south) are part of a protected region which has been administered by the Icelandic forest service since the early 1900s. Soils at Skorradalur are developed in glacial till composed of poorly sorted basaltic material. Soil thickness in the vegetated plots ranges from 100 to 150 cm thick. The barren area of Thvergil exhibits soil thickness from 0 to 150 cm, while the barren area on Dragafell has no soil or till. All areas contain sporadic mosses and lichens.

The relationships between the processes controling solutes in the study areas are expressed by the following equation (Moulton & Berner 1998):

$$F_{weathering} = F_{stream} - F_{input} + F_{tree\ storage} + F_{litter} \quad (1)$$

The stream term is given by the water chemistry of streams draining the area, the input term is derived from data on precipitation corrected for evapo-transpiration by normalization to Cl⁻, and the tree storage term is calculated from data on tree composition, size, age, and density. The litter storage term is neglected because it is small relative to the tree storage term. Storage in soils is also neglected.

3 RESULTS

The results of flux and tree storage calculations are shown in Table 1. The negative fluxes of K^+ and SO_4^{2-} reflect a sink for these solutes (uptake by trees?) without concomitant contribution by rock weathering. The principle source of these ions is input from precipitation and aerosols.

Ratios of vegetated to bare, no soil, weathering fluxes are shown in Table 2. The difference in element mobility between nutrient cations (K^+, Ca^{+2}, and Mg^{+2}), and bicarbonate, Na^+, and Si may be the result of inefficient scavenging of nutrient cations lost by leaching from vegetation and litter.

4 DISCUSSION

Gymnosperms and angiosperms are expected to affect the underlying soil chemistry and weathering regime differently because of differences in growth rates and effects on soil chemistry.

The native Icelandic downy birch grows more slowly than the imported conifers at Skorradalur; 33 yr old conifers have about twice the dry mass of birch of the same age. This is not representative of forests in other parts of the world where conifers typically grow more slowly than angiosperms.

Conifers aggressively leach the soil by secreting tannins and fatty acids which bind with nutrients and are flushed out of the soil (Salisbury and Ross, 1992). Angiosperms such as birch trees generally cannot compete in such a low-nutrient environment against the conifers, which have low nutrient requirements and nutrient-scavenging mycorrhizae. The high acidity and leaching associated with conifers suggest that they might have a more pronounced silicate-weathering effect on a per mass basis than birches. However, evergreen conifers moderate nutrient cycling more efficiently than deciduous angiosperms (Day & Monk 1977). Evergreens lose leaves continuously throughout the year, while deciduous trees drop their leaves at one time and may suffer severe ion loss from the litter by runoff or percolation beyond the rooting zone when they are not actively acquiring nutrients (Thomas and Grigal, 1976). Also, evergreen litter releases nutrients more slowly upon decay than deciduous litter, allowing for a more gradual addition of ions to the soil through litter decay (Monk, 1966, 1970). The recycling efficiency of the deciduous trees should therefore be less than that of the evergreens, and might result in greater loss of nutrient cations in runoff.

Weathering fluxes normalized to biomass are shown in Table 3. These data indicate that the birch forests weather more silicate rock per unit dry mass than the conifer forest. This is in agreement with the previous study of Likens et al. (1977) which showed that on a world-wide basis, temperate-region angiosperm forests lose four times more K and Ca and three times more Mg annually than conifer forests, while Na losses are similar.

In addition to the effect plants have on the multimillion-year carbon cycle, the presence of vegetation affects the short term carbon cycle, which involves the transfer of CO_2 between the atmosphere, oceans, soils, and biosphere, and operates on the time scale of years to tens of thousands of years. The oceans and the terrestrial biosphere plus soils are both much larger carbon reservoirs than the atmosphere, and exchange carbon rapidly with the atmosphere on the order of years to hundreds of years.

Estimates of carbon storage in biomass and leakage from the Skorradalur forests are shown in Table 4. The leakage of carbon out of the system is calculated as the difference between the HCO_3^- weathering fluxes of the vegetated and unvegetated (no soil) areas. The results of the present study suggest that the addition of HCO_3^- to the hydrosphere via secretion of acids and weathering by plants is an additional and perpetual (for the lifetime of the forest) short-term mechanism by which plants remove CO_2 from the atmosphere to the soil and ocean reservoirs. However, the large uncertainties in these estimates suggest the need for rigorous study of the HCO_3^- flux from forests before assigning numbers to mandates calling for the planting of forests as a sink for C emitted by anthropogenic sources, such as was suggested at the Kyoto conference on global warming in 1998.

The results presented here represent a minimum weathering effect by the trees because of the slow growth rate of trees in Iceland. Forests in warmer

Table 1. Results from stream flux and tree storage calculations.

	Stream Flux (mol/yr ha)							Tree Storage (mol/yr ha)		
	SO_4^{-2}	HCO_3^-	Na^+	K^+	Mg^{+2}	Ca^{+2}	Si	K	Mg	Ca
Bare, soil	-49	353	138	-53	49	127	320	n.a.*	n.a.	n.a.
Bare, no soil	-8	362	109	-19	39	96	287	n.a.	n.a.	n.a.
S. Birch	-98	911	109	-86	158	268	646	8	4	22
N. Birch	-55	985	158	-48	167	252	679	8	5	21
Conifer	-63	999	217	-59	164	277	756	13	7	35

*n.a. = not applicable

Table 2. Ratios of vegetated to unvegetated (no soil) weathering fluxes.

	HCO_3^-	Na^+	K^+	Mg^{+2}	Ca^{+2}	Si
S. Birch/Bare, no soil	3	1	4	4	3	2
N. Birch/Bare, no soil	3	1	2	4	3	2
Conifer/Bare, no soil	3	2	2	4	3	3

Table 3. Weathering fluxes of forested watersheds, normalized to biomass.

	Biomass	Weathering flux (mol/(yr kg) x 10^3)					
	(Kg/ha)	HCO_3^-	Na^+	K^+	Mg^{2+}	Ca^{2+}	Si
S. Birch	1679	543	65	-50	96	173	385
N. Birch	1698	580	93	-23	101	161	400
Conifer	2620	381	83	-18	65	119	289

Table 4. Carbon storage in biomass and leakage from forests at Skorradalur.

	C storage (mol/(yr•ha))	C leakage (mol/(yr•ha))
South birch	1398 ± 182	549 ± 566
North birch	1927 ± 289	623 ± 623
Conifer	2975 ± 446	637 ± 461

climates or with substrates other than basalt may differ from the estimates for carbon storage and leakage presented here because of differences in growth rates of the trees. Faster-growing trees would become larger during their life-span and thus store more carbon, and their higher nutrient turnover might result in greater leakage of nutrients and carbon from the ecosystem.

5 CONCLUSIONS

The results of this investigation support the suggestion that the rise of deeply-rooted vascular plants on the continents during the Devonian exerted a major control on atmospheric CO_2 by at least a four-fold enhancement of the weathering of calcium-magnesium-silicates. Of particular importance is the leakage of nutrient cations from vegetated areas and concomitant replacement by weathering, especially in the case of deciduous angiosperms.

These results also suggest a perpetual short term sink for atmospheric C as leakage of HCO_3^- from the forest ecosystem to rivers and streams, in addition to the finite sink of C in forest biomass. This is of importance in consideration of protocol which calls for CO_2 emmiters to plant forests in order to provide an equivalent sink for C which they have contributed to the atmosphere.

REFERENCES

Berner, R. A. 1998. The carbon cycle and CO_2 over Phanerozoic time: the role of land plants: serial B, *Royal Society of London Philosophical Transactions* 353: 75-82.

Day, F. P. & C. D. Monk 1977. Seasonal nutrient dynamics in the vegetation on a southern Appalachian watershed: *American Journal of Botany* 64: 1126-1139.

Drever, J. I. 1994. The effect of land plants on weathering rates of silicate minerals: *Geochimica et Cosmochimica Acta* 58: 2325-2332.

Drever, J. I. 1997. *The geochemistry of natural waters*, 3rd edition: Englewood Cliffs, New Jersey: Prentice-Hall, 436 p.

Garrels, R. M. 1967. Genesis of some ground waters from igneous rocks, *in* , P. H Abelson.(ed), *Researches in geochemistry* 2. New York: Wiley.

Gíslason, S. R., S. Arnórsson & H. Ármannsson 1996. Chemical weathering of basalt in southwest Iceland: Effects of runoff, age of rocks and vegetative/glacial cover: *American Journal of Science* 296: 837-907.

Griffiths, R. P., J. E. Baham & B. A. Caldwell 1994. Soil solution chemistry of ectomycorrhizal mats in forest soil: *Soil Biology and Biochemistry* 26: 331-337.

Knoll, M. A. & W. C. James 1987. Effect of the advent and diversification of vascular land plants on mineral weathering through geologic time: *Geology* 15:. 1099-1102.

Likens, G. E., F. H. Bormann, R. S. Pierce, J. S. Eaton & N. M. Johnson 1977. *Biogeochemistry of a forested ecosystem*: New York: Springer-Verlag.

Monk, C. D. 1966. An ecological significance of evergreenness: *Ecology* 47: 504-505.

Monk, C. D. 1970. Leaf decomposition and loss of ^{45}Ca from deciduous and evergreen trees: *American Midland Naturalist* 86: 379-384.

Moulton, K. L. & R. A. Berner 1998. Quantification of the effect of plants on weathering: Studies in Iceland. *Geology* 26: 895-898.

Salisbury, F. B. & C. W. Ross 1992. *Plant Physiology*, 4th edition: Belmont, California: Wadsworth Publishing Company.

Shukla, J. & Y. Mintz 1982. Influence of land-surface evapo-transpiration on the Earth's climate: *Science* 215: 1498-1501.

Thomas, W. A. & D. F. Grigal 1976. Phosphorus conservation by evergreenness of mountain laurel: *Oikos* 27: 19-26.

Geochemistry of the Earth's Surface, Ármannsson (ed.)© 1999 Balkema, Rotterdam, ISBN 90 5809 073 6

Salt accumulations in ancient Antarctic soils

D.S. Sheppard, K.M. Rogers, I.J. Graham & R.G. Ditchburn
Institute of Geological and Nuclear Sciences, Lower Hutt, New Zealand

G.G.C. Claridge & I.B. Campbell
Institute of Geological and Nuclear Sciences, Lower Hutt, New Zealand & Land and Soil Consultancy Services, Stoke, New Zealand

ABSTRACT: The anions in salts in ancient soils are derived from atmospheric fallout over the last 15 million years or less. The salt compositions are related to altitude and distance from open sea, and so a climatic record may be preserved in these profiles. ^{10}Be dating techniques have been developed and are confirming the ages derived by other methods. The isotopic compositions of nitrate are most unusual and may reflect a unique pathway from a distant open ocean source.

1 OLD ANTARCTIC SOILS

The Antarctic continent is almost completely ice-covered with a very small ice-free area. The largest such area is in the McMurdo Sound region and has an area of about 5000 m^2 (Figure1). The ground remains free of ice or snow cover because of the extreme aridity of the area. On the exposed ground weathering and soil formation can occur (Campbell & Claridge 1987)

According to Barrett (1996) continental deglaciation began in Antarctica between 34 and 15 million years ago and ice sheets have covered the continent continuously since then, although over the last 2 million years there have been fluctuations in accord with world-wide glaciations. Much of the present day topography of Antarctica was cut during the initial stages of the onset of ice cover and many upland areas have remained free of ice cover since that time (Prentice et al. 1998).

Tills emplaced by retreating glaciers in upland valleys 1100 to 2000 m above the larger valleys occupied or formerly occupied by glaciers draining the inland ice sheet have been largely unaffected by recent (< 2 Ma) changes in the levels of these glaciers, although at times ice has flooded into the mouths of the valleys as the volume of the continental ice-sheet has increased. This invasion may be likened to the movement of the tracks of a tracked vehicle which rolls over the surface without disturbing the material. As the retreat of the ice is largely by ablation i.e. no liquid phase is necessarily involved, rock material is deposited on top of pre-existing tills and soluble salts in the pre-existing tills are not necessarily mobilised by any liquid water phase.

The ages of soils can be estimated using the extent of weathering (Campbell & Claridge 1975): other techniques for attributing ages rely on inclusion of datable local volcanic ashes (e.g. Marchant et al. 1993), salt content and geomorphic relationships to other features (e.g. moraine ridges).

In this paper we will discuss one soil profile taken from a small high altitude valley in the Quartermain Mountains (Figure 1). Site 829 was excavated to about

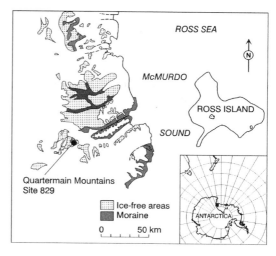

Figure 1. Location of the sampling site.

1.2 m depth and sampled largely by measured interval. The soil at this site is considered, on the basis of weathering, to be > 10 Ma old and contains at least two paleosol surfaces (at c. 65 cm and 100 cm depths).

Salts are found in the soils in large quantities. The anionic component of the salts are believed to be dominantly atmospherically derived, and of a composition which varies with altitude and distance from open sea. They thus may serve as indicators of past environments and processes at the atmospheric/land surface interface. In order to determine if they can be used in this way, we need an improved understanding of the sources of the salts, their ages of emplacement and that of the soils in which they exist and an appreciation of the mechanisms by which they are remobilised within the soils.

2 SALT COMPOSITIONS

2.1 Chemical

The salts are dominantly Na and Ca chlorides, sulphates and nitrates, with $NaNO_3$ increasingly dominant inland (Campbell & Claridge 1987).

The dominant salt anion concentrations in the soil profile 829 are shown in Figure2. There is wide variation in concentrations throughout the profile, the dominating features being the high levels at about 10 cm depth, and the steady decline in total anion concentration with depth. The anion concentrations show peaks between 30 to 45 cm depth and 50 to 70 cm, as well as that at 10 cm. The paleosol surface at 65 cm may be reflected in the lower anion peak: there is no evidence in the anion concentrations of the lower paleosol but any associated salt peak may be deeper than the profile excavated. A combination of factors such as movement of the salts within the soil, coarse sampling intervals, and sampling bias would contribute to failure to identify salt accumulations associated with specific paleosol surfaces.

Anion ratios vary widely, particularly those involving nitrate. The question to be addressed is whether the distribution reflects the compositions at the time of deposition from the atmosphere, or subsequent redistributive processes. That the salts are redistributed is suggested by the shape of the anion distribution profile.

2.2 Isotopic compositions

The isotopic compositions of the water-soluble nitrate and sulphate have been measured and the results are shown on Figure 3. Immediately apparent is that both species have quite abnormal compositions nitrate $\delta^{15}N$ at -12 to -20 per mil is more negative than for any other natural measured nitrates except for other Antarctic soils and biota

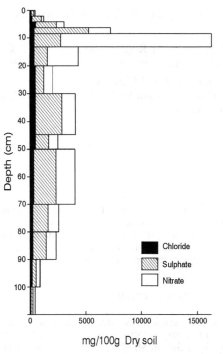

Figure 2. Major anion composition of soil in profile of site 829.

(Wada et al. 1981). The composition is significantly different from that of Chilean and Californian desert nitrate deposits (Böhlke et al. 1997) and northern hemisphere precipitation. The $\delta^{18}O$ in the nitrate has a similar composition to that of atmospheric oxygen, which suggests that the nitrate is produced by the interaction of marine-derived proteins and subsequent atmospheric oxidation, as proposed by Claridge & Campbell (1968). The different range of the compositions, compared with those for Atacama desert nitrates, may be a function of greater travel distances which provides opportunities for reactions to take place.

The sulphate isotopic composition similarly is quite different from that in other environments. Sea-spray sulphate $\delta^{18}O$ is approximately 20 per mil higher than found in the soils. While biological activity could reduce this difference, it is unlikely to be significant in these conditions. Antarctic precipitation has a $\delta^{18}O$ of –20 to –40 per mil which suggests some interaction between seawater sulphate and local snow. Lyon (1978) has shown highly depleted $\delta^{18}O$ (-33 per mil) in the water of crystallisation of gypsum in the nearby Miers Valley lake deposits: however the interchange of oxygen between water and sulphate molecules is an extremely slow process even at elevated temperatures.

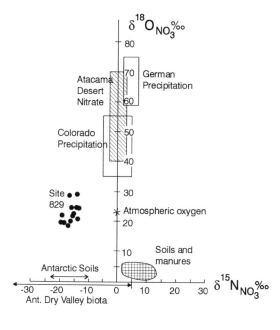

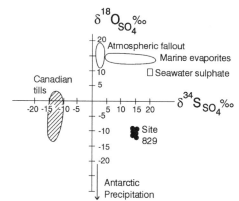

Figure 3. Nitrate and sulphate isotopic compositions compared with possible sources.

The sulphate $\delta^{34}S$ is similar to the values for marine sulphur and other Antarctic evaporites.

3 COSMOGENIC ISOTOPE DATING

Two methods of dating geological materials using the radioisotope ^{10}Be are possible – that using the ^{10}Be in exposed rock surfaces (surface exposure dating) and that using ^{10}Be derived from atmospheric fallout. The former has given ages in Antarctica which are too young, as the technique

does not always adequately allow for the degree of weathering and other surface disturbances such as the rolling over of the rocks.

Much greater quantities of ^{10}Be are produced in the upper atmosphere as compared to surface rocks (Lal & Peters 1967) and this is rapidly removed from the atmosphere by precipitation, probably in soluble form (McHargue & Damon 1991).

We have analysed archived and recently collected samples from a number of locations in the Ross Sea Region for salts and ^{10}Be (Graham et al. in prep.). Determining the age of the soils from the measurements requires assumptions about ^{10}Be

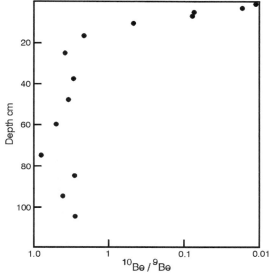

Figure 4. Variation of $^{10}Be/^9Be$ in soils in the site 829 profile.

fluxes and redistributive processes within the soils. For the site 829 profile the initial slope of $^{10}Be/^9Be$ (Figure 4) best defines the age of the uppermost surface of > 11.3 Ma, quite consistent with inferred geomorphic ages of 11 Ma (Sugden et al. 1995).

In addition to the derived age, the ^{10}Be distribution within the profiles indicates that active downward transport of atmospherically derived salts within the soil profiles.

4 CONCLUSIONS

Chemical, stable and cosmogenic isotope measurements strongly indicate that salts are redistributed downwards in the Antarctic soils. ^{10}Be dating techniques have been adapted and confirm the ages derived by other methods. The isotopic compositions of nitrate in one site are most unusual and can be interpreted to indicate transport from a distant open ocean source. The sulphur in sulphate is clearly from sea-salt, but the oxygen isotopic composition of the sulphate is not so clearly related to this source and may reflect interaction with local precipitation.

Data from a wider range of sites is required to better determine the processes and sources of salts in the soils.

REFERENCES

Barrett, P.J. 1996. Antarctic paleoenvironment through Cenozoic times – a review. *Terra Antarctica* 3: 103-119.

Böhlke, J.K., G.E. Ericksen, & K. Revesz 1997. Stable isotope evidence for an atmospheric origin of desert nitrate deposits in northern Chile and southern California, U.S.A. *Chem. Geol.* 136: 135-152.

Campbell, I.B. & G.G.C. Claridge 1975. Morphology and age relationships of Antarctic soils. In Suggate P. & Cresswell, M.M. (eds) *Quaternary studies.* N.Z. Royal Society Bull. 13: 83-88.

Campbell, I.B. & G.G.C. Claridge 1987. *Antarctica: soils, weathering processes and environment.* Elsevier: Amsterdam.

Claridge, G.G.C. & I.B. Campbell 1968. The origin of nitrate deposits. *Nature* 217 428-430.

Graham, I.J., R.G. Ditchburn, G.C.C. Claridge, D.S. Sheppard, & N.E. Whitehead, (in prep.) Dating Antarctic soils with atmospherically-produced [10]Be.

Lal, D. & B. Peters B. 1967. Cosmic-ray produced radioactivity on the earth. In: *Handbook der Physik 4/2.* Berlin: Springer Verlag.

Lyon, G.L. 1978. The stable isotope geochemistry of gypsum, Miers Valley, Antarctica. In Robinson, B.W. (ed), *Stable Isotopes in the Earth Sciences.* DSIR Bull. 220: 97-103.

Marchant, D.R., C.C. Swisher, D.R. Lux, D.P. West & G.H. Denton 1993. Pliocene paleoclimate and East Antarctic ice-sheet history from surficial ash deposits. *Science* 260: 667-670.

McHargue, L.R.& P.E.Damon 1991. The global beryllium-10 cycle. *Rev. Geophys.* 29: 141-158.

Sugden, D.E., G.H. Denton & D.R. Marchant 1995. Landscape evolution of the Dry Valleys, Transantarctic Mountains: tectonic implications. *J.Geophys.Res.*100: 9949-9967.

Prentice, M.L., K. Kleman & A.J. Stroeven 1998. The composite glacial erosional landscape of the northern McMurdo Dry Valleys: implications for Antarctic Tertiary glacial history. In Priscu, J.C. (ed.) *Ecosystem Dynamics in a Polar desert: the McMurdo Dry Valleys, Antarctica.* Am. Geophys. Union Antarctic Research Series 72: 1-38.

Wada, E., R. Shibata, & T. Torii 1981. [15]N abundance in Antarctica: origin of soil nitrogen and ecological implications. *Nature* 292: 327-329.

Geochemistry of the Earth's Surface, Ármannsson (ed.)© 1999 Balkema, Rotterdam, ISBN 90 5809 073 6

Sulfur isotope study of the source materials of the aeolian dust of inland Asia

S. Yabuki – *Institute of Physical and Chemical Research, Wako, Japan*

A. Okada – *Japan Science and Technology, Kawaguchi, Japan*

A. Ueda – *Mitsubishi Material Corporation, Ohmiya, Japan*

Q. Chang – *Kumamoto University, Japan*

ABSTRACT: In saline lands around the desert areas in Xinjiang, NW China, saline soils contain significant amount of sulfate salt minerals. We investigate the sulfur isotope ratios of the salt materials to study the salt accumulation in the environment. In Xinjiang, sulfate salts are accumulated on the ground surface owing to the very arid condition, thus a large amount of sulfur-bearing aeolian dust are transferred from the ground into the atmosphere due to the frequent dust storm events. We suggest that the sulfur isotope ratios of the air-borne dust particles can be used as an indicator to trace the origin of aeolian dust of inland Asia.

1 INTRODUCTION

The dust storm is a common phenomenon that occurs frequently in arid and semi-arid lands. Xinjiang, NW China is one of the major frequent dust storm outbreak regions of inland Asia. Dust storms generated most frequently in spring in and around the desert areas in Tarim and Zhungar Basins. Particles ranging from submicron to micron in size are blown up to the upper atmosphere, and are transported long range to Japan and Central and North Pacific regions by the strong west winds. Such wind-blown particles surely have strong effects not only on formation of soil and crop growth but also on ocean sedimentation, air pollution and climate change. The global annual input of mineral dust to the atmosphere is estimated to be 1000 – 2000 Mt/year (Duce 1995), and the annual input of sulfate sulfur to the atmosphere from soil dust is estimated to be 3 – 20 Mt sulfur /year (Andreae 1990). In Xinjiang, NW China, sulfate salts have accumulated on the ground surface owing to the extreme arid climate, and thus a large amount of sulfur-bearing aeolian dust are transferred from the ground to the upper atmosphere due to the frequent dust storm events. Therefore the sulfur isotope ratio of air-borne dust particles is expected to be a useful indicator to investigate the source of aeolian dust.

2 SAMPLING SITE

Salt material and soil samples were collected in and around the saline lands at the margins of the desert areas in Xinjiang, NW China (Fig. 1) during 1987 –

1994 (Yabuki et al. 1996). Tarim Basin is an arid, closed basin occupied mostly by Taklimakan Desert, the largest sand desert in China. The annual precipitation is <50mm and evaporation is from 2000 to 3000mm. Zhungar Basin is surrounded by Tianshan and Artai Mountains. The basin is opened toward northwest and moist currents come from this direction. Therefore the climate of Zhungar Basin is relatively moderate compared with other desert areas in Xinjiang. Turpan Basin is an intermontane basin within the eastern Tianshan Mountains, known as one of the most dried and high temperature region in China, with less than 20mm of annual precipitation and 3000mm evaporation. A salt lake called Aidinghu is located at the deepest part, 154m below sea level.

3 RESULTS AND DISCUSSION

3-1 *Regional characteristics of sulfur isotope ratios of salt materials in Xinjiang*

Sulfur isotopic composition ($\delta^{34}S$) of salt materials collected in Xinjiang have wide range from -30.3 to +26.1‰, with most frequent values between +5 and +10‰ (Fig. 2). The regional characteristics of the sulfur isotope data are summarized as follows.
1. In southern Zhungar Basin area $\delta^{34}S$ values, mostly ranging from +1.6 to +4.9‰, are inclined to be lower than those of Tarim Basin and Turpan Basin. This is probably due to the fact that the salt precipitates in these areas are strongly influenced by the hydrological and geological conditions of the northern part of the Tianshan Mountains.

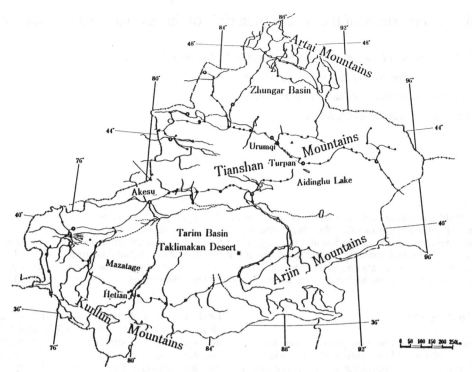

Figure 1.   A geographic map of Xinjiang area including sampling sites.

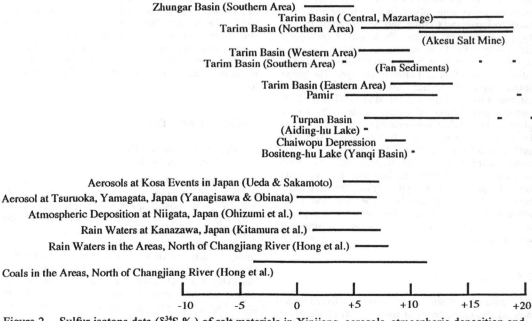

Figure 2.   Sulfur isotope data (δ^{34}S ‰) of salt materials in Xinjiang, aerosols, atmospheric deposition and rain waters in Japan, rain waters and coals in the areas, north of Changjiang River.

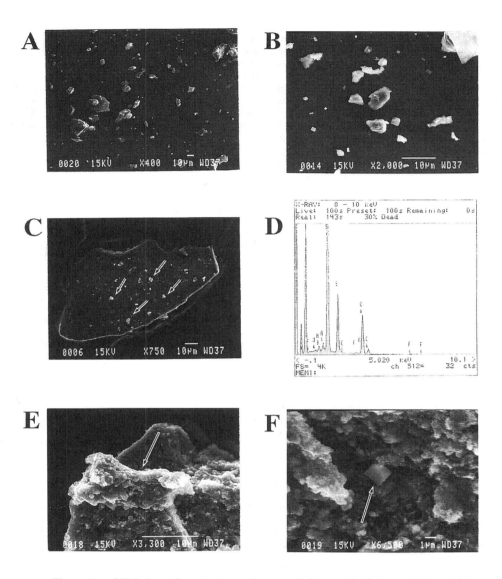

Figure 3. SEM (scanning electron microscope) image of air-borne dust particles trapped on the adhesive tapes on the slide glass.

A: Dust particles collected at Hetian, southern Tarim Basin.

B: A magnified view indicates the presence of minute single grains of salt, euhedral in shape, Hetian.

C: A quartz grain bearing sub-micron calcium sulfate particles on the surface (arrows), southern Tarim Basin.

D: The EDX analysis indicates the particles in Photo C are $CaSO_4$.

E: A magnified view of Na_2SO_4 aggregate (arrow), Ruoqiang, eastern Tarim Basin.

F: Occurrence of micron-sized NaCl crystal (arrow) in the hollow of the Na_2SO_4 aggregate, Ruoqiang.

2. Diluvial-alluvial fan sediments at southern Tarim Basin which extends from the northern foot of the Kunlun Mountains has relatively constant $\delta^{34}S$ values from +8.4 to +10.1‰.

3. Both of salt deposit at Akesu Salt Mine located at the northern margin of Taklimakan Desert and the gypsum-anhydrite deposit at Mazartage at central Tarim Basin have high $\delta^{34}S$ values between +10.7 and +17.8‰. Geological history of Xinjiang area indicates that the west part of the Tarim Basin experienced a marine transgression in the Cretaceous to early Tertiary (Tang et al. 1989). Both the salt deposits at Akesu and Mazartage Mountains were formed due to the transgression during the period. Both salt deposits are distributed in early Tertiary strata, and the $\delta^{34}S$ value of salt samples almost agrees with the sulfur isotope curve of marine evaporites at same geological period (Claypool et al. 1980).

4. In Turpan Basin, the $\delta^{34}S$ values vary from site to site in the basin, but finally approach +6.0‰ at Aidinghu salt lake, into which the brines in the basin flow.

3-2 *Salt mineral species*

Our investigation of atmospheric dust particles in the areas around Taklimakan desert showed that minute Na- and Ca- bearing sulfate salts, probably corresponding to thenardite (Na_2SO_4) and gypsum ($CaSO_4 \cdot 2H_2O$), and halite (NaCl) crystals are present not only as individual grains but also as micronsized to submicron-sized grains on the surface of silicate and silica dust particles in the atmosphere (Fig. 3). In the dust aerosol samples collected in Taklimakan Desert, the presence of halite is also confirmed by the electron microscopic study (Okada et al. 1997). These facts suggest that the sulfur-species, which accumulate on the ground surface in Xinjiang, could become significant sources of sulfur in the aerosol.

Ueda and Sakamoto (1989) measured sulfur isotope ratios of the "Kosa" aerosol sampled in Japan at the time of "Kosa" phenomena, and obtained $\delta^{34}S$ values of the non-sea-salt sulfate components ranging from +4.1 to +7.6‰. As seen in Fig. 2, it is interesting that these values partly overlap with the range found in the $\delta^{34}S$ distribution of salt samples from Xinjiang. It is suggested that the mixing of sulfur species emitted by human activity influences the aerosols of Asian continental origin. $\delta^{34}S$ values of the aerosol particles which collected in Japan are generally inclined to be lower compared with those of salt samples from Xinjiang area. This could be because aeolian dust including "Kosa" sampled near the surface of the ground are more or less influenced by anthropogenic pollution materials during deposition onto the ground. The recent remote sensing investigations of the Asian dusts using lidar and satel- lite observations have revealed that the dust particles are blown up high in the troposphere, 2000 – 8000 meters above sea level. These facts indicate that the aeolian dust particles in the upper troposphere are hardly influenced by the anthropogenic during long-range transportation. It is promising to trace the dust-supplying sources by sulfur isotope studies of atmospheric dust particles collected at high altitude.

REFERENCES

Andreae, M.O. 1990. Ocean-atmosphere interaction in the global biogeochemical sulfur cycle. *Mar. Chem.* 30: 1-29.

Claypool, G. E. et al. 1980. The age curves of sulfur and oxygen isotopes in marine sulfate and their mutual interpretation. *Chem. Geol.* 28:199-260

Duce, R.A. 1995. Sources, distribution, and fluxes of mineral aerosols and their relationship to climate. In R. J. Charlson, and J. Hentzenberg, (ed), *Aerosol forcing of climate*: 43-72. New York: John Wiley & Sons.

Hong, Y. et al. 1992. Compositional characteristics of coal in China and..: fractionation of sulfur isotopes during coal-burning process. In *Institute of Geochemistry, Chinese Academy of Sciences, Development in Geochemistry*: 241-250. Beijing: Seismological Press,

Kitamura, M. et al. 1993. An estimation of the origin of sulfate ion in rain water in view of sulfur isotopic variations. *Chikyukagaku (Geo-chemistry)* 27:109-118.

Ohizumi, T. et al. 1991. Source of sulfur in the atmospheric deposits in view of sulfur isotopic variations –A case study in Niigata Prefecture, Japan. *Nippon Kagakukaishi* 44: 675-681.

Okada, A. et al. 1997. Salt efflorescent materials in saline lands of Xinjiang, China. *J. Arid Land Studies* 7: 53-67.

Ueda, A. & K. Sakamoto 1989. Isotope ratio analysis – Sulfur isotopes. In *Research Association of Air Pollution, Report of the Investigation on the "Kosa" Aerosol*: 30-34

Yabuki, S. et al. 1996. The behaviors of ions from soluble salts in inland waters around the desert area, Xinjiang, China. *J. Arid Land Studies* 5: 195-216.

Yanagisawa, F. & H. Obinata 1997. Sulfur isotope ratio of non-sea-salt sulfate in dry deposition in Tsuruoka and Yamagata, Yamagata-Prefecture, Japan. *J. Arid Land Studies* 7:119-126.

Geochemistry of the Earth's Surface, Ármannsson (ed.)© 1999 Balkema, Rotterdam, ISBN 90 5809 073 6

Sulfur and strontium isotope study of some gypsum samples from southern Taiwan

Hsueh-Wen Yeh & Chen-Hong Chen
Institute of Earth Science, Academia Sinica, Taipei, Taiwan

Kai-Shuan Shea
Central Geological Survey, MOEA, Taipei, Taiwan

ABSTRACT: Gypsum is widespread in Pleistocene clastic sediments of southern Taiwan. The sulfur isotope data indicate that the gypsum samples are not of evaporite but terrestrial origin. The strontium isotope data reveal additional source(s), besides the Pleistocene seawater, of the strontium and the calcium. The probable modes of formation of the gypsum include a consequence of reverse diagenesis, of acid rain impact or of a combination of the above.

1 INTRODUCTION:

Gypsum has been found in many outcrops of Pleistocene, with one exception of Pliocene, rocks in southern Taiwan. The outcrops scatter randomly but somewhat evenly within an elongated (c. 50 by 120 km), northwest to southeast trending strip in Focthill Providence of Taiwan (Ho 1986). The northern half of the strip constitutes the area of this study (Figure1).

The occurrence of gypsum in clastic sediments other than as a member of an evaporite is not very common (Blatt et al. 1972, Pettijohn 1975, Tucker 1981). It implies unusual elevation of activity product (Ca^{++}) (SO_4^{--}) and thus an unique geological process and /or an environmental setting (Engelhardt 1977, Tucker 1981, Blatt et al. 1972,Krauskopf & Bird 1995, Garrels & Christ 1965, Drever 1982). This study is an attempt to gain some insight into the mode(s) of formation of the mineral. The possible modes of origins include as a member of an evaporite, a consequence of weathering product of sedimentary pyrite and other sulfide and of acid rain; our preliminary S-isotope data, however, rule out the likelihood of evaporitic origin. We are updating the S-isotope results and presenting newly acquired Sr-isotope data

2 BACKGROUND:

Gypsum occurs almost exclusively in either muddy sandstone or sandy mudstone and either along the bedding planes or the joints. The size of the gypsum crystal varies but is mostly within 3 to 15 cm long. Physically the crystal ranges from semi-transparent to transparent (selenite).

Geologically, the sediments are Pleistocene and Pliocene marine sandstones and shales. Their sedimentary environments are either coastal or of inner continental shelf. The sediments, especially within and adjacent to the outcrops, contain abundant calcareous fossils of marine fauna.

3 SAMPLES AND METHODS:

Aliquots of mineral sample containing a few crystal aggregate, a single crystal or a few pieces of a broken crystal, depending on the situation, taken from each of the large bags of gypsum samples brought back from the outcrops were used for the study. The mineral samples were pulverized in an agate mortar and pestle. The powder samples were then used in S- and Sr-isotope measurements.

For S-isotope analysis, the powder samples were first processed to quantitatively produce SO_2 samples according to Breit et al. (1985) and Bailey & Smith (1972). The $^{34}S/^{32}S$ ratio of the SO_2 samples were then measured with IRMS and are reported as $\delta^{34}S_{CDT}$ values with reproducibility better than 0.2‰ (Yeh & Shea 1995).

For Sr-isotope analysis, the powder samples were dissolved and Sr separated for $^{87}Sr/^{86}Sr$ determination following the procedures of Lan et al. (1986). $^{87}Sr/^{86}Sr$ ratios of the Sr samples were determined in a thermal ionization MS. All the ratios have been corrected for mass fractionation by normalizing $^{88}Sr/^{86}Sr$ to 0.1194. The $^{87}Sr/^{86}Sr$ ratio of NBS-987 was 0.710239±38 at the time.

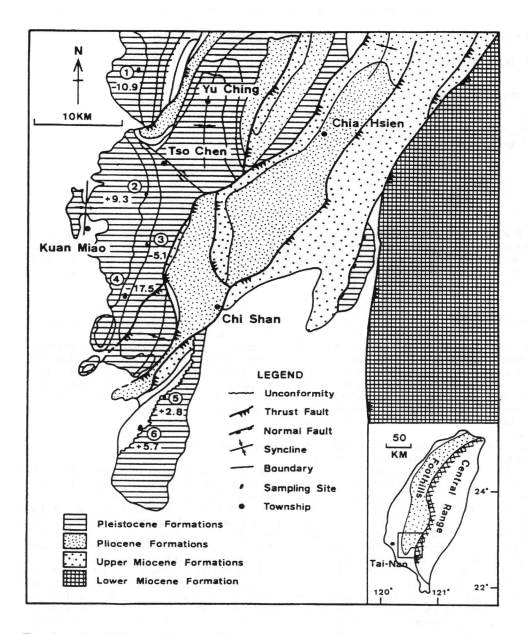

Figure 1.   General Geology of the area (modified after Lin, 1991) of this study and the sampling sites: (1) Liu Shuang Kng, (2) Wu Shu Lin, (3) Chu Chueh Hu Shan, (4) Ta Kang Shan Gulf Course, (5) Hsin I Gulf Course, (6) Kao Hsiung Kung Hsueh Yuan. The number beside the sampling sites are the δ^{34}SCDT values. See Table 1 for more information.

70

Table 1. δ^{34}SCDT and ^{87}Sr/^{86}Sr and some relevant information of the gypsum samples of southern Taiwan

Location	Stratigraphy	Sample I.D.	δ^{34}S$_{CDT}$(‰)	^{87}Sr/^{86}Sr
Liu Shuang Kng	Liu-Shuang Ts'eng (Pleistocene Sandstone/Mudstone)	PP-G-1	-10.9±0.0(2)	0.710166±14
Wu Shu Lin	Liu-Shuang Ts'eng (Pleistocene Sandstone/Mudstone)	PP-G-2	9.3±0.2(2)	0.71730±25
Chu Chueh Hu Shan	Erh-Chung-Ch'i Ts'eng(Pleistocene Sandstone/Mudstone)	PP-G-3	-5.1	0.70978±16
Ta Kang Shan Gulf Course	Erh-Chung-Ch'i Ts'eng(Pleistocene Sandstone/Mudstone)	PP-G-4	-17.5	0.710635±15
Hsin I Gulf Course	Lin-K'ou Conglomerate (Pleistocene)	PP-G-5	+2.8±0.0(2)	0.711862±17
Kao Hsiung Kung Hsueh Yuan	Nan-Shih-Lung Sandstone (Pliocene)	PP-G-6	+5.7±0.0(3)	0.709815±15

4 RESULTS AND DISCUSSION:

The results of this study are shown in Table 1 and Figure 1. δ^{34}S$_{CDT}$ values of the samples analyzed in this study range from –17.5 to +12.0 ‰. For comparison, δ^{34}S$_{CDT}$ value of seawater of Pliocene to modern ocean is quite uniform and around +20 ‰ (Sasaki 1972, Claypool et al. 1980, Longinelli 1989) and the values of SO$_4^{--}$ in many world's major rivers are mainly between +2 to +10‰ and subjected to temporal fluctuation (Hitchon & Krouse 1972, Longinelli & Cortecci 1970, Longinelli & Edmond 1983, Rabinorich & Grienko 1979, Schwarz & Cortecci 1974). It is clear that the gypsum samples are neither a member of evaporite nor a precipitate from seawater of the last one to two million years. The possible sources of the sulfur include the relevant sedimentary pyrite and related sulfide and the acid rain.

^{87}Sr/^{86}Sr ratios of the gypsum samples range from 0.70978 to 0.71862. The range of the ratios of the Pliocene to Modern seawater is between 0.7090 to 0.7092 (DePaolo & Ingram 1985). The results indicate that the unaltered contemporary seawater could not be the only source of Sr of the gypsum samples. This is consistent with the S-isotope data in

that they both rule out the possibility of the gypsum samples being a member of a marine evaporite. It can be argued that although major proportion of Ca and Sr of the samples probably derived from the calcareous fossils in the relevant rock formations, significant amount also came from older sediments of Taiwan.

5 SUMMARY AND CONCLUSIONS:

The S- as well as Sr-isotope data of the gypsum samples of southern Taiwan provide the conclusive evidence that the mineral is terrestrial and not marine. The sulfur of the gypsum samples is probably either derived from the relevant sedimentary sulfide such as pyrite etc. or from pollutant in acid rain. The source of Ca and Sr is probably mainly the ambient calcareous fossils.

We suggest that the formation of the gypsum follows a consecutive sequence of probable events. The initial event was the arrival of SO$_4^{--}$ solution during a wet "season". The solution then percolated through the sediments and dissolving some of the calcareous fossils on the way. The Ca concentrated SO$_4^{--}$ solution was stored in the sediments that is

71

porous enough to store the solution but not permeable enough to allow free flow. Finally, the solution was subjected to evaporation through the capillary systems and to precipitate gypsum crystal during dry "season".

REFERENCES

Ault, W.V. & A.L. Kulp 1972. Isotope geochemistry of sulfur. *Geochim. Cosmochim. Acta* 16:201-235.

Bailey, S.A. & J.W. Smith 1972. Improved method for the preparation of sulfur dioxide from barium sulfate for isotope ratio studies. *Analytical Chemistry* 44(8):1542-1543.

Blatt, H. & G. Middleton & R. Murray 1972. Evaporites and native sulfur. In: *Origin of Sedimentary Rocks*: 501-526. New Jersey :Prentice - Hall, Inc, Englewood Cliffs.

Breit, G.N. & E.C. Simmons & M.B. Goldhaber 1985. Dissolution of barite for the analysis of strontium isotopes and other chemical and isotopic variations using aqueous sodium carbonate. *Chem. Geol.* 52(3-4):333-336.

Claypool, G.E. & W.T. Hosler & I.R. Kaplan & H. Sakai & I. Zak 1980. The age curves of sulfur and oxygen isotopes in marine sulfate and their mutual interpretation. *Chem. Geol.* 26:199-260.

Engelhardt, v. W. 1977. *The origin of sediments and sedimentary rocks*. New York: John Wiley & Sons,

Drever, J.I. 1982. *The Geochemistry of natural waters* New Jersey: Prentice-Hall, Inc., Englewood Cliffs.

DePaolo, D.J. & B.L. Ingram. 1985. High resolution stratigraphy with strontium isotopes. *Science* 227:938-941.

Garrels, R.M. & C.L. Christ 1965. *Solutions, Minerals, and Equilibria*. Freeman, Cooper & Company.

Hitchon, B. & H.R. Krouse 1972. Hydrogeochemis-try of surface waters of the Mackenzie River drainage basin. *Geochim. Cosmochim. Acta* 36:1337-1357.

Ho, C.S. 1986. *An Introduction to the Geology of Taiwan: Explanatory Text of the Geological Map of Taiwan.* MOEA, Central Geological Survey.

Krauskopf, K.B. & D.K. Bird 1995. *Introduction to geochemistry*. New York: McGraw-Hill.

Lan, C.Y. & J. J.S. Shen & T. Lee 1986. Rb-Sr isotopic study of andesites from Lu-Tao, Lan-Hsu, and Hsiao-Lan-Hsu: Eruption ages and isotopic heterogeneity. *Bulletin of the Institute of Earth Sciences, Academisa Sinica* 6:211-226.

Lin, T.S. 1991. *Phases and Evolution of Sedimentary Environment of clastic Sediments of southwest Foothills, Taiwan.* Master thesis (in Chinese), National Taiwan University.

Longinelli, A. 1989. Oxygen-18 and sulphur-34 in dissolved oceanic sulphate and phosphate. In: P. Fritz & J.C. Gontes(eds), The Marine Environ-ment A: *Handbook of environmental Isotope Geochemistry*, 3:219-380.

Longinelli, A. & G. Cortecci 1970. Isotopic abundance of oxygen and sulfur in sulfate ions from river water. *Earth Planet. Sci. Lett.*7:307-308.

Longinelli, A. & J.U. Edmond 1983. Isotope geochemistry of the Amazon basin: a reconnaissance. *J.Geo. Res.* 88(c6):3703-3717.

Pettijohn, F.J. 1975. *Sedimentary Rock.* 2nd ed., New York: Harper & Row..

Rabinovic A.L. & V.A. Grineuko 1979. Sulphate sulfur isotope ratios for USSR river water. *Geochem. Int.*3:68-79.

Sasaki, A. 1972. Variation in sulphur isotopic composition of oceanic sulphate. 24th *Int. Geol. Congr., Montreal* 10:342-345.

Schwarcz, H. & G. Cortecci 1974. Isotopic analyses of spring and stream water sulfate from Italian Alps and Apeunines. *Chem. Geol.* 13:285-294.

Tucker, M.E. 1981. *Evaporites*: 158-173. New York: John Wiley & Sons Inc.

Yeh, H.W. & K.S. Shea 1995. Sulfur isotope study of Gypsum of Pleistocene and Pliocene Formations of Tai Nan - Kuo Hsiung Area: Preliminary results. *Bull. Central. Geol. Surv.* 10:135-144(in Chinese).

2 Chemical weathering and climate, river catchment studies

Geochemistry of the Earth's Surface, Ármannsson (ed.)© 1999 Balkema, Rotterdam, ISBN 90 5809 073 6

Invited lecture: Intensities and fluxes of global silicate weathering deduced from large river study

J. Gaillardet & C.J. Allègre
Laboratoire de Géochimie et Cosmochimie, URA CNRS 1758, IPGP, Université Paris, France
B. Dupré
Laboratoire de Géochimie, OMP, UMR-CNRS 5563, Toulouse, France

ABSTRACT: The materials transported by the largest rivers give us information concerning the fluxes and intensities of silicate weathering. We compiled the most recent data on river geochemistry for both soluble and solid phases and explored the coupled relationships of silicate weathering fluxes, weathering intensities, climate and relief.

1 INTRODUCTION

The global Carbon cycle issue considerably improved our understanding of river geochemistry over the last decade, because it is necessary to establish global laws of (silicate) weathering rates and associated CO_2 consumption. The most interesting advantage of river catchment studies is that they lead to averaged informations. A first group of studies (pioneered by Meybeck 1986) focused on « small » streams draining one type of rock in order to reduce the natural complexity. The second group of studies favored a more global approach by exploring the chemical composition of the largest river systems. Examples of contributions are those of Stallard (1980) for the Amazon, Krishnaswami et al. (1992), Sarin et al. (1989), Galy & France-Lanord (1999) for the Ganges-Brahmaputra system, Négrel et al. (1993) for the Congo, Edmond et al. (1995) and Edmond et al. (1996) for the Orinoco, Huh et al. (1998) for the Siberian rivers. Recently Meybeck & Ragu (1997) compiled the water quality of the 550 largest rivers.
Most of the large catchment studies that have been published so far have reported major solutes concentrations. Simultaneously, Strontium isotopic ratios became a classic tool to distinguish between the contribution of the different lithologies.
In contrast, very few studies reported trace element concentrations in river waters and tried to use them as weathering proxies. In addition, the geochemistry of the solid products of weathering, transported in rivers as suspended sediments and bottom sands, retained little attention. A pioneering work has been published by Martin & Meybeck (1979), who reported the chemical composition of the suspended sediments of the largest rivers and attempted the first geochemical budget between solids and solutes inherited from continental weathering.

Here we compile the chemical compositions of the largest rivers for both soluble and solid phases and apply to these data the mass budget models that we have previously developed for the Congo and Amazon river basins (Négrel et al. 1993, Gaillardet et al., 1995).

2 DISSOLVED LOAD: CHEMICAL WEATHERING AND CO_2 CONSUMPTION FLUXES.

It is now well admitted that the chemical composition of a river water depends on the proportions of the different rocks outcropping within the drainage basin. In particular, all large drainage basins contain carbonates that dissolve very easily and supply most of the Ca and alkalinity for the river. In order to estimate the proportion of Ca and alkalinity which is derived from the weathering of silicates, the contribution of carbonate dissolution has to be subtracted. The use of chemical elemental ratios and Sr isotopic ratios, at the scale of each basin allow us to demonstrate the mixing between the main groups of lithologies and allow us to determine the proportions of each source, for each element. It is clear that such a calculation requires the knowledge of the endmembers (for both elemental ratios and Sr isotopic ratios). Such a knowledge relies on the study of monolithological streams. For example, it appears that at a global scale, the Ca/Na ratio is a good tracer of the carbonate vs. silicate contribution. Typical Ca/Na ratios close to 0.4 for waters draining granites and associated rocks contrast with higher Ca/Na values (Ca/Na = 100) in the case of rivers draining pure carbonates.

Using this method, the contribution of silicates can be deduced from the dissolved load of each river and the weathering rates of silicates and associated CO_2 consumption rates can be estimated (Gaillardet et al. 1999a).

The relationships between these rates and the environmental factors are of special interest. Among the parameters that have been tested, three seem to have a major influence on weathering fluxes : runoff, temperature and mechanical denudation. The relationship of silicate weathering rates with temperature, deduced from this study is shown in Figure 1. There is no general relationship between temperature and silicate weathering rates. However, if some rivers such as the rivers of the arid areas or the Amazon, Orinoco and Congo rivers are ignored, then a trend of increasing rates with temperature emerges. We therefore conclude that the chemistry of large rivers is consistent with a positive action of temperature on weathering rates, but that temperature is not the over-riding parameter. As observed by Edmond et al. (1995) in Guyana, the rivers of the tropical zone have weathering rates that are low with regards to their mean annual temperature.

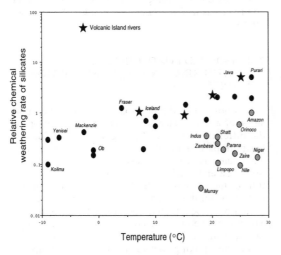

Figure 1 : General relationship between silicate weathering rates and temperature. The results for volcanic island rivers are from Louvat (1997).

The clear positive correlation between silicate weathering rates and physical denudation (as measured by river sediment yields) may give us one possible explanation of this paradox. Lowland tropical rivers have very low sediments yields, leading to the development of thick soils. In such areas, the shielding effect of soils prevents weathering fluxes to be important in spite of optimum climatic conditions.

More generally, this correlation speaks in favor for a global coupling between chemical and mechanical weathering at the surface of the Earth.

3 SUSPENDED SEDIMENTS: GLOBAL WEATHERING INTENSITIES

The comparison between the chemical composition of suspended sediments and the Upper Continental Crust shows that the most soluble elements are depleted because they are released during chemical weathering (Gaillardet et al. 1999b). Exceptions are Ca and Sr which are often enriched in suspended sediments, due to the presence of calcite. These features are observed in all large rivers, but the amplitude of the soluble depletion varies from one river to another. By computing (for example) the Na depletion, we compared the degree of weathering of the suspended sediment from the largest rivers. The most weathered sediments are produced in the tropics, while the least weathered sediments are made in the subarctic regions, the high mountainous areas and on volcanic islands (independent of the climate) . There is therefore no observed relationship between weathering intensity and climate, except that low temperature areas do not produce highly weathered sediments. At a global scale, for a given range of temperature, there is an inverse relationship between the degree of weathering of river sediments and sediment yields. The highest denudation rates are associated with the lowest degree of silicate weathering. The reason for this inverse relationship is probably that the residence time of water in soils is short (and then water-rock interaction is weak) when mechanical denudation is intense.

4 A MASS BUDGET PROBLEM

The important question is then: is the degree of sediment weathering compatible with the river dissolved load derived from silicate weathering? The answer to this question is basically a mass balance problem.

Under steady state conditions, the amount of any element present initially in the pristine bedrock is equal to the amount as solutes plus the amount transported by particulates in the river. The two major unknowns to solve this mass budget are the amount of suspended sediments transported annually into the river (mechanical denudation) and the concentration of the element in the pristine bedrock. Using the mean Upper Continental Crust composition, we calculate sediment yields that are generally lower than those determined by multi-year sampling surveys. If we assume that present day measured sediment yields are not fair estimates of mechanical denudation, and that mechanical denudation is higher because of sediment storage within the drainage basin, then the discrepancy between measured and calculated sediment yields is much greater.

One way to reconcile the measured sediment yields with those predicted by the mass balance, is to assume that the pristine bedrocks are depleted in soluble elements with regards to the mean Upper Continental Crust. This is possible if a significant part of the drainage basin is composed of sedimentary rocks, that were affected by previous weathering/sedimentation cycles. Thus, the proportion of recycled rocks vs. pristine rocks (e.g. granites) is an important parameter that should retain attention for weathering and CO_2 consumption studies.

5 CONCLUSIONS

The study of large rivers allows us an integrated approach to estimate silicate weathering over time and space. We insist on the need to develop new tracers, in particular isotopic tracers, such as Os, B, U series desequilibrium, to progress in our understanding of the global weathering processes and rates.

REFERENCES

Edmond J. M., M. R. Palmer , C. I. Measures, E. T Brown, and Y Huh 1996. Fluvial geochemistry of the northeastern Andes and its fordeep in the drainage of the Orinoco in Columbia and Venezuela. *Geochim. Cosmochim. Acta* 60(16): 2949-2976

Edmond J. M., M. R. Palmer , C. I. Measures, B Grant, and R. F. Stallard 1995. The fluvial geochemistry and denudation rate of the Guyana Shield in Venezuela, Columbia and Brazil. *Geochim. Cosmochim. Acta* 59: 3301-3325.

Gaillardet J., B. Dupré, and C. J. Allègre 1995. A global geochemical mass budget applied to the Congo Basin rivers: erosion rates and continental crust composition. *Geochim. Cosmochim. Acta* 59(17), 3469-3485.

Gaillardet J., B. Dupré, P. Louvat, and C. J. Allègre 1999a. Global silicate weathering and CO2 consumption rates deduced from the chemistry of large rivers. *Chem. Geol.*:in press.

Gaillardet J., B. Dupré and C. J. Allègre 1999b. Geochemistry of large river suspended sediment: global silicate weathering or recycling tracer? *Geochim. Cosmochim. Acta* : in press

Galy A. & C. France-Lanord 1999. Processes of weathering in the Gangs-Brahmaputra basin and the riverine alkalinity budget. *Chem. Geol* : in press.

Huh Y., G. Panteleyev , D. Babich, A. Zaitsev, and J. M. Edmond 1998. The fluvial geochemistry of the rivers of Eastern Siberia: II. Tributaries of the Lena, Omoloy, Yana, Indigirka, Kolima and Anadyr draining the collisional/accretionary zone of the Verkhoyansk and Cherskiy ranges. *Geochimi. Cosmochim. Acta* 62(12): 2053-2075.

Krishnaswami S., J. R. Trivedi, M. M. Sarin, R. Ramesh, and K. K. Sharma 1992. Sr isotopes and Rb in the Ganga-Brahmaputra river system: Weathering in the Himalaya, fluxes to the Bay of Bengal and contributions to the evolution of oceanic [87]Sr/[86]Sr. *Earth Planet. Sci. Lett.* 109: 243-253.

Louvat P. 1997. *Etude géochimique de l'érosion fluviale d'îles volcaniques à l'aide des bilans d'éléments majeurs et traces.* Ph. D. thesis, University of Paris 7.

Martin J. M. & M. Meybeck 1979 Elemental mass balance of materiel carried by world major rivers. *Mar. Chem.* 7(2): 173-206.

Meybeck M. & A. Ragu 1997. *River discharges to the Oceans: An assessement of suspended solids, major ions and nutrients.* Nairobi: UNEP in press.

Négrel P., C. J. Allègre , B. Dupré and E. Lewin 1993 Erosion sources determined by inversion of major and trace element ratios in river water: The Congo Basin case. *Earth Planet. Sci. Lett.* 120: 59-76.

Sarin M. M., S. Krishnaswami, K. Dilli, B. L. K. Somayajulu, and W. S. Moore 1989 Major ion chemistry of the Ganga-Brahmaputra river system: Weathering processes and fluxes to the Bay of Bengal. *Geochim. Cosmochim. Acta* 53: 997-1009.

Stallard, R. F. 1980 *Major elements geochemistry of the Amazon River system.*, Ph. D. Thesis. Cambridge: MIT.

Geochemistry of the Earth's Surface, Ármannsson (ed.)© 1999 Balkema, Rotterdam, ISBN 90 5809 073 6

Invited lecture: The effect of climate on chemical weathering of silicate rocks

A. F. White, T. D. Bullen, D. V. Vivit & M. S. Schulz
US Geological Survey, Menlo Park, Calif., USA

A. E. Blum
US Geological Survey, Boulder, Colo., USA

ABSTRACT: Climate, principally temperature and precipitation, has been proposed as a linkage by which the rates of surficial weathering of silicates closely balance rates of atmospheric CO_2 production, thus promoting stable climatic conditions that permit life on earth. Such a feedback mechanism however is not universally accepted particularly in relationship to natural weathering systems on the scale of watersheds and river basins. The present paper reviews the available literature related to the effect of climate on chemical fluxes in such systems. It discusses new experimental data related to the magnitude of temperature effect on weathering of multi-mineralic granitoid rocks and presents additional data and revised activation energies that describe temperature-dependent silica fluxes in a global distribution of small watersheds.

1 INTRODUCTION

The term "weathering" implies that chemical weathering is strongly affected by climate, principally by moisture and temperature. The intensive interest in past and present global climate change has renewed efforts to quantitatively understand feedback mechanisms between climate and chemical weathering. Berner & Berner (1997) and others have proposed a direct linkage in which any increase in atmospheric CO_2 from sources such as volcanism is counter-balanced by increased CO_2 consumption by more rapid silicate weathering under increased greenhouse temperatures. In contrast, diminished atmospheric CO_2 is counter-balanced by decreased weathering rates caused by lower temperatures.

This linkage between climate, continental silicate weathering, and atmospheric CO_2 is not universally accepted. For example Francois & Walker (1992) suggested that low temperature seafloor-basalt alteration and not terrestrial silicate weathering exerts the dominant long-term control on atmospheric CO_2. Bickle (1996) and Edmond & Huh (1997) concluded that weathering rates controlled by tectonics, rather than climate, principally influence atmospheric CO_2.

2 TEMPERATURE EFFECTS

An important issue in this ongoing controversy is the impact of climate on quantitative weathering rates of silicate rocks. The effects of temperature on weathering rates are experimentally established for silicate minerals such as feldspars (Blum & Stillings 1995). Temperature effects on other important silicate phases such as biotite and hornblende are not well established. In addition, complexities associated with temperature effects on weathering of common multi-mineralic silicate rocks have not been investigated.

Direct observations of temperature differences on natural weathering processes have proven to be elusive and surrogate weathering studies comparing spatially separated climatic regimes are employed. The utility of such comparisons depends on the ability to separate out the effect of temperature from other variables influencing chemical weathering including precipitation, geomorphology, vegetation and lithology. This ability decreases as the scale of the weathering process increases. This explains why comparison of solute concentrations and fluxes in large-scale river systems most often fail to detect a temperature effect (Huh et al. 1998).

3 WATERSHED WEATHERING

Comparison of smaller watershed weathering environments have been more successful in documenting temperature impacts due to an increased ability to separate out other non-weathering variables (Velbel 1993, Louvat 1997). White & Blum (1995) tabulated average chemical fluxes from a global distribution of 68 small upland watersheds underlain by granitoid rocks and observed positive correlations between increasing Si and Na fluxes and increasing temperature and precipitation based on the expression

$$Q_{i,w} = (a_i P)\exp\left[\frac{E_a}{R}\left(\frac{1}{T_o} - \frac{1}{T}\right)\right] \quad (1)$$

where $Q_{i,w}$ is the average flux of species i (mol ha^{-1} yr^{-1}), P is average annual precipitation (mm yr^{-1}) and a_i is magnitude of the precipitation dependence. The exponential temperature function is essentially the Arrhenius expression in which E_a is the apparent activation energy and R is the gas constant. T represents a specific watershed temperature and T_o is a reference temperature taken to be 5°C, the mean average annual temperature of the watershed database. The above relationship implies a climatic reinforcing effect for watersheds in which

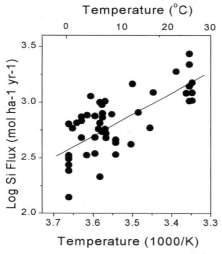

Figure 1. Relationship between annual silica discharge fluxes and temperature for a global distribution of small watersheds underlain by granitoid rocks. Line is optimized fit to model (Equation 1).

weathering rates at both high temperature and precipitation, as in the tropics, are very rapid compared to watersheds in more temperate climates. The original database was weighted heavily in favor of temperate climates based on extensive watershed research conducted principally in North American and Europe. Additional Si flux data have since become available (20 watersheds). Of particular significance are fluxes for a number of tropical granitoid watersheds in mountainous areas of Malaysia.

This new temperature, precipitation and Si flux data are combined with the previous data (White & Blum 1995) and Si fluxes are plotted as the log inverse of absolute temperature in Figure 1. Equation 1 is numerically optimized based on the expanded data set for values of a_1 and E_a. The apparent activation energy for Si discharge fluxes from granitoid watersheds is calculated to be 51 kJ mol^{-1}.

4 EXPERIMENTAL WEATHERING

The present study further investigates the temperature effects on granitoid weathering based on long-term experimental dissolution of granitoid rocks from several of the watersheds included in the data based described above. Pairs of fresh and weathered samples were obtained from two alpine watersheds; the upper Merced River in Yosemite National Park, Colorado USA and the Loch Vale watershed in Rocky Mountain National Park, Colorado, USA. Additional samples were collected from the subtropical Panola Mountain Research Watershed near Atlanta Georgia, USA, and the tropical Rio Icacos watershed in the Luquillo Mountains of Puerto Rico.

White et al. (1999) describe sample preparation and the initial experimental setup. The 0.25 to 0.85-mm size fraction were packed into glass columns through which distilled/deionized water, saturated with a 5% CO_2/95% air, was introduced at flow rates of 10 ml/hr. Effluents from the columns were collected and analyzed for chemistry and Sr isotopes.

Experiments at ambient temperature 22 ± 1 °C were run for 1.5 years. For non-ambient conditions, the columns were jacketed with plastic tubing and insulation and the temperature regulated with circulating water baths. The column experiments were first decreased to a temperature range of 5 ±1 °C for 100 days, subsequently increased to 17±1 °C and finally increased to 35 ±1 °C.

Effluent Si, Na, and K increased exponentially with temperature. Figure 2 shows an example of this effect for Si release from fresh and weathered Panola granodiorite. The temperature trends for effluent Si, Na, and K are parallel to sub-parallel for fresh and weathered paired granitoids indicating that the temperature sensitivities are similar whether or not the granitoid has undergone natural weathering. This implies that the net activation energies associated with the breakdown of the silicate structure remains constant.

However, at any given temperature, the effluent concentrations from the weathered rocks are lower than from the fresh rocks. No systematic differences

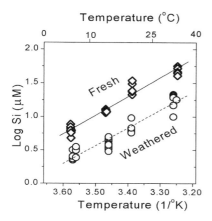

Figure 2. Temperature dependence of effluent Si released from fresh and naturally weathered Panola granodiorite during column experiments

in bulk rock chemistries are apparent between any of the fresh and weathered granitoid pairs. These solute concentration decreases from the weathered granitoids must correlate with diminished mineral reactivity, which is not dependent on temperature effects.

4 APPARENT ACTIVATION ENERGIES

The temperature effect on weathering is commonly characterized as the change in the ratio of reaction rates r and r_o (mol m^{-2} s^{-1}) over a temperature interval defined by $T - T_o$ ($^\circ$K) such that (Brady & Carroll, 1994)

$$\frac{r_T}{r_o} = \exp\left[\frac{E_a}{R}\left(\frac{1}{T_o} - \frac{1}{T}\right)\right] \quad (2)$$

where E_a is the activation energy. The ratios of effluent concentrations C and C_o characterized in the present study, are equivalent to the ratio of reaction ratios in Equation 2 (i. e., $r_T/r_o = C_T/C_o$) if the mineral surface areas and fluid flow rates are assumed constant.

The average Si activation energies for the fresh and weathered granitoids are respectively 53 ± 5 kJ mol^{-1} and 57 ± 5 kJ mol^{-1} (Table 4). Activation energies for Na are only slightly higher (57 ± 15 kJ mol^{-1} and 59 ± 8 kJ mol^{-1} respectively). These values are similar to those reported for experimental plagioclase dissolution (60 kJ mol^{-1}, Blum & Stillings 1995) and for watershed Si fluxes based on Equation 1 (51 kJ mol^{-1}).

The average activation energies for K release from the fresh and weathered granitoid columns (27 ± 9 kJ

mol^{-1} and 32 ± 11 kJ mol^{-1} respectively) are much lower than for Si and Na. High Rb/K ratios in the effluents are indicative of biotite dissolution. The lower bonding energy for interlayer K in biotite relative to tetrahedral Na in plagioclase is expected to result in faster K release rates as evidenced by high K/Na effluent ratios. However, the lower activation energy of K relative to Na is expected to diminish the selective loss of K with increasing temperature. This effect is shown by the progressive decreases in the K/Na ratios in effluents from the fresh granitoids (Figure 3).

This temperature trend is consistent with high K/Na ratios in glacial runoff commonly ascribed to the relatively rapid initial weathering of biotite induced by exposure of fresh surfaces through grinding and physical erosion (Blum & Erel 1997). However, differences in the activation energies between biotite and plagioclase implies that temperature differences between glacial and warmer non-glacial enviroments are of comparable importance in controlling observed K/Na solute ratios.

Effluent Ca, Mg and Sr do not exhibit positive correlations with temperature. Ca concentrations are controlled in part by diminishing trace amounts of calcite and possible retrograde calcite solubility. The close correlation between Ca, Mg and Sr concentrations implies additional control by ion exchange.

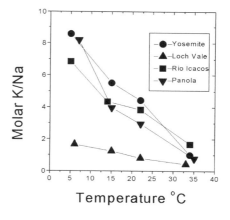

Figure 3. Decreases in K/Na ratios in effluents with increasing temperature.

CONCLUSIONS

Results presented in the present study provide strong support for significant temperature effects on rates of chemical weathering of granitoid rocks both under experimental and natural conditions. The strong structural bonds associated silicate minerals such as plagioclase produce relatively high activation energies for release of species such as Na and Si (60 kJ mol^{-1}) which make chemical weathering of granitoid rocks temperature sensitive. The Arrhenius temperature fit to a coupled temperature-precipitation model produces activation energies for Si (51 kJ mol^{-1}) that are similar to values produced by the experimental weathering of the fresh and weathered granitoids (53 and 57 kJ mol^{-1} respectively).

Experimental K release rates are more rapid but less temperature sensitive than Na and Si. Lower activation energies for weathered and fresh granitoids (27 and 32 kJ mol^{-1}) indicate that both these effects are attributed to K loss from interlayer sites in biotite. Differences in mineral temperature sensitivities are reflected in decreases in effluent K/Na ratios with temperature, an effect that has been previously observed in comparisons of glacial versus more temperate watersheds.

REFERENCES

Berner R. A. & E. K. Berner 1997. Silicate weathering and climate. In W. F. Ruddiman (ed) *Tectonic Uplift and Climate Change*: 353-364. New York: Plenum Press.

Bickle M. J. 1996. Metamorphic decarbonation, silicate weathreing and the long term carbon cycle. *Terra Nova* 8: 270-276.

Blum A. E. & L. L. Stillings 1995. Feldspar dissolution kinetics. In A. F. White and S. L. Brantley (eds), *Chemical Weathering Rates of Silicate Minerals 31*: 291-351. Mineralogical Society of America.

Blum J. D. & Y. Erel 1997. Rb-Sr isotope systematics of a granitic soil chronosequence: The importance of biotite weathering. *Geochimica Cosmochimica Acta* 61: 3193-3204.

Brady P. V. and S. A. Carroll 1994. Direct effects of CO_2 and temperature on silicate weathering: possible implications for climate control. *Geochemica Cosmochimica Acta* 58(7): 1853-1863.

Edmond J. M. & Y. Huh 1997. Chemical weathering yields from basement and and orogenic terrains in hot and cold climates. In W. F. Ruddiman (ed), *Tectonic Uplift and Climate Change*: 329-351. New York: Plenum Press.

Francois L. M. & J. C. G. Walker 1992. Modelling of the Phanerzoic carbon cycle and climate: constraints from the 87Sr/86Sr isotopic ratio of seawater. *Amer. J. Science* 292: 81-135.

Huh Y., M. Tsol, A. Zaitsev & J. Edmond 1998b. The fluvial geochemistry of the rivers of Eastern Siberia: I. Tributaries of the Lena River draining the sedimentary platform of the Siberian Craton. *Geochimica Cosmochimica Acta* 62: 1657-1676.

Louvat P. 1997. *Etude geochimique de l'erosion fluviale d'iles volcaniques a l'aide des bilans d'elements majeurs et traces*. PhD Thesis, Institutde Physique du Globe de Paris.

Velbel M. C. 1993. Temperature dependence of silicate weathering in nature: how strong a feedback on long-term accumulation of atmospheric CO2 and global greenhouse warming. *Geology* 21: 1059-1062.

White A. F. & A. E. Blum 1995. Effects of climate on chemical weathering rates in watersheds. *Geochimica Cosmochimica Acta* 59: 1729-1747.

White A. F., T. D. Bullen, D. V.Vivit & M. S. Schulz 1999. The role of disseminated calcite in the chemical weathering of granitoid rocks. *Geochimica Cosmochimica Acta*, in press.

Geochemistry of the Earth's Surface, Ármannsson (ed.) © 1999 Balkema, Rotterdam, ISBN 90 5809 073 6

Features of chemical composition of Kamchatka rivers

V.A. Chudaeva – *Pacific Institute of Geography, Far East Branch Russian Acad. of Sciences, Vladivostok, Russia*

O.V. Chudaev – *Far East Geological Institute, Far East Branch Russian Acad. of Sciences, Vladivostok, Russia*

G.A. Karpov – *Inst. of Volcanology, Far East Branch Russ. Acad. of Sciences, Petropavlovsk-Kamchatsky, Russia*

W.M. Edmunds & P. Shand – *British Geological Survey, Wallingford, UK*

ABSTRACT: Chemical composition of the rivers of the Paratunka basin have been studied in order to evaluate any influence of hydrothermal activity. Rivers of the Kronotsky reserve, a major hydrothermal area, are used for comparison. It was found that in areas with increased hydrothermal activity of Kamchatka the chemical composition of river water can be affected.

1 INTRODUCTION

Kamchatka is a unique region of Russia, where active volcanic and hydrothermal activity is observed. Directly or indirectly this affects the chemical composition of surface waters of the region. In this paper attention is given to the chemistry of the river water in the Paratunka Basin. Within the catchment of the river hydrothermal manifestations are widely distributed, occurring as springs on the valley sides, as baseflow but are also found in boreholes up to 1000 and more meters depth and may add to this flow.

Rivers of the Kronotsky reserve with well known hydrothermal activity in the Geyser valley and elsewhere were used for comparison.

2 METHODS

The main sampling of the Paratunka River was conducted in July 1996 with additional sampling in August 1996 and September 1997 (Figure 1). A range of parameters was measured at each site including pH, specific electric conductance (SEC), dissolved oxygen (DO), and bicarbonate (HCO$_3$). Major, minor and trace cations and sulphate were analyzed using ICP-OES and ICP-MS. Anion concentrations were determined using automated colorimetry.

3 RESULTS AND DISCUSSION

The three main tributaries of the Paratunka (the Poperechnaya, Karimshina and Bistraya, including Right and Left Bistraya) were tested, as well as a smaller river, (R. Khaikova) which is directly influenced by hydrothermal springs.

3.1 *The Karimshina Rriver*

The average water discharge of this river is about 4.6 m^3 sec^{-1}. The catchment area (Figure 1) is mostly composed of acid tuffs. Despite high water level in July 1996 some hydrothermal influence of alkaline warm waters was nevertheless measurable geochemically in the river although the temperature of water remained constant. Over a distance of two kilometers an increase of pH from 6.2 up to 7.6, was observed together with some increase in major ion concentrations: Na, K, and HCO$_3$ increased from 20-31 mg l^{-1}, SO$_4$ from 6.8-8.6 mg l^{-1}, and Si from 3.63 to 4.35 mg l^{-1}. Over a short distance, where the manifestations of thermal springs below water level are known, the contents of Li increase from 2.6 to 6.3 µg l^{-1}. A similar increase is found for B from 14 to 70 µg l^{-1}, Al from 21 to 51 µg l^{-1}, Ga from < 0.04 to 0.07 µg l^{-1} and Cs from 0.05 to 0.36 µg l^{-1}. Over the same distance Ba, La, Y, Rb increased slightly, together with Mo, from 0.78 to 1.24 µg l^{-1}. Over the same distance no increase of heavy metals (Cu, Zn, Pb, Cd, Fe, Mn) is observed.

For the period of low flow (August, 1997) in the lower section of Karimshina river an increase in SEC was found, also Na (1.2 times), SO$_4$ (1.4 times), but HCO$_3$ decreased 1.4 times. The Sr, Ba, Li, and B contents, also increased. Over the same period the concentrations of some heavy metals (Fe, Mn, Cu) also increased slightly. The concentration of Zn-increased 4 times, Co from < 0.03 up to 1.4 µg

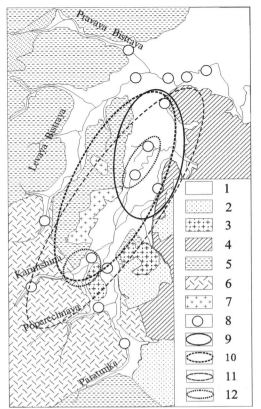

Figure 1. Geological setting and the main features of hydrogeochemical distribution.1-alluvium; 2-andesite-basalt, Q_{1-4}; 3-liparite, N_2; 4-basalt-tuff, N_{1-2}; 5- sedimentary rock, N_1; 6-acid tuff, N_1; 7-granite, N_1; 8-sampling points; 9-area with SEC > 70 µS/cm, Na > 5 mg l^{-1}, K > 0.5 mg l^{-1}, Cl> 2 mg l^{-1}; 10-B >10 µg l^{-1}; 11-Li >5 µg l^{-1}; 12- Rb > 0.5 µg l^{-1}

l^{-1}, Ni from < 0.4 up to 5.7 µg l^{-1}, Pb from 0.18 up to 21 µg l^{-1}, Cd from < 0.03 up to 2.5 µg l^{-1}, and also Y from 0.03 up to 1.6 µg l^{-1}, La from 0.02 up to 7.5 µg l^{-1}, Cr from 0.38 up to 3.7 µg l^{-1}, V from < 0.1 to 2.7 µg l^{-1}, Zr from 0.37 to 2.8 µg l^{-1}. Some increase in P takes place, most likely, due to seasonal dynamics of the nutrients. As found by the Hydrological Survey (Russian Hydrological Survey, 1995), the maximum amount of PO_4 in the Paratunka river is observed in July (0.025-0.030 mg l^{-1}).

3.2 The Poperechnaya River

In this valley no thermal springs were found. The Poperechnaya and Karimshina Rivers both drain mostly acid tuffs (N_1 Figure 1). The Ca composition of the Poperechnaya River is appreciably higher than in the Karimshina River, but Na and K contents are lower. The SO_4 concentration is twice as lower. Moreover, in the Poperechnaya River Li is twice as low, and B is 4-5 times lower. Mo Rb, Cs, Cu, Zn,

Sb, Y concentrations are 2-3 times lower than in the Karimshina River. Similar concentrations are found for Al, Sr, Ba, Cd, Pb in both rivers.

3.3 The Bistraya River

This river discharges about 22.1 m^3 sec^{-1} into the Paratunka. The Bistraya consists of two large tributaries: Right Bistraya (Right B) and Left Bistraya (Left B), the upper part of their catchment is composed of acid tuffs and the greater part of it of volcanic and sedimentary rock.

Hydrothermal springs are not found in the catchment, therefore the chemical composition is controlled only by natural processes of rock leaching.

The water of the Left B was relatively acid during the period of sampling (pH 6.69-6.66). The ratios of major ions in the upper and lower sections were similar. In comparison with the Karimchina River the Na, K contents of the Bistraya are lower, but Ca and Mg contents are respectively 1.5 and 2 times higher. The SO_4 content is 1.7 times higher, while the Cl content is lower.

The Right B has higher pH (7.27) and lower SO_4 content, but higher HCO_3 than the Karimchina After the confluence of the Right B and Left B an average composition for all main components is found. The Fe, Mn, and F contents did not differ from those in the Karimchina river and the Li content of Bistraya River was an order lower. The Left B, (draining the same rock massif, as the Karimshina River) had Li concentrations 3-4 times higher than the Right B and Bistraya River. The Li content decreased downstream.

The boron content in the Bistraya River and its tributaries (Right B and Left B) is an order of magnitude lower than that of the Karimchina River. The Left B contained more Al, possibly related to the lower pH. For trace elements some increase of Co, Cu, Zn, Mn Cd, Rb in the Left B were observed. Concentrations of Sr, and Ba in the Left B were higher than in the Karimshina River.

Sampling of the Bistraya River over different periods of the summer season have shown a rather stable major ion composition, as well as near constant Sr and Ba.

3.4 The Khaikova river

The Khaikova river has a small catchment area - 13.3 km^2 and a discharge of 0.1 m^3 sec^{-1} (Vaskovsky, 1973). Hydrothermal waters directly affect the river. The temperature of river water in the period of sampling increased by up to 11.5^{0}C. On the diagram (Figure 2) it is clearly distinct from all the other Paratunka waters. High concentrations of many elements such as: SO_4 (47.5 mg l^{-1}), Cl (14.5 mg l^{-1}),

Table 1. Some element concentrations in Kamchatka rivers

*	Si, mg l⁻¹	Al, μg l⁻¹	Sr, μg l⁻¹	Li, μg l⁻¹	B, μg l⁻¹	Cu, μg l⁻¹	Zn, μg l⁻¹	Mo, μg l⁻¹
1	3.6-4.3	8.9-50.6	28.7-32.5	2.6-6.1	14-70	0.24-1.16	1.2-4.7	0.78-1.24
2	4.6-4.9	8-13.6	31.7-46.3	0.17-0.34	4.2-7.2	0.28-0.69	1.7-5	0.48-1.1
3	4.1	23.8	33.9	2.02	11	0.13	<1	0.59
4	5-8.7	13.2-23.3	18.1-37.1	0.65-5.8	15.9-62.5	0.34-0.97	1.7-9.2	0.37-1.1
5	10.8	14.5	175	9.16	220	0.33	2.92	1.6
6	0.1-11.7	17.9-89.2	0.8-8.6	0.2-3	0.8-33.3	<0.7-2.3	0.9-2.9	<0.2-10.5
7	20.4	4232.3	1060.	4.8	75.2	46.0	262	1189.4
8	23.2	42.8	110	4.6	22.2	5.6	5.4	8.11
9	6.7-30.1	46.4-73.2	14.7-57.1	<0.6-1.2	4.3-1281	<0.5-2	1.8-8.8	0.4-5.7

* 1- Karimshina; 2- Bistraja; 3- Poperechnaja; 4- Paratunka; 5- Khaikova; 6-9 - Kronotsky reserve: 6- background rivers; 7- Kislaja; 8- Travertinovaja; 9- rivers of the Geyser valley

Na (30.6 mg l⁻¹), B (up to 220 μg l⁻¹), Sr (175 μg l⁻¹), Ni (0.56 μg l⁻¹), Ge (0.4 μg l⁻¹), Rb (1.35 μg l⁻¹), Ba (5.7 μg l⁻¹). The composition of these waters is close to the upper Paratunka hydrothermal springs.

4 DISCUSSION

Considering all results, an increase in the concentrations of Na (> 5 mg l⁻¹), K (> 0.5 mg l⁻¹), SEC (> 70 μS cm⁻¹), Cl (> 2 mg l⁻¹) takes place in the central part of the Paratunka valley, coincident with the main hydrothermal area recognised by (Manukhin & Vorojeikina (1976). This is also clearly shown by the increased concentration of Li (> 5 μg l⁻¹), B (> 40 μg l⁻¹), Rb (> 0.5 μg l⁻¹).

Thus, the composition of river waters of the Paratunka basin is initially controlled by leaching of catchment rocks. However, direct inputs of thermal waters as springs also takes place.

Discharge of the Paratunka river into the sea (in the Avacha Bay), is included in the sum of the Paratunka River, Mikija (34 m³/cek) and the Bistraya River (22 m³/cek), corresponding to a total load of 109 t yr⁻¹ chemical load. In the lower section of the Paratunka river (Nikolaevka) a marine influence is also added. (Figure 1). Total dissolved solids concentrations in the Bistraya and Poperechnaya rivers, are both close to 57 mg l⁻¹. Taking these as the background values the input of hydrothermal waters in the Paratunka River amounts to about 12% of the chemical load. Rivers of other regions of Kamchatka (for example, in the Kronotsky reserve) represent a rather different water composition (Figure 2). Background waters (five rivers have been tested) have slightly alkaline pH, low SEC and no increased contents of chemical components. Rivers connected with the Valley of Geysers (the Geysernaya River and its tributaries) are under the influence of acid hydrothermal solutions and have pH from 5 to 2.4 with

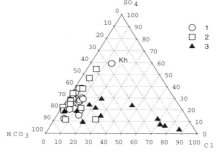
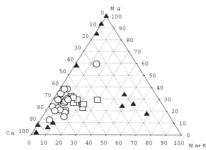

Figure 2. Diagram of main ions of Kamchatka rivers. 1- authors data on Paratunka basin; 2- data of Hydrology Survey; 3- Kronotsky reserve; Kh- Khaikova river.

predominance of SO₄ (NaSO₄ composition) and increased contents of heavy metals and other trace Geysernaya (in upper part) rivers, pH 7.4-7.6, with prevailing (Na+K)Cl composition. For the latter rivers an increased content of Li, B (> 0.1 mg l⁻¹) is typical.elements (Fe, Mn, Cu, Zn, Cd, Ni, Mo, Pb, Co, V, Y, F, Zr). The other types of river water in this area are alkaline warm waters in the Travertinovaya river, with pH 7.4-7.6 and prevailing Ca(+Mg)HCO₃ composition and increased contents of Fe, Mn, Co, Ni, Mo, Cr, V and the water of the Vodopadnaya and Thus, the chemical composition of surface waters in Kamchatka in areas of increased hydrothermal activity can be influenced by the direct input of thermal (or other mineral waters).

ACKNOWLEDGEMENTS

The authors thank the INTAS foundation for financial support through project 94-1592.

REFERENCES

Russian Hydrological Suvey. 1996. *Annual data on quality of surface waters*. 1996. Part 1. V1(20). Issue 18. *Kamchatka.*. (In Russian).

Manukhin U.F. & L.A. Vorojeikina 1976. Hydrogeology of Paratunka hydrothermal system and conditions of it formation. In V.M. Sugrobov (ed.) *Hydrothermal systems and thermal fields of Kamchatka*: 143-177. Vladivostok (In Russian).

Vaskovsky, M.G. (ed) Resources *of surface water USSR*. 1973. Vol 20. Kamchatka. (In Russian).

Geochemistry of the Earth's Surface, Ármannsson (ed.)© 1999 Balkema, Rotterdam, ISBN 90 5809 073 6

Bedrock and climatic controls on the cationic composition of karst waters

Ian J. Fairchild, Anna F. Tooth & Yiming Huang – *Department of Earth Sciences, University of Keele, UK*

Andrea Borsato & Silvia Frisia – *Museo Tridentino, Trento, Italy*

Frank McDermott – *Department of Geology, University College Dublin, Belfield, Ireland*

Baruch Spiro – *NERC Isotope Geosciences Laboratory, Keyworth, UK*

ABSTRACT: A combination of field data collection, experimental dissolution and growth studies, and theoretical modelling of calcite and dolomite dissolution leads to an overall assessment of the controls on the Mg/Ca and Sr/Ca compositions of karst waters and speleothems. Palaeohydrology is a key variable. Relatively prolonged contact of water with soil and karst zones (relatively drier climatic conditions) tends to lead to enhanced Mg/Ca and Sr/Ca ratios in cave waters due to some combination of : 1) enhanced selective leaching of Mg and Sr in soils, 2) increased prior precipitation of calcite due to degassing within the aquifer above a precipitation site, and 3) enhanced dolomite dissolution (increased Mg/Ca). Individual drip sites and speleothems show individual sensitivities and thresholds to change so that modern calibration of behaviour is desirable in interpretation of palaeohydrology.

1 INTRODUCTION

Shallow karst cave environmental systems have several advantages as recorders of palaeoclimate data (Gascoyne 1992), including the slow semi-continuous growth of speleothems (especially stalagmites) and the equation of cave temperature with the mean annual temperature of the overlying ground. Attention has focused mainly on applications of oxygen isotope geothermometry (e.g. Bar-Matthews et al. 1997, McDermott et al. 1999) and improved methods of speleothem dating by U-series techniques. However cave hydrology is also strongly influenced by the water budget to the karstic system. We have previously presented (Fairchild et al. 1996) preliminary data suggesting that changes in water budget are related to hydrochemical trends, with particular reference to two parameters (Mg/Ca and Sr/Ca) that are preserved in speleothems. Roberts et al. (1998) have also demonstrated the preservation of annual variations in both Mg/Ca and Sr/Ca in an ancient speleothem. We now have more extensive data on hydrochemical changes at individual drip sites in several caves, experimental data on trace element partitioning during speleothem growth, and element yields during leaching. Combined with further theoretical analyses of the dissolution behaviour of dolomite and calcite, we are closer to an overall understanding of the controls on the Mg-Sr-Ca chemistry of karst waters in relation to the karst bedrocks and can state with more confidence how chemical variations reflect

hydrological conditions (Fairchild et al. in review). The implication is that trace element patterns in favourable cases will yield palaeohydrological information, particularly on the seasonal water balance.

2 OVERVIEW OF MG/CA and SR/CA RATIOS OF CAVE WATERS

At each of several sites, the range of karstic bedrock compositions have been compared with that of karst waters. In each case (e.g. Figure 1) the cave waters are enriched in Sr and Mg in relation to the bedrock mixing line and at individual locations display covariations in Sr/Ca and Mg/Ca. We therefore need an explanation of the enrichment and the covariations. More subtly, the differential disssolution kinetics of dolomite and calcite will also have resulted in displacement of leachate compositions along bedrock mixing line from their source bedrocks.

3 DIFFERENTIAL DISSOLUTION OF DOLOMITE AND CALCITE

The kinetics of calcite dissolution are relatively well understood, but dolomite dissolution is less well known except far from equilibrium Dolomite dissolution kinetics is known to vary with composition and crystallinity (Busenberg &

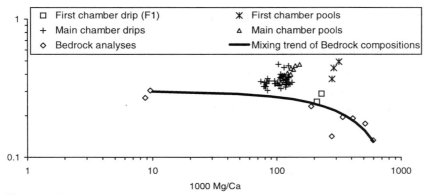

Figure 1: Sr/Ca-Mg/Ca crossplot comparing bedrock compositions with cave water data from the alpine Ernesto cave McDermott et al. in press), NE Italy

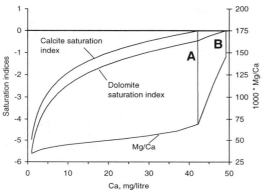

Figure 2. Modelled dissolution of dolomite and calcite: see text

Plummer 1982) which limits the precision of our treatment. Nevertheless, Chou et al. (1989) present comparative data for dolomite and calcite dissolution kinetics which will serve to illustrate the general principles. Their expressions were used to calculate the changing solution composition during the progressive dissolution of calcite-dolomite mixtures towards equilibrium in an open system, that is one with a P_{CO_2} fixed by equilibrium with (soil) gas. Figure 2 represents the results of a computation utilizing Chou et al's (1989) rate expressions in which progressive dissolution of equal surface areas of dolomite and Mg-free calcite is modelled at a fixed P_{CO_2} of $10^{-2.6}$, leading (at line A) to a calcite-saturated water with the Mg and Ca concentration similar to that of the main Ernesto chamber. The effects of dolomite dissolution become relatively more important (higher Mg/Ca) as dissolution proceeds. Once calcite is saturated, only dolomite can dissolve, and the ultimate possible cumulative ratio of calcite to dolomite dissolved is independent

of kinetics. Figure 2 indicates that the Mg/Ca ratio rises significantly between calcite saturation (line A) and dolomite saturation (line B).

Figure 3 summarizes the results of similar calculations across a range of calcite-dolomite bedrock compositions illustrating both the drop in Mg/Ca in calcite-saturated leachates compared with bedrocks due to preferential calcite dissolution and how this is most pronounced under lower P_{CO_2} conditions. A general agreement is found when these predicted compositions are compared with those of experimental leachates. At dolomite saturation, Mg/Ca ratios increase (not shown in figure), but are still short of the original values in the bedrock.

It can be concluded that at relatively short reaction times, a key process is preferential calcite dissolution over dolomite. This leads to leachate compositions being systematically lower in Mg/Ca with respect to bedrocks, although Mg/Ca ratios may increase in particularly slow-flowing waters in the shallow karstic aquifers. Other processes are required however, to cause the aqueous trace element to calcium ratios to deviate from those of the bedrocks.

4 CALCITE PRECIPITATION

The partition coefficients of Mg and Sr in calcite are $\ll 1$. Hence Mg/Ca and Sr/Ca increase in solutions which have precipitated calcite. Modelling shows that precipitation of calcite induced by degassing of CO_2 to log P_{CO_2} values around -0.5 lower than that of the primary leachates in the soil zone and epikarst, will lead to a several-fold increase in solute Mg/Ca and Sr/Ca. Precipitation of calcite by the common ion effect following dolomite dissolution during prolonged water-rock contact is much less effective. Also many shallow karst waters also have

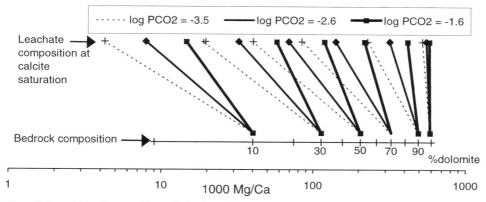

Figure 3. Comparison of compositions of bedrocks and leachates at calcite saturation at different PCO$_2$ values calculated using experimental data of Chou et al (1989).

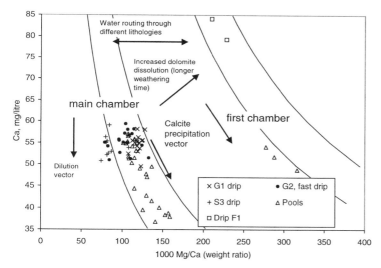

Figure 4. Crossplot of Ca and Mg/Ca from sites in the Ernesto cave

residence times of hours to weeks (Bottrell & Atkinson 1992): too short to approach dolomite saturation.

Calcite precipitation leads to a characteristic trend of data on plots of Ca versus either Mg/Ca or Sr/Ca. For example figure 4 illustrates that in the Grotta di Ernesto the composition of pool waters can be similar to those of dripwaters, but tend to lower Ca values (in the winter), resulting from calcite precipitation within the chamber. At other sites similar types of behaviour are shown by drip waters, pointing to the occurrence of calcite precipitation above the chamber roof induced by degassing into air pockets during dry conditions.

5 SELECTIVE LEACHING

Selective leaching of Mg and Sr relative to Ca from soil or aquifer materials can arise by non-congruent dissolution of calcite, by contribution of Mg and Sr from silicate or other phases, or by calcite precipitation in the soil zone in response to evaporation or freezing. It can be argued that all these effects will be enhanced when water availability is less. Amongst the sites shown in Figure 4, that from drip G2 appears to show this effect in that Mg/Ca is proportional to Sr/Ca and Mg/Ca is inversely proportional to drip rate. Leachates of weathered Ernesto bedrocks also showed the effect (Fairchild et al. 1996).

6 DISCUSSION

Figure 4 illustrates the way in which the effects of different processes on cave water chemistry may be distinguished in principle. However, the behaviour of different drip sites is highly variable. Some show virtually no change in either drip rate or chemical composition or both. Others cease flowing altogether in dry conditions. Nevertheless, drips feeding speleothems are commonly responsive to hydrological conditions in ways that should leave a permanent record in the deposit. Our experimental studies on calcite growth also yield similar results for Sr and Mg partition coefficients as those inferred from field sites, although the value for Sr is lower, at a given growth rate than literature values from experiments on solutions with higher ionic strengths Like Roberts et al. (1998), we also observe seasonal fluctuations in trace element to calcium ratios in ion microprobe scans across seasonally banded material. These should be interpretable in palaeohydrological terms.

7 CONCLUSIONS

The Mg/Ca and Sr/Ca of karst waters is controlled by:
a) the composition of the karstic aquifer and soil carbonates
b) differential dissolution of dolomite and calcite, with Mg/Ca increasing with water residence time
c) (in some cases) selective leaching of Mg and Sr relative to Ca, especially during drier climatic conditions.
d) (in some cases) enhanced calcite precipitation leading to increased Mg/Ca and Sr/Ca, arguably especially during dry weather conditions.

Under favourable circumstances cave waters and speleothems will show seasonal and longer term fluctuations in Mg/Ca and Sr/Ca reflecting palaeohydrological conditions.

ACKNOWLEDGEMENTS

This research was supported by NERC grant GR3/10801 and EU grant CT94-0509CV5V.

REFERENCES

Bar-Matthews, M., A. Ayalon, & A. Kaufman, 1997. Late Quaternary paleoclimate in the eastern Mediterranean region from stable isotope analysis of speleothems at Soreq Cave, Israel, *Quaternary Research* 47: 155-168.

Bottrell S.H. & T.C. Atkinson, 1992. Tracer study of flow and storage in the unsaturated zone of a karstic limestone aquifer. In: H. Hötzl and A. Werner, (eds*) Tracer Hydrology*: 207-211. Rotterdam: Balkema.

Busenberg, E. & L.N. Plummer, 1982. The kinetics of dissolution of dolomite in CO_2-H_2O systems at 1.5 to 65°C and 0 to 1 atm PCO_2. *American. Journal of Science* 282: 45-78.

Chou, K., R.M. Garrels, & R. Wollast, 1989. Comparative study of the kinetics and mechanisms of dissolution of carbonate minerals, *Chem. Geol.* 78: 269-282.

Fairchild, I.J., A.F. Tooth, Y. Huang, A. Borsato, S. Frisia, & F. McDermott, 1996. Spatial and temporal variations in water and stalactite chemistry in currently active caves: a precursor to interpretations of past climate In S. Bottrell (ed),*Proceedings of the fourth International Symposium on the Geochemistry of the Earth's Surface, Ilkley, Yorkshire, July 1996* :229-233,. Leeds: University of Leeds.

Fairchild, I.J., A. Borsato, A.F. Tooth, F. McDermott, S. Frisia, C.J. Hawkesworth, Y. Huang, and B. Spiro, (submitted). Controls on trace element (Sr-Mg) compositions of carbonate cave waters: implications for speleothem climatic records. *Earth and Planetary Science Letters*

Gascoyne, M., 1992. Palaeoclimate determination from cave calcite deposits, *Quaternary . Science. Res.* 11: 609-632.

McDermott, F., S. Frisia, Y. Huang, A. Longinelli, B. Spiro, T.H.E. Heaton, C.J. Hawkesworth, A. Borsato, E. Keppens, I.J. Fairchild, K. van der Borg, S. Verheyden. & E. Selmo, 1999. Holocene climate variability in Europe: evidence from $\delta^{18}O$ and textural variations in speleothems. *Quaternary Science Reviews*, xx:

Roberts, M.S., P.L. Smart, & A. Baker, 1998. Annual trace element variations in a Holocene speleothem, *Earth & Planetary Science Letters* 154: 237-246.

Geochemistry of the Earth's Surface, Ármannsson (ed.) © 1999 Balkema, Rotterdam, ISBN 90 5809 073 6

Chemical weathering of volcanogenic sediments, Skeidarársandur, SE Iceland

Ian J. Fairchild, Anna F. Tooth & Andrew J. Russell
Department of Earth Sciences, University of Keele, UK

ABSTRACT: Our hydrochemical investigations in the Skeiðarársandur outwash plain in SE Iceland are aimed at an understanding of low-temperature chemical weathering processes in a terrain of reactive volcanogenic sediments. The area forms a seaward-sloping 20-km long proglacial aquifer predominantly composed of sands and gravels resulting from jökulhlaup flood events. Seaward increases in alkalinity and electroconductivity result from a combination of enhanced chemical weathering, increased salinity of rainfall and evaporation. Close to the glacier margin, springs yield waters with PCO_2 lower than the atmosphere, some saturated with calcite, whereas PCO_2 higher than the atmosphere characterizes the distal sandur. Although dissolution of volcanic glass is important, a pronounced co-variation of Ca^{2+} and SO_4^{2-} in waters around 1 km from the ice front is consistent with coupled oxidation of pyrite and dissolution of calcite.

1 INTRODUCTION

Glacial environments are important contributors to global geochemical fluxes (Anderson et al. 1997). Young volcanic terrains with glassy lithologies should be significantly more reactive during chemical weathering than other silicate areas. In the Iceland region, this factor is combined with a high atmospheric precipitation enhancing the total flux to the oceans. Extensive studies on dissolution of volcanic glass at earth surface temperatures have been made (Gislason & Eugster 1987a, b, Crovisier et al. 1992). Dissolution is congruent at high water-rock ratios, but becomes increasingly incongruent in respect of different elements at higher solute concentrations. A study in SE Iceland by Raiswell & Thomas (1984) suggests that pyrite and calcite dissolution need also to be considered. In sedimentary and metamorphic terrains, species resulting from dissolution of pyrite and calcite dominate the solute load, owing to the rapid dissolution kinetics of these minerals (Raiswell 1984, Fairchild et al. 1994, Tranter et al. 1993).

There is increasing interest in the role of proglacial zones as aquifers (Boulton et al. 1995) and Sigurðsson (1990) has established the presence of distinctive glacier-fed aquifer zones in many parts of Iceland. The sandar along parts of the southern coastline are dominated by highly permeable sands and gravels resulting from periodic jökulhlaup (glacier outburst flood) events stimulated by

volcanic events. Skeiðarársandur is the longest active sandur in the world, stretching 20 km from ice margin to the coast, and is underlain by a minimum of 100 m of sediments. Meltwater and atmospheric precipitation is routed both via several major rivers and through groundwater. Hydrological studies by Bahr (1997) illustrated that the groundwater flows seawards and that the water table is close to the surface throughout the lower sandur where evaporation can be significant. There is also significantly more vegetation in this zone than in the ice-proximal part of the sandur. Bahr (1997) attributed solute Na and Cl to marine aerosol and solute Ca, Mg, K and silica to hydrolysis of volcanic glasses. Our study is aimed ultimately towards quantifying the nature and magnitude of low-temperature weathering fluxes of elements and isotopes in the sandur setting: initial elemental results are presented here.

2 METHODS

One hundred water samples were collected during July-August 1998 from streams, springs and shallow pits dug into the sandur surface. In-situ temperature was recorded, and pH and electroconductivity (automatically corrected to 25 °C) measured on-site. Waters were filtered immediately and separate acidified and unacidified aliquots were collected for cation and anion analysis respectively. Eighteen rainwater samples were collected from a permanent rain gauge between 18th July and 18th August, on each day when rain fell. Ionic species were determined in the laboratory by ion chromatography except for Sr^{2+} which was determined by Graphite Furnace Atomic Absorption Spectrometry. Alkalinity was determined by Gran titration using.6N acid in a Hach digital titrator in a field laboratory

Mean ionic balances (= (cations-anions)/(anions + cations)×100) were +4.5% for surface waters with a standard deviation of 4.3%, excluding a few outlying results which are not presented here. Electroconductivities calculated from the ionic compositions averaged 101% of the measured electroconductivity, being relatively lower where ionic balances were more positive (and vice-versa) suggesting that the main source of analytical uncertainty is in the alkalinity determination. This is supported by the small difference between cations and anions in rainwater samples (mean ionic

balance-1.1 ± 6.4 %) where alkalinity determination was inapplicable

3 RESULTS

Figure 1 and Table 1 illustrate the increase in both electroconductivity and alkalinity towards the coast. The hot springs are offset from the trend because of their high sulphate content. Hydrothermal sources contribute negligibly to the total mass balance in the system (except during jökulhlaup events, Steinthórsson. & Óskarsson 1983, Bjornsson & Kristmannsdóttir 1984).

The rise in alkalinity reflects increased weathering loads whereas the electroconductivity also includes an increased marine aerosol component near the coast. This is shown in Figure 2 by the higher chloride contents in the coastal and lower sandur (sample groups 1 and 2), previous

Table 1. Mean values of chemical parameters in different sample groups. SI_{cc} (calcite saturation index) and −log PCO_2 calculated using PHREEQM model. Solute concentrations in μeq/litre.

1. Coast (latitude 63° 47' 53'' N and longitude 17° 16' 35'' W).

2. Lower sandur (63° 52' N and longitude 17° 17' 33'' W).

3 and 4."Twin peaks" area (63° 59' N and longitude 17° 24' W) around 1 km from the ice margin, near the western side of the southern margin of the Skeiðarárjökull glacier. This site is laterally displaced to the west from a direct flowline to sites 1 and 2. 3. Springs, streams, lakes and groundwater fed by rainwater falling on a 19th century moraine system. 4. Springs, streams and groundwater fed both by rainfall and by glacier meltwater.

5. Rivers: samples of major rivers draining Skeiðarársandur.

6. Glacier margin and melted ice: meltwaters from various locations on southern margin of Skeiðarárjökull glacier.

7. Hot springs, near SE margin of Skeiðarárjökull glacier. Saturation index calculated using T = 10° C.

8. Rainwater, Bolti farmhouse, Skaftafell, near SE margin of Skeiðarárjökull glacier.

Sample group	1	2	3	4	5	6	7	8
n	9	7	29	28	6	18	3	18
Distance to sea (km)	1	8	19	19	15-20	20-21	22	21
T (°C)	14.1	14.5	11	8.1	6.3	5.4	30	n.d.
EC (μS/cm)	168	267	68	60	23	55	690	n.d.
-logPCO$_2$	2.9	2.8	4.0	4.3	3.5	3.7	3.6	n.d.
SI$_{cc}$	-1.0	-1.2	-1.0	-0.9	-1.8	-3.6	+0.2	n.d.
Ca^{2+}	560	600	240	230	230	110	1080	20
Mg^{2+}	230	230	80	70	50	30	70	10
Na$^+$	550	560	230	210	210	80	504	10
K$^+$	40	30	10	10	10	10	14	20
Sr^{2+}	0.6	0.6	0.3	0.2	0.3	0.1	13	n.d
Cl$^-$	470	760	140	110	70	30	3400	200
SO$_4$$^{2-}$	300	280	180	170	130	50	5200	100
NO$_3$$^-$	<10	30	10	10	<10	20	<10	<10
F$^-$	3	<3	<3	<3	<3	<3	65	<3
Alk-alinity (meq)	1100	900	400	400	400	200	1800	(0)

EC studies having shown that negligible chloride is derived from rock weathering. Some lower sandur samples lie close to the marine aerosol line; these presumably reflect a high proportion of local marine aerosol-enriched rainwater input, perhaps coupled with the effects of evaporation. The calcium and alkalinity contents of these samples are however similar to that of other samples.

Significant Na yields are noted in the Twin Peaks area and locally in water pools having opportunities for prolonged water-rock contact at the ice margin. A source of volcanic glass is provided by tephra which is particularly abundant on the glacier surface on the western glacier margin which is now rapidly melting back following a surge in 1991. Glassy basalts are also present in the reworked sands and gravels on the sandur. Overall in the data set there are crude correlations of Na with Ca, K and alkalinity (even when corrected for aerosol inputs), consistent with an origin of these species by carbonation of glass. On the other hand, Mg and Sr correlate better with Ca than with Na and a possible significant source of these species in carbonate should be considered.

The tightest correlation between solute parameters is provided by Ca and SO_4 (Figure 3) in the Twin Peaks area, both locally-derived and glacier-derived surface waters and groundwaters showing the same pattern. The slope of the covariation is around 2:1 in equivalent units. This slope is consistent with a coupled dissolution of pyrite and calcite (Tranter et al. 1993). In the Twin Peaks area, fine pebbles were counted (n = 800) at several localities and 32 samples of the different lithologies were thin-sectioned. Calcite is present as an accessory phase in some amygdaloidal basalts and breccias, whereas pyrite is found in rare altered rhyolites.

Figure 4 illustrates that samples in the Twin Peaks area display a trend from calcite-saturated samples, with P_{CO_2} values lower than 10^{-5} to samples with P_{CO_2} values approaching that of the atmosphere. There is no correlation between SO_4 and $-\log P_{CO_2}$. Preliminary modelling suggests that the various samples have experienced various

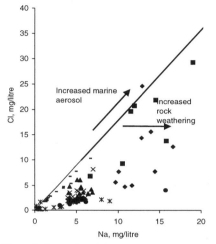

Figure 2. Cross-plot of Cl and Na with line indicating relative proportions in marine aerosol. Additional symbol: short horizontal lines are rain samples

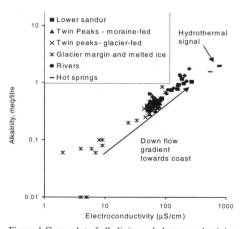

Figure 1 Cross-plot of alkalinity and electroconductivity

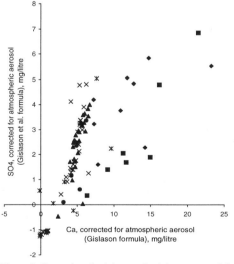

Figure 3. Cross-plot of sulphate and calcium corrected for total atmospheric input using the equations generated by Gislason et al. (1996).

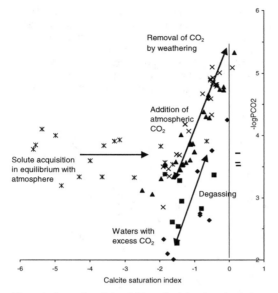

Figure 4. Cross-plot of –log PCO_2 and calcite saturation index.

degrees of consumption of CO_2 by calcite (and glass) weathering, and enrichment of CO_2 both from the atmosphere and by protons produced by pyrite oxidation.

By contrast, the glacier margin samples all have PCO_2 values close to atmospheric ($10^{-3.5}$) and hence have stayed in close contact with the atmosphere throughout. In the lower sandur PCO_2 values exceed atmospheric values which is tentatively attributed to introduction from soil zone plant respiration.

4 CONCLUSIONS

Preliminary analysis of the data indicates that the higher ionic loads encountered in lower parts of the sandur are due to a combination of evaporation (Bahr 1997), increased mineral weathering, and increased input of atmospheric aerosols. PCO_2 values in this zone are typically higher than atmospheric whereas they are below atmospheric close to the glacier. Volcanic glass weathering leads to increases in Na and K and some proportion of the Ca, other cation, and alkalinity loads.

In an area 1 km from the western ice front, both locally-derived rain waters and glacier-derived waters generate similar weathering trends in which a tight covariation of calcium and sulphate occurs, attributed to coupled oxidation of pyrite and dissolution of calcite. Calcite is also a potential source of significant Mg and Sr. This needs to be borne in mind when assessing the magnitude and significance of weathering fluxes, in particular of Sr isotopes.

ACKNOWLEDGEMENTS

This research is supported by a grant from Earthwatch International to Russell, Knight and Fairchild at Keele University. The dedication of numerous Earthwatch volunteers and the help of personnel in the Skaftafell National Park Warden's office and the Bolti farmhouse is gratefully acknowledged.

REFERENCES

Anderson, S.P., J.I. Drever, & N.F. Humphrey, 1997. Chemical weathering in glacial environments, *Geology* 25: 399-402.
Bahr, T. 1997. *Hydrogeologische Untersuchungen im Skeiðarasandur (Südisland)*. Müncher Geologische Hefte Reihe B, 3, xiv + 142 pp.
Bjornsson, H. & H. Kristmannsdóttir, 1984. The Grímsvötn geothermal area, Vatnajökull, Iceland. *Jökull* 34: 25-50.
Boulton, G.S., P.E. Caban, & K. van Gijssel, 1995. Groundwater flow beneath ice sheets: Part I – large scale patterns. *Quaternary Science Reviews* 14: 545-562.
Crovisier, J.-L., J. Honnorez,, B. Fritz, & J.-C. Petit, 1992. Dissolution of subglacial volcanic glasses from Iceland: laboratory study and modelling. Applied Geochemistry Suppl. Issue No. 1: 55-81.
Fairchild, I.J., L. Bradby, M. Sharp, & J.-L. Tison, 1994. Hydrochemistry of carbonate terrains in Alpine glacial settings. *Earth Surface Processes and Landforms* 19: 33-54.
Gislason, S.R. & H.P. Eugster, 1987a. Meteoric water-basalt interactions. I: A laboratory study. *Geochimica Comochimica Acta* 51: 2827-2840.
Gislason, S.R. & H.P. Eugster, 1987b. Meteoric water-basalt interactions. II: A field study in N.E. Iceland. *Geochimica Cosmochimica Acta* 51: 2841-2855.
Gislason, S.R., S. Arnorsson, & H. Armannsson, 1996. Chemical weathering of basalts in southwest Iceland: effect of runoff, age of rocks and vegetative/glacial cover. *American Journal of Science* 296: 837-907.
Raiswell, R. 1984. Chemical models of solute acquisition in glacial melt waters. *Journal of Glaciology* 30: 49-57.
Raiswell, R. & A.G. Thomas, 1984. Solute acquisition in glacial meltwater I. Fjallsjökull (south-east Iceland): bulk meltwater with closed system characteristics. *Journal of Glaciology* 30: 35-43.
Sigurðsson, F. 1990 Groundwater from glacial areas in Iceland. *Jökull*, 40: 119-146.
Steinthórsson, S. & N. Óskarsson, 1983. Chemical monitoring of Jökulhlaup water in Skeiðara and the geothermal system in Grímsvötn, Iceland. *Jökull* 33: 73-85.
Tranter, M., G. Brown, R. Raiswell, M. Sharp & A. Gurnell, 1993. A conceptual model of solute acquisition by Alpine glacial meltwaters. *Journal of Glaciology* 39: 573-581.

Geochemistry of the Earth's Surface, Ármannsson (ed.) © 1999 Balkema, Rotterdam, ISBN 90 5809 073 6

A leaching column study of bone meal amendments to metal polluted soil

M. E. Hodson & É. Valsami-Jones
Department of Mineralogy, Natural History Museum, London, UK

ABSTRACT: Many metal phosphates are extremely stable under the range of Eh and pH conditions encountered in the natural environment. It has been suggested that, for metal contaminated soils, the formation of metal phosphates could effectively "lock up" the metals *in situ* so that they would no longer pose a pollution problem. In this paper we report a leaching column investigation into the suitability of bone meal as a phosphorus source for the formation of metal phosphates and remediation of metal contaminated soils. Bone meal amendments reduced the release of metal ions into soil leachate, reduced the concentration of metal ions in pore waters, increased soil pH, reduced metal availability as assessed by DTPA and $CaCl_2$ extraction and reduced toxicity of soil leachate to soil organisms as assessed by bioassays. These results suggest that bone meal could be a suitable remediation treatment for metal contaminated soil.

1 BACKGROUND

The presence of high concentrations of metals in soils and waters associated with industrial, mining and waste disposal activities is a common and growing environmental problem. Many metals are known to be particularly harmful to human health and the remediation of metal contaminated land is essential in order to improve its environmental quality.

Phosphates of many metals (e.g. Pb, Zn, Cu) have exceptionally low solubilities with solubility coefficients = 10^{-60} – 10^{-80} (Nriagu, 1984). In addition they are stable over almost the entire range of Eh and pH encountered in the natural surface environment. Therefore, the formation of these stable phosphates in contaminated soils would immobilise metals resulting in reduced leaching of heavy metals into groundwater and reduced bioavailability to plant and animal life.

A variety of phosphate amendments, including soluble phosphate (K_2HPO_4) and rock phosphate (apatite) have been tested in laboratory and field trials (e.g. Cotter–Howells and Caporn, 1996, Ma *et al.*, 1993) and problems related to the rate of release of phosphate have been encountered (too fast or too slow respectively). Recent experiments (Valsami – Jones *et al.*, 1998) have shown that the solubility of synthetic apatite is greater than that of rock phosphate and, due to the structural similarity of synthetic apatite and bone meal it has been suggested that bone meal may provide a readily available, cost effective natural phosphate source for the remediation of soils contaminated with certain metals. This paper reports the results of a series of leaching column experiments designed to investigate the potential of bone meal as a remediation treatment for metal contaminated land.

2 EXPERIMENTAL MATERIALS

The soil used in the experiments was a metal contaminated made ground of pH = 3.1 from the former coke works site at Lambton, Sunderland in the U.K (Table 1). The site is typical of many former industrial areas throughout the UK and as such was deemed a suitable source for material to test the suitability of bone meal as a remediation treatment. The soil was sieved to remove the > 2 mm fraction and to homogenize the material. Commercially available bone meal was sieved to produce a 90 – 500 µm fraction.

Table 1. Soil metal content and mineralogy.

Metal content (ppm) after removal of > 2mm fraction	Zn	Ni	Cu	Pb
HF digest	93.6	35.5	86.3	233.0
HNO_3 – $HClO_4$ digest	85.7	31.0	77.2	222.0

Mineralogy (weight %) after removal of > 2mm fraction					
Quartz	Gypsum	Fe-oxide	Jarosite	Kaolinite	Illite-smectite
14	4	6	14	9	50

3 EXPERIMENTAL PROCEDURE

Bone meal was mixed with subsets of the soil in the proportions 1 g bone meal to 10, 50, 100 and 200 g soil. Triplicate sets of leaching columns containing 200 g of bone meal : soil mixture, together with control columns containing just soil were set up. The columns were irrigated twice daily at a rate equivalent to 0.9 m yr^{-1}, which is approximately the average annual precipitation rate in Sunderland, using a solution with a similar composition to rainfall in the Sunderland area. Leachate was collected from the columns over a period of 3144 hours, except for from the 1:10 bone meal : soil columns which were terminated after 1968 hours (and are consequently not referred to in the tables below). Leachate was acidified when it was collected and filtered through 0.2 μm filters prior to analysis by ICP – AES. The pH of unacidified leachate was measured after 1080 hours and this leachate was used for bioassay experiments to assess the toxicity of the leachate to the common soil organisms *Colpoda steinii*, *Escherichia coli* and *Pseudomonas fluorescens* (Forge *et al.*, 1993; Paton *et al.*, 1995). After the experiment ended the pore water of the soil was extracted by centrifugation for metal content analysis. The material from the columns was air dried. The pH of the soil was measured (after McLean, 1982). Metal availability in the samples to plants and soil organisms was assessed by DTPA and 0.01 M $CaCl_2$ extractions respectively (Norvell 1984; Pickering, 1986). The 1:10 bone meal to soil mixture was examined before and after the experiments using analytical SEM and XRD for metal phosphates.

4 RESULTS

4.1 *Leachate composition*

Zn, Ni and Cu were detectable in the leachate from the columns. Concentrations of these metals in the leachate decreased over time (e.g. Fig. 1).

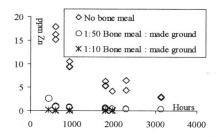

Figure 1. Change in concentration of Zn in leachate over time.

Table 2. Total cumulative release of metals from leaching columns over duration of experiment (μg of metal per g soil). Figures are means of triplicate columns. < DL = below detection levels.

Bone meal : soil	Released metals		
	Zn	Ni	Cu
Control	17.1	1.5	0.4
1:200	14.0	1.5	0.3
1:100	8.6	1.1	0.2
1:50	1.5	0.4	0.1

Concentrations of metal in the leachate decreased with increasing concentrations of bone meal in the soil and, after 3144 hours the columns containing the most bone meal had released the least metals (Table 2). At the end of the experiment Zn and Ni concentrations in the leachate from the 1:50 bone meal : soil treatments were below the maximum Environmental Quality Standards (EQS) for freshwater as laid out in the EC Dangerous Substances Directive (76/464/EEC) (0.5 ppm for Zn and 0.2 ppm for Ni) and were close to the permitted Cu level (0.02 ppm). Concentrations of these elements in the leachates from the other columns were higher than maximum permitted EQS values. The actual permitted EQS values decrease with decreasing pH and the maximum values might not be appropriate for solutions of the pH of the leachates reported here. pH was highest in the leachate from the columns containing the most bone meal. For example, after 1080 hours the pH of the leachate from the control columns was 3.75 whilst that from the columns containing the 1:50 bone meal : soil treatments was 4.04.

4.2 *Leachate toxicity*

After 1080 hours the leachate from the control columns and the columns containing the 1:100 and 1:200 bone meal : soil treatments was toxic to the *C. steinii*, killing them on contact and greatly reduced the growth rate of *E. coli* and *P. fluorescens*. The leachate from the columns containing the 1:50 bone meal : soil treatment was less toxic, did not kill the soil organisms and did not reduce growth rate so much.

4.3 *Pore water composition*

The concentration of metals in the pore waters generally decreased with increasing bone meal concentration (Table 3).

4.4 *Soil pH*

The pH of the soil increased with increasing bone meal concentration. At the start of the experiment

Table 3. Soil pore water composition (ppm) after the end of the experiments.

Bone meal : soil	Zn	Ni	Cu
Control	1.75	0.16	0.09
1:200	0.90	0.08	< DL
1:100	0.71	0.09	< DL
1:50	0.29	0.08	0.07

the pH of the soil was 3.1. At the end of the experiment the pH of the soil in all the experiments had increased. The pH of the soil in the control experiment had increased to 4.2. The pH of the bone meal : soil mixtures was 4.2, 4.4, 5.0 and 6.3 for the 1:200, 1:100, 1:50 and 1:10 bone meal : soil mixtures respectively.

4.5 *SEM and XRD observations*

Small Ni, Ti, Zn, Cu, Fe particles were detected using the SEM in the bone meal. At the end of the experiments small concentrations of metal bearing phosphate phases were detected in the 1:10 bone meal : soil mixture. These contained Ca, Zn, Pb and Cu but were too small to characterise more fully. No metal phosphate phases were detected using XRD due to their low concentration.

4.6 *Chemical extractions*

Generally the concentration of DTPA extractable metals increased with increasing bone meal concentration but the concentration of $CaCl_2$ extractable metals decreased (Tables 4a and 5a). Taking into account both the metals released from the columns during the experiments and the extraction of metals, apart from for Cu and DTPA

Table 4a. DTPA extractable metals in bone meal : soil mixtures at end of experiment (μg of metal per g of solid).

Bone meal : soil	Extracted metals		
	Zn	Ni	Cu
Control	2.5	0.2	3.4
1:200	2.6	0.2	3.7
1:100	3.3	0.2	4.7
1:50	3.9	0.3	4.9

Table 4b. Total release of metals from leaching columns (μg of metal per g of solid) including release by leaching and DTPA extraction.

Bone meal : soil	Extracted metals		
	Zn	Ni	Cu
Control	19.6	1.7	3.8
1:200	16.6	1.7	4.0
1:100	11.9	1.3	4.9
1:50	5.4	0.7	5.0

Table 5a. $CaCl_2$ extractable metals in bone meal : soil mixtures at end of experiment (μg of metal per g of solid).

Bone meal : soil	Extracted metals		
	Zn	Ni	Cu
Control	2.0	0.2	0.1
1:200	2.0	0.2	0.1
1:100	1.7	0.2	0.1
1:50	0.1	0.1	< DL

Table 5b. Total release of metals from leaching columns (μg of metal per g of solid) including release by leaching and $CaCl_2$ extraction.

Bone meal : soil	Extracted metals		
	Zn	Ni	Cu
Control	19.1	1.7	0.5
1:200	16.0	1.7	0.4
1:100	10.3	1.3	0.3
1:50	1.6	0.5	0.1

there was a decrease in the total release of metals from the soil with increasing bone meal concentrations (Tables 4b and 5b).

5 DISCUSSION

The bone meal additions to the soil reduced the concentration of metals passing into solution. This could have been due to adsorption of the metals onto the bone meal, the formation of metal phosphates or the precipitation of other metal phases in response to the pH rise of the soil. SEM observations indicate that at least some metal phosphates formed. The decrease in $CaCl_2$ extractable metals with increasing bone meal indicates that some of the reduction in metal concentration in the leachate was due to the formation of metal bearing phases rather than the adsorption of the metals onto the bone meal.

The increase in DTPA extractable metals with increasing bone meal content means that either 1) the bone meal was a source of DTPA extractable metal contaminants and / or 2) some of the metals which were immobilised by the bone meal were immobilised in a DTPA extractable form. Analysis of the bone meal by ICP – AES and analytical SEM indicates that it contains 127 ppm Zn, 3 ppm Ni and 11 ppm Cu in the form of metal splinters, probably derived during the bone crushing process from metal jaw crushers. In addition DTPA extractions show that of this metal 10 μg Zn, 0.4 μg Ni and 2 μg Cu are extractable per gram of bone meal. Whilst this does show that the bone meal is a source of metal contaminants, the concentration of DTPA extractable metal is insufficient by itself to explain the increase in DTPA extractable metals seen with increasing bone meal concentration. Thus some of the metals which were immobilised and prevented from passing into solution during the leaching column experiments must have been fixed in a DTPA extractable form. However, some of the increase in DTPA extractable metals will have been due to the fact that, in the experiments with lower concentrations of bone meal, a large proportion of potentially DTPA extractable metals had already passed into solution. If it is assumed that all the readily soluble metal which was lost from the column during the experiment would have been in a DTPA extractable form then, taking into account both the amount of metals passing into solution over the course of the leaching column experiments and the concentration of DTPA extractable metals from the bone meal : soil mixtures (Table 4b) it is seen that the bone meal amendments reduced the concentration of plant available Zn and Ni in the soil. The increase in total release of Cu with bone meal concentration indicates that the bone meal and soil reacted to convert insoluble Cu present in the soil into a still insoluble, but DTPA extractable form. The presence of metal contaminants in the bone meal could also explain the high Cu concentration in the pore waters of the 1 : 50 bone meal : soil treatments.

Both high levels of metals and low pH can have a toxic effect on *C. steinii* and *P. fluorescens* soil organisms so it is not possible to say whether the decrease in the toxicity of the leachate with increasing bone meal amendments was due to the rise in pH or the fall in metal concentrations. However, the response of *E. coli* is independent of pH between pH of 3 and 11 (Palmer *et al.*, 1998) so the reduction in leachate toxicity to this organism is probably due to the reduction in metal concentrations in the leachate.

6 CONCLUSION

The leaching column experiments carried out here indicate that bone meal amendments have the potential to be a useful remediation treatment for metal contaminated land.

1. Despite metal impurities in the bone meal, bone meal amendments resulted in immobilisation of metals in the soil. At least some of the metals were fixed as metal phosphates.
2. Bone meal amendments reduced the concentration of metals in solutions draining through the soil and increased solution pH. This would help remediate streams draining from metal contaminated land.
3. Bone meal amendments generally reduced metal concentrations in the soil pore water and reduced metal availability to plants and soil organisms (DTPA and CaCl₂ extractions).

Further experiments are now in progress to assess the sustainability of the treatment, the applicability of the treatment to less acidic soils and the applicability of the treatment to more heavily contaminated soils prior to larger scale field testing.

ACKNOWLEDGEMENTS

This work is funded by a grant from The BOC Foundation and by the Environment Agency. Meryl Batchelder carried out the XRD analysis.

REFERENCES

Cotter–Howells, J.D. and S. Caporn 1996. Remediation of contaminated land by formation of heavy metal phosphates. *Ap. Geochem.* 11:335–342.

Forge, T.A., M.L. Berrow, J.F. Darbyshire and A. Warren 1993. Protozoan bioassays of soil amended with sewage sludge and heavy metals, using the common soil ciliate *Colpoda steinii*. *Biol. Fertil. Soils* 16:282-286.

Ma, Q.Y., S.J. Traina, T.J. Logan and J.A. Ryan 1993. In-situ lead immobilisation by apatite. *Environ. Sci. Technol.* 27:1803-1810.

McLean, E.O. 1982. Soil pH and lime requirement. In A.L. Page, R.H. Miller and D.R. Keeney (eds) *Methods of soil analysis part 2. Chemical and microbiological properties*: 199-224. Wisconsin, Soil. Sci. Soc. Am.

Norvell W.A. (1984) Comparison of chelating agents as extractants for metals in diverse soil material. *Soil Sci. Soc. Am. J.* 48:1285–1292.

Nriagu J.O. 1984. Formation and stability of base metal phosphates in soils and sediments. In J.O. Nriagu and P.B. Moore (eds) *Phosphate minerals*: 318 – 329. London: Springer – Verlag.

Palmer G., R. McFadzean, K. Killham, A. Sinclair and G.I. Paton 1998. Use of lux-based biosensors for rapid diagnosis of pollutants in arable soils. *Chemosphere* 36:2683-2697.

Paton, G.I., C.D. Campbell, L.A. Glover and K. Killham 1995. Assessment of bioavailability of heavy metals using *lux* modified constructs of *Pseudomonas fluorescens*. *Lett. Appl. Microbiol.* 20:52-56.

Pickering W.F. 1986. Metal ion speciation – soils and sediments (A review) *Ore Geol Revs.* 1:83-146.

Valsami – Jones É, K.V. Ragnarsdottir, A. Putnis, D. Bosbach, A.J. Kemp and G.Cressey 1998. The dissolution of apatite in thepresence of aqueous metal cations at pH 2 – 7. *Chem. Geol.* 151:215-233.

Geochemistry of the Earth's Surface, Ármannsson (ed.) © 1999 Balkema, Rotterdam, ISBN 90 5809 073 6

Holocene sediment-borne chemical storage in the Yorkshire Ouse basin, UK

K.A. Hudson-Edwards
Department of Geology, Birkbeck College, University of London, UK

M.G. Macklin
Institute of Geography and Earth Sciences, University of Wales, Aberystwyth, UK

M.P. Taylor
School of Geography, University of Oxford, UK

ABSTRACT: Floodplain overbank sediments are often used to evaluate the influence of environmental change on sediment and chemical fluxes within river basins. This paper presents the results of an investigation of sediment-borne chemical storage during the Holocene in seven floodplain reaches in the Yorkshire Ouse basin in North-East England. Storage has been greatest in the last 1000 years, probably due to climatic influences, increased anthropogenic activity (mining, industry, agriculture), and the increased delivery of fine-grained sediment to the catchment, especially during the High Middle Ages.

1 INTRODUCTION

Overbank sediment profiles record both the natural and anthropogenic geochemical evolution of floodplains, and have been used as sampling and mapping media for the assessment of heavy metal pollution (e.g. Macklin et al. 1994; Bölviken et al. 1996). Few attempts have been made, however, to use overbank sediments as media for assessing the rate and amount of long-term chemical storage in floodplains, particularly on a basin-wide scale. This is surprising, considering the unavailability of Holocene-aged water (normally used to evaluate chemical fluxes in river basins, e.g. Dupré et al. 1996), and the relatively good preservation of overbank sequences. In order to put present-day chemical fluxes (e.g. Neal et al. 1997) into context, it is important to quantify and explain past chemical fluxes. Overbank sediments provide a window into past sediment-borne chemical storage, and thereby fluxes, within river basins, provided several reaches within the basin are studied (Macklin et al. 1994).

The Yorkshire Ouse basin in north-east England is an ideal study site for long-term river catchment studies because because the human (Tinsley, 1975, Campbell, 1990) and Holocene sedimentation histories (Gaunt et al. 1971, Catt, 1987, Gaunt, 1994, Taylor & Macklin, 1997, Howard & Macklin, 1998) are reasonably well understood. This paper presents the results of a study on sediment-borne chemical floodplain storage during the Holocene in the Yorkshire Ouse basin. This research is part of the UK Natural Environment Research Council's Land-Ocean Interaction Study programme which is aimed at evaluating the influence of environmental change on sediment and chemical fluxes from river basins to east coast estuaries and coastal zones.

2 STUDY AREA

The major rivers of the Yorkshire Ouse basin include the Swale, Nidd, Ure, Ouse, Wharfe, Aire and Derwent, which have a total catchment area exceeding 8000 km². The first six rivers rise on Pennine uplands (to over 700 m) and drain in a south-easterly direction, and the River Derwent rises on the North York Moors and flows in a southerly direction, into the Humber Estuary.

The Yorkshire Ouse basin is underlain, from west to east, by Carboniferous to Cretaceous sedimentary rocks which include limestones, shales, sandstones, thin coals, marls, mudstones, and chalks (Gaunt, 1994). A wide variety of Late Quaternary glacial, periglacial, lacustrine, fluvial, aeolian and estuarine deposits also occur in valleys and in river interfluves, and are typically thicker and more extensive in the eastern lowland part of the basin near the Humber Estuary (Gaunt, 1994).

3 METHODOLOGY

Seven one to two km long floodplain reaches on the Rivers Swale, Ure, Nidd, Ouse, Wharfe, Aire and Derwent, located at Catterick, Myton, Kirk Hammerton, York, Tadcaster, Beal and Wressle, respectively, were examined in detail for this study. At

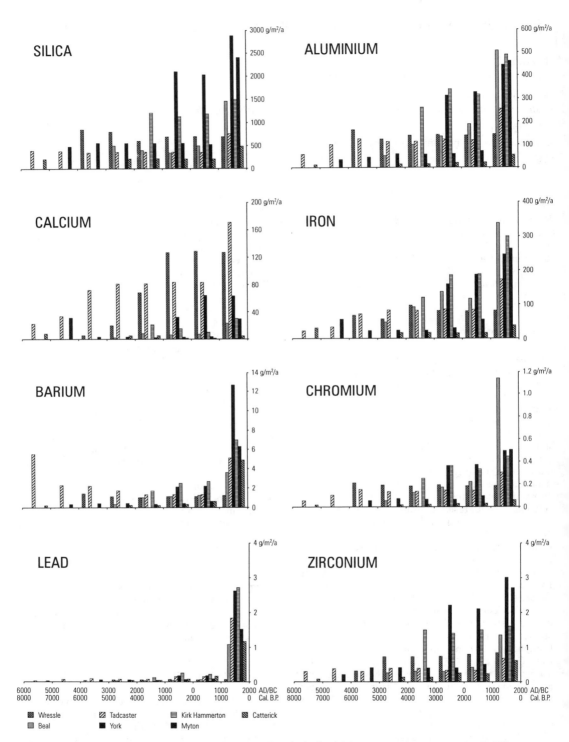

Figure 1. Storage of selected major and trace elements in Ouse basin floodplain reaches over time intervals of 1000 years, expressed in g of metal per m² per a.

each of the reaches, valley floor morphology was mapped onto aerial photographs at a scale of 1:10,000, and representative floodplain and terrace units exposed and sampled both by mechanical excavator and percussion drilling. Terrace and floodplain surfaces, palaeochannels, trench and drill core heights were determined to an accuracy of 0.01 m using an electronic measuring device (EDM), with reference to the nearest accessible Ordnance Survey benchmark.

Sediment samples were collected from distinct sedimentological horizons of 0.06-0.73 m (average 0.33 m) thickness from representative trench sections and cores from each of the floodplain reaches. Thicker sedimentological units (> 0.5 m) were subdivided into smaller samples. The samples were air dried, disaggregated and sieved to pass through a 2 mm aperture. They were then reduced to powder in an agate mill and pressed to produce pellets of c. 12 g mass for determination of a range of major oxides and trace elements by high-precision X-ray flourescence (XRF, Philips PW2400 and PW1480/10) at the facilities of the British Geological Survey. Calibrations were verified by analysis of Reference Materials and a stable silica glass disc, containing all the analytes of interest, was analysed regularly as a quality control standard with the data plotted on QC charts to monitor performance.

Age control was provided by radiometric (^{14}C) dating of organic material (wood, bone, peat) incorporated within alluvial deposits. All ^{14}C dates were calibrated using the Oxcal Program, v2.18 (Ramsey, 1995), and the results are reported as calibrated years before present (cal. BP) at the $\pm 2\sigma$ level for analytical significance.

Estimates of the amounts of sediment-borne major and trace elements stored per annum in the Yorkshire Ouse floodplains over 1000 year intervals for the last 8000 years were calculated according to a scheme described in Hudson-Edwards et al. (in press). These are expressed as a storage index (in g/m^2/annum) which can be compared from reach to reach.

4 RESULTS AND DISCUSSION

The chemical storage indices (Fig. 1) reflect several characteristics and processes within the Ouse catchment. These include the nature of the underlying bedrock and Quaternary deposits, the amount of weathering and erosion which have taken place, the geomorphology of the reach, and the nature and amount of anthropogenic activity in the catchment.

The composition of the alluvial source material is indicated by the storage indices. For example, the large Ca storage indices at the Tadcaster and Wressle reaches (Fig. 1) reflect the limestone and chalk bedrock, respectively, which underlie these reaches. Relatively high Si and Zr storage at York reflects the sandy, quartz- and zircon-rich nature of the alluvial sediment there, and high Al, Fe and K (not shown) storage, particularly at Kirk Hammerton, reflect clay mineral-rich alluvial material. Both of these types of alluvium are derived mainly from the Carboniferous sandstone and shale lithologies of the Yorkshire Dales. The large chemical storage indices at Kirk Hammerton may also be due to the unconfined nature of the reach, providing room for sediment and therefore, chemical storage. This is also true of the reach at Beal. In contrast, the reaches at York and Tadcaster are narrow and confined by large Quaternary deposits, restricting the amount and rate of sediment deposition.

For many elements (e.g. Si, Al, Ca, Fe, Cr, Zr) the storage indices have been steadily increasing since the early Holocene. One of the major factors responsible for these increases may be the timing and rate of alluviation. Macklin et al. (in press) have summarised eight major phases of river alluviation and incision in the Ouse catchment, between c. 3700-3380 cal. BC, 2320-1850 cal. BC, 900-430 cal. BC, cal. AD 645-775, cal. AD 1015-1290 and cal. AD 1420-1645. The first of these (c. 3700-3380 cal. BC) occurred mainly in the piedmont and lowland parts of the Ouse catchment, and is reflected in notable increases in Al and Fe storage at Beal and Wressle. Al and Fe are associated with clay and Fe oxides minerals, both of which are abundant in the fine-grained Ouse floodplains sediments of this age.

Many of the major phases of river alluviation coincide with inferred climate shifts identified by Smith (1985) and Chiverell (1998), and may thus be due to increased flooding and related sediment mobilisation and re-deposition. This is particularly true of the third phase of river alluviation between c. 900-430 cal. BC (Macklin et al. in press), and is reflected in relatively high amounts of chemical storage at all of the Ouse reaches (Fig. 1). This storage could also have been exacerbated by woodland clearance from c. 1850 cal. BC until 100 cal. BC (Tinsley, 1975, Atherden, 1976).

The indices in Figure 1 indicate that for all of the elements, the period of greatest storage has been the last 1000 years. This can probably be intimately linked to sedimentological factors in that the most rapid phases of fine-grained alluviation in the Ouse basin during the Holocene. This is recorded at Myton and Beal between cal. AD 1015 and 1290 (Macklin et al. in press), and is probably due to intense agricultural activity in the Yorkshire Dales during the High Middle Ages. Coincidental shifts to a wetter climate at this time (Chiverrell, 1998) may have accelerated sediment delivery and deposition within the catchment. Walling et al. (1998) have also recorded considerable present-day floodplain storage within the Ouse catchment in the rivers Swale, Ure and Ouse.

There is a very sharp rise in Pb storage in all of the reaches, and Cr storage in the Beal reach in the last 1000 years. This is also shown by Zn and in some cases, Cu, and can be attributed to intensive heavy metal mining in the Yorkshire Dales (Dunham and Wilson, 1985) and industrial activity in the Aire catchment (Dawson, 1997), especially during the last 250 to 300 years (cf. Hudson-Edwards et al. in press).

5 CONCLUSIONS

Sediment-borne chemical storage in the Yorkshire Ouse basin has increased consistently throughout the Holocene, and can be attributed to sedimentary, land-use, mining and industrial activity, and climatic factors. This information is critical in assessing present-day storage, although it should be noted that the floodplain overbank sediments used to evaluate the chemical storage provide only a window into alluvial, rather than incision events, and thus do not record the material which may have once been stored in the floodplain and subsequently eroded.

REFERENCES

Atherden, M.A. 1976. Late Quaternary vegetational history of the North York Moors, III, Fen Bogs. *Journal of Biogeography* 3: 115-124

Bölviken, B., Bogen, J., Demetriades, A., De Vos, W., Ebbing, J., Hindel, R., Langedal, R., Locutura, J., O'Connor, P., Ottesen, R.T., Pulkkinen, E., Salminen, R., Schermann, O., Swennen, R, Van der Sluys, J. & Volden, T. 1996. Regional geochemical mapping of Western Europe towards the year 2000. *Journal of Geochemical Exploration* 56: 141-166.

Campbell, B.M.S. 1990. People and land in the Middle Ages, 1066-1500. In R.A. Dodgshon & R.A. Butlin (eds) *An Historical Geography of England and Wales*: 69-121. London: Academic Press.

Catt, J.A. 1987. The Quaternary of East Yorkshire and adjacent areas. In S. Ellis (ed) *East Yorkshire Field Guide*: 1-14. Oxford: Quaternary Research Association.

Chiverrell, R.C. 1998. *Moorland vegetation history and climate change on the North York moors during the last 2000 years*. PhD thesis, University of Leeds.

Dawson, E.J. 1997. *The dispersal, storage and remobilisation of heavy metals in the River Aire contaminated by urban and industrial wastes*. PhD thesis, University of Leeds.

Dunham, K.C. & Wilson, A.A. 1985. Geology of the Northern Pennine Orefield, Volume 2 Stainmore to Craven. *Economic Memoir of the British Geological Survey*, Sheets 40, 41, 50, and parts of 31, 32, 51, 60, and 61.

Dupré, B., Gaillardet, J., Rousseau, D. & Allègre, C.J. 1996. Major and trace elements of river-born material: The Congo Basin. *Geochimica et Cosmochimica Acta* 60: 1301-1321.

Gaunt, G.D., Jarvis, R.A. & Mathews, B. 1971. The late Weichselian sequence in the Vale of York. *Proceedings of the Yorkshire Geological Society* 38: 281-284.

Gaunt, G.D. 1994. Geology of the country around Goole, Doncaster and the Isle of Eaxholme. *Memoirs of the British*

Geological Survey, sheets 79 and 88 (England and Wales). London: HMSO.

Howard, A.J. & Macklin, M.G. (eds) 1998. *The Quaternary of the Eastern Yorkshire Dales: Field Guide, The Holocene Alluvial Record*. London: Quaternary Research Association.

Hudson-Edwards, K.A., Macklin, M.G. & Taylor, M.P. in press. 2000 years of sediment-borne heavy metal storage in the Yorkshire Ouse basin, NE England, UK. *Hydrological Processes*.

Macklin, M.G., Ridgway, J., Passmore, D.G. & Rumsby, B.T. 1994. The use of overbank sediment for geochemical mapping and contamination assessment: results from selected English and Welsh floodplains. *Applied Geochemistry* 9: 689-700.

Macklin, M.G., Taylor, M.P., Hudson-Edwards, K.A. & Howard, A.J. in press. Holocene environmental change in the Yorkshire Ouse basin and its influence on river dynamics and sediment fluxes to the coastal zone. *Geological Society Special Publication*.

Neal, C., Robson, A. J., Jeffery, H. A., Harrow, M. L., Neal, M., Smith, C. J. & Jarvie, H. P. 1997. Trace element interrelationships for Humber rivers: inferences for hydrological and chemical controls. *Science of the Total Environment* 194/195: 321-343.

Ramsey, C.B. 1995. *Oxcal Program v. 2.18*. Oxford: Oxford Radiocarbon Accelerator Unit, Oxford University.

Smith, B. 1985. *A palaeoecological study of raised mires in the Humberhead Levels*. PhD thesis, University of Wales.

Taylor, M.P. & Macklin, M.G. 1997. Holocene alluvial architecture and valley floor development on the River Swale, Catterick, North Yorkshire, U.K. *Proceedings of the Yorkshire Geological Society* 51: 317-327.

Tinsley, H.M. 1975. The former woodland of the Nidderdale Moors (Yorkshire) and the role of early man in its decline. *Journal of Ecology* 63: 1-26.

Walling, D.E., Owens, P.N. & Leeds, G.J.L. 1998. The role of channel and floodplain storage in the suspended sediment budget of the River Ouse, Yorkshire, UK. *Geomorphology* 22:225-242.

Geochemistry of the Earth's Surface, Ármannsson (ed.)© 1999 Balkema, Rotterdam, ISBN 90 5809 073 6

The geochemistry of disseminated calcite in high Himalayan silicate rocks

A. D. Jacobson & J. D. Blum
Department of Earth Sciences, Dartmouth College, Hanover, N.H., USA

ABSTRACT: Mass-balance calculations of stream chemistry in the Raikhot watershed located within the High Himalayan Crystalline Series (HHCS) of northern Pakistan lead Blum et al. (1998) to suggest that the dissolved flux of radiogenic Sr is derived dominantly from the dissolution of trace amounts of calcite contained within HHCS rocks. Here, we confirm the presence of disseminated calcite in silicate rocks obtained from the Raikhot watershed and other nearby locations. In the Raikhot watershed, molar $Ca/(1000*Sr)$ and $^{87}Sr/^{86}Sr$ ratios of disseminated calcite range from 1.33 to 5.19 and 0.794039 to 0.930619, respectively. Elsewhere in the region, disseminated calcite $Ca/(1000*Sr)$ and $^{87}Sr/^{86}Sr$ ratios range up to 3.69 and 0.771244, respectively. These findings demonstrate that disseminated calcite with highly radiogenic Sr is a ubiquitous trace phase within HHCS silicate rocks. Thus, the dissolved Sr flux leaving the HHCS cannot be used to directly estimate rates of silicate weathering and atmospheric CO_2 consumption.

1 INTRODUCTION

The composition of stream water in catchments underlain by igneous rock is typically attributed to the chemical weathering of silicate minerals (e.g. Garrels & Mackenzie 1967). However, more recent studies both in the field (e.g. Drever & Hurcomb 1986; Mast et al. 1990; Blum et al. 1998; Horton et al. 1999) and in the laboratory (e.g. White et al. 1999) indicate that excess Ca observed in streams draining granitoid watersheds may be due to the weathering of disseminated calcite crystallized during magmatic cooling or precipitated during hydrothermal alteration. Moreover, ongoing studies suggest the presence of calcite within granitoid rocks is more widespread than previously thought (White et al. 1999). Research addressing the consequences of hydrothermal fluid flow has noted varying degrees of Sr isotope equilibration between coexisting silicate and carbonate bands in high-grade metamorphic rocks (e.g. Bickle et al. 1995; Gazis et al. 1998). In particular, Blum et al. (1998) proposed that isotopically re-equilibrated calcite existed within the silicate rocks of the high Himalaya, although they did not measure this endmember directly. Given that carbonate weathering does not influence long term atmospheric CO_2 concentrations, the presence of disseminated calcite implies that geochemcial mass balance studies which seek to relate the solute and Sr isotope chemistry of streams draining silicate terrains to changes in climate may overestimate the contribution provided by silicate weathering.

This study investigated the occurrence and geochemistry of disseminated calcite found within fresh silicate rocks obtained from the Nanga Parbat massif located within the High Himalayan Crystalline Series (HHCS) of northern Pakistan. We examined 11 polished thin sections, 5 from the Raikhot watershed and 6 from a wider area within the massif, with optical microscopy and cathodoluminescence (CL) in order to identify the presence of disseminated calcite. The use of CL to identify carbonate minerals is well documented (e.g. Marshall 1988) and is often more useful than optical microscopy which cannot always provide the resolution necessary to identify trace phases or complex internal structures. We further analyzed the chemistry of the calcite by pulverizing rock samples and sequentially leaching them in deionized water and 4N acetic acid. The water leach removes ions easily susceptible to hydration as a result of the pulverizing process and simulates the weathering of fresh rock surfaces generated by glacial grinding in the alpine environment of the HHCS. Further leaching of the water-rinsed residue in 4N acetic acid preferentially dissolves the carbonate component of the samples and allows for the analysis of the chemistry and Sr isotope ratios of the disseminated calcite. The specific goal of this research was to test the assertions of Blum et al. (1998) that stream water chemistry in the HHCS is strongly influenced by the dissolution of a calcite endmember possessing molar $Ca/(1000*Sr)$ ratios near 4.5 and $^{87}Sr/^{86}Sr$ ratios equivalent to 0.82. In addition, the results of this study are relevant to larger debates surrounding the use of the marine Sr isotope record as a proxy for silicate weathering and atmospheric CO_2 consumption in models that attempt to relate Himalayan uplift to changes in global climate (e.g. Raymo & Ruddiman 1992; Berner 1994; Goddéris & François 1995).

2 GEOLOGIC SETTING

The ~8125 m Nanga Parbat massif is located in the extreme northwestern portion of the Himalayan syntaxis, and the ~300 km² Raikhot watershed is found at an average elevation of ~4000 m on the northern side of the mountain. The bedrock of the massif, considered to be representative of much of the HHCS, stretches for ~2000 km across the length of the Himalayan Mountains and primarily consists of high-grade quartzofeldspathic biotite gneiss and schist intruded by small leucogranitic dikes and plutons. Narrow lenses of marble as well as calcsilicate and amphibolite schists comprise ~1% of the outcrop exposure area within the Raikhot catchment. The rocks of the Nanga Parbat massif have metasedimentary protoliths that were primarily pelitic sedimentary rocks with ages of ca. 1.8 Ga (Zeitler et al. 1993). Because these rocks have been metamorphosed and rapidly uplifted within the last 10 m.y. (Zeitler & Chamberlain 1991; Zeitler et al. 1993), biotite and feldspars have nearly concordant Ar-Ar and Rb-Sr ages (e.g. Zeitler et al. 1993; Gazis et al. 1998). In addition, the Nanga Parbat massif is characterized by high rates of physical denudation, and much of the landscape has been significantly affected by glaciation.

3 METHODS

3.1 Leaching

Approximately 5 to 10 grams of rock were crushed to a fine powder in a shatterbox, and 1 gram subsamples were weighed into 15 ml polypropylene centrifuge tubes. The samples were leached with 10 ml of deionized water for 6 h, centrifuged at 2500 rpm for 20 minutes, and the solutions were passed through 0.45 μm filters into teflon beakers. The carbonate fraction was dissolved by leaching the resulting residue with 10 ml of ultrapure 4N acetic acid for 6 hours. The samples were again centrifuged and filtered. All samples were then taken to dryness and re-dissolved in 8 ml of 5% HNO_3.

3.2 Elemental chemistry and Sr isotope analysis

Water leaches were analyzed for Ba, Ca, Fe, K, Mg, Mn, Na, Si, and Sr using a Finnigan ELEMENT inductively coupled plasma mass spectrometer (ICP-MS). Acetic acid leaches were analyzed for Ba, Ca, Fe, K, Mg, Mn, Na, Si, and Sr using a Spectro inductively coupled plasma optical emission spectrometer (ICP-OES). For all samples, Rb was measured using the ICP-MS. Analyses of SRM 1643d and USGS QLO-1 standards were analyzed to test the accuracy and precision of the methods. Measured and reported values agreed within ± 5% for the ICP-MS and ± 10% for the ICP-OES.

A pure Sr fraction was separated from each sample by eluting an aliquot through a quartz cation exchange column packed with Eichrom Sr-Spec resin. Approximately 75 ng of Sr was loaded onto a W filament with Ta_2O_5 powder and H_3PO_4, and the $^{87}Sr/^{86}Sr$ ratios were measured on a Finnigan 262 multiple-collector thermal ionization mass spectrometer. The $^{86}Sr/^{88}Sr$ ratios were normalized to 0.1194. The ^{87}Rb isobaric interference was monitored by measuring ^{85}Rb during each analysis, and a small correction was made to the measured $^{87}Sr/^{86}Sr$ ratios. Total procedural blanks were less than 20 pg for Sr and 1 pg for Rb. Analyses of SRM-987 yielded a mean $^{87}Sr/^{86}Sr$ ratio of 0.712068 ± 11 (±2σ, n=15) during the period of analysis.

4 RESULTS AND DISCUSSION

4.1 The occurrence of calcite

Calcite was identified in roughly half of the polished thin sections using optical microscopy. Cathodoluminescence of the same slides clearly showed the presence of calcite in all samples. The calcite, which luminesces bright orange relative to the darker hues of most co-existing silicate minerals, occurs as disseminated blebs within individual silicate grains as well as between grain boundaries and along fractures within the rock.

4.2 Correction for silicate dissolution

The objective of the progressive leaching procedure was to dissolve the disseminated calcite component within silicate rocks and measure the $Ca/(1000*Sr)$ ratio and the Sr isotope composition of the dissolution products. However, the acetic acid leaching procedure also dissolved some silicate material in addition to calcite, as evidenced by significant concentrations of K, Na, Rb, and Si in the solutions. To determine the composition of the calcite, we had to estimate the Ca and Sr contribution from silicate dissolution and subtract this from the measured concentrations. Thus, we assumed that Na in the acetic acid leach was derived solely from plagioclase dissolution and subtracted Ca and Sr in proportion to the Ca/Na (0.61) and Sr/Na (2.73) molar ratios of plagioclase in a typical Nanga Parbat gneiss. We view this as a reasonable methodology given that plagioclase is the dominant source of Ca, Sr, and Na in these Nanga Parbat rock samples. Other silicate minerals that contain significant concentrations of these elements - such hornblende, titanite, and apatite - occur in very low abundance relative to plagioclase. The corrected Ca concentrations were also used to calculate the total amount of disseminated calcite in the silicate rock samples. This correction was only applied to the major element data, as it does not significantly affect the measured $^{87}Sr/^{86}Sr$ ratios. Even though silicate rocks from the Nanga Parbat massif exhibit a wide range of Sr isotope ratios (0.7721 – 1.0642), major silicate minerals within each hand sample studied possess nearly identical Sr isotope ratios as a result of very recent thermal re-equilibration during metamorphism (Gazis et al. 1998). Our observation of similar Sr isotope ratios measured in the water and acetic acid leaches further supports this assumption.

4.3 Ca/(1000*Sr) and $^{87}Sr/^{86}Sr$ ratios

Raikhot watershed rocks contain between 0.03 and 0.28 wt% disseminated calcite with molar Ca/(1000*Sr) and $^{87}Sr/^{86}Sr$ ratios ranging from 1.33 to 5.19 and 0.794039 to 0.930619, respectively (Table 1). Elsewhere in the Nanga Parbat massif, disseminated calcite represents between 0.59 and 13.1 wt% of the total rock sample, with Ca/(1000*Sr) and $^{87}Sr/^{86}Sr$ ratios varying between 0.50 to 3.69 and 0.709979 to 0.768697, respectively. Figure 1 shows the Ca/(1000*Sr) and $^{87}Sr/^{86}Sr$ ratios for disseminated calcite within Raikhot watershed rocks. In addition, analyses of the carbonate and silicate fractions of individual rock samples as well as a sample of Raikhot riverbed sand are shown in Figure 1. The riverbed sand fractions are thought to be representative of the average carbonate and silicate endmembers for the watershed (Blum et al. 1998). Figure 1 also includes data for three types of water samples collected in the Raikhot watershed. Clear streams drain unglaciated granite and gneiss catchments and therefore contain no glacial silt. Field and laboratory filtered symbols refer to samples of silt laden glacial waters, either filtered in the field immediately following collection or filtered in the laboratory six months after collection. The riverbed sand carbonate and disseminated calcite endmembers define a mixing relationship that encompasses all water samples and shows little influence from the weathering of bulk silicate material. Clear

Table 1. Disseminated calcite content and composition in silicate rocks.

Sample	CaCO$_3$ (wt%)*	Ca/(1000Sr) *,**	$^{87}Sr/^{86}Sr$
Raikhot Watershed			
S6D2	0.10	4.43	0.845463
S6D5	0.08	5.19	0.827996
S6D6	0.28	2.18	0.842707
S9E1	0.08	0.98	0.794039
S9E6	0.03	1.33	0.930619
Nanga Parbat Region			
FNS8 BI3A	0.59	2.25	0.768697
CHURIT	0.83	3.69	0.753551
KC4	3.63	2.80	0.723742
RUS23	0.82	2.97	0.771244
66/27B	13.1	1.94	0.709979
5/30C	0.80	0.50	0.715757

* Calculated after correction for silicate dissolution
** Molar ratio

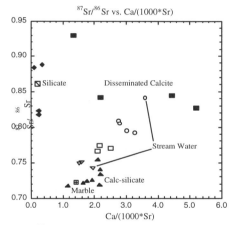

Figure 1. $^{87}Sr/^{86}Sr$ plotted versus Ca/(1000*Sr) ratio for Raikhot watershed disseminated calcite, water samples, rocks, and riverbed sand. Disseminated calcite (■), Carbonate rocks (π), Riverbed sand carbonate (⊞), Silicate rocks (▣), Riverbed sand silicate (◪), Clear streams (O), Field-filtered water (□), Lab-filtered water (σ).

large effect disseminated calcite weathering. Due to the homogenization of all rock types by glacial processes, field-filtered glacial waters show a significant contribution from small marble outcrops denuded by the Raikhot glacier. Lab-filtered glacial waters plot very close to the riverbed sand carbonate endmember for the watershed signifying marble dissolution during the time between collection and filtration of the suspended sediment in the bottle. These findings attest to the high reactivity of all carbonates in the watershed, but particularly demonstrate that the weathering of disseminated calcite with high $Ca/(1000*Sr)$ and $^{87}Sr/^{86}Sr$ ratios is the predominant source of dissolved Ca and Sr to the clear streams draining areas without appreciable marble, even though the calcite represents only ~0.1 wt% of the total silicate rock. Moreover, these observations agree well with Blum et al. (1998) who proposed the existence of a disseminated calcite mixing endmember in the Raikhot watershed possessing a $Ca/(1000*Sr)$ ratio over 4.5 and a $^{87}Sr/^{86}Sr$ ratio greater than 0.82.

5 CONCLUSIONS AND IMPLICATIONS

In summary, this research further emphasizes the importance of characterizing trace amounts of calcite present within silicate rocks. Given that disseminated calcite may be a more prevalent component of igneous rocks than previously recognized (e.g. White et al. 1999), calculations relating the solute and Sr isotope chemistry of streams draining predominantly silicate terrains to changes in climate may overestimate contributions provided by silicate weathering. Here we demonstrate that finely disseminated calcite is a ubiquitous trace phase within HHCS silicate rocks that has achieved Sr isotope equilibration with co-existing silicate minerals. Consistent with Blum et al. (1998), we suggest that the modern dissolved flux of radiogenic Sr emanating from the Pakistan HHCS is largely a function of disseminated calcite and minor meta-sedimentary marble weathering and hence is not a simple proxy for silicate weathering and atmospheric CO_2 consumption. Furthermore, we add the observation that the continuous exposure of fresh disseminated calcite in the HHCS by glacial and tectonic processes may provide a mechanism for sustaining a riverine flux of radiogenic Sr on timescales long enough to influence the evolution of the marine Sr isotope record.

REFERENCES

Berner, R.A. 1994. GEOCARB II: A revised model for atmospheric CO_2 over Phanerozoic time. *Amer. J. Sci.* 294: 56-91.

Bickle, M.J., Chapman, H.J., Wickham, S.M., & Peters, M.T. 1995. Strontium and oxygen isotope profiles across marble-silicate contacts, Lizzies Basin, East Humbolt Range, Nevada: Constraints on metamorphic permeability contrasts and fluid flow. *Contributions to Mineralogy and Petrology* 121: 400-413.

Blum, J.D., Gazis, C.A., Jacobson, A.D., & Chamberlain, C.P. 1998. Carbonate versus silicate weathering in the Raikhot watershed within the High Himalayan Crystalline Series. *Geol.* 26: 411-414.

Drever, J.I. & Hurcomb, D.R. 1986. Neutralization of atmospheric acidity by chemical weathering in an alpine drainage basin in the North Cascade Mountains. *Geol.* 14: 221-224.

Gazis, C.A., Blum, J.D., Chamberlain, C.P., & Poage, M. 1998. Isotope systematics of granites and gneisses of the Nanga Parbat Massif, Pakistan Himalaya. *Am. J. Sci.* 298: 673-698.

Goddéris, Y. & François, L.M. 1995. The Cenozoic evolution of the strontium and carbon cycles: Relative importance of continental erosion and mantle exchanges. *Chem. Geol.* 126: 169-190.

Horton, T., Chamberlain, C.P., Fantle, M., & Blum, J.D. 1999. Chemical weathering and lithologic controls of water chemistry in a high-elevation river system: Clark's Fork of the Yellowstone River, Wyoming and Montana. *Water Resources Research* In press.

Marshall, D.J. 1988. *Cathodoluminescence of geological materials.* Boston: Unwin Hyman.

Mast, M.A., Drever, J.I., & Baron, J. 1990. Chemical weathering in the Loch Vale watershed Rocky Mountain National Park, Colorado. *Water Resources Research* 26: 2971-2978.

Garrels, R.M. & Mackenzie, F.T. 1967. Origin of the chemical composition of some springs and lakes. In R.F. Gould (ed.), *Equilibrium concepts in natural water systems*: Washington, D.C., American Chemical Society, Advances in Chemistry Series. 67: 222-242.

Raymo, M.E. & Ruddiman, W.F. 1992. Tectonic forcing of late Cenozoic climate. *Nature* 359: 117-122.

White, A.F., Bullen, T.D, Vivit, D.V., & Schulz, M.S. 1999. The role of disseminated calcite in the chemical weathering of granitoid rocks. *Geochimica et Cosmochimica Acta.* In press.

Zeitler, P.K., Chamberlain, C.P., & Smith, H.A. 1993. Synchronous anatexis, metamorphism, and rapid denudation at Nanga Parbat (Pakistan Himalaya). *Geol.* 21: 347-350.

Zeitler, P.K. & Chamberlain, C.P. 1991. Petrogenetic and tectonic significance of young leucogranites from the northwestern Himalaya, Pakistan. *Tectonics* 10(4): 729-741.

Geochemistry of the Earth's Surface, Ármannsson (ed.)© 1999 Balkema, Rotterdam, ISBN 90 5809 073 6

REE in river sediments: Partitioning into residual and labile fractions

L. Leleyter & J. L. Probst – *Centre de Géochimie de la Surface, EOST (CNRS/ULP), Strasbourg, France*

P. Depetris – *Department of Chemistry, University of Cordoba, Argentina*

S. Haida – *Department of Geology, University Ibn Tofail, Kénitra, Morocco*

J. Mortatti – *University of São Paulo, CENA, Piracicaba, Brazil*

ABSTRACT: To assess rare earth element (REE) distribution, fractionation during weathering processes and river transport, a sequential 7-steps extraction procedure was applied on 36 river suspended matter and bottom sediments collected from different rivers in Argentina, Brazil, France and Morocco. The results show that labile REE fractions are mainly linked to carbonates, iron oxides or organic matter. Moreover the leached REE do not behave as a coherent group : middle REE (specially europium) are mainly bound to carbonates, whereas light REE (except cerium) are preferentially linked to organic matter and heavy REE and cerium are associated with iron oxides. The REE distribution patterns change in the different labile fractions and consequently, the distribution pattern of the total sediment sample can be modified according to the relative abundance of the different labile fractions.

1 INTRODUCTION

The rare earth elements (REE, La through to Lu) are a coherent geochemical group characterized by a single oxidation state : REE (III) except for cerium and europium. During the last few decades, REE have become important geochemical tracers in order to understand and describe the chemical evolution of the earth continental crust (McLennan 1989, Goldstein & Jacobsen 1988). Moreover REE have been used as analogues for actinide elements in studies related to radioactive waste disposal in order to demonstrate their general immobility in weathering environments (Wood 1990).

Despite this, many studies have indicated that REE may be significantly mobilized during weathering and that REE may behave non-conservatively (Sholkovitz et al. 1994). However, little is know about REE distribution and fractionation during weathering and river transport. Accordingly, the purpose of this study is to determine the speciation of REE in river sediments. Sequential extraction procedures are probably the most useful tool to study the speciation of particular elements in order to determine their origin, their transport and their fate in the environment.

River sediments are considered to consist of two major phases : the residual fraction and the labile one. REE from leachable fractions are very sensitive to changes in the physico-chemical properties of the aquatic environment (such as salinity, pH, redox potential, and concentration of chelators) and are participating in liquid-solid adsorption-desorption, surface complexation and co-precipitation with different solid fractions like clay minerals, carbonates, oxides and organic matter. To assess and to identify the different fractions concerned with the REE transport and the fractionation of REE in river sediments, a sequential extraction procedure was applied on 36 river sediment samples (suspended matters and bottom sediments).

2 MATERIALS AND METHODS

2.1 Sampling

In order to have a global view of REE speciation, 36 river sediments from different countries were collected from oxidizing environment. They were selected according to their different chemical and mineralogical characteristics : the Patogonian rivers, in Argentina (Rio Chico, Rio Coyle, Rio Colorado and Rio Deseado) whose sediments are rich in clay minerals, the Piracicaba river, in Brazil, whose sediments are rich in iron oxides, the Ill river, in France, whose sediments are rich in organic matter and the Sebou river, in Morocco, whose sediments are rich in carbonates. Then all the samples were rapidly dried at 40°C in a stove and stored at 4°C in polypropylene bottles to avoid any bacterial activity.

Table 1. Procedure for the 7-step sequential extraction (Leleyter & Probst 1999).

	fraction	reagent	time	pH
S1	evaporitic	10 ml water	0.5h	5.7
S2	exchangeable	10 ml 1M Mg(NO$_3$)$_2$	2 h	5.0
S3	carbonates	10 ml 1M NaOAc	5 h	4.5
S4	manganese oxides	10 ml 0.1M NH$_2$OH HCl	0.5h	3.5
S5a	amorphous iron oxides	10 ml {0.2M (NH$_4$)$_2$C$_2$O$_4$ – 0.2M H$_2$C$_2$O$_4$} in the dark	4 h	3.0
S5b	crystalline iron oxides	10 ml {0.2M (NH$_4$)$_2$C$_2$O$_4$ – 0.2M H$_2$C$_2$O$_4$ - 0.1M C$_6$H$_8$O$_6$} T=80°C	0.5h	2.3
S6	organic matter	1) 3ml 0,02M HNO$_3$ and 8 ml 35% H$_2$O$_2$, T= 85°C. 2) 5 ml 3.2 M NH$_4$OAc	5 h 0.5h	2.0
S7	labile	sum of all the labile fractions		

2.2 Extraction procedure

All the samples were leached by a chemical extraction procedure (Leleyter & Probst 1998) that takes place in a watertight container with continuous agitation to increase the interaction surface between the reagent and the sediment. The quantity of reagent used refers to 1g sediment (dry weight at 100°C) of the original sample used for the initial extraction. After each reaction, the residue was filtered through a 0.45 µm pore size filter and washed with 20 ml of distilled water (Table 1).

This procedure serves to dissolve selectively and efficiently all the chemical constituents of the river sediments, which can be affected by changes in physico-chemical conditions, in the following order : elements dissolved by water (S1), exchangeable elements (S2), elements bound to carbonates (S3), manganese oxides (S4), amorphous iron oxides (S5a), crystalline iron oxides (S5b) and organic matter (S6). The labile fraction of the sample (S7) is the sum of all the leached fractions. The total initial sample concentration of REE was determined by digesting the sample by an alkaline-melting attack.

2.3 Analysis

REE concentrations were determined by Inductively Coupled Plasma Mass Spectrometry (ICP-MS) Fisons VG-Plasma Quad. The detection limit is 0.01 µg/l and the standard deviation is less 5%. The reported REE distribution patterns are PAAS (Post Archean Australian Average Shale) normalized to smooth the distributions (McLennan 1989).

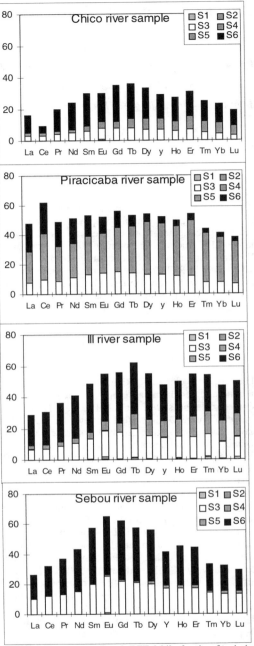

Figure 1. Percentage of each REE labile fraction for 4 river sediments (% of the initial sample). The residual fraction represents the difference between the initial sample and the sum of all labile fractions. S1 to S6, see text and Table 1.

3 RESULTS AND DISCUSSION

3.1 REE speciation in river sediments

The results obtained on the 36 river sediment samples show that the REE occur most located in carbonates, iron oxides, organic matter and in the residual fraction (Figure 1). The dominate fraction depends essentially on the initial composition of the sediment. Indeed for the Patagonian river samples, REE are relatively immobile (Figure 1, Chico river for example). Whereas leached REE from the Piracicaba river samples are mainly associated with iron oxides, the leached REE from the Ill river samples are mainly linked with organic matter and, at last, the leached REE from Sebou river samples comes from carbonates principally (Figure 1).

The labile REE do not behave as a coherent group. They do not have the same preferential scavenged fractions. An unexpected result, obtained in this study, is that the fractionation of the REE between the different fractions of the sediment is independent of the type of sediment involved.

The middle REE coprecipitate always mainly with carbonates (specially europium), whereas organic matter is mainly associated with light or middle REE (except cerium). On the contrary, iron oxides adsorb always preferentially heavy REE and cerium. Thus it is to be expected that the light and heavy REE mobility depends on the relative abundance of carbonates, oxides and organic matter in the initial sample (Figure 1) which is partly controlled by the physico-chemical composition of the river water.

3.2 REE distribution patterns

As labile REE do not behave as a coherent group, the abundance of carbonates, oxides or organic matter in the initial sample can provoke some changes in the PAAS normalized REE distribution patterns.

Thus the abundance of carbonates in the total sample can cause a positive europium anomaly, as seen for the Sebou river sediment (Figure 2.A) (Eu* = 1.12; with Eu* = $Eu_{normalized}/[Sm_{normalized} * Gd_{normalized}]^{1/2}$; McLennan 1989), whereas the residual fraction of the sediment presents a negative one (Figure 2.A, Eu* = 0.88). This affinity of europium toward carbonates is due to the similarity of the ionic radius (Shannon, 1976) of Eu^{2+} (r = 1.25 Å) and Sr^{2+} (r = 1.26 Å) which can provoke some Sr substitutions in carbonates (Brookins 1989).

In the same way, abundance of organic matter in the initial sediment can cause a light and middle REE enrichment and sometimes a negative cerium anomaly in the total sample, as seen for the Coyle river sediment (Ce* = 0.32; with Ce* = $Ce_{normalized} / [2 La_{normalized} + Nd_{normalized}]$; Elderfield et al. 1990) (Figure 2.B), whereas the

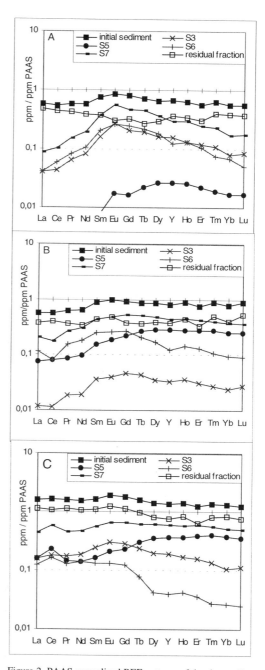

Figure 2. PAAS normalized REE patterns of the river sediment sample, and its different fractions concerned by REE transport (Sx : see Table 1). Above : Sebou river sediment; middle : Coyle river sediment and below : Piracicaba river sediment.

residual fraction of this sample present a positive one (Ce* = 0.37).

At last, abundance of oxides, which are the third fraction concerned by the transport of labile REE, can explain the positive cerium anomaly of the total sample (Ce* = 0.35), as for the Piracicaba river sediment whose the residual fraction present a negative one (Ce = 0.32) (Figure 2.C). The affinity of cerium towards oxides is well documented (Shokovitz et al. 1994) and is probably due to the oxidative scavenging of Ce(IV).

These results show that the PAAS normalized REE distribution pattern of the initial sample could be really different from that of the residual material (Figure 2).

4 CONCLUSION

This study points out that REE distribution pattern of river sediments is not only dependent on the continental crust signature but also on the relative abundance of REE labile fractions associated to carbonates, iron oxides and organic matter in river sediment which are controlled by the physico-chemical parameters of the aquatic environment and which can provoke important changes in the normalized REE distribution patterns. These results must be taken into account in the study of the erosion of continental crust using river transport of particulate material, and to characterize the REE signature of the river sediments discharged into the oceans.

Moreover, these results could be useful tool to chose a disposal site for sediments, in order to minimize mobility of trace element toward their disposal site. pH and redox potential are the 2 primordial parameters to be controlled to verify the stability of iron oxides and of organic matter.

REFERENCES

Brookins, D.G. 1989. Aqueous geochemistry of rare earth elements. In : *Geochemistry and mineralogy of rare earth elements*. 21 : 201-223.

Elderfield, H., R. Upstill-Goddard & E.R. Sholkovitz 1990. The rare earth elements in rivers, estuaries and coastal seas and their significance to the composition of ocean water. *Geochim. Cosmochim. Acta.* 54 : 971-991.

Goldstein, S.J. & S.B. Jacobsen 1988. Rare earth elements in rivers waters. *Earth Planet. Sci. Lett.* 89 : 35-47.

Leleyter, L. & J.L. Probst 1998. A new sequential extraction procedure for the speciation of particulate trace elements in river sediments. *Int. J. Environ. Anal. Chem.* 73 : 109-128

McLennan, S.M. 1989. Rare earth elements in sedimentary rocks : influence of the provenance and sedimentary process. In : *Geochemistry and Mineralogy of rare earth elements*. 21 : 169-200

Shannon, R.D. 1976. Revised effective ionic radii and systematic studies of interatomic distances in halides and chalcogenides. *Acta Crystallogr., Sect A.* 32 : 299-308.

Sholkovitz, E.R., W.M. Landing & B.L. Lewis 1994. Ocean particle chemistry: the fractionation of rare earth elements between suspended particles and seawater, *Geochim. Cosmochim. Acta.* 58 : 1567-1579

Wood, S.A. 1990. The aqueous geochemistry of rare-earth elements and yttrium. *Chem. Geol.* 82 : 159-186.

Geochemistry of the Earth's Surface, Ármannsson (ed.)© 1999 Balkema, Rotterdam, ISBN 90 5809 073 6

Chemical and mechanical erosion of major Icelandic rivers: Geochemical budgets

P. Louvat & C. J. Allègre
Laboratoire de Géochimie et Cosmochimie, IPGP, Paris, France

S. R. Gíslason
Science Institute, University of Iceland, Reykjavík, Iceland

ABSTRACT: We present a coupled geochemical study of the dissolved and suspended material carried by most of the biggest rivers of Iceland. Chemical erosion rates have been estimated after subtraction of the atmospheric and hydrothermal inputs from the chemical compositions of the dissolved river loads. Rates of atmospheric CO_2 consumption by chemical weathering have also been determined. Mechanical erosion rates were calculated using the steady-state model of erosion, which establishes chemical mass budgets between the pristine rocks of the drainage basins and the dissolved and solid erosion products. As reliable estimates of measured sediment loads are available, it is possible to compare calculated and measured sediment loads and to discuss the efficiency of this steady-state model of erosion. Chemical and mechanical erosion are both very high, underlining the high erodability of basalt, even in a cold climate, and the influence of glaciers on riverine sediment transport.

1 INTRODUCTION

In the general attempt to link erosion and climate, the study of riverine erosion in Iceland provides important information. Despite the relatively small size of Icelandic rivers, their basaltic catchments are good candidates for intense cold climate weathering. Because of its sparse vegetation and almost absence of pollution Iceland is an ideal experimental field for defining pristine water-basalt weathering interactions. Furthermore, the study of Icelandic rivers is important for estimating the global riverine sediment supply to the oceans. In fact, glaciers greatly enhance riverine mechanical erosion in Iceland and render it comparable to that of many areas in South-East Asia.

This study expands previous work on chemical weathering in Iceland (Gíslason et al. 1996) and adds coupled geochemical study of both chemical and mechanical erosion rates. Such coupled studies have already been undertaken by the IPGP group in the Earth's two biggest basins: the Congo and the Amazon (Gaillardet et al. 1995, Gaillardet et al. 1997, Dupré et al. 1996) and in that of smaller basaltic catchments of Réunion Island and the Azores (Louvat & Allègre 1997, Louvat & Allègre 1998). In all these previous studies, mechanical erosion rates were estimated using the steady-state model of erosion first evoked by Martin & Meybeck (1979).

The present study offers an opportunity to test the steady-state model by comparing the calculated mechanical erosion rates to that calculated from the measured suspended loads and discharge of the rivers (Pálsson & Vigfússon 1996).

2 SAMPLING AND ANALYSIS

Eleven of the largest Icelandic rivers (Hvítá-West, Blanda, Austari Jökulsá, Skjálfandafljót, Jökulsá á Fjöllum, Jökulsá á Brú, Jökulsá í Fljótsdal, Hverfisfljót, Skaftá , Thjórsá and Hvítá-South) were sampled for their dissolved and solid contents (water, suspended sediments and beach sands). Constituents smaller than 0.2 μm were defined as dissolved solids. Dissolved loads of seven smaller rivers and one hydrothermal spring (Geysir) were also sampled to improve the study. Temperature, pH and alkalinity were measured at the sampling sites. The major elements, Na, K, Ca, Mg, were analysed by HPLC as well as dissolved Cl and SO_4. The trace elements, B, Li, Sr, Rb, Ba, REE, Th and were analysed by ICP-MS. Details of the analytical procedures and results are given in Louvat et al. (1999).

3 RATES OF CHEMICAL EROSION AND ATMOSPHERIC CO_2 CONSUMPTION.

3.1 Determination of the chemical erosion rates

The dissolved constituents of the rivers represent a mixture of the atmospheric, weathering and hydrothermal/volcanic sources. The chemical composition of the rivers and that of the different end-member sources makes it possible to decipher the part in the mixed information that only stems from basalt weathering. This was done by considering that Cl and B both arise mostly from atmospheric and hydrothermal inputs but to a much lesser degree from rock weathering. The Cl/B ratios for atmospheric and hydrothermal inputs are very distinct (Arnórsson & Andrésdóttir 1995); thus, it is possible to quantify the proportions of Cl and B arising from each of the two sources. Knowing the chemical compositions of these two end-members, oceanic-type rain and well studied hydrothermal sources in Iceland, the weathering part of the dissolved load was simply calculated by subtracting for each chemical species the concentrations originating from atmospheric and hydrothermal sources.

The atmospheric source correction was applied to all the rivers but the geothermal correction was only necessary for six rivers with particularly high concentrations of SO_4, B, Li and HCO_3. The hydrothermal discharges are often negligible compared to the river discharges; however, their dissolved loads may be up to 1000 times higher. Thus the resulting dissolved constituents contribution of these hydrothermal springs to the rivers may be substantial.

The chemical erosion rates were calculated by summing the concentrations of the major elements of the dissolved load excluding the HCO_3 concentrations, which arise directly and indirectly from atmospheric CO_2.

3.2 Weathering rates and atmospheric CO_2 consumption

Chemical erosion rates of the eleven major rivers of Iceland under study ranged between 17 and 80 $t/km^2/yr$. Including the seven smaller rivers of the study, the range was extended to 110 $t/km^2/yr$ (Figure 1). For the same eleven rivers, the rates of atmospheric CO_2 consumption during the weathering ranged between 0.35 and 1.20 × 10^6 $mol/km^2/yr$ and 0.16 to 2.10 × 10^6 $mol/km^2/yr$ when including the seven smaller rivers (Figure 1). As the degassing CO_2 flux of Iceland is 0.27 × 10^6 $mol/km^2/yr$ (Gíslason et al., 1996), we can consider in a first approximation that chemical weathering in

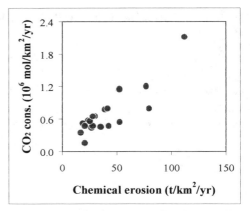

Figure 1. Chemical erosion versus resulting atmospheric CO_2 consumption

Iceland acts as a sink for atmospheric CO_2 in the global CO_2 budget. These rates are relatively high if compared to the worldwide estimations of silicate weathering rates; 26 $t/km^2/yr$ for average chemical weathering and 0.3 × 10^6 $mol/km^2/yr$ for average atmospheric CO_2 consumption (Berner & Berner 1996), especially considering the relatively cold climate of Iceland. The chemical erosion rates are about half of those determined for Réunion Island in a tropical climate and equivalent to those for Sao Miguel, Azores, in a temperate climate (Louvat & Allègre 1997, Louvat & Allègre 1998).

The average ages for the rocks of each drainage basin were estimated from geological maps. There was a good inverse correlation between age of the rocks and the chemical erosion rates: the younger the basalt, the more intense the chemical weathering (Figure 2).

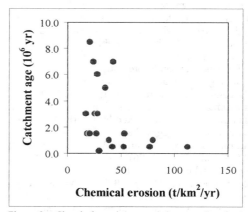

Figure 2. Chemical erosion versus the age of rocks in the catchment areas.

There was also a conspicuous correlation between chemical erosion rates and runoff, and between chemical erosion rates and the proportions of glaciers in the catchments areas. As runoff was used to calculate the chemical erosion rates (in t/km²/yr, rate = TDS × runoff), such a correlation underscores the fact that despite considerable variation in runoff the TDS (Total Dissolved Solids) of the Icelandic rivers, stemming from weathering, is not very variable: 22 ± 6 mg/l, excluding the Ytri Rangá River. The correlation between chemical erosion rates and proportion of glacier area rather reflects the correlation between the runoff and the proportion of glaciers in the catchments since runoff tends to be high in glacier terrain.

4 MECHANICAL EROSION RATES

4.1 The steady state model of erosion

The mechanical erosion rates were determined using the steady-state model of erosion. This model is based on the chemical mass budget between the initially pristine rocks of the watershed and the erosion products (Gaillardet et al. 1995, Gaillardet et al. 1997, Louvat & Allègre 1997, Louvat & Allègre 1998).

Over a sufficiently long period of time, the dissolved and solid erosion products issued from a given mass of pristine rock will have been washed out of the watershed. If the chemical compositions of the dissolved and solid erosion products are sufficiently constant through time, then any sampling of these materials plus a good estimation of hydrologic cycles of the watersheds allow the assessment of mechanical erosion rates. The validity of this model can be tested by a simple comparison of the calculated and measured erosion rates, in so far as measurements of sediment loads can be trusted.

One delicate step for the calculation is to evaluate a mean chemical composition for the pristine rocks of the watershed. Rock geochemical studies are useful but do not give average compositions. Average chemical compositions can be best re-estimated from the river beach sands and suspended loads which come from the whole drainage basin. This is done using the most insoluble rock elements like REE or Th.

Once the average chemical composition of the pristine rock has been defined, it is easy to compare it to the chemical compositions of the erosion products in order to define the weathering state of the solid erosion products. As chemical erosion rates are easily calculated (see 3.1), the river's solid loads are calculated from the chemical mass balance equations between pristine rock and dissolved and solid erosion products.

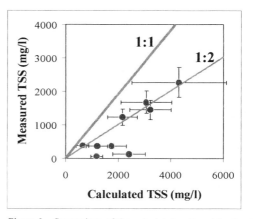

Figure 3. Comparison of the calculated sediment load using the steady-state model of erosion and the measured sediment load of the Icelandic rivers (Total Suspended Sediment, TSS).

4.2 Calculated and measured sediment loads

Measuring river sediment loads is often problematic, as sediment concentrations are very variable during hydrologic cycles. Estimating reliable mechanical erosion rates from sediment load measurements is thus very difficult, especially for regions with highly variable discharge (Walling & Webb 1996). Iceland is unfortunately one of these regions but there is fortunately a good hydrological survey of its rivers done by the National Energy Authority of Iceland (Pálsson & Vigfússon, 1996). Thus, estimations of the measured sediment loads on a multi-annual basis are available for many rivers (nine in eleven of this study).

Total suspended sediment concentration (TSS) predicted by the steady state model range between 650 and 4300 mg/l, with errors between 20% and 40% (Figure 3). These values were always higher than the measured TSS, 60 to 2300 mg/l (Figure 3). For five of the nine rivers, all of them glacier-fed, there was a good linear correlation between the calculated and measured TSS, with the calculated to the measured ratio being close to 2. For the four remaining rivers, the calculated TSS was undoubtedly very much higher than the measured TSS. Of these four rivers, one is now dammed (Blanda River) and one other (Austari Jökulsá River) is braided in its upper part, where sediments may accumulate.

Considering the lowest possible calculated and the highest possible measured TSS (with a 20% error for the measured values), the equilibrium was realized for four of the nine rivers (Jökulsá á Brú, Jökulsá í Fljótsdal, Hverfisfljót and Skaftá Rivers).

The differences between the calculated (inferred from the weathered state of the suspended material) and measured sediment loadsn of the rivers may be explained by accumulation areas (braided rivers) in the glacial river catchments, by accumulation in lakes or by dam construction. Wind transport of sediments is also a concern for many Icelandic catchments but its effect may well be a loss of sediments in one catchment as a supply in another.

4.3 Factors controlling mechanical erosion rates

Mechanical erosion rates deduced from the calculated TSS values ranged between 960 and 16500 t/km²/yr. The highest rate was for the Hverfisfljót River, mainly fed by the Vatnajökull Glacier. These mechanical erosion rates were very high and reflected the dominant influence of glaciers for sediment transportation in Icelandic rivers.

In order to emphasize how high these rates are, we may compare them to the worldwide average estimation of Milliman & Syvitski (1992): 230 t/km²/yr. Icelandic mechanical erosion rates are otherwise equivalent to those determined for Reunion Island and about 2 to 100 times higher than those of the Azores (Louvat & Allègre1997, Louvat & Allègre 1998). The Icelandic rivers rank among the maximum reported values of mean annual specific yields, together with New Zealand, New Guinea or Java (Walling & Webb, 1996).

A correlation between the mechanical erosion rates and the proportions of glaciers in the drainage basins is shown in Figure 4. The effect of glaciers upon mechanical erosion in Iceland is striking. We also observed a correlation between the mechanical erosion rates and the mean age of the drainage basins: the mechanical erosion rates were always higher for young rocks than for old ones. However, there was no obvious correlation between mechanical erosion rates and the runoff or surface area of the basins.

5 CONCLUSION

The chemical erosion rates, rates of atmospheric CO_2 consumption and mechanical erosion rates calculated by this coupled geochemical study of the dissolved and solid loads of Icelandic rivers are among the highest worldwide estimates. The basaltic lithology provided easily weatherable material; young glassy basalt is more prone to weathering than the more crystalline basalt of Tertiary age, and finally, glaciers act as crushers, enhancing sediment production.

Relatively small rivers with high erosion rates, such as the Icelandic rivers, have to be taken into consideration in the global budget of dissolved and solid river material supplied to the oceans, and in the CO_2 cycle. And finally, small monolithic catchments provide a way to study the fundamental mechanisms of erosion.

REFERENCES

Arnórsson S. & A. Andrésdóttir 1995. Processes controlling the distribution of boron and chloride in natural waters in Iceland. *Geochim. Cosmochim. Acta* 59: 4125-4146.

Berner E. K. & R. A. Berner 1996. *Global environment: water, air, and geochemical cycles.* New Jersey: Prentice Hall.

Dupré B., J. Gaillardet, D. Rousseau & C. J. Allègre 1996. Major and trace elements of river-borne material: The Congo Basin. *Geochim. Cosmochim. Acta* 60: 1301-1321.

Gaillardet J., B. Dupré & C. J. Allègre 1995. A global geochemical mass budget applied to the Congo Basin rivers: Erosion rates and continental crust composition. *Geochim. Cosmochim. Acta* 59: 3469-3485.

Gaillardet J., B. Dupré, C. J. Allègre & P. Négrel 1997. Chemical and physical denudation in the Amazon River Basin. *Chem. Geol.* 142: 141-173.

Gíslason S. R., S. Arnórsson & H. Ármannsson 1996. Chemical weathering of basalt in SW Iceland: effects of runoff, age of rocks and vegetative/glacial cover. *Am. J. Sci.* 296: 837-907.

Louvat P., S. R. Gislason & C. J. Allègre 1999. Chemical and mechanical erosion rates of Iceland as deduced from dissolved and suspended material in rivers. *Am. J. Science,* in prep

Louvat P. & C. J. Allègre 1997. Present denudation rates at Réunion Island determined by river geochemistry: basalt weathering and mass budget between chemical and mechanical erosions. *Geochim. Cosmochim. Acta* 61: 3645-3669.

Louvat P. & C. J. Allègre 1998. Riverine erosion rates on Sao Miguel volcanic island, Azores archipelago. *Chem. Geol.* 148: 177-200.

Milliman J. D. & J. P. M. Syvitski 1992. Geomorphic/tectonic control of sediment discharge to the ocean: the importance of small mountainous rivers. *J. Geol.* 100: 525-544.

Martin J. M. & M. Meybeck 1979. Element mass-balance of material carried by major world rivers. *Mar. Chem.* 7: 173-206.

Pálsson S. & G. H. Vigfússon 1996. *Results of suspended load and discharge measurements 1963-1995*: OS-96032/VOD-05 B. Reykjavik: National Energy Authority.

Walling D. E. & B. W. Webb 1996. Erosion and sediment yield: a global overview. In: *Erosion and Sediment Yield: Global and Regional Perspectives* (Proceedings of the Exeter Symposium, July 1996), IAHS publ. n°236: 4-18.

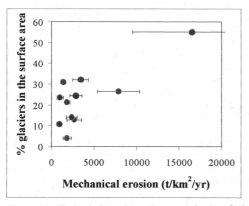

Figure 4. Mechanical erosion rate versus the % of glacial cover of the catchments.

Geochemistry of the Earth's Surface, Ármannsson (ed.)© 1999 Balkema, Rotterdam, ISBN 90 5809 073 6

Silicate weathering rates in the Mackenzie river basin, North-West Territories, Canada

R. Millot, J. Gaillardet & C. J. Allègre
Laboratoire de Géochimie et Cosmochimie, Institut de Physique du Globe de Paris, Université Paris, France

B. Dupré
Laboratoire de Géochimie, OMP, Toulouse, France

ABSTRACT: We present in this study the first Sr isotopic systematics for the Mackenzie river basin. The coupled use of Sr isotopic ratios and major element ratios in the dissolved load allows us to confirm the dominant control of lithology on weathering rates, and to extract the silicate weathering component from other sources. The three main results are : Firstly, high Sr isotopic ratios (0.7320) are found to be associated with high Ca/Na ratios (20 to 30), such a feature was only observed in the Himalayas. Secondly, the silicate weathering and CO_2 consumption rates are calculated for the Mackenzie basin and are higher for the Amazon basin, especially for the Andean tributaries. Finally, Na concentrations (up to 300 µmol/l) delivered by silicate weathering are found in organic-rich rivers draining sedimentary rocks in the lowlands.

1 INTRODUCTION

The role of climate and dissolved organic matter on global silicate weathering is still poorly known. The Mackenzie river basin is one of the largest basins under subarctic climatic conditions, and one of the most organic-rich river systems in the world. An extensive study of the surface waters of the Mackenzie basin has been carried out by Reeder et al. (1972), without focusing on the silicate weathering processes. The main problem associated with the determination of silicate weathering rates from river catchment studies, is that the dissolved load of large rivers is mostly controlled by lithology (e.g. Négrel et al. 1993, Edmond et al. 1996, Gaillardet et al. 1999). In order to compare silicate weathering rates under cold (arctic and subarctic basin of the Mackenzie river) and hot climate (Amazon river basin), the contribution of silicate rocks must be extracted from the total dissolved solids load (TDS).

As shown previously by Négrel et al. (1993) and Gaillardet et al. (1997), the coupled use of Sr isotopic ratios and major element ratios allows to separate the silicate contribution from other sources, and calculate CO_2 consumption and silicate weathering rates.

Finally, the role of dissolved organic matter in weathering processes can be addressed in the Mackenzie river basin, which is characterized by both low dissolved organic carbon (DOC) rivers (highland rivers) and DOC rich rivers (black rivers) of the lowlands). Due to the weakness of organic matter degradation in the Mackenzie basin, organic-

rich rivers are very abundant in the lowlands. It has been recently demonstrated that organic-rich environments allow very intense hydrolysis of silicate (Viers et al. 1997). Our aim is to constrain the influence of organic matter on weathering processes in the subarctic catchment of the Mackenzie river.

2 GEOGRAPHICAL AND GEOLOGICAL SETTING OF THE MACKENZIE BASIN

The Mackenzie river catchment is located in the Yukon, British Columbia, Alberta and Northwest Territories in western Canada. The Mackenzie basin is the second largest basin in North America after the Mississippi, it drains an area of 1787 km^2 (Meybeck & Ragu 1997). The runoff is low and ranges from 160 to 172 mm/year (Meybeck & Ragu 1997) and the mean annual temperature is close to -4°C (Summerfield & Hulton 1994). The vegetation cover is mainly tundra and boreal forest, and only a small part of the total area of the basin is located under temperate climate.

The morphology of the basin is characterized by three principal zones. The first one is situated in the western part of the basin and corresponds to the cordilleran mountain system (Rocky and Mackenzie mountains). The second one is the interior platform and is characterized by a large carbonate lowland, where swampy zones are abundant. The third one is the Canadian Shield in the eastern part of the basin.

These three morphological systems correspond to three geological zones. The headwaters of the basin

(Rocky and Mackenzie mountains) are characterized by mixed lithologies, mainly alumino-silicates with some dolomitic limestones. The lowlands (interior platform) are characterized by sedimentary rocks from Cambrian to Cretaceous and Tertiary (marine, non marine limestones and evaporites). The Canadian Shield (Slave Province), mostly comprises old crystalline rocks. According to Reeder et al. (1972) most of the surface of the basin is sedimentary (limestones and clastic rocks : 68%, evaporitic rocks : 2.5% and igneous and metamorphic rocks : 29.2%). We have sampled the Mackenzie river basin during an high flow stage (summer 1996). Samples from the main tributaries were collected (Peel, Liard, Hay, Peace, Slave and Athabasca rivers) as well as some typical small rivers for each of the morphological zones.

3 STRONTIUM ISOTOPIC AND MAJOR ELEMENT RATIOS SYSTEMATICS

We present the first Sr isotopic systematics for the dissolved load transported in the Mackenzie River basin. The sedimentary part of the basin is characterized by $^{87}Sr/^{86}Sr$ ratios ranging from 0.7079 to 0.7120, while the Canadian Shield displays higher values close to 0.7424 and the Rocky and Mackenzie Mountains exhibit values ranging from 0.7116 to 0.7336. The samples from the Canadian Shield present very high $^{87}Sr/^{86}Sr$ values (0.7411 to 0.7424) with regard to the rest of the basin, but the discharge of these rivers is low, and the flux of radiogenic Sr is very dilute. Two samples of monolithological rivers draining only saltrocks present $^{87}Sr/^{86}Sr$ ratios close to 0.7086. The Mackenzie river (at Inuvik) present $^{87}Sr/^{86}Sr$ ratio of 0.7113, which is in good agreement with the literature (Wadleigh et al. 1985, Palmer & Edmond 1989). The Sr isotopic systematics as well as major element ratios confirm the dominant control of lithology for river dissolved load, with higher concentrations resulting from carbonates and evaporites.

Plotted in a $^{87}Sr/^{86}Sr$ vs. Ca/Na diagram (Figure 1), the rivers of the Mackenzie basin define a singular trend of increasing $^{87}Sr/^{86}Sr$ with increasing Ca/Na ratio. This type of relationship is not classically observed in other large river systems like the Congo, the Orinoco and the Amazon river system (Négrel et al. 1993, Edmond et al. 1996, Gaillardet et al. 1997), where trends of decreasing Sr isotopic ratios with increasing Ca/Na ratios are observed.

The rivers characterized by a high Ca/Na ratio (20 to 30) and a high $^{87}Sr/^{86}Sr$ ratio (0.7320) are the headwaters of the mountainous tributaries of the Mackenzie river. The same feature have been observed in the rivers draining the High Himalayan Crystalline Series (HHC) in northern Pakistan (Blum

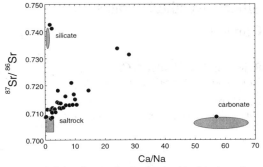

Figure 1 : Mixing diagram for the dissolved load in the main tributaries of the Mackenzie river basin.

et al. 1998). For the Mackenzie basin, it is unlikely that these rivers define a silicate endmember, because their drainage basin are mostly sedimentary rocks. This particular endmember could be a mixture between silicate (with low Ca/Na ratio and high $^{87}Sr/^{86}Sr$ ratio) and carbonate. It could also be a radiogenic carbonate endmember (Galy et al. 1999), or a calcite vein component, like in the HHC (Blum et al. 1998).

4 CO₂ CONSUMPTION AND SILICATE WEATHERING RATES : COMPARISON WITH THE AMAZON RIVER BASIN

On a geological time scale, only silicate weathering acts as a sink for atmospheric CO_2 (Berner et al. 1983). Here, we use major elements ratios to estimate the percentage of each element coming from silicate weathering and to calculate CO_2 consumption and silicate weathering rates. The fraction of each major element delivered by silicate weathering is calculated after correction for the dissolved load coming from other lithologies.

We find that 50, 49, 26 and 51×10^3 mol of $CO_2/km^2/year$, are consumed by silicate weathering in the Mackenzie, Liard, Peace and Athabasca rivers respectively. These values are two times higher than those given by Edmond et al. (1996) for silicate weathering based on Si concentrations.

Furthermore the Total Dissolved Solids (TDS) load produced by silicate weathering for the main tributaries of the Mackenzie river basin are compared with the values for the Amazon river basin (Gaillardet et al. 1997). For the tributaries draining the Guyana and Brazilian shield (e.g. Rio Madeira, Urucara, Rio Trombetas and Rio Tapajos), the silicate TDS values are one to two times higher than those of the rivers draining the Mackenzie lowlands. For the Andean tributaries of the Amazon river basin, the

TDS values are four to five times higher than those of the Mackenzie river basin. This result reemphazises the idea that silicate weathering rates are sensitive to temperature on a global scale. Moreover, in the Mackenzie basin, the lowest silicate weathering rates are observed in highlands, this is exactly the opposite in the Amazon basin.

5 INFLUENCE OF DISSOLVED ORGANIC MATTER

Finally, the role of organic matter in weathering processes can be addressed in the Mackenzie river basin, which gives examples of both low dissolved organic carbon (DOC) rivers (highland rivers) and DOC rich rivers (black rivers of the lowlands). The DOC values for the black rivers of the Mackenzie river basin are similar to the values for the organic-rich waters in Cameroon (Viers et al. 1997).

The relationship between La, Th and Nd concentrations of the dissolved load and dissolved organic carbon (Figure 2) shows that these insoluble elements are mainly complexed by humic acids of the dissolved organic matter.

In contrast, we find a positive correlation between sodium or alkalinity derived from silicate weathering and DOC concentrations (Figure 3), which is not a complexation by DOC, as shown by ultrafiltration experiments.

In organic-rich rivers, Al (insoluble element) is solubilized by complexation with dissolved organic matter. This complexation causes the destruction of the alumino-silicate lattice, and the solubilization of Na.

Na contents derived from silicate weathering are very high in the Mackenzie river basin (up to 300 μmol/l), especially for black rivers in the lowlands. These values are higher than those of many large river basins like Amazon, Changjiang, Zaire and Orinoco, where Na contents delivered by silicate weathering are respectively 22, 53, 54 and 42 μmol/l (Gaillardet et al. 1999).

These results suggest that under the subarctic conditions of the Mackenzie basin, dissolved organic matter may be a key parameter controlling the weathering processes and possibly the silicate weathering rates.

6 CONCLUSIONS

The comparison of river chemistries for the Mackenzie and Amazon river basin offers the possibility of comparing silicate weathering rates under

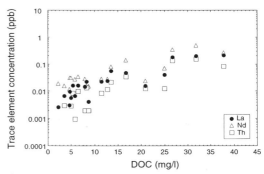

Figure 2 : Relationship between La, Th and Nd concentrations of the dissolved load (filtration at 0.2 μm) and dissolved organic carbon (DOC).

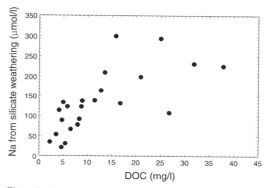

Figure 3 : Na contents delivered by silicate weathering plotted as a function of the content of dissolved organic carbon (DOC).

cold and hot climate.

Based on mixing diagrams and major element ratios, we have estimated the contribution of each element delivered by the various lithologies, and especially the contribution of the dissolved load coming from silicate weathering for the calculation of CO_2 consumption and silicate weathering rates.

Thus, silicate weathering rates values are obviously higher for the Andean tributaries of the Amazon river basin than for the Mackenzie river basin. This result suggests that silicate weathering rates are sensitive to temperature on a global scale. Moreover, dissolved organic matter may be a key parameter controlling the weathering processes in the Mackenzie river basin.

REFERENCES

Berner, R.A., Lasaga, A.C. & R.M. Garrels 1983. The carbonate-silicate geochemical cycle and its effects on atmospheric carbon dioxide over the past 100 millions years. *American Journal of Science* 283 : 641-683.

Blum, J.D., Gazis, C.A., Jacobson, A.D. & C.P. Chamberlain 1998. Carbonate versus silicate weathering in the Raikhot watershed within the High Himalayan Crystalline Series. *Geology* 26 : 411-414.

Edmond, J.M., Palmer, M.R., Measures, C.I., Brown, E.T. & Y. Huh 1996. Fluvial geochemistry of the eastern slope of the northestearn Andes and its foredeep in the drainage of the Orinoco in Colombia and Venezuela. *Geochimica et Cosmochimica Acta* 60 : 2949-2976.

Gaillardet, J., Dupré, B., Allègre, C.J. & P. Négrel 1997. Chemical and Physical denudation in the Amazon River Basin. *Chemical Geology* (142) : 141-172.

Gaillardet J., Dupré, B., Louvat, P., & C.J. Allègre 1999. Global silicate weathering and CO_2 consumption rates deduced from the chemistry of large rivers. *Chemical Geology*, in press.

Galy, A., France-Lanord, C. & L.A. Derry 1999. The strontium isotopic budget of himalayan rivers in Nepal and Bangladesh. *Geochimica et Cosmochimica Acta.*, submitted.

Meybeck, M. & A. Ragu 1997. *River discharges to the Oceans : An assessement of suspended solids, major ions and nutrients.* Nairobi: UNEP

Négrel, P., Allègre, C.J., Dupré, B. & E. Lewin 1993. Erosion sources determined by inversion of major and trace element ratios and strontium isotopic in river water : The Congo basin case. *Earth and Planetary Science Letters* 92 : 59-76.

Palmer, M.R. & J.M. Edmond 1989. The strontium isotope budget of the modern ocean. *Earth and Planetary Science Letters* 92 : 11-26.

Reeder, S.W., Hitchon, B. & A.A. Levinson 1972. Hydrogeochemistry of the surface waters of the Mackenzie River drainage basin, Canada-I. Factors controlling inorganic composition. *Geochimica et Cosmochimica Acta* 36 : 825-865.

Summerfield, M.A. & N.J. Hulton 1994. Natural controls of fluvial denudation rates in major world drainage basins. *Jounal of Geophysical Research* 99 : 13,871-13,883.

Viers, J., Dupré, B., Polvé, M., Schott, J., Dandurand, J.L. & J.J. Braun 1997. Chemical weathering in the drainage basin of a tropical watershed (Nsimi-Zoetele site, Cameroon) : comparison between organic-poor and organic-rich waters. *Chemical Geology* 140 : 181-206.

Wadleigh, M.A., Veizer, J. & C. Brooks 1985. Strontium and its isotopes in Canadian rivers : Fluxes and global implications. *Geochimica et Cosmochimica Acta* 49 : 1727-1736.

Geochemistry of the Earth's Surface, Ármannsson (ed.)© 1999 Balkema, Rotterdam, ISBN 90 5809 073 6

Water circulation in a karstified area: Sr, Pb isotope constraints

E. Petelet
Laboratoire Hydrosciences, Case 057, Université Montpellier II, France (Presently: ANTEA, Montpellier)

J-M. Luck & D. Ben Othman
Laboratoire Hydrosciences, Case 057, Université Montpellier II, France

P. Négrel
BRGM, Direction de la Recherche, Orléans, France

ABSTRACT: This paper presents geochemical constraints (major and trace elements), especially radiogenic Sr and Pb isotopes, on the study of water circulation in high and low flows in a highly karstified area (Hérault valley, S. France). Sr isotopes coupled with major and trace elements allow the understanding of the water dynamics of karstic springs : water origins and circulation, surficial - groundwater relations according to hydrological conditions. Sr isotopes coupled with Pb ones make it possible to clearly point out underground water circulation differences not revealed by Sr isotopes alone. The mean chemical erosion rate, based on Ca fluxes, is about 50 to 68 μm/y in this carbonated environment.

1. INTRODUCTION

The chemical study of water circulation in karstic systems is often based on major element concentrations, pH, water conductivity and calculated parameters such as pCO_2, saturation indexes for calcite or dolomite. More recently, geochemical tools (trace elements and radiogenic isotopes) classically used for large watershed studies in order to determine erosion rates and, since about 10 years, to trace the origin and remobilization of chemical elements (Goldstein & Jacobsen 1987; Negrel et al. 1993), are now also used in small scale studies to identify spring water origins and movements especially in karstified carbonated areas (Ben Othman et al. 1997, Luck & Ben Othman 1998).

Rock and soil erosion by rainwater and rivers are the major geological processes which modify the earth's surface morphology (Berner & Berner 1987). One way to estimate the current erosion rates of watersheds is to study of the loads transported by rivers. In this paper we only present the dissolved load (< 0.2 μm) which results from the chemical weathering processes.

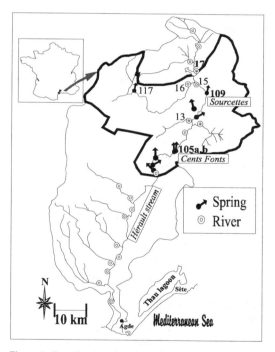

Figure 1. Sampling location sites (rivers and springs) in the Hérault watershed. Only the middle part is treated in this paper.

2. SITE STUDIED AND ANALYTICAL METHODS

The Hérault basin covers 2500 km² and can be divided into 3 sub-basins according to contrasting lithologies : the Paleozoic granite, schists and dolomites to the north (300 km²), the Jurassic and Cretaceous highly karstified limestones in the middle part (700 km²) and the Tertiary-Quaternary alluvial plain downstream to the south

(1500 km²) (Figure1). In this paper we will focus on the middle part. Six sampling series have been performed in various hydrological conditions (high, low flows and flood) along the Hérault stream, on its main tributaries and 16 karstic springs in the middle part.

We present analyses of the dissolved load. Major and trace elements (especially alkalis, alkali-earths and heavy metals) have been analyzed by Capillary Ion Electrophoresis and ICP-MS respectively. Sr and Pb isotopes were measured, after chemical separation, with a VG Sector Mass Spectrometer. Pb isotopes of karstic spring waters with very low Pb contents (about 5 ng) have been measured with an ICP-MS with a magnetic sector (Plasma 54®, Fisons Instruments®) and corrected for the blank of the bottle.

3. WATER CIRCULATION

3.1. *The Sourcettes spring (109)*

This spring is located in the upper part of the karstic zone and results from water infiltrations at the limit of the Palaeozoic basement and the Mesozoic cover. The signature of water entering the system is assimilated to the one of surface water (17) draining the Palaeozoic basement (Figure 2).

During the underground circulation the Sr isotopic composition of water decreases to less radiogenic values because of the dissolution of Sr contained in the Jurassic carbonated surrounding rocks.

In low flows, the Sourcettes spring Sr isotopic compositions are more radiogenic than in high flows, similar to the one of the Vis river (16, squares) and the Foux spring (117). The Foux spring drains a carbonated aquifer whose base level is composed of Liassic marls. Marls contain Rb which induces higher $^{87}Sr/^{86}Sr$ in water draining the base of the aquifer. The Vis river (16) is mainly fed by the Foux (17) but also by other springs which drain Liassic marls too.

In high flows, spring water has two origins : 1- water infiltration at the limit of the Palaeozoic basement and the Mesozoic cover, 2- rain water infiltrated directly at the surface of the karstic cover which flushes an older water (few days to few weeks) temporally stored in the pores of the karstic aquifer. This water has had a longer contact with the surrounding carbonate which allowed a greater dissolution of the Sr of the rock ($^{87}Sr/^{86}Sr \sim 0.708$). High flow Sourcettes samples plot on a mixing line between the Liassic marls and Carbonates endmembers reflecting the influence of both water origins. The sample which plots near to the carbonate endmember corresponds to the highest

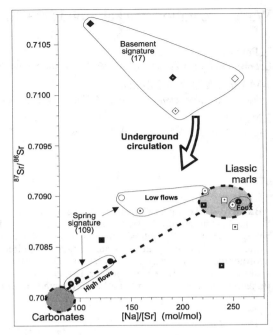

Figure 2. Sourcettes spring signature evolution according to hydrological conditions in a $^{87}Sr/^{86}Sr$ vs. Na/Sr diagram.

flood of the six sampling series. At the opposite, the minimum flow sample plots close to the Liassic marls endmember. This reveals the various proportions of water of both origins in the different hydrological conditions.

Sr concentration is constant in high and low flows, while Na increases in low flows reflecting the more important proportion of water originating from the Palaeozoic basement (granite and schists).

3.2 *The Cent Fonts Springs (105a and 105b)*

These springs are located a few meters from each other. A coloration experiment (Paloc 1972) has shown a relation with the Buège interrupted stream (13). For the study of these springs, we have coupled two isotopic tools : Sr and Pb radiogenic isotopes.

From the infiltration to the resurgence, the water Sr isotopic composition decreases slightly to 0.7083 due to Sr dissolved from the Jurassic surrounding carbonates ($^{87}Sr/^{86}Sr < 0.7076$; Koepnick et al. 1990) (Figure3). Both springs (105a and 105b) present the same Sr isotopic signature and Rb/Sr chemical ratio. The Rb/Sr ratio is constant for the infiltration (13) and the resurgence (105a,b). These results suggest that the waters have the same origin and the same underground circulation.

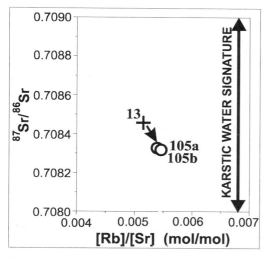

Figure 3. Evolution of the geochemical signature of water during underground circulation in a $^{87}Sr/^{86}Sr$ vs. Rb/Sr diagram.

Lead isotopes have been measured in the dissolved load of the karstic spring waters. Results have been reported in a $^{207}Pb/^{204}Pb$ vs. $^{206}Pb/^{204}Pb$ diagram (Figure 4) with local rock endmembers analysed in previous studies, and Pb-Zn mineralizations of the Malines district located in the upper part of the Hérault watershed.

Springs 105a and 105b present clearly different Pb isotopic signatures. 105a plots in the Oxfordian carbonates domain and presents the lowest Pb concentration (0.04 nmol/l), whereas 105b plots in the Malines mineralizations domain and has a slightly higher Pb content (0.25 nmol/l). These differences of Pb signatures suggest different underground circulations, at least part of the way. 105a waters seem to drain the surrounding upper Jurassic carbonates and acquire their Pb signature with low Pb concentrations. 105b waters seem to drain some trace of disseminated mineralizations with the same Pb isotopic signature as the Malines ones. This example well illustrates the high sensibility of the Pb isotopic tracer even at very low Pb concentrations. Note that the 105b sample also presents the same Pb signature as the Triassic and Bathonian endmembers, but the higher Pb concentration in this sample suggests a contribution from the Malines mineralizations.

Pb isotopic compositions coupled with Sr ones and major and trace elements, give complementary information on water origin and circulation in a carbonated environment. They suggest water circulations at different levels or ways of the aquifer not revealed by Sr isotopes, major and trace elements.

4 CHEMICAL EROSION RATE

Chemical erosion rate of the Mesozoic cover in the middle part of the Hérault watershed has been

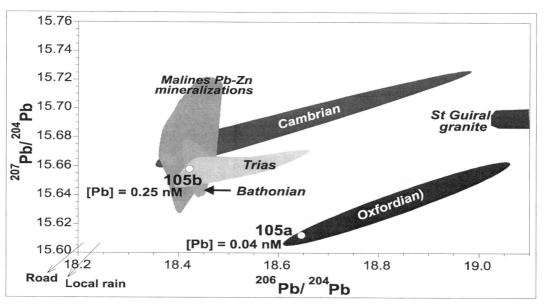

Figure 4. Pb isotopic signature of karstic springs 105a and 105b with local rocks endmembers.

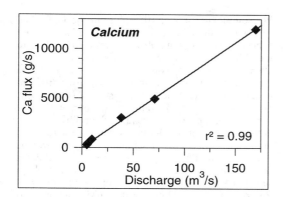

Figure 5. Ca flux-discharge relation in the Mesozoic sub-basin.

calculated according to Ca fluxes - discharge relations based on the six sampling series at the outlet of this sub-basin. This relation has been established by subtracting the Ca flux entering the karstic zone, originating from the basement in the upper part of the basin (Figure 5).

The mean annual discharge at the outlet of the karstic zone is 15.5 m^3/s. We assimilate the limestones to CaCO$_3$ (40 % Ca) with a mean rock density of 2.5. In these conditions, the mean annual Ca flux is 49.5 t/km²/y corresponding to a rock dissolution of 50 µm/y.

The same calculation has been done on the Vis (16) sub-catchment, part of the karstic zone. It gives a chemical erosion rate of 68 µm/y.

These estimations of the chemical erosion rate are in good agreement with previous calculations (61 µm/y) based on Ca concentrations in karstic springs in the same basin (Paloc 1972).

Various attempts at calculations have been make with trace element fluxes, especially Sr. Results cannot be validated because of the heterogeneity of Sr concentrations at local Jurassic levels (50 to 750 ppm; Ben Othman et al. 1997).

5. CONCLUSION

This study has make it possible to identify a dynamical scheme of water circulation in a karstic area with coupled geochemical tools (major, trace elements and Sr isotopes). Sr isotopic composition variations in contrasted hydrological conditions reveal the presence of marly argillaceous levels in aquifers and water storage in the pores of the surrounding carbonates. We have also shown surficial - underground relations especially with Sr isotopes and trace element ratios. Geochemical tracers make it possible to confirm underground circulations identified by colorimetric experiments.

Pb isotopic compositions have made it possible to identify Pb origins in the karstic area of the Hérault watershed : 1- Pb issuing from the Jurassic surrounding rocks and 2- Pb originating in the dissolution of disseminated Pb mineralizations. Pb isotopic compositions are a very sensible tracer even with very low Pb concentrations in samples. They give complementary information on the karstic underground circulation of waters which have the same Sr isotopic signature.

Chemical erosion rate based on Ca fluxes at the outlet of the Mesozoic cover is about 50 µm/y.

REFERENCES

Ben Othman, D., Luck, J.M. & Tournoud, M.G., 1997. Geochemistry and water dynamics: application to short time-scale flood phenomena in a small Mediterranean catchment. I- Alkalis, alkali-earths and Sr isotopes. *Chem. Geol.* 140 : 9-28.

Berner, E.K. & Berner, R.A., 1987. *The global water cycle. Geochemistry and environment.* Englewoods Cliffs, New Jersey : Prentice-Hall.

Goldstein, S.J. and Jacobsen, S.B., 1987. The Nd and Sr isotopic systematics of river-water dissolved material: Implications for the sources of Nd and Sr in seawater. *Chem. Geol. (Isot. Geosci. Sect.)* 66 : 245-272

Koepnick, R.B., Denison, R.E., Burke, W.H., Hetherington, E.A. & Dahl, D.A., 1990. Construction of the Triassic and Jurassic portion of the Phanerozoic curve of seawater ^{87}Sr/^{86}Sr. *Chem. Geol. (Isot. Geosci. Sect.)* 80 : 327-349.

Luck, J.-M. & Ben Othman, D., 1998. Geochemistry and water dynamics. II. Trace metals and Pb-Sr isotopes as tracers of water movements and erosion processes. *Chem. Geol* 150 : 263-282.

Négrel, P., Allègre, C.J., Dupré, B. & Lewin, E., 1993. Erosion sources determined by inversion of major and trace element ratios and strontium isotopic ratios in river water: The Congo Basin case. *Earth Planet. Sci. Lett.* 120 : 59-76.

Paloc, H., 1972. *Atlas hydrogéologique du Languedoc-Roussillon.* CERGA-BRGM.

Geochemistry of the Earth's Surface, Ármannsson (ed.)© 1999 Balkema, Rotterdam, ISBN 90 5809 073 6

Dissolved trace elements in a seasonally snow-covered catchment

A. M. Shiller
Department of Marine Science, University of Southern Mississippi, Stennis Space Center, Miss., USA

ABSTRACT: Over two years of weekly data have been collected on the stream concentrations of a number of dissolved trace elements in a seasonally snow-covered alpine/subalpine catchment in the Rocky Mountains, USA. In general, the trace element concentrations show significant seasonal changes in these dissolved trace elements and the changes appear to be linked to the seasonal snow cover and its effects. However, not all trace elements vary in the same way. For instance, Mn and Mo concentrations appear to be controlled by flushing and ventilation of an alpine lake whereas U follows the flushing of DOC from the soils. Furthermore, ratios of alkali metals are found to have the potential to reveal seasonal changes in weathering and/or vegetative processes.

1 INTRODUCTION

There are three basic approaches that researchers have taken in studies of dissolved trace elements in fluvial systems. The first might be termed the "universal approach" in which the emphasis is on using laboratory studies to obtain an understanding of the most basic factors that control the general concentration levels of these elements. The second can be termed the "global approach" in which the dissolved concentrations of a given trace element in a wide variety of rivers are compared with potential controlling factors such as pH, rock type, or indicators of weathering regime. The third approach can be termed the "local approach" in which the goal is understanding the local temporal variability in dissolved trace elements in one system and the factors that lead to that variability.

There have been surprisingly few studies of fluvial dissolved trace elements utilizing the local approach. This has resulted in part from the problems of contamination that have plagued national and regional monitoring programs as well as the reluctance of scientific funding agencies to sponsor research projects that are typically viewed as being only monitoring. However, time series data on dissolved constituents in fluvial systems can be an important means of identifying or eliminating possible mechanisms controlling the concentrations of those constituents. Additionally, in a seasonally snow-covered catchment, there is the possibility that spring melt processes will mimic the processes that occur over longer timescales during global climate change.

Previous work in my lab has examined the seasonal variability of dissolved trace elements in the lower Mississippi River (Shiller 1997), that is, a major flood plain river. In the work described here, I examine seasonal variability of dissolved trace elements in a small, alpine/subalpine, seasonally snow-covered catchment. A brief comparison is made between these two very different fluvial systems.

2 SITE DESCRIPTION AND METHODS

The Loch Vale watershed is a 660 ha alpine-subalpine watershed located in Rocky Mountain National Park, about 80 km northwest of Denver, Colorado. A detailed description of the watershed can be found in Baron (1992) and Clow et al. (1997) and only the most pertinent details are summarized here. Over 80% of the watershed consists of bare rock surfaces and talus slopes. Most of the rest of this northeast facing watershed consists of alpine tundra and meadow with only the lower 6% having a mature fir/spruce/pine forest. Bedrock consists mainly of Precambrian gneiss intruded by the Precambrian Silver Plume granite which also underlies the lower 20% of the basin. True soils cover only about ~10-15% of the basin area and are coarse, immature, and acidic.

Weekly samples were collected at the outlet of the Loch at an elevation of 3050 m. The Loch is the lowest of three small alpine/subalpine lakes (and a glacial tarn) in the basin The trace element sampling began in February 1996 and continues now. Major element data (including pH and DOC) plus discharge and water temperature were provided by J. Baron (USGS). Collection and analysis of the major element samples is described by Campbell et al. (1995). Samples for trace element analysis were collected using acid-washed all-polyethylene/ polypropylene syringes and filtered through acid-washed 0.45 µm polyethylene syringe filters and stored in acid-washed 15 ml polyethylene bottles. Trace element samples were then acidified to pH ~2 with quartz sub-boiling distilled 6 N HCl. Prior to analysis the samples were diluted in half with 0.1 N HNO_3 and then analyzed by quadrupole ICP-MS.

3 RESULTS AND DISCUSSION

Major elements have a concentration variability in Loch outflow waters significantly determined by seasonal dilution with snow melt; i.e., higher concentrations during winter, decreasing in spring and gradually increasing in summer and fall. DOC shows an increase during early spring which appears to be due to flushing of soil waters (Baron et al. 1991; Denning et al. 1991). There is also a smaller peak in DOC in the fall which is probably related to die-off of seasonal vegetation. For pH, there does appear to be a ~0.5 unit decrease in the spring and an increase in the summer though there are also longer term changes of a greater magnitude.

Mean concentrations for the dissolved trace elements are: Mn = 44 nM, Mo = 3 nM, Cu = 10 nM, Rb = 4 nM, Ba = 30 nM, Li = 160 nM, and U = 0.5 nM. However, the data show significant seasonal changes in these dissolved trace element concentrations and the changes appear to be linked to the seasonal snow cover and its effects. There are four types of seasonal patterns shown by these trace elements. Alkali and alkaline earth metals such as Rb and Ba shown a seasonal dilution pattern similar to the major elements. Manganese shows a pattern similar to this, though with a very abrupt drop in concentration associated with the breakup of ice cover on the Loch. Uranium, and to a lesser extent copper, show a seasonal pattern similar to that of DOC. Molybdenum is the only element determined so far that increases in concentration at the time of the spring melt, continues to increase slowly through the summer, and then decreases as ice cover forms on the Loch.

The variability of Mn and Mo is compatible with the redox behaviors of these elements (i.e., mobilization of Mn and removal of Mo during reducing conditions) as well as with the longer wintertime residence time of water in the Loch. It is interesting to note that these two elements show a similar seasonal variability in the lower Mississippi River. However, in that system the temperature effect on microbial oxidation is a more likely cause of the Mn variability (Shiller 1997). Mo variability in both systems, however, probably results from removal into anoxic sediments during the winter versus a lack of removal during summer when lakes are better flushed and ventilated.

The approximately three-fold increase in U associated with the early spring DOC maximum suggests that most of the mobilization of U in this system is associated with organic matter. Cu also appears to be so affected, though its early spring increase is only about 30% of its mean concentration. In contrast, in the flood plain river (lower Mississippi) DOC varies little as does Cu. U concentrations do vary seasonally in the lower Mississippi, however this may result from seasonal changes in the mixing ratio of the major tributaries.

For the alkali metals (including the major elements Na and K), it is interesting to examine the seasonal variation of their ratios. An early spring increase is observed in the K/Na ratio which could result from the soil flush of the decayed remnants of the previous summer's seasonal plant growth. Alternatively, it is possible that the seasonal change in the K/Na ratio results from preferential weathering of K-rich primary minerals freshly exposed by wintertime physical erosion followed by the spring melt. Because Rb, like K, is taken up by plants, the relative constancy of the Rb/K ratio tends to support the vegetative interpretation. Interestingly, the Na/Li ratio is also variable, mainly because of only minimal concentration variation in Li. Why this is so is uncertain, though possibilities include the slow incorporation of Li into clays.

4 CONCLUSIONS

The two years of data from the Loch Vale watershed demonstrate significant seasonal concentration variability for a number of trace elements. In general this variability results from the seasonal snow cover and its effects on runoff, flush of soils, and the redox state and residence time of lake waters. Comparison with dissolved trace element variability in the lower Mississippi River shows similar patterns for some elements (e.g., Mn and Mo) and different for others (e.g., U). However, the Mn data suggest that similar seasonal concentration variations in these two systems do not necessarily result from the same processes. A more detailed study of alkali metal ratios would have the promise of separating seasonal vegetative effects from seasonal weathering effects. To address these questions soil water, spring, and vegetation samples are being collected.

REFERENCES

Baron, J. 1992. *Biogeochemistry of a subalpine ecosystem.* New York: Springer-Verlag.

Baron, J., D. McKnight & A.S. Denning 1991. Sources of dissolved and particulate organic material in Loch Vale Watershed, Rocky Mountain National Park, Colorado, USA. *Biogeochem.* 15: 89-110.

Campbell, D.H., D.W. Clow, G.P. Ingersoll, M.A. Mast, N.E. Spahr & J.T. Turk, 1995. Processes controlling the chemistry of two snowmelt- dominated streams in the Rocky Mountains. *Water Resour. Res.* 31: 2811-2821.

Clow, D.W., M.A. Mast, T.D. Bullen & J.T. Turk 1997. Strontium 87/strontium 86 as a tracer of mineral weathering reactions and calcium sources in an alpine/ subalpine watershed, Loch Vale, Colorado. *Water Resour. Res.* 33: 1335-1351.

Denning, A.S., J. Baron, M.A. Mast & M.A. Arthur 1991. Hydrologic pathways and chemical composition of runoff during snowmelt in Loch Vale Watershed, Rocky Mountain National Park, Colorado, USA. *Wat. Air Soil Pollut.* 55: 107- 123.

Shiller, A.M. 1997. Dissolved trace elements in the Mississippi River: seasonal, interannual, and decadal variability, *Geochim. Cosmochim. Acta* 61: 4321-4330.

Geochemistry of the Earth's Surface, Ármannsson (ed.)© 1999 Balkema, Rotterdam, ISBN 90 5809 073 6

Effect of lithology on silicate weathering rates

A.S.Taylor & A.C.Lasaga
Department of Geology, Yale University, New Haven, Conn., USA

J.D.Blum
Department of Earth Sciences, Dartmouth College, Hanover, N.H., USA

ABSTRACT: Present day and long-term average weathering rates have been measured for a range of silicate lithologies exposed in adjacent drainage basins. The exposure of all the basins to the same weathering environment allows for direct comparison of weathering rates as a function of lithology while all other climatic variables are constant. On average, basalt weathers twice as fast as granite and tonalite both in modern weathering environments and over the past 11,000 years. Basalt will have an even more pronounced effect on atmospheric CO_2 concentrations. We estimate that CO_2 consumption resulting from basalt weathering is almost 3 times greater than during the weathering of more siliceous rock-types

1 INTRODUCTION

Chemical weathering of silicate minerals is one of the most important sinks for atmospheric CO_2 on geologic time scales. Variations in average global weathering rates play a central role in controlling atmospheric CO_2 concentrations. Understanding the factors that control chemical weathering rates is therefore critical to predictions of a variety of climatic variables in the geologic past. Chemical weathering rates are a function of an array of factors including temperature and precipitation (White & Blum 1995), river and groundwater discharge (Bluth & Kump 1994; Amiotte Suchet & Probst 1993), vegetative cover (Moulton & Berner 1998), glacial history (Taylor & Blum 1995), tectonic environment, and lithology (Meybeck 1987). While Meybeck's (1987) survey of mono-lithologic drainage basins in France vastly improved our understanding of the relative weathering rates of a variety of lithologies, no effort was made to control the large number of other variables that affect silicate weathering rates. Because of the infrequency with which they are exposed in the same locality, direct comparison of basalt and granite weathering rates in the field has difficult. As a result, assumptions are typically made to try to correct for climatic variables when comparing the weathering rates of these two end-member silicates. We have measured modern weathering rates of several silicate lithologies in a location where all other variables are relatively constant. In addition, we determined the relative long-term average weathering rates of these rock-types in order to develop a more complete understanding of the effect of lithology on dissolution rates.

2 RESULTS

Near McCall Idaho, USA there is a series of adjacent drainage basins that consist entirely either of the Columbia River basalt, the Idaho Batholith granite or an intermediate tonalite. The geographic proximity of these basins allowed us to determine weathering rates as a function of lithology in an environment where all other variables (climate, geomorphology, vegetation, glacial history, atmospheric deposition) were essentially constant. We considered both present-day weathering rates and long-term average weathering rates by measuring solute fluxes in streams draining the basins and cation release from soil profiles. Samples were taken during late summer field seasons in order to minimize the contribution of snow to the stream fluxes.

We determined present day weathering rates (R_{PD}) by measuring the fluxes of dissolved base cations and Si in springs and streams draining several basins of each lithology. In general the solute fluxes were significantly greater in the streams draining the basalt than in those draining the more siliceous terranes. On average, Si fluxes out of the basaltic basins were 2.1-2.4 times the fluxes of the tonalite and the flux granite. Base cation fluxes were also higher in the basaltic streams by a factor of 2.9. To compare the weathering rates more directly, we normalized the stream ratios to the rock compositions. The

Table 1. Relative weathering rates of silicate lithologies

Measure of weathering rate	Ratios	
	Basalt/Granite	Basalt/Tonalite
Present Day (R_{PD})		
Si	2.4	2.1
Base Cations	2.9	2.9
Long-term average (R_{LT})	2.1	1.8
CO_2 consumption (R_{CO2})	2.8	2.9

normalized basalt/granite Si flux ratio was 3.3 while the normalized base cation flux ratio was 2.0.

Similar relative weathering rates were observed when we measured long term averages (R_{LT}). Base cation depletion profiles were determined for 3 soil profiles on each of the 3 lithologies. In our depletion profile calculations, Ti was assumed to be immobile. The long-term average weathering rates were calculated based on the assumption that the most recent glaciation in this area (11,000 years ago) exposed fresh material to the weathering environment. The long-term base cation depletion rates of the basalt are also 1.8-2.0 times that of the tonalite and the granite. The agreement between weathering rates calculated using solute fluxes and soil depletion profiles suggests that the basalt/granite weathering rate ratio is reasonably constant with time.

While the ratio of basalt/granite weathering rates may be roughly 2 for both modern conditions and long-term averages, basalt weathering consumes more than twice the CO_2 that granite weathering does because it has higher Ca and Mg concentrations. We calculated the present day CO_2 consumption (R_{CO2}) rates for each lithology based on the dissolved Ca and Mg fluxes. This calculation agreed well with CO_2 consumption rates calculated based on bicarbonate fluxes measured in the streams and we estimated that basalt weathering consumes 3 times more CO_2 than granite weathering.

By measuring weathering rates of several silicate lithologies in a location where other weathering variables could be considered constant, we were able to more accurately determine the relative weathering rates of these rock-types.

REFERENCES

Amiotte Suchet P. & J. L. Probst 1993. Modelling of atmospheric CO_2 consumption by chemical weathering of rocks: Applications to the Garonne, Congo and Amazon basins. *Chem. Geol.* 107: 205-210.

Bluth G. J. S. & L. R. Kump 1994. Lithologic and climatologic controls of river chemistry. *Geochim. Cosmochim. Acta* 58: 2341-2359.

Meybeck, M. 1987. Global chemical weathering of surficial rocks estimated from river dissolved loads. *Am. J. Sci.* 287: 401-428.

Moulton K. M. & R. A. Berner 1998. Quantification of the effect of plants on weathering; studies in Iceland. *Geology* 26(10): 895-898.

Taylor A. S. & J. D. Blum 1995. Relation between soil age and silicate weathering rates determined from the chemical evolution of a glacial chronosequence. *Geology* 23(11): 979-982.

White A. F. & A. E. Blum 1995. Effects of climate on chemical weathering in watersheds. *Geochim. Cosmochim. Acta* 59(9): 1729-1747.

Geochemistry of the Earth's Surface, Ármannsson (ed.) © 1999 Balkema, Rotterdam, ISBN 90 5809 073 6

Weathering of granite – A field study in SW Sweden

P.Torssander, H.Strandh & C-M.Mörth
Department of Geology and Geochemistry, Stockholm University, Sweden

G.Åberg
Department of Environmental Technology, Institute for Energy Technology, Kjeller, Norway

ABSTRACT: A large field study on weathering of granite surfaces is undertaken in the Tanum area in SW Sweden. Numerous rock carvings of cultural interest have been found in the area. In this study the weathering and weathering rates are investigated by mass balances and element ratios of major-, trace elements and isotopes (Sr and S). Samples were collected from atmospheric deposition and runoff from two confined ponds on the granite surface. The main objective is to investigate if a roof cover could serve as a preservation technique. Sr isotope ratios are enriched in the runoff as well as many trace elements. The chemistry expressed as the release rate of Si indicate so far that the roof might decrease the weathering rates.

1 INTRODUCTION

Increased weathering rates of natural stone in buildings and cultural heritage have become an acute problem as a response to acid deposition. In Tanum, SW Sweden, there are numerous rock carvings of high preservation interest from the Swedish Bronze Age. During recent decades these rock carvings deteriorate, vanish and disappear at a high rate.

The area is severely affected by acid rain, and the proximity to the North Sea expose the rock surfaces to high salt concentrations and give the area a relatively mild climate. During winter snow covers are seldom stable and the temperature is often around the freezing point of water. These are the worst possible conditions for physical weathering which easily can overcome the chemical weathering that initializes the process.

Many ideas for the preservation of rock carvings have been and are currently applied in Scandinavia: These ideas include roof covered rock surfaces combined with water cleaning at known intervals and burial with different types of material such as soil, peat as well as artificial materials. The purpose of this intense chemical study is to determine the sources and sinks of elements so that the rock weathering rates can be calculated. The weathering rates can later be used to evaluate different preserving techniques. The objective of this study is to determine the rate controlling mechanism of granite rock surfaces.

2 METHODOLOGICAL APPROACH

Chemical weathering of the rock surface is one source for the elements found in runoff waters. The chemical weathering can be calculated with mass balances if other sources are identified and subtracted. To separate between the sources; throughfall, rain and bulk deposition are collected as measurement of the input, while runoff from the rock surface is collected from two small about 2 m^2 large confined ponds on the rock surface. Chemical analysis is made of all these waters for major- and trace elements and isotopes (Sr and S). The rock surface pond is sampled during rainfall, while the rock surface pond under roof is sampled during washing with 2L de-ionized water on a weekly basis. The chemical study was initiated in early 1998.

One m long drill cores (7 cm in diameter) were sampled during the roof construction in late 1997. The mineralogical composition of the rock was determined by visual inspection in hand-specimen and microscopic analysis and the specific surface area of the rock will be determined through laser micro mapping and BET analysis of small surface rock pieces.

In order to determine the weathering rate, it is necessary to have knowledge about the ponds areas and water balance quite exactly and desirable to know the surface temperature and time the surfaces stay wet etc. At an early stage in our investigation a computer-logged system equipped with a mobile telephone was installed at the site. All to ensure a continuous recording of temperature and relative humidity in the air and on the rock surfaces below and outside the roof, and similarly, continuous determination of rainfall and con-

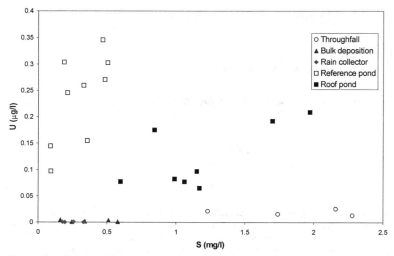

Figure1. U vs S for compiled Tanum data.

ductivity of runoff water. Data is logged every ten minutes and collected daily. An attempt is also made to determine the physical weathering through collection of rock fragments in the ponds and collection of small rock particles in filtered samples from the runoff. These are analysed mineralogically and chemically.

3 RESULTS AND DISCUSSION

The granite rock has a rather heterogeneous modal composition of quartz, plagioclase, potassic feldspar and minor biotite. It has a typical chemistry for a granite with more than 70 % SiO_2 and rather high alkali con-

centrations (up to 7% K_2O). It has a weathered surface down to about 4-5 cm depth.

The results from the water chemistry are shown in Figures 1-4. Only preliminary results exist so far from this study. Comparisons of ICP-MS (HR) analysis between deposition and runoff samples show that the trace elements U, Nd, La, Ga, Ce, Mo, Pr, Sm, Sc, Sr, Ti, Y, and Zr are enriched in runoff. Other elements like N, S and Cl are enriched in the atmospheric deposition. Figure 1 is plot of U vs S where this can be seen clearly, highest sulfur concentrations occur in the runoff from the washed pond (roof pond where S is dry deposited). This is due to the small amounts of water used for wetting and leaching the roof pond surface

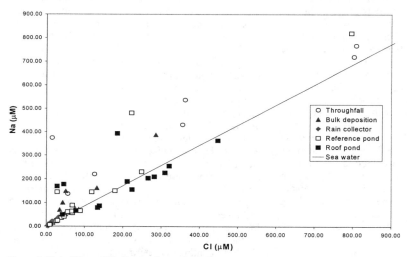

Figure 2. Plot of Na vs Cl indicating the marine influence.

130

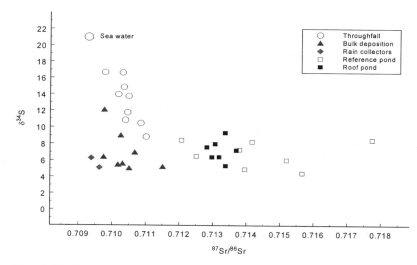

Figure 3. Plot showing the influence of sea water vs weathering as a source of sulfur and strontium.

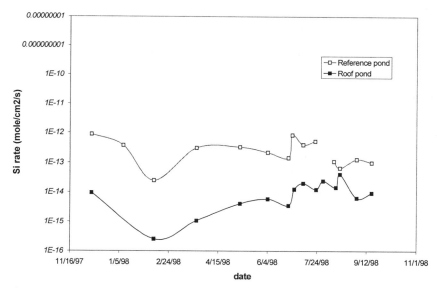

Figure 4. Preliminary results of weathering rates based on silica release. The graph shows the decreased chemical weathering in pond under roof.

(the difference between the ponds were pronounced by an unusually rainy summer in 1998). Analyses of dissolved and particulate faces show that 0-50% of the trace elements are in particulate form. Blanks from de-ionized water show that the contribution to the trace element contents is less than 1%. There is a strong correlation between Na and Mg versus Cl in the water samples, indicating a large influence of the coastal proximity (Figure 2). This implies that it might cause problems to use Na and Mg for mass balances.

Strontium isotope ratios ($^{87}Sr/^{86}Sr$) vary in the fresh rock minerals from around 0.7792 up to ratios for whole rock of around 0.7913. The $^{87}Sr/^{86}Sr$ ratios in water from the rain collectors is very similar to the seawater Sr ratio, while bulk deposition and throughfall show slight increases of the ratio. Hence, the Sr isotope ratio can be used to estimate the dry deposition. The samples from the ponds and the roof exhibit higher $^{87}Sr/^{86}Sr$ ratios up to around 0.718 as a cause of weathering input (Figure 3). Extremely good correlation is found between Ca and Sr, concurring that Sr isotope ratios appears to be a promising tool for the evaluation of weathering data, previously used in catchment studies (Åberg et al. 1989). The δ^{34}S values in Figure 3 plot along the vertical axis; highest δ^{34}S in the deposition is found in throughfall, large variation in bulk deposition and the lowest δ^{34}S values in the rain collectors. The runoff from the ponds has generally high S concentrations and variable δ^{34}S between +4‰ and +9‰. Basically, this indicates two sources of S: an anthropogenic source and sea salt, previously noted in the nearby located Lake Gårdsjön area (Torssander & Mörth 1998).

ICP-MS analyses of the filters show that an important part of the weathering products are small rock fragments contributing to the total weathering rate. As a first approximation, chemical weathering rates have been estimated without correction for dust fall and specific surface area. The maximum chemical weathering rate (release rate of Si) varies between $8*10^{-13}$ and $2*10^{-14}$ mole/cm^2/s with a mean of $3*10^{-13}$ mole/cm^2/s (Figure 4), which is about 100 times faster than experimental data (Schweda 1990). The rate is positively correlated with the rock surface temperature. Also, the runoff pH from the pond under the roof varies between 5.5 and 6.5 while the pH in runoff from the pond outside the roof is between 4 and 5 similar to the atmospheric deposition. The rock surface under the roof show a mean maximum release rate of Si at $1*10^{-14}$ mole/cm^2/s, indicating that a roof might be an alternative preservation technique against chemical weathering. However, during the second half of 1998 the rainfall at the site amounted to 500 mm to be compared with the 20 mm that is poured over the roof covered area. The experiment will therefore be modified so that the addition of deionised water to the roof covered pond will equal the amount of rainfall to ensure larger comparability.

4 CONCLUSIONS

Trace elements, S- and Sr isotopes form a promising tool that could be used to evaluate weathering data. Preliminary results indicate that the release rate of Si from a granite rock surface could be reduced with the aid of a roof and cleaning of the rock surface with deionized water.

ACKNOWLEDGEMENT

B. Boström, K. Hajnal and H. Siegmund are thanked for laboratory and mass spectrometrical work. This project is part of the Interreg. II program and jointly funded by EC and the Swedish Central Board for National Antiquities.

REFERENCES

Åberg G., G. Jacks & P. J. Hamilton 1989. Weathering rates and ^{87}Sr/^{86}Sr ratios: an isotopic approach. *J. Hydrol.* 109: 65-78.

Schweda, P. 1990. Kinetics and mechanisms of alkali feldspar dissolution at low temperatures. *Meddelanden Stockh. Univ. Inst. Geol. Geok.* p. 100.

Torssander P. & Mörth C-M., 1998. Sulfur dynamics in the roof experiment at Lake Gårdsjön deduced from sulfur and oxygen isotope ratios in sulfate, In H Hultberg & Skeffington R (eds): *Experimental reversal of acid rain effects: The Gårdsjön Roof Project*: 185-206. New York: John Wiley & Sons.

3 Environmental geochemistry of the terrestrial environment and its effect on health

Geochemistry of the Earth's Surface, Ármannsson (ed.) © 1999 Balkema, Rotterdam, ISBN 90 5809 073 6

Invited lecture: Soils and soil erosion in Iceland

Ólafur Arnalds
Agricultural Research Institute, Reykjavík, Iceland

ABSTRACT: The soils of Iceland are Andosols that draw their major characteristics from their volcanic parent materials. Rapid chemical weathering results in the formation of allophane and imogolite clay minerals together with poorly crystalline ferrihydrate. Organic wetland soils also show strong andic influences due to steady input of eolian materials and occational tephra deposition events. Large areas in Iceland have desert-like barren surfaces. Much of the desert landscape has formed as a result of intense erosion over the past 11 centuries, since the arrival of man to Iceland. The volcanic characteristics of the soils and the intense erosion processes shape the physical and chemical environment of the surface. Erosion and desertification have recently been surveyed in all of Iceland and a new soil map of the country is being prepared.

1 INTRODUCTION

Iceland is an island in the North Atlantic Ocean, between 63° and 66° northern latitudes. The climate is humid cold temperate to low arctic. Permafrost is nearly absent. The island is mountainous with lowland areas and river plains along the coastline. Rainfall varies between 600 and 2000 mm yr^{-1} in lowland areas.

Three factors exert major influence upon the Icelandic soil environment 1) volcanic activity and the volcanic nature of the soil parent materials; 2) cold maritime climate with intensive cryogenic processes; and 3) extremely active soil erosion, that has created vast unstable "desert" areas; the source of steady eolian sedimentation in the country.

Icelandic soils began to form when the Pleistocene glacier retreated about 10,000 years ago, but many surfaces have since then been disturbed by volcanic tephra deposition, lavaflows, soil erosion, solifluction and landslides, sand encroachment, and glacio-fluvial flooding. Surfaces of undisturbed sites continue to rise due to frequent deposition of volcanic and eolian materials.

2 ICELANDIC SOILS

Soils that form in volcanic ejecta exhibit unique soil properties, so-called andic soil properties, and are therefore recognized at the order level in international soil classification systems as Andisols (US Soil Taxonomy) or Andosols (FAO soil classification). Andosols are common in the volcanic regions of the world, including Japan, New Zealand, and along the western part of the Americas. Andosols constitute about 1% of the world soils. Maeda et al. (1977), Wada (1985) and Shoji et al. (1993) provided summaries of the properties of Andosols. Andosols generally contain high levels of carbon, have high water holding capacity, and low bulk density. The volcanic parent materials have high surface area and weather rapidly to form such clay materials as allophane and imogolite (Wada 1985). The low contents of lattice clay minerals and the physical properties of allophane explain why Andosols often lack cohesion and are extremely susceptible to disturbance, especially when saturated by water.

All Icelandic soils exhibit andic soil properties because of their volcanic origin which is usually basaltic in composition. The parent materials weather rapidly, both volcanic glass and crystalline materials. This is reflected in high measured chemical weathering rates (Gíslason et al. 1996). The soils are traditionally divided into three major groups (Jóhannesson 1960, Arnalds 1999): Typical Andosols of freely drained sites, organic soils in wetland positions (Histosols/Andosols), and the soils of barren deserts (Regosols, Arenosols, Leptosols, Gleysols). Some of the common soil properties of

Table 1. Selected properties of major soil types in Iceland, common range of values.

Type	Area km^2	Depth[#] m	OC[&] %	pH	Clay %	CEC meq/100 g	15 bar water ret. %
Andosol	43,770	0.5-1.5	2-10	5.5-6.5	15-40	10-40	30-60
Histosol	8660	0.5-5	5-30	4.5-5.5	?	30-60	60-120
Regosol	18,600	0.2-0.5	0.5-1.5	6.5-7.0	5-15	5-15	5-15
Arenosol	15,090	0.1-2	0.2-1.0	6.5-7.0	1-5	2-10	1-10

#: Characteristic depth of the soil profile
&: C content throughout profile for Histosols and Andosols, top 10-20 cm for Regosols and Arenosols.

these three major soil types are summarized in Table 1. The Icelandic Agricultural Research Institute is preparing a soil map of Iceland that will be published later this year. Areal numbers for soil types in this paper are derived from this soil map.

Soils that form in eolian and tephra materials under vegetative cover at freely drained sites are typical Andosols (dominant soils on about 43,800 km^2), often with abundance of allophanic clay minerals (up to 30%), ferrihydrates and some imogolite (Wada et al. 1992). Distinct volcanic tephra layers are common. The age of most tephra layers has been determined, allowing for quantification of eolian sedimentation rates for different time intervals. Some of the layers are light colored rhyolitic tephra giving the profiles colorful appearances. Jóhannesson (1960), Helgason (1963, 1968) and Arnalds et al. (1995) provided summaries of the charactersitics of the Andosols of freely drained areas.

The Icelandic wetland soils (organic soils, dominate about 8900 km^2) have lower organic contents than would be expected due to the steady influx of eolian materials. They classify either as Histosols or Andosols. Their characteristics is a mixture of andic and organic properties. The eolian influx is most rapid near glacio-fluvial flood plains, interior deserts and active volanoes. The organic content of the Histosols rises and pH decreases with reduced eolian influx rates. Gudmundsson (1978) and Ólafsson (1974) have provided discussion on Icelandic wetland soils.

Soils of barren landscapes are the third major soil type, comprising a variety of soils, which classify as Regosols, Gleysols, Arenosols, and Leptosols according to the FAO legend, depending on factors such as the local geology and landscape position. The desert soils dominate about 40,000 km^2, but also occur within the Andosol areas. All of these soils exhibit andic properties and are mostly classified as Andisols according to the US system. They are characterized by low organic content but high contents of poorly weathered volcanic glass and basaltic crystalline materials. Water retention is low

and most precipitation enters the ground water system except during spring snow melt, when runoff is rapid. Allophane can be up to 15% in some of the old glacial till surfaces (Regosols), but very low in recently deposited tephra and on active flood plains. The balsaltic origin of the materials color the surfaces dark or black. The desert soils are infertile and generally have 0.5 - 5% vegetation cover (Arnalds et al. 1987). They have an immense potential for sequestering carbon when revegetated, because Andosols can accumulate large quantities of carbon.

Cryoturbation is intense in Iceland for several reasons. During winter, the temperature tends to stay near 0°C creating stationary freezing front. Hydraulic conductivity is often high and freeze-thaw events are so frequent that it is often referred to as the "Icelandic cycle" in geomorphology of cold areas. Most slopes have well developed solifluction features and hummocks develop on freely drained Andosols (e.g. Schunke & Zoltai 1988). Barren surfaces often develop patterned ground. Permafrost is rare, however, but some palsa areas are found in the central highland areas (Thórhallsdóttir 1997).

3 SOIL EROSION AND EOLIAN FLUX

Some of the properties of andic soils have important implications for erosion. The Icelandic Andosols contain almost no phyllosilicate minerals, such as smectite, that provide cohesion, and the mineralogy is dominated by allophane, imogolite and poorly crystallised ferrihydrite (Wada et al. 1992). The formation of silt-sized aggregates, which are susceptible to wind erosion, is favoured. The physical characteristics of Andosols include high infiltration rates and hydraulic conductivity, but also high wind erosion susceptibility. They have extremely high water retention, liquid limit and plastic limit, but a low plasticity index (Maeda et al. 1977). These properties are characteristic of Icelandic Andosols and they contribute to the high susceptibility of these soils to frost heave,

Table 2. The Icelandic erosion classification system (erosion forms)[#].

Erosion froms associated with erosion of of Andosols/Histosols	Deserts erosion forms
Rofabards	Melar (lag gravel, till surfaces)
Advancing erosion fronts (sand encroachment)	Lavafield surfaces
Isolated spots	Sandur (bare sand, sand sources)
	Sandy lavafields
Isolated spots and solifluction features on slopes	Sandy melar (sandy lag gravel)
Water channels	Scree slopes
Landslides	Andosol remnants

[#]: The erosion forms are devieded into two overall categories: erosion forms associated with vegetated Andosol and Histosol systems, and erosion forms on barren (desert) surfaces.

landsliding, and transport by rain-splash and running water, intensified by periodic cold spells and tephra deposition (Arnalds 1990)

Soil erosion processes in Iceland are extremely active, but also quite varied. They include erosion by wind and water, landslides and cryogenic processes. The Agricultural Research Institute (ARI) and the Soil Conservation Service (SCS) conducted a national survey of erosion and desertification in the scale 1:100,000 over the period 1991-1996 (Arnalds et al. 1997). The assessment of soil erosion in Iceland is based on a classification of erosion forms that can be identified in the landscape. The Icelandic erosion classification system is presented in Table 2. Erosion severity is estimated for each of the erosion forms on a scale from zero to five, five being considered extremely severe erosion. The ARI-SCS erosion database is made of about 18,000 polygons. Each polygon is characterised by one or more erosion forms.

Some of the most striking erosional features of Icelandic landscapes are the "rofabards", which are erosional escarpments where Andosols are being truncated from the surface and barren desert is left behind (Arnalds 1999). Other means of erosion are, however, equally important in destroying vegetated Andosol ecosystems. Sand encroachment is an especially severe problem in many areas. The barren desert surfaces are unstable and subject to erosion by wind and water.

It is well established that a large portion of the Icelandic deserts were vegetated at the time of settlement, about 1125 years ago. The evidence for this include historical records, Sagas, annals, old farm surveys, old place names, relict areas and current vegetation remnants, pollen analyses, and soils buried under sand (e.g. Einarsson 1963, Thórarinsson 1961, 1981, Arnalds 1987, Hallsdóttir 1995, Kristinsson 1995, Gísladóttir 1998). Major change with accelerated erosion occurred at the time of settlement and is indicated by accelerated eolian deposition rates soon after settlement 874 AD (e.g. Thórarinsson 1961, Haraldsson 1981). There is no documented evidence for similar massive country-wide erosion in Iceland prior to settlement. The causes of the severe erosion have therefore traditionally been attributed to the the use of the land.

It should however be stressed, that other causes than land use are also important. Additional factors include a cooling trend that began 2500 BP, and also growing sources for eolian sand associated with the advancement of glaciers. The cooler climate resulted in larger glaciers that caused an increased number of melt water floods from sub-glacial thermal areas and volcanoes. This results in larger unstable sandy areas at glacial margins ("sandar"). Increased eolian processes on these unstable surfaces cause steady influx of eolian materials that make the Andosol mantle continually thicker and therefore more unstable than it was before (Arnalds 1999). It can be argued that the combination of these factors caused a "snowball effect" that greatly accelerated erosion rates, but the cause/process relationships still remain to be documented adequately.

REFERENCES

Arnalds, A. 1987. Ecosystem disturbance and recovery in Iceland. *Arctic and Alpine Res.* 19: 508-513.

Arnalds, O. 1990. *Characterization and erosion of Andisols in Iceland.* Ph.D. Dissertation, College Station, Texas:Texas A&M University.

Arnalds, Ó. 1999. The Icelandic "rofabard" soil erosion features. *Earth Surface Processes and Landforms.* In press.

Arnalds, Ó., C.T. Hallmark & L.P Wilding 1995. Andisols from four different regions of Iceland. *Soil Sci. Soc. of Am. J.* 59:161-169.

Arnalds, Ó., E.F. Thórarinsdottir, S. Metúsalemsson, Á. Jónsson, E. Grétarsson & A. Árnason 1997. *Soil Erosion in Iceland.* Reykjavik: Icelandic SCS and the Agricultural Research Institute. (In Icelandic, available in English late 1999).

Einarsson, Th. 1963. Pollen analytical studies on the vegetation and climate history of Iceland in Late and Post-Glacial times. In A. Löve &. D. Löve (eds), *North Atlantic biota and their history:* 355-365. Oxford: Pergamon Press.

Gísladóttir, G. 1998. *Environmental Characterisation and Change in South-western Iceland.* Stockholm: Department of Physical Geography, Stockholm University, Dissertation Series 10.

Gíslason, R.S., S. Arnórsson & H. Ármannsson 1996. Chemical weathering of basalt in Southwest Iceland: effects of runoff, age of rocks and vegetative/glacial cover. *Am. J. Sci.* 296:837-907.

Gudmundsson, Th. 1978. *Pedological studies of Icelandic peat soils.* PhD. Thesis. Aberdeen: University of Aberdeen.

Hallsdóttir, M. 1995. On the pre-settlement history of Icelandic vegetation. *Icel. Agric. Sci.* 9:17-29.

Haraldsson, H. 1981. The Markafljot sandur area, southern Iceland. Sedimental, petrological and stratigraphical studies. *Striae* 15:1-60.

Helgason, B. 1963. Basaltic soils of South-west Iceland. I. *J. of Soil Sci.* 14:64-72.

Helgason, B. 1968. Basaltic soils of South-west Iceland. II. *J. of Soil Sci.* 19:127-134.

Jóhannesson, B. 1960. *The Soils of Iceland.* Reykjavík: University Research Instiute (Agricultural Research Institute).

Kristinsson, H. 1995. Post-settlement history of Icelandic forests. *Icel. Agric. Sci.* 9:31-35.

Maeda, T., Takenaka, H. & Warketin, B.P. 1977. Physical properties of allophane soils. *Adv. in Agr.* 29:229-264.

Ólafsson, S. 1974. *Fysiske og fysisk-kemiske studier af Islanske jordtyper. Licentiatafhaldling.* Copenhagen: Hydroteknisk Laboratorium, Den Kongelige Veterinar Landbohöjskole. (In Danish).

Schunke, E. &S.C.Zoltai 1988. Earth hummocks (thufur). In M.J.Clark (ed), *Advances in Periglacial Geomorphology* 231-245. New York:Wiley.

Shoji, S., M. Nanzyo & R.A. Dahlgren 1993. *Volcanic Ash Soils. Devel. in Soil Sci. 21.* Amsterdam: Elsevier.

Thorarinsson, S. 1961. Wind erosion in Iceland. A tephrocronological study. *Icel. Forestry Soc. Yearb.* 1961:17-54. In Icelandic, extended English summary.

Thórarinsson, S. 1981. The application of tephrochronology in Iceland. In S. Self & R.S.J. Sparks (eds.), *Tephra Studies:* 109-134. Reidel: Dordrecht.

Thórhallsdóttir, Th. E. 1997. Tundra Ecosystems of Iceland. In F.E. Wiegolaski (ed), *Polar and Alpine Tundra. Ecosystems of the World 3:* 85-96. New York: Elsevier.

Wada, K. 1985. The distinctive properties of Andosols. *Adv. in Soil Sci.* 2:173-229.

Wada, K., O. Arnalds, Y. Kakuto, L.P. Wilding, & C.T. Hallmark 1992. Clay minerals in four soils formed in eolian and tephra materials in Iceland. *Geoderma* 52:351-365.

Geochemistry of the Earth's Surface, Ármannsson (ed.)© 1999 Balkema, Rotterdam, ISBN 90 5809 073 6

Invited lecture: Consilience and conciliation: The need for less human impact and more humane impact

D. Kirk Nordstrom

US Geological Survey, Boulder, Colo., USA

ABSTRACT: "For the material world at least, the momentum is overwhelmingly toward conceptual unity. Disciplinary boundaries with the natural sciences are disappearing, to be replaced by shifting hybrid domains in which consilience is implicit" (Wilson 1998). This statement is more profound and deeper than our usual pronouncements for more holistic environmental perceptions — it calls for the unity of all knowledge. Combining consilience with conciliation is the means by which we can hope to attain sustainability, minimize degradation of our environment, and heal relationships between individuals, between communities, and between ourselves and nature.

1 INTRODUCTION

The theme of this paper is simple: we must reduce deleterious human impact on our environment and increase beneficial humane impact. To achieve this goal, we must promote *consilience* and *conciliation*. We must find ways to link causal explanations of phenomena across discipline boundaries (consilience) and to link the goodwill and cooperation of all affected people (conciliation). Our future depends on our willingness and ability to unify knowledge, to communicate it honestly and effectively to society, and to equitably facilitate the cooperation of everyone to manage resources and waste in a sustainable manner. This is not a new message but one that must be told over and over again until enough people take it seriously enough to make a sustainable difference. Human survival depends on it.

The remainder of this paper describes some geochemical problems that result from human impact and the role of consilience and conciliation in achieving reasonable and practical goals..

2 ACID MINE EFFLUENT: TO PLUG OR NOT TO PLUG

Acid mine drainage discharging from a mine portal is like a person bleeding from a deep cut. If a flesh wound is left unattended it can result in death. If a mine portal discharging acid mine waters is left unattended, normal aquatic life will die and stay dead. If the portal is plugged the discharge will stop,

at least temporarily, just as direct pressure applied to a person's wound will temporarily stop the bleeding. Whether this temporary fix can be permanent or not depends on some important factors. We know that if a person experiences a cut, it is important to know the depth of the cut, whether any organs were damaged, whether any major veins or arteries were cut, whether any bones were broken, nerves severed, and so on. Obviously if any of these things had occurred, much more than a few stitches or direct pressure will have to be applied to save that life. After major surgery on the earth, however, the practice has been to walk away, or to install a plug and then walk away. Why? Perception has been limited to simple hydraulics with no links to hydrogeology, geochemistry, microbiology, and mineralogy. Mine plugging has been utilized without considering such questions as — What are the consequences? How long does a plug last? How do the ground-water conditions change? When the ground water stabilizes, what will be the piezometric surface? How will that affect remediation? What will be the chemistry of the mine pool? How will that affect remediation? Will that mine pool leak? If so, where and how fast? When the mine pool stabilizes and the recharge and discharge are in balance, where will be the discharge? At that time will discharge of contaminants be comparable to pre-plugging discharge or not?

Enough mines have now been plugged that some of the answers are available. The short answers are: (a) the consequences of mine plugging can be deleterious, even disastrous, (b) a plug can last a few weeks or it can last a few decades, or possibly longer but it is not immortal, (c) ground-water conditions will usually change tremendously and quickly, (d) the ground-water table will often rise many tens of meters in large hard-rock mines, (e) large ground-water mounds usually produce seeps and springs, outflows from other portals, adits, shafts, faults, fissures, fractures, and drillholes, (f) the resultant mine pool can be very acid with high concentrations of dissolved metals if the mine is sulfidic in a non-carbonate crystalline rock, (g) new ground-water mounds will move and cause fracture-flow in a crystalline bedrock if they are above the existing water table, (h) leakage rates can vary enormously depending on the hydrogeology, structure, and related site conditions, (i) metal loads can be less than or comparable to pre-plugging loads.

At Iron Mountain Superfund site, California, the importance of some of these questions were recognized by the U.S. Environmental Protection Agency and given some serious thought. The largest point source at Iron Mountain of both acid effluent (pH values of 0.0 - 1.1) and dissolved metals is the Richmond Mine. It was a prime location for a set of mine plugs. An underground investigation, followed by some geochemical computations, indicated that plugging the Richmond would lead to a 600,000 m^3 mine pool containing water of pH about 1 or less, with many grams per liter of dissolved metals, sitting in a bedrock with little or no neutralization capacity, dominated by fracture-flow, placed at or near the top of the ground-water table, and probably discharging to the nearby streams in a few decades or less (Nordstrom & Alpers 1999). This scenario was one of several reasons why mine plugging was not an acceptable remediation strategy for Iron Mountain.

Other mines in the area to the north of Iron Mountain have been plugged and some continue to be successful at eliminating discharges, others leak and produce comparable loads to those existing before plugging. There are insufficient investigations to say why some have worked and others have not. Mine plugging is still a common remedy for which the consequences are not carefully considered. Engineering practices have not been adequately linked with hydrology and geochemistry and incorporated them into the design plan. Post-remedial monitoring is also rarely designed to effectively determine the actual consequences of plugging, but this situation is changing. More and more frequently hydrologists and geochemists are being consulted, but it has taken a long time and many costly mistakes have been made.

3 THE MINE FLOOD DISASTER AT AZNALCOLLAR, SPAIN

On the evening of April 24, 1998 and on into the next day, an impoundment dam holding several million tons of tailings, pyrite stockpile, and acid water (pH = 2) failed and flooded down the Guadiamar River toward Doñana National Park, one of Europe's most important wildlife refuges. About 26 km^2 of farmland was covered with sulfides (details at www.csic.es and in a special issue of *Science of the Total Environment*). Several features of the impoundment were questionable. The impoundment was constructed near the headwaters of the Guadiamar, which has a very direct path south to Doñana Park. It was constructed on a base of marl (typically carbonate and clays) which would partially dissolve in acid waters and might not have the necessary shear strength to endure much physical or chemical stress. At the very least, there should have been some backup system or wastewater management contingency in the event of leakage or failure. Again, a disconnect is apparent between engineering practice and hydrogeochemical knowledge.

4 GROUND-WATER WITHDRAWAL

Many surface waters have been polluted to the point where they are no longer usable, especially for drinking purposes. In many arid regions, the available surface-water supply has been fully utilized, often to the detriment of ecosystems. These conditions have encouraged the extraction of ground water for both drinking and agriculture. Ground water exists most everywhere, but at least five important questions should be addressed before utilization begins: (1) At what depth is the ground-water table? (2) How permeable is the aquifer containing the ground water? (3) What are the potential consequences of ground-water withdrawal? (4) What is the water balance and the recharge rate relative to discharge? and (5) What is the quality of the water? These questions seem obvious, but are not always adequately considered. Consequently, there have been numerous examples of subsidence (Texas, Nevada, and California have classic examples), of ground-water contamination, and of disappearing water supplies.

4.1 *Arsenic poisoning from ground water usage*

Consider the fact that numerous aquifers used for drinking water have been found to contain high concentrations of arsenic, after (not before) symptoms of arsenic poisoning have been found among residents. Notable occurrences of high arsenic ground waters used for drinking are known in Mexico, Argentina, Chile, Taiwan, Mongolia, India, and Bangladesh. In the United States both anthropogenic and natural occurrences of high arsenic ground waters have been found in nearly every state. The world's largest mass poisoning is presently occurring in the Bengal delta (West Bengal, India and Bangladesh) where many thousands are stricken and many millions are at serious risk from excessive arsenic intake (Das et al. 1994, Bagla & Kaiser 1996, Nickson et al. 1998). The obvious disconnect in this instance, was the lack of just a few complete major and trace element analyses for the ground waters from the early shallow wells that were built. It would seem that you would want to know what was in the water before you let millions of people drink it!

From a geochemical viewpoint, one would not have expected arsenic in the deltaic plain of the Ganges-Brahmaputra-Megna river system. The question of whether it is anthropogenic or not does not have a simple answer. It is generally agreed that the arsenic itself is of natural origin, but its mobility in the shallow ground water may well be anthropogenic. There are three main hypotheses that have been put forward to explain its mobility: (1) oxidation of arsenian pyrite by lowering of the ground water table through well usage (Chowdhury et al. 1998), (2) reductive dissolution of iron oxyhydroxides causing desorption of arsenic (Bhattacharya et al. 1997, Smedley et al. 1998, Nordstrom 1998) and (3) competitive desorption of arsenic by overutilization of phosphate fertilizers and by the high dissolved organic carbon content. Only the second hypothesis of reductive dissolution might be considered a natural process. All of these processes are probably occurring to some extent depending on hydrology, stratigraphy, and land usage. Appropriate connections between hydrology, geochemistry, and engineering practice could have avoided the present disaster.

The conciliation of governing and affected parties in both India and Bangladesh is a greater problem than investigating the extent and cause of arsenic mobility. Several government and non-government agencies have responsibilities with respect to water supplies and water quality but they have shown little ability to coordinate and collaborate on such a large-scale disaster. In Bangladesh, a central clearinghouse for ground-water data is needed but agreement cannot be easily reached on who should do it and where. Meanwhile millions of people need to be educated on what wells are safe for drinking and cooking and which are not. Modern water supply, treatment, and distribution systems for Bangladesh are not feasible. Educational barriers, political barriers, and social and religious barriers all have to be dealt with while proper investigations to determine what wells have safe levels of arsenic and how to best treat water for 3 million wells in such an economically poor country with more than 120 million people is a challenge on a greater scale than anything before attempted. This situation presents a tremendous challenge to consilience and conciliation.

5 RADIOACTIVE CONTAMINATION

In the rush to maintain military superiority, "superpowers" developed a tremendous nuclear arsenal, contaminated the atmosphere through nuclear bomb testing, downplayed the effects of radiation dosages, and created huge stockpiles of radioactive waste (some of which are under water in the Arctic Ocean). More radioactive waste was created by the construction of more than 430 nuclear power plants around the world. The waste generated during the life cycle of a 30-year nuclear power plant must be safely disposed or stored for many thousands of years. This hardly seems a fair trade-off for future generations no matter how safe we say it is today. When the long-term economics are all calculated, it doesn't appear to be a viable energy source. Meanwhile we experienced the Three Mile Island disaster, the Chernobyl disaster, the Mayak disaster, and dozens of smaller radioactive contaminant releases all over the world. Former East Germany and Russia have radioactive and non-radioactive wastelands of unparalleled proportions. These events occurred because an insufficient number of people understood the importance of strict safety requirements and strict engineering design and operation. The health risks were often underestimated or ignored. Insufficient links existed between production of radionuclides and the dangers of their disposal and dispersion in the environment. We shall be paying for this ignorance for a very long time.

6 CONCLUSIONS

Untold millions of people world-wide have suffered or died because of insufficient attention to consilient thinking between engineering practice and hydrogeochemical knowledge. The lack of education, even among the "educated," is frightening. On environmental issues I offer a few suggestions in the best interests of consilience and conciliation:

1. Avoid picturing any argument or situation as a simple either/or, black or white. Only uninteresting non-issues are truly that simple.

2. Make sure the overall objective, goal, and purpose of the issue is understood and agreed upon by all interested and affected parties. Avoid details here, the overall goal should be the focus.

3. Get the right science in at the beginning and get it right. Don't consider the science after the fact, as damage control, because by then it is usually too late.

4. An accurate, balanced, and informative synthesis of information should be one of the main goals.

5. Be honest about what you know and what you don't know.

6. Communication promotes trust and helps to resolve opposing viewpoints. Get the participation right and get the right participation. Include national oversight with local facilitation and management.

7. Any organization responsible for risk decisions should develop capabilities that conform to sound risk characterization.

8. Expand your useful geochemical knowledge into hydrology, microbiology, toxicology, ecology, and public policy.

REFERENCES

Bagla, P. & J. Kaiser 1996. India's spreading health crisis draws global arsenic experts. *Science* 274: 174-175.

Bhattacharya, P., D. Chatterjee & G. Jacks 1997. Occurrence of arsenic-contaminated groundwater in alluvial aquifers from delta plains, eastern India: Options for safe drinking water supply. *Water Resour. Devel.* 13(1): 79-92.

Chowdhury, T.R., G.K Basu, G. Samanta, C.R. Chanda, B.K. Mandal, R.K. Dhar, B.K. Biswas, D. Lodh, S.L Ray & D. Chakraborti 1998. Borehole sediment analysis probable source and mechanism of arsenic release to groundwater in affected districts of West Bengal, India. In *Intl. Conf. Arsenic Pollution Ground Water Bangladesh: Causes, Effects, Remedies*, abstracts, Feb. 8-12, 1998: 157-158. Dhaka, Bangladesh.

Das, D., A. Chatterjee, G. Samanta, B. Mandal, T.R. Chowdhury, G., Samanta, P.P Chowdhury, C. Chanda, G. Basu, D. Lodh, S. Nandi, T. Chakraborty, S. Mandal, S.M. Bhattacharya & D. Chakraborti 1994. Arsenic contamination in groundwater in six districts of west Bengal, India: the biggest arsenic calamity in the world. *Analyst* 119: 168N-170N.

Nickson, R., J. McArthur, W. Burgess, K.M. Ahmed, P. Ravenscroft & M. Rahman 1998. Arsenic poisoning of Bangladesh groundwater. *Nature* 395: 338.

Nordstrom, D.K., 1998. Geochemistry of the arsenic-pyrite connection and biogeochemical processes leading to arsenic mobility in natural waters. In *Intl. Conf. Arsenic Pollution Ground Water Bangladesh: Causes, Effects, Remedies*, abstracts, Feb. 8-12, 1998, , 148-149. Dhaka: Bangladesh.

Nordstrom, D.K. & C.N. Alpers 1999. Negative pH, efflorescent mineralogy, and consequences for environmental restoration at the Iron Mountain Superfund site, California. *Proc. Natl. Acad. Sci.* 96: 3455-3462.

Smedley, P.L., D.G. Kinniburgh & M. Hussain 1998. Arsenic distribution and mobility in groundwaters from Chapai Nawabganj, western Bangladesh. In *Intl. Conf. Arsenic Pollution Ground Water Bangladesh: Causes, Effects, Remedies*, abstracts, Feb. 8-12, 1998, Dhaka, Bangladesh, 147.

Wilson, E.O., 1998. *Consilience: the unity of knowledge*. New York: Alfred A. Knopf.

Geochemistry of the Earth's Surface, Ármannsson (ed.)© 1999 Balkema, Rotterdam, ISBN 90 5809 073 6

Invited lecture: Budgets of Pb and Cd in soils and critical time to reach a harmful level

Tomas Paces
Czech Geological Survey, Praha, Czech Republic

ABSTRACT: Heavy metals in soils can be hazardous to ecosystems. Budgets of Pb and Cd in agricultural soils are presented. The budgets are not in a steady state. They change with time. This makes the concentrations of the heavy metals a function of time. The time function is derived. A critical time is set when the concentrations of the metals reach a level critical to an ecosystem. The predicted temporal trends in Cd and Pb concentrations in agricultural soils of polluted central Europe (Czech Republic) are compared to the temporal trends in much less polluted forest soils in south Sweden. While the critical concentration of Cd is reached in about sixty years in the Czech agricultural soils, the data from Sweden indicate that Cd is being depleted from the forest soils so that the critical upper level will never be reached. A critical concentration of lead in soil will be reached in about one thousand years in the Czech Republic, while the critical concentration will never be reached in the Swedish forest soils at present atmospheric deposition and agricultural inputs.

1 INTRODUCTION

Terrestrial ecosystems may react adversely to high man-generated inputs of heavy metals (Pb, Cd, Hg). On the other hand, insufficient supply of some of the metals can be also ecologically harmful (Mo). It is therefore desirable to define such inputs of heavy metals that would yield steady-state concentrations in soils not yet harmful to the ecosystems. Such theoretical input is called a "critical load" (Nilsson and Grennfelt 1988, de Vries and Bakker 1998, de Vries, Bakker and Sverdrup 1998, Paces 1998). The critical load is a quantitative estimate of a pollutant input below which harmful effects on specified sensitive elements of the environment do not occur. In a case of some nutrients, critical load can be even a low limit above which a harmful effect does not occur. The critical load is therefore a calculated input that depends on the present knowledge of toxicity of the pollutant in ecosystem. The critical load depends also on inputs, outputs, and reservoirs or pools of the heavy metal. Such system can be characterised by a budget defined by a general mass balance equation:

$$\Sigma \text{ Input (t)} - \Sigma \text{ Output (t)} = \Delta \text{ (t)} \qquad (1)$$

Input (t), Output (t) and Δ (t) are functions of time. The units commonly used to present the flux of heavy metals are $g.ha^{-1}.a^{-1}$. The budget, $\Delta(t)$, is positive if the heavy metal accumulates in the pool or negative if the heavy metal is depleted.

Metals in terrestrial ecosystems come from dissolution of minerals, atmospheric deposition, dispersed inputs of agricultural chemicals and from industrial and domestic point sources of pollution. After entering the ecosystem trace metals are distributed among soil components, living matter and ground and surface water. Runoff of water from the system and removal of biologically fixed metals by harvesting crop or lumbering timber are the major outputs. We will not consider the point sources of the pollution because the critical load is a concept useful to characterise the limiting load of dispersed atmospheric and agricultural pollution. A portion of metals in soil is irreversibly fixed in newly formed minerals. The easily soluble or dissolved portion of metals is assumed to be biologically available for plants. The biologically available part is defined somewhat arbitrarily by a chosen extraction method for chemical analysis. (Mehlich, 1978, 1984, Wolf 1982, Lindsay and Norvell 1978, Jones, 1990).

The concentration of heavy metal in soil is a function of time because it depends on the budget $\Delta(t)$. Under pristine conditions the concentration of heavy metal reaches a steady state favourable value for a given ecosystem because it is the result of the ecosystem metabolism. After a disturbance of the pristine conditions due to an increased input of trace metals, the concentrations in soil increase. The increasing concentrations are approaching critical levels harmful to the ecosystem. An important environmental information is the time span before the concentrations reach their critical levels. This time is called "critical time" (Paces, 1998). This information can be environmentally more significant that the critical load itself because it tells us whether the present inputs are imminently dangerous or are such that they will not endanger the ecosystem in foreseeable future.

2 DEFINITON OF CRITICAL LOAD AND CRITICAL TIME

The simple mass balance mode of soil system is illustrated in Figure 1.

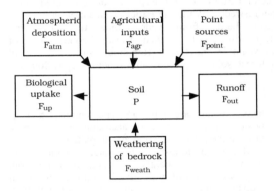

Figure 1 A black box model for calculation of critical load and critical time of heavy metals in soil. Symbols are explained in the text.

The mass balance equation for the system is

$$F_{atm} + F_{agr} + F_{wea} - F_{up} - F_{run} = dP/dt \qquad (2)$$

F_x is flux [g.ha^{-1}.yr^{-1}] of the metal by a transport mechanism x.
The fluxes considered in the model are:
F_{atm} - input of metal by atmospheric deposition
F_{agr} - input of metal by application of agro-chemicals
F_{wea} - input of metal by chemical weathering of bedrock
F_{up} - output of metal by biological uptake
F_{run} - output of metal by runoff of water
P is the biologically available pool of the metal in soil
t is time;

$$P = c_s.\vartheta \qquad (3)$$

$$\vartheta = 10.\sigma. h(1-p/100) \qquad (4)$$

c_s is the concentration of the metal in dry soil [mg.kg^{-1}] determined as a biologically active fraction
p is porosity of soil in percent of *in situ* volume [dimensionless]
σ is density of dry soil [kg.m^{-3}]
h is thickness of biologically active soil [m]
constant 10 transfers units of P to g.ha^{-1}.

The annual inputs are considered to be constant with time. The outputs are assumed to be proportional to the size of the pool:

$$F_{up} = k_{up}.P \qquad (5)$$

$$F_{run} = k_{run}.P \qquad (6)$$

k_{up} and k_{run} are the first-order rate constants [yr^{-1}].
After substitution to the mass balance equation, the change in the pool is

$$dP/dt = F_{atm} + F_{agr} + F_{wea} - (k_{up} + k_{run})P \qquad (7)$$

Integrating from an initial pool P_o at t_o :

$$P = \frac{F}{k} - \frac{F - kP_o}{k} e^{-k(t-t_o)} \qquad (8)$$

$$F = F_{atm} + F_{agr} + F_{wea} \qquad (9)$$

and

$$k = k_{up} + k_{run}. \qquad (10)$$

Expressed in terms of concentrations when the initial condition is given by present state of soil ($t_o = 0$, $P_o = \vartheta.c_s^o$ where c_s^o is the present concentration of heavy metal in soil.

$$c_s = \frac{F}{\vartheta.k} \left(1 - e^{-k.t}\right) + c_s^o e^{-k.t} \qquad (11)$$

The critical time when the soil reaches the critical concentration $c_{s\ critical}$ is

$$t_{critical} = -\frac{1}{k} \ln \frac{F - \vartheta k.c_{s\ critical}}{F - \vartheta k.c_s^o} \qquad (12)$$

The critical load of trace metals $(F_{atm} + F_{agr})_{critical}$ is calculated from the mass balance equation where size of pool was converted to the concentration in soil:

$$\left(F_{atm} + F_{agr} \right)_{critical} = c_{s\,critical} \cdot \vartheta \left(k_{up} + k_{run} \right) - F_{wea} \quad (13)$$

The "exceedance" (*EX*) of the critical load is the difference between the calculated value and the real atmospheric and agricultural load:

$$EX = \left(F_{atm} + F_{agr} \right)_{critical} - \left(F_{atm} + F_{agr} \right) \quad (14)$$

3 CRITICAL LOAD, CRITICAL TIME AND EXCEEDANCE FOR Pb AND Cd

Lead and cadmium are of environmental concern. The data representing agricultural land in the Bohemian Moravian Upland in the Czech Republic are summarised in Table 1. The calculated critical loads, critical times and the exceedance of critical loads by the present atmospheric and agricultural inputs are given in Table 2.

Table 1. Data representing average characteristics of the biogeochemical mass balance of Pb and Cd in agricultural catchments with crystalline bedrock in the Czech Republic (F_{atm}, F_{agr} and F_wp - Benes 1994, Paces 1998).

	Units	Cd	Pb
F_{atm}	g.ha^{-1}.yr^{-1}	9.3	181
F_{agr}	g.ha^{-1}.yr^{-1}	5.6	14.8
F_{wea}	g.ha^{-1}.yr^{-1}	0.053	6.1
F_{up}	g.ha^{-1}.yr^{-1}	2.6	26
F_{run}	g.ha^{-1}.yr^{-1}	0.12	0.67
k_{up}	yr^{-1}	5.6 e-3	3.5 e-4
k_{run}	yr^{-1}	2.6 e-4	9.0 e-6

Table 2 Calculated critical load ($F_{atm}+F_{agr}$)$_{crit}$, and critical times ($t_{critical}$), and exceedance (*EX*)

	Units	Cd	Pb
($F_{atm}+F_{agr}$)$_{crit}$	g.ha^{-1}.yr^{-1}	6.4	73
$EX = (F_{atm}+F_{agr})$ $(F_{atm}+F_{agr})_{crit}$	g.ha^{-1}.yr^{-1}	8.6	123
$t_{critical}$	yr	61	980

The exceedance values of atmospheric and agricultural inputs of Pb and Cd are positive numbers. The critical concentrations in soil will be reach after 61 and 980 years. The critical load, critical time and the exceedance are dependent on our choice of the critical concentration of the trace metal in soil $c_{s\ critical}$. The data in Table 2 were calculated for critical concentrations defined by the Czech state norms for soils. They are 0.4 and 80 mg.kg^{-1} for Cd and Pb respectively (Vrubel et al., 1996). On the other hand, the time function does not depend on the critical concentration. It depends on our measurements of atmospheric, agricultural, hydrological and geochemical and on physical and chemical properties of soils. The graphs of the time functions for Cd and Pb are presented in Figure 2.

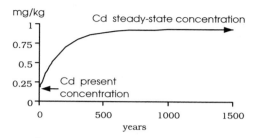

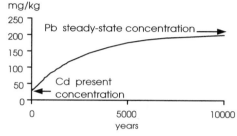

Figure 2 The temporal evolution of Cd and Pb in agricultural soils corresponding to the data in Table 1

The relations between the critical concentrations and the critical times are shown in Figure 3.

The definition of critical load by equation (13) indicates that the faster is the release of heavy metals in soil pool by weathering, the lower is the critical load and consequently the more sensitive is the region to dispersed pollution by heavy metals. On the other hand, the higher are the concentrations of heavy metals in runoff, the higher is the rate constant k_{run} and the higher is the critical load. flux of water due to runoff purifies the soil pool. this leads us to a conclusion that the more polluted water is by heavy metals the less sensitive the soil seems to be with respect to pollution by the metals. therefore, the pollution of soils and pollution of runoff water work against each other. the calculation of critical loads for soils without considering the pollution of runoff water can lead to incorrect environmental interpretations.

The temporal evolution of Cd and Pb in agricultural soils in central Europe is compared to evolution (Figure 4). calculated for forest soils in southern Sweden using data by Begkvist (1986). A much lower input of Cd and Pb from atmosphere and no agricultural input were measured in the Swedish forests. Therefore, cadmium is depleted from the soil profile (the exceedance is a negative number) and lead will never reach the critical concentration of 80 mg.kg^{-1} because the steady state concentration is only 35 mg.kg^{-1}.

mg/kg

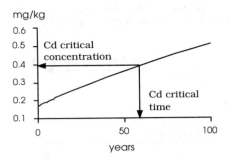

mg/kg

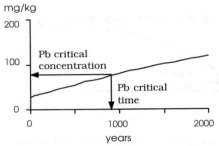

Figure 3 Critical times corresponding to chosen critical concentrations of Cd and Pb in soil

mg/kg

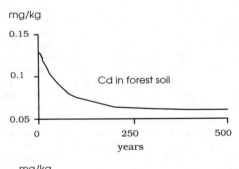

mg/kg

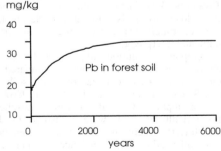

Figure 4. The temporal evolution of Cd and Pb in forest soils calculated with data by Bergkvist (1986)

4 CONCLUSIONS

A simple mass balance approximately characterises temporal trends of trace metal concentrations in soils. This model enables calculation of a critical time when a metal in soil will reach its environmentally critical concentration. The mass balance determines the critical load of heavy metals from atmosphere and from agricultural dispersed inputs in soils.

The method of calculation is based on a very simplified hypothetical model in which the rate of runoff and biological uptake of the metal is proportional to the pool of the biologically available metal is soil. This assumption of the first-order kinetics is an oversimplification of the complex soil - water - plant system. Therefore the calculated characteristics are only approximate indicators of the behaviour of trace metals in soil systems.

Data obtained in a typical Czech agricultural countryside (Paces, 1998) and data from Swedish forest soils (Bergkvist, 1986) were used to test the model. The results indicate that the present agricultural and atmospheric inputs of Pb and Cd are higher than their critical loads. The critical concentrations in soil, 0.4 and 80 mg.kg^{-1} for Cd and Pb respectively, will be reached after 61 and 980 years. The Swedish forest soils do not seem to be in danger of the metal pollution. These results depend mainly on the choice of critical concentrations. Therefore the calculated critical loads have just a formal meaning. On the other hand, the time function of the pool of the heavy metal concentrations in soil indicates the temporal trend that depends on *in situ* determined characteristics of environmental fluxes and properties of local soils.

REFERENCES

Benes S. 1994. *Obsahy a bilance prvku ve sferach zivotniho prostredi, II. cast*. Praha: Ministerstvo zemedelstvi Ceske republiky (in Czech).

Bergkvist B. 1986. *Metal fluxes in spruce and beech forest ecosystems of south Sweden*. Lund: Dept. of Ecology, Plant Ecology, University of Lund, Sweden.

de Vries W. & D.J. Bakker 1998. Manual for calculating critical loads of heavy metals for terrestrial ecosystems, soils and surface waters. *Report 166*. Wageningen: DLO Winand Staring Centre, Den Helder, The Netherlands: TNO Institute of Environmental Sciences.

de Vries W., D.J. Bakker & H.U. Sverdrup 1998. Manual for calculating critical loads of heavy metals for terrestrial ecosystems, soils and surface waters. *Report 165*. Wageningen: DLO Winand Staring Centre, Den Helder, The Netherlands: TNO Institute of Environmental Sciences.

Jones J.B. 1990. Universal Soil Extractants, their composition and use. *Comun. Soil Sci Plant Anal.* 21: 1091-1101.

Lindsay W.L. & W.A. Norvell 1978. Development of a DTPA soil test for zinc, iron, manganese and copper. *Soil Sci. Amer. J.* 42: 421-428.

Mehlich A. 1978. New extract for soil test evaluation of phosphorus, potassium, magnesium, calcium, sodium, manganese and zinc. *Commun. Soil Sci. Plant Anal.* 9: 477 - 492.

Mehlich A. 1984. Mehlich No. 3 soil test extractant: A modification of Mehlich No. 2. *Commun. Soil Sci. Plant Anal.* 15: 1409 - 1416.

Nilsson J. & P. Grennfelt (eds.) 1988. Critical loads for sulfur and nitrogen. *Report from a Workshop at Skokloster, Sweden, March 1988. Miljo rapport 1988*.15. Copenhagen:Nordic Council of Ministers.

Paces T. 1998. Critical loads of trace metals in soils: A method of calculation. *Water, Air, and Soil Pollution.* 105: 451-458, Dordrecht.

Vrubel J., T. Paces, G. Cejková, C. Kantor, V. Majer M. Sanka, P Trojasek,. & M. Zapletal 1996. *Critical leads of heavy metals for soils in Czech Republic* - Draft. Prague: Ekotoxa, Opava and Czech Geological Survey,.

Wolf B. 1982. An improved universal extracting solution and its use for diagnosis in soil fertility. *Commun. Soil Sci. Plant Anal.* 13: 1005 - 1033

Geochemistry of the Earth's Surface, Ármannsson (ed.)© 1999 Balkema, Rotterdam, ISBN 90 5809 073 6

Invited lecture: Pesticide toxicology and behaviour in the natural environment

K. Vala Ragnarsdóttir

Department of Earth Sciences, University of Bristol, UK

ABSTRACT: Organophosphate pesticides (OPs) have recently been linked with many degenerative diseases, calling for an examination of their toxicology and transport in the environment. OPs are generally regarded as "safe", due to "fast" degradation. However, environmental rates vary significantly as a function of microbial composition, pH, temperature, and availability of sunlight. Under laboratory conditions (25°C; pH 7) biodegradation is about one order of magnitude faster than chemical hydrolysis which in turn is ten times faster than photolysis. However, soil microbial biomass needs an adaptation period for mutation. In groundwater (pH 6; 5°C) hydrolysis of an OP that has a half life of 10 days in the laboratory increases to one year, suggesting that OPs can persist in the environment for long periods of time. OPs are systemic and are found throughout the whole body of all food stuffs for which limited rate data are available.

1 INTRODUCTION

In the 1990s more than 25,000 t/yr. of pesticides were used in Great Britain (MAFF/SOFAD 1995) and over 50,000 t/yr. of pesticide products are used in the USA. At present there are about 450 active ingredients approved for use in the UK (Eke et al. 1996) and 600 are approved for use in the US (OTA 1990). Western Europe and the US report that 2-3 kg/ha of chemicals are used to protect plants, and the world-wide annual average is 0.3 kg/ha (Ananyeva et al. 1992). It is therefore not surprising that the global agrochemical market was valued at US $30 billion in 1995 (Pesticide Trust 1996). At the present time organophosphates (OPs) are the most widely used pesticides in the world (Pesticide Trust 1996). As early as the late 1950s hundreds of fatal OP poisoning cases were reported annually (Carson 1962). Recent estimates suggest that pesticides account for more than 20,000 fatalities yearly, and that most of these occur in developing countries (Forget 1991). OPs are not only used for agricultural purposes. They are also used in gardening and homes for pest control. Residues of OPs are also commonly found in food stuffs such as vegetables, fruit, cereal, meat, and milk, as well as in drinking water. The medical and toxicological literature now shows that OPs have a role to play in the development of many modern day degenerative diseases. It is because of these observations that the toxicology and environmental behaviour of OPs is reviewed in this article.

2 ORGANOPHOSPHATE TOXICOLOGY

OPs possess the common characteristic of being excellent inhibitors of the enzyme acetylcholine esterase (AChE), which catalyses the break-down of the neurotransmitter acetylcholine in mammals and insects (e.g. Vijverberg & van den Bercken 1990). This occurs by the formation of a covalent bond between the phosphorylating group of the OPs with an OH group at the active site of the enzyme (e.g. Manahan 1992). OPs also interfere with four other enzymes, including neuropathy target esterase (NTE), serum cholinesterase (ChE), serum paroxonase (PON) and serum arylesterase (Mutch et al. 1992, Ehrich 1995). These enzymes affect the central nervous system, which ultimately control the function of the whole body. OPs are primarily used as systemic insecticides, i.e. the OPs are translocated throughout plant and animal tissues.

The neurotoxicological effects described above are indeed the reason why these pesticides were developed. While the acute toxicity and lethal doses needed to kill living organisms are well known, little is known about continued low level exposure to OPs. However, in the past decade the medical literature increasingly reports cases of OP linkage with many modern day degenerative diseases (Ragnarsdottir 1999), including Creutzfeld Jacobs disease (CJD) (King 1996), Gulf War Syndrome (Hom et al. 1997), neurological diseases such as Parkinson's disease and peripheral neuropathy (Marrs 1995), Multiple Sclerosis and motor neuron disease (Purdey 1998), sensory neuropathy (Stephens et al. 1995), polyneuropathy (McConnell et al. 1994), and chronic neurological sequelae (Steenland et al. 1994). Other OP related diseases include cancerous lymphomas (Newcombe et al. 1994), intermediate syndrome (Marrs 1995), immune dysfunction (Thomas 1995), asthma (O'Malley 1997), chronic fatigue syndrome or myalgic encephalomyelitis (ME) (Behan 1996), farmers' flu (Stephens et al. 1995), multiple chemical sensitivity (Bell et al. 1992), induced hypothermia (Gordon & Rowsey 1998), sleep disturbance (Bell et al. 1996), systemic illness, skin diseases and eye injury (Weinbaum et al. 1995), Saku disease (Dementi 1994), and mental retardation (Weiss 1997). Mental disorders linked with OPs include psychiatric disorder (Stephens et al. 1995), delayed psycho neurodegenerative syndrome (Purdey 1998), and schizophreniform- and depressive psychosis (Marrs 1995). Last but not least, the transfer of OPs through a human placenta has also been demonstrated and their teratogenic effects have been manifested by severe birth defects (Sherman 1995). OPs have also been shown to be mutagenic, causing genetic toxicity to humans (Flessel et al. 1993). Of high importance also is that OPs have been reported to be linked to the Bovine Spongiform Encephalopathy (BSE or Mad Cows disease) crisis in the U.K. and Switzerland (King 1996, Axelrad 1997, Purdey 1996a,b, 1998). These results suggest that the OPs cause permanent damage to the central nervous- and autoimmune system, as well as having detrimental effects on cell reproduction.

3 OP FATE IN THE ENVIRONMENT

OPs are found in the environment due to diverse agricultural use. OPs have a relatively high solubility (10-10,000 ppm) due to their polar nature. It is therefore no surprise that OPs have been reported in soils (Weber 1992), river water and sediments (Winger & Laiser 1998, Domagalski et al. 1997), estuary water and sediments (Domagalski & Kuivila 1993), and even the atmosphere (Plimmer 1992).

In the natural environment OPs are primarily affected by adsorption, microbial degradation, chemical hydrolysis, and photolysis. When OPs are applied to fields they are more effective for pest control if the soil is dry than wet, allowing adsorption to clay and organic material surfaces (Valsaraj and Thibodeaux 1992).

Bacteria break down OPs through various metabolic activities including hydrolytic and oxidation/reduction processes. Biodegradation is enhanced through repeated use of OPs (e.g. Niemcsyk & Chapman 1987), indicating that that gene modification occurs in the microbial flora, evolving genes required for OP degradation. Of note is that OPs have been found to be present in soils 16 years after application (Stewart et al. 1971), suggesting that we do not understand all of the factors involved in OP breakdown/stabilisation.

In groundwaters, OPs have been shown to have concentrations greater than 2 ppm below pastures where sheep dipping solutions were poured for disposal (Blackmore & Clark 1994). This value is four orders of magnitude higher than drinking water limits (0.1 ppb). Half lives for the hydrolysis of an OP in groundwater are increased to one year at 5°C and pH of 6 for an OP with a laboratory half life of 10 days (pH 7; 25°C) (Ragnarsdottir 1999), showing that OPs can persist in water for long periods of time.

In surface waters OPs are affected by biodegradation, hydrolysis and photolysis. Release of OPs into the River Rhine allowed the relative rates of these processes to be tested (Schnoor et al. 1992). It was shown that for disulfoton (solubility 25 ppm) the degradation was consistent with combined laboratory half lives for biodegradation (0.57 days), hydrolysis (1.3 days), and photolysis (40 days).

Reflecting the high levels of pesticides used in the western world, the UK spends £2 million/yr. on monitoring pesticide levels in food and the water industry spends £880 million/yr. on drinking water monitoring and treatment (Pesticide Trust 1996).

4 OPs IN FOOD

Pesticides are applied onto fields as often as 25 times over the growing season. Due to the systemic nature of OPs, they are distributed throughout the whole body of roots, leaves and fruit. Therefore, washing of foodstuffs is not of much use for lowering concentrations of OPs. Indeed concentrations of hundreds of ppm are frequently measured in fruit, vegetables and cereals (National Consumer Council 1998) which, interestingly, have a maximum residue levels (MRL) of 10 ppm (6 orders of magnitude higher than drinking water limits). Due to these high MRLs it has been estimated that most humans obtain over 90% of their pesticide intake from foodstuffs.

OPs are used in many countries to control mosquitoes (whole cities and counties are sprayed from the air in USA), sheep dipping (against ticks), cow dressing (against warble fly), dog and cat collars and sprays (against ticks and fleas), shampoo for adults and children (against lice), and for pest control in domestic gardening, houses, stalls and pet shops. Due to the frequent direct use of OPs on farm animals OPs are found in both milk and meat (Coulibali & Smith 1993, Di Muccio et al. 1996).

5 CONCLUSIONS

Human beings and animals are continuously exposed to OPs in their homes, food, and water. Due to the neurotoxicological, immune, and cell damage caused by these pesticides that are outlined above, it follows that new guidelines for the use of OPs in agriculture and in homes is urgently needed.

REFERENCES

Ananyeva, N.D., N.N. Naumova, J. Rogers, & W.C. Steen 1992. Microbial transformation of selected organic chemicals in natural aquatic systems. In J.L. Schnoor (ed.), *Fate of pesticides and chemicals in the environment*: 275-294. New York: John Wiley.

Axelrad, J. 1997. An autoimmune response causes transmissible spongiform encephalopathies. *Medic. Hypothes.* 50: 259-264.

Behan, P.O. 1996. Chronic fatigue syndrome as a delayed reaction to chronic low-dose organophosphate exposure. *J. Nutrit. Envir. Medic.* 6: 431-350.

Bell, I.R., C.S. Miller & G.E. Schwartz 1992. An olfactory-limbic model of multiple chemical-sensitivity syndrome – possible relationships to kindling and affective spectrum disorders. *Biol. Psychiatry* 32: 218-242.

Bell, I.R., R.R. Bootzin, C. Ritenbaugh, J.K. Wyatt, G. DeGiovanni, T. Kulinovich, J.L. Anthony, T.F. Kuo, S.P. Rider, J.M. Peterson, G.E. Schwartz & K.A. Johnson 1996. A polysomnographic study of sleep disturbance in community elderly with self-reported environmental chemical odor intolerance. *Biol. Psychiatry* 40: 123-133.

Blackmore, J. & L. Clark 1994. *The disposal of sheep dip waste. The effects on water quality.* National River Authority R&D Report 11.

Carson, R. 1962. *Silent spring.* New York: Fawcett Crest Books.

Coulibali, K. & J.S. Smith 1993. Effect of pH and cooking temperature on the stability of organophosphate pesticides in beef muscle. *J. Agric. Food Chem.* 42: 2035-2039.

Dementi, B. 1994. Ocular effects of organophosphates – A historical perspective of Saku disease. *J. Appl. Toxicol.* 14: 119-129.

Di Muccio, A., P. Pelosi, I. Camoni, D.A. Bargini, R. Dummarco, T. Generali & A. Ausili 1996. Selective, solid-matrix dispersion extraction of organophosphate pesticide residues from milk. *J. Chromatogr. A* 754: 497-506.

Domagalski, J.L. & K.M. Kuivila 1993 Distribution of pesticides and organic contaminants between water and suspended sediments, San-Francisco Bay, California. Estuaries 16: 416-426.

Domagalski, J.L. , N.M. Dubrovsky, & C.R. Kratzer 1997 Pesticides in the San Joaquin River, California: Inputs from dormant sprayed orchards. J. Envir. Qual. 26: 454-465.

Ehrich, M. 1995. Using neuroblastoma cell-lines to address differential specificity to organophosphates. *Clinic. Experim. Pharmacol. Phsysiol.* 22: 291-292.

Eke, K.R., A.D. Barnden & D.J. Tester 1996. Impact of agricultural pesticides on water quality. In R.E. Hester and R.M. Harrison (eds.), *Aricultural chemicals and the environment.* Iss. Environm. Sci. Technol. 5: 43-56.

Flessel, P., P.J.E. Quintana & K. Hooper 1993. Genetic toxicity of malathion – a review. *Environm. Molec. Mutagen.* 22: 2-17.

Forget, G. 1991. Pesticides and the third-world. *J. Toxicol. Environm. Health* 32: 11-31.

Gordon, C.J. & P.J. Rowsey 1998. Poisons and fever. *Clinic. Experim. Pharmacol.* 25: 145-149.

Hom, J., R.W. Haley & T.L. Kurt 1997. Neuropsychological correlates of Gulf War Syndrome. *Arch. Clinic. Neurophyschol.* 12: 531-544.

King, J.W. 1996. *Bovine spongiform encephalopathy and organophosphates.* Mathematica Press.

MAFF/SOFAD, 1995. *Arable farm crops in Great Britain 1994.* Pesticide Usage Survey Group.

Manahan, S.E. 1992. *Toxicological chemistry*, 2nd ed. London: Lewis.

Marrs, T. 1995. Organophosphate sheep dips and human health. An overview of the delayed effects of organophosphates. *Seminar for farmers, medical practitioners and policy makers*, June 2, 8-10.

McConnell, R., M. Keifer & L. Rosenstock 1994. Elevated quantitative vibrotactile threshold among workers previously poisoned with methamidophos and other organophosphate pesticides. *Amer. J. Indust. Medic.* 25: 325-334.

Mutch, E., P.G. Blain & F.M. Williams 1992. Individual variations in enzymes controlling organophosphate toxicity in man. *Human Experim. Toxicol.* 11: 109-116.

National Consumer Council 1998. *Farm policies and our food: The need for change.* PD 11/B2/98.

Newcombe, D.S., A.M. Saboori & A.H. Esa 1994. Chronic organophosphorous exposure – biomarkers in the detection of immune dysfunction and the development of lymphomas. *Amer. Chem. Soc. Symp. Ser.* 546: 197-212.

Niemcsyk, H.D. & R.A. Chapman, 1987. Evidence of enhanced degradation of isofenphos in turfgrass, thatch and soil. *J. Econ. Entomol.* 80: 880-882.

O'Malley, M. 1997. Clinical evaluation of pesticide exposure and poisonings. *Lancet* 349: 1161-1166.

OTA, 1990. U.S.A. Congress Office of Technology Assessment (OTA). *Neurotoxicology: Identifying and controlling poisons in the nervous system.* Van Norstand Reinhold.

Pesticides Trust Review 1996. The Pesticides Trust, London.

Plimmer, J.R. 1992 Dissipation of pesticides in the environment. In J.L. Schnoor (ed.), *Fate of pesticides and chemicals in the environment*: 79-91. New York: John Wiley.

Purdey, M. 1996a. The UK epidemic of BSE: Slow virus or chronic pesticide-initiated modification of the prion protein? 1. Mechanisms for a chemically induced pathogenesis/ transmissibility. *Medic. Hypothes.* 46: 429-443.

Purdey, M. 1996b. The UK epidemic of BSE: Slow virus or chronic pesticide-initiated modification of the prion protein. 2. An epidemiological perspective. *Medic. Hypothes.* 46: 445-454.

Purdey, M. 1998. High-dose exposure to systemic phosmet insecticide modifies the phosphatidylinositol anchor on the prion protein: The origins of new variant transmissible spongiform encephalopathies. *Medic. Hypothes.* 50: 91-111.

Ragnarsdottir, K.V. 1999. Environmental fate and toxicity of organophosphate pesticides. *J. Geol. Soc.*, in press.

Schnoor, J.L., D.J. Mossman, V.A. Borzilov, M.A. Novitsky, O.I. Voszhennikov & A.K. Gerasimenko 1992. Mathematical model for chemical spills and disturbed source runoff to large rivers. In J.L. Schnoor (ed.), *Fate of pesticides and chemicals in the environment*: 275-294. New York: John Wiley.

Sherman, J.D. 1995. Chlorpyrifos (Dursban)-associated birth defects: A proposed syndrome, report of four cases, an discussion of the toxicology. *Intern. J. Occup. Medic. Toxicol.* 4: 417-431.

Steenland, K., B. Jenkins, R.G. Ames, M. O'Malley, D. Chrislip & J. Russo 1994. Chronic neurological sequelae to organophosphate pesticide poisoning. *Amer. J. Publ. Health* 84: 731-736.

Stephens, R., A. Spurgeon, J. Beach, I. Valcert, H. Berry, L. Levy & J.M. Harrington 1995. *An investigation into the possible chronic neurophsychological and neurological effects of occupational exposure to organophosphates in sheep farmers.* HSE Contract Research Report 74.

Stewart, R., D. Chrisholm & M.T. Ragab 1971. Long term persistence of parathion in soil. *Nature* 229: 47-48.

Thomas, P.T. 1995. Pesticide-induced immunotoxicity: Are the Great Lakes residents at risk? *Envir. Health Perspect.* 103: 55-61.

Valsaraj, K.T. & L.J. Thibodeaux 1992. Equilibrium adsorption of chemical vapors onto surface soils: Model predictions and experimental data. In J.L. Schnoor (ed.), *Fate of pesticides and chemicals in the environment*: 275-294. New York: John Wiley.

Vijverberg, H.P.M. & J. van den Bercken 1990. Neurotoxicological effects and the mode of action of pyrethroid insecticides. *Crit. Rew. Toxicol.* 21: 105-121.

Weber, J.B. 1992 Interaction of ortanic pesticidws with particulate matter in aquatic and soil systems. In J.L. Schnoor (ed.), *Fate of pesticides and chemicals in the environment*: 275-294. New York: John Wiley.

Weinbaum, Z., M.B. Schenker, M.A. O'Malley, E.B. Gold & S.J. Samuels 1995. Determinants of disability in illness related to agricultural use of organophosphates (OPs) in California. *Amer. J. Industr. Medic.* 28: 257-274.

Weiss, B. 1997 Pesticides as a source of develompnental disabiltities. *Mental Retard. Developm. Disabil. Res. Rev.* 3: 246-256.

Winger , P.V. & P.J. Laiser 1998 Toxicity of sediment collected upriver and down river of major cities along the Mississippi River. *Arch. Envir. Contam. Toxicol.* 35: 213-217.

Geochemistry of the Earth's Surface, Ármannsson (ed.)© 1999 Balkema, Rotterdam, ISBN 90 5809 073 6

Environmental impact on soil sulfur by mining activities – Sulfur isotope evidence

E. Carlsson, P. Torssander & C-M. Mörth
Department of Geology and Geochemistry, Stockholm University, Sweden

ABSTRACT: For about thousand years the area around the Falun copper mine, South Central Sweden, has been exposed to acid deposition. The main purpose of this investigation was to examine the spatial effects from the mining activities using stable S isotopes. The soil S concentration and S isotope composition were analyzed for six podzole profiles at increased distance from the Falun mining area. The $\delta^{34}S$ values ranged from +1.3 ‰ to +7.6 ‰ and showed an increase in $\delta^{34}S$ values with increased distance from the mining site for both total sulfur and inorganic sulfate and for all depths. The results partly reflects the spatial effect from mining activities and that soil profiles close to the mining site still hold a signal from deposited ore sulfur.

1 INTRODUCTION

During almost thousand years the environment in the Falun area, South Central Sweden has been affected by anthropogenic S emissions. The dominating S source has been SO_2; a product from oxidized sulfide ore during mining processes in the Falun copper mine. The S deposition from the mining activities has decreased considerably since the beginning of the 20th century and has later been replaced by long range transported S from fossil fuel burning. Good estimates of the $\delta^{34}S$ values of these two dominating S sources in the area exist from previous studies. Therefore, by using stable sulfur isotopes as tracers, the soils in the Falun area provide an opportunity to study soil S dynamics and acidification reversals. In a previous study we have shown that soils close to the mining site still hold a signal from deposited ore sulfur (Carlsson et al. 1999). In this study we have examined soil profiles at increased distance from the mining site in order to study the spatial effects of S deposition in the area.

2 STUDY AREA

The investigated area is situated in South Central Sweden (Figure 1). The bedrock consists of sulfide bearing leptite formations and primorogenic granites, with minor intrusions of gabbro and diorite. The effects of the S emission from mining activities have probably become enhanced by the fact that the city of Falun is situated in a valley, with more than 100-m difference in elevation between the mountains and the base of the valley. The valley has a northeast-southwest trend, channelling the wind accordingly. The largest SO_2 emissions during the active period of the mining business took place between the 17th and the 19th century with a maximum during the middle of the 17th century. During this maximum, the SO_2 emission is estimated to have been almost 40,000 tons SO_2 year^{-1} corresponding to an air concentration of at least 200 $\mu g\ SO_2\ m^{-3}$ around Falun (Löfgren & Harsbo 1994).

3 METHOD

Six podzol profiles at distances from 2 to 40 km in the prevailing wind direction from the mining site were sampled during November 1997 (Figure 1). The profiles were sampled with depth in O-, E- and B-horizons and analyzed for S isotope ratios and S concentrations of total sulfur (organic sulfur + inorganic sulfur) and inorganic (adsorbed + dissolved) sulfate using HNO_3 and Br_2 (Krouse & Tabatabai 1986) and $NaHCO_3$ (Van Stempvoort et al. 1990). The dissolved SO_4^{2-} obtained from both extraction methods was converted to $BaSO_4$ for subsequent isotope analysis. Sulfur isotope measurements were made at Swedish Museum of Natural History on a rebuilt MM 602 and at Stockholm University on a Finnigan Delta+. The S isotope composition is defined as a deviation in ‰ of the ratio $^{34}S/^{32}S$ between a sample and a standard expressed in the conventional $\delta^{34}S$ notation relative to Cañon Diablo Troilite (CDT). The precision of the isotope measurements is within ± 0.2 ‰.

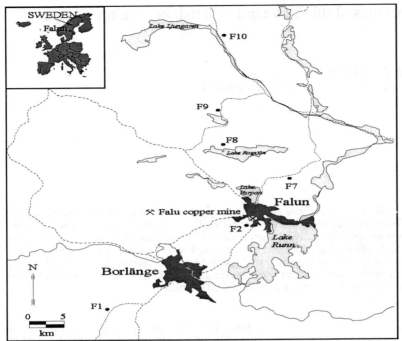

Figure 1. Map showing samplings sites in the Falun area.

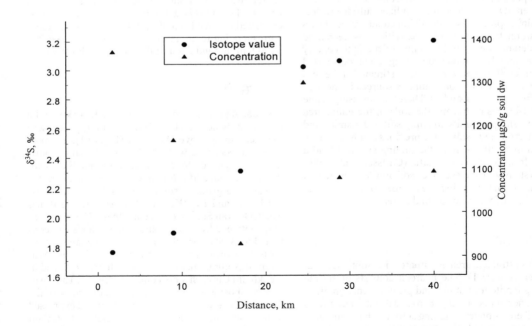

Figure 2. O-horizon concentration and isotope ratios vs distance from the Falun copper mine

The absence of reduced inorganic sulfur in soil samples was tested with distillation with Cr-solution (Canfield et al. 1986)

4 RESULTS AND DISCUSSION

In the soil profiles $\delta^{34}S$ values ranged from +1.3 ‰ to +7.6 ‰ and showed an increase in $\delta^{34}S$ values with increased distance from the mining site for both total sulfur and inorganic sulfate (Figure 2). The increase in $\delta^{34}S$ values for total S with distance was found to be larger in the deeper soil horizons than in the topsoils. Multiple linear regression also showed good correlations for $\delta^{34}S$ vs depth and distance (Figure 3). For the sulfur concentration no clear trend with increased distance from the mining site could be seen either for organic or inorganic sulfur, which in part can be explained by differences in the sulfate adsorption capacity for the profiles. With depth all profiles exhibited decreasing total S contents.

By previous studies we can assume the Falun ore to hold a $\delta^{34}S$ value of about 0 ‰. Studies of anthropogenic atmospheric deposition in Sweden have reported fairly consistent $\delta^{34}S$ values between +4‰ to +6‰. Measurements of bulk deposition at Buskbäcken (80 km S of Falun) showed $\delta^{34}S$ mean values of + 5.6 ‰, indicating that the present day $\delta^{34}S$ of deposition in Falun is around +6‰. Mixtures of fossil fuel S and ore S result in deposition $\delta^{34}S$ values between these two end members. This means that the atmospheric S deposition will have proportionally lower $\delta^{34}S$ values, dependent upon the amount of that is derived from the ore. The homogeneous $\delta^{34}S$ values in the profiles close to the mining site therefore imply that the soil sulfur still mainly originate from deposited ore sulfur with minor influences of the heavier modern deposition.

The correlation of $\delta^{34}S$ values with distance and depth imply that underlaying factors such as deposition and turnover time of S in soils can explain the increase in $\delta^{34}S$ values at greater distances and soil depths. Several studies have show that a large part of the organic sulfur circulation takes place in the upper soil horizons meanwhile organic sulfur in lower horizons can be preserved for a long time (Mitchell et al. 1989, Torssander & Mörth 1998). As the O-horizon sulfur pools are large any change to heavier $\delta^{34}S$ values due to S deposition from fossil fuel burning will take long time, i.e. the turnover time may be long for organic S in the B-horizons. The present day sulfate deposition in the Falun region is around 12 mmol/m². As the soil sulfur pools are about 1500 mmol/m² on average the turnover time is in the order of 125 years. In reality the exchange of sulfur takes longer time as only a smaller part of the organic sulfur is replaced each year. The replacement of organic sulfur has been shown to be only a part of the yearly sulfur deposition (Van Stempvoort et al. 1992, Torssander & Mörth 1998). This means that the sulfur isotope signature will be preserved for a very long time. At the greatest distance and in the lowest soil horizons the $\delta^{34}S$ values for total S may therefore indicate that this sulfur originates from even older deposition than ore sulfur. For profile F10 the very low sulfur concentrations at the greatest depth may also indicate that any influence on $\delta^{34}S$ values from another S sources such as the bedrock S may be possible. However, these findings must be further examined before more conclusions can be made.

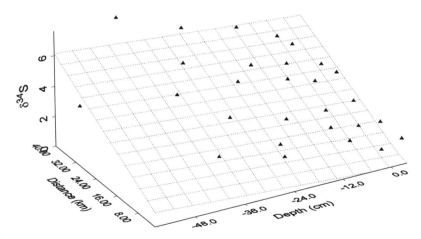

Figure 3. $\delta^{34}S$ of total sulfur vs sample depth and distance for all samples. $R^2=0.79$.

CONCLUSIONS

The homogeneous δ³⁴S values in the profiles close to the mining site imply that the soil sulfur still mainly originate from historically deposited ore sulfur. The higher δ³⁴S values in the O-horizons with increased distance from the point source indicate decreased influences of deposition of ore sulfur and increased influence from long range transported fossil fuel sulfur.

REFERENCES

Canfield, D.E., R. Raiswell, J.T. Westrich, C.V. Reaves, C.V. & R.A. Berner 1986. The use of Chromium reduction in the analysis of reduced inorganic sulfur in sediments and shales. *Chem. Geol.* 54:149-155

Carlsson E., P. Torssander, C-M. Mörth & M. Kusakabe 1999. Historical atmospheric deposition in a Swedish mining area traced by S isotope ratios in soils. *Water, Air and Soil Pollution* 10/1-2:103-118

Krouse, H. R. & M.A. Tabatabai 1986. Stable sulfur isotopes. In M. A. Tabatabai (ed.), *Sulfur in agriculture*: 160-205, Amer. Soc. Agron.- Soil Sci. Soc. Amer. Madison, Wisconsin.

Löfgren, S. & K. Harsbo 1994. *Potentiella luftföroreningar och torrdeposition av svaveldoixid i Falun sedan 1200-talet,* Preliminär rapport, SLU, Uppsala (in Swedish).

Mitchell, M.J., C.T. Driscoll, R.D. Fuller, M.B. David & G.E. Likens 1989. Effect of whole-tree harvesting on the sulfur dynamics of a forest soil. *Soil Science Society of America Journal* 53:933-940.

Torssander, P. & C-M. Mörth 1998. Sulfur dynamics in the roof experiment at Lake Gårdsjön deduced from sulfur and oxygen isotope ratios in sulfate. In Hultberg, H. and Skeffington, R (eds), *Experimental reversal of acid rain effects: The Gårdsjön roof project*: 185-206, John Wiley & Sons, New York.

Van Stempvoort, D. R., E.J. Reardon & P. Fritz 1990. Fractionation of S and oxygen isotopes in sulfate by soil sorption, *Geochim. et Cosmochim. Acta* 54:2817-2826.

Van Stempvoort, D.R., P. Fritz, E.J. Reardon 1992. Sulfate dynamics in upland forest soils, central and southern Ontario, Canada: stable isotope evidence. *Applied Geochemistry* 7:159-175.

Geochemistry of the Earth's Surface, Ármannsson (ed.) © 1999 Balkema, Rotterdam, ISBN 90 5809 073 6

Plant-microbe interactions in extreme environments of lignite mining lakes

A.Chabbi

Department of Soil Protection and Recultivation, Brandenburg Technical University, Cottbus, Germany

ABSTRACT: Bulbul rush (*Juncus bulbosus*) initiates the plant colonization in acidic mining lakes in the Lusatian mining district. The extreme and hostile site conditions in mining lakes suggest the presence of adaptive mechanisms that enable bulbul rush to survive in those ecosystems. Scanning electron microscopy shows that the iron plaques around the root are characterized by the presence of a mineral-free space between the root and the sand grains. This unusual microenvironment is inhabited by colonies of micro-organisms. Chemical analyses revealed that the iron plaque contains several trace concentrations of root exudates. The results suggest that there are interactions of the microbial component embedded on the root surface and the root exudates beneath the iron plaque (i.e. mineral free space). This may be the mechanism by which bulbul rush avoid inorganic carbon deficiency and phosphorus limitation in man made-ecosystems.

1 INTRODUCTION

Acidic mining lakes resulting from coal mining operations in the eastern part of the Federal Republic of Germany have been regarded as environmental disasters. But for researchers, these man-made ecosystems offer a remarkable research opportunity (Charles 1998). In the Lusatian area alone, more than 100 lakes of various sizes contain water having a pH between 2.5 and 3.5 and concentrations of dissolved iron, manganese and aluminum at levels highly toxic or lethal to most plant species. Despite the extreme conditions, bulbul rush (*Juncus bulbosus*) was recognized as a pioneer species of the littoral areas of these lakes (Pietsch 1979). The presence of an adaptive mechanism has been long speculated but remains an unresolved question. A recent study (Chabbi 1999a) has shown that the plant responds to toxic compounds through a sophisticated ecophysiological adaptation. This report explores the effective interactions of root excreted compounds and micro-organisms beneath an iron plaque formed on roots as key processes enabling the plant to survive in the extreme environment of acidic mining lakes.

2 MATERIAL AND METHODS

Our investigations were carried out in Senftenberg See (lake SFB) and Koyne-Plessa (Lakes numbers 108 and 109) of the Lusatian mining district in the eastern part of the Federal Republic of Germany. These lakes are the final result of lignite mining that lasted for decades. The water levels fluctuate depending largely on ground water. The substrate is mainly Pleistocene-Sand with little Tertiary material (Senftenberg See) and Tertiary material rich in pyrite at Koyne-Plessa district. *Juncus bulbosus,* floating and submersed stands, is the dominant macrophyte species in the littorals of these lakes.

On August 5, 1997, turgid and structurally intact living roots were collected at the sampling sites from both the sediment of acid lignite mine sediment rich in iron (roots with iron plaque formation) and the sediment of gravel pit poor in iron (control roots without iron plaque formation). Root collection were carefully carried out with a stainless steel shovel, keeping the root and soil intact. Afterwards the roots were placed in plastic bags, transported to the lab and stored overnight at 4°C. Root and soil were separated using de-ionised water. Root material with iron plaques was used for scanning electron microscopy (SEM) investigations. The trace elements of root exudates in iron plaques were determined by chemical analysis. For SEM fresh roots segment with iron plaques (10 mm distance to apex) were fixed with 2.5% glutaraldehyde in 0.1 M cacodylate buffer, pH 7.4 at 4°C overnight. After gently washing with buffer solution, segments were postfixed with 1% OsO_4 for 2 hr, dehydrated with acetone and embedded in epoxy resin (Spurr 1969). After polymerization at 70°C for 24 hr, specimens were cut

157

in 4 mm sections, mounted on aluminum specimen mounts with epoxy resin. After gently grinding the specimens were polished and coated with carbon. The specimens were investigated with a SEM (ZEISS DSM 962) at 20 kV with a working distance of 25 mm using a backscattering electron detector and an energy dispersive X-ray detector (Oxford Instruments, Link ISIS). For trace elements of root exudates, iron plaque was carefully separated from fresh root segment to expel iron plaque solution using a specially equipped cooling centrifuge, keeping sample temperature at 4°C, at 12 000 rev. min^{-1}, for 1 hr. The amounts of exudates have been determined by ion chromatograph following the procedure described by Shen et al. 1996.

3 RESULTS AND DISCUSSION

Juncus bulbosus was characterized as having an extensive plaque surrounding the roots (Figure 1)

bar: 200 μm

Figure 1: Root of *Juncus bulbosus* with iron plaque (left) clearly visible as an reddish brown precipitate and root without iron (White root, control) (right).

which is known to serve as a protective mechanism against the entry of phytotoxic levels of reduced elements into the root cells (Chabbi et al. 1998). In spite of the well-designed research reported on iron plaque formation, the specific nature of the processes involved in the protective mechanisms within species is not clear (Mendelssohn et al. 1995). Transverse sections of oxidized root channels show that the root is covered by quartz and iron oxide, mainly Goethite (Chabbi 1999a). Between the root and the sand grains exists a mineral-free space (Fig. 2a-b) which is inhabited by colonies of micro-organisms (Fig. 2c-d). The microbial component is quite interesting since observations show that the bacteria are true rhizobacteria attached to or embedded on the root surface beneath the iron plaque. This is in contrast to typical "iron associated bacteria" that one would expect to be literally coated with iron (Brock & Gustafson 1976).

The investigation of iron plaque (Figure 3)

revealed that this material contained several concentrations (nano- to micromolar) of malate, citrate, glucose and glycine which likely increased with plant stress (Hale et al. 1978, Hale & Moore 1979, Curl & Trelove 1986) and thereby provided micro-organisms with available substrates for metabolism.

These exudates may also function as complexing ligands (Francis et al. 1992) for toxic metal immobility/and or in sequestering nutrients such as phosphorus for uptake by the plant (Jones et al. 1996). The feedback involving stress, quantity of exudates, and microbial establishment in this unusual rhizosphere environment may have beneficial effects that promote survival and plant growth in acidic mining lakes. The discovery of this unique microenvironment challenges current ideas about inorganic carbon assimilation and phosphorus uptake in such ecosystems. Much evidence in the literature indicated that carbon dioxide controls the growth dynamics of *Juncus bulbosus* (Svedäng 1992) and its competitive ability increases due to the shift in the carbon dioxide cycle towards free CO_2 the in water that has been acidified by aerial deposition.

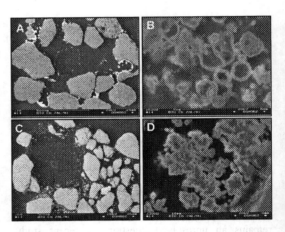

Figure 2. SEM of *Juncus bulbosus* root with iron plaque (10 mm distance to apex) taken from lake SFB (a) and lake 108 (c). Visible free room between the surface root (center) and mineral component. Figures (b) and (d) represents microbial component between surface root and mineral component in Figures (a) and (c).

The amount of inorganic carbon in extremely acidic lakes is assumed to be a limiting factor for primary biological production (Goldman et al. 1974). We found a very low concentration of CO_2 around the leaves of *Juncus* plants (Chabbi, unpul. material). At the some lakes, a previous study (Nixdorf et al.

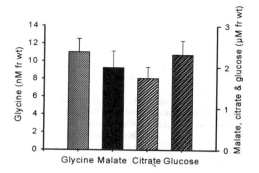

Figure 3. Trace of root exudates extracted from iron plaque surrounding the roots of *Juncus bulbosus*.

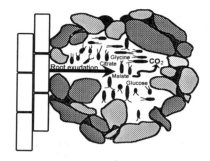

Figure 4. Iron plaque formation, root exudates and microbial interactions.

1998) reported that the concentration of DIC is very low and is often below the level of detection (0.5 mg C L^{-1}). A subsequent study (Kapfer 1998) documented that DIC produced by microbial activities (e.g. sulfate reduction, denitrification, respiration) in the sediment is rapidly lost to the atmosphere. We believe that the micro-organisms likely metabolize root exudates to different extents and thereby cause an increase in carbon dioxide released in the rooting medium, (Figure 4) which is presumed to be a substantial part of photosynthetically-fixed CO_2 evolved from *Juncus bulbosus* roots. Furthermore the formation of iron plaque around root surfaces may prevent the loss of this inorganic carbon from the system. So far net CO_2 assimilation may remove the acidity produced in the Fe^{2+} oxidation and in the cation-anion intake imbalance from the micro-space as follows:

$$HCO_3^- + H^+ \rightarrow CO_2 + H_2O \qquad (1)$$

A reaction that may cause an increase in pH in the micro-space.

Several reports in the literature suggest that iron plaque on roots can influence phosphorus uptake by the affected plant (Christensen & Wigand 1998). The tissue - phosphorus content of the plant was at a level for optimal nutrition (0.02 & 0.03 mmol g^{-1} d wt; Chabbi 1999b) even though iron plaque (e.g. goethites) typically exhibit a high capacity for phosphorus binding (Schwertmann & Taylor 1977). The P status in plant tissue was further approved by using the model of Verhoeven et al. (1996) (figure 5) which shows that most of our values are above the critical level for either N and P. The plant also does not support mycorrhizal associations that could mediate the process of solubilization of P bound to metals within the plaques. Bacterial metabolites,

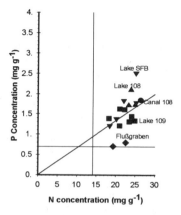

Figure 5. Nitrogen (N) and phosphorus concentrations in the shoots of *Juncus bulbosus* (mg g^{-1} d wt). The horizontal and vertical lines indicate the critical P and N concentrations respectively. The line through the origin represents an N:P of 15 according to Verhoeven et al. (1996).

other than siderophores (i.e., phytochelators), however, are likely important in P solubilization and uptake patterns. To our knowledge, evidence on the occurrence of P-solubilizing bacteria and their interaction with plaque in this unique extreme habitat has not been previously reported. It seems that Juncus plant through the bacterial metabolites is able to increase the bioavailability of phosphorus. The specific mineral free space beneath the iron plaque may foster a favorable zone for pH buffering capacity, rhizobacteria establishment on root surfaces, nutrient uptake and carbon source for plant metabolism. We believe that *J. bulbosus* growth in acid mine sediments is not a situation of directly living in and tolerating acidic milieu, rather the plant is capable of forming a favorable growth environment by achieving an alkaline micro-medium within an acidic milieu. The specific nature of these interactions may help explain why *Juncus bulbosus* is a primary plant colonizer in acid lignite mining lakes in spite of extreme and unfavorable growth conditions.

ACKNOWLEDGEMENTS

I gratefully acknowledge Dr. W. Wiehe for providing the SEM. This work was supported by the German Ministry of Research and Education (BMBF) and LMBV GmbH.

REFERENCES

Brock, T.D. & J. Gustafson 1976. Ferric iron re duction by sulfur-and iron-oxidizing bacteria. *Appl. Environ. Microbiol.* 32:567-571.

Chabbi, A., W. Pietsch, W. Wiehe, & R.F. Hüttl 1998. *Juncus bulbosus* L.: Strategies of survival under extreme phytotoxic conditions in acid mine lakes in the Lusatian mining district, Germany. *Inter. J. of Ecol. & Envir. Sc.* 24:271-292.

Chabbi, A. 1999a. Extreme environmental conditions in mining lignite lakes and ecophysiological responses of *Juncus bulbosus* L. *New Phytol.* In press.

Chabbi, A. 1999b. Element concentrations in tissues of *Juncus bulbosus* L. and sediments of the Lusatian lignite mining lakes. *Plant and Soil*. In press.

Charles, D. 1998. Wasteworld. *New Scientist* 157:32-35.

Christensen, K. K. & C. Wigand 1998. Formation of root plaques and their influence on tissue phosphorus content in *Lobelia dortmanna Aquat. Bot.* 61:111-122.

Curl, E.A., B. Trelove 1986. The rhizosphere. *Adv. Seri Agric. Sc.* 15:9-54.

Francis, A.J., C.J. Dodge, & J.B. Gillow 1992. Biodegradation of metal citrate complexes and implications for toxic metal mobility. *Nature* 356:140-142.

Goldman, J.C., W.J. Oswald, D. Jenkins 1974. The kinetics of inorganic carbon-limited algal growth. *J., Water Pollut. Cont. Fed.* 46(3)554-574.

Hale, M.G., L.D. Moore, & G.J. Griffin 1978. Roots exudates and exudation. In Y.R. Dommergues, & SV. Krupa, (eds.) *Interaction between non-pathogenic soil micro-organisms and plants*: 163-204. Elsevier Publisching Co., Amsterdam.

Hale, M.G. & L.D. Moore 1979. Factors affecting root exudation II:1970-1978. *Adv. Agron.* 31:93-124.

Jones, D.L., A.M. Prabowo & L.V. Kochian 1996. Kinetics of malate transport and decomposition in acid soils and isolated bacterial populations: The effect of microorganisms on root exudation of malate under Al stress. *Plant and soil* 182:239-247.

Kapfer M., 1998. Assessment of colonization and primary production of microphytobentos in the littoral of acidic mining lakes in Lusatia, Germany. *Wat. Air and soil Pollut.*, 108:331-340.

Mendelssohn, I.A., B.A. Kleiss, & J.S. Wakeley 1995. Factors controlling the formation of oxidized root channels: a review. *Wetlands*, 15(1)37-46.

Nixdorf, B., K. Wollmann & R. Denke 1998. Ecological potentials for planctonic development

Svedäng, M.U. 1992. Carbon dioxide as a factor regulating the growth dynamics of *Juncus bulbosus*. *Aquatic Botany* 42:231-240.

Verhoeven, J.T.A., and food web interactions in extremely acidic mining lakes in Lusatia. In Geller W. H. Klapper W. Salomons (eds.), *Acidic Mining Lakes*: 147-167. Springer.

Pietsch, W. 1979. Zur hydrochemischen Situation der Tagebauseen des Lausitzer Braunkohlen-Reviers. Arch Naturschutz Landschftsforsch Berl. 19:97-115.

Schwertmann, U. and R.M. Taylor 1977. Iron oxides. In B. Dixson & S.B. Weeb (eds), *Minerals in soil environments*: 145-179. Soil Sci. Soc. Am. Madison, WI, USA

Shen, Y., V. Obuseng, L. Grönberg & J.Å. Jönsson 1996. Liquid membrane enrichment for the ion chromatographic determination of carboxylic acids in soil samples. *J. chromatogr.* A 725:189-197.

Spurr AR. 1969. A low-viscosity epoxy resin embedding medium for electron microscopy. *Journal Ultrastructure Research* 26:31-43.W. Koerselman & A.F.M. Meuleman 1996. Nitrogen- or phosphorus-limited growth in herbaceous, wet vegetation: relations with atmospheric inputs and management regimes. *TREE* 11:494-497.

Geochemistry of the Earth's Surface, Ármannsson (ed.)© 1999 Balkema, Rotterdam, ISBN 90 5809 073 6

Heavy metal contaminated waters, Coeur d'Alene Mining District, USA

V.E.Chamberlain & S.Gill
University of Idaho, Moscow, Idaho, USA

B.Williams
Idaho Water Resources Research Institute, Moscow, Idaho, USA

ABSTRACT: Geochemical analyses of water from 20 ground water and 17 surface water sites in the Cataldo Flats region of the Coeur d'Alene Mining District reveal that severe contamination of the natural water system has occurred. This is due to the percolation of precipitation through mining spoils, which were dredged from the Coeur d'Alene River and deposited on its floodplain at Cataldo, approximately 60 years ago. In ground water, concentrations of heavy metals in micrograms per millilitre range up to 0.39 for Pb, 150 for Zn and 0.43 for Cd. In surface water, concentrations range up to 0.009 for Pb, 7.3 for Zn and 0.012 for Cd.

1 INTRODUCTION

Mining, for silver, lead and zinc has taken place in the Coeur d'Alene region of northern Idaho, U.S.A., for over one hundred years, and the disposal of mining and milling waste has always been a problem in the narrow valleys of the Coeur d'Alene River and its tributaries. The first tailings ponds were constructed in 1968, and until that time, all waste was allowed to enter the drainage system and be carried downstream. Large quantities of these wastes became lodged in a bend in the Coeur d'Alene River at Cataldo (Figure 1), which was also a landing stage for paddle boats linking the mining district with the nearest town, Coeur d'Alene. Because of the necessity of keeping the Cataldo dock open and the river flowing, these spoils, together with natural alluvium, were, between 1880 and 1940, frequently dredged from the river and deposited on the adjacent flood plain, named the Cataldo Flats.

These dredge spoils, now up to 13 m thick, cover an area of approximately one sq. km. to the northwest of the Old Mission State Park and on both sides of the freeway, Interstate I-90. The dredge spoils are bounded to the west and east by wetlands (consisting of shallow lakes and marshes), to the north by hills and to the south by the Coeur d'Alene River. The wetlands are replenished by the south flowing Hayden Creek, which drains the hills forming the northern boundary to the Flats. The western wetlands in turn are drained into the river via Coffee Creek. Iron-stained seeps can be seen draining out of the spoils southward into the river.

Latour Creek flows into the Coeur d'Alene River from the south.

The rocks outcropping north and south of the Cataldo Flats flood plain are primarily Middle Precambrian Belt Supergroup metasediments, similar to those in the Mining District, though barren here. Heavy metal concentrations in the creeks (Hayden and Latour) draining these hills are negligible, being consistently below the detection limits of our instruments (0.0015 micrograms per millilitre for Pb, 0.0025 for Zn and 0.0023 for Cd)

Previous studies of the geochemistry of these spoils was conducted over twenty years ago (Galbraith et al. 1972, Mink et al. 1971). More recently, Paulson (1994) has investigated the fate of metals from mine waste in the lower Coeur d'Alene River water.

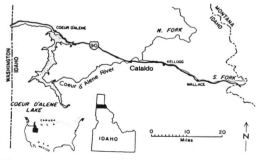

Figure 1. Map showing the location of Cataldo in the Lower Coeur d'Alene River valley, Idaho, USA.

2. SAMPLE SITES

A total of thirty-seven sample sites were chosen in the Cataldo Flats area, seventeen surface water sites and twenty ground water sites. Surface water sites included four sites in the Coeur d'Alene River (RV01-RV04) one upstream from, one downstream from and two within the Cataldo Flats. Other surface water sites comprised two seeps within the dredge pile (WT02 and WT07), eight sites in the wetland lakes in the Flats (WT01, WT03, WT04, WT05, WT05B, WT06, WT08 and WT09), one site in each of the creeks (Hayden, CK01, and Latour, CK02) and one site in Coffee Creek (CC01).

The ground water sites were chosen to monitor water flowing westward from the dredge pile to the wetlands and southward from the dredge pile to the river. Piezometers were installed at locations PZ01 - PZ20 as shown in Figure 2. Manual installation of these piezometers afforded least disturbance to recently seeded areas as well as being economical. Each hole was hand-augered until the sediments encountered were too wet to be removed, and then a steel casing with a disposable HDPE point was driven approximately two meters deeper into the sediments. This extra depth was intended to allow for water table fluctuations, but even with this precaution, many wells went dry at the end of the summer. The piezometers, constructed of ¾" PVC,

with a four foot slotted interval in the lower end wrapped with screen, were inserted into the steel casings. In most cases the saturated sediments moved in against the PVC and screen material as the casing was removed, but sand was added if the screened interval was not in intimate contact with the saturated sediments. Bentonite pellets were placed above the saturated interval up to the annulus in order to prevent any preferential downward movement of water. At each site, core samples were taken for analysis.

At location PZ03-PZ04, a dry layer of tight clay was encountered below a perched water table of saturated sediments. In this locality, two piezometers were inserted, one (PZ03) was completed below the clay layer, and the other (PZ04) above it. Global Positioning System (GPS) was used to survey the location and elevation of each well. Surface water level datum was established at the wetland culvert (location WT03) and at the Coffee Creek Bridge (CC01).

Water elevations in the piezometers were taken at monthly intervals, and these together with pump tests were used to determine the hydrological gradients at the site. Water-table contours for January 16 1998 are shown in Figures 2-4. There are steep hydraulic gradients within the spoils southward towards the river and less steep gradients to the west. The dredge spoils themselves were sampled in a selected area near the center of the pile, north of the

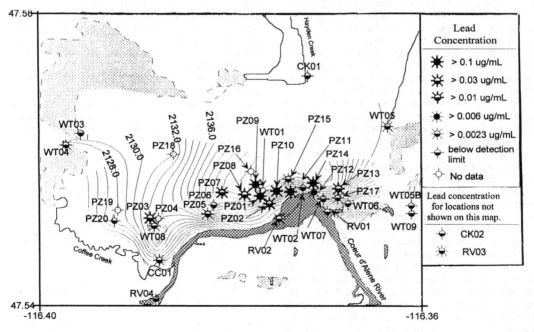

Figure 2. Pb concentrations in ground and surface waters at Cataldo Flats

freeway I-90 and in a location free of all vegetation. In an area approximately 80 m by 50 m, eighteen samples were collected on a grid at approximately 16 m intervals and at approximately 30 cm depth.

3. SAMPLE COLLECTION AND ANALYSIS

Water samples from piezometers, standing surface water, seeps, creeks and the river were collected monthly from June through September, 1997, and again in January, March and June 1998. At each site, water was collected and filtered through a 0.45 micron Millipore filter into a pre-cleaned high-density polyethylene (HDPE) bottle. It was then acidified with Seastar ultra pure nitric acid to pH < 2 and kept in a cooler or refrigerator until analyzed.

Temperature, conductivity, pH, and Eh of each sample were measured in the field, immediately after collection, using a YSI model 3500 multifunction water quality monitor. All chemical analyses were performed at the University of Idaho's Analytical Laboratory, using Environmental Protection Agency (EPA) methodology. Cd and Zn were determined by EPA method 200.7, using a Perkin Elmer P40 inductively coupled plasma atomic emission spectrometer. Low detection limits by this method are 0.0023 µg/ml for Cd and 0.0025 µg/ml for Zn. Pb was determined by graphite furnace atomic absorption spectrometry, using a Perkin Elmer 5100 spectrometer and EPA method 239.2. Low detection limits for Pb are 0.0029 µg/ml.

Heavy metal concentrations in the sediments were determined by EPA method 3050, with dry samples being dissolved in HNO_3 (nitric acid) and then refluxed in hot concentrated HCl (hydrochloric acid) before being filtered and diluted for analysis by atomic absorption spectrometry. A Perkin Elmer 603 flame atomic absorption spectrometer was utilized to measure concentrations of Pb (EPA method 7420), Zn (EPA method 7950), and Cd (EPA method 7130). Standards, blanks and replicate analyses followed EPA procedures.

4. RESULTS

Average heavy metal concentrations in water samples are shown in Figures 2-4, Figure 2 showing Pb concentrations, Figure 3, Zn concentrations and Figure 4, Cd concentrations. Heavy metal concentrations in the spoils were extremely high and somewhat heterogeneous, varying randomly by 10% to 50% from one site to the next only 16 m distant. Concentrations of Pb range from 1168 ppm to 2746 ppm, Zn from 0 to 2775 ppm, and Cd from 4.6 to 13.6 ppm.

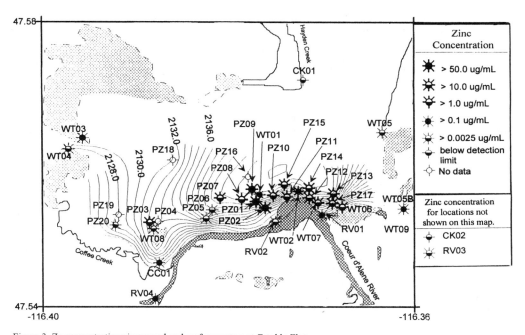

Figure 3. Zn concentrations in ground and surface waters at Cataldo Flats

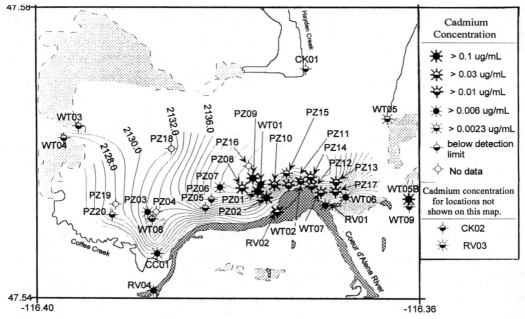

Figure 4. Cd concentrations in Ground and surface waters at Cataldo Flats

5. CONCLUSIONS

Concentrations of heavy metals in ground water within the dredge spoils are high and locally variable. The simplest interpretation is that the deepest and/or most contaminated spoils occur near the center of the pile. In this area, precipitation percolating through the spoils is dissolving metal oxides and sulfides.

Galbraith et al. (1972) described a process of oxidation and reduction of metal sulfides through the action of sulfide-oxidizing and sulfate-reducing microorganisms, the latter being destroyed at depth due to a decrease in pH. However, in this study, all attempts to correlate heavy metal concentrations with pH, Eh, alkalinity, sulphate concentrations or mineralogy failed. It is concluded that the system is too large and complex to be easily modelled. (Benjamin 1978, Benner et al. 1995, Charlet et al. 1993, Davis & Leckie 1978, Filipek et al. 1981, Hem 1976). Thus, it was not possible to determine with any certainty the exact chemical and bio-chemical processes operating at any given point in the system. Our chemical analysis of the spoils has shown their extreme heterogeneity with respect to heavy metal content, and at least some of the varying concentrations in the ground water are a direct result of the varying concentrations in the spoils themselves

REFERENCES

Benjamin, M.M. 1978 *Effects of Competing Metals and Complexing Ligands on Trace Metal Adsorption at the Oxide/Solution Interface*. Ph. D. dissertation Stanford University 228 p.

Benner, S.G., E.W. Smart & J.N. Moore 1995. Metal behavior during surface-groundwater interaction, Silver Bow Creek, Montana. *Envir. Sci. Tech.* 29:1789-1795.

Charlet, L., N. Dise & W. Stumm 1993. Sulfate adsorption on a variable charge soil and on reference minerals. *Ag. Ecosys. Envir.* 47:87-102.

Davis, J.A. & J.O. Leckie 1978. Surface ionization and Complexation at the Oxide/Water Interface. *Science* 67:90-102.

Filipek, L.H., T.T. Chao & R.H. Carpenter 1981. Factors affecting the partitioning of Cu, Zn and Pb in boulder coatings and stream sediments in the vicinity of a polymetallic sulfide deposit. *Chem. Geol.* 33:45-64.

Galbraith, J.H., R.E. Williams, & P.L. Siems 1972. Migration and Leaching of Metals from Old Mine Tailings Deposits. *Ground Water* 10(3):14-32.

Hem, J.D. 1976.Geochemical controls on Lead concentrations in stream water and sediments. *Geochim. Cosmochim. Acta.* 40:599-609.

Mink, L.L., R.E. Williams & A.T. Wallace 1971. *Effect of industrial and domestic effluents on the water quality of the Coeur d'Alene River Basin.* Idaho Bureau of Mines and Geology pamphlet 149.

Paulson, A.J. 1994. *Fate of Metals in Surface Waters of the Coeur d'Alene Basin, Idaho.* United States Department of the interior, Bureau of Mines Report of Investigations 9620

Geochemistry of the Earth's Surface, Ármannsson (ed.) © 1999 Balkema, Rotterdam, ISBN 90 5809 073 6

Hydrochemistry of Khanka (Xangkai) lake and its ecological problems

V.A.Chudaeva
Pacific Institute of Geography, Far East Branch Russian Academy of Science, Vladivostok, Russia

G.I.Semikina
Hydrological Survey, Vladivostok, Russia

ABSTRACT: Khanka lake (boundary with China) is the largest fresh water reservoir on the territory of the Far East of Russia and all East Asia. The high pollution load at the end of the 80-ies and beginning the 90-ies years has coincided in time with a phase of natural falling of the level of the lake and has resulted in a high level of water pollution of various components, including pesticides derived from farmlands, especially rice fields and some municipal and industrial pollutants as well. In 1990's the degree of water pollution became moderate with local increases

1 INTRODUCTION

Khanka lake (boundary with China, the Chinese name Xangkai) is the largest freshwater reservoir the territory of Far East Russia and all East Asia (Figure 1), and is unique in species of plants and animals, i.s. birds and fishes. Its catchment area is 16890 km^2, including Russia territory of 15370 km^2. The water area for the average lake level is 4070 km^2, in Russia -3030 km^2); average volume of water - 18.5 km^3. Khanka Lake is rather shallow with maximum depth up to 5-6 m. The pollution of the lake is connected, firstly with agricultural activity and secondly with municipal and industrial discharge from Spassk-sity (Chudaeva 1996).

2 DISCUSSION

2.1 *Fluctuation of water level*

Cyclic changes with a period from 19 to 29 years in long-term fluctuation of a water level of the lake since 1880's till the present time were found. The falling (lower) of water level in the Khanka lake since the 1960's was aided by the using of water for growing rice.

2.2 *Currents*

The main currents on the surface are caused by wind effects (Figure 2), but they are very unstable. Nearshore currents caused by inflow of river waters and outflow through the Sungach River are present

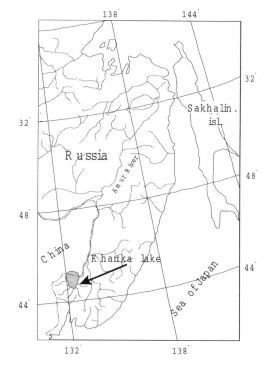

Figure 1. Location of Khanka lake

also. These currents can influence the distribution of chemical substances in the water mass as shownby studies of dissolved matter in the

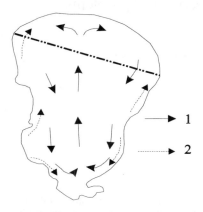

Figure 2. The main surface (1) and nearshore (2) currents directions.

lake. The main directions of nearshore movement of debris material discharged into the lake are also due to these processes (Korotky et al. 1979)

2.3 *Water balance of lake*

The natural water balance of the Khanka lake may be defined by the equation $A + B + C + D - A_1 - B_1 - C_1 - D_1 = E$, where A, A_1 - meteoric water coming on the surface of the lake and evaporation; B, B_1 - river water inputs to lake and output of the lake water through the Sungach River; C, C_1 - underground inflow into the lake and underground outflow from the lake; E- change of water volume of the lake during the time considered.

The average amount of meteoric water entering the lake surface is 567 mm or 72.9 m³/sec, and evaporation from the surface about 584 mm or 75.1 m³/sec (Vaskovsky 1978).

The annual river discharge jn the area of the Khanka basin was investigated by the Hydrological Survey in six rivers with a catchment area 11319 km² from the total -15370 km². Data reveals that the average annual discharge of the rivers is 61.3 m³/sec. The discharge of water from Chinese territory does not change the magnitude.

The underground inflow to and outflow from the lake practically is not taken into account. The possibility of some underground discharge is indicated by some changes in the major ion composition of the lake water in wintertime.

2.4 *Suspended matter*

A specific feature of the Khanka lake is a high suspended matter concentration, which now most of all is connected with ploughing of the land. Concentrations of suspended matter in the lake

water have changed from 20 up to more then 300 mg/l during the last few years. The maximum concentration was found in the summertime and connected with wind mixing of shallow water masses. In wintertime suspended concentrations were the lowest.

The fine size-fraction suspended material (Figure 3) cannot be fast deposited and can easily be lifted during a wave mixing. For this reason the suspended solid contents of the lake, as a rule, is above the concentrations of suspended solids in the inflow rivers. The concentrations of suspended matter in the Sungach River running out of the lake is similar to that of the lake and exceeds the contents of the rivers of the catchment.

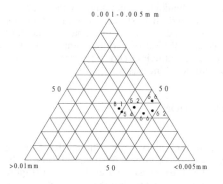

Figure 3. Size-fraction composition of suspended solids of the Khanka lake.

2.5 *Major ions*

The chemical composition of the lake water is characterized by small fluctuations of salts in time. The TDS of the lake water varies mostly within 60-120 mg/l. The composition of the waters is usually $Ca-HCO_3$.

Total discharge of all dissolved matter by the rivers into the Khanka lake is evaluated as 2519 x 10³ t/y, and major ions as 1736 x 10³ t/y or 89.3 t/km²/y. The total discharge of organic matter is 617 x 10³t/y or 31.8 t/km²/y. The biggest amount of organic matter carried by the Ilistaya River - 176.6 x 10³ t/y, and the smallest - by the Melgunovka River -75.7 x 10³ t/y. Data agrees with the distribution of organic substances in deposits of Khanka lake (Korotky et al. 1979). The increase in organic matter was found near the mouth of the Ilistaya River.

2.6 *Heavy metal*

Four dissolved heavy metals were analyzed by the atomic absorption method and these concentrations order was usually Zn > Fe > Cu, Mn.

166

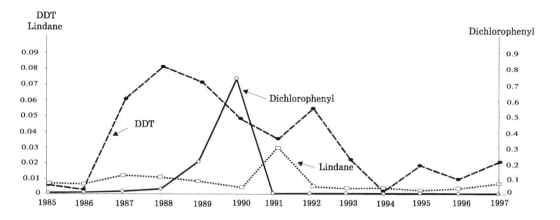

Figure.4. The changes of pesticides contents in the Khanka Lake in time.

Other element contents were below detectable limits. In suspended matter the order of heavy metal concentration differs from the dissolved concentrations : Fe > Mn > Zn > Pb,Cr > Cu,Ni > Cd. Some seasonal changes in element contents both in river input and in the lake were found. It is more common for iron and manganese, as they are the most sensitive to O_2 contents. Some other elements followed Fe and Mn. Some mobile (sorbed) forms differ for different elements, as was found by the treatment of suspensions and sediments of the Khanka lake with successively stronger agents (Chudaeva 1991).

From rivers to the lake an increase of suspended contents of elements in the water (mg/l) takes place. In the lake sediments the higher contents of heavy metal contents are connected with fine-size fractions.

Strong heavy metals pollution of the lake was not found except some places (close to Spassk-sity).

2.7 Pesticides

Use of pesticides in agriculture and especially in rice fields during the years has been far from rational. Enormous volumes of the water from lake were carried out to the rice fields, and then returned back to the lake with pesticides and components of fertilizers which have been the main polluting factor the of ecosystem of the lake.

The high pollution load has coincided on time with a phase of natural falling of the lake level (Chudaeva & Semikina 1988). That resulted at the end of 1980's and the beginning of 1990's in a high level of water pollution of various components, including pesticides (Figure 4).

The decrease of the pesticide contents in the lake water in the 1990's is a result of decreasing rice growing, a consequence of both economic recession, and also natural rise of the water level of the lake.

The presence of pesticides in the water affects fishes and other water species. Our study in 1986 has established a presence of some pesticides in fish and mussels in concentrations exceeding Russian permitted level for food. Their presence was fixed later too.

2.8 Nutrients

The river water carried away exposed topsoil and nutrients. Rather high concentrations of nitrogen were found in all rivers of the Khanka basin and the lake water. The main form ($N-NH_4$) was 3-5 times above the Russian permitted level (0.39 mg/l), and in two especially polluted rivers the level was much higher. But $N-NO_3$ contents were not very high (from < 0.01 mg/l up to 0.8 mg/l).

The P (PO_4) contents changed from < 0.01mg/l to 0.19 mg/l. Comparison of the $P-(PO_4)$ contents in 60'sh - 70's - 80's indicate an increase in concentrations in some rivers and the lake for this period. During field studies it was found that dissolved and suspended concentrations of P in the lake, were as a rule higher than in streams. Only in two rivers, Bolshie Usachi and Spasovka were total(dissolved and suspended) concentrations of P higher than in the lake: > 2 g/l.

Calculation the input of N and P in the lake and discharge through the Sungach River shows that a large part of the nutrients stays in the lake. It seems, that high suspended concentrations have prevented increases in the algae population. In the 1990's

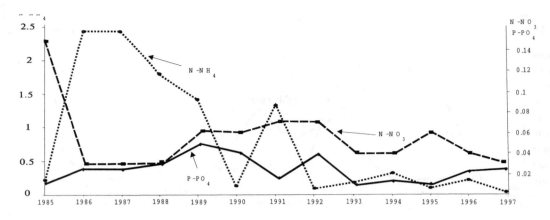

Figure 5. The change of nutrient contents in the Khanka lake with time

concentrations of nitrogen and phosphorus in the water of the lake have decreased (Figure 5).

3 CONCLUSION

In connection with the decrease of pollutants and natural rising of the water level of the lake in the 1990's, the degree of water pollution, as a whole, is moderate with local increases.

The pollution is, first of all, connected with pesticides and nutrients arriving from farmlands, especially rice fields. Some municipal and industrial pollution inputs to the Khanka lake have been observed too. Heavy metal pollution of the Khanka lake is not abnormal, except at a few localized sites.

REFERENCES

Chudaeva V.A. 1991. Dynamic of heavy metals concentrations in Khanka lake and rivers of the catchment. In P.V.Ivashov (ed) *Biogeochemical distribution of chemical elements in Far East ecosistems:* 100-122. Vladivostok. (In Russian).

Chudaeva V.A. 1996. Some ecological problems of the Khanka Lake in the Russian Far East. In B.Nath, I.Lang, E.Mezaros, J.P.Robinson & L.Hens (eds), *Environmental pollution.* Proceedings of ICEP.3. V.1: 373-380. London: European Center of Pollution Research.

Chudaeva V.A. & G.I. Semikina 1988. Some data on organochlorine pesticides in the ecosystem of the Khanka lake. *Problems of water resources of the FarEastern Economic region:* 144-145. Vladivostok. (In Russian).

Korotky A.M., M.A. Mikhailov, I.V.Kitaev & V.B.Kurnosov. 1979. *Lithology and geochemistry of the modern lake sediments of the humid zone.* Moskow: Nauka. (In Russian).

Vaskovskv M.G. 1978. *Hydrological regime of the Khanka Lake.* Leningrad: Gidrometeoizdat (In Russian).

Geochemistry of the Earth's Surface, Ármannsson (ed.) © 1999 Balkema, Rotterdam, ISBN 90 5809 073 6

The environmental impact of heavy metals from sewage sludge in Ferralsols

S. Cornu – *SESCPF, INRA, Orléans, France*

C. Neal & M. Neal – *Institute of Hydrology, Wallingford, UK*

P. Whitehead – *Department of Geography, Reading, UK*

J. Sigolo – *Instituto de Geociencias, São Paulo, Brazil*

P. Vachier – *INRA, Grignon, France*

ABSTRACT: The application of sewage sludge over an acidic Ferralsol was simulated in a soil column experiment in order to estimate the hazard of such a practice on soil and water quality. Drainage and runoff water were sampled daily over two months and analyzed for Fe, Ni, Cu, Pb, TOC. Results show a clear influence of the sludge on the water quality, especially at the start of the experiment. The presence of sludge increases the leaching of metals (Cu, Ni, Pb) and TOC from the soil, mainly in solution. The mobility sequence is Ni>Cu>Pb in soils and sludge. No soil chemistry alteration due to the sludge application was recorded on this time scale.

1 INTRODUCTION

Through its economic activities, man generates a large amount of waste, which has to be disposed of, either by recycling or by accumulation in the environment. Sewage sludge is rich in N and P rendering it of interest to agriculture. However, this residue is enriched in heavy metals, which are toxic to plants and animals. Its application in the fields may induce risks of ground water pollution, plant toxicity and heavy metals being passed on up through the food chain.

Although mobility of heavy metals in sewage sludge-amended soils has been much investigated during the last decade in Europe and in the United States, few studies have been conducted on the mobility of heavy metals in acidic soils such as ferrallitic soils (pH = 4), that have a very small cation exchange capacity. These soils are, however, common and found around some of the most populated cities of the world.

We studied heavy metals mobility in ferrallitic soils in the São Paulo vicinity (Brazil). Application of sewage sludge on a Ferralsol was simulated in a soil column in order to estimate the hazard of such a practice to soil and water quality.

2 MATERIALS AND METHODS

The sampled soil is located 50 km to the southwest of the city of São Paulo (Brazil) near the town of Suzano. This soil developed from a mica-schist bedrock is composed of three main horizons: a brownish coloured, A-horizon, a yellow B-horizon, and a very thick, reddish coloured, C-horizon.

Table 1: Description of the columns

Column number	Total length (cm)	Column composition			
		Sludge[1]	A	B	C[2]
1	20	x			
2	17		x		
3	17	x	x		
4	35		x	x	
5	35	x	x	x	
6	60		x	x	x
7	60		x	x	x

1. São Paulo sewage sludge was load on the soil column top at 74 t/ha/y dry matter.
2. The C-horizon was artificially located under the 35 first centimetres of the soil.

An experiment was performed on unsaturated and undisturbed soil columns. Seven columns were built as described in the table 1 and on the figure 1. Chemical and mineralogical analyses were performed on soil and sludge samples before and after the experiment.

Daily rainfall events were simulated on the top of the soil column over a two months period. Runoff and drainage water were collected daily and analysed for Ni, Cu, Pb, Fe (ICP-OES and ICP-MS), TOC (TOC analyser). Blank samples were also run.

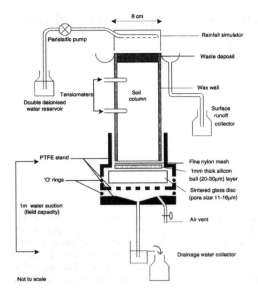

Figure 1: Column setup

Table 2: Concentration in heavy metals within the sludge before and after the experiment (ppm)

Element	Before	After experiment		
		Mean	SD	Balance
Cu	498	590	40	32 - 52
Pb	122	151	10	19 - 39
Ni	195	182	16	-29 - 2

Table 3: Range of heavy metal content for the different soil horizons

Horizon	Depth cm	Cu ppm	Pb ppm	Ni ppm
A	0-7	37-44	31-19	23-26
	7-14	23-27	27-19	15-16
	14-20	19-25	94-16	12-16
B	20-24	22-20	22-21	18-15
	24-30	19-19	10-25	14-15
	30-37	18-21	51-23	14-18
C	120-127	22-24	16-14	14-15
	127-135	28-25	25-9	15-15
	135-145	22-27	27-11	15-19

Some samples were ultra-filtered a through pre-washed Millipore biomax apparatus. The ultrafiltration cuts used were 10 and 5 kDa.

3 CHARACTERISATION OF THE SOLIDS

Heavy metal concentrations of the sludge are reported in table 2. During the experiment, the sludge underwent a decomposition process inducing a weight loss ranging from 3 to 14 %. At the end of the experiment, the sludge was enriched from 6-10% in Cu and 15-30 % in Pb. Concentrations of Ni slightly decreased, indicating a loss of Ni from the sludge of about 20%.

The mineralogy of the three soil horizons were mainly composed of kaolinite, quartz, goethite, muscovite and some gibbsite with pH values ranging from 4.5 to 4. Cu and Ni were enhanced in the topsoil, mostly in the first 7 centimetres (Table 3). Pb content was highly variable. About 15 mg Cu, 10 mg Pb and 10 mg Ni were found within the upper 7 centimetres of the soil columns. Thus the metal input due to the sludge application represents about one third, one tenth and one fourth of the initial topsoil content for Cu, Pb and Ni respectively. Heavy metal input was too small to significantly enhance soil metal content. This explains why no chemical change was recorded in the soil at the end of the experiment.

4 CHEMICAL EXPORT

4.1 Surface runoff waters

Despite a slight enrichment in TOC, Cu, Ni and Pb of the runoff water after sludge application, the concentrations remained low. It seems thus that even at the start of the experiment, when concentrations were maximal, sludge spreading on a horizontal surface does not expose surface water to a pollution risk.

4.2 Drainage water

- The sludge (column 1)

Sludge water samples were not analysed for Pb and TOC. Cu, Ni and Fe were rapidly flushed at the start of the experiment. This is due to the leaching of the soluble and exchangeable elements and probably to a rapid decomposition of organic matter. Soluble organic compounds that carry heavy metals are thus released (Minnich et al. 1987, Alloway & Jackson 1991, Chaney & Ryan 1993).

At the end of the experiment the loss was less than 5% of the total metal content, except in the case of Ni those loss was 15% of the sludge initial content. These results were in agreement with those obtained from the solid analyses. 10 % of Ni contained in the sludge was in soluble or exchangeable form (Lake et al. 1984) and was easily released. As the sludge has undergone processes of decomposition (3 to 14% of weight losses) during the experiment, other forms of Ni may also be released from the sludge.

In water, 90 % of Ni and Fe were found in the <5 kDa fraction, while 30 % of Cu is found in particles and 5 % in colloids. According to previous studies (Legret 1993), Ni is the metal most associated with the soluble phase, while Cu is more associated with organic matter (Karapanagiotis et al. 1991). Baham & Sposito (1986) showed that Cu-organic complexes have a high molecular weight. Organic matter also contributes to Ni mobility (Karapanagiosti et al. 1991).

• In soils

In the presence of sludge, drainage water concentrations of Ni, TOC, Cu and Pb increased, while Fe concentrations decreased.

Element speciation in water showed that, for the control treatment, more than 60 % Ni and Cu were transported in dissolved form, while 50% of Pb and Fe were mobilised in colloidal form. Pb mobility was thus probably due to Fe colloids. In the presence of sludge, the dissolved contribution increased for all the studied heavy metals.

These results indicated a clear impact of the sludge on drainage water quality, especially at the start of the experiment, when the sludge is fresh.

However, in all cases, the total loss represents less than a few percent of the initial column content.

Table 4 reports ratios of column losses. These ratios were calculated as:

$$(N+1)/N = (E_{N+1} - E_N)/ E_N \times 100 \qquad (1)$$

where E_{N+1} and E_N are the element lost from horizon N+1 and N respectively.

Table 4: Ratio of elemental losses for the control treatment and the sludge application case. Results expressed in %.

Case	Ratio	Fe	TOC	Cu	Ni	Pb
Polluted	A/sludge	71	nd	-73	-99	nd
	B/A	-78	-55	-86	-37	-73
	C/B	nd	93	1338	514	-56
Natural	B/A	-96	-53	-90	-75	-67
	C/B	575	20	-100	25	100

nd : not determined

Cu and Ni coming from the sludge are retained in the A-horizon. Crouzet & Bourg (1992) showed in a column experiment study that Cu was more strongly retained than Ni on soil solid phases. On the other hand, the A-horizon releases Fe.

From the A- to the B-horizons, losses in Cu, Ni, TOC, Pb and Fe decrease in both treatments. This decrease might be due to TOC fixation within the B-horizon. All metals associated with TOC are then retained.

From the B- to the C-horizon, losses of most elements increase. During the experiment, columns 6 and 7 underwent a saturation stage. This may have induced Fe oxi-hydroxides reduction and associated Ni and Pb release, and then large element losses for these two columns. Indeed (Salim et al. 1996)

showed that these two elements were closely associated with iron oxi-hydroxides at background level in clay rich material.

Table 5: Polluted versus control losses calculated as (P-N)/Nx100 where P and N are the element lost from the polluted and unpolluted horizon respectively. (%)

Horizon	Fe	TOC	Cu	Ni	Pb
A	-94	99	217	372	12900
B	-100	121	500	400	1150
C	-56	276	nd	4200	0

Sludge spreading increases the loss of Cu, Ni, TOC, and Pb (Table 5), while it decreases that of Fe. This might be due to a stabilising effect of TOC contained in the sludge drainage water, while in the control experiment deionised water has an aggressive effect on the soil.

• Mobility sequences of the studied elements

The element mobility sequence was Ni>Cu>Fe for the sludge and Ni>Cu, TOC>Pb>Fe at the start of the experiment and Ni>Pb, Cu, TOC>F for the soils. This sequences were in agreement with work on element speciation in sludge (Legret 1993) and on element affinities for clay and iron oxi-hydroxides (Alloway 1990).

In addition, Cu and TOC mobility were similar suggesting a link between these two elements. Pb mobility changes with time, probably due to the dissolution of a solid phase during the experiment.

5 CONCLUSION

There was a clear impact of sludge when spread on the Ferralsol:
• runoff and drainage water were enriched in TOC, Cu, Ni and Pb, inducing higher export fluxes from the polluted columns;
• elements in water were to a greater extent in soluble form in the presence of sludge.
• However, no soil chemistry alteration was recorded.

Cu, Ni and Pb released by the sludge are partially retained within the A and B-horizons, probably due to TOC fixation. On the other hand, the C-horizon released these elements. A reduction process was thought to be responsible for this release.

In all cases, Ni was the most mobile element, Pb being the least mobile. This was due to the form of these elements within the sludge and the soil. While a large part of Ni was probably in soluble and exchangeable form, Cu was more closely associated with organic compounds and Pb with iron.

6 ACKNOWLEDGEMENTS

This work has been undertaken with the financial support of a Lavoisier fellowship of the French foreigh affair ministary, and with the contribution of the Department of Geography of Reading, the Institute of Hydrology of Wallingford and The Institute of Geosciences of São Paulo.

REFERENCES

Alloway, B.J., 1990. Soil processes and the behaviour of metals. in *Heavy metals in soils*. ed B.J. Alloway. Blackie, J.Wiley & Sons, Inc. 7-28.

Alloway, B.J., & A.P. Jackson, 1991. The behaviour of heavy metals in sewage sludge amended soils. *Sci. Tot. Env.* 100: 151-176.

Baham, J. & G. Sposito, 1986. Proton and metal complexation by water soluble ligands extracted from anaerobically digested sewage sludge. *J. Environ. Qual.* 15: 239-244.

Chaney, R.L., & J.A. Ryan, 1993. Heavy metals and toxic organic polluants in MSW-compost: Research results on phytoavaibility, bioavaibility, fate, etc. In: H.A.J., Hoiting & H.M., Keener (eds.). *Science and engineering of composting: design, environmental, microbiological and utilisation aspects*. Renaissance Publ. Workington, Ohio: 451-506.

Crouzet, C., & A. Bourg, 1992. Relargage de métaux lourds lors de différentes conditions d'épendage de boues d'épuration sur un sol agricole. BRGM internal report R35927, 61 p.

Karapanagiotis, N.K., R.M. Sterritt, & J.N. Lester, 1991. Heavy metal complexation in sludge-amended soil. The role of organic matter in metal retention. *Environ. Technol.* 12: 1107-1116.

Lake, D.L.P., W.W. Kirk, & J.N. Lester, 1984. Fractionation, characterization and speciation of heavy metals in sewage sludge and sludge-amended soils: a review. *J. Environ. Qual.* 13(2): 175-183.

Legret, M. 1993. Speciation of heavy metals in sewage sludge and sludge amended soil. *Int. J. Env. Anal. Chem.* 51: 161-165.

Minnich, M.M., M.B. McBride, & R.L. Chaney, 1987. Copper activity in soil solution : II. Relation to copper accumulation in young snapbeans. *Soil Sci. Soc. Am. J.* 51: 573-578.

Salim, I.A., C.J. Miller, & J.L. Howard, 1996. Sorption isotherm-sequential extraction analysis of heavy metal retention in landfill liners. *Soil Sci. Soc. Am. J.* 60: 107-114.

Geochemistry of the Earth's Surface, Ármannsson (ed.) © 1999 Balkema, Rotterdam, ISBN 90 5809 073 6

REE mobility associated to acid mine drainage: Investigations in the Libiola area, northern Italy

E. Dinelli, G. Cortecci, F. Lucchini & M. Fabbri
Dipartimento di Scienze della Terra e Geologico-Ambientali, Università di Bologna, Italy

M. D'Orazio
Dipartimento di Scienze della Terra, Università di Pisa, Italy

ABSTRACT: High contents of rare earth elements were recorded in acidic waters in the surroundings of the Libiola Fe-Cu mine area, their PAAS-normalized patterns closely resembles those of basaltic rocks outcropping in the area. No fractionation occurs suggesting that intense water/rock interaction took place extensively. Their dissolved concentrations depend on water pH (being lower at higher pH), and their aqueous mobility seems to be controlled by iron-ochre precipitation associated to alkaline geochemical barrier.

1 INTRODUCTION

The importance of REE and other trace elements related to them, such as Y and Sc is generally based on their low mobility. However, under particular environmental conditions, the stability of REE-bearing phases might be lowered. Acid drainage areas represent context were this phenomenon could occur. We investigated REE distribution in stream waters in the surroundings of a pyrite-chalcopyrite mine in northern Italy.

2 STUDY AREA AND GEOLOGICAL SETTING

The study area is located in the drainage basin of the Gromolo river, near the village of Libiola (GE), about 10 km NE from the town of Sestri Levante (Fig. 1). The Libiola mine area is situated in the Mesozoic ophiolitic rocks of the Internal Ligurides outcropping in the Eastern Liguria (Brigo and Ferrario 1974, Zuffardi 1977). It is hosted in pillow and brecciated basalts, and partially involves also tectonically overlying serpentinites (Ferrario and Garuti 1980). Pyrite (FeS_2), chalcopyrite ($CuFeS_2$) and sphalerite (ZnS) are the main ore forming minerals, hosted in a quartz and/or calcite gangue.

The mine was already known in the Copper Age (about 2500 y B.C.), economic exploitation started in the XVII century and ended in 1965. Mining operations were active both in open pit and underground shafts. Nowadays several open galleries and waste dumps testify the past mining activities.

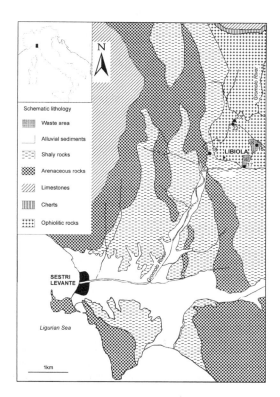

Figure 1. Location map, water sampling sites and schematic lithology of the Libiola mine area

Schematic lithology

▦ Waste area

☐ Alluvial sediments

⋮⋮ Shaly rocks

▨ Arenaceous rocks

▨ Limestones

▥ Cherts

⋮⋮ Ophiolitic rocks

LIBIOLA

SESTRI LEVANTE

Gromolo River

Ligurian Sea

1km

Table 1. Major chemistry of samples from the Libiola mine area.

	pH	Eh (mV)	Cond (µScm⁻²)	HCO₃ (mg/l)	Cl	SO₄	Na	K	Ca	Mg	Fe	Al	Mn	Zn	Cu (mg/l)	Cd (µg/l)	Ni	Co	Cr	Rb	Sr	V
18	2.5	466	7670	nr	nr	5100	24.9	1.4	340	306	891	265	9.9	34	221	130	6400	4140	2540	3	1184	16
22	2.6	600	5290	nr	nr	4500	32.2	9.2	469	344	358	120	6	20	80	85	480	564	450	4.9	1819	2
16	5.8	220	1800	21.3	17.8	1300	12	5.5	345	146	0.12	nr	1.3	3.7	5	9.2	230	210	0.5	25.8	1400	<0.05
17	7.4	62	2100	33.6	14.2	1400	7.1	3.6	265	202	0.14	nr	1	3.6	0.97	10	510	201	nr	4.9	2811	<0.05
23	8.3	193	297	88.5	10.7	150	1.9	0.5	50	15	0.1	nr	nr	nr	nr	26	18	7.7	0.2	48	2

3 MATERIALS AND METHODS

Samples of water and rivershed sediments were collected between 1997 and 1998 in the surroundings of the Libiola mine area in order to evaluate the environmental impact of the mining waste. We discuss only the water samples on which REE analyses were performed.

Major cations were analysed by AAS, alkalinity by titration, chloride by argentometry and sulfate by turbidimetry. Trace elements were measured by AAS, REE, Y and Sc by ICP MS on filtered samples (0.45µm). Electrical conductivity, pH, Eh and temperature were measured directly in the field.

Rivershed sediments were sampled from the active river bed and sieved in the field, retaining the fraction < 180 µm for analysis. Major and trace elements were determined by X-ray fluorescence spectrometry on pressed powder pellets. The mineralogy of the stream sediments has been evaluated by X-ray diffraction using both Cu and Co sources.

4 CHEMICAL DATA

4.1 Major element composition

Unaffected waters in the area are represented by sample 23 (Table 1), a Ca-Mg-carbonate water with low-dissolved metal contents, which contrast markedly with acid drainage originating from the mine area (e.g sample 18 and 22, Table 1). These waters are characterized by high sulfate contents as well as Ca, Mg, Al, Fe Cu and Zn and high concentrations of other dissolved trace elements, such as Cr, Co, Ni, Cd. Sample 16 was collected within the waste area and represent the only site were visible changes in drainage features were observed. In wet periods acidic waters with typical ochreous precipitates were observed, whereas in the dry season higher pH and blue precipitates occurred. Shift in the Fe/Cu ratio towards lower values is typical of this site. Sample 17 is a water sample in contact with a blue Cu-Zn-Al hydrate sulfate recipitate and has peculiar chemical features such as high pH and low dissolved metal concentrations.

4.2 REE distribution

REE distribution are particularly high in the acidic water samples. They reach values not unusual for stream waters (e.g. McLennan 1989), but are several times higher than the local background, for almost all REEs are below analytical detection limits (sample 23).

A possible origin for these elements is related to the intense leaching effects of acidic solutions on basaltic rocks, extensively outcropping in the area (Fig. 2). The PAAS-normalized REE contents of the more acidic samples is parallel to the range observed in the basaltic rocks of the Internal Ligurides (Ottonello et al. 1982), characterized by an increasing pattern from La to Eu and a rather flat trend for MREE and HREE. Sample 16 records a similar pattern, with minor negative spikes for Tb and Tm, whereas a slightly more complex patterns occurs in sample 17, that shows a tendency towards depletion in HREE. High concentration of Y (541 ppm Y) has been observed in the blue precipitate at this sampling site, which might involve also the HREE and explain their relative depletion in the water samples. Sample 17 also shows an anomalous La/Ce ratio that is specular to the one observed in the blue precipitate (57 ppm La, 180 ppm Ce).

Yttrium displays variations similar to REE, with high dissolved concentrations in the acidic waters, progressively decreasing in the other samples (as a reference can be used the concentration of 40 ng/l Y reported by Reimann and de Caritat 1998). As concerns Sc, its dissolved concentrations are rather high in the acidic waters, but are also significant in the other water samples (Reimann and de Caritat 1998, reported 4 ng/l Sc in a world mean stream water) suggesting that other phases (mostly Fe-Mg minerals) then those REE-bearing, are involved in dissolution reactions.

4.3 Stream sediments composition

The downstream variations in stream sediment composition along the Gromolo river (Fig. 3) report comparable distribution for Fe_2O_3, Cu, Sc, Ce and Y, with spikes dowstream of the confluences of the acidic waters 18 and 22.

Table 2. Dissolved REE, Sc and Y concentrations (ng/l)

sample	18	22	16	17	23
La	51	17.4	2.37	4.7	1.9
Ce	86	50	3.8	1.81	<0.01
Pr	15.3	9	0.51	0.19	<0.01
Nd	81.4	47.8	3.4	1.44	<0.01
Sm	25.6	15.6	0.76	0.15	<0.01
Eu	7.7	4.3	0.18	0.04	<0.01
Gd	34.2	20	0.95	0.29	<0.01
Tb	6.3	3.6	0.09	0.04	<0.01
Dy	39.4	22.9	1.2	0.32	<0.01
Ho	8.2	4.5	0.19	<0.01	<0.01
Er	19.2	10.7	0.52	0.06	<0.01
Tm	2.73	1.46	0.02	<0.01	<0.01
Yb	15	8.2	0.27	0.01	<0.01
Lu	2.6	1.12	<0.01	<0.01	<0.01
Sc	181	183	22	23	16
Y	209	121	8.5	3.5	<0.02

The relative magnitude of the peak anomaly in the stream sediments is roughly proportional to the quantities dissolved in the acidic waters: the highest changes occurring for Y (5 times background values) and the lowermost observed for Ce (twice the background values).

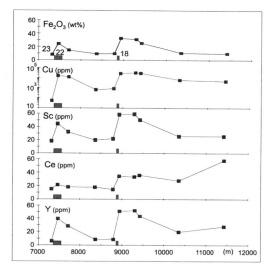

Figure 3. Downstream variations along the Gromolo river for Fe_2O_3, Cu, Sc, Ce and Y.

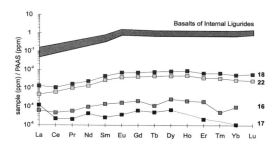

Figure 2. REE patterns, normalized to PAAS (McLennan, 1989), for the Libiola mine area water samples. The field of basaltic rocks of the Internal Ligurides (data from Ottonello et al. 1982) is reported for comparison.

These sites can be considered alkaline geochemical barriers (Perel'man 1986), which are accompanied by the precipitation of fine-grained iron ochres (ferrihydrite and schwertmannite). These secondary phases are likely to remove, either by adsorption or coprecipitation processes, many of the elements dissolved in the acid drainage, including also REE and elements related to them such as Y and Sc. The net result is a dramatic decrease in dissolved trace REE concentrations, coupled to an increase in the solid phase values.

The peaks observed downstream of the acidic emissions are evident for all the elements, even if some complication may arise in the final section of the Gromolo river, cutting through sandstones and shales (e.g. Ce downstream trend).

5. CONCLUSIVE REMARKS

Although only preliminary, our observations indicated that in the Libiola mine area acidic mine drainage strongly interacts with rocks, mostly basalts. REE data suggest that no selective dissolution of REE-bearing phases occurred, since the dissolved REE pattern closely matches those reported for basaltic rocks. These data could be used to evaluate the intensity of water/rock interaction given the assumption that no other rock types were strongly involved.

The REE mobility is however influenced, as that of many other elements, by the flocculation of iron ochres along the Gromolo river. Further studies on the solid phase speciation are running in order to better constrain the mechanisms of precipitation. Moreover studies on the stability of ochreous precipitates (e.g. Dinelli et al. 1998) should bettern constrain the environmental impact of the mining wastes.

REFERENCES

Brigo, L. & A. Ferrario 1974. Le mineralizzazioni nelle ofioliti della Liguria orientale. *Rend. Soc. It. Min. Petrol.* 30: 305-316.

Dinelli, E., N. Morandi & F. Tateo 1998. Fine-grained weathering products in waste disposal from two sulphide mines in the northern Apennines. Italy. *Clay Minerals* 33: 423-433.

Ferrario, A. & G. Garuti 1980. Copper deposits in the basal breccia and volcano-sedimentary sequences of the Eastern Ligurian Ophiolites (Italy). *Min. Deposita* 15: 291-303.

McLennan, S.M. 1989. Rare earth elements in sedimentary rocks: influence of provenance and sedimentary processe. In B.R. Lipin & G.A. McKay (eds) *Geochemistry and mineralogy of rare earth elements.* Rev. Mineral. 21: 169-200.

Ottonello, G., J.L. Joron & G.B. Piccardo 1982. rare Earth and 3d transition element geochemistry of peridotitic rocks: II. Ligurian peridotites and associated basalts. *J. Petrol.* 25: 373-393.

Perel'man, A.I. 1986. Geochemical barriers: theory and practical applications. *Applied Geochem.* 1: 669-680.

Reimann, C. & P. de Caritat 1998. *Chemical elements in the environment.* Berlin: Springer-Verlag.

Zuffardi, P. 1977. Ore/mineral deposits related to the Mesozoic ophiolites in Italy. In H.J. Schneider (ed) *Mineral deposits of the alpine epoch in Europe:* 314-323.

Geochemistry of the Earth's Surface, Ármannsson (ed.)© 1999 Balkema, Rotterdam, ISBN 90 5809 073 6

Dissolution of Aznalcóllar sulphide sludge in O_2-bearing water

C. Domènech & C. Ayora
Instituto de Ciencias de la Terra 'Jaume Almera', CSIC, Barcelona, Spain

J. de Pablo
Departamento de Enginyeria Química, Universidad Politècnica de Catalunya, Barcelona, Spain

ABSTRACT: On April 25th, 1998, the wall of the tailing dam of Aznalcóllar pyrite mine (SW of Spain) failed. As a consequence, 4×10^6 m^3 of sludge covered a farmland and riverflat extension of nearly 4×10^7 m^2 in a layer of 5 cm to 1 m thick. The oxidation rate of the sludge due to $O_2(aq)$ has been studied by means of laboratory flow-through experiments. The rate values obtained for the different sulphides (pyrite, chalcopyrite, sphalerite and galena) were all between 1.0×10^{-9} and 8.5×10^{-9} mols m^{-2} s^{-1}. The behaviour of trace elements, such as As, Cd, Co, Sb, Mn and Tl seems to be explained by the congruent dissolution of the sulphides, whereas Ni probably belongs to a silicate phase.

1 INTRODUCTION

The finely crushed mill tailing and water resulting from the processing plants of massive sulphide mines are stored in dams. On April 25th, 1998, the tailing dam of the Aznalcóllar pyrite mine (70km north of Doñana National Park, SW Spain) collapsed and the valleys of the Agrio and Guadiamar rivers were flooded with nearly 4×10^6 m^3 of sulphide sludge. As a consequence, 4×10^7 m^2 of riverflats and farmlands were covered with a blanket of sulphide sludge ranging from 5 cm to 1 m thick. Moreover, 2×10^6 m^3 of metal bearing acidic waters were retained within the channelized lower course of the Guadiamar river, located in the marshes of the Guadalquivir river.

The bulk of the sulphide sludge was removed during the three months following the failure. However, a percentage of this sludge (0.1 to 10 % wt) remains in the unsaturated zone of the soil and represents a potential source of future pollution. Therefore, it is important to know the chemical behaviour of these sulphides in order to foresee the rate of release of contaminants into the soil, runoff and groundwater.

The pyrite dissolution rate, both by $O_2(aq)$ and Fe^{3+}, is well documented (Nicholson 1994). In the case of sphalerite, galena and chalcopyrite, only the oxidation rate due to Fe^{3+} has been previously measured (Rimstidt et al. 1994). However, only Fe^{2+}-sulphates were observed in Aznalcóllar forming crusts during the dry summer season following the spill. Therefore, sulphide processes due to O_2 dissolved in water were initially considered.

This is the first part of a study focused on obtaining the rate of dissolution of the Aznalcóllar sludge. The work presented here gives us the maximum rate of dissolution and the relative ratios of release of the potential contaminants (Zn, Cd, Cu, Pb, As, Tl). Nevertheless, limiting factors, such as oxygen availability and the formation of sulphate crusts, may decrease the overall sludge dissolution rate in the unsaturated zone.

2 SAMPLE CHARACTERIZATION

A few days after the dam failure, the sulphide sludge used in the experiments was collected from the flooded Agrio river, in front of the collapsed wall.

Chemically, it is a solid rich in Fe (36.44%) and S (39.75%) with significant amounts of potentially toxic elements such as Zn (9448μg/g), Pb (7141μg/g), As (5223μg/g), Cu (1968μg/g), Sb (474μg/g), Cd (31μg/g), Ag (29μg/g), Bi (47μg/g).

The mineralogical composition of the slurry was determined by XRD and confirmed by SEM-EDS. It is mainly composed of pyrite (76%wt) and contains lesser amounts of quartz (8%), gypsum (6%) and clays (10%). A few grains of chalcopyrite, galena and sphalerite were detected by SEM-EDS. Therefore, it was assumed that several contaminant elements were major constituents of sulphides present in minor amounts (Cu in chalcopyrite, Zn in sphalerite and Pb in galena). The rest of the contaminants were considered to be trace constituents of pyrite and the above minor sulphides.

The granulometric analyses of the sludge showed a great range of sizes between 330 and 0.5 µm. Of the grains 50% had a diameter below 11.6 µm and 90% were below 46.0 (Querol et al. 1998).

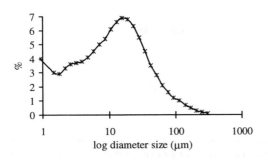

Figure 1. Distribution of grain sizes of the Aznalcóllar sludge (Querol et al. 1998)

To facilitate a quantitative discussion about the results obtained in the dissolution experiments, a geometrical approximation of the reactive surface area was carried out. Considering only the most abundant metals (Fe, Cu, Zn, Pb) and their concentration in the sludge, the sulphide was divided into four minerals, and considering spherical particles and the sludge size distribution, the area of each mineral (pyrite, sphalerite, galena and chalcopyrite) could be calculated. A surface area of 1.06×10^{-3}, 3.90×10^{-4}, 3.21×10^{-4} and 0.043 m^2 g^{-1} of sludge was obtained for sphalerite, chalcopyrite, galena and pyrite respectively.

Due to the objectives of this work, the solid sample was not subjected to any kind of pretreatment. It was only dried in order to facilitate its handling and weighing.

3 THEORETICAL BACKGROUND AND EXPERIMENTAL DESIGN

A dissolution/precipitation reaction is a complex process and the experimental rate values have been proven to depend on a variety of factors (Lasaga et al. 1994)

$$rate = k_o e^{\frac{-Ea}{RT}} a_{H^+}^{n_{H^+}} \Pi a_i^{n_i} f(\Delta G_r) \qquad (3.1)$$

where k_o is the rate constant (mol m^{-2} s) of the reaction at the temperature of reference. The dependence on pH of the dissolution/precipitation rate is included explicitly in the $a_{H^+}^{n_{H^+}}$ term and other possible catalytic/inhibitory effects by other species in solution (such as $O_2(aq)$), in the $a_i^{n_i}$ term. The term

$f(\Delta G_r)$ accounts for the important variations of the dissolution/precipitation rate with deviation of the solution from equilibrium ($f(\Delta G_r)=0$).

In a flow-through experiment the dissolution rate of a mineral leads to the following mass balance expression for the concentration of the ith solute in a reactor cell:

$$\frac{dc_i}{dt} = rate \; v_i \frac{A}{V} - \frac{q}{V}\left(c_i - c_i^o\right) \qquad i=1...Nc \qquad (3.2)$$

where v_i is the stoichiometric coefficient of the ith solute in the mineral, c_i is the total concentration (mol m^{-3}) of solute i in the cell (and output solution), V is the volume of the cell (m^3), A is the total reactive surface area of the mineral (m^2), q is the fluid flux through the system (m^3 s^{-1}), and c_i^o is the concentration of solute in the input solution.

In a flow-through experiment the value of the dissolution rate is obtained from the difference between the output and input solutions when the steady state is reached. If q, c_i and c_i^o are known, the mineral surface reactive area and the stoichiometric coefficient of the element in the mineral, the rate can be obtained from:

$$\frac{rate}{v_i} = \frac{q(c_i - c_i^o)}{A} \qquad i=1...Nc \qquad (3.3)$$

A flow-through experimental design was used to carry out the experiments. The input solution, prepared with analytical grade Merck reagents and Millipore MQ water, was kept in a bottle and equilibrated with a known O_2 partial pressure. From the bottle, the input solution was sent to the reactor where the sample was located by means of a peristaltic pump (Gilson) which controlled the flow (2% error). The solution entered the reactor and after reacting, flowed out of the system. The solution circulated in teflon tubes except in the peristaltic pump, where tygon tubes were used to avoid deformation with time. The experiments were carried out at room temperature (22°C).

The reactor was made of metacryllate. The stirrer was separated from the sample with a 1.2 µm white Rawp Millipore filter to avoid variations in reactive surface area due to stirrer erosion. Before leaving the reactor, sample was filtered with a 0.45 µm white Hawp Millipore filter.

The pH was measured in the output solution with a Crison combined glass electrode.

Samples were taken every 2-3 days. Total iron was measured both by UV-VIS spectrophotometry following the method developed by Gibbs (1976) and ICP-AES. The amount of V, Cr, Cd, Co, Cu, Ni, Pb, As, Sb and Tl was determined by ICP-MS, and the concentrations of Zn, S and Mn and major cations were determined by ICP-AES.

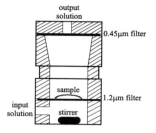

Figure 2. Scheme of the flow-through reactor used.

4 RESULTS AND DISCUSSION

The concentration of an element at steady state depends linearly on the flow at which the input solution circulates (Eq. 3.3). Therefore, a set of experiments at different water flow rates were carried out to check this behaviour. Finally, a flow of 5.0×10^{-4} cm^3 s^{-1} was chosen because it allowed enough trace metal concentration to be quantified.

Experiments were performed at variable pH and O$_2$ partial pressure. HCl and HNO$_3$ solutions were used to fix the initial pH. A pH of 3 was initially chosen, although experiments at pH 5 and 4 were also done.

Steady state was clearly achieved for most elements after 300 hours. As expected, the most abundant elements in solution were Fe and S, with concentrations of around 10^{-4} kmol m^{-3}. Tl and sometimes Cd and Co, had very low concentrations and in some experiments they were not detected.

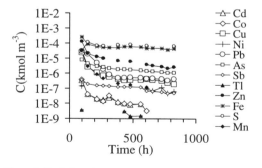

Figure 3. Evolution of the concentration of elements over time. A 10^{-3} kmol m^{-3} HNO$_3$ solution (pH 3) equilibrated with 21% O$_2$(g) was used as input solution and 0.2 g of sulphide sludge was initially used. The flow was 5.0×10^{-4} cm^3 s^{-1}.

The dissolution rate for each of the following minerals: pyrite, sphalerite, chalcopyrite and galena, was calculated using Equation 3.3 and the concentrations of Fe, Zn, Cu and Pb respectively. Results can be observed in Figure 4.

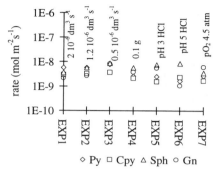

◇ Py □ Cpy △ Sph ○ Gn

Figure 4. Dissolution rates for pyrite, chalcopyrite, sphalerite and galena obtained in experiments made at pH 3, 21% O$_2$ and at 5.0×10^{-4} cm^3 s^{-1}, except when indicated. Estimated errors are of the order 20%.

A pyrite O$_2$ dissolution rate equation (Nicholson 1994) gives values of 2.3×10^{-10} mol m^{-2} s^{-1} for pH 3 and 21% O$_2$(g). The difference from the experimental value obtained here could be due to several factors related to the experimental conditions used. The sample used here was not pretreated, while samples reported by Nicholson were subjected to a complete cleaning process. Moreover, the experiments were not done in a saturated medium and as said before, the rate in those cases should be lower than in a saturated medium. No O$_2$(aq) dissolution rates were found in the literature for other sulphides.

From results obtained, no clear dependence of the sludge dissolution on pH and O$_2$(g) was observed. Nicholson et al. (1988) showed that significant variations in the pyrite dissolution rate at neutral pH can only be observed when O$_2$ partial pressures are lower than 0.05 atm. Moreover, according to this author, the oxidation rate depends to a slight extent on pH ([H$^+$]$^{-0.11}$).

In order to describe the trace metal behaviour a new variable called *FIAP*, Fraction of Inventory in Aqueous Phase (Grambow et al. 1997), was defined. $FIAP_i$ is defined as the fraction of element i that has been dissolved with respect to the amount of element i present initially in the solid phase:

$$FIAP_i = \frac{c_i V M_i}{m f_i} \quad (4.1)$$

where c_i is the concentration of element i in solution (mol m^{-3}), V, the unit volume (m^3), m, the mass of solid used in the experiment (g), f_i, the fraction of element i in the solid and M_i, the molecular weight of element i (g mol^{-1}). *FIAP* is independent on the mass (and area) used in the experiment.

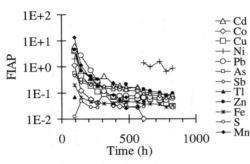

Figure 5. Evolution of *FIAP* of elements over time. Values obtained from experiment described in Figure 3.

Although some trace elements were not detected in all experiments, As, Co, Cd, Sb, and Mn showed *FIAP* values similar to the major sulphide-forming metals, suggesting congruent dissolution of the sulphides as can be seen in Figure 5. Tl, however, showed lower *FIAP* values, although the results for it were not conclusive. Due to similar dissolution rates for all the sulphides, no relation of each trace to a particular sulphide phase could be concluded. Although arsenopyrite was described in minor amounts in the Aznalcóllar deposit (Sierra 1984), the amount of As in solution was consistent with a content of about 0.5 wt % As in pyrite, analysed by Electron Microprobe. Ni had *FIAP* values significantly different from the rest of metals, suggesting that it is related to silicates and oxides rather than to sulphides.

5 CONCLUSIONS

Sulphide sludge dissolution rates in oxygenated aqueous solution were determined. Pyrite, chalcopyrite, sphalerite and galena showed similar rates of (1.6 to 8.0)×10^{-9} mol m^{-2} s^{-1}, (1.6 to 3.6)× 10^{-9} mol m^{-2} s^{-1}, (3.5 to 8.5)×10^{-9} mol m^{-2} s^{-1} and (1.0 to 8.0)×10^{-9} mol m^{-2} s^{-1}, respectively. This fact allowed us to treat the sludge as one solid without making distinctions between minerals.

With the experiments carried out to present, no clear dependence of the sludge dissolution on pH and O$_2$ partial pressure was found. More experiments regarding this aspect are in progress.

Most of the trace metals analysed appeared to dissolve congruently, although their relationship to particular sulphide phases was not established. As

seemed to come from pyrite dissolution. On the other hand, Ni did not appear to be related to the dissolution of sulphides.

ACKNOWLEDGEMENTS

This paper was supported by the Spanish Government CICYT AMB96-1101-C02 and FD97-0765 contracts and by the Instituto Tecnológico y Geominero de España.

REFERENCES

Gibbs, C.R. 1976. Characterization and application of ferrozine iron reagent as ferrous iron indicator. *Anal.Chem.* 48:1197-1200.

Grambow, B., Loida, A., Dressier, P., Geckeis, H., Gago, J., Casas, I., de Pablo, J., Giménez, J. & Torrero, M.E. 1997. Chemical reaction of fabricated and high burn-up spent UO$_2$ fuel with saline brines. Final report. *Nuclear Science and Technology.EUR 17111.*

Lasaga, A., Soler, J.M., Ganor, J., Burch, T.E. & Nagy, K.L. 1994. Chemical weathering rate laws and global geochemical cycles. *Geochim. Cosmochim. Acta.* 58:2361-2386.

Nicholson, R.V. 1994. Iron-sulfide oxidation mechanisms: laboratory studies. *Mineral. Assoc. Canada.* 22:163-183.

Nicholson, R.V., Gillham, R.W. & Reardon, E.J. 1988. Pyrite oxidation in carbonated-buffered solution: 1. Experimental kinetics. *Geochim. Cosmochim. Acta.* 52:1077-1085.

Querol, X., Alastuey, A., Plana, F. & Ayora, C. 1998. *Caracterización físico-química y mineralógica de muestras ambientales de Aznalcóllar-Guadiamar.* Informe técnico del Instituto de Ciencias de la Tierra "Jaume Almera"-CSIC.(www.ija.csic.es).

Rimstidt, J.D., Chermak, J.A. & Gagen, P.M. 1994. Rates of reaction of galena, sphalerite, chalcopyrite and arsenopyrite with Fe(III) in acidic solutions. In C.N. Alpers & D.W. Blowes, eds. *Environmental Geochemistry of Sulfide Oxidation.* Am. Chem. Soc. Symp. Series 550: 2-13.

Sierra, J. 1984. Geología, mineralogía y metalogenia del yacimiento de Aznalcóllar (Segunda parte: Mineralogía y Sucesión Mineral). *Boletín Geológico y Minero.*XCV:553-568.

Geochemistry of the Earth's Surface, Ármannsson (ed.) © 1999 Balkema, Rotterdam, ISBN 90 5809 073 6

High arsenic-containing soils in SW England and human exposure assessment

M.E.Farago & P.Kavanagh

T.H.Huxley School of Environment, Earth Science and Engineering, Imperial College of Science, Technology and Medicine, Royal School of Mines, London, UK

ABSTRACT: Soils in SW England are heavily contaminated by arsenic, which has been dispersed by past mining and smelting activities. Contaminated soils are used for agricultural purposes, and in some areas houses are close to contaminated wastes. From an assessment of arsenic in hair and urine we conclude that such residents may have an arsenic intake exceeding WHO recommendations.

1 INTRODUCTION

The sources of contamination of waters, soils and dusts leading to arsenic into the food chain, may be derived from:

    (i) naturally occurring As-rich minerals or rocks, e.g. pyrite;

    (ii) mining and smelting of sulphide minerals and the resulting wastes;

    (iii) the burning of coal with a high As content

This investigation has been carried out in part of the UK highly contaminated with waste from the mining and smelting processes.

The south-western peninsula of England (the South West) consists of the counties of Cornwall to the west, and Devon. The River Tamar forms the boundary between the two counties. This area is extensively contaminated with heavy metals arising from centuries of mining activity in the region. From about 1860 to 1900, this region was the world's major producer of arsenic. The principal minerals of economic importance were arsenopyrite ($FeAsS$), chalcopyrite ($CuFeS_2$) and galena (PbS). Other local ores were casterite (SnO_2) and stannite ($CuSnS_4$). Mining and smelting activities have left a legacy of contaminated land, with As- and Cu-rich mine tailings and other wastes. Further extensive areas of land were contaminated with fallout from the smelting process and over the area some 700 km^2 of land are affected (Abrahams and Thornton 1987). Most of the contaminated area is agricultural with villages and small towns; urban development has sometimes taken place on contaminated land. Sources of arsenic in the

region and some aspects of the exposure of local populations have been discussed (Thornton 1994; Mitchell & Barr 1995; Farago *et al.* 1997; Kavanagh *et al.* 1998; Kavanagh 1998).

The area under investigation is in the Tamar Valley. On the east (Devon side lies the abandoned Devon Great Consols Mine, where mining and smelting of the ores was carried out. On the other (Cornwall) side of the river is the village of Gunnislake, which is also in close proximity to abandoned waste sites. There is small number of houses on the Devon Great Consols Mine, close to the abandoned waste tips. These were investigated, together houses from Gunnislake and Cargreen villages. The latter was taken as a control area, being further down the river and away from past mining activities. Arsenic exposure was assessed from the measurement of arsenic concentrations in hair and urine from populations in Gunnislake, Devon Great Consols and Cargreen.

2 RESULTS AND DISCUSSION

2.1 Arsenic in soils and dusts

Soil samples (0-15 cm) were taken on a grid at 500 m intervals and were collected from sites as near as possible to the junction of the eastings and northings of the O.S. map for the Devon Great Consols, Gunnislake and Cargreen. Collection of samples on this regular interval grid was conducted in order to produce a sub-regional geochemical database for As and its associated elements for this area. Data for As are shown in Table 1.

Table 1. Summary statistics of concentrations (μg/g) of As in top soils from the Tamar Valley area

	Range	GM	Mean	N
Devon Great Consols (DGC)	173-52600	2557	9894	21
Study area (excluding GDC)	50-26485	282	819	118

A second survey was carried out to measure concentrations in garden soils and house dusts from residences in the area. The results are shown in Table 2. It can be seen that the concentrations of arsenic are much higher in the areas near the mine wastes than in the control area of Cargreen.

"Normal" values for England and Wales are expected to be in the range 2-55 μg/g. The UK ICRCL (Interdepartmental Committee on the redevelopment of Contaminated Land 1987) suggest the following "trigger" values for arsenic, above which further investigation should be carried out, for domestic gardens and allotments 10 μg/g, and for playing fields and open spaces 40 μg/g. The Netherlands' guideline threshold value for arsenic is 30 μg/g and the action threshold value is 50 μg/g. It can be seen that these values are of little use in a consideration of soils near the Tamar Valley.

2.2 Arsenic in hair samples

Results for two separate hair assays are shown in Tables 3 and 4. The concentrations of total As in scalp hair have been shown in this present study to increase with increasing soil As concentrations. These results follow a trend seen in the urinary As results. However the data can be used to observe the relative increase from the As concentration in the unwashed control hair and that of Gunnislake and also the relative decrease arising from the washing of the hair samples.

The results of As determinations in unwashed hair reflect the concentration of the As present externally on the hair and also shafts As present in the hair structure attached to sulphurhydryl groups. The washing of hair should remove the external As

contamination and one would expect a lower concentration value for the hair. The data suggest increased arsenic exposure in the Gunnislake population compared with the control group.

2.3 Arsenic in urine

Arsenic is methylated in the human liver as a detoxifying mechanism (Vahter 1994). In the process inorganic arsenic entering the body is reduced to As(III), the substrate for methylating enzymes which produce the less toxic metabolites (Buchet & Lauwerys 1994). The methylated metabolites (MMAA, monomethylarsonic acid and DMAA, dimethylarsinic acid) are considered to be less reactive than inorganic arsenic with tissue components and are excreted in the urine more readily than inorganic arsenic (Buchet et al. 1981; Vahter et al. 1984). It is the As(III) species that is most reactive with tissue components (Vahter & Marafante 1983) and consequently, factors that effect the methylation process may thus affect arsenic toxicity.

First void urine samples were collected from residents from the study area. Data in Tables 5 and 6. show the concentrations of "total" urinary arsenic, As_T which is the sum of inorganic arsenic (As(III) + As(V)) and the methylated species DMAA and MMAA; any arsenobetaine, from the consumption of seafoods was excluded. Thus this "total" arsenic in the urine represents the uptake from all sources except seafood. Seafood normally represents about 70-75% of the arsenic intake, but the arsenic from this source is in the form of the inert organoarsenic, arsenobetaine, which is thought to pass through the body without being metabolised. Thus urine analysis, when the chemical species of the arsenic is also determined, is a more reliable estimate of environmental exposure than the determination of the total of all arsenic species in urine, since this may be dominated by seafood intake. The results are presented per g creatinine, thus correction for dilution effects in the raw urine.

Table 2. As (μg/g) in garden soils and housedusts in the Tamar Valley

		Soils			Dusts	
Site	N	GM[*]	Range	N	Mean	Range
Gunnislake	71	365	120-1695	9	217	33-1160[**]
DGC	15[***]	4499	345-52600	13	1167	24-3740
Cargreen	18	37	16-198	4	49	20-114

[*]Geometric mean; [**]Outlying value of 16700 μg/g ignored; [***]Some samples contain mine wastes

Table 3. Concentrations of As in unwashed hair samples

µg/g	Cargreen	Cargreen	D G C
GM	0.83	1.32	2.14
Median	0.62	1.37	1.92
Min	0.08	0.28	0.93
Max	6.96	8.46	4.18
N	5	15	6

Table 4. Concentrations of As in washed hair samples

µg/g	Cargreen	Cargreen
M	0.24	1.71
Median	0.19	0.23
N	5	13

Table 5. Total As (As_i + DMAA + MMAA) concentrations in urine in children

Total As µg/g creatinine	Cargreen Children N= 4	Gunnislake + DGC Children N= 8
Mean	5.32	16.36
SD	0.07	17.71
Median	5.32	11.41
Min	5.27	2.65
Max	5.37	58.95

Table 6. Total As (As_i + DMAA + MMAA) concentrations in urine in adults

Total As µg/g creatinine	Cargreen Adults N= 3	Gunnislake + DGC Adults N= 16
Mean	3.28	12.4
SD	1.01	12.50
Median	2.95	8.28
Min	2.48	5.5
Max	4.41	47.3

Although the numbers are relatively small and so statistical evaluation of these differences is precluded, the data do show trends in the exposure levels.

Table 7. Arsenic concentrations in drinking water

As µg/l	1st drawn	2nd drawn
Mean	0.5	0.5
Minimum	0.2	0.2
Maximum	1.9	1.0
Median	0.4	0.5
N	12	12

The data are separated into two groups: adults and children for each area. Differences in "total" As concentrations (As_i + DMAA + MMAA) were observed. Totals are higher in the contaminated areas, than in Cargreen, and higher in children.

2.4 Exposure routes

Possible exposure routes are through ingestion of As in food, water and soil/dust; and by inhalation. A survey of drinking water supplied to homes of the participating residents was carried out. This included the first drawn water in the morning and a sample from later in the day.

It can be seen that none of samples reached the current health guideline limit of 50 µg/L or the recently set provisional limit of 10 µg/L.

We have estimated that the daily ingested intake of inorganic arsenic by children from soils and dusts, from hand-to-mouth activities is 50 mg of soil and 50 mg of dust, leading to a mean intake for the contaminated area of Gunnislake of 29 µg/day from these sources alone. A further intake of 11.16 µg/day of inorganic arsenic is estimated from diet (the assumptions underlying these figures are set out in Farago et al. 1997). These lead to an estimated daily intake of around 39 11.16 µg/day. The mean "total arsenic" in the urine (i.e. As_i + DMAA + MMAA derived from the intake of inorganic arsenic) for the 8 children from Gunnislake and Devon Great Consols was 16 µg/g creatinine. Assuming that the average creatinine output for the children is 1 g, the mean daily output is 16 µg, which represents approximately 50% of the daily intake, and leads to a daily intake of 32 µg/day, in reasonable agreement with the previous estimate. Assuming a body weight (bw) of 13 kg , then the mean daily intake is 2.5 µg/day/kg bw, in excess of the WHO recommendation of 2.0 µg/day/kg bw. For the maximum value of (Table 6) of 59 µg/g creatinine, similar calculations lead to a value of 9.5 µg/day/kg bw far in excess of the WHO recommended value. The comparable exposure value for children form Cargreen is 0.85 µg/day/kg bw, below the WHO recommended limit.

The concentrations of urinary arsenic and its organo-metabolites in the populations were assessed in a pilot study and the results are shown in Table 8. From that we conclude that populations in both Gunnislake and Devon Great Consols are chronically exposed to inorganic arsenic, since inorganic arsenic appears in the urine. The data also indicate that chronic exposure results from soil and dust ingestion of arsenic in a partially available form, since dust and soil appear to be the only significant exposure route·

Considering the individuals in our control population, only one had detectable inorganic arsenic in the urine and none had MMAA in the urine. These results are very similar to those reported by Johnson & Farmer (1989) for an unexposed population of 40 adults from Glasgow, where the geometric mean of As_T was 4.4 µg/g creatinine.

Table 6. Concentration ranges of arsenic species detected in urine samples
From Kavanagh *et al.* 1998.

μg/g creatinine	Cargreen (n=7)	Gunnislake (n=17)	Devon G C (n=7)
$As_T(As_i + DMAA + MMAA)$			
range	2.5- 32.7 (2.5-5.3)*	2.7 - 58..9	5.1-17.6
geometric mean	5.4 (4.0)*	10.5	10.8
median	4.7 (4.5)*	9.2	10.0
Arsenite (As III)			
range	BDL**-0.6(BDL-0.6)*	BDL- 8.5	0.6-1.8
median	BDL (BDL)*	1.7	0.9
Number detected	1	14	7
Arsenate (As V)	BDL ** (BDL)*	BDL -2.95	BDL - 2.06
range		0.9	1.34
median	BDL (BDL)	13	6
Number detected			
DMAA			
range	2.5-32.7 (2.5-5.4)*	1.9 –54.3	3.3 - 15.5
median	4.7 (4.2)*	5.6	8.5
Number detected	7	17	7
MMAA			
range	BDL **	BDL - 3.8	BDL -0.9
median	BDL (BDL)*	0.3	0.7
Number detected		2	2

* indicates that statistics based on data with outlier omitted;
**where As in urine was below detection limit (BDL) of 0.5 μg l⁻¹, the value was taken as zero i.e. not detected, ND = 0

REFERENCES

Abrahams, P. and Thornton, I., 1987, Distribution and extent of land contaminated by arsenic and associated metals in mining regions of south west England. *Transactions of the Institute of Mining and Metallurgy (Sheet B: Applied Earth Science,* 6: B1-B8.

Buchet, J.P. and Lauwerys, R. 1994, Inorganic arsenic metabolism in humans. In: *Arsenic; Exposure and Health,* W.R. Chappell, C.O., Abernathy and C. Cothern (eds), , Northwood, Middlesex: Science Technology Letters 1994: 181-190.

Buchet, J.P., Lauwerys, R. and Roels, H., 1981, Urinary excretion of inorganic arsenic and its metabolites after repeated ingestion of sodium meta-arsenite by volunteers. *International Archives of Occupational Environmental Health* 48:111-118.

Farago, M.E., Thornton, I., Kavanagh, P., Elliott, P. and Leonardi, G., 1997, Health aspects of human exposure to high arsenic concentrations in soil in south-west England, In: *Arsenic; exposure and health effects,* C.O. Abernathy, W.R. R.L. Calderon, W.R Chappell, (eds), London: Chapman and Hall: 191-209.

Johnson, L.R. and Farmer, J.G., 1989, Urinary arsenic concentrations and speciation in Cornwall residents. *Environmental Geochemistry and Health* 11: 39-44.

Kavanagh, P., 1998, *Impacts of high arsenic concentrations in south west England.* Ph.D. thesis, Imperial College, University of London.

Kavanagh, P., Farago, M.E., Thornton, I., Goessler, W., Kuehnelt, D., Schlagenhaufen, C. and Irgolic, K.J., 1998, Urinary Arsenic Species in Devon and Cornwall Residents, UK. *The Analyst* 123(1): 27-30.

Mitchell, P. and Barr D., 1995, The nature and significance of public exposure to arsenic: a review of its relevance to South West England, . *Environmental Geochemistry and Health* 17(2): 57-82.

Thornton, I., 1994, Sources and pathways of arsenic in south-west England; health implications, in *Arsenic; exposure and health effects,* W.R. Chappell, C.O. Abernathy and C.R. Cothern (eds.), Northwood,:Science and Technology Letters: 61-70.

Vahter, M.E., 1994, Species differences in the metabolism of arsenic, in *Arsenic; Exposure and Health,* W.R., Chappell, C.O., Abernathy, and C.R. Cothern (eds), Northwood: Science and Technology Letters: 171-180

Vahter, M.E., Marafante, E. and Dencker, L., 1983, Intracellular interaction and metabolic fate of arsenite and arsenate in mice and rabbits. *Chemical Biological Interactions* 47: 29-44.

Vahter, M.E., Marafante, E. and Dencker, L., 1984, Tissue distribution and retention of [74]As-dimethylarsinic acid in mice and rats. *Archives of Environmental Contamination and Toxicology* 13: 259-264.

Geochemistry of the Earth's Surface, Ármannsson (ed.) © 1999 Balkema, Rotterdam, ISBN 90 5809 073 6

Tracing the impact of irrigation on water quality using chemistry and isotopes

J. Guttman & J. Kronfeld
Department of Geophysics and Planetary Sciences, Tel-Aviv University, Israel

ABSTRACT: The Pleistocene aquifer of the Western Galilee Coastal Plain has been exhibiting increased salinity over the past forty years. Isotopic and chemical analyses show that intrusion from the adjacent sea is not responsible: rather, the recycling of irrigation water that has been leaching agricultural lands which receive continuous deposition of sea spray and have been extensively artificially fertilized is causing a continuous rise in chloride and nitrate.

1 INTRODUCTION

Along the Western Galilee Coastal Plain groundwater is exploited for domestic and agricultural purposes, especially during the summer growing season. Rainfall is restricted solely to the winter season.

Over the past forty years many wells have exhibited increases in salinity concomitant with increasing nitrate values primarily in the wells that exploit the phreatic Pleistocene aquifer. The city of Naharriya's water well #6 can be taken as an example. In 1957 the water that was pumped contained 36 mg/l Cl^- and 14 mg/l NO_3, rising in 1973 to 53 mg/l Cl^- and 21 mg/l NO_3. By 1993 the chloride had risen to 290 mg/l and the nitrate to 25 mg/l. In wells situated within agricultural communities the rise has often been greater still, with chloride levels rising from approximately 100 mg/l forty years ago, to 300-1500 mg/l Cl^- today. In these cases likewise, the nitrate increase has been equally dramatic, reaching concentrations that frequently reach or exceed the Israeli permissible level of 70 mg/l NO_3. The purpose of the study is to identify the source/process of the water quality degradation. Thus, a suite of isotopic techniques (hydrogen, carbon, oxygen, nitrogen, strontium and uranium) was employed to supplement the standard geochemical study.

2 GEOGRAPHY AND HYDROLOGY

The study area is situated in the westernmost region of the Upper Galilee. It extends from the cliffs of Rosh Hanikra, near the Lebanese border in the north, southwards approximately 15 km to Kibbutz Shomerat. In the west the border is determined by the seashore. The foothills of the Galilee are the borders to the east. These hill rise up to an elevation of 150 meters. They are composed of chalks, limestone and dolomites ranging from Cenomanian to Eocene in age. Intensive agriculture is carried out over a stretch of land lying between the hills and the sea, which is covered by an alluvial soil and a fossil dune ridge of late Pleistocene age. At its widest point this strip is 6-7 km in the south, while narrowing to only 3 km at Betzet in the north. Two aquifers of importance are exploited in the region. The Judea Group aquifer of Cenomanian-Turonian age is comprised of marine carbonates. Lying over the Judea Group, and nestled between the carbonate hills and the sea, is the sandy Pleistocene aquifer. The thinner phreatic Pleistocene aquifer is only 15-40 m in thickness, 20 m on average. Through out much of the area there is hydrologic separation between the two aquifers. However, in the region to the east of the city of Naharriya, the intervening aquiclude is absent (the Ben-Ami "window"). Thus hydrologic interconnection between the two aquifers can take place in this region.

The water exploited from the Judea Group aquifer is of higher quality. It is of calcium carbonate composition. The composition of the ground water in the Pleistocene aquifer has changed over time from a calcium carbonate water type, when exploitation of the aquifer first began, to a primarily sodium chloride type. The Pleistocene aquifer is today used primarily for irrigating the fields. As the agricultural demand for irrigation water has increased, water has been piped into the area. The imported irrigation water comes from large springs, such as the Kabri and Keren springs wells located along the foothills to the south , which issue from the carbonate aquifer. The irrigation water, which is spread over the fields, percolates rapidly into the underlying sandy aquifer. It is subsequently recycled by the Pleistocene aquifer exploitation wells. Thus, while the abstraction of the water from the Pleistocene aquifer has been strong, there has been no corresponding fall in the water levels over the years. Both the Judea Group and the Pleistocene aquifers drain to the sea.

3 METHODS AND RESULTS

Representative wells from both the Judea Group aquifer and from the Pleistocene aquifer were collected in the field from actively pumping wells. The isotopes of hydrogen, oxygen, carbon, nitrogen, strontium, and uranium were analyzed in the laboratory. Chemical analyses were mostly taken from the archives of the Hydrological Services in Jerusalem. The stable isotopes in the water from the Judea Group aquifer are similar to that of the isotopic composition of the local precipitation (Gat and Dansgaard, 1972). The isotopes of hydrogen and oxygen in the water of the Pleistocene aquifer are more variable and generally enriched (-18 to -28 ‰ (SMOW) for deuterium and -4.3 to -6.0 ‰ (SMOW) for ^{18}O) compared to the water of the Judea Group (approximately -27 to –28 ‰ or deuterium and -6.0 ‰ for ^{18}O).

This indicates that the Pleistocene aquifer water has undergone evaporation. The Pleistocene aquifer contains younger water as evidenced by the presence of tritium which is lacking in the Judea Group water. Likewise, the Pleistocene aquifer water has higher amounts of radiocarbon (above 70 pMC (percent modern carbon) compared to a ^{14}C activity in the range of 52-58 pMC for the Judea Group aquifer water). The nitrate concentrations in the Judea Group are the lowest. Moreover, it is seen that there is a inverse trend which correlates the lowest ^{15}N values (+4.5 to +5.7 ‰ (air)) to the highest nitrate concentrations. The ^{15}N of the fertilizer used in the area is approximately +1 ‰. The ^{15}N background values appear to be in the range of +6 to + 8 ‰. Likewise the lowest uranium concentrations (in the Judea Group) are associated with the highest $^{234}U/^{238}U$ activity ratios, while the highest uranium concentrations are associated with low $^{234}U/^{238}U$ activity ratios. The phosphate fertilizer that is used in the area is rich in uranium (140 ppm U) with a low activity ratio of 1.0 (secular equilibrium). The $^{87}Sr/^{86}Sr$ ratios of the Pleistocene aquifer water fall between the modern sea water value and the imported carbonate aquifer water values.

4 DISCUSSION

The process affecting the degradation of the Pleistocene aquifer is not due to the intrusion of the immediately adjacent sea. Rather, it appears that the rapid recycling of the irrigation water is dissolving salts from the fertilization of the fields and from the continuously depositing sea-spray. The high to excessive amounts of fertilizer use in particular contribute large amounts of nitrate. The quality of the Pleistocene aquifer will continue to degrade as long as there is the recycling of the irrigation water that dissolves sea spray- derived salts and large amounts of fertilizer derived nitrates from the top of the soil profile.

REFERENCE

Gat, J.R. & W. Dansgaard, 1972 Stable isotopes survey of the fresh water occurrences in Israel and the Jordan Rift Valley. *J. Hydrol.* 16: 177-211.

Geochemistry of the Earth's Surface, Ármannsson (ed.) © *1999 Balkema, Rotterdam, ISBN 90 5809 073 6*

Gadolinium in aquatic systems as indicator for sewage water contamination

A. Knappe, Chr. Sommer-von Jarmersted & A. Pekdeger
FR Rohstoff- und Umweltgeologie, Freie Universität Berlin, Germany

M. Bau & P. Dulski
GeoForschungsZentrum Potsdam, Germany

ABSTRACT: We used the rare earth element (REE) Gadolinium to trace the influence of treated waste water on surface water quality and bank infiltrates. Positive Gd anomalies in shale normalized REE patterns (REE_{SN}) of rivers in Berlin area were discovered in previous studies and traced back to sewage plants. The source of Gd is gadopentetic acid, $Gd(DTPA)^{2-}$, which is applied as a contrast agent in magnetic resonance imaging. Systematic sampling shows an increase of anthropogenic Gd along the river systems in Berlin. Samples of rivers upstream and uninfluenced groundwater do not show Gd anomalies. In the municipal area the Gd anomalies of rivers increase significantly behind outlets of sewage plants. Gd anomalies are also detected in drinking water wells being influenced by bank infiltrates.

1 INTRODUCTION

Public water supply of Berlin municipal area originates from groundwater and bank infiltration in more or less similar proportions. Therefore river water quantity and quality are of great importance for Berlin. The production wells are generally situated close to the surface waters which are more or less influenced by clear water effluents from sewage plants. Due to groundwater exploitation the groundwater level decreases below surface water level and leads to bank infiltration. During low water times the proportion of clear water in surface water may be up to 65 % (Sommer-von Jarmersted 1998).

This study is focused on positive Gd anomalies which seem to be an excellent tracer to determine sewage load in surface and groundwater. Results from Dulski & Bau (1996) were taken up to investigate if positive Gd anomalies in surface and groundwater can be used as a tracer to quantify the proportions of bank storage and to qualify water conditions. The positive Gd anomalies are of anthropogenic origin and result from the application of gadopentetic acid $Gd(DTPA)^{2-}$, used in magnetic resonance imaging (MRI) as a paramagnetic contrast agent.

The organic aqueous Gd(III) complex is very stable and is excreted very fast by the human body. Therefore it becomes part of the waste water. However sewage plants are not able to eliminate the complex and so it will be reintroduced into surface water with clear water effluents. The measured concentrations of Gd in surface water are not known to

be dangerous for health. Gutierrez et al. (1997) used the Gd(DTPA) complex in tracer tests and showed that it can be used as ideal tracer.

To get the anomaly factors it was necessary to analyse the whole group of rare earth elements (REE). By interpolation of Gd^*_{SN} with its neighbour elements Sm and Tb the size of anomaly can be quantified by:

$$(Gd / Gd^*)_{SN} = Gd_{SN} / (0,33\ Sm_{SN} + 0,67\ Tb_{SN})$$

Normalisation was carried out with Post Archaean American Shale (PAAS). Only $(Gd / Gd^*)_{SN}$ ratios above 2 will be interpreted as significant Gd_{SN} anomalies.

Examples of the geogenic patterns of REE distribution in our investigation area are given in Figure 1.

2 SAMPLING, PRECONCENTRATION AND ANALYSES

Besides water production wells (3 samples), observation wells (12 samples) and clear water effluents of Berlin sewage plants (16 samples) particularly the surface waters (34 samples) were examined systematically to get an idea of distribution and amount of anthropogenic Gd (Fig.2). Samples of Upper Havel, Upper Spree and observation wells out of bank storage areas gave us the geogenic background.

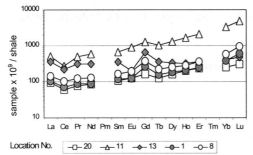

Figure 1. Shale normalised REE patterns of the rivers Upper Spree (8), Dahme (1), Upper Havel (13) and observation wells from uninfluenced groundwater (11,20). These samples represent the geogenic patterns of REE distribution.

Sample volumes of 1-2 l were immediately filtered first through 0,65μm and secondly through 0,2 μm. They were acidified with 0,6 M HCl until a final pH between 1,9 and 2,1 was reached. A Tm-spike was added to determine the recovery rate of the rare earth elements. With a modified method based on Shabani et al. (1992) the REE were preconcentrated by passing the water samples through C_{18} exchange columns (Millipore corp. USA). The loaded cartridges were purged with 15 ml 0,1 M of HCl to remove remaining matrix elements and then eluated

with 6 M of HCl. This solution was evaporated to incipient dryness, again taken up with 1 ml of 5 M HNO_3 an filled up to a volume of 10 ml. The elements Ru and Re were added to correct the instrumental drift. Measuring the REE was carried out with ICP-MS as described by Dulski (1994).

Additionally filtration experiments were carried out to distinguish between (1) truly dissolved or colloidal REE and (2) dissolved plus particle-bound (acid-soluble) REE, using samples from Upper Spree and Lower Teltow Channel. One part of the samples was treated as described above to determine the sum of truly dissolved and colloidal REE. The other part was first acidified and spiked and then filtered after a period of 24 h.

Boron as further indicator for sewage load was additionally analysed using a photometric method (Spectroquant, Merck).

3 RESULTS

3.1 *Sewage plants*

To get a general idea of the average input of anthropogenic Gd, samples were taken at the clear water effluent of all Berlin sewage plants (Tab. 1, Fig. 2).

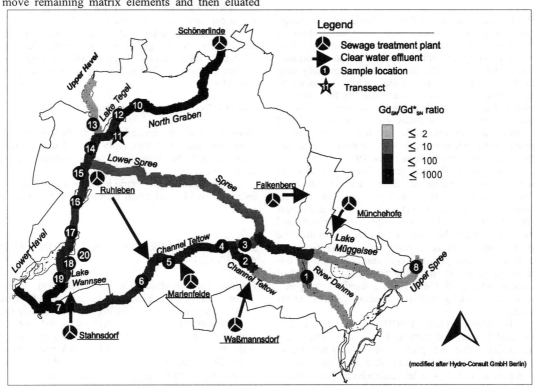

Figure 2. Sketch of Berlin, the river systems and locations of some sampling points. The "contamination map" shows increasing $(Gd/Gd^*)_{SN}$ ratios in the rivers during their passage through the municipal.

Table 1. Daily capacities and measured outputs of anthropogenic (1) Gd of Berlin sewage treatment plants. (The value of Ruhleben is the average of a sampling over a period of 10 days.

Sewage plant	m³/day	Gd[1] g/day*	Gd[1] g/Year*
Ruhleben	233.000	115	41932
Waßmanns.	95.000	8	2857
Schönerlinde	90.000	32	11760
Falkenberg	87.000	9	3399
Marienfelde	62.000	2	642
Münchehofe	57.000	5	1733
Stahnsdorf	40.000	5	1786
total	664.000	176	64109

*estimated amounts

In the largest plant 'Ruhleben` with an average capacity of 233.000 m³/day samples were taken over a period of 10 days. The $(Gd/Gd^*)_{SN}$ ratio varies from 207 to 2014 (corresponding Gd concentrations from 116 to 1160 ng/l). To get the anthropogenic component we subtracted the interpolated Gd* from measured Gd. If we take the ten days average concentration of Gd and capacity m³/day and assume it to be the average for every day, a daily input of 115 g anthropogenic Gd can be calculated for this sewage plant. The yearly input of all sewage plants is estimated to be 64 kg (Tab. 1).

3.2 Surface (river) waters

The rivers Spree, Dahme and Havel do not show Gd_{SN} anomalies before they reach the city zone (Fig. 2). Their $(Gd/Gd^*)_{SN}$ ratios vary from 1,3 to 1,8 representing the 'natural' input for Berlin in this study. The influence of the municipal area leads to increasing REE concentrations in general and to $(Gd/Gd^*)_{SN}$ ratios especially. Particulary after locations where clear water from sewage plants reaches the rivers, concentrations augmented dramatically (Fig. 2, 3 and 4). Along the Teltow Channel there are several clear water effluents. The effects of these inputs are very significant. The $(Gd/Gd^*)_{SN}$ ratio increases here up to three orders of magnitude. A comparison with Boron as a further indicator for sewage influence shows the same increasing pattern like Gd but to a less degree (Fig. 4).

3.3 Observation- and pumping wells

A transect at Lake Tegel built by 11 observation wells of variable depth and a production well shows different Gd_{SN} anomalies from shore to land side (Fig. 5). All the observation wells and the production well are affected by bank filtration processes.

They show $(Gd/Gd^*)_{SN}$ ratios > 2. Only one observation well behind the production well is not influenced by the bank infiltration and shows the geogenic background. However there is no clear trend of concentration distribution from shore to produc-

tion well. The interpretation of these results is - due to the complicity of interacting factors - still difficult. Presumably responsible for this are: (1) seasonal variations of input into surface water, (2) the heterogeneity of the aquifer and (3) varying groundwater flow velocities.

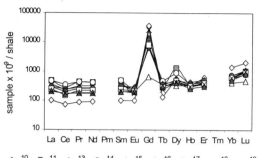

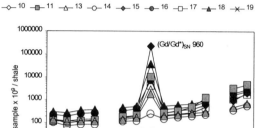

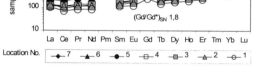

Figure 3. Increasing $(Gd/Gd^*)_{SN}$ ratios can be seen after every sewage plant outlet in the different river systems of (a) Lake Tegel-Havel, (b) Teltow Channel. Gradually increasing ratios up to three orders of magnitude reflect the strong influence of sewage plants on Teltow Channel.

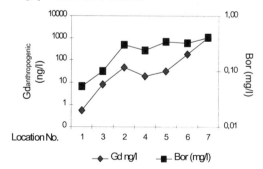

Figure 4. Comparison of Boron and anthropogenic Gd concentrations as indicators for sewage influence. They show the same increasing pattern but Gd is a much more sensitive indicator (note the different scales).

3.4 Filtration experiments

Two different samples (influenced and uninfluenced by sewage water) were compared regarding their

189

REE patterns in 'non-filtrated' (A-sample acidified prior to filtration) and 'filtrated' (B-sample filtrated before acidification) samples (Fig. 6).

Contrary to other REE the absolute Gd concentrations are the same in filtrated and unfiltrated samples taken at the end of Teltow Channel. This means that anthropogenic Gd occurs in form of a very stable aqueous complex which is not considerably particle reactive and stays in solution. The REE pattern of this sample shows an immense Gd_{SN} anomaly (>900) in the filtered sample.

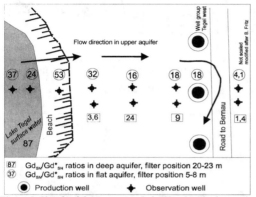

Figure 5. Sketch of the transect at Lake Tegel with observation wells of different depths and distances from the shore and a production well. Different Gd_{SN} anomalies show the influence of bank infiltration.

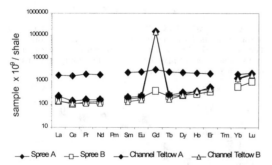

Figure 6. REE patterns of surface water influenced and uninfluenced by sewage water.(A-sample acidified prior to filtration; B-sample filtrated before acidification)

The REE patterns of the differently treated sample of the Upper Spree (not influenced by sewage water) do not show distinct Gd_{SN} anomalies. Due to much higher concentrations of REE, including Gd, in suspended material, total Gd exceeds the dissolved Gd in concentration.

4 DISCUSSION AND CONCLUSION

In urban areas the REE Gadolinium is a very useful tool for determining the sewage water influence on surface and groundwater. Gd as a sewage component occurs in form of a very stable aqueous complex.

A detailed sampling campaign gave a first overview of the input, distribution and contamination of Berlin waters by anthropogenic Gadolinium. The investigation proved that under influent conditions Gd_{SN} anomalies will reach groundwater and production wells. Presumably the Gd complex passes from surface water to production wells without reacting with the underground.

The proportion of bank infiltration may be determined by using Gd. However this quantification requires long time series of Gd measurements. They will give us a better understanding of the time dependence of anthropogenic Gd concentrations in bank infiltrates. Nevertheless detailed knowledge about the aquifer conditions and the hydrogeological situation must be taken into account.

The next objective is to set up further column- and batch-experiments in order to get a better understanding of the mobility of the Gd complex in the aquifer.

5 REFERENCES

Bau, M. & Dulski, P. 1996. Anthropogenic origin of positive gadolinium anomalies in river waters. *Earth and Planet. Sci. Lett.* 143: 245-255.

Dulski, P. 1994. Interferences of oxode, hydroxide and chloride analyte species in the determination of rare earth elements in geological samples by inductively coupled plasma-mass spectrometry. *Fresenius´Journal of Analytical Chemistry* 350: 194-203.

Gutiérrez, M., Guimerà, J., Yllera de Llano, A., Hernández Benitez, A., Humm, J. & Saltink, M. 1997. Tracer test at El Berrocal site. *Journal of Contaminant Hydrology* 26: 179-188.

Shabani, M. B., Akagi, T. & Masuda, A. 1992. Preconcentration of trace rare-earth elements in seawater by complexation with bis (2-ethylhexyl) hydrogen phosphate and 2-ethylhexyl dihydrogen phosphate adsorbed on a C18 cartridge and determination by inductively coupled plasma mass spectrometry. *Anal. Chem.* 64: 737-743.

Sommer-von Jarmersted, C. 1998. Erkundung des nutzbaren Süßwasserspeichers für Berlin – Reichen die Grundwasservorräte zur Trinkwasserversorgung der Stadt? Die Verbindung zu den Oberflächenwässern und die Höhe des Uferfiltrates. In Senatsverwaltung für Stadtentwicklung, Umweltschutz und Technologie (ed.), *Zukunft Wasser* 72-76. Berlin

Geochemistry of the Earth's Surface, Ármannsson (ed.)© 1999 Balkema, Rotterdam, ISBN 90 5809 073 6

Technogenic heavy metals and the soil regional background

S. I. Kovalev, I. N. Malikova, V. M. Tsibulchik, B. L. Scherbov & V. D. Strakhovenko
United Institute of Geology, Geophysics and Mineralogy, Novosibirsk, Russia

ABSTRACT: Data on Cd contents in soils of southern Siberian regions characterized by various levels of technogenic influence are reported. Maps of Cd distribution in sod and humus horizons of the Altay region are presented. An estimation of the technogenic constituent of Cd regional geochemical background in the soils of the region is made using the comparison of the ratios of Cd content in sod to content in humus horizon.

1 INTRODUCTION

The global character of antropogenic pollution of the environment involves some difficulties in the determination of the natural concentration levels for many toxic elements in soils and other ecosystems. The amount of technogenic emission of some heavy metals, in particular Cd, Pb, Hg, considerably exceeds their natural levels (Nriagu 1979) resulting in the constant accumulation of pollutants in the environment. The most intensive heavy metal accumulation in soils takes place near technogenic emission sources. After a sufficiently long operating time such a source the polluted areas can become extended to dozens of kilometers away, acquiring regional character.

Atmospheric migration is the main way of pollutant dissipation in terrestrial ecosystems. In this case is the sod horizon that is primarily subjected to pollution. This horizon of virgin soils has some properties (structure, composition, sorption capacity, etc), which sharply differentiate it from the lower humus horizon and enable it to be considered as an independent object for ecogeochemical studies. Nevertheless the sod horizon has not received proper attention in environment studies up till now. A method of ecogeochemistry study of sod horizon as a separate source of the information about atmospheric fallout of heavy metals on the soil surface is proposed.

2 AREAS AND METHOD OF INVESTIGATION

Investigations cover southern Siberian territories: Altay, Novosibirsk and Kemerovo regions,

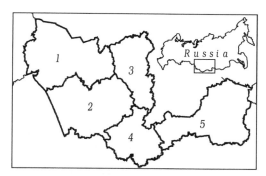

Figure 1. Map of Russia and location of the study regions. 1 –Novosibirsk region, 2 – Altay region, 3 - Kemerovo region, 4 – Republic of Altay, 5 – Republic of Tuva.

Republics of Altay and Tuva. (Figure 1) The virgin soil samples were taken by manual digging from 40 cm depth.

Sampling was conducted according to soil genetic horizons. Toxic elements were analyzed in Analytical Center of United Institute of Geology, Geophysics and Mineralogy of Siberian Branch of Russian Academy of Science. Cd, Pb, Cu, Zn, Mn, Cr, Ni, Co, Sb were determined using atomic-absorption analysis. Hg was determined by a "cold vapor" atomic-absorption method with gold amalgamation.

3 RESULTS

The territories investigated differ essentially from each other by the level of technogenic influence. The Republics of Altay and Tuva are the most ecologically uncontaminated of the investigated regions.

Soils from these regions are marked by the least average Cd contents (Table 1). In the investigated southern parts of Novosibirsk region large industrial enterprises are also absent. In the Altay region large

Table 1. Average Cd contents in soils of some southern Siberian regions (ppm)

Region	A_S	A
Novosibirsk region	0.18±0.02 (34)	0.11±0.02 (34)
Altay region	0.16±0.01 (443)	0.10±0.01 (556)
Kemerovo region:		
West	0.30±0.06 (28)	0.11±0.02 (28)
East	0.23±0.05 (26)	0.15±0.04 (26)
Republic of Altay	0.13±0.02 (104)	0.09±0.01 (100)
Republic of Tuva	0.13±0.05 (26)	0.16±0.05 (26)

Here and in Table 2. A_S – sod horizon, A – humus horizon. In parentheses - amount of samples

enterprises of metallurgical and mechanical engineering industries are situated in south-western part. The cities of Barnaul and Biysk are significant sources of technogenic contamination. The territory of the Altay region is polluted by the industrial enterprises, located in the north-eastern territory of Kazakhstan. Ore deposits, mainly, nonferrous are numerous in South-west Altay. Thus, the increased heavy metal contents in the soils of this territory can be enhanced not only big antropogenic but also natural factors. Revealing the antropogenic component of heavy metal contents in the ore deposits areas is more complicated than in the background territories. The most intensive technogenic influence on the environment is characteristic of the western part of Kemerovo region where one of the largest coal basins of Russia is located, and chemical and metallurgical industries are developed.

The heavy metal distribution in soils of the Altay region was studied most extensively. For this territory the proposed approach is illustrated by the example of Cd (Figure 2).

About 50 % of the area of the Altay region is characterized predominantly by Cd concentrations of less then 0.15 ppm in sod horizon (Zone I). Concentrations from 0.15 to 0.30 ppm prevail in the wide band, extending through the central part of the region in the north-eastern direction (Zone II). The direction of this band coincides with the prevailing wind direction in the region. The maximum Cd contents in the sod horizon (> 0.30 ppm) are founded in the south-western part of the Altay region (Zone III), where large technogenic (the cities of Rubtsovsk, Zmeinogorsk) and natural (nonferrous deposits) sources of pollution are located.

In the humus horizon a concentration less than 0.15 ppm predominates in the greater part of the Altay region. The zone with concentrations from 0.15 to 0.3 ppm is considerably smaller than in the sod horizon. Concentrations > 0.30 ppm are found in the south-western part of the region, as well as in the sod horizon.

4 DISCUSSION

The more extensive area of increased Cd contents in the sod horizon than in the humus one, as well as of its coincidence with the prevailing wind direction indicate antropogenic pollution in this territory by industrial enterprises in South-west Altay. The average Cd concentrations in the soil and humus horizons within of boundaries of each of the allocated zones increase distinctly from zone I to III (Table 2).

For approximate evaluation of the technogenic contribution to Cd content in the sod horizon the ratio of Cd concentration in sod to the concentration in the humus horizon (I_A) was used. The increase of the ratio testifies that atmospheric Cd input increases from zone I to III (Table 3). In zone I, with the lowest Cd concentrations in upper soil horizons, Cd accumulates in sod horizon, obviously, only by plant

Table 2. Average Cd contents in soils of Altay region (ppm).

Zone	A_S	A
I	0.10±0.01 (72)	0.10±0.01 (139)
II	0.16±0.01 (173)	0.15±0.01 (143)
III	0.31±0.03 (99)	0.21±0.05 (18)

uptake and global atmospheric deposition. In zones II and III, technogenic emission from the sources in the south-east part of the region is added to these factors. Taking the average I_A value in zone I to be 100%, the technogenic regional atmospheric Cd deposition constitutes an increase of 107% in the sod horizon of zone II by 21%, and zone III by In the soils of Republics of Altay and Tuva and in the eastern part of Kemerovo region I_A values are close to those in zone I of the Altay region. In the western part of the Kemerovo region this ratio is close to that observed in zone III. The maximum I_A value is characteristic of the soils near the zinc smelter factory in the Kemerovo region. The calculations concern a part of the technogenic Cd concentrated only in the sod horizon. The method proposed disregards the technogenic Cd migrated to the humus horizon. Taking into consideration that the sod horizon has a high sorption capacity and that Cd is slightly mobile at pH values that characterize the greater part of the investigated soils (6.5 - 7.5) (Kabata-Pendias & Pendias 1989) it is likely that the estimates obtained account for the major portion of the technogenic Cd input to the soils from regional sources.

Table 3. Average ratios of Cd content in sod horizon to content in humus horizon.

Region	I_A	N
Novosibirsk region	1.7±0.2	35
Altay region:		
I	1.4±0.1	61
II	1.7±0.1	158
III	2.9±0.2	89
Kemerovo region:		
West	2.7±0.5	28
East	1.4±0.2	30
KE-32*	5.9	1
Republic of Altay	1.2±0.2	40
Republic of Tuva	1.3±0.1	28

* - Near zinc smelter factory. N – amount of samples.

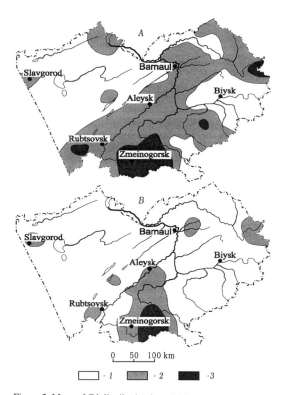

0   50   100 km

☐ - 1   ▨ - 2   ■ - 3

Figure 2. Maps of Cd distribution in sod (A) and humus (B) soil horizons of Altay region. Cd contents : *1* – less than 0.15 ppm, *2* – from 0.15 to 0.30 ppm, *3* – greater than 0.30 ppm.

5 CONCLUSION

1 The sod horizon of virgin soils is an important source of information about the atmospheric input of antropogenic pollutants. The comparison of the ratios of the heavy metal contents in the sod and humus horizons in the territories distinguished by antropogenic influence make possible to reveal the technogenic heavy metals constituent in soils on the background of their increased natural contents.

2 Among the investigated regions the least Cd pollution of soils is characteristic of the Republics of Altay and Tuva. The western part of the Kemerovo region and the south-west of the Altay region are distinguished by greatest technogenic influence on the soils.

3 The atmospheric Cd emission from industrial enterprises located on the south-west of the Altay region is an important factor of the modern lateral Cd distribution in the upper soil horizon.

ACKNOWLEDGEMENT

This work was founded by grant 97-05-65201 from Russian Foundation for Basic Research.

REFERENCES

Kabata-Pendias, A. & Pendias, H. 1989. *Microelements in the Soils and Plans.* Moscow: Mir.
Nriagu, J.O. 1979. Global Invertory of Natural and Antropogenic Emissions of Trace Metals to Atmosphere. *Nature* 279: 409-411

Geochemistry of the Earth's Surface, Ármannsson (ed.) © 1999 Balkema, Rotterdam, ISBN 90 5809 073 6

Metal redistribution within the sulfide tailings body

E. V. Lazareva, S. B. Bortnikova, U. P. Kolmogorov, A. G. Kireev & V. G. Tsimbalist
United Institute of Geology, Geophysics and Mineralogy SB RAS, Novosibirsk, Russia

ABSTRACT: This paper considers distribution and redistribution features of antimony, zinc, copper, arsenic, lead and cadmium within a technogenic body (tailings impoundment) containing cyanide solutions. Initial metal distribution within this technogenic body is determined by gravitation differentiation of their minerals. The metal redistribution is caused by dissolution of the initial minerals both at ore processing (technological stage) and during the storage period (inside the technogenic body). Extraction of metals from the solution occurs due to formation of separate mineral phases and their sorption by secondary ferriferous phases and organic substance. A detrital horizon of buried soil is the main barrier, which prevents dissemination of dissolved metals beyond the technogenic body. However, despite the ability of the organic substance to absorb metals effectively, some amount of it penetrates into the host soil.

1 INTRODUCTION

The man actively changes the earth landscape by mining industry, which exposes and lifts great rock masses and forms technogenic bodies such as rock piles and tailings impoundment. The formation rate of technogenic landscapes exceeds significantly the rate of geological processes. Simultaneously, the balance in the ore regions is being disturbed. Understanding of the processes occurring within the technogenic bodies is a key to the prediction of results of interaction between these bodies and environment. The paper presented demonstrates investigation results for one of numerous technogenic bodies situated in the Kemerovo region (Russia). We examined the redistribution of heavy metals (Sb, As, Zn, Cu, Pb and Cd) within a technogenic body during its formation and development of the oxidation zone.

Komsomolsk gold-extracting plant processes gold-quartz-arsenic-pyrite ores of the Komsomolsk deposit and the matte of settling melting of antimony concentrate from the Kadamzhay plant by means of cyanidation method. The amount of processed matte was not higher than 0.5% of the total mass, but due to the high contents of antimony (8.6 % Sb), tailings were significantly enriched by antimony (Bortnikova et al. 1998). Since 1964, industrial wastes were added to the tailings impoundment without preliminary neutralization. As a result the Komsomolsk technogenic body was formed. Now it is a natural ravine filled with wastes of cyanide conversion.

Quaternary sediments are the host rocks for this technogenic body and include soil with vegetation residua, loams and eluvial deposits. Native rocks are presented by diorites.

2 METHODS OF INVESTIGATION

Solid tailings were sampled along lateral and vertical profiles. Structural anomalies and accumulation points of secondary mineral phases were thoroughly studied. At the edges of the body, vertical cuts reached underlying soils. Averaged substance samples were collected layer by layer over the whole height of the stripped vertical column. Samples from the prospecting pit wall were taken into special containers for thin slides and polished sections. Analyses of the waste solid substance for the metal content were carried out by the X-ray-fluorescent method. Mineral composition of samples was determined by the X-ray analysis (executive T.N. Grigorieva), heavy concentrate analysis, and microscopic studying of thin and polished sections (using electronic microscope JSM-35 of "JEOL", operator S.V. Letov). Composition of separate minerals was determined using automatic roentgen-spectral microanalyzer "CAMEBAX MICRO" with four crystal-diffraction spectrometers (operator L. N. Pospelova).

During two field seasons (1997-98), we measured a total concentration of dissolved metal species. During plant operation (1997) and inactive periods (1998) the following components were studied: solutions of the "fresh" pulp, collection pond and pore waters. The samples were filtered through the 0.45 µm filters. Unstable parameters were determined directly under field conditions. The samples were conserved by chemically pure HNO_3. Metal contents were determined by the AAA methods (executive N. V. Androsova).

3 RESULTS AND DISCUSSION

3.1 Initial distribution of metals

Initial minerals of the Komsomolsk technogenic body are as follows: quartz, plagioclase, micas, pyrite, pyrrhotite, magnetite, ilmenite, arsenopyrite, sphalerite, and galena. Antimony is mainly presented by Sb-Fe-Si, Sb-S-0 phases and elementary antimony (98 mass % Sb). Granulometric and gravitation fractionation is clearly seen within the body, which is typical for the flooded impoundment (Robertson 1994). The eastern part, adjacent to the pulp line, is enriched by the sand fraction, whose content reaches up to 5 % (average content is 1 %). The western part is far from the pulp line and is predominantly composed of silt. Lenses of up to 10 mm in thickness are observed in the sand. They are formed by the accumulation of heavy metals in basins of ripples on the alluvium surface.

Initial metal distribution within the technogenic body is determined by granulometric differentiation of their minerals. Elevated contents of arsenic, zinc and copper are related to the lenses of gravitational enrichment. Silt fraction of tailings is enriched by antimony and lead due to light Sb-Fe-Si phases (Table 1). An increase of antimony content in gravitation lenses is caused by incompletely dissolved heavy phases (Sb-S-0 phases and elementary antimony).

3.2 Solution composition.

The composition of pore and surface solutions of the Komsomolsk technogenic body is caused by the method of material treatment. These solutions are of a weak alkaline character. The main salt composition can be described by the following components: SO_4^{2-}, HCO_3^-, F^-, Cl^-, Na^+, Ca^{2+}, K^+ and Mg^{2+}. Ions SO_4^{2-}, Na^+ and Ca^{2+} are the predominant ones spatially during inactive periods of the plant operation accompanied by drying the collection pond. According to Table 2, considerable amounts of antimony (1090 µg/l), copper (2800 µg/l) and zinc (47800 µg/l) are released into solution during the substance treatment. Zinc dust, used in the technological process, is the main source of zinc (1, 2).

$$2Au + 4NaCN + 2H_2O + O_2 =$$
$$2Au(CN)_2^- + 2NaOH + H_2O \quad (1)$$
$$2Au(CN)_2^- + Zn = Zn(CN)_4^{2-} + 2Au \quad (2)$$

Sphalerite dissolves together with chalcopyrite, which is responsible for the copper in solution of the fresh pulp. Both minerals dissolve easily in the technological solutions (Scott & Ingles 1987). Due to formation of insoluble cyanide complexes, the contents of copper and zinc in solution decrease drastically after the pulp is supplied into the tailings impoundment. Similar behavior is typical for cadmium.

It is assumed that antimony transfers into solution by the following reactions:

$$Sb_2O_3 + 2NaOH + 3H_2O = 2NaSb(OH)_4 \quad (3)$$
$$Sb_2O_5 + 2NaOH + 4H_2O = 2NaSb(OH)_6 \quad (4)$$

This process starts at the technologic stage and continues within the technogenic body, and antimony accumulates in solution. Its content increases especially sharply due to shrinkage (Table 1, 1998) simultaneously with an increase in Na^+ concentration.

Arsenic begins to transfer into solution at the technological stage, and gradually increases in solution within the technogenic body. Concentration of dissolved lead is not high in

Table 1. Composition of Sb-Fe-Si phases, wt. %.

K	Ca	As	Al	Si	Sb	Pb	Fe
1.08	1.59	0.098	3.67	15.45	11.6	1.90	6.83
0.52	3.22	0.906	3.09	9.33	20.4	1.34	7.00

Table 2. Concentration of metals and CN⁻ in the solutions of Komsomolsky body, µg/l.

Year		pH	CN⁻	Sb	As	Zn	Cu	Pb	Cd	Fe
1997	"Fresh" pulp water	10.2	530	1090	63	47800	2800	6.9	0.35	660
	Surface water	n.d.	273	4780	91	21	10	<1	0.05	<10
	Pore water	n.d.	25	5750	500	23	6	14.0	0.05	360
1998	Surface water	8.6	111	11800	180	440	20	<2.5	0.15	180
	Pore water (№9)	6.3	96	31000	340	<50	<1	<2.5	0.08	420
	Pore water (№4)	8.2	206	11125	100	430	4.3	<2.5	0.25	250

Note: n.d. – not detected.

different water types (1997) and becomes lower with the time (1998).

3.3 Mineralogical peculiarities of secondary compounds of the oxidation zone.

Signs of the formation of an oxidation zone are observed in the sand material of the Komsomolsk body down to the depth of 0.8 m. Initial minerals (pyrrhotite, pyrite, and arsenopyrite) are inter substituted there by secondary minerals, or they demonstrate the traces of intensive dissolution (sphalerite, and Sb-S-0 phases). Secondary minerals are concentrated in the gravitation lenses. The following compounds are established there: Iron hydroxides form a zonal accumulations and substitution rims over pyrite and pyrrhotite; gypsum is precipitated as metacrystals; Fe-Ca hydrous arsenate replaces arsenopyrite (the phase of a similar composition was described by Paktunc et al. 1998); and mineral X. The latter is shown as aggregates of hexagonal tabular crystals. Spatial manifestations of mineral X are related to the upper boundary of the gravitation lenses. The complete determination of this mineral is difficult. According to the diagram from Vink (1996), for Eh-pH parameters of pore solutions of the Komsomolsk technogenic body (Eh=0.38, pH=8), antimony should exist in the form of an antimonate-ion. On the basis of composition and formation conditions, we can assume that mineral X can be an iron antimonate. It is obvious from Table 3 that secondary compounds contain antimony, arsenic, lead, and zinc as isomorphic admixture. Therefore, also mentioned metals are retained in the oxidation zone by the formation of separate mineral phases and sorption on secondary ferriferous phases. During dry periods, white blooms are formed on the surface of the technogenic body.

3.4 Redistribution of metals in solid

While studying the cross-section of the Komsomolsk technogenic body, we revealed that metal redistribution is already taking place. According to Table 4, in the mean, contents of antimony, zinc, lead and cadmium are lowest in the sand of oxidation zone, in spite of numerous gravitation

These blooms are predominantly made of gypsum and thenardite.lenses with increased contents of metals. Metal contents of the silt layers of oxidation zone are slightly higher than in the sand material. Perhaps, this is caused by weak permeability of the former and, as a consequence, less intensive dissolving and removal processes. A drastic increase in contents of antimony, copper and zinc is observed at the boundary of the technogenic body in detrital material of the buried soil. The detailed study of a

Table 3. Composition of the secondary minerals and phases, wt. %.

Ca	As	Si	Sb	Pb	S	Zn	Fe
Alteration rim over pyrite							
0.50	0.68	0.75	1.43	0.80	0.28	1.11	44.8
Replacement product of pyrrhotite							
1.22	1.97	1.38	2.24	0.17	0.74	0.76	45.7
Ca-Fe arsenate-hydrate							
2.76	17.33	0.51	0.43	1.43	3.49	0.26	34.7
2.72	11.40	0.89	0.80	0	1.72	0.36	34.8
Zonal iron (oxy)hydroxides							
1.90	0.62	3.62	1.97	0.19	0.33	0.68	41.4
0.59	0.85	0.48	2.90	0.13	0.28	1.18	49.4
Mineral X							
1.03	0.99	0.92	11.41	1.01	0.24	0.72	41.2

vertical cross-section "silt - detrital horizon - underlying soil" at the periphery of technogenic body, where metal contents in tailings are significantly lower, demonstrates that the detrital horizon retains all the metals under consideration. We assume that metals concentrate in the detrital horizon as a result (1) of neutralization of technologic solutions in the presence of organic carbon (Bernardin 1973) and (2) high sorption capacity of organic substance. It should be noted that the highest concentrations in the detrital horizon are determined for antimony, zinc and copper, which reach high concentrations in solutions at any stage of formation and existence of the technogenic body. Metal contents in the host soil decrease sharply even at the depth of 15 cm from the boundary of the technogenic body.

4 CONCLUSIONS

Initial metal distribution within the Komsomolsk technogenic body is caused by gravitation differentiation of phases during body formation.

Zinc, copper and antimony are actively dissolved at the technological stage (their concentrations are 47800, 2800 and 1090 µg/l accordingly). Within the technogenic body contents of zinc and copper decrease drastically (21 and 10 µg/l), and antimony keeps accumulating, especially in pore solutions (up to 31000 µg/l). The main arsenic accumulation occurs in the pore solution of the technogenic body (up to 500 µg/l).

Metal precipitation from solutions in the oxidation zone occurs due to formation of separate minerals and sorption on secondary ferriferous phases, the most abundant of which are iron hydroxides.

All metals concentrate in the detrital horizon of buried soil as a result of (1) neutralization of technological solutions in the presence of organic carbon and (2) high sorption capacity of organic

Table 4. Metal contents in various substance of the Komsomolsk technogenic body and underlying soils.

n/cm	Sb %	As %	Zn %	Cu %	Pb %	Cd ppm	Fe %
	Sand layers + lenses of gravitation enrichment, oxidation zone						
9/80	0.002-0.1	0.17-1.26	0.03-0.156	0.005-0.017	0.015-0.044	0.2-9.4	5.9-15.8
	0.027	0.38	0.06	0.008	0.024	1.2	6.8
	Lens of gravitation enrichment						
1/1	0.1	1.26	0.156	0.017	0.044	9.4	15.8
	Silt layers, oxidation zone						
6/40	0.0054-0.087	0.19-0.52	0.068-0.125	0.002-0.013	0.021-0.062	0.5-5.7	6.7-8.3
	0.05	0.28	0.086	0.008	0.036	2.0	7.5
	Not-oxidized part						
5/66	0.023-0.10	0.32-0.45	0.085-0.173	0.009-0.013	0.027-0.091	2.5-4.9	6.5-7.6
	0.075	0.4	0.144	0.012	0.065	4.2	7.2
	Detrital horizon						
3/8	0.065-0.37	0.15-0.34	0.149-0.918	0.008-0.297	0.047-0.114	1.8-2.9	7.5-8.7
	0.3	0.31	0.365	0.082	0.093	2.6	8.3
	Underlying soils						
6/52	0.0003-0.22	0.004-0.067	0.018-0.116	0.005-0.039	0.004-0.009	0.1	5.6-7.2
	0.0023	0.008	0.027	0.008	0.004		5.9

Notes: n/cm – number of analyses/interval, min-max/weighted average.

substance. Antimony, zinc and copper reach the highest concentrations there (0.3, 0.37 and 0.08 %) because these elements are the most movable under conditions of the Komsomolsk technogenic body. Despite the retaining role of the detrital horizon, metals are supplied into the host rocks, i.e. they are removed to the environment.

ACKNOWLEDGEMENTS

This work was financially supported by the UIGGM, grant "Mintex", RFBR # 97-05-65185 and by the grant of the Ministry of Higher Education of Russia for basic research in the field of search and exploration of useful mineral deposits.

REFERENCES

Bernardin F. E. 1973. Cyanide detoxification using adsorption and catalytic oxidation on granular activated carbon. *Jornal WPCF*. 45 (2): 221-231.

Bortnikova S.B., E. V. Lazareva, P. Serbin, E. I. Hozhina 1998. Transformation of metal speciation under an action of the aquatic vegetation in flooded tailings (wastes of gold-ore cyanidation). In Atak, Onal & Celik (eds), *Innovations in Mineral and Coal Processing* (Proceedings of the 7th International Mineral Processing Symposium Istanbul, Turkey, 15-17 September 1998): 793-797 Rotterdam: Balkema.

Paktunc, D.J., J. Szymansky, K. Lastra, J. H. G. Iaflamme, V. Enns & E. Soprovich 1998. Assessment of potential arsenic mobilization from the Ketza river mine tailings, Yukon, Canada. *Proceedings of Waste Characterization and Treatment Symposium, Orlando, Florida, 1998.*

Robertson W. D. 1994. The physical hydrogeology of mill-tailings impoundments In J. L. Jambor & D. W. Blowes (eds), *Short course handbook on environmental geochemistry of sulfide mine-wastes*: 1-17. Waterloo, Ontario, May.

Scott J. D. & J. Ingles 1987. State of the art processes for the treatment of gold mill effluents. *Mining, Mineral, and Metalurgical Processes Division, Industrial Programs Branch, Environment Canada*, Ottawa, Ontario March.

Vink, B.W. 1996. Stability relations of antimony and arsenic compounds in the light of revisec and extended Eh-pH diagrams. *Chemical Geology* 130(1-2): 21-30.

Geochemistry of the Earth's Surface, Ármannsson (ed.)© 1999 Balkema, Rotterdam, ISBN 90 5809 073 6

Pb, Zn and Cu isotopic variations and trace elements in rain

J. M. Luck & D. Ben Othman
UMR 5569 Hydrosciences (cc 057), Université Montpellier, France

F. Albarède & P. Telouk
Géologie, Ecole Normale Supérieur de Lyon, France

ABSTRACT : Pb isotopes are reported for rains from southern France, as well as trace elements and, for the first time, Cu and Zn isotopes (expressed as $\delta^{65}Cu$ and $\delta^{66}Zn$). Isotopic variations for transition metals have recently been observed in ores, sediments etc. The purpose here was to establish whether such isotopic variations would also show up in surface samples, whether they would correlate with well-known Pb isotope variations and could then be used as new tracers for processes and sources of metals. Zn and Cu have also been measured in potential endmembers (natural : rocks, and anthropogenic : chemicals etc.). Local carbonate rocks exhibit positive $\delta^{65}Cu$ and $\delta^{66}Zn$, as observed for low-temperature rocks. In rains, isotopic variations do show up : in particular Zn isotopes correlate with Zn content and may be interpreted for the moment as a simple binary mixture.

1. INTRODUCTION

It is known that certain metals have been released in the atmosphere and hydrosphere in much larger quantities than natural levels since the development of human activities. Even in Roman times, lead (Pb) was disseminated in the environment and may be found today in ice cores. Fluxes of metals such as Cd, Pb, Zn due to anthropogenic activities greatly exceed (10-100 times) those linked to natural erosion, even in areas remote from cities or industries. Lead is an ideal tracer thanks to the natural variations of its isotopes in rocks and mineralisations used by man : measuring them in various kinds of samples (aerosols, rivers, oceans etc.) allows the calculation of the proportions of the various end-members. For the other metals, these proportions could only be addressed by establishing tentative correlations between their levels and Pb isotopic variations, until recently. However, this approach has strong limitations due to the different chemical behaviour of elements such as Cu, Zn, Cd, Pb etc., particularly in surface environments, where numerous parameters are involved : pH, Eh, ligands etc. The design of sophisticated machines (ICP-Magnetic Sector with Multi-Collector) capable of measuring the isotopic composition of elements other than radiogenic ones with a great precision has revealed the existence of natural isotopic variations for elements such as Mo, Cu, Zn etc. (e.g. Halliday et al. 1995, Marechal et al. 1999). Although small -

of the order of a few permil - these variations open up the possibility to use a large spectrum of elements as potential tracers of sources or processes in the future. We report here Cu and Zn isotopic data obtained for the first time for rains, and try to correlate them with major, trace elements and Pb isotopes.

2. SAMPLING AND ANALYTICAL METHODS

Major, trace elements and Zn, Cu and Pb isotopes have been measured in various rains of essentially marine (Mediterranean) origin, collected in two neighbouring locations in South France (Languedoc-Roussillon). One of them ("Montpellier") is close to the Sea, the other ("Roujan") somewhat more inland (50 km). Both areas are mostly covered with vineyards and only a few industries (cements, leisure ports, chemicals) are clustered around the large marine Harbour at Sète. Several rains were collected over the period 1996-99 and analysed as described in Luck and Ben Othman (1998) and Ben Othman et al. (1997). Two heavy rain events were sampled sequentially (3 subsets). All samples were collected in plastic funnels, one of them being covered with teflon to minimize metal contamination. Samples were filtered through 0.2 μm acid-cleaned PVDF filters in the laboratory under class 100 US laminar flow, and stored in acid-cleaned PP bottles. Sequential samples were stored in teflon-lined bottles. Both dissolved and particulate subsets were analyzed for trace elements and isotopes. Particulates from the sequential rains were furthermore subjected to leaching experiments by the procedure of Hamelin et al. (1989) : 2 times HBr 0.5N, then HF-HNO$_3$, in order to better constrain the various origins of metals on aerosols and particulates.

A new chemical technique has been designed that allows the separation of Cu, Zn, Pb and Cd from a single sample on an AG1X4 anion-exchange resin, with a further separation and purification of Cu from major elements on an AG-MP resin using the procedure developed by Maréchal et al. (1999). Because isotopic fractionation effects for Cu through column chemistry have been described by Maréchal et al. (1999), we carefully checked this possibility for all four metals by regularly analysing different mixtures of standards, including major elements (cations and anions).

Major elements were determined by Capillary Ion Electrophoresis. Trace elements were determined, either directly (dissolved) or after acid dissolution or leaching (particulate) by ICP-MS. Isotopic composition for Pb was mostly determined by Solid Source Mass Spectrometry. Low level Pb samples and Cu and Zn isotopic compositions were determined with a VG Sector 54 ICPMS with multi-collector at the ENS of Lyon. The data were corrected for mass discrimination by normalizing to added JMC Zn and NIST Cu standards. Data are expressed as $\delta^{65}Cu$ and $\delta^{66}Zn$ (with respect to ^{64}Zn) in permil (‰). Precision and reproducibility for most samples are ±0.02‰.

3. RESULTS AND DISCUSSION

3.1 *Trace elements*

For some trace elements (U, Ni), our samples show levels similar to those found in remote countries (e.g. Congo (Freydier et al. 1997)). Some other metals typically associated with human activities (Cu, Zn, Cd, Pb) show strong enrichments. In the sequential events, several metals are correlated (e.g. Pb-Cd; Ni-Cu). The [Pb]/[Cu] ratios are within the range determined for other rains from the Western Mediterranean (Chester et al. 1997).

3.2 *Sr and Pb isotopes*

Sampling locations at Montpellier are close to the sea. Thus as expected, they show the influence of seaspray aerosols in their high [Na], [Mg], [Cl] contents, but the low values of $^{87}Sr/^{86}Sr$ (0.7083-0.7089) combined with high [Ca] contents show also the incorporation of locally-derived carbonate

aerosols. Continental rains show more radiogenic values for Sr (0.7098), although not as high by far as in the Massif Central (Négrel and Roy1998) which are strongly influenced by granitic terrain aerosols.

Lead isotopes (Figure 1) for continental rains also show the influence of radiogenic Pb derived from rocks. Meanwhile marine rains show less radiogenic values : they define a domain strongly tilted in the $^{206}Pb/^{204}Pb$-$^{207}Pb/^{204}Pb$ diagram with $^{206}Pb/^{204}Pb$ around 17.7-17.8. This domain is parallel to, but lower than industrial data from the litterature (Petit et al. 1984) Such a feature has also been observed for other rain samples over Paris (Roy1996), even inside a single event : this can probably be related to the presence in the rain of at least two distinct sources of Pb with different isotopic signatures, being scavenged at different rates. Surprisingly, the values for our area are only slightly less radiogenic than those for the center of Paris. Finally it seems remarkable that the slope for all these is also parallel to that for Water Treatment Plants in the area suggesting similar fractionation and maybe similar (industrial ?) sources although not in the same proportions.

3.3 *Zn and Cu isotopes*

Zn and Cu were analysed for their isotopes in the dissolved and particulate loads of rains. Moreover we added isotopic data obtained on some of the various natural and anthropogenic endmembers already caracterized by Pb isotopes : carbonate bedrock samples, which are dominant in this area; anthropogenic products, like Cu-enriched chemicals, which have been used for decades in the catchment; and Water Treatment Plant effluents. The value for a mussel's flesh (known to be a bioindicator of

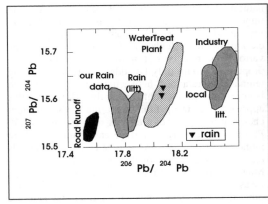

Figure 1. (after Luck & Ben Othman 1998) Lead isotopic compositions for rains from South France (vertically hatched) compared to Paris rain (horizontally hatched) and various anthropogenic endmembers.

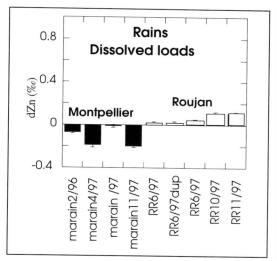

Figure 2. Histogram of Zn isotopic composition (expressed as δ^{66}Zn) for rains from Montpellier ("Marain", filled) and Roujan ("RR", open) areas (note that δ^{66}Zn values are relative to ENS shelf standard, and that relative variations are important).

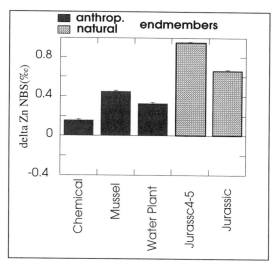

Figure 3. Histogram of δ^{66}Zn values for some anthropogenic (chemical, Plant effluent) and natural endmembers (rocks).

environmental metallic concentrations) from the nearby Thau lagoon is also reported : however its value is difficult to interpret since we cannot eliminate possible kinetic effects during absorption. Data are plotted as a histogram in Figure 2 and compared to endmembers in Figure 3. Variations in

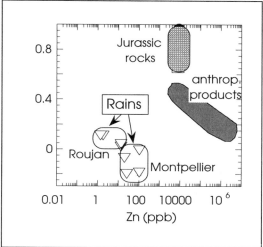

Figure 4. Plot of δ^{66}Zn vs. Zn content in rains and endmembers.

rains are small (+/- 0.2 ‰) but resolvable considering precision (0.01‰). Duplicate analyses including chemical procedure show excellent agreement (0.01‰). First, we notice that the two locations are characterized by distinct isotopic ranges (Figure 2), the inland samples having for example higher δ^{66}Zn. Jurassic carbonates show even higher delta values, while the anthropogenic products give delta values lower than the natural endmembers (Figure 3). The rains sampled in the agricultural area (Roujan) have values close to that of a largely used Cu-enriched chemical product. On the other hand, Montpellier (mostly marine) rains have distinctly lower isotopic values. When coupled with higher Cu and Zn levels, this requires one or more additional endmember, yet to be identified.

4 CONCLUSIONS

Elements (Zn, Cu) other than radiogenic ones exhibit isotopic variations in rains : these variations are linked to those reported by other authors in rocks, sediments and ores, which we have also confirmed in the various endmembers from our study area. The observed Zn and Cu isotopic variations seem to be linked to the metal contents. Combined with Pb isotopes, these variations therefore establish new possibilities in using these tracers in assessing metal origins, proportions and/or processes in the environment.

REFERENCES

Ben Othman, D, J-M. Luck & M.G. Tournoud 1997. Geochemistry and water dynamics : application to short-time scale flood phenomena in a small Mediterranean catchment. I- alkalis, alkali-earths and Sr isotopes. *Chem. Geol.* 140:9-28.

Chester, R., M. Nimmo., & P..A. Corcoran.,1997. Rain water-aerosol trace metal relationships at Cap Ferrat: a coastal site in the western Mediterranean *Mar. Chem.* 58 : 293-312.

Freydier, R., B. Dupré., & J.P. Lacaux. 1997. Precipitation chemistry in intertropical Africa *Atmosph. Environ.* 32: 749-765

Halliday,A.N., D.C. Lee, J.C. Christense, A.J. Walder, P.A. Freedman, C.E. Jones, C.M. Hall, W. Yi & D. Teagle 1995. Recent developments in inductively coupled plasma magnetic sector multiple collector mass spectrometry. *Intl. J. Mass Spec. Ion Proc.* 146/147: 21-33.

Hamelin,B., F.E. Grousset., P.E. Biscaye & A. Zindler 1989. Lead Isotopes in Trade Wind Aerosols at Barbados : The influence of European Emissions Over the North Atlantic. *J. Geophys. Res.* 94 : 16243-16250.

Luck,J-M. & D. Ben Othman 1998. Geochemistry and water dynamics : II- Metals and Pb isotope constraints on water movements and erosion processes. *Chem. Geol.* 150: 263-282

Maréchal, C., F. Albarede & P. Telouk 1999 Precise analysis of Cu and Zn isotopic compositions by Plasma-source mass spectrometry *Chem. Geol.* (accepted)

Négrel, P. and S. Roy 1998. Chemistry of rainwater in the massif Central (France). An Sr and Pb isotope and major element study.*Anal. Chem. (submitted),*.

Petit, D., J.P. Menessier., & L. Lamberts 1984. Stable lead isotopes in pond sediments as tracer of past and present atmospheric lead pollution in Belgium. *Atmos. Env.* 18 : 1189-1193.

Roy , S. 1996. *Utilisation des isotopes du Pb et Sr comme traceurs des apports anthropiques et naturels dans les précipitations et les rivières du bassin de Paris.*Thesis, Univ. Paris 7.

Geochemistry of the Earth's Surface, Ármannsson (ed.)© 1999 Balkema, Rotterdam, ISBN 90 5809 073 6

Environmental study of waste dumps at the Pb-Ag Bottino mine, Apuane Alps, Italy

I. Mascaro, M. Benvenuti, F. Corsini, P. Costagliola, M. Ferrari, P. Parrini & G. Tanelli
Dipartimento di Scienze della Terra, Firenze, Italy

S. Da Pelo & P. Lattanzi
Dipartimento di Scienze della Terra, Cagliari, Italy

ABSTRACT: The abandoned Bottino Pb(Zn)-Ag mining district (Apuane Alps, northwestern Tuscany) has a long mining history dating back at least to the eleventh century AD. Abandoned mining wastes represent a potential source of heavy metal pollution. A study of mineralogy and chemistry of the waste dumps, and of the chemistry of stream, mine drainage and waste seepage waters indicates that alteration phenomena occurred, until now, under near neutral conditions, and with no significant pollution of surficial waters. However, some waste heaps may present some hazard in the long run.

1 INTRODUCTION

The Bottino Pb(Zn)-Ag deposit (Apuane Alps, northwestern Tuscany) has a long mining history dating back at least to the eleventh century AD, reaching a maximum in the past century, until final closure in 1967. Total production is estimated in the order of 4000 t of Pb, 600 t of Zn, and 22 t of Ag (Lattanzi et al., 1994). As very little action was undertaken to minimize the environmental impact during and after exploitation, many waste piles are left scattered over the deposit area, and drainage from mining tunnels freely runs into the stream network. We describe the mineralogy and chemistry of the waste dumps in the area, and we document the supergene weathering reactions affecting the primary minerals, with the ultimate goal of predicting the environmental impact of the past mining activities. An actual estimate of the present impact was obtained by reconnaissance analyses of stream, mine drainage and mining waste drainage waters in the area.

2 GENERAL FEATURES OF THE STUDY AREA

2.1 *Background geology*

The Bottino deposit consists of a set of NW-SE striking, steeply dipping veins, entirely hosted within siliciclastic metasediments ("Filladi inferiori") and metavolcanic (rhyolitic) rocks ("Porfiroidi" and "Scisti porfirici") of the Paleozoic basement of Apuane Alps. Mineralization is composed mainly of galena, sphalerite, pyrite, pyrrhotite, chalcopyrite, and a variety of sulphosalts, in a quartz-carbonate (chiefly siderite-dolomite) gangue, and in strict association with a tourmaline-rich rock. According to Benvenuti et al. (1989) and Lattanzi et al. (1994), ore deposition occurred synchronously with Tertiary regional metamorphism, possibly by (re)mobilization of preexisting Palaeozoic deposits. The deposit area extends at the surface between ca. 700 and 300 m above sea level (asl) over the steep northern slope of Monte Ornato. Surface waters form a very short, torrent-like network (Bottino creek) flowing to the Vezza river.

2.2 *The waste dumps*

Our field work led to recognition of eleven waste bodies. Seven bodies, offering the best exposures, were chosen for detailed work (Table 1). Overall, they cover an area of approximately 5300 square meters; their thickness is of 2-5 m. The probable total volume of waste material in the area may lie somewhere between 20,000 and 50,000 cubic meters. The waste bodies typically lie next to Bottino creek, over steep slopes (about 40°) at higher elevations (500-700 m asl), and gentler slopes (10°-20°) down to about 200 m asl. Most dumps consist of coarse-grained (pebble, gravel, coarse-sand) waste rock from excavation; tailings from mineral processing (jigging and handpicking) also

Table 1. Characteristics of waste dumps in the Bottino area

Waste types and ages	N	Fe[1]	S[1]	As[1]	Cu[1]	Pb[1]	Zn[1]	Cd[1]	pH	Primary Minerals[2]	Secondary minerals[2]
jigging (LAV I) 1919-1929	4	8.58 ±1.11	1.22 ±0,68	146 ±105	392 ±205	0.88 ±0.45	0.90 ±0.41	96 ±41	6.3 ±0.8	qz, wh-mi, clay, dol, sid, sp, ga, (py, asp, ccp, po, tetr, calc)	prl, goe, lp,cer, (ang, Au, Fe-sulp, Fe-Mn-ox, clay)
hand picking (LAV II) 1919-?	4	7.78 ±1.30	0,96 ±1.14	266 ±221	384 ±401	0.82 ±0,84	0.76 ±1.14	45 ±64	6.2 ±1.0	qz, wh-mi, clay, ab, ga, sid (dol, sp, py, po, tetr, calc)	lp, prl, clay, cer, (goe, ang, Fe-Mn-ox, Fe-sulp)
excavation (La Rocca) 1919-?	4	3.83 ±0.22	0.75 ±0,38	470 ±225	50 ±21	0.22 ±0,20	0.03 ±0.01	4 ±3	6.7 ±0.5	qz, wh-mi, clay, tourm, dol, py, sid, (ga, asp, ccp, po)	clay, prl (goe, cer, Fe-sulp, Mn-Fe-Al-ox, Fe-Pb-ars)
excavation (Redola) ?-1593	4	9.80 ±2.44	0,69 ±0,17	86 ±77	442 ±121	0.47 ±0.17	0.34 ±0.04	30 ±6	6.8 ±0.4	qz, wh-mi, clay, dol, py, sid, ga, tourm, (sp, ccp, po, tetr, calc)	goe, prl, cer, lp (Fe-sulp, Fe-ox, clay)
excavation (Paoli) 1840-1868	14	7.62 ±2.20	0.98 ±1.16	75 ±45	177 ±94	0.62 ±0.87	0.57 ±0.77	75 ±119	6.8 ±1.3	qz, wh-mi, ab, clay, sid, dol, sp, ga, (tourm, ccp, po, tetr, py, calc)	goe, cer, prl, (Au, Pb-ars, Fe-sulp, Fe-Mn-ox, clay)
excavation (Casello) ?-1593	8	6,25 ±1.48	2.30 ±3.10	402 ±354	123 ±75	0.47 ±0.30	1.10 ±1.94	119 ±204	5.4 ±0.8	qz, wh-mi, clay, sp, sid, dol, ga, py (ccp, po, tetr)	goe, prl, cer,
excavation (Nuova) 1840-1868?	6	6.17 ±1.29	0,50 ±0.30	46 ±22	80 ±47	0.06 ±0.06	0.10 ±0.07	7 ±7	6.7 ±0.8	qz, wh-mi, clay, ab, dol, sid, py, (sp, ga, calc, ccp, po)	goe, prl, lp

N= number of samples; [1]Mean values and standard deviation (in italics) expressed in wt% except for As, Cd, and Cu in ppm; [2]Mineral abbrevations: ab: albite, ang: anglesite, ars: arsenate, asp: arsenopyrite, Au: native gold, calc: calcite, ccp: chalcopyrite, cer: cerussite, clay: clay minerals, dol: dolomite, ga: galena, goe: goethite, lp: lepidocrocite, ox: oxyhydroxide, po: pyrrhotite, prl: pyrolusite, py: pyrite, qz: quartz, sid: siderite, sp: sphalerite, sulp: sulphate, tetr: tetrahedrite, wh-mi: white micas

occur. Soil development over the dumps is moderate, and mostly confined at the borders.

3 SAMPLING AND ANALYTICAL METHODS

Samples of waste dumps were collected at the surface, or at shallow depths (5-30 cm). Mineralogy of the samples was determined by a combination of optical (binocular, transmitted and reflected light) microscopy, X-ray diffraction (XRD), and SEM/EDS observation. Mineral chemistry of selected phases was determined by means of SEM/EDS, EPMA/WDS. The concentration of Pb, Zn, Cu, Cd, As and Sb in the bulk samples was determined by X-ray fluorescence.

Water samples include five stream waters, one tunnel drainage, and four seepage waters from a waste heap (LAV1). After measuring pH, conductivity and temperature in the field, samples were filtered (0.45 µm), and analyzed for major and selected trace ions by a combination of techniques.

4 WASTE DUMPS

4.1 Mineralogy

The mineralogy of the waste dumps is reported in Table (1). We distinguish between primary phases, i.e., minerals originally present in the mineralization and/or in the country rocks, irrespective of their origin, and secondary phases that developed in situ by supergene processes. Among primary phases, quartz and white mica (muscovite-paragonite) are ubiquitous, and quite abundant. Other common silicates include chlorite, albite, and tourmaline. Carbonates are mainly represented by terms of the siderite-magnesite and dolomite-ankerite solid solutions; calcite is comparatively rare. The most abundant sulphides are sphalerite, galena, and pyrite; chalcopyrite, pyrrhotite, arsenopyrite, Ag-bearing tetrahedrite, meneghinite, and boulangerite also occur. Clay minerals occur as primary (illite and chlorite) and as secondary phases (kaolinite, montmorillonite, vermiculite).

The main secondary minerals are goethite, lepidocrocite, pyrolusite, and cerussite. Other phases comprise: anglesite, pyrochroite, ferrihydrite, Fe-sulphates, a Fe-phosphate, a Zn carbonate (?), a Zn-Ag-Fe sulphate (?), a Fe-Pb arsenate (?), and native gold. Fe-oxyhydroxides, Mn-oxides and Fe-Mn-Al - hydroxides mostly occur as amorphous material, containing in places appreciable amounts of other metals (up to 20 wt. Zn, 25% Pb, 7% Cu, 2% Ni, 1% Co). These may result from: a) submicroscopic admixtures of separate phases, or b) isomorphogenous substitution in the lattice, or c) surface adsorption, or any combination of a, b, c. The fine-grained, porous nature of the Bottino material makes very difficult a clearcut distinction among these alternatives.

4.2 Chemistry

Table 1 reports the pH values and the contents of several metals in the waste dumps. Most pH values fall within one unity from neutrality; however, at Casello they are consistently lower, reflecting a comparative scarcity of carbonates. Metal contents are variable; some elements, such as Zn and Cd, show significantly positive correlation coefficients with S, suggesting that they are still bound in sulphides. The high correlation between Fe and Mn reflects the occurrence of these metals in oxides-hydroxides.

4.3 Supergene processes

At Bottino, supergene alteration of primary phases in the dumps may occur in two ways: development in situ of pseudomorphic replacement of secondary phases, and/or leaching and dissolution; elements released in this process may be eventually reprecipitated in secondary phases. A relative sequence of alterability of sulphides was established based on the observed extent of dissolution textures

and/or development of coatings by secondary phases: pyrrhotite (most altered) > galena-sphalerite > (chalcopyrite, arsenopyrite)-pyrite (least altered). Alteration of pyrrhotite and galena is mainly of the replacement type. In the case of pyrrhotite, this leads to almost complete pseudomorphosis by goethite and/or lepidocrocite. Cerussite (and much rarer anglesite) form coatings of galena grains.

Iron-bearing carbonates and sphalerite show more pronounced dissolution textures, accompanied however by development of Fe (Mn) oxyhydroxides rims. Additional reactions of minor relevance for acidity control/metal release are those involving less abundant sulphides (chalcopyrite, arsenopyrite, sulphosalts), carbonates (dolomite and calcite) and alumo-silicates. Most of Fe released from primary minerals during alteration appears precipitated in situ as Fe oxyhydroxides, but the minor occurrence of iron sulphates and Fe oxyhydroxides deposited onto Fe-devoid minerals indicates some iron transport.

5 WATER CHEMISTRY

All Bottino creek waters and mining waste drainage waters are characterized by low salinity values (conductivity 72-190 µS/cm), and neutral pH (6.8 – 7.4). The tunnel drainage water (810 µS/cm, pH=7) is moderately more saline, but still of neutral pH. All samples have a dominant sulphate-chloride character and low contents of heavy metals, often below the instrumental (ICP-AES) detection limits. Slightly higher contents of Zn occur in mining waste seepage waters (300-450 ppb compared to the range of 30-300 ppb of the other samples). The substantially unpolluted and neutral pH character of all waters collected suggests a very limited dispersion of heavy metals and of possibly acidic solutions away from the mining waste heaps. This is in agreement by the very scarce amount of metal-rich particulates observed in the bottom sediments.

6 CONCLUSIONS

At Bottino, the abandoned wastes piles contain variable amounts of heavy metals that are of potential concern for environment. A comparison of mineralogical data and pH measured in waste material indicates that the alteration phenomena occurred, until now, under near neutral conditions and with no significant pollution of surficial waters. Mineralogical data show that supergene alteration of primary phases in the dumps occurs in two ways: development in situ of pseudomorphic replacement of secondary phases, and/or leaching and dissolution; elements released in this process may be eventually reprecipitated in secondary phases. The observed neutral pH values have favoured a relatively efficient fixation of heavy metals in phases, like Fe- and Mn-hydroxides and cerussite, stable in such conditions, limiting in this way the pollutants dispersion.

In the Bottino area, the main factors controlling the alteration of minerals and the pollutant transport seem to be: 1) the dominant content of sulphides like galena and sphalerite, characterized by oxidation under near neutral conditions, in comparison with a lesser amount of pyrite and pyrrothite; 2) the relative abundance of buffering phases like carbonates and alumo-silicates; 3) the very coarse grain size of waste materials that, limiting the total extension of sulphide crystal surfaces available for interaction with oxidizing agents (O_2, etc.), has strongly slowed down the oxidation processes. Moreover, the coarse grain size of waste materials, and the steep morphology of the area, restrict both the contact time between minerals and oxygenated drainage waters and the pedogenic development.

At present, therefore, the abandoned waste piles apparently pose no serious threats to the environment. However, in the long run we must consider the possibility of increasing acidity production for those waste bodies (eg. La Rocca, Casello, Redola), that are particularly rich of Fe sulphides in comparison with the amount of buffering phases.

7 ACKNOWLEDGMENTS

We thank A. Buccianti for statistical analyses. The research was supported by MURST grants (G. Tanelli), Università di Firenze funds (F. Corsini), and by CNR through the Centro Studi per la Minerogenesi e la Geochimica Applicata.

REFERENCES

Benvenuti, M., P. Lattanzi & G. Tanelli 1989. Tourmalinite-associated Pb-Zn-Ag mineralization at Bottino, Apuane Alps, Italy: Geologic Setting, Mineral Texture, and sulfide chemistry. *Economic Geology* 84: 1277-1292.

Lattanzi P., M. Benvenuti, P. Costagliola & G. Tanelli 1994. An overview of recent research on metallogeny of Tuscany, with special reference to Apuane Alps. *Mem. Soc. Geol. It.* 48: 613-625.

Geochemistry of the Earth's Surface, Ármannsson (ed.) © 1999 Balkema, Rotterdam, ISBN 90 5809 073 6

Zinc and cadmium behaviour in tailings

L. P. Mazeina & D. U. Bessonov
Novosibirsk State University, Russia

S. B. Bortnikova
United Institute of Geology, Geophysics and Mineralogy, Novosibirsk, Russia

ABSTRACTS: The relative behaviour of zinc and cadmium in flooded tailings was investigated. Field experiments and laboratory analyses show the alteration of Zn/Cd ratio in solid substances. Variation of such ratios depends upon the leaching of cadmium from sphalerite and its oxidative dissolution. These processes are reflected in the water composition of a collection pond. Calculation of zinc and cadmium speciation in water shows primary ionic and sulphate forms for cadmium and carbonate form for zinc. The water is saturated with respect to $Zn(OH)_2$, ZnO, Zn_2SiO_4, $ZnCO_3$; pore water is saturated with respect to CdS, ZnS, that is, these compounds are able to become a suspended phase. Presence of ionic forms of zinc and cadmium leads to the assimilation of these metals by water plants. Zinc is a more dangerous element in the system but according to the prognosis, cadmium is most dangerous element, because its high mobility causes the transportation of cadmium to distant places.

1 INTRODUCTION

It is known that mining wastes are the source of large amount of toxic elements. Macrocomponents (sulphide minerals) are well known but microcomponents, which are present as admixture (As - in pyrite, cadmium – in sphalerite and so on), can be dangerous. Admixture concentration is initially one thousandth or one millionth of that of the mineral-forming elements. In spite of this admixture elements can contribute a significant part to the geochemical background of the processes due to easier leaching from mineral structure (Perel'man, 1972) and due to features of speciation. The problem of mining waste storage is still actual in the world and interest to this problem increases more and more every year.

The investigative object is the Salagaev tailings (Russia, Southwest Siberia, Kemerovo region). The Salagaev is the active tailings impoundment, measuring 1400×900 m, which had been filling since 1967 by wastes of lead-zinc recovery plant. At presents, amount of material composes about $30×10^6$ tons here. It is situated within the natural ravine and

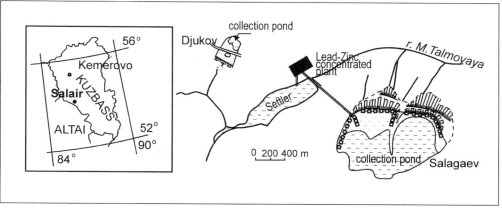

Figure 1. The scheme of the hydrosystem around plant.

its depth is 25m at the north end. By means of the high water table, the entire tailings impoundment is essentially covered with water. Sulfide content does not exceed 1-2% and the main mineral is pyrite. The remainder of the mass includes debris from hosted rocks of different composition (quartzites, limestones, quartz-sericite schits) and mineral grains (quartz, barite, carbonate). Sulphides of lead, zinc and copper are represented by small monomineral grains of galena, sphalerite, chalcopyrite, tennantite and tetrahedrite. Content of sphalerite (main source of zinc and cadmium) varies from 0.1 to 0.5%.

As wee see from Figure 1 drainage from Salagaev discharges into the M. Talmovaya River. Taking into account large resources of heavy metals in Salagaev tailings (Table 1) there is a potential danger of metal dispersion from the storage area into the environment. The main objective of this investigation is to compare the behaviour of zinc and cadmium as geochemicaly similar elements so in the tailings as outside.

Table 1. Heavy metal concentration of Salagaev tailings.

Metal	Content, ton
Zinc	136250
Lead	49750
Copper	7750
Cadmium	250

2 METHOD AND INSTRUMENTS

Our investigation consisted of three part: field experiments and sampling, laboratory analyses and thermodynamic calculation of metal speciation.

During the fieldwork, sampling of the following components was carried out.

1. Solid tailings from the exposed part over the surface of the impoundment and through prospecting pits.

2. Water from the collection pond and the river (both in its and lower reaches) was sampled. During the sampling pH and temperature of water were measured. Samples were filtered in the field by 0.45 μm filter, then divided into portions. In the first portion, the anion composition was determined. The second portion was conversed by adding chemically pure HNO_3 for later analysis for heavy metal contents.

3. The sediments, mainly the uppermost part (approximately to 5-7 cm depth) were studied. Sediments were sampled from vertical cross-sections of about 0.8-1m depth by a sampler that made possibility extraction of a column without disturbing the material. Columns were divided into visually different layers, which were placed in a gas impermeable dishware for laboratory analysis.

Laboratory investigations entailed the following steps.

1. Heavy metal contents in the conserved portions of water were determined. Pore water was pressed from samples of sediment with 100 atm pressure.

2. Solid tailings were analysed by atomic absorption method (AAM).

3. The sediments were dried, cleaned from admixture of coarse-grained manerial, and after milling they were analysed by analysis RFA (X-Rey fluorescence) method.

The next step of the investigation consisted of thermodynamic calculations using the program WATEQ4F (Ball et al. 1987). This program calculates element speciation in solution(through the corresponding constants, which are collected in a large database) and in solid phase (through the saturation index) from the concentration of macro- and micro- water components. We also use the zinc and cadmium concentration ratio to compare their behaviour. This ratio gives (according to Nordstrom & Ball 1985) quite a clear estimate their relative geochemical behaviour in different components.

3 RESULTS AND DISCUSSION

The complicated composition of concentrated ores (fine sulphide grains presence of intergrowths and so on) causes considerable losses of useful components including sphalerite. The partioning of zinc between ore and working products is shown in Figure 2. The diagram shows the high zinc concentration in the storage waste, the zinc and cadmium content of the bottom sediment is higher. We suggest that there are two main reasons for zinc and cadmium concentrations increasing in sediment in comparison with the average concentration in the tailings. The first reason is that metal content is higher in the silt fraction of waste, which is a basic part of bottom sediment, than in the medium- and coarse-grained fraction (Bortnikova et al. 1996). The second on is that the dissolved metal species rise by vertical diffusion of solution and therefore metals are redeposited in upper layers of bottom sediment.

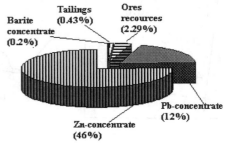

Figure 2. Percentage of Zn in different components.

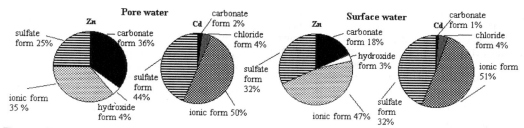

Figure 3. Zinc and cadmium speciation in pore and surface water.

Zinc and cadmium contents exceed backrgound level by more than one order of magnitude but it is not higher than standard levels for sewage water (for cadmium). But their concentrations increase greatly in pore water of bottom sediment (see Tables 2,3). The reason is the constant interaction with bottom sediment, which is primary, represented by original pulp. It is to be noted that the zinc content of the water increases 32 times and the cadmium content 13 times. But in the river, where drainage discharges the cadmium increase is much greater (44 times that of collection pond water and 571 times that of the background concentration); at the same time zinc concentration increases 6 and 44 times respectively. The ionic speciation of metals release from solid waste is compatible with the solid waste is compatible with the sulphide oxidation.

$$MeS + 2O_2 + H_2O = Me(H_2O)_6^{2+} + SO_4^{2-} \quad (1)$$

Thermodynamic calculation shows characteristic speciation of these metals in water: surface and pore (see Figure 3). Zinc species to free ion (Zn^{2+}) and carbonate ($ZnCO_{3aq}^{0}$ and $Zn(CO_3)_2^{2-}$), at the same time cadmium is present as Cd^{2+} and $CdSO_{4aq}^{0}$. The different metal speciation varies in pore water because of changes in physico-chemical conditions and the metal concentration tend to be higher but not

considerably. The zinc carbonate species share alters at the expense of a decreasing share of the free ion. Cadmium metal speciation in pore water is practically the same as in surface water of the tailings impoundment. Tailings solution is saturated with respect to the following minerals. Surface water in the collection pond is saturated with respect to ZnO (cinkite), $Zn(OH)_2$, $ZnCO_3*H_2O$, Zn_2SiO_4 (willemite), $ZnCO_3$ (smitsonite); in the pore water - CdS (grinockite), ZnS (sphalerite). These phases can form a suspension and redeposit under these conditions extracting metals from solution.

The changes in the Zn/Cd ratio are shown in Tables 2 and 3. The Zn/Cd ratio in sphalerite is 290. We take that value as an equilibrium value for this system. Moreover, Zn/Cd ratio in the water is very close to 290. The Zn/Cd ratio increases in the ore in comparison with sphalerite itself, because of the zinc admixture of other sulphides. This ratio increases in the waste and the bottom sediment of the collection pond. But this ratio is significantly lower in different types of water than in the solid substance; it changes in this collection pond water in comparison with background level and also in pore water. If wee see the changes in this ratio we can observe its decrease in the chain: ore –waste – bottom sediment – water, due to the increasing share of cadmium and therefore increasing migration rate from ore to waste, from waste to bottom sediment, from bottom

Table 2. Zn and Cd content (%) and their ratio in solid substance

	Sphalerite	Ore	Tailings	Bottom sediment
Zn	66.85	1.5	0.2-0.4	0.35-1.00
Cd	0.23	0.0006	0.0003-0.0004	0.00051-0.0016
Zn/Cd	290	2500	667-1000	274-660

Table 3. Zinc and cadmium content and their ratio in water (mg/l)

	Max. allowable concentration	Background	Pore water of sediment	Surface layer of the pond	Lower reaches of the Talmovaya River
Zn	10	10	1100-6200	147-392	1.45
Cd	5	0.14	10.3-58	1.3-2.4	0.0046
Zn/Cd		71	69-338	93-236	315.2

sediment to pore water, and, in the end, to surface water. This regularity correlates with the general tendency of elemental behaviour under supergenic conditions (Perel'man 1972). It means that element migration intensity (P_x) in the geochemical environment studied increases with its concentration (b_x).

$$P_x = \frac{1}{b_x}\frac{db_x}{dt} = \frac{d\ln b_x}{dt} \qquad (2)$$

And in the end, it has to be noted that metals are predominantly assimilated by water plants in ionic form (it was established (Bortnikova et. al., in press) that cadmium and zinc are most easily of all metals assimilated by the macrophites best of all metals for which ionic form is main).

4 CONCLUSIONS

Thus, we can say that the Salagaev region and territories close to the river are influenced by a high technogenic load.

Zinc is dangerous element in the derived system as its concentration level considerably exceeds the maximum permitted concentration in some places. Zinc compounds can reach saturation level in the solution and thus it can be preserved in the bottom sediment. But from the prognosis point of view cadmium is more dangerous as there are no clear ways to detoxify it and therefore cadmium distributes further a field in nature reservoirs. Therefore there is need to find ways to decrease cadmium mobility because zinc toxicity we can regulate by water component controlling.

ACKNOWLEDGEMENTS

This investigation was supported by Grant of RFBR #97-05-65181 and Grant of UIGGM SB RAS "Mintex".

We want to express our thanks to E.Lazareva for help us to design the figures.

REFERENCES

Ball, J. W., Nordstrom, D. K. & Zachmann, D. W. 1987. WATEQ4F . *U.S. Geological Survey Open-File Report* 87-50.

Bortnikova, S.B., Gaskova O.L., Airijants, A.A., Kolonin G.R. & Tzimbalist V.G. 1997. Influence of carbonate minerals on contemporary oxidation processes in tailings of Pb-Zn ore recovery. *Society of Economic Geologists. Special Publication* 4: 640-653.

Bortnikova, S.B., Hozhina E.I., Fazlullin S.M. & Androsova, N.V. 1999 Heavy metals in aquatic vegetation in mining region. *Proceedings of GeoEnv'97, Turkey, September 1-5,* in press.

Nordstrom, D.K. & Ball, J.W., 1985. Toxic element composition of acid mine water from sulfide ore deposits. In R. Fernandez-Rebio (ed), *Mine Water-Proceedings of the Second international Congress, Granada, Spain, 1985.* Granada: International Mine Water Association.

Perel'man A.I., 1972. *Geochemistry of elements in the gipergenezis zone.* Moscow:" Nedra".

Perel'man A.I., 1966. *Landscape Geochemistry.* Moscow: "High school".

Geochemistry of the Earth's Surface, Ármannsson (ed.) © 1999 Balkema, Rotterdam, ISBN 90 5809 073 6

Relevance of self-sealing processes in pyrite sinters for heavy metal mobility

J. Müller & K.-P. Seiler

GSF, Institute of Hydrology, Neuherberg, Germany

ABSTRACT: Pyrite sinters as industrial waste are widespread deposits in Germany, and usually allow free access of precipitation and drainage of seepage water and might endanger ground water quality because of its high content of heavy metals (Cd, Cu, Ni, Pb, Zn, As, Sb). The pyrite sinters studied were of different ages (present, 10, 40 and 60 year), originated from the Bohemian Massif and were processed in the same industrial manner. The mode of deposition and therefore the weathering conditions differed. Long-term elution-experiments, mineralogical studies, field observations on ground waters and a sequential extraction have been executed to highlight leaching behaviour and assesment. The increasing portion of strongly fixed and non-leachable heavy metals with increasing age as well as mineralogical studies indicate weathering, co-precipitation and recrystallization processes immobilizing most of the heavy metals. These „self-sealing" processes do not endanger ground water systems, but could be accelerated by conditioning activities and might replace remediation.

1 INTRODUCTION

Pyrite sinters result from roasting processes to produce sulfuric acid and cellulose. Just like the raw material pyrite, the pyrite sinters are usually enriched in heavy metals like Cd, Cu, Ni, Pb, Zn and in the semi-metals As and Sb. The method of pyrite roasting for sulfuric acid production was a widespread method in Germany, especially in Bavaria and lead to a lot of deposits in open gravel pits or the material was used for road surfaces and railway embankments. Lately, the pyrite sinters have been used as further raw material for copper smelting.

Most of the deposits allow free access and drainage of precipitation and might endanger ground water quality via seepage. The assesment of ground water systems below such deposits is required because of the high content of heavy metals in such pyrite sinters.

Within the scope of a research project four pyrite sinters of different deposition age (present, 10, 40 and 60 year) were examined (Table 1).

The raw material of the pyrite sinters originated mostly from the Bohemian Massif or the former Yugoslavia and they were all processed in the same industrial manner. The mode of deposition differed from slope to horizontal layered or railway embankment deposition (Table 1) and therefore different weathering conditions and processes might have prevailed.

Table 1. Material properties and chemical composition.

Material	unit	Pyrite sinter	Pyrite sinter	Pyrite sinter	Pyrite sinter
Age of deposition	[year]	present	10	40	60
Mode of deposition		slope	layered	railway embankment	layered
Origin of pyrites		Bor	Bor	Meggen/Waldsassen	
Spec. Gravity	[g/cm³]	4.4	2.8	4.0	2.6
Dry matter	[%]	94	90	87	76
Chemical composition					
(Extraction with aqua regia following DIN 38 414-S 7)					
Al	[mg/kg]	4500	7300	8500	33,000
As	[mg/kg]	500	230	1250	270
Cd	[mg/kg]	26	7.3	41	5.1
Co	[mg/kg]	71	1.8	17	1.1
Cr	[mg/kg]	56	20	9.9	74
Cu	[mg/kg]	1450	970	4100	97
Fe	[mg/kg]	640,000	99,000	530,000	76,000
Hg	[mg/kg]	2.1	4.9	3.9	0.6
Mg	[mg/kg]	210	1600	130	4,900
Mn	[mg/kg]	1700	200	320	350
Ni	[mg/kg]	67	21	32	83
Pb	[mg/kg]	3550	1100	6250	1,200
Sb	[mg/kg]	50	50	290	32
Zn	[mg/kg]	4800	900	5100	920

2 METHODS

In addition to long-term elution-experiments (Schimetschek 1997, Seiler 1997) a sequential extraction has been executed to highlight the bonding and mobility of heavy metals in the pyrite sinters in detail than water extraction and aqua regia extraction could do (Müller 1998).

The long-term elution-experiments were designed as column-experiments under saturated conditions. The properties of the columns are listed in Table 2. The eluent was rain water and the experiments has taken 2 year. The elution rate simulated 1200 mm/year precipitation which means one order of ground water recharge higher than in the area of investigation (Kelheim, Germany).

Table 2. Properties of the column-elution-experiments.

Material	Unit	Pyrite sinter	Pyrite sinter	Pyrite sinter	Pyrite sinter
Age of deposition	[year]	present	10	40	60
Solid : eluent		1 : 4.5	1 : 4.6	1 : 4.1	1 :7.8
Exchange coefficient	[V_i/V_{eff}]	11.7	16	12.4	15.8
Hydraulic conductivity	[m/s] · 10^{-6}	6.2	4.2	3.5	1.9
Effective porosity	[%]	51	35	48	40

A sequential extraction method developed especially for iron rich materials does not exist so that a well known and established extraction sequence (Zeien & Brümmer 1989) was applied apart from step I and VIII which followed DIN 38 414 (1984, Table 3).

Table 3. Sequential extraction method after Zeien & Brümmer (1989), apart from step I and VIII which followed DIN 38 414.

Step	Character of bonding and mobility	Extraction agent	pH-value
I	Water soluble fraction (DIN 38 414-S 4, 1984)	$H_2O_{Milli-Q}$	7.0
II	Mobile fraction water soluble and exchangeable heavy metals (non-specific adsorbed) and easy soluble metal-organic complexes	Ammonium nitrate	7.0
III	Easy available fraction specific adsorbed, near sur-face occluded and heavy metals fixed to CaCO$_3$; metal-organic complexes with low bonding capacity	Ammonium acetate	6.0
IV	Fraction fixed to manganese-oxides	Hydroxylamine-hydrochloride + Ammonium acetate	5.5
V	Fraction fixed to organic substances	Ammonium-EDTA	4.6
VI	Fraction fixed to crypto-crystalline iron-oxides	Ammonium oxalate	3.25
VII	Fraction fixed to well-crystalline iron-oxides	Ascorbic acid (buffered with oxalate)	3.25
VIII	Aqua regio soluble fraction: (DIN 38 414-S 7, 1983)	HNO$_3$/HCL: 1:3	< 1

Neither of the sequential extraction methods characterize mobility and bonding conditions of heavy metals satisfactorily; they represent a compromise of understanding. The described mobilities and bondings are consequently of operational character and not of exact chemical definition. The mineralogical studies included thin cut analysis, X-ray diffraction analysis, raster-electron-microscopy analysis and energy-dispersive-x-ray analysis (Schimetschek 1997).

The chemicals used in this study were all of p.a. quality and the water used for dissolution and dilution was a MILLIPORE Milli-Q water with electric conductivity of < 18.2 MΩ.

3 RESULTS AND DISCUSSION

The results of the elution-experiments are calculated as mobility of heavy metals in [%] of total amount summed up after an elution time of 2 year (Figure 1). The unit [%] allows a direct comparison with the results of the sequential extraction which are shown in Figure 2; the different portions of mobilities and bonding are given in [%] of total amount.

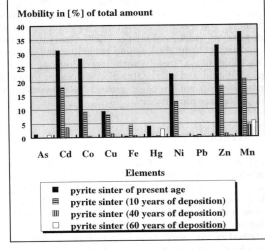

Figure 1. Mobility of heavy metals in [%] of total amount as result from the column-elution-experiments.

The results of the elution-experiments point out, that the mobilities of heavy metals decrease significantly with increasing age of deposition. Only the present material and the 10 year old material have relevant portions of mobile heavy metals. The mobility of As, Hg and Pb is rather small in all materials. The decrease of mobility with increasing age of deposition is characterized by a clear-cut between the 10 and 40 year old material. This is astonishing because the 40 year old material is characterized by heavy metals as much as those of the present material but has a leaching behaviour like the 60 year old material. In contrast, the heavy metals of the 10 year old material are as mobile as the present one although the amount of heavy metals is still smaller. The immobility of most of the heavy metals could be characterized by the sequential extraction because these portions of heavy metals are fixed to iron-oxides or are only soluble in aqua regia. The portions of water soluble heavy metals correlate very well with the mobile elements characterized by the elution-experiments, especially the elements As, Cd, Cu, Ni and Zn. In the cases of Hg, Mn and Pb the results of the sequential extraction do not correlate with the results of the elution-experiments concer-ning the mobile fractions. The elution-experiments seem to leach more than the water soluble fraction of these elements.

The development of heavy metal bonding from the present material up to the 40 year old material shows a significant increase of the heavy metal fraction fixed to crypto-crystalline iron-oxides, especially for As, Cd, Co, Cu and Zn.

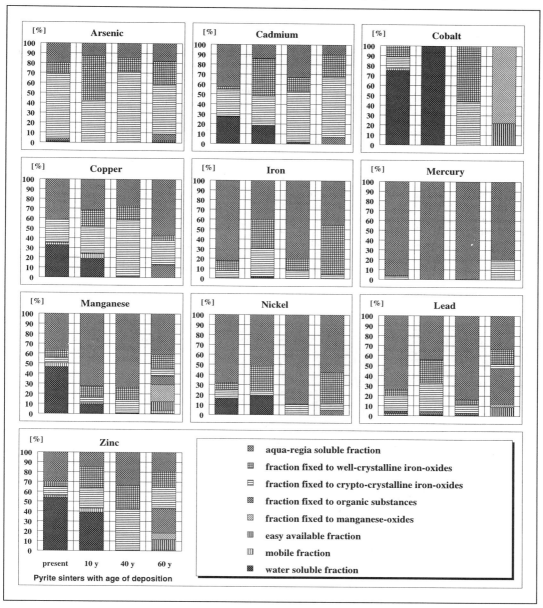

Figure 2. Distribution of bonding and mobilities of heavy metals from the pyrite sinters with different deposition age after sequential extraction described in Table 3; calculated in [%] of each fraction from the total amount given in Table 1.

213

The mineralogical characterization of the pyrite sinters (Table 4) shows a small decrease for hematite and magnetite with age of deposition. But the appearance of typical weathering minerals such as bassanite, ettringite and jarosite is much more interesting if weathering processes during deposition are considered.

Table 4. Mineralogical characterization of the pyrite sinters.

Material	Pyrite sinter	Pyrite sinter	Pyrite sinter	Pyrite sinter
Age of deposition	present	10 year	40 year	60 year
	hematite	hematite	hematite	quartz
	magnetite	magnetite	magnetite	gibbsite
	quartz	quartz	quartz	**bassanite**
	gibbsite	gibbsite	gibbsite	**jarosite**
	bassanite	**bassanite**	**bassanite**	hematite
		ettringite	biotite	magnetite

The limited mobility of heavy metals with increasing age of deposition especially the 40 year old material is due to processes that have taken place during deposition because the initial material from processing was in all cases identical. The results of the mineralogical studies indicate recrystallization processes which lead to typical weathering minerals like jarosite, bassanite and ettringite. Thin-cut analysis (Figure 3) show such recrystallization products in the form of „coatings" around e. g. gibbsite minerals or in other cases gibbsite relics. Gibbsite relics could still be oberved in the 60 year old material because they have been incorporated by such coatings and therefore sealed against further leaching.

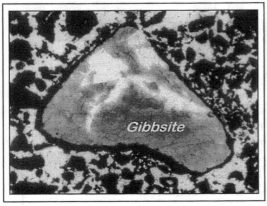

Figure 3. Thin-cut photo of a pyrite sinter with a gibbsite mineral incorporated by coatings of recrystallization products.

The 40 year old pyrite sinter shows the lowest amount of leaching in relation to its total amount of heavy metals. This material was used as railway embankment and was strongly compressed before deposition. Therefore only a small portion of rain water could infiltrate into the hydophobic material so that weathering under dry conditions took place.

For that reason, it is possible that besides recrystallisation processes surface sorption and diffusion into mineral failings e. g. of the iron-oxides could be responsible for the reduced amount of leaching of the heavy metals.

4 CONCLUSIONS

It is obvious, that the „coatings" also coprecipitated heavy metals and therefore immobilized them. These „self-sealing" processes lower the contamination potential of such pyrite sinters and minimize danger to ground water. This has been proved by respective ground water studies in carbonate-gravel waters (Schimetschek 1997, Seiler 1997).

If the processes of „self-sealing" were understood in detail, it would be possible to condition pyrite sinters in future during deposition and thus avoid ground water contamination as well as remediation activities.

REFERENCES

DIN 38 414-S 7 1983. *Schlamm und Sedimente (Aufschluß mit Königswasser zur nachfolgenden Bestimmung des säurelöslichen Anteils von Metallen); Baugrund. Erkundung durch Schürfe und Bohrungen sowie Entnahme von Proben.* Deutsches Institut für Normung e.V. (ed), Weinheim: VHC.

DIN 38 414-S 4 1984. *Schlamm und Sedimente (Bestimmung der Eluierbarkeit mit Wasser). - Deutsche Einheitsverfahren zur Wasser-, Abwasser- und Schlamm-Untersuchung.* Deutsches Institut für Normung e.V. (ed), Weinheim: VHC.

Müller J. 1998. Der Einfluß von Alterungsprozessen bei der Lagerung von Pyritabbränden auf das Auslaugungsverhalten und den Bindungscharakter von Schwermetallen. *Mitt. Dtsch. Bodenkundl. Ges.* 88: 217-220.

Schimetschek J. 1997. *Die Umweltrelevanz von Pyritabbränden für das unterirdische Wasser im Raum Kelheim.* Ph. D. Thesis. Munich: University of Munich.

Seiler K.-P. 1997. *Die Umweltrelevanz des Stoffaustrags von Produktionsresten und Sonderabfällen am Beispiel von Schwermetallen im Boden und unterirdischen Wasser des Industriestandortes Kelheim.* Neuherberg: Final report BSTMLU ADC-Nr. 66790.

Zeien H. & G. Brümmer 1989. Chemische Extraktionen zur Bestimmung von Schwermetallbindungsformen in Böden. *Mitt. Dtsch. Bodenkundl. Ges.* 59(1): 505-510.

Geochemistry of the Earth's Surface, Ármannsson (ed.) © 1999 Balkema, Rotterdam, ISBN 90 5809 073 6

The Tundra Nentsy: Health and environmental radiation on the territory they inhabit

L. P. Osipova, O. L. Posukh, A. V. Ponomareva & V. G. Matveeva
Institute of Cytology and Genetics, Siberian Branch of the Russian Academy of Sciences, Novosibirsk, Russia

B. L. Shcherbov, V. D. Strakhovenko & F. V. Sukhorukov
Analytical Center of the United Institute of Geology, Geophysics, and Mineralogy, Siberian Department of the Russian Academy of Sciences, Novosibirsk, Russia

ABSTRACT: Long-term investigation of the Tundra Nentsy population inhabiting the lower part of the Pur river basin (Pur raion, Yamal-Nentsy autonomous district) has demonstrated an increased frequency of pronounced chromosome aberrations, secondary immunodeficiency diseases, and the changes in blood indices typical of the populations affected by radiation. We have analyzed the long-lived radionuclides in the components of biogeocenosis and recorded an essential radioactive pollution of the territory inhabited by the aboriginal population of the Russian Extreme North. A special attention was focused on the ecological chain "lichen-reindeer-human". The proximity of the Severnyi Nuclear Test Site to the region under study allowed us to consider the test site as the chief source of radioactive contamination even after cessation of nuclear explosions on Novaya Zemlya.

1 RESULTS OF MEDICO-GENETIC STUDIES

1.1 *Object of investigation*

The population of Tundra Nentsy inhabiting a settlement of Samburg and the adjacent tundra (Pur raion, Yamal-Nentsy autonomous district) is rather small (about 2000), mostly nomadic, and occupies an area of about 100,000 km² (Figure 1).

Figure 1. The region investigated.

The Tundra Nentsy population has been well studied from the population and demographic genetic standpoint (Osipova et al. 1996, Posukh et al. 1996); this allows us to carry out systematic medico-genetic and family studies. According to demographic criteria, the group in question belongs to a growing type populations, ethnically uniform to a large extent, with a preserved "genetic core", and a small fraction of migrants both inbound and outbound. Their basic activities are reindeer breeding, fishing, and hunting. Therefore, their food chains are relatively simple: "lichen-reindeer-human" and "water-fish-human".

1.2 *Cytogenetic analysis*

Blood samples were taken from 170 Tundra Nentsy, virtually healthy by the time of examination. The peripheral lymphocytes stimulated with phytohemagglutinin were used for chromosome analysis. The lymphocytes were cultivated conventionally over 48h and fixed. The entire range of structural chromosome and chromatid aberrations, used conventionally to assess mutagenic effects, were considered. On the average, 100 metaphases were analyzed in each individual. The majority of the people examined was born and is living constantly in Pur raion. A group of children born after 1980 (N=56) was considered separately.

The data obtained were compared with the available published data, indicating that, depending on ecological situation and natural background radiation, the basic background frequency of chromosome aberrations (CA) varied in a range of 0-1.5%; the frequency of radiation markers (rings and dicentrics), 0.05-0.21%. The average frequency of these radiation markers is 0.078% (Pohl-Ruling 1990); in Russia, 0.05-0.1% (Bochkov 1993).

The results of cytogenetic analysis are listed in Table 1.

Table 1. Chromosome aberrations (CA) in the population of Tundra Nentsy

	N	Number of meta-phases	Total number of aberrant cells; frequency (confidence interval)	Cells containing rings and dicentrics; frequency (confidence interval)
Entire sample	170	18,406	588 3.19% (2.9%-3.4%)	83 0.45% (0.36%-0.55%)
Adult population	114	12,310	444 3.61% (3.3%-3.9%)	70 0.57% (0.44%-0.71%)
Children	56	6096	144 2.36% (2.0%-2.8%)	13 0.21% (0.11%-0.34%)

The frequencies of CA (3.19%) and rings-dicentrics (0.45%) over the entire sample (N=170) are significant ($p<0.001$) and exceed the upper conventional level of these parameters (1.5 and 0.1%, respectively). The adult population displays even higher values: 3.61 and 0.57%, respectively.

The total CA frequency in the children group (N=56) amounted to 2.36%, exceeding significantly ($p<0.05$) the control level. Rings and dicentrics were recorded in 13 children of the 56 examined (23%). The mean frequency of radiation markers in children from a settlement of Samburg amounted to 0.21%, exceeding twofold the upper level of the conventional control range. Similar cytogenetic analysis was performed in the children living, together with their parents, in Altai kray near the Semipalatinsk Nuclear Test Site and exhibiting multiple developmental defects (MDD) and atypical neonatal jaundice (AJ). The total CA frequency in children with MDD and AJ amounted to 2.8 and 2.7%, respectively, and up to 40% aberrations were of chromosome type (disruptions, translocations, rings, etc.) (Matveeva et al. 1993). The chromosome instability recorded in children from Samburg was similar to that in the children with MDD and AJ living in the radioactively contaminated region in Altai.

The entire sample studied was divided into three groups depending on the total CA frequency. Group I comprised the individuals displaying the control CA level (CA ≤ 1.5%); group II, with a moderate increase in CA frequency (1.5% < CA ≤ 3%); and group III, with high CA frequency values (CA > 3%) (Figure 2a). Only 38% of the people examined

fell into group I (CA ≤ 1.5%); 27% displayed a moderately increased CA frequency; and 35%, at least a twofold increase relative to the control level. In addition, the number of individuals displaying radioactive markers (rings and dicentrics) amounted to 9% in group I relative to the number of individuals in the group; 39%, in group II; and 53%, in group III. Thus, the trend of the increase in the frequency of radioactive markers with the increase in the total CA frequency is evident in the groups examined.

The data on adult and children groups are shown separately in Figures 2b and 2c. Note a positive dynamics among children: the group displaying a control level of chromosome aberrations is exceeding the group with an increased chromosome instability almost twofold. This suggests an extreme importance of the follow-on monitoring of younger children population.

a)
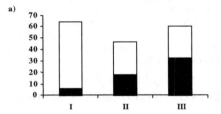

b)
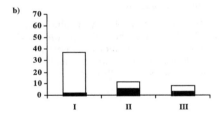

c)
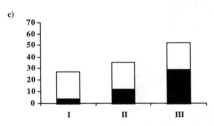

Figure 2. Distribution of the sample studied with respect to the total frequency of chromosome aberrations (CA): (a) the entire sample of Tundra Nentsy (N=170); (b) children (N=56); and (c) the adult part of the sample (N=114). The number of individuals examined is on the ordinate; the groups differing in CA, on the abscissa: (I) CA ≤ 1.5%; (II) 1.5% < CA ≤ 3%; and (III) CA > 3%. The fraction of individuals displaying rings and dicentrics are marked black.

1.3 Other examinations within epidemiological screening

The Tundra Nentsy population was examined therapeutically and immunologically to reveal secondary immunodeficiency diseases (SIDD). It appeared that the individuals with SIDD displayed the ring and dicentric chromosome aberrations 1.7-fold more frequently than the people merely composing the risk group of chronic diseases. Peripheral blood indices, studied in 90 randomly chosen individuals, have demonstrated certain changes recorded earlier in populations inhabiting the regions affected by radiation. Among the individuals examined, 17% displayed maximally pronounced deviations from the "regional norm", that is, reduced hemoglobin content and leukocyte number, and erythrocyte shape abnormalities, while only 24%, normal blood indices.

Preliminary analysis of the morbidity and mortality rates of the Samburg tundra population employing the statistical data of the Pur raion hospital demonstrated a drastic increase in the mortality rate from the cancer diseases, previously not typical of the aboriginal population in this area; of them, 50% were the cancer of digestive organs.

2 RESULTS OF RADIOECOLOGICAL STUDIES

The sample of biogeocenosis components included forest litters, mosses, lichens, peats, and lake sediments as well as certain samples of plants eaten by reindeers in summer. Venison, the main food of Tundra Nentsy, was sampled in northern and southern reindeer herds. ^{137}Cs was determined in the analytical Center of the United Institute of Geology, Geophysics, and Mineralogy, Siberian Branch of the Russian Academy of Sciences (Novosibirsk, Russia) by γ-spectrophotometry; ^{90}Sr, by β-radiometry. All the data listed are recalculated relative to the air-dried mass with the exception of various samples of reindeer organs analyzed in a natural state.

The papers on "lichen-reindeer-human" food chain, both Russian and foreign, are numerous; however, they were published as early as in 60-70s (Liden 1961, Nizhnikov et al. 1969, Holleman et al. 1971, Moiseev & Ramzaev 1975, etc.). The current radioactive contamination of the North is poorly known.

^{137}Cs was detected at all the 94 observation sites. Its activity (Table 2) in the biogeocenosis components of the same type varies in a wide range. These variations result undoubtedly from the uneven radioactive fallouts, recorded by all researches dealing with radioecological problems. However, the role of plant species range plays a conspicuous role too. For example, the ^{137}Cs activity in different lichen species

collected from an area of 10 m × 10 m ranges 58-144 Bq/kg.

Our data on ^{137}Cs activity in lichens are inconsistent with the data indicating a lesser contamination of Russian northern territories compared with southern regions (Troitskaya et al. 1971). For example, the ^{137}Cs activity recorded in the lichen *Cladina stellaris* in Pur raion amounted to 118.2 Bq/kg, whereas the same lichen species from the Altai region affected by the Semipalatinsk Nuclear Test Site exhibited only 38 Bq/kg. The same observation is true for the forest litter, the substrate of lichen growth (320 and 130 Bq/kg, respectively). However, the lake sediments in the North are less contaminated with ^{137}Cs compared with those in Altai. Unfortunately, the insufficient data on the lakes of the region studied prevents the correct comparison.

Table 2. The range of ^{137}• s activity in the biogeocenosis components in Pur raion

Components	^{137}Cs, Bq/kg mean value (range)
Forest litter	320 (19-610)
Moss	89.2 (2-738)
Lichen	118.2 (9-372)
Peat	130 (60-200)
Lake sediments	59 (33-102)
Cep (*Boletus edulis*)	169.5 (120-219)
Birch and alder leaves	64
Horsetail	41.7 (8-76)
Venison	162.1 (66-315)
Reindeer liver	71.4 (30-131)
Kidneys	163.9 (46-370)
Lungs	54.0 (30-82)
Heart	74.8 (47-99)
Bone marrow	14.7 (2-38)
Bone tissue	22.4 (5-57)

Note a considerable ^{137}Cs activity in venison (Table 2). Although, the mean values fall below the permissible level, the continuous consumption of venison and other reindeer organs is a chronic source of internal irradiation of the Tundra Nentsy. Moreover, such activities as 1200 Bq/kg, recorded in a sample of jerked venison (representing the major part of the summer ration of aboriginal population) should be considered as extremely dangerous.

On the whole, the venison from the northern herds (Samburg tundra) is less radiocesium-contaminated than in the southern herds (88.3 and 203.5 Bq/kg, respectively). For comparison: ^{137}Cs was not detected in the mutton and beef samples from Altai.

Note also a more pronounced ^{137}Cs accumulation in organs of young reindeers compared to old individuals. For example, the meat of a nine-month calf exhibited 256 Bq/kg of ^{137}Cs vs. 141 Bq/kg in a six-year bull from the same herd; in liver, 131 and 51 Bq/kg, respectively; in kidneys, 370 and 141 Bq/kg;

and in bones, 43 and 23 Bq/kg, respectively. And it is not a singular example.

As for ^{90}Sr, its activity in the soft reindeer tissues is negligible, varying from 0.25 to 1 Bq/kg in the northern herds. In bone tissue, the activity is essentially higher: 54.5 to 103.4 Bq/kg. The data on southern herds are absent.

3 CONCLUSION

Analysis of individual components of the Pur raion biogeocenosis has demonstrated their considerable contamination with artificial radionuclides. Of special importance is the increased radiocesium contamination of lichens in Pur raion. The aboriginal reindeer-breeding populations consume ^{137}Cs mainly through venison, their major food. Thus, the incorporated radiocesium may be a source of continuous internal irradiation, negatively affecting the health of humans, whose nutrient chain starts from lichen.

The scale of the recorded cytogenetic abnormalities, secondary immunodeficiency diseases, and specific changes of the general blood indices along with their similarity to these parameters recorded in the population of the radioactivity-contaminated regions suggest that radiation factor, which is a consequence of nuclear tests in the Severnyi Nuclear Test Site (the island of Novaya Zemlya) is presently contributing to the health deterioration of the aboriginal population of the North.

This work was funded by the Integrative grant of SB RAS (1997-1998) and the RFBR grant No. 97-05-65235.

REFERENCES

Bochkov, N.P. 1993. An analytic survey of cytogenetic studies following the Chernobyl accident. *Vestnik RAMN* 6: 51-56.

Holleman, D.F., Luik, J.R., Whicher, F.W. 1971. Transfer of radiocesium from lichen to reindeer. *Health Phys.* 21(5): 657-666.

Liden, K. 1961. Cs^{137} burdens in Swedish Laplanders and reindeer. *Acta Radiolog.* 56: 64-65.

Matveeva, V.G., Sablina, O.V., Eremina, V.R., et al. 1993. Cytogenetics of innate pathology in Altai populations in the regions of radioactive contamination. In: *Genetic Effects of Anthropogenic Environmental Factors,* No. 1: 5-17, Novosibirsk: ICG SB RAS.

Moiseev, A.A., Ramzaev, P.V. 1975. *Cesium-137 in the Biosphere.* Moscow: Atomizdat.

Nizhnikov, A.I., Nevstrueva, M.A., Ramzaev, P.V., et al. 1969. *Cesium-137 in the chain lichen-reindeer-human in the USSR Extreme North (1962-1968).* Moscow: Atomizdat.

Osipova, L.P., Posukh, O.L., Ivakin, E.A., Kryukov, Yu.A., Karafet, T.M. 1996. Gene pool of the aboriginal population of the Samburg Tundra. *Genetika* 32(6): 830-836.

Pohl-Ruling, J. 1990. Chromosome aberrations of blood lymphocytes induced by low-level doses of ionizing radiation, In: G.Obe (ed.), *Advances in Mutagenesis Research.* 2: 155-190. Berlin: Springer-Verlag.

Posukh, O.L., Osipova, L.P., Kryukov, Yu.A., Ivakin, E.A. 1996. Genetic-demographic analysis of the population of native inhabitants of the Samburg Tundra. *Genetika* 32(6): 715-721.

Troitskaya, M.N., Ramzaev, P.V., Moiseev, A.A., et al. 1971. Radioecology of Extreme North landscapes. In: *Modern Problems of Radiobiology.* 2: 325-353. Moscow: Atomizdat.

Geochemistry of the Earth's Surface, Ármannsson (ed.)© 1999 Balkema, Rotterdam, ISBN 90 5809 073 6

Mercury levels in soils in the vicinity of a crematorium in North London, United Kingdom

S. Panyametheekul, M. E. Farago & J. Rieuwerts
T.H. Huxley School of Environment, Earth Sciences and Engineering, Imperial College of Science, Technology and Medicine, London, UK

ABSTRACT: Mercury has been known since prehistoric times and was used by the Greeks and Romans. After the well-documented outbreak of Minamata Disease in Japan in the 1950s, there has been much concern over mercury contamination of water. However, there is a paucity of research regarding mercury in soils surrounding crematoria or atmospheric discharges from crematoria stacks. Nowadays, the widespread use of mercury in amalgam fillings and the increasing numbers of cremations should be considered as a possible cause of mercury contamination. This study attempts to estimate the dispersion of mercury in the plume of a crematorium by reference to the levels of mercury in the surrounding soils. Soil samples for the analytical and experimental work were taken from the vicinity of selected crematorium. Total mercury concentrations were measured using Inductively Coupled Plasma-Atomic Emission Spectrometry (ICP-AES). The results indicated significantly higher levels of mercury (maximum 0.69 mg kg^{-1}) in the vicinity of the crematorium compared to background levels in the region, which was range between 0.008-0.30 mg kg^{-1}

1 INTRODUCTION

Mercury, Hg, belongs to group II-B of the Periodic Table, and has an atomic weight of 200.59. It was named quicksilver by Aristotle over two thousand years ago (Suzuki *et al.* 1991). In previous centuries it was widely employed in medicine and is still used in some applications, for example dental filling.

Even though the use of amalgam fillings has been banned in Germany for pregnant women, small children, and patients with kidney problems, the UK and US dental associations continue to use amalgam.

There is no proven risk to patients from amalgam and there is a lack of suitable alternatives (Phillips *et al.* 1994). The increasing use of cremation as a method of corpse disposal, coupled with the fact that dental amalgam consists of 50% metallic mercury, implies that a significant amount of mercury should be released into the environment every year via this route (DoE/ERM 1996; Maloney *et al.* 1998; Nieschmidt and Kim 1997; Phillips *et al.* 1994). As demonstrated in Figure 1, the pathway of amalgam mercury can be traced from individual cavities to stack emissions and possible hazards.

There is a lack of investigation to consider how much mercury is emitted into the environment and in particular, how much is deposited on to soils when human bodies are cremated. In addition, there is no current limit on emissions of metals from crematoria in the UK, other than the guidance that coffins made from lead and zinc should not be cremated (DoE/ERM 1996; Chamberlain 1993).

The aim of this study was to determine whether mercury concentrations in soils around a crematorium in north London could be considered sufficient to cause environmental problems. A secondary aim was to investigate relationships between mercury concentrations in soil and other parameters, for instance pH, loss-on-ignition (LOI), and other element concentrations.

2 EXPERIMENTAL

2.1 *Study area*

The crematorium chosen for investigation was a crematorium in North London, UK. The sampling points in the vicinity of crematorium are shown in Figure 2.

2.2 *Soil collection*

For measuring metal elements in soils, soil samples are generally collected in Kraft bags. In contrast, for the analysis of mercury in soils, keeping the samples in the Kraft bags with sealed polyethylene bags is essential to avoid absorption of mercury from air and

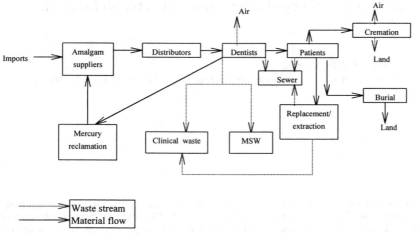

Figure 1 A profile of mercury from amalgam (DoE/ERM, 1996)

revolatilisation back to the atmosphere from soil samples (Davies 1976; Jones *et al*. 1995).

2.3 *Soil sampling point*

The soil samples were collected along a transect which crossed the stack of the crematorium. With the intention of determining the dispersion of mercury from the stack by reference to the levels of mercury in soils around the area of the crematorium, samples taken followed the main wind direction (from southwest to northeast).

In order to yield a representative sample, each soil sample comprised sub-samples from 4-5 separate points within a 1 m² area.

2.4 *Sample collection*

Fifty topsoil (0-15 cm) samples were taken using a stainless steel trowel, placed in paper bags which were transferred to zip-lock polyethylene bags. The samples were frozen to preserve their chemical integrity until analysis (Jones *et al*. 1995; Pereira *et al*. 1998).

2.5 *Extraction and analysis*

Samples, in Kraft bags, were air-dried at room temperature for 2-3 days. The dried samples were ground in a porcelain pestle and mortar, which were acetone-washed between each sample. Ground samples were sieved through a 2-mm nylon sieve to eliminate coarse material. The sieved samples were each separated into two zip-lock plastic bags. One bag was used for measurement of pH, loss-on-ignition (LOI), and mercury concentration and the other for crushing in a TEMA swing-mill. The lat-

ter fraction was analyzed for a suite of elements; Li, Na, K, Be, Mg, Ca, Sr, Ba, Al, La, Ti, V, Cr, Mo, Mn, Fe, Co, Ni, Cu, Ag, Zn, Cd, Pb, and P. Mercury and all other element concentrations in soils were determined by Inductively Coupled Plasma-Atomic Emissions Spectrometry (ICP-AES)

3 RESULTS AND DISCUSSION

The spatial distribution of mercury concentrations around the vicinity of the crematorium is shown in Figure 2. Results are also summarized in Table 1.

In order to elucidate data obtained, some estimate of normal or background level of mercury is needed Several studies have reported typical concentration for soils in the UK. A figure of 0.25 mg kg⁻¹ was reported by DoE/ERM (1996). Yaron (1996) stated on average level of 0.1 mg kg⁻¹ and 15 mg kg⁻¹ for contaminated soils. A range of 0.01-0.3 mg kg⁻¹ was quoted by Petts (1997). Thornton (1991) reported concentrations of 0.008-0.19 mg kg⁻¹. As seen in Table 1, the range of mercury concentrations measured in the recent study is 0.04-0.69 mg kg⁻¹. From a comparison of these figures, it appears that there are some elevated levels of mercury to be found in the vicinity of the stack. In addition, the results of a recent study of Maloney *et al*. (1998) suggest that more than 60 per cent of staff at crematoria had higher concentrations of mercury in their hair than controls. However, Figure 2 indicates that the cremation may contribute to the loading of mercury in the vicinity of the crematoria. This study perhaps needs further research to enhance the understanding of mercury deposition.

In order to examine inter-relationships between metal concentrations and other soil parameters, cor-

Grid Reference

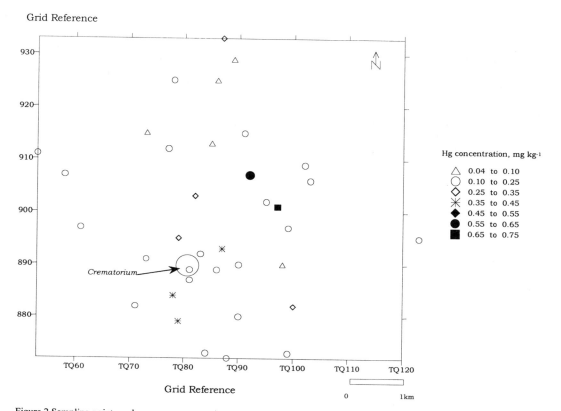

Figure 2 Sampling points and mercury concentration around a crematorium in North London, UK.

Table 1 Summary data for soil samples at crematorium site, n=50

| | pH | LOI(%) | Metal content in dry soil (mg kg⁻¹) | | | | | | | | | | | |
			Hg	Al	Cr	Mo	Mn	Fe	Co	Ni	Cu	Zn	Cd	Pb
Geometric mean	5.67	15.5	0.16	194	30.2	2.34	327.5	177	7.9	18.4	25.1	105	0.44	100.3
Maximum	7.54	83.1	0.69	397	49.8	4.4	240	345	35	48	90.2	427	1.6	538
Minimum	4.01	6.2	0.04	301	6.6	0.8	63.9	291	1.8	7	8.62	30.5	0.2	31.2

relation coefficients were calculated. The results demonstrate that mercury concentration in soil is significantly correlated (+) with soil levels of lead and copper. This relationship is likely to be an indicator of a source of Hg, Pb and Cu pollution, such as a crematorium. It is hoped that future work will clarify this point.

4 CONCLUSIONS

There is no current limit on emissions of metals from crematoria in the UK. However, an in-creasing number of cremations could be associated with a significant amount of mercury released into the environment.

Even though a great deal of toxic effects from mercury released during cremation have been considered, there is not a large body of research regarding mercury in soils surrounding crematoria in the UK. According to Mills (1990), in the UK, approximately 11 kg of mercury is being released each year from unfiltered crematorium chimneys. A recent investigation of mercury in soils around a crematorium in Brighton (Nunn, 1996) has shown that the maximum level of mercury in the area was

7.3 mg kg^{-1}. A study by Phillips *et al.* (1994) found maximum levels of mercury in soil of 1.32 mg kg^{-1} around another British crematoria. Maximum mercury concentrations recorded in the present work are 0.69 mg kg^{-1}. In general, mercury concentrations around the crematorium appear elevated compared to background levels. However, the maximum concentrations presented are lower than those observed around other crematoria in the UK. Nevertheless this research indicates that mercury levels in the vicinity of crematoria are of growing concern.

ACKNOWLEDGEMENTS

SP acknowledges a Studentship from the Royal Thai Government. The technical advice of Barry J. Coles is greatly appreciated. The authors would also like to acknowledge the contribution of Alban Doyle for experimental assistance.

REFERENCES

Chamberlain, C.T. (1993). Experiences with cremators under the Environmental Protection Act in the United Kingdom. *Pharos International*. 59(3): 102-103.

Davies, B.E. (1976). 'Mercury content of soils in Western Britain with special reference to contamination from base metal mining'. *Geoderma* 16(3): 183-192.

DoE/ERM (1996). *Mercury in the UK*. Final report prepared by Environmental Resources Management (ERM) on behalf of the Department of Environment (DoE).

Jones, R.D., M.E. Jacobson, R. Jaffe, J. West-Thomas, C. Arfstrom, & A. Alli (1995). Method development and sample processing of water, soil, and tissue for the analysis of total and organic mercury by cold vapour atomic fluorescence spectrometry. *Water, Air and Soil Pollution* 80: 1285-1294.

Maloney, S.R., C.A. Phillips, & A. Mills (1998). 'Mercury in the hair of crematoria workers'. *The Lancet*. 352(Nov. 14): 1602.

Mills, A. (1990). Mercury and crematorium chimneys. *Nature* 346: 615

Nieschmidt, A.K. & N.D. Kim (1997). Effects of mercury release from amalgam dental restorations during cremation on soil mercury levels of three New Zealand crematoria. *Bull. Environ. Contam. Toxicol* 58: 744-751.

Nunn, A. (1996). *An investigation into the concentration of mercury found in the soil in the vicinity of emissions from Brighton Crematoria*. Unpublished B.Sc. Dissertation. University of Sussex.

Pereira, M.E., A.C. Duarte, G.E. Millward, S.N. Abreu, & C. Vale (1998). An estimation of industrial mercury stored in sediments of a confined area of the lagoon of Aveiro (Portugal). *Wat. Sci. Tech.* 37(6-7): 125-130.

Petts, J. (1994). Incineration as a waste management option. In: Hester, R.E. & R.M. Harrison (eds.): *Waste incineration and the environment*, Royal Society of Chemistry.

Phillips, C.A., T. Gladding, & S. Maloney (1994). Clouds with a quicksilver lining. *Chem. Brit.* (August): 646.

Suzuki, T., N. Imura, & T.W. Clarkson (eds.) (1991). *Advances in mercury toxicology*. New York: Plenum Press.

Thornton, I. (1991). *Metals in the global environment: facts and misconceptions*. ICME.

Geochemistry of the Earth's Surface, Ármannsson (ed.)© 1999 Balkema, Rotterdam, ISBN 90 5809 073 6

Critical loads of metals in UK soils: An overview of current research

J. S. Rieuwerts, M. E. Farago & I. Thornton – *Geochemistry and Health Group, T. H. Huxley School, Imperial College, London, UK*

M. R. Ashmore – *Department of Environmental Science, University of Bradford, UK*

D. Fowler & E. Nemitz – *Institute of Terrestrial Ecology, Penicuik, UK*

J. Hall & D. Kodz – *Institute of Terrestrial Ecology, Huntingdon, UK*

A. Lawlor & E. Tipping – *Institute of Freshwater Ecology, Ambleside, UK*

ABSTRACT: A consortium of several research institutions in the United Kingdom is currently investigating the potential of the critical loads concept for assessing impacts, on terrestrial and freshwater ecology, of the long-range transboundary transport of several metals (Cd, Cu, Pb and Zn). Preliminary measurements of these metals in wet deposition, soils and freshwaters have been made and are continuing throughout the project. Preliminary investigations have borne several key findings, although data for estimating national critical loads will be obtained as the project progresses. Future work will involve verification of models and their integration into a GIS framework and subsequent production of national critical loads maps, highlighting susceptible areas.

1 INTRODUCTION

The critical loads concept, originally developed for acid deposition (Critical Loads Advisory Group 1994), is now being evaluated as a tool to assess the impacts of long-range atmospheric transport of metals. This project is focussed on UK uplands, and aims to extend our understanding for metals of deposition processes, soil mobility and bioavailability, and pathways into freshwaters, integrating the models developed in a GIS framework to critically evaluate the critical loads approach for metals. Here we report progress to date.

2 DEPOSITION

To quantify wet and dry deposition to UK upland areas, knowledge is required of: the air concentration field; average particle size; washout coefficients (w); and size-dependent deposition velocity (V_d). At a Scottish moorland site, wet deposition and air concentrations below 2.5 μm ($PM_{2.5}$) and 10 μm (PM_{10}) have been measured over the past three years and a first deposition estimate has been derived (Table 1). From the ratio of concentration in rain and air, washout coefficients can be estimated, which decrease strongly with decreasing particle size, expressed by the ratio $PM_{2.5}/PM_{10}$.

The deposition velocity can be inferred from micrometeorological size-segregated particle flux measurements, as was done at a Dutch heathland. The results show a clear diurnal cycle and an increase of V_d with particle diameter (Dp). These results will be used in deposition models to derive the yearly deposition of heavy metals at a 5 km x 5 km grid.

Table 1. Dry deposition estimates based on measured mean air concentrations ($PM_{2.5}$ / PM_{10} – $PM_{2.5}$) and V_{ds} parameterizations as well as measured wet deposition.

Aerosol species	Average concentration	Estimated ave. V_{ds}	Dry deposition	Wet deposition
	ng m^{-3}	mm s^{-1}	g ha^{-1} yr^{-1}	g ha^{-1} yr^{-1}
Zn	5.98/4.55	1.5/6.9	8.6	66
Pb	8.42/3.79	1.5/5.3	6.3	14
Cd	0.06/0.07	1.5/6.9	0.067	0.29
Cu	0.23/0.48	1.5/6.9	0.96	9.4
Ni	0.22/0.89	1.5/6.9	0.77	0.076
Cr	0.16/0.19	1.5/6.9	0.22	0.82

The air concentration field will be inferred from an emissions inventory and the results will be compared with direct or indirect measurements of the deposition on a coarser resolution.

3 SOIL ANALYSIS AND BIOAVAILABILITY MODELLING

Soil samples from six soil series in N. Wales were analysed for total metals, metals extractable by water and 0.01 M $CaCl_2$, and a range of soil parameters. Regression analysis shows that pH is the strongest predictor of Cd and Zn extractability by water and $CaCl_2$ (e.g. Fig.1) and for Pb extractability by $CaCl_2$

(Fig.2). Water-extractable Cu and Pb was not correlated with pH: the strongest predictor in both these cases was dissolved organic carbon, indicating organic complexation of free ions.

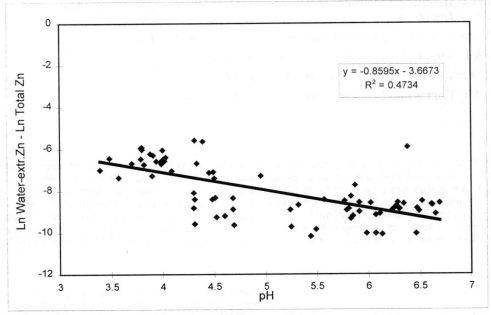

Figure 1. Relationship between pH and water-extractable Zn.

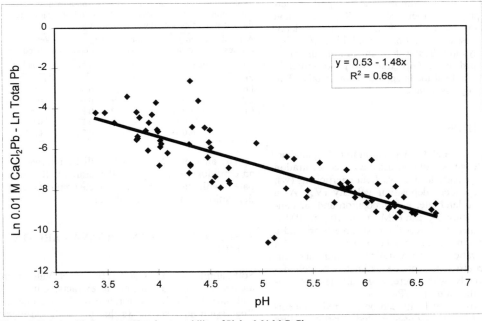

Figure 2. Relationship between pH and extractability of Pb by 0.01 M CaCl₂.

Figures 1 and 2 illustrate the generally observed increase in metal extractability with lower pH. Although the R^2 value recorded with pH could be increased slightly with the addition of further soil parameters, no one variable was consistent in this respect. Empirical models, derived from statistical analysis of the data, will be verified with further field work and will used to predict extractable metal concentrations in other upland areas of the UK.

4 DYNAMIC MODELLING

We are monitoring the atmospheric deposition of Cu, Zn, Cd and Pb to two upland catchments: Great Dun Fell, N.Pennines and River Duddon, Lake District. Rain collectors have been deployed every two weeks from April 1998 and bulk metal concentrations measured by ICP-MS. Weekly stream water chemistry is measured. Particular attention is given to reducing contamination during sampling and subsequent handling, and measurements are controlled by a QA/QC scheme. Soil extractions have also been performed to investigate the behaviour of metals in soils. Soils have been leached with acids of varying strengths and batch titrations have been used to in-vestigate metal behaviour with changes in pH, a major controlling factor.

Mean concentrations for metal deposition to the catchments are given in Table 2. Values are similar for the two sites and of the same order of magnitude to those found in other studies. Figure 3 shows metal concentrations for one stream draining the River Duddon catchment. Metals are largely in the dissolved phase due to low SPM concentrations for these upland waters. Future work will involve application of the CHemistry of the Uplands Model to simulate metal hydrochemistry.

Table 2. Volume weighted mean metal concentrations for bulk rainwater collected at Cockley Beck and Great Dun Fell: two upland catchments. Sampling period: April-October 1998. The values are means from two 14 cm funnel collectors at each site.

Element	Cockley Beck	Great Dun Fell
	$\mu g\ l^{-1}$	$\mu g\ l^{-1}$
Ni	0.30	0.29
Cu	0.56	0.72
Zn	3.0	2.9
Cd	0.05	0.05
Pb	0.87	1.0

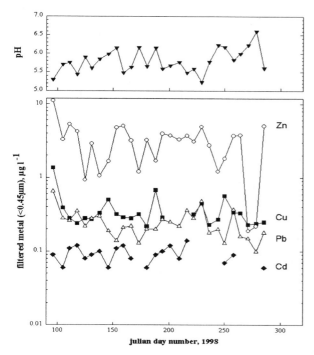

Figure 3. Trace metal time series for Troughton Gill, a stream draining the River Duddon catchment. The lowest pH values are observed at high flow. There are good correlations between Zn and pH (r = -0.63) and Pb and pH (r = -0.74). The missing values for Cd are below ICP-MS detection limits (0.06 $\mu g\ l^{-1}$)

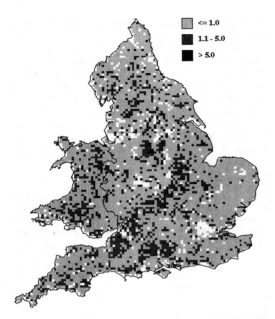

Figure 4. Concentrations of cadmium in topsoils (0-15cm) in England and Wales (mg kg^{-1}).

5 GIS MODELLING

We have assembled the various datasets required for calculating critical loads in the UK, such as the top-soil metal concentration data illustrated (Fig. 4): other key datasets include stream sediment metal concentrations and data on soil series, solid and drift geology, run-off and meteorology. Initially, we are using a steady state critical loads model devised in the Netherlands. A sensitivity analysis has been carried out to identify those variables in the model which have the most significant influence on the calculated critical load. The results indicate the relative importance of three terms; water flux through soil; critical metal concentration in soil for biological effects; and the partition coefficient between total metal and soil solution metal concentrations. Root uptake and weathering can be excluded from the steady state model due to their negligible influence. Consequently this highlights the areas in which work should be undertaken to ensure the greatest accuracy possible of input variables comprising these terms.

6 CONCLUSIONS

Preliminary measurements of several metals, in wet deposition, soils and freshwaters, have been made and are continuing throughout the project. Dry deposition estimates have also been made based on at-

mospheric measurements of $PM_{2.5}$ and PM_{10} and size-dependent deposition velocity. Key results for estimating critical loads will be obtained in the later stages of the project but several key findings have been made from the preliminary measurements obtained to date. Dry deposition measurements have indicated a diurnal cycle and an increase of deposition velocity with particle diameter. Metals measured in rainwater are similar to those found in other studies. Analysis of soils has demonstrated the importance of pH as an influence on the solubility of Cd, Cu and, in some cases, Pb. Dissolved organic material appears to be a more important influence on Cu solubility. Metals in some of the upland streamwaters measured appear to occur largely in the dissolved phase. Preliminary GIS work has shown the importance (on calculated critical load) of soil metal concentration, the soil to soil solution partition coefficient and soil water flux.

7 FUTURE WORK

Dry deposition data currently being obtained will be used to model the yearly deposition of heavy metals. Predictive models derived for estimating metal concentrations in soil solution will be verified with further field work and will be used to predict soluble metal concentrations in other upland areas of the UK Future work on dynamic modelling will involve application of the CHemistry of the Uplands Model to simulate metal hydrochemistry. Verified models for the different components of the research will be integrated into a GIS framework and national critical loads maps will be produced, highlighting areas susceptible to further metal loading.

8 ACKNOWLEDGEMENTS

The authors acknowledge funding from the Department of the Environment, Transport and the Regions; National Environment Research Council; Environment Agency and Rio Tinto.

REFERENCES

Critical Loads Advisory Group 1994. *Critical Loads of Acidity in the United Kingdom*. Summary report for the Department of the Environment.

Geochemistry of the Earth's Surface, Ármannsson (ed.) © 1999 Balkema, Rotterdam, ISBN 90 5809 073 6

The influence of different soil parent material on cadmium biopurification

M. Svetina
ERICO Velenje, Slovenia

ABSTRACT: In the Šalek Valley (Slovenia) a study of Cd in soil and plant materials on three different geological bedrocks (calcareous, sedimentary and andesitic rocks) was made. An alternative assessment method called biopurification was used to estimate Cd uptake into the food chain. Biopurification is expressed by the Cd and Zn ratio in plants and soils. It was established that soils formed from calcareous rock are more advantageous with respect to uptake from sedimentary and andesitic rock, because they limit Cd bioaccumulation in the plants growing on these types of soil.

1 INTRODUCTION

The abundance of Cd in limestones and dolomites is around 0.035 µg/g, in magmatic and sedimentary rocks it does not exceed 0.3 µg/g, and this metal is likely to be concentrated in argillaceous and shale deposits. Cd is strongly associated with Zn in its geochemistry. Cd exibits a higher mobility than Zn in acidic environments. The main factor determining the Cd content of soil is the chemical composition of the parent rock. The average contents of Cd in soils lie between 0.07 and 1.1 µg/g. Higher values than 0.5 µg/g reflect anthropogenic impact on the Cd status in top soils (Kabata-Pendias & Pendias 1986).

The uptake of Cd into plants generally depends upon the availability of the metal in soil solution. The soil pH and composition, particularly the nature of the soil clays, the organic matter content, and the soil Cd level all affect this availability (WHO 1992). Cd contents in plants depend on sampling time (physiological age of the plant) and on plant species (Sillanpää & Jansson 1992). Due to incompatible units in different materials, it is difficult to assess the impact of cadmium on the environment. An alternative metod is called biopurification; this recognizes that the flow of the non-nutrient Cd through successive trophic levels follows a pathway similar to that of nutrients such as Ca or Zn. Both Cd and Zn should be measured at each trophic level (material) and expressed as the molar ratio of these two elements (WHO 1992).

The present study was performed in order to establish the influence of different parent rock on Cd biopurification from soil to plants. Fifteen study areas in three of most most typical geological formations in the polluted Šalek Valley were chosen

(Svetina & Pirc 1998). Five of them are on calcareous bedrock, five on sandstone and five on andesitic tuff. Identical reference areas representing clean areas were chosen for each geological formation.

2 METHODS

To obtain a more complete picture of Cd behaviour in soil, this study is based on two separate analyses, soil analysis and plant analysis.

2.1 *Soil analysis*

The soil samples were taken from 5 to 10 cm soil depth in the pedological profile. The samples were stored in plastic bags, air-dried, sieved through a sieve with 2 mm mesh size and pulverized in an agate mill. Prior to analysis the samples were extracted by aqua regia digestion. Cd and Zn were analysed by the ICP analytical technique in the Acme Analytical Laboratories. The detection limit for Cd in soil samples was 0.01 and for Zn 1 µg/g. To provide quality control the NIST standard reference materials 2704 (Buffalo River Sediment) was analysed together with the soil samples.

2.2 *Plant analysis*

The plants were sampled by separately collecting the typical meadow plants (*Plantago lanceolata, Trifolium pratense, Rumex acetosa, Dactylis glomerata, Taraxacum officinale and Achillea millefolium*). Fodder, namely a composite sample, includes all 6 plant species (5 mg of each species).

Samples were oven-dried at 30°C and milled in an agate mill. Prior to analyses samples were decomposed with nitric acid in a microwave oven. Cd and Zn were determined by the AAS analytical technique in the ERICo laboratory. The detection limit for Cd in plant samples was 0.01 and for Zn 10 µg/g. To provide quality control the NIST standard reference material 1515 (Apple Leaves) was analysed together with the plant samples.

3 RESULTS

Tables 1 to 6 summarise the Cd and Zn contents with the Zn/Cd molar ratio in the last column. Results are presented separately for each sampling material (soil and plant) and each parent material (calcareous rock, sedimentary rock and andesitic tuff). Tables 1-3 give results for soil analysis and Tables 4-6 results for plant analysis. In the last two rows of the tables the average values and standard deviations for the samples are presented.

Table 1. Cd and Zn contents of soils on calcareous rock.

Study area	Cd*	Zn*	Zn/Cd
Šalek Valley 1	1.66	103	106
Šalek Valley 2	1.70	121	122
Šalek Valley 3	1.22	69	97
Šalek Valley 4	1.96	198	174
Šalek Valley 5	1.32	85	110
Reference area	1.42	159	193
Average	1.55	123	134
Standard deviation	0.28	48	39

*µg/g

Table 2. Cd and Zn contents of soils on sedimentary rock.

Study area	Cd*	Zn*	Zn/Cd
Šalek Valley 6	0.50	78	269
Šalek Valley 7	0.95	89	160
Šalek Valley 8	4.19	84	35
Šalek Valley 9	0.43	81	324
Šalek Valley 10	0.57	97	292
Reference area	0.17	56	565
Average	1.14	81	274
Standard deviation	1.50	14	177

*µg/g

Table 3. Cd and Zn contents of soils on andesitic tuff.

Study area	Cd*	Zn*	Zn/Cd
Šalek Valley 11	1.09	102	160
Šalek Valley 12	0.50	88	302
Šalek Valley 13	0.84	113	232
Šalek Valley 14	0.70	83	204
Šalek Valley 15	0.28	54	330
Reference area	0.28	96	588
Average	0.62	89	303
Standard deviation	0.32	20	153

*µg/g

Table 4. Cd and Zn contents of plants on calcareous rock.

Study area	Cd*	Zn*	Zn/Cd
Šalek Valley 1	0.54	41.9	133
Šalek Valley 2	0.49	56.7	199
Šalek Valley 3	0.48	57.6	206
Šalek Valley 4	0.21	34.9	286
Šalek Valley 5	0.84	53.6	110
Reference area	0.47	49.7	182
Average	0.51	49.1	186
Standard deviation	0.20	8.9	62

*µg/g

Table 5. Cd and Zn contents of plants on sedimentary rock.

Study area	Cd*	Zn*	Zn/Cd
Šalek Valley 6	1.25	60.0	83
Šalek Valley 7	0.58	35.7	106
Šalek Valley 8	1.37	46.8	59
Šalek Valley 9	0.55	37.9	118
Šalek Valley 10	0.85	54.6	110
Reference area	0.68	42.7	108
Average	0.88	46.3	97
Standard deviation	0.35	9.5	22

*µg/g

4 DISCUSSION

There is a general problem in expressing changes in content of Cd between different environmental spheres due to different sampling materials. The biopurification

Table 6. Cd and Zn contents of plants on andesitic tuff.

Study area	Cd*	Zn*	Zn/Cd
Šalek Valley 11	0.83	35.0	72
Šalek Valley 12	1.01	54.7	93
Šalek Valley 13	0.83	51.1	106
Šalek Valley 14	1.19	42.6	62
Šalek Valley 15	1.50	76.1	87
Reference area	0.96	46.9	84
Average	1.05	51.1	84
Standard deviation	0.26	14.1	15.5

*µg/g

method is an alternative method, suggested for estimation of uptake of Cd into the food chain (WHO 1992). Biopurification is expressed by a biopurification index (BPI). BPI is the molar ratio between Cd and Zn at the higher trophic level and the lower level. In our case the higher level is plant material and the lower level is soil material. The process could be described as bioaccumulation and in the following discussion this term will be used. A BPI value between 0 and 1 means biopurification of Cd in the bioaccumulation process and a BPI higher than 1 means Cd enrichment at the higher level relative to the lower level.

In Figure 1 the BPI in three different parent materials is shown graphically, expressed

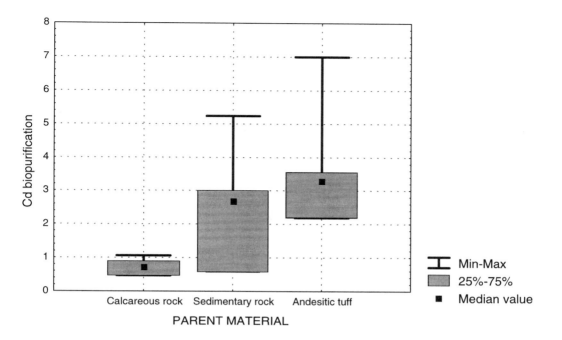

Figure 1. Plot of Cd biopurificiation in soils formed from different parent rock materials.

statistically by median, minimum and maximum values. The graph illustrates the quite different BPI of Cd in different soil types. BPI is the highest in soils formed from andesitic tuff (3.2), somewhat lower in soils formed from sedimentary rock (2.7), and the lowest in soils formed from calcareous rocks (0.7). The distribution of the BPI is quite uniform in calcareous rocks and more dispersed in sedimentary rocks and andesitic tuffs. The highest BPI (7.0) was calculated for andestic tuff. From the ecological point of view, this means that soils formed from calcareous rocks are more advantageous than those formed from sedimentary rock and andesitic tuff. Irrespective of the Cd content in the soil, plants grown in soil with calcareous parent material accumulate two to three times less Cd than those grown in soils formed from sedimentary and andesitic tuff parent materials.

5 CONCLUSIONS

In this study an alternative assessment method was used to estimate Cd bioaccumulation in plants overlying different types of soil.

It was established that soils formed on calcareous parent material are more advantageous than soils formed from sedimentary and andesitic tuff parent materials with respect to Cd uptake into the food chain.

ACKNOWLEDGEMENT

The author is highly indebted to J. Flis, V. Ro• i• and A. Glinšek for their help. Acknowledgement is made to the Thermal Power Plant Šoštanj and the Coal Mine Velenje for financial support.

REFERENCES

Kabata-Pendias, A. & H. Pendias 1986. *Trace Elements in Soils and Plants*: 92-109. Florida: CRC Press.
Sillanpää, M. & H. Jansson 1992. Status of cadmium, lead, cobalt and selenium on soils and plants of thirty countries. *FAO Soils Bulletin*. 65:8-15.. Forssa: FINNIDA.
Svetina, M. & S. Pirc 1998. Cadmium enrichment in the soil of the Šalek Valley, Slovenia. *Terra Nova*. 9:664. Strasbourg: EUG.
World Health Organization (WHO) 1992. Cadmium - Environmental Aspects. *Environmental Health Criteria*. 135:42-58. Geneva: IPCS.

Geochemistry of the Earth's Surface, Ármannsson (ed.)© 1999 Balkema, Rotterdam, ISBN 90 5809 073 6

Impact of organic waste disposal on phosphorus losses during simulated storm events

H. Vanden Bossche
Géosciences Rennes, Université de Rennes I, France

G. Bourrié, C. Gascuel & F. Trolard
INRA, Sols et Agronomie, Rennes, France

ABSTRACT: Rainstorm simulations performed on an agricultural field gave evidence for a drastic increase of dissolved phosphorus losses on soils supplied with sewage sludge. The dissolved P contents obtained (0.8 to 0.5 mg P / L) are the result of a desorption process. They are far over the eutrophication threshold and might therefore damage the quality of surface water. On the other hand the effect of this organic amendment on soil structure led to a significant decrease in erosion and thus in total P losses. Thus, the negative impact of sludge on water quality, may often be hidden by the dominance of particulate-bound phosphorus in the average budget.

1 INTRODUCTION

Eutrophication of continental surface waters has been mainly attributed to elevated nutrient concentrations. Phosphorus is known to be the limiting factor for algal growth, owing to C and N availability from the atmosphere (Fisher et al. 1995, Pietiläinen 1997).

Large concentrations of phosphates in surface water were primarily ascribed to sewage-derived phosphorus inputs. But the persistency of eutrophication in areas where point sources have been curtailed now turns the focus of attention to agricultural losses (Tunney et al. 1997). Little information is available on P loss from agricultural soils. But it appears to be linked to cultural practices and P over-fertilization (Sharpley et al. 1995). Moreover, surface runoff and drainage water are considered to be the main pathways for water contamination.

Among organic fertilizers likely to enhance P surplus, sewage sludge, as a product of wastewater treatment, deserves specific interest. It presents a high P content, especially in regions subject to eutrophication where water treatment plants are required to remove P from effluents (which results in P enrichment in the sludge). Previous monitoring of the quality of natural runoff water on a field supplied with sludge, did not show any noticeable effect of sludge on the average phosphorus losses during the year following application (Vanden Bossche & Bourrié 1998). Hence, the purpose of this study is to assess the impact of sludge disposal during simulated critical storm events.

2 MATERIALS AND METHODS

Rainfall simulations were conducted on a maize field in Brittany. The place, located in western France presents a humid and temperate oceanic climate (mean annual rainfall of 760 mm). The soil chosen for this investigation is a loamy acid brown soil, slightly leached and developed on shale. Its initial P content was high (0.9 g P / kg dry matter). The slope of the field surface is 4%.

The sludge applied was liquid (solid load of 23 g / L). It had been produced by a biological treatment plant, fitted out with biological P removal.

The device used to simulate the rainfall was sheltered by a tent to avoid wind effect. It allowed to produce a constant rain on a 4 m × 2 m plot (Asseline & Valentin 1978, Casenave & Valentin 1989). Three rain applications of equal intensity (40 mm / h) and duration (30 min) were performed.

The surface runoff water (SRW) was collected in the spacing between two maize rows running downslope. The device used to recover the sheet flow has been designed by Cros-Cayot (1996) and Gascuel-Odoux et al. (1996). Two 4 m × 0.75 m plots, limited by maize plants, were monitored at the same time. One of them was supplied with sludge at the rate of 12 L / m², corresponding to 7.3 g P / m² or 168 kg P_2O_5 / ha. The sludge was disposed on the field five days before simulation. The soil of both plots was tilled before the first simulation.

Samples were collected every minute at the first half of the rain and then every third minute. The particulate and dissolved fraction were separated through 0.2 μm filtration. The dissolved molybdate reactive phosphorus (DMRP) was analysed by colorimetry (Murphy & Riley 1962). The total P content was determined on an aliquot after an acid digestion.

3 RESULTS AND DISCUSSION

3.1 *Sludge and runoff intensity*

Sludge application had a significant effect on runoff magnitude. This result is illustrated by Figure 1a. It shows that during the third simulation (with a well saturated soil) the runoff intensity increased to 28 mm / h on the reference row while it only rose up to 17 mm / h on the row provided with sludge.

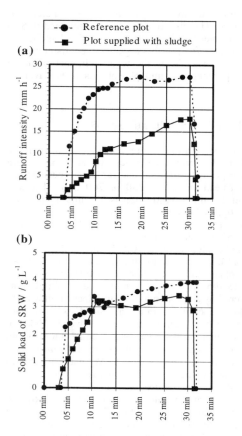

Figure 1. Runoff intensity (a) and solid load (b) during the third rainfall simulation

This difference can be ascribed to the consequence of sludge application on soil structure. It improves the stability of soil macro-aggregates that remained preserved even after the third rain. This can be at-

tributed to the cement effect of sludge organic matter.

On the plot supplied with sludge, flow pattern is controlled by puddles connection. Evidence for this is given by the successive steps of the hydrogram in Figure 1a. On contrary the soil roughness of the reference plot quickly decreases during the two first rain events (not shown here). The formation of a crust leads to the rapid contribution of the entire surface to sheet flow. Thus the runoff intensity reaches a high value at the beginning of the rain and then rises slowly as soil saturation increases.

The magnitude of erosion is affected by the runoff intensity. As shown on Figure 1b, SRW solid load increases more slowly on the row supplied with sludge. It reaches a stage after 10 min, when runoff intensity exceeds 10 mm / h. But it stays 15% below the solid load of the reference row until the end of the rainfall, as an effect of the lower flow (hence lower energy).

3.2 *Dissolved phosphorus release*

Sludge application results in a drastic increase of the SRW dissolved phosphorus content. This appears on Figure 2, which clearly illustrates a "flush effect" on the row supplied with sludge, with the highest concentrations at the beginning of the rain. The flush of a labile pool of phosphorus is also illustrated by a decrease of the average P content from the first to the third rain (from 0.73 mg P / L to 0.53 mg P / L).

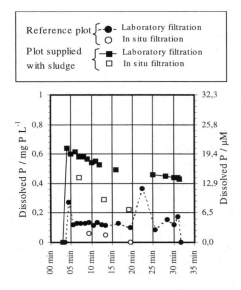

Figure 2. Dissolved phosphorus content of the runoff water, during the third rain

The acquisition of the DMRP content of the runoff water is interpreted as the consequence of a desorption process. The lower contents of samples filtered in the field show that filtration interrupted this reaction before completion of a steady state. This result is consistent with the study of desorption kinetics from the soil (without sludge) in a batch process that gives an equilibration time of 4 hours at 20° C. The same experiment conducted with the sludge (without soil) shows a very fast release (<10 min). Therefore, the relatively low equilibration time observed in the runoff water samples is an evidence that desorption is controlled by soil particles. This means that part of the sludge labile phosphorus has already been transferred on soil particulate matter, in the five days preceding the rain simulation.

3.3 Particulate phosphorus losses

Despite the difference in DMRP content, and owing to the role of particulate-bound phosphorus, the total P losses from the reference row were higher than the losses from the row supplied with sludge. This result is illustrated on Figure 3.

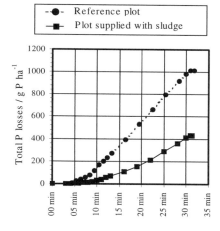

Figure 3. Total P losses during the third rain.

Particulate-bound phosphorus accounts for more than 95 % of the total P losses. The average P content of the sediment appears to be the same on each row (2,5 mg P/ g of dry matter). Therefore, erosion intensity is the main factor controlling P losses. In this respect, sludge disposal appears to have a beneficial impact on P losses, since its consequence on soil surface structure minimizes erosion.

It is to be noted that, although the sludge application increases the labile particulate-bound phosphorus in the soil, it hardly changes the total P content.

4 CONCLUSIONS

Rain simulations brought to the fore a relevant impact of sludge disposal (as a result of human activity), on the geochemistry of soil and surface water. It leads to an increase of labile phosphorus adsorbed on soil particles. Consequently, it causes an augmentation of dissolved phosphorus in runoff water. But average P losses budgets are likely to hide this harmful result, since they seldom take in account critical storm events. They would mainly point out the beneficial effect on the structure of soil surface and the decrease of total P transfer.

REFERENCES

Asseline, J. & C. Valentin 1978. Construction et mise au point d'un infiltromètre à aspersion. *Cahiers de l'ORSTOM - Série hydrologie* XV(4): 321-349.

Casenave, A. & C. Valentin 1989. *Les états de surfaces de la zone sahélienne*. Editions de l'ORSTOM.

Cros-Cayot, S. 1996. *Distribution spatiale des transferts de surface à l'échelle du versant. Contexte armoricain*. Doctorat - thèse de troisième cycle. ENSAR.

Fisher, T. R., J. M. Melack, J. U. Groobelaar & R. Howarth 1995. Nutrient limitation of phytoplankton and eutrophication of inland, estuarine and marine waters. In H Tiessen (ed), *Phosphorus in the global environment*: 171 - 199. New York: John Wiley & Sons Ltd.

Gascuel-Odoux, C., S. Cros-Cayot & P. Durand 1996. Spatial variations of sheet flow and sediment transport on an agricultural field. *Earth Surface Processes and Landforms*. 21: 843-851.

Murphy, J. & J. P. Riley 1962. A modified single solution method for determination of phosphates in natural waters. *Anal. Chim. Acta*. 27: 31-36.

Pietiläinen, O. P. 1997. Agricultural phosphorus load and phosphorus as a limiting factor for algal growth in finnish lakes and rivers. In H. Tunney, O. T. Carton, P. C. Brookes, & A. E. Johnston (eds) *Phosphorus loss from soil to water*. 354-356. CAB Internationnal.

Sharpley, A. N., M. J. Hedley, E. Sibbessen, A. Hillbricht-Illowska, W. A. House & L. Ryszkowski 1995. Phosphorus transfers from terrestrial to aquatic ecosystems. In H Tiessen (ed.) *Phosphorus in the global environment*. 171 - 199. New York: John Wiley & Sons Ltd.

Tunney, H., O. T. Carton, P. C. Brookes & A. E. Johnston 1997. *Phosphorus loss from soil to water*: 461. Cab Internationnal.

Vanden Bossche, H. & G. Bourrié 1998. Phosphorus rich organic amendments and release of soil phosphorus to runoff. *Mineralogical magazine*. 62A(3): 1584-1585.

Geochemistry of the Earth's Surface, Ármannsson (ed.) © 1999 Balkema, Rotterdam, ISBN 90 5809 073 6

Geochemistry and remediation of phosphate mining and wastes, Florida, USA

J. M. Whitmer
University of Kansas, Lawrence, Kans., USA

J. A. Saunders
Auburn University, Ala., USA

ABSTRACT: Phosphate mining has been a major industry in Florida for over 50 years and the environmental consequences are just now being scrutinized. The phosphate ore and byproduct gypsum are enriched in uranium, metals, and rare earth elements. Due to the high uranium content, the byproduct gypsum has no industrial use and is stockpiled at the surface. The groundwater near and beneath the gypsum piles contains the same contaminants as the gypsum, including uranium decay-series elements. The high sulfate content of the groundwater suggests the possibility that the stimulation of naturally occurring sulfate-reducing bacteria can be used to partially remediate the groundwater. Reaction-path modeling of representative groundwater shows that sulfate reduction does have positive effects on the geochemistry of the water.

1 INTRODUCTION

The Central Florida phosphate district (USA) produces approximately 25% of the world's supply of phosphate fertilizer (Burnett et al. 1995). The ore is mined from the phosphate-rich horizons in the Miocene Hawthorne Formation by large-scale open pit operations, which are rapidly reclaimed to pasture land or freshwater lakes. The principal ore mineral is francolite, which is a carbonate-rich apatite (Van Kauwenbergh & McClellan 1990). After the coarser sand tailings are removed, the remaining calcium-phosphate "rock" or ore is dissolved in sulfuric acid. The addition of ammonia initiates the precipitation of diammonium phosphate so it can be removed and used to produce the valuable commodity, fertilizer. The remaining solution, enriched in Ca from the Ca-PO_4 ore and SO_4 from the sulfuric acid, is directed toward ponds where it evaporates to form gypsum. The gypsum builds up over time and these "gypstacks" may reach 30 m in height and cover several square kilometers. Approximately 30 million tonnes of this gypsum are produced per year, and about 1 billion tonnes are currently stored above ground in Florida presenting the potential for further contamination (Burnett et al. 1995).

No suitable industrial use has been found for this byproduct gypsum because it contains significant levels of uranium and U-decay series elements (Burnett et al. 1995). The gypstack focused on in this study was produced before the current regulations, which prohibit mixing gypsum with mining spoil piles, were in place. The heterogeneous nature of the gypstack has led to a relatively high permeability allowing transport of contaminated water into the shallow aquifer. The primary environmental concern is the potential for continued contamination and further spreading through the aquifer. The groundwater in the shallow aquifer locally has elevated levels of sulfate, radionuclides, and heavy metals that have been leached from the gypstack (Florida State University 1998). In this paper, the geochemistry of one gypstack is documented and compared to the trace-element composition of the present-day and historic phosphate ores from central Florida and an innovative remediation for contaminated ground water at the site is proposed.

2 HYDROGEOLOGY AND GEOCHEMISTRY

The gypstack in the study area covers approximately 1.5 square miles and there are monitoring wells positioned around the perimeter (Figure 1a). On a regional basis, groundwater primarily flows to the west and south within the phosphate district (Southeastern Geological Society 1977). Near the study area there is a general radial component of the flow around the gypstack because the water tends to follow topography with a pronounced plume to the northwest moving toward a small stream (Figure 1a). Waters leach sulfate, metals, and radionuclides from the gypsum, seep through the underlying sands, and mix with shallow groundwaters at the site (Figure 1b). The phosphatic rich ore unit serves as a semi-confining unit at the base of the shallow aquifer, and also generally separates it from the underlying intermediate aquifer (Figure 1b). Underlying the intermediate aquifer is the principal water-bearing unit in the region, the Floridan aquifer (Southeastern Geological Society 1977). The shallow and intermediate aquifers both have been affected by contaminants in the study area (Florida State University 1998).

The gypstack under scrutiny was deposited before the current regulations, requiring liners and allowing no mixture of byproduct gypsum and sand tailings, were adopted, creating a more permeable environment than around the more recent gypstacks. At the site, cracks in the gypsum due to shrinkage also contribute to this increased permeability. The contaminants are leached out of the cracks in the gypsum and the permeable mine tailing zones and allow the contaminated water to more easily percolate through to the shallow aquifer (Figure 1b).

The dissolved species, including Ca, SO_4, Fe, As, rare earth elements (REE), and Ra-226 in the groundwater are similar to those present in the gypsum and the phosphate ore. Because the shallow groundwater at the site is close in proximity to both the remaining phosphatic ores and the gypsum, both can impact the trace-element geochemistry of groundwater. Most likely the contamination is a combination of both sources, but the high levels of dissolved Ca and SO_4 are indicative of leaching from the highly soluble gypsum.

The concentration of heavy metals, radionuclides, and some REEs were analyzed by instrumental neutron activation (Table 1). The concentrations of various constituents in the gypsum and phosphate can be compared, as well as an evaluation of the geochemistry of ores mined at different time periods.

Phosphate ore samples typically contain 31-37% Ca, whereas gypsum samples contain about 18-24% (Table 1). In the phosphate ore samples Ca content correlates with P_2O_5 content due to a general 1:1 atomic ratio of Ca to PO_4 in apatite. Samples with the highest Ca content generally have the highest U and REE contents, as expected (Table 1).

The sand tailings (sample 6) generally have the lowest amount of nearly all analyzed constituents.

The different age of mining of phosphate samples showed little variation in most constituents (samples

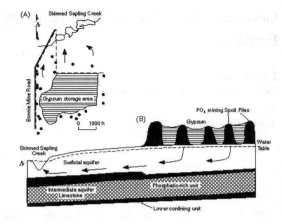

Figure 1a. Plan map showing the gypstack and monitoring wells with arrows showing movement of the plume toward the creek.

Figure 1b. Diagrammatic cross section showing the gypstack with the mixture of sand tailings and gypsum and the general interpreted underlying geology.

2-5). These samples represent district-wide composites and were obtained from the sample archives of the Florida Institute of Phosphate Research.

Comparison of the geochemistry of the gypsum and the phosphate samples generally reveals that trace elements are more concentrated in the phosphate ores. For example, phosphate ores contain about 100 ppm U on average, whereas the sand tailings contain about 20 ppm and the gypsum samples have <5 ppm (Table 1). Similarly, phosphate ores contain 200-300 ppm La + Ce + Nd, sand tailings approximately 100 ppm, and gypsum <100ppm. Other rare earth elements follow this general trend (Table 1). Fe, As, Th, Co, Ni, and Cr are also more abundant in the phosphate ores than in the sand tailings or gypsum. The highest grade and only non-composite ore sample (sample 1, Table 1) had the highest Se content (21 ppm).

From this study, it is not clear if trace element depletions in the gypsum relative to phosphate ores is due to removal of trace elements during fertilizer production or leaching of these trace elements from the gypstack over about 50 years, or by both. However, high levels of Ca and SO_4 in the shallow groundwater indicate that gypsum leaching could have removed significant amounts of trace elements and this was confirmed by laboratory leaching experiments (Burnett et al. 1995).

Table 1. Geochemistry of phosphate ores and gypsum byproduct

Sample	1	2	3	4	5	6	7	8	9
Constituents in ppm									
As	5	5	5	7	7	<2	<2	<2	<1
Ca%	37.0	34.1	34.4	31.4	31.7	7.2	21.8	18.1	24.2
Co	5	3	4	6	8	2	<1	<1	1.8
Cr	51	51	50	50	45	20	<2	4	3.2
Fe	0.54	0.82	0.75	0.68	0.79	0.34	0.12	0.11	0.067
Na	0.13	0.33	0.36	0.30	0.42	0.08	0.03	0.03	292
Sb	4.8	1.7	1.8	1.6	1.4	5.5	0.2	<0.2	0.1
Se	21	<3	<3	<3	<3	4	<3	<3	<0.5
Th	3.6	7.9	7.0	6.3	7.3	4.6	1.1	1.6	0.9
U	138	94.6	79.1	82.4	117	19.8	1.9	3.0	3.8
Zn	188	92	126	85	85	47	<40	<40	13
La	64.1	91.5	88.1	87.7	88.3	24.4	30.9	17.8	28
Ce	82	121	108	115	131	39	43	26	39
Nd	50	80	63	74	80	18	26	11	21
Sm	8.6	12.8	11.6	11.3	12.0	3.9	4.6	2.4	3.93
Eu	2.4	3.5	3.1	3.2	3.4	1.0	1.1	0.5	0.99
Tb	1.9	2.6	2.4	2.3	2.2	0.6	0.7	<0.5	0.6
Yb	6.7	9.8	9.2	9.1	8.1	3.1	2.2	2.1	1.91
Lu	1.08	1.49	1.34	1.53	1.19	0.55	0.29	0.17	0.27

Concentrations are in ppm (except Ca)

Sample 1: Clear Springs, Hopewell Mine   Sample 2: composite of central Florida ores, 1994
Sample 3: IMC-180, 1960's ore   Sample 4: composite ore, NBS 120C
Sample 5: composite ore from 1998 for district   Sample 6: 1993 sand tailings
Samples 7, 8, 9: gypsum samples randomly collected at the surface

3 REMEDIATION

The EPA has recently mandated remediation of the groundwater in this area. Up to this point nothing has been done to clean up the contamination, and considering the extent of the mining industry in this area, the contamination has not migrated very far. Remediation using pump and treat methods can be expensive, time consuming, and not completely effective. Due to the nature of the contaminants and the environmental conditions present, it has been proposed that stimulating naturally occurring sulfate-reducing bacteria (SRB) should theoretically be a feasible remediation technology at this specific site (Whitmer & Saunders 1998). Studies by others (Burnett et al. 1995) have shown that SRB are present within the gypstack at the site and can locally lead to gypsum dissolution. The innovative process proposed here involves utilizing the bacteria's metabolic processes to remove the sulfate, metals, and radionuclides from the groundwater.

Modeling of representative water from the site with the USGS reaction-path model PHREEQC (Parkhurst 1995) indicates that the contaminants can be removed from the groundwater to reclaim the water to within or near federal and state drinking water standards. The sulfate reduction results in beneficial environmental effects such as the increase of pH and alkalinity and the overall decrease in TDS. These effects are facilitated by the precipitation of the contaminating species as insoluble sulfide and oxide phases. The model indicates that the sulfate concentration decreases as the H_2S concentration increases due to the coupled reduction of sulfate and oxidation of organic carbon (Figure 2). This reaction produces the H_2S and bicarbonate, which increases alkalinity. Model results suggest that the aqueous iron is completely removed from solution by $P_{CH4}=-11$ as it precipitates as pyrite. In the model, increasing concentration of methane is used to simulate increasing bacterial sulfate reduction, shown toward the right of the graph (Figure 2). Methane is used as a proxy for any source of organic carbon, and it is the only organic compound for which there is thermodynamic data in the model.

The two common concerns with using the SRB for remediation are the possibility of reoxidation of the minerals precipitated during remediation and clogging the aquifer from precipitation of authigenic pyrite and calcite. The reoxidation of the contaminants that precipitate as their respective oxide and sulfide mineral phases should not pose a problem because the mineral phases are relatively insoluble under the water table and this process, if it occurs, should be slow due to the low solubility of oxygen in groundwater.

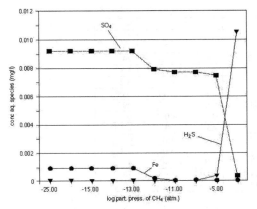

Figure 2. Plot of model results of aqueous species concentrations vs. log partial pressure of methane.

Calculations show that the porosity of the aquifer will not be affected by the sulfate reduction because all model predicted minerals must precipitate to decrease the porosity by just 1%. Overall, the positive environmental effects of this technology greatly outweigh the small possibility of reoxidation or clogging the aquifer. The model results show that the sulfate reduction theoretically can remediate the water to required drinking water standards with the possible exception of calcium and sulfate. To stimulate biogenic sulfate reduction at the site, water soluble organic carbon will be injected into the sulfate rich groundwater using wells (Saunders 1998).

4   CONCLUSIONS

1. Trace elements are more concentrated in the ore than the byproduct gypsum or sand tailings.
2. High levels of calcium and sulfate suggest leaching of the contaminants from the soluble gypsum but it is not clear if the trace element depletions in the gypsum is due to their concentration in the fertilizer during production or leaching from the gypsum over time.
3. Remediation of the groundwater at the site by stimulating the naturally occurring SRB should be a feasible solution and reaction path modeling supports this conclusion.

REFERENCES

Burnett, W.C., M. Schultz & C. Hull 1995. Behavior of radionuclides during ammonocarbonation of phosphogypsum. Publication No. 05-040-111, Florida Institute of Phosphate Research: 1-5.

Florida State University 1998. How does phosphogypsum storage affect groundwaters? Publication No. 05-042-142, Florida Institute of Phosphate Research.

Parkhurst, D.L. 1995. Users guide to PHREEQC—A computer program for speciation, reaction-path, advective-transport, and inverse calculations. U.S.G.S. Water Supply Paper 95-4227: 143.

Saunders, J.A. 1998. In situ bioremediation of contaminated groundwater: U.S. Patent No. 5,833,855.

Southeastern Geological Society 1977. Environment of the Central Florida Phsophate District, 21st Field Conference, Publication No. 19,:11-12.

Whitmer, J.M. & J.A. Saunders 1998. Use of sulfate-reducing bacteria in bioremediation of groundwater contaminated by phosphate mining and extraction: Predictions from reaction-path modeling: in press.

Van Kauwenbergh, S.J. & G.H. McClellan 1990. Comparative geology and mineralogy of the southeastern United States and Togo phosphorites, In A.J.G. Notholt and I. Jarvis (eds), *Phosphorite Research and Development:* 139-155. Geological Society Special Publication No. 52.

4 Organic geochemistry

Geochemistry of the Earth's Surface, Ármannsson (ed.) © 1999 Balkema, Rotterdam, ISBN 90 5809 073 6

Invited lecture: Atmosphere O_2 control by a 'mineral conveyor belt' linking the continents and ocean

J. I. Hedges & R. G. Keil
School of Oceanography, University of Washington, Seattle, Wash., USA

C. Lee
Marine Sciences Research Center, State University of New York, Stony Brook, N.Y., USA

S. G. Wakeham
Skidaway Institute of Oceanography, Savannah, Ga., USA

ABSTRACT: One of the most fascinating questions in geochemistry is how the global cycles of O_2 and organic carbon have maintained conditions suitable for multicellular life during multiple deposition/weathering cycles of sedimentary rock over the last half billion years. To sustain the required precarious balance, ~0.1% of the organic matter photosynthesized by plants must now circumvent heterotrophic recycling to be preserved in marine sediments. Preservation of this minute fraction frees up just enough O_2 to oxidize fossil organic matter in sedimentary rocks uplifted on the continents, thereby maintaining a constant atmospheric O_2 concentration. Recent biogeochemical studies suggest that only organic materials intimately associated with mineral surfaces survive destruction. Organic substances protectively escorted to coastal marine sediments by a continuous stream of weathered minerals appear to harness the global oxygen and carbon cycles to tectonic control.

1 INTRODUCTION

Formation of organic matter and molecular oxygen during photosynthesis,

$$CO_2 + H_2O \Rightarrow H_2CO + O_2 \qquad (1)$$

and recombination of these energy-rich products during respiration, inextricably link the global cycles of organic carbon (OC) and molecular oxygen (Berner & Canfield 1989). The net balance of these two reactions over geologic time has set the redox state of the Earth's surface, thereby directing the chemical and biological evolution of the planet (Holland 1984).

The global balance between photosynthesis and respiration is dynamic and delicate, as indicated in numeric models by the need for negative feedback mechanism for O_2 regulation (Berner & Canfield 1989). Presently, ~0.1 x 10^{15} g of OC are buried annually in marine sediments accumulating along upper continental margins, the only setting where appreciable long-term OC preservation presently occurs (Berner 1989). The biochemical and isotopic compositions of the buried organic remains, equivalent to about 0.1% of annual global primary production, indicate a modern marine origin, as opposed to recycled fossil material. Because of the 1/1 stoichiometry of equation (1), each mole of OC buried leaves one "excess" mole of O_2 in the atmosphere. The present-day global OC burial rate annually releases ~0.3 x 10^{15} g of O_2. This influx, if uncompensated, would double the amount of O_2 in today's atmosphere (1,200,000 g) in about 4,000,000 years, leading to runaway forest fires and O_2 toxicity (Berner 1989). Because ~ 1/1000 carbons in biomass is now preserved, the production side of the global O_2 cycle is particularly volatile.

An amount of O_2 equivalent to that released by sedimentary OC preservation must be removed to maintain a constant atmospheric reservoir. The only known mechanism for withdrawing such large amounts of atmospheric O_2 over geologic time is by oxidizing fossil OC (and reduced minerals) exposed in weathering continental rocks, the bulk of which are "recycled" sedimentary material (Hunt 1996). This tectonically driven cycle is closed on a 10^8 year time scale, when OC buried in marine sediments is uplifted in continental rocks and oxidized.

Because uplift of sedimentary rocks occurs sporadically, removal rates of atmosphere O_2 vary appreciably over geologic time. Burial of OC in marine sediments, therefore, must continuously shift to release just the right amount of O_2 to compensate for the changing terrestrial sink. This ever-shifting balance must be continuously maintained between two physically separate environments characterized by drastically different physical conditions, chemistry and biota. In spite of perils on both sides of imbalance, the half-billion-year fossil record of multicellular life on Earth attests to the remarkable stability of the global O_2 cycle over a period more than a hundred times longer than is required for a catastrophic fluctuation to occur.

We speculate here that this precarious balance has been maintained over the Phanerozoic by protective association of finite amounts of organic matter per unit of mineral surface area, and near-conservative transfer of mineral surface area from weathering continental rocks to ocean margin sediments. This transport process appears to be fine-tuned by negative feedback control between atmospheric O_2 concentrations and the length of time that sedimenting organic materials are exposure to severe oxic degradation in the ocean. A gauntlet of biological and abiotic oxidation processes work together to ruthlessly destroy essentially all types of organic matter - including such recalcitrant forms as kerogen, pollen, coal, and char - that either do not ride the "mineral conveyer belt", or are not buried rapidly enough (within $\sim 10^2$-10^3 years) below the oxic horizon within marine sediments.

2 THE CONVEYER BELT MODEL

2.1 Protective mineral association

It has long been recognized that OC concentrations are higher in finer-grain marine sediments. In addition, most sedimentary organic matter cannot be physically separated from coexisting minerals by density fractionation of hydrodynamic sorting (e.g. Keil et al. 1994a). These phenomena can be explained by intimate association of organic materials with mineral surfaces, whose ratio to mass increases with smaller average particle size. Mayer (1994) convincingly demonstrated that weight percentages of OC (%OC) increase in proportion to the average surface area (SA) of continental shelf sediments, with typical OC "loadings" of 0.5-1.0 mg OC/m^2. Since this pioneering work, the same range has been observed for bulk sediments from the upper Washington State margin (Keil et al. 1994a, Hedges et al. submitted) and the Amazon River (Keil et al. 1997). Although higher organic loadings have been measured in sediments accumulating beneath anoxic ocean bottom waters (Hedges & Keil 1995), such zones presently account for only ~1% of total OC preservation (Berner 1989).

Evidence that organic matter associated with mineral surfaces is protected from complete mineralization comes from both field and laboratory observations. For example, the trend that most non-

deltaic shelf sediments contain 0.5-1.0 mg OC/m^2 surface area seems more than a coincidence. In addition, Mayer (1994) has demonstrated that the diagenetically active upper few centimeters of many marine sediments exhibit higher values which decrease asymptotically at depth into the "average" loading range. Moreover, "old" sedimentary organic materials can be remineralized within weeks when dissolved from mineral surfaces and then exposed to oxic conditions (Keil et al. 1994b).

2.2 Severe oxic degradation

The pattern that incrementally depositing deep-sea sediments (underlying roughly 80% of the ocean surface) exhibit low organic loadings on the order of 0.05-0.1 mg OC/m^2 indicates that sorptive protection must give way to OC removal mechanisms somewhere along deeper continental margins (Hedges & Keil 1995). The reasons that pelagic sediments account for < 5% of OC preservation (Berner 1989) are numerous and include lower productivity, greater particulate OC loses in deeper water columns, slower rates of burial to diagenetically inactive sedimentary depths, more extensive bioturbation and longer periods of exposure to oxic conditions. Although these variables are inextricably linked, recent studies of oxidized deep-sea turbidites demonstrate that long term exposure to oxic conditions is sufficient alone to produce the low concentrations that characterize organic matter in pelagic sediments.

The most studied example of oxic degradation is the relict f-turbidite that was emplaced approximately 140,000 years ago in the Madeira Abyssal Plain off the northwest coast of Africa (Cowie et al. 1995). This initially homogeneous 4-m-thick deposit was oxidized by diffusing O_2 to a depth of ~0.5 m, over a time span of ~10^5 years. This extended exposure decreased the %OC of the upper turbidite from concentrations near 0.9% (as is seen in the unoxidized deeper section) to values of 0.1-0.2%. The organic matter in the oxidized surface horizon not only occurs within the same low %OC range that typifies open ocean sediments, but also exhibits greatly elevated levels (> 30 mole%) of non-protein amino acids which are essentially unique to such highly degraded deposits (Cowie et al. 1995). Oxidation was so severe under these conditions that essentially all pollen (Keil et al. 1994c) in the surface section of this turbidite also was destroyed. Even after 140,000 years in the presence of porewater sulfate the lower portion of the f-turbidite compositionally resembles sediments depositing today on the adjacent African continental margin.

Almost all particles comprising ocean sediments will have some history of exposure to oxic condition, either within the water column or surface marine sediments. It has been demonstrated in fact that the burial efficiency of sedimentary OC is inversely related to the O_2 exposure time (OET) during their deposition (Hartnett et al. 1998). The average OET of continental margins sediments accumulating under oxic bottom waters can be estimated by dividing the depth of O_2 penetration into surface sediment porewaters by the average sediment accumulation rate at the site (Hedges & Keil 1995). Although local productivity, bottom water O_2 concentrations, bioturbation, irrigation and sediment accumulate rate all affect OET, this parameter integrates these variables into one value that appears to represent the combined effect of the processes directly affecting OC preservation.

This contention is supported (Figure 1) by a recent study in which OET values were calculated for sediments accumulating along the Washington State margin (Hedges et al. submitted). In this setting, the surface-normalized concentrations of OC decrease exponentially offshore from loadings typical of upper continental margin deposits to values < 0.4 mg OC/m^2 on the lower continental slope. This OC decrease corresponds to an increase in OET from nearshore values of decades to offshore values of centuries to a thousand years (Figure 1). All diagenetic indicators, including percentages of physically degraded pollen and of nonprotein amino acids, also increased consistently offshore and with increasing OET. The lack of discernable downcore increases in the same parameters indicates that most degradation occurs near the sediment/water interface, as would be expected for an oxic effect in sediments where O_2 penetrates no deeper than 4 cm.

2.3 The mineral conveyer belt link

The previous evidence for protective association of organic matter with mineral surfaces, and for severe

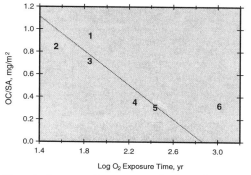

Figure 1. Average surface area-normalized organic carbon loadings for modern sediments from the Washington State margin plotted versus the log of the oxygen exposure time calculated for this region (site numbers increase offshore).

The Mineral Conveyer Belt

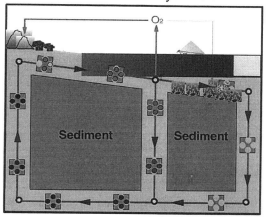

Figure 2. A cartoon of the hypothetical "mineral conveyer belt." The two cycles represents the conservative passage of mineral surface area (boxes with holes) and its associated organics (balls in holes) through the tectonic (anoxic) and weathering (oxic) cycles. The outer loop represents deposition below the "OC compensation depth" where long term O_2 exposure removes most organic matter from mineral "carriers".

oxidation of essentially all organic substances exposed to oxic conditions, sets the basis for the mineral conveyer belt hypothesis illustrated in Figure 2. One of the controlling influences in this model is continuity of surface area during continental uplift/weathering cycles. Because ~80% of all continental rocks are sedimentary, and can be expected to break down to approximately the same size particles from which they were formed, the amount of surface area generated at any time should be proportional to the amount of rock then being uplifted and weathered. Increased O_2 uptake by a greater mass of weathering rock, will generate proportionately more mineral surface area, which is then available to transport proportionately more organic matter to upper continental margin sediments where the bulk of these remobilized particles eventually are deposited. At such nearshore sites, OETs less than 100 years should allow the accumulating particles to protect and bury an "average" complement (0.5-1.0 mg OC/m^2) of associated organic matter. As a result, the molar equivalent of additional O_2 will be released to compensate for the increased kerogen-weathering sink. This scenario should work well because most of the organic matter being oxidized on the continents resides in shales (Hunt 1996), which are particularly prone to weather back to fine clay minerals that represent most of the OC preservation potential of marine deposits. Because %OC values in depositing fine grained continental margin sediments and ancient shales are both near 1% (Hunt 1996), the subsurface (anoxic) component of OC transfer by sediment subsidence and rock uplift appears to be relatively conservative (Figure 2). Conservation of mass and surface area during weathering/deposition cycles can thus constrain atmospheric O_2 imbalances, even during periods of unusual tectonic activity.

To obtain the high degree of atmospheric O_2 stability needed over geologic time, the passive constraints of mass throughput likely require fine-tuning by an active negative-feedback mechanism. As illustrated in Figure 2, the pervasiveness of O_2 gas and the constraints of slow oxic degradation on the efficiency of sedimentary organic matter preservation may provide the additional needed control. If in fact the abrupt "break point" along many continental margins between organic rich and poor sediments corresponds to exponentially increasing OET (Figure 1), then this horizon may correspond to the organic counterpart of the calcite composition depth. In the organic case, this transition point would mark where oxic degradation increases to the rate where essentially no organic matter preservation is possible. Other factors being equal, an increase in atmospheric O_2 concentration will lead to greater saturation concentrations of O_2 in surface ocean waters and to higher remnant O_2 levels in ocean bottom waters. Elevated deep-water concentrations will drive greater diffusive fluxes of O_2 and deepen the penetration of oxic conditions into surface marine sediments. As O_2 penetration depths increase linearly, the preservation rates of oxygen-sensitive organic matter in the underlying sediments will decrease exponentially (Hedges & Keil 1995). In response, the "OC compensation depth" will move inshore globally to a new balance position that results in less organic matter preservation, and a decreased rate of O_2 release to the atmosphere. The resulting readjustment in atmospheric O_2 content will be on the time scale of million of years, and hence rapid enough to compensate for tectonically-driven perturbations.

The superposition of a sensitive negative O_2 feedback mechanism on mass balance constrains would still not be sufficient, if organic materials unassociated with mineral surfaces were to accumulate shoreward of the OCCD, or if a major form of reduced carbon were to resist slow oxic degradation further offshore. The observation that kerogen is supplanted by slowly cycling soil humus on land, and that these recalcitrant materials are largely replaced on discharged mineral surfaces by recently-formed marine organic matter (Keil et al. 1997), indicates that severe mechanisms for organic matter remineralization occur in all oxic settings.

The world appears to be able to host multicellular life only by ruthless destroying essentially all forms of organic matter that are not protected by minerals or saved by rapid transfer to anoxic sediments.

REFERENCES

Berner, R.A. 1989. Biogeochemical cycles of carbon and sulphur and their effect on atmospheric oxygen over Phanerozoic time. *Palaeogeography, Palaeoclimatology, Palaeoecology* 75: 97-122.

Berner, R.A. & D.E. Canfield 1989. A new model for atmospheric oxygen over Phanerozoic time. *American Journal of Science* 289: 333-361.

Cowie, G.L., J.I. Hedges, F.G. Prahl & G.J. De Lange 1995. Elemental and major biochemical changes across an oxidation front in a relict turbidite: A clear-cut O_2 effect. *Geochimica et Cosmochimica Acta* 59: 33-46.

Hartnett, H.E., R.G. Keil, J.I. Hedges & A.H. Devol 1998. Influence of oxygen exposure time on organic carbon preservation in continental margin sediments. *Nature* 391: 572-575.

Hedges, J.I. & R.G. Keil 1995. Sedimentary organic matter preservation: An assessment and speculative synthesis. *Marine Chemistry* 49: 81-115.

Hedges, J.I., F.S. Hu, A.H. Devol, H.E. Hartnett, & R.G. Keil submitted. Sedimentary organic matter preservation: A test for selective oxic degradation. *American Journal of Science*.

Holland, H.D. 1984. *The Chemical Evolution of the Atmosphere and Oceans*. Princeton Univ. Press.

Hunt, J.M. 1996. *Petroleum Geochemistry and Geology*. Freeman, New York.

Keil, R.G., E.C. Tsamakis, C.B. Fuh, J.C. Giddings & J.I. Hedges 1994a. Mineralogical and textural controls on the organic composition of coastal marine sediments: Hydrodynamic separation using SPITT-fractionation. *Geochimica et Cosmochimica Acta* 58: 879-893.

Keil, R.G., D.B. Montluçon, F.G. Prahl & J.I. Hedges 1994b. Sorptive preservation of labile organic matter in marine sediments. *Nature* 370: 549-552.

Keil, R.G., F.S. Hu, E.C. Tsamakes & J.I. Hedges 1994c. Pollen in marine sediments as an indicator of oxidation of organic matter. *Nature* 369: 639-641.

Keil, R.G., L.M. Mayer, P.D. Quay, J.E. Richey & J.I. Hedges 1997. Losses of organic matter from riverine particles in deltas. *Geochimica et Cosmochimica Acta* 61: 1507-1511.

Mayer, L.M. 1994. Surface area control of organic carbon accumulation in continental shelf sediments. *Geochimica et Cosmochimica Acta* 58: 1271-1284.

Geochemistry of the Earth's Surface, Ármannsson (ed.) © 1999 Balkema, Rotterdam, ISBN 90 5809 073 6

Invited lecture: Geochemistry of hydrothermal ecosystems

Everett L. Shock, Jan P. Amend & Gavin W. Chan
GEOPIG, Department of Earth and Planetary Sciences, Washington University, St. Louis, Mo., USA

ABSTRACT: Microbially-mediated redox reactions involving organic and inorganic constituents of hot springs and hydrothermal systems can be integrated into geochemical models of hydrothermal processes through thermodynamic analysis. Incomplete oxidation of organic compounds coupled to sulfate reduction, for example, can be quantified by considering the effects of stoichiometry on the overall Gibbs free energy of reaction. Analyses of organic acids and other compounds in hydrothermal fluids from Vulcano, Yellowstone, Iceland, and elsewhere permit evaluation of the energy available for heterotrophic and autotrophic organisms, and estimates of the biomass supported by geochemical processes. This approach also allows the prediction of unknown metabolic processes, and assists in the development of growth media suited to hydrothermal environments.

1 HOT LIFE AND GEOCHEMISTRY

Hot springs and hydrothermal systems host microorganisms that possess greater genetic diversity than that represented by all visible life on Earth. From an evolutionary perspective, there may be a connection between genetic diversity and the number of niches available. In a hydrothermal system, niches closely correspond to sources of metabolic energy, and the number of heterotrophic and autotrophic reactions that can supply energy is enormous. Advances in theoretical geochemistry (Shock 1995, Shock et al. 1997, Amend & Helgeson 1997a, 1997b, Prapaipong et al. 1999) allow us to evaluate the energetics of microbially-mediated geochemical processes involving organic and inorganic compounds found in hydrothermal ecosystems.

By placing reactions conducted by microbes in the same geochemical models that illustrate the consequences of water/rock/organic reactions, likely sources for metabolic energy from the many redox and pH dependent reactions that link the inorganic and organic constituents of geologic materials can be explored (McCollom & Shock 1997, Shock et al., 1998, Amend & Shock 1998). This approach permits a quantitative assessment of known sources of metabolic energy, but many other likely candidate reactions can be revealed through theoretical calculations that combine thermodynamic predictions with analytical measurements of natural systems on organic compounds and elements in various redox states (Amend et al. 1998).

The biologically-mediated reactions that supply the greatest amount of energy are oxidation-reduction reactions that are thermodynamically favored but kinetically inhibited (iron oxidation, sulfur reduction, etc.) Heterotrophs often couple these reactions with organic transformations. As an example, sulfate reduction is coupled with organic compound oxidation by many hyperthermophilic microbes. The energy that can be obtained from these reactions is highly variable and extremely sensitive to changes in temperature, pressure, pH, fugacities of gases, and activities of dissolved ions and organic compounds.

2 HETEROTROPHIC REACTIONS

Archaeoglobus fulgidus, a member of the hyperthermophilic Archaea, is capable of dissimilatory sulfate reduction coupled to formic acid oxidation via:

$$4\,HCOOH + SO_4^{-2} + 2\,H^+ \rightarrow 4\,CO_2 + H_2S + 4\,H_2O \qquad (1)$$

The standard state Gibbs free energy for this reaction (ΔG°_r) shifts from -326.1 kJ mol^{-1} at 25°C to -366.5 kJ mol^{-1} at 83°C, the optimal temperature for growth in the lab (Stetter 1988). Of course, the disequilibrium that *A. fulgidus* uses as its energy source is revealed by the overall Gibbs free energy of reaction (ΔG_r) and not by the standard state value. The familiar expression:

$$\Delta G_r = \Delta G°_r + RT \ln Q_r, \qquad (2)$$

where R stands for the gas constant and Q_r represents the activity product, links standard state calculations with the analytical measurements required to evaluate the activity product and yields the quantities of available energy. We combined chemical analyses of hot spring fluids (Amend et al. 1998, Gurrieri et al. 1999) with the calculated value of $\Delta G°_r$. The value of ΔG_r in a 98°C hot spring at Vulcano is -227 kJ per mole of sulfate reduced. In other words, 227 kJ can be liberated from this exergonic catabolic process to drive otherwise energy-demanding anabolic reactions.

A. *fulgidus* is also capable of dissimilatory sulfate reduction coupled to lactic acid oxidation via

$$2\,C_3H_6O_3 + 3\,SO_4^{-2} + 6\,H^+ \rightarrow$$
$$6\,CO_2 + 3\,H_2S + 6\,H_2O. \qquad (3)$$

This reaction yields energy in numerous hydrothermal environments that can support many types of sulfate reducers. For example, at 98°C in a hot spring in the Baia di Levante, Vulcano, where the pH = 3.3, $f CO_2 = 2.8 \times 10^{-2}$, $f H_2S = 4.7 \times 10^{-4}$ and $aSO_4^{-2} = 0.144$ (Gurrieri et al. 1999) there is -315.4 kJ (mol lactic acid)$^{-1}$ available, assuming the lactic acid concentration is similar to that of acetic acid determined by Amend et al. (1998).

3 INCOMPLETE OXIDATION

The stoichiometries of the coupled redox reactions pursued by heterotrophs are adjustable depending on the concentrations of various compounds available in a hydrothermal ecosystem. At Vulcano, the hot spring fluids are acidic, so the organic acids will be predominantly associated. In addition, the extent of oxidation of lactic acid is variable depending on the amounts of CO_2 and acetic acid produced. As an example, a generalized version of reaction (3), written to include acetic acid is given by:

$$2x\,C_3H_6O_3 + y\,SO_4^{-2} + 2y\,H^+ \rightarrow$$
$$(3x-y)\,C_2H_4O_2 + 2y\,CO_2 + 2y\,H_2O + y\,H_2S \qquad (4)$$

where x and y are stoichiometric variables that indicate the completeness of oxidation. In coupled redox reactions, electrons are transferred from the oxidation half-reaction (lactic acid to CO_2) to the reduction half-reaction (SO_4^{-2} to H_2S). The amount of energy per unit of transferred electrons is related to the extent of lactic acid oxidation. During complete oxidation, every carbon in lactic acid is oxidized to CO_2, and 12 moles of electron are released. In incomplete oxidation, some carbons from lactic acid are sequestered in acetic acid, and

thus, less CO_2 is produced. Concomitantly, the reduction of one mole of SO_4^{-2} to H_2S accepts 8 moles of electron. Figure 1 illustrates that the more completely lactic acid is oxidized to CO_2, the more energy is released. But complete oxidation requires more electron acceptors, and so the energy released per mole of SO_4^{-2} is less. In an environment where sulfate is limiting, complete oxidation of lactic acid is energetically expensive, and incomplete oxidation yields more energy.

4 AUTOTROPHIC REACTIONS

Autotrophic methanogens produce methane from CO_2 and hydrogen through reactions like

$$CO_2 + 4\,H_2 \rightarrow CH_4 + 3\,H_2O. \qquad (5)$$

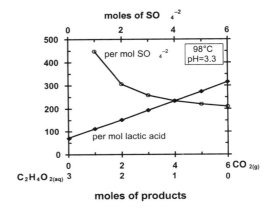

Figure 1. Overall Gibbs free energy as a function of the stoichiometry of the lactic acid oxidation/sulfate reduction (reaction (4) in the text).

Geochemical calculations demonstrate that this reaction yields metabolic energy when submarine hydrothermal fluids carrying H_2 and CO_2 mix with seawater carrying bicarbonate at temperatures 40°C (McCollom and Shock 1997). The metabolic systems of thermophilic autotrophic methanogens also generate organic compounds through analogous reactions in which methane is replaced on the right with many small organic compounds such as carboxylic and amino acids. Many of these reactions are thermodynamically favored during mixing at submarine hydrothermal systems (Shock & Schulte 1998, Amend & Shock 1998). As a consequence, these autotrophs may be able to generate organic compounds and gain metabolic energy at the same time. With this in mind, it may not be surprising that some of these organisms populate the deepest and shortest branches of the phylogenetic tree.

One representative of the deeply-branching, hyperthermophilic Bacteria is *Thermocrinis ruber*, a new genus belonging to the order *Aquificales* (Huber et al. 1998). *T. ruber* is a rod-shaped, motile, and nonsporulating bacterium that forms pink filamentous streamers in its natural habitat of continental hot springs. It is also an aerobic-microaerophilic facultative chemolithoautotroph that oxidizes elemental sulfur, molecular hydrogen, or thiosulfate in the presence of oxygen. In the lab, growth occurs at temperatures between 44° and 89°C (optimal T = 80°C); pH = 7, 8.5. The overall energy-yielding reaction used by *T. ruber* can be depicted by the following aerobic sulfur oxidation reaction

$$2 S^0 + 3 O_{2\,(aq)} + 2 H_2O \rightarrow 2 SO_4^{-2} + 4 H^+ \,. \quad (6)$$

Thermocrinis ruber was isolated from Octopus Spring, which is located in the Lower Geyser Basin of Yellowstone National Park. Thomas Brock (1978), one of the pioneer microbiologists who studied and reported on the high-temperature communities in this hot spring, provides the following data. At the source, the temperature of Octopus Spring was measured to be 91.2°C and the pH 8.3 (other data in Table 1).

Table 1. Octopus Spring Composition (in mg/l)

Na^+	321	Cl^-	256
Fe	0.01	SO_4^{-2}	23
NH_4^+	0.00	H_2S	0.2
NO_3^-	0.35	SiO_2	254

Pink-filament-forming communities were sampled by Huber et al. (1998) in the rapidly flowing, fully oxygenated upper effluent channel; temperatures were 82°-88°C and pH = 8.0. We calculated the amount of dissolved O_2 based on atmospheric equilibrium. For the temperature, pressure and composition of Octopus Spring, reaction (6) yields about 625.1 kJ of energy for every mole of S^0 oxidized (Figure 2).

Huber et al. (1998) have cultured *T. ruber* at 85°C under 300 kPa of a N_2-H_2-O_2 (94:3:3) gas phase in an optimized medium containing the inorganic components listed in Table 2.

Table 2. Optimized Growth Medium (in mg/l)

Na^+	386	Cl^-	169
Fe	0.22	SO_4^{-2}	32.6
NH_4^+	0.25	H_2S	0
NO_3^-	0.22	SiO_2	0

We calculated the activity of dissolved O_2 in these experiments based on the composition of the growth medium including the gas mixture. For a given temperature and pH, compositions in the growth medium supply slightly less energy (610.8 kJ per mol S^0) than those in the natural environment (Figure 2).

In this case, the difference in geochemical bioenergetics between natural and culture conditions is not great. However, this is unusual for laboratory studies of hyperthermophiles based on what is reported in the literature. How do geochemical environments support microorganisms? What amounts of geochemical constituents are necessary to support a given biomass? Where are ecosystems located given certain geochemical constraints? What is the community structure of these ecosystems?

All these questions can be addressed if a few additional measurements are made in the laboratory (or in nature!) to further quantify growth yield and biomass production. According to the culture condition for *T. ruber* in the laboratory, 10 ml of growth medium contains 5 mg of S^0, supporting a reported concentration of 10^7 cells/ml. The energy required to support this amount of biomass production could be evaluated if the amount of S^0, O_2, or SO_4^{-2} was measured subsequent to cultivation. For example, if 1 mg of S^0 was consumed during the growth experiment (which would correspond to consumption of 84.7 mole % of the headspace O_2),

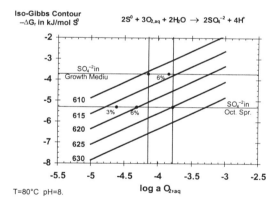

Figure 2. Activities of sulfate are plotted against activities of O_2, and constant values of overall Gibbs free energy are shown as contour lines; T=80°C, pH=8.3. Solid circles represent different aerobic-microaerophilic conditions. The energy supply between natural and culture conditions differs by about 14.3 kJ per mol of sulfur. In Octopus Spring, the equilibrium activity of O_2 is $10^{-65.4}$ ($10^{-64.4}$ in the growth medium) which is equal to a concentration of $O_2=10^{-66.1}$ mol/kg. At concentrations of O_2 higher than this value, the system is out of equilibrium and can release energy for biosynthetic processes if sulfur oxidation serves as a catabolic reaction.

then 190 joules from sulfur oxidation can support 10^9 cells (a *Gigabug*). This type of information would allow us to use geochemical constraints to evaluate biomass production in hydrothermal ecosystems.

5 THE SUBSURFACE BIOSPHERE

In natural settings, both heterotrophic and autotrophic microorganisms coexist and interact with one another Therefore, one may envision that autotrophs (like, methanogens) harness the geochemical energy of inorganic disequilibrium and produce organic compounds, which are consumed by heterotrophs (like, sulfate reducers). Both types of microbes mediate oxidation-reduction reactions that are thermodynamically favored but kinetically sluggish, and both inorganic and organic compounds can be sources of energy if they are out of equilibrium with their surrounding geochemical environment.

If geochemical energy is the key, then the biosphere is not confined to the Earth's surface. Hydrothermal ecosystems found at Vulcano, Yellowstone, Iceland, and elsewhere represent just the tip of the Earth's invisible, subsurface biosphere, which may extend as deep as geochemical energy is available. Recently, numerous scientists have reported on abundant microbial communities in deep granitic aquifers (Szewzyk et al. 1994, Sheriff & Brown 1995, Pedersen 1997), basaltic aquifer (Stevens & McKinley 1995, Stevens 1997), geothermal basin (Love et al. 1993, Byers et al. 1998), deep terrestrial subsurface sediments (Fliermans & Balkwill 1989; Onstott et al. 1998), deep marine sediments (Parkes et al. 1994, Wellsbury et al. 1997), and oil reservoirs (Stetter et al. 1993, L'Haridon et al. 1995).

With suggestions that life may thrive at temperatures of at least 150°C (Daniel 1992, Segerer et al. 1993), and possible evidence for biological processes at 169°C (Cragg and Parkes 1994), a huge number of geochemical processes and a huge volume of the subsurface may fall in the biosphere, and the potential for genetic diversity may vastly surpass that which is currently known. Studies of the physiology and evolution of subsurface life will benefit greatly from experimental and theoretical efforts that quantify the supporting geochemical processes. In this sense, the exploration of global genetic diversity becomes a geochemical problem, and efforts to quantify geochemical processes will be more successful upon including biologically-mediated reactions.

REFERENCES

Amend J.P. & H.C. Helgeson 1997a. Group additivity equations of state for calculating the standard molal thermodynamic properties of aqueous organic species at elevated temperatures and pressures. *Geochim. Cosmochim. Acta* 61: 11-46.

Amend J.P. & H.C. Helgeson 1997b. Calculation of the standard molal thermodynamic properties of aqueous biomolecules at elevated temperatures and pressures I. L-α Amino acids. *J. Chem. Soc. Faraday Trans.* 93: 1927-1941.

Amend J.P. & E.L. Shock 1998. Energetics of amino acid synthesis in hydrothermal ecosystems. *Science* 281: 1659-1662.

Amend J.P., A.C. Amend, and M. Valenza 1998. Determination of volatile fatty acids in the hot springs of Vulcano, Aeolian Islands, Italy. *Org. Geochem.* 11: 699-705.

Brock T.D. 1978. *Thermophilic microorganisms and life at high temperatures.* New York: Springer-Verlag.

Byers H.K., E. Stackebrandt, C. Hayward, & L.L. Blackall 1998. Molecular investigation of a microbial mat associated with the Great Artesian Basin. *FEMS Microbiol. Ecol.* 25: 391-403.

Cragg B.A. & R.J.Parkes 1994. Bacterial profiles in hydrothermally active deep sediment layers from Middle Valley (N.E. Pacific) sites 857 and 858. *Proc. Ocean Drilling Program, Sci. Res.* 139: 509-516.

Daniel R.M. 1992. Modern life at high temperatures. *Orig. Life Evol. Biosph.* 22: 33-42.

Fliermans C.B. & D.L. Balkwill 1989. Microbial life in deep terrestrial subsurfaces. *BioScience* 39: 370-377.

Gurrieri S., H.C. Helgeson, J.P. Amend, & K. Danti 1999. Biogeochemistry of the geothermal system in the Aeolian Islands: Authigenic phase relations in the hot springs of Vulcano, Italy. *Extremophiles* (submitted).

Huber R., W. Eder, S. Heldwein, G. Wanner, H. Huber, R. Rachel, & K.O. Stetter 1998. *Thermocrinis ruber* gen. nov., sp. nov., a pink-filament-forming hyperthermophilic bacterium isolated from Yellowstone National Park. *Appl. Environ. Microbiol.* 64: 3576-3583.

L'Haridon S., A-L. Reysenbach, P. Glénat, D. Prieur, & C. Jeanthon 1995. Hot subterranean biosphere in a continental oil reservoir. *Nature* 377: 223-224.

Love C..A., B.K.C. Patel, R.D. Nichols, & E. Stackebrandt 1993. *Desulfotomaculum australicum*, sp. nov., a thermophilic sulfate-reducing bacterium isolated from the Great Artesian Basin of Australia. *Syst. Appl. Microbiol.* 16: 244-251.

McCollom T.M. & E.L. Shock 1997. Geochemical constraints on chemolithoautotrophic metabolism by microorganisms in seafloor hydrothermal systems. *Geochim. Cosmochim. Acta* 61: 4375-4391.

Onstott T.C., T.J. Phelps, F.S. Colwell, D. Ringelberg, D.C. White, & D.R. Boone 1998. Observations pertaining to the origin and ecology of microorganisms recovered from the deep subsurface of Taylorsville Basin, Virginia. *Geomicrobiology J.* 15: 353-385.

Parkes R.J., B.A. Craig, S.J. Bale, J.M. Getliff, K. Goodman, P.A. Rochelle, J.C. Fry, A.J. Weightman, & S.M.

Harvey 1994. Deep bacterial biosphere in Pacific Ocean
sediments. *Nature* 371: 410-413.

Pedersen, K. 1997. Microbial life in deep granitic rock.
FEMS Microbiol. Rev. 20: 399-414.

Prapaipong P., E.L. Shock, & C.M. Koretsky 1999. Metal-
organic complexes in geochemical processes:
Temperature dependence of standard partial molal
thermodynamic properties of aqueous complexes between
metal cations and dicarboxylate ligands. *Geochim.
Cosmochim. Acta* (in press).

Segerer A.H., S. Burggraf, G. Fiala, G. Huber, R. Huber, U.
Pley and K.O. Stetter 1993. Life in hot springs and
hydrothermal vents. *Orig. Life Evol. Biosph.* 23: 77-90.

Sheriff B.L. & D.A. Brown 1995. Microbial geochemistry of
granite. *Geol. Soc. Amer.: Abst. Prog.* 27: A-185.

Shock E.L. 1995. Organic acids in hydrothermal solutions:
Standard molal thermodynamic properties of carboxylic
acids, and estimates of dissociation constants at high
temperatures and pressures. *Am. J. Sci.* 295: 496-580.

Shock E.L. & M.D. Schulte 1998. Organic synthesis during
fluid mixing in hydrothermal systems. *J. Geophys. Res.*
103: 28513-28527.

Shock E.L., D.C. Sassani, M. Willis, & D.A. Sverjensky
1997. Inorganic species in geologic fluids:
Correlations among standard molal thermodynamic
properties of aqueous ions and hydroxide complexes.
Geochim. Cosmochim. Acta 61: 907-950.

Shock E.L., T. McCollom, & M.D. Schulte 1998. The
emergence of metabolism from within hydrothermal
systems. In Wiegel and Adams (eds.), *Thermophiles: the
keys to molecular evolution and the origin of life?*: 59-76.
London: Taylor & Francis.

Stetter K.O. 1988. *Archaeoglobus fulgidus* gen. nov., sp.
nov.:a new taxon of extremely thermophilic
archaebacteria. *Syst. Appl. Microbiol.* 10: 172-173.

Stetter K.O., R. Huber, E. Blöchl, M. Kurr, R.D. Eden, M.
Fielder, H. Cash, & I. Vance 1993. Hyperthermophilic
archaea are thriving in deep North Sea and Alaskan oil
reservoirs. *Nature* 365: 743-745.

Stevens T. 1997. Lithoautotrophy in the subsurface. *FEMS
Microbiol. Rev.* 20: 327-337.

Stevens T.O. & J.P. McKinley 1995. Lithoautotrophic
microbial ecosystems in deep basalt aquifers. *Nature*
270: 450-454.

Szewzyk U., R. Szewzyk, & T.-A. Stenström 1994.
Thermophilic, anaerobic bacteria isolated from a deep
borehole in granite in Sweden. *Proc. Natl. Acad. Sci.
USA* 91: 1810-1813.

Wellsbury P., K. Goodman, T. Barth, B.A. Cragg, S.P.
Barnes, & R.J. Parkes 1997. Deep marine biosphere
fuelled by increasing organic matter availability during
burial and heating. *Nature* 388: 573-576.

Geochemistry of the Earth's Surface, Ármannsson (ed.) © 1999 Balkema, Rotterdam, ISBN 90 5809 073 6

Adsorption of marine pore water organic matter to montmorillonite

T.S. Arnarson & R.G. Keil
School of Oceanography, University of Washington, Seattle, Wash., USA

ABSTRACT: Adsorption of marine pore water natural organic matter to clean montmorillonite clay surfaces was found to be linear with respect to concentration, depend on the molecular weight of the compounds, and be affected by the pH of the solution and the type of ions in the solution. The pore water NOM is heterogeneous by nature, with multiple compounds and varying chemical groups, resulting in complex adsorption mechanisms. The data indicate several possible mechanisms: electrostatic interactions, cation bridging, ligand exchange and hydrophobic expulsion.

1 INTRODUCTION

Adsorption of natural organic matter (NOM) to particle surfaces is very important to the fate of organic matter in the environment. The observed strong relationship between organic carbon content and surface area in coastal sediments may be a consequence of adsorbed molecules being less accessible for bacteria to degrade (Mayer 1994). In order to understand the importance of adsorption for the preservation of organic matter in marine sediments it is important to determine the basic mechanisms of pore water organic matter adsorption to particle surfaces.

In this study the adsorption behavior of marine pore water organic matter on montmorillonite is investigated, and preliminary results are presented.

2 MATERIALS AND METHODS

The NOM sample used in these experiments was collected by box core from Liberty Bay (Puget Sound, WA). The bulk sediment was centrifuged to isolate the pore water, which was fractionated into two organic fractions with an ultrafilter using a 1000 D nominal cutoff.

The solid phase used in the adsorption experiments was Na-montmorillonite from Clay Spur, WY (API #26; Ward's Scientific, Rochester, NY). The <2 μm size fraction was isolated by gravity settling, and consequently heated at 450 °C overnight to remove any organic carbon.

A batch adsorption method was used to measure the adsorption of NOM to montmorillonite (described in Thimsen & Keil 1998). However, in this study a greater solid to solution ratio of 280 g/l (equivalent to a porosity of 0.9) was used. Thimsen & Keil (1998) used a 2 g/l solid to solution ratio.

3 RESULTS AND DISCUSSION

The method employed in the experiments uses a solid to solution ratio that is similar to what is typically found in marine surface sediments (Hedges & Keil 1995). Due to the high solid to solution ratio, it is possible to study the adsorption over a DOC concentration range that is also similar to sedimentary pore water values; it is not necessary to use abnormally high concentrations to get measurable adsorption (eg. Thimsen & Keil 1998).

Adsorption isotherms for both pore water NOM fractions are approximately linear over the examined concentration range (Fig. 1). The higher molecular weight fraction exhibits greater adsorption over the entire range, as can be seen by its greater partition coefficient (K_d; Table 1). This observation is consistent with results for adsorption of fresh water NOM on hydrous oxide surfaces (Davis & Gloor 1981; Gu et al. 1995) and of sewage sludge on montmorillonite (Baham & Sposito 1994). The reasons for greater adsorption for larger molecules can be that the larger molecules can either bind with more binding sites at the surface or that they are more hydrophobic and are therefore more surface reactive (Stumm & Morgan 1996).

Adsorption of NOM depends on the pH of the solution (Fig. 2). The amount of NOM adsorbed increases with decreasing pH. This has also been

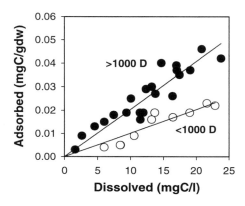

Figure 1. Adsorption isotherms for both size fractions of Liberty Bay pore water NOM.

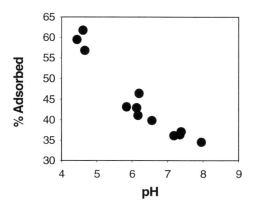

Figure 2. pH dependence of adsorption of Liberty Bay >1000 D fraction on montmorillonite. Initial concentration 10 mgC/l at all pH values; % Adsorbed is the % carbon that was removed from solution by adsorption.

Table 1. The partition coefficients of the two NOM size fractions (± 1 SD)

Sample	K_d*(lkg^{-1})
>1000 D	2.0 ± 0.1
<1000 D	1.0 ± 0.1

*Determined by linear regressions of the data in Figure 1.

found for adsorption of fresh water NOM on hydrous oxide surfaces (Gu et al. 1995) and of soil fulvic acid on montmorillonite (Schnitzer & Kodama 1966). There are two possible reasons for increased adsorption at lower pH. On the one hand this may be due to electrostatic interactions; at low pH's both the organic compounds and the clay surfaces are protonated, but at higher pH's both lose protons and become negatively charged, resulting in repelling electrostatic forces and reduced adsorption (Rashid et al. 1972). On the other hand, this may be due to formation of complexes of organic acids and hydroxyl groups on the clay edges (ligand exchange); the formation of these complexes is favored at lower pH's closer to the pK_a of the organic acids involved (Stumm & Morgan 1996).

Varying the type and concentration of ions in solution affects the adsorption of the NOM (Fig. 3). Adsorption generally increases with increasing ionic strength of solution. This has been observed by other investigators, e.g. for humic acids on montmorillonite and other clay types (Zhou et al. 1994). The reason for the effect of ionic strength is thought to be decreased repulsion between negatively charged organic molecules and clay surfaces in solutions with more ions (i.e. of higher ionicstrength) that can mask the opposite charges and hence, favor adsorption (Rashid et al. 1972).

The amount of NOM adsorbed at a given ionic strength for different ions decreases in the order $CaCl_2$ > NaCl > Seawater > Na_2SO_4. Increased adsorption with Ca^{2+} present in solution is consistent with results for adsorption of humic acids on clay surfaces showing that di- and trivalent cations increase their adsorption (Theng 1979). The reason for this may be that the Ca^{2+} ions are capable of forming bridges between the organic molecules and the clay surface, which results in stronger bonds and

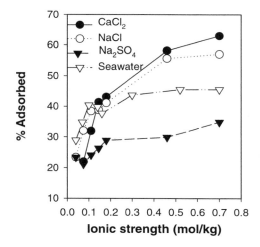

Figure 3. Effect of different ions and their ionic strength on adsorption. Initial concentration 5 mgC/l for all ions and seawater.

greater adsorption (Theng 1979). In addition, the observed decreased adsorption with SO_4^{2-} present (relative to the other ions) is consistent with results of Gu et al. (1994) for fresh water NOM on iron oxide surfaces. This is thought to be due to competition between the SO_4^{2-} and organic ligands for hydroxyl sites on the surface (ligand exchange; Gu et al. 1994).

4 CONCLUSIONS

1. The adsorption method used allows estimation of partition coefficients of NOM at concentrations that are within the range found in marine sediments.

2. The pore water NOM – montmorillonite system is quite complex, due to the heterogeneity of the pore water NOM (multiple compounds with varying functional groups). The results indicate several possible mechanisms of adsorption: electrostatic interactions, cation bridging, ligand exchange and hydrophobic expulsion.

ACKNOWLEDGEMENTS

We are thankful to Crystal Thimsen for help with field work. We thank John Hedges and Allan Devol for ongoing discussion of the project. We would also like to thank Allan Devol for providing ship-board opportunities for collecting samples. This research was supported by US-NSF grant OCE9711792.

REFERENCES

Baham, J. & G. Sposito 1994. Adsorption of dissolved organic carbon extracted from sewage sludge on montmorillonite and kaolinite in the presence of metal ions. *Journal of Environmental Quality* 23: 147-153.

Davis, J.A. & R. Gloor 1981. Adsorption of dissolved organics in lake water by aluminum oxide. Effect of molecular weight. *Environmental Science & Technology* 15: 1223-1229.

Gu, B., J. Schmitt, Z. Chen, L. Liang & J.F. McCarthy 1994. Adsorption and desorption of natural organic matter on iron oxide: mechan-isms and models. *Environmental Science and Technology* 28: 38-46.

Gu, B., J. Schmitt, Z. Chen, L. Liang & J.F. McCarthy 1995. Adsorption and desorption of different organic matter fractions on iron oxide. *Geochimica et Cosmochimica Acta* 59: 219-229.

Hedges, J.I. & R.G. Keil 1995. Sedimentary organic matter preservation: an assessment and speculat-ive synthesis. *Marine Chemistry* 49: 81-115.

Mayer, L.M. 1994. Surface area control of organic carbon accumulation in continental shelf sediments. *Geochimica et Cosmochimica Acta* 58. 1271-1284.

Rashid, M.A., D.E. Buckley & K.R. Robertson 1972. Interactions of a marine humic acid with clay minerals and a natural sediment. *Geoderma* 8: 11-27.

Schnitzer, M. & H. Kodama 1966. Montmorillonite: effect of pH on its adsorption of a soil humic compound. *Science* 153: 70-71.

Stumm, W. & J.J. Morgan 1996. *Aquatic Chemistry*. New York: Wiley-Interscience.

Theng, B.K.G. 1979. *Formation and properties of clay-polymer complexes*. Amsterdam: Elsevier Scientific Publishing Company.

Thimsen, C.A. & R.G. Keil 1998. Interactions between sedimentary dissolved organic matter and mineral surfaces. *Marine Chemistry* 62: 65-76.

Zhou, J.L., S. Rowland, R.F.C. Mantoura & J. Braven 1994. The formation of humic coatings on mineral particles under simulated estuarine conditions - a mechanistic study. *Water Research* 28: 571-579.

Geochemistry of the Earth's Surface, Ármannsson (ed.) © 1999 Balkema, Rotterdam, ISBN 90 5809 073 6

Molecular selectivity of dissolved organic matter sorption to sediments

Anthony K. Aufdenkampe, John I. Hedges & Jeffrey E. Richey
School of Oceanography, University of Washington, Seattle, Wash., USA

Alex V. Krusche
Centro de Energia Nuclear na Agricultura, Universidade de São Paulo, Piracicaba, Brazil

ABSTRACT: A consistent observation in the Amazon River Basin and elsewhere is that suspended fine particulate organic matter (POM) is compositionally distinct from co-existing dissolved organic matter (DOM). This paper presents experimental results showing that these compositional patterns are the outcome of selective partitioning of nitrogen-rich DOM components onto mineral surfaces. In every case, nitrogen was preferentially taken into the particulate fraction relative to the "parent" DOM, as were total hydrolyzable amino acids with respect to total organic carbon and nitrogen. Particulate amino acid compositional patterns also indicated preferential sorption of basic amino acids with positively charged nitrogen side chains. In short, all the various organic nitrogen compositional patterns of the Amazon Basin could be recreated in a beaker. While conjectured from POM/DOM compositional patterns among river samples, this is the first direct evidence for preferential uptake of naturally-occurring nitrogenous DOM by suspended riverine minerals.

1 INTRODUCTION

Riverine transport of organic matter (OM) from land to sea represents a major link in the global cycles of bioactive elements, which modulate the biosphere over geological time (Berner, 1989; Meybeck, 1993). These terrestrial OM losses support significant heterotrophic activity within rivers, estuaries and marine systems alike (Kaplan and Newbold, 1993; Hedges *et al.*, 1997; Mayer *et al.*, 1998). Thus, understanding the mechanisms that determine the forms and compositions of riverine organic matter fluxes is important for both regional and global scales.

In the Amazon and other major rivers of the world, about 90% of transported organic matter is either sorbed to fine minerals or has remained dissolved, yet the compositions of these two fractions are quite different (Meybeck, 1982 & 1993; Keil *et al.*, 1997). Relative to co-existing dissolved organic matter (DOM), fine particulate organic matter (POM) consistently has lower carbon to nitrogen ratios, higher total amino acid concentrations and higher ratios of basic to acidic amino acids (Figure 1a) (Hedges *et al.*, 1994). Because recent findings demonstrate the importance of sorption to OM preservation (Mayer, 1994a&b; Keil *et al.*, 1994a&b; and Hedges and Keil, 1995), selective partitioning of organic nitrogen into the particulate phase could play a critical role in

preferential retention of nitrogen by terrestrial ecosystems.

This study examines whether preferential sorption of nitrogenous DOM components to mineral surfaces determines OM compositional trends in the Amazon Basin and other river systems. Whereas previous observational results implicate preferential sorption, particulate and dissolved OM in rivers have vastly different histories and residence times. Thus to test the hypothesis, changes in DOM and POM compositions during sorption were directly observed in a set of nine laboratory experiments, in which natural suspended river sediments and organic-free kaolinite were mixed with various natural DOM samples from the Peruvian Amazon.

2 METHODS

2.1 Sample Collection

All samples were collected during the October 1996 CAMREX (Carbon in the Amazon River Experiment) expedition to Amazon River source basins in Peru. Samples spanned all major environments found in the Amazon watershed—from 4000+ m Andean "Altiplano" grasslands, to cobbled "mesoscale" rivers in the 300-500 m foothills, and finally to the typical lowland Amazon mainstem at 50 m elevation near Iquitos.

255

2.2 Sorptive Partitioning Experiment

Individual partitioning experiments were conducted in the field by mixing natural pre-filtered DOM with sediment minerals at representative riverine concentrations (~300 mg sediment/L). DOM samples were collected from two rivers and one wetland. In addition, two leaf litter leachates were created by immersing litter from dominant grass communities in distilled water for 24 hours. Suspended sediments from two rivers were isolated by settling and decantation, and organic-free kaolinite was obtained by peroxide oxidation of a commercial mineral source. These dissolved and particulate sample sets each represent continuums of increasing freshness: the DOM suite follows a decreasing history of exposure to mineral surfaces and bacterial degradation, and the sediments represent end-members of equilibration with ambient natural organic substances. In all, nine different combinations of DOM and sediments were mixed.

Mixtures were then incubated for 24 hours under live conditions (ie. not sterile) at ambient temperatures of 30°C. Incubations were terminated by filtration onto glass fiber (Whatman GF/F) filters and subsequent preservation of both the filtered minerals and the resultant DOM for organic analysis in the laboratory. Sub-samples of both the DOM and the POM, before and after mixing, were analyzed for total organic carbon and nitrogen and hydrolyzable amino acids. After analysis, sorbed components were calculated as the difference between the initial and final concentrations.

2.3 Organic Analyses

Particulate carbon and nitrogen were analyzed directly on the GF/F filter with a Leeman Labs CE440 Elemental Analyzer. Dissolved organic carbon (DOC) concentrations were measured with a modified high temperature combustion MQ Scientific 1001 DOC Analyzer, and dissolved organic nitrogen was measured as the change in nitrate concentration after high intensity UV oxidation. Individual amino acid concentrations were determined after acid hydrolysis and derivitization with o-phthaldialdehyde (OPA) by reverse-phase HPLC as per the method of Cowie and Hedges (1992).

3 RESULTS

3.1 Carbon and nitrogen sorption

Appreciable sorption of DOM to sediments occurred in all nine mixing sets (Table 1), with newly sorbed organic matter ranging from 0.1 to 3 weight percent

POC (mg OC/100 mg sediment). In general, carbon lost from the DOM pool was gained by the POM pool, with additional DOM losses likely due to respiration during the experiment. Nitrogen mass balance was also good. The surface area of the kaolinite was 10.7 m^2/g; therefore a change of one weight percent POC corresponds to roughly one mg OC/m^2 surface area. Sorption of river and wetland DOM to kaolinite occurred below the range of 0.5-1.0 mg OC/m^2 that is typical of most riverine and coastal margin sediments (Hedges and Keil, 1995; Keil et al., 1997). Diluted leachates sorbed within this typical range, and concentrated leachates sorbed at higher levels commonly found in sediments from anoxic environments. The extent of sorption generally increased with increasing DOM freshness.

Nitrogen was preferentially partitioned in all cases into the POM fraction relative to the initial "parent" DOM (Figure 1b). Despite a wide range of organic carbon to nitrogen ratios (10-31 mol OC/mol N) for the initial DOM, organic matter transferred into the particulate fraction had C/N ratios in the narrow range of 3-7. This pattern of bulk nitrogen enrichment on particles is a dominant characteristic of organic matter in the Amazon River (Figure 1a) and in other rivers around the world (Meybeck, 1982 & 1993), and is commonly observed in fine soil fractions (Oades, 1988).

Table 1. Carbon mass balance for partitioning experiments. All organic carbon concentrations are normalized to 100 mg sediment in the incubation; thus being directly comparable.

Source for DOM	Sediment	Initial DOC mg/100mg	Δ DOC mg OC/100mg sed.	Δ POC
river 1	kaolinite	1.3	-0.02	0.08
river 2	kaolinite	2.5	-0.08	0.08
wetland	kaolinite	3.7	-0.24	0.18
wetland	river 2	7.3	0.13	0.18
leachate 1	kaolinite	64.6	-4.9	2.5
leachate 1	river 1	29.3	-3.5	3.0
dil. leach. 1	kaolinite	2.9	-1.1	0.7
dil. leach. 2	kaolinite	8.3	-2.9	1.2
leachate 2	kaolinite	72.3	-2.0	2.8

3.2 Amino acids

The percent of total organic carbon occurring as amino acids (%TaaC) indicated selective enrichment of amino acids into POM during sorption to all sediments (Figure 1b). This trend also holds when total hydrolyzable amino acids were normalized to total organic nitrogen (%TaaN). As seen previously, the degree of fractionation of DOM during sorption seemed to be related to its freshness, with leachates

sorbing POM with ~40 %TaaC. For reference, bars representing individual experiments in Figure 1b are given in the same order as found in Table 1.

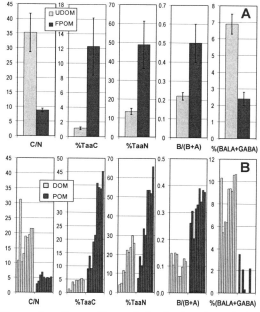

Figure 1. Composition of dissolved and particulate organic matter for A) the Amazon River and major tributaries (mean and standard deviation of nine samples) and B) the nine partitioning experiments presented in this study, in which natural DOM was sorbed to sediments Parameters are defined in text.

Molecular amino acid analysis also demonstrates distinct compositional trends. Of the 20 amino acids found in protein, 15 can be quantified by the OPA method along with four additional non-protein amino acids. The method is sensitive enough that differences in the relative concentrations of individual amino acids could be discerned between initial DOM samples and the newly sorbed POM. These differences generally corresponded to the side-chain functionality of individual amino acids. Basic amino acids, with positively charged nitrogenous side chains, were strongly enriched in the sorbed POM relative to the initial DOM. This enrichment is represented by the amino acid parameter B/(B+A), which is the ratio of basic amino acids to the sum of basic and acidic amino acids. Amino acids with hydrophobic side chains were also enriched in the POM, but to a lesser extent. The extent of selective partitioning of basic and hydrophobic amino acid groups onto particles during the experiment is consistent with observed differences in DOM and POM amino acid compositions in the Amazon River (Figure 1).

Non-protein amino acids had mostly undetectable concentrations in POM despite having high concentrations in DOM. Beta-alanine and gama-aminobutyric acid are the two most abundant non-protein amino acids, whose sum normalized to total amino acid yield, %(BALA+GABA), is a commonly used amino acid parameter. All partitioning experiments showed a pattern of higher %(BALA+GABA) for DOM than POM, which is similar, yet more pronounced, than partitioning patterns in Amazon Basin samples (Figure 1).

4 DISCUSSION

The sorptive partitioning experiments clearly show nitrogen enrichment onto mineral surfaces at three different levels—bulk elemental enrichment, amino acid macromolecular enrichment and nitrogen functional group enrichment—mimicking compositional patterns observed in the Amazon River and it's major tributaries. In the experiment this enrichment is relative to a single, characterized DOM source. However, the mechanism leading to the preferential uptake of these components by mineral surfaces, which must be important in natural and simulated systems alike, is still unknown.

Most clay minerals, including kaolinite, have a net negative surface charge. Because basic amino acids have a net positive charge and acidic amino acids a net negative charge at riverine pH, using electrostatic mechanisms to explain the POM compositions is logical. This type of preferential sorption has been observed for mixtures of dissolved free amino acids (Hedges and Hare, 1987), and for melanoidin polymers synthesized from the condensation of glucose with acidic, neutral and basic amino acids (Hedges, 1978). However, in natural systems most amino acids are bound together in proteins and other biomacromolecules, making a purely electrostatic mechanism more difficult to invoke. A similar argument holds for explanations of hydrophobic amino acid enrichment. Despite requiring macromolecular sorting, physicochemical sorptive mechanisms could play an important role in determining POM compositions.

The total extent of sorption and the selectivity for nitrogen components seemed to be a function of the freshness of the initial DOM. This brings to question the role that microorganisms may play in these processes. Because incubations lasted 24 hours at 30°C, numerous generations of microbial colonies should have been able to leave behind biofilms and necromass. Whereas microbial biomass alone could not account for the concentrations seen, bacterial exopolymers and cell wall remains could be quantitatively and qualitatively important (Costerton et al., 1987;

Hedges and Oades, 1997). Additional experiments are now being conducted to determine the relative extents to which microbially mediated processes versus purely abiotic sorptive processes contribute to the observed selective partitioning into POM.

The non-protein amino acid results in this study give insight to previous interpretations. Because non-protein amino acids are thought to be the products of bacterial degradation, the %(BALA+GABA) parameter has been used as an indicator of diagenetic alteration of sedimentary organic matter (Cowie and Hedges, 1994). Results from partitioning experiments show that sorptive processes are important in determining differences in %(BALA+GABA) between phases; thus non-protein amino acids can not be interpreted solely as a diagenetic parameter when comparing two phases, such as DOM and POM.

5 CONCLUSION

This study demonstrates that all the various organic nitrogen compositional patterns of the Amazon Basin can be recreated in a beaker by sorbing natural dissolved organic matter to mineral surfaces under representative conditions. Additional studies are underway to investigate the relative importance of abiotic, physicochemical mechanisms versus microbially mediated processes in determining these compositions.

REFERENCES

Berner, R. A. 1989. Biogeochemical cycles of carbon and sulphur and their effect on atmospheric oxygen over Phanerozoic time. *Palaeogeography, Palaeoclimatology, Palaeoecology* 75: 97-122.

Costerton, W. J., K.-J. Cheng, G. G. Geesey, T. I. Ladd, J. C. Nickel, M. Dasgupta and T. J. Marrie. 1987. Bacterial biofilms in nature and disease. *Annual Review of Microbiology* 41: 435-464.

Cowie, G. L. and J. I. Hedges. 1992. Improved amino acid quantification in environmental samples: charge-matched recovery standards and reduced analysis time. *Marine Chemistry* 37: 223-238.

Cowie, G. L. and J. I. Hedges. 1994. Biochemical indicators of diagenetic alteration in natural organic matter mixtures. *Nature* 369: 304-307.

Hedges, J. I. 1978. The formation and clay mineral reactions of melanoidins. *Geochimica et Cosmochimica Acta* 42: 69-76.

Hedges, J. I. and J. M. Oades. 1997. Comparative organic geochemistries of soils and marine sediments. *Organic Geochemistry* 27: 319-361.

Hedges, J. I. and P. E. Hare. 1987. Amino acid adsorption by clay minerals in distilled water. *Geochimica et Cosmochimica Acta* 51: 255-259.

Hedges, J. I. and R. G. Keil. 1995. Sedimentary organic matter preservation: An assessment and speculative synthesis. *Marine Chemistry* 49: 81-115.

Hedges, J. I., G. L. Cowie, J. E. Richey, P. D. Quay, R. Benner, M. Strom and B. R. Forsberg. 1994. Origins and processing of organic matter in the Amazon River indicated by carbohydrates and amino acids. *Limnology and Oceanography* 39: 743-761.

Hedges, J. I., R. G. Keil and R. Benner. 1997. What happens to terrestrial organic matter in the ocean. *Organic Geochemistry* 27: 195-212.

Keil, R. G., D. B. Montluçon, F. G. Prahl and J. I. Hedges. 1994. Sorptive preservation of labile organic matter in marine sediments. *Nature* 370: 549-552.

Keil, R. G., E. C. Tsamakis, C. B. Fuh, J. C. Giddings and J. I. Hedges. 1994. Mineralogical and textural controls on the organic composition of coastal marine sediments: Hydrodynamic separation using SPITT-fractionation. *Geochimica et Cosmochimica Acta* 58: 879-893.

Keil, R. G., L. M. Mayer, P. D. Quay, J. E. Richey and J. I. Hedges. 1997. Losses of organic matter from riverine particles in deltas. *Geochimica et Cosmochimica Acta* 61: 1507-1511.

Mayer, L. M. 1994. Relationships between mineral surfaces and organic carbon concentrations in soils and sediments. *Chemical Geology* 114: 347-363.

Mayer, L. M. 1994. Surface area control of organic carbon accumulation in continental shelf sediments. *Geochimica et Cosmochimica Acta* 58: 1271-1284.

Mayer, L. M., R. G. Keil, S. A. Macko, S. B. Joye, K. C. Ruttenberg and R. C. Aller. 1998. Importance of suspended particulates in riverine delivery of bioavailable nitrogen to coastal zones. *Global Biogeochemical Cycles* 12: 573-579.

Meybeck, M. 1982. Carbon, Nitrogen, and Phosphorus Transport by World Rivers. *American Journal of Science* 282: 401-450.

Meybeck, M. 1993. C, N, P and S in rivers: from sources to global inputs. *Interactions of C, N, P and S Biogeochemical Cycles*. R. Wollast, F. T. Mackenzie and L. Choo, Springer-Verlag: 163-193.

Oades, J. M. 1988. The retention of organic matter in soils. *Biogeochemistry* 5: 35-70.

Geochemistry of the Earth's Surface, Ármannsson (ed.)© 1999 Balkema, Rotterdam, ISBN 90 5809 073 6

Early diagenesis of amino acids in high organic content marine sediments

R.G. Keil

School of Oceanography, University of Washington, Seattle, Wash., USA

ABSTRACT: Marine sediments containing high amounts of organic matter (>5%) are not enriched in amino acids relative to other sediments, but the molecular composition of amino acids in organic matter-rich systems suggests less extensive alteration of source materials. During degradation, sediments become enriched in the amino acids glycine, threonine, serine and arginine relative to bulk planktonic sources. Isoleucine, leucine and glutamic acid are depleted. Compositions of amino acids in high-carbon sediments can be explained simply as a mixture of planktonic-derived protein, amino acids in diatom frustules, and glycine-rich bacterial peptidoglycan in approximately 0.50 – 0.25 – 0.25 proportions. These results suggest that although degradative losses of amino acids are large, extensive diagenetic alteration of remaining amino acids is minimal.

1 INTRODUCTION

Extensive effort has gone into investigating early diagenesis in marine sediments because of the important role sediments play in modulating organic matter burial and in recording global change, (e.g. Hedges & Keil 1995). Amino acids represent a substantial portion of the organic matter input to continental margin sediments (Lee 1988) and amino acids extracted from a bulk sedimentary system can provide information about the conditions of deposition and the extent to which the deposit has undergone decay (Henrichs 1993).

Amino acid concentrations and compositions in marine sediments change during diagenesis, and these changes have been related to progress along a diagenetic continuum between fresh material (e.g. phytoplankton) and organic-depleted materials (e.g. oxidized deep sea turbidites; Cowie & Hedges 1994, Dauwe & Middelburg 1998, Whelan & Emeis 1992). Over a wide variety of time scales, amino acids have been shown to be degraded faster than bulk organic matter (Henrichs 1993). This results in a depletion of amino acids with respect to the total organic matter of a sample.

Much of the information about amino acid diagenesis has been gathered from degradation experiments (e.g. Nguyen & Harvey 1997), sampling of water column particles using sediment traps (e.g. Lee 1988, Lee & Cronin 1982, Lee & Wakeham 1988) or examination of diagenetic signals down cores (e.g. Burdige & Martens 1988,

Haugen & Lichtentaler 1991, Henrichs 1987, Henrichs et al. 1984). Despite a large number of individual studies, there have been few previous attempts to broadly compile sedimentary data for gross, system-level evaluation of amino acid diagenesis under a wide range of depositional conditions (Dauwe & Middelburg 1998, Keil et al. in press). Here I present amino acid concentration and compositional data for a wide range of marine sediments, and discuss how the compositions and quantities of amino acids in high organic matter sediments (>5% OC) that are deposited under oxygen-deficient conditions relate to known source and diagenetic information. Sediments with high organic carbon contents are unique because they potentially provide a diagenetic bridge between fresh materials and more degraded systems (e.g. Tyson 1987). However, the validity of the supposition that organic-rich sediments indeed contain less degraded organic matter is debatable (Pederson et al. 1992).

2 METHODS

I have compiled data from twenty-four articles (Keil et al. in press). Three of the works include data for sediments enriched in organic matter (Table 1). Many articles in the literature could not be used for this compilation because the data were incomplete; most commonly no compositional data are reported. The data are clustered into five groups; plankton

259

Table 1. Literature sources of amino acid compositions in high organic matter sediments used in this compilation.

Region	Reference
Peru Upwelling Zone	*Henrichs et al. 1984*
Pettaquamscutt River Estuary	*Henrichs 1987*
Mexican Margin	*Keil et al. submitted.*

Data for organic carbon and total nitrogen from Henrichs and Farrington, 1984.

samples, sinking particles, high carbon sediments deposited under oxygen-deficient water masses, shallow marine sediments (3000 m or less; most are from <300 m), and open ocean sediments (slowly accumulating sediments and oxidized turbidite samples). Two additional groups, diatom frustule and bacterial peptidoglycan, were used to reconstruct a hypothetical marine sediment sample (Hecky et al. 1973, Shemesh et al. 1993, Cowie & Hedges 1996, Koch 1990).

3 RESULTS AND DISCUSSION

Marine organisms and sinking particles are uniformly enriched in amino acids relative to sediments (Figure 1). On average, 52 ± 23% of the organic carbon in marine organisms is chromatographically-identifiable amino acid carbon, and 77 ± 16% of the nitrogen is amino acid nitrogen (Keil et al. in press). Subsequent depletion of amino acids relative to bulk carbon is an ongoing process, but two stages of degradation can markedly alter the bulk amino acid content of organic matter; conversion from plankton to sinking particle (Lee & Wakeham 1988) and incorporation of sinking particle into the sediment (Wakeham et al. 1997).

In contrast to either plankton sources or sinking particles, marine sediments are depleted in amino acids by a factor of two or more (Figure 1), with open ocean sediment being depleted by orders of magnitude (Whelan 1977). Continental margin sediments contain on average only 12 ± 10% of their carbon and 30 ± 12% of their nitrogen in amino acid form. Carbon depocenters such as the Mexican margin (Keil et al. submitted) or the Peru upwelling region (Henrichs & Farrington 1984), which can contain as much as 30% organic matter, have been hypothesized to contain organic matter that is less degraded and more similar in composition to source materials (Tyson 1987). Based on bulk amino acid content (Figure 1), these sediments are clearly not enriched in amino acids relative to other sediments, as one might expect if the organic matter were less degraded. However, as discussed next, the amino acid composition of high organic content sediments

is different (i.e. less degraded) than that of many other sediments.

At the molecular level, glycine, serine, threonine and arginine are preferentially present in sediment relative to sources (Figure 2). Three of these amino acids (glycine, serine and threonine) are present in excess abundance in silacious diatom frustules and their associated organic matrix (Hecky et al. 1973), suggesting that this material might be preferentially preserved in marine sediments. Sedimentary enrichment of glycine could potentially support the idea that bacterial biomass contributes significantly to the amino acid pool. Bacterial peptidoglycan can be replete in glycine, which in pentameric form acts

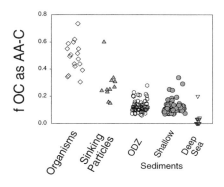

Figure 1. The fraction of organic carbon that is amino acid carbon as a function of sample type. ODZ= organic carbon-rich sediments underlying oxygen deficient zones.

as a bridge peptide (Koch 1990). This peptide bridge is most abundant in Gram+ bacteria, although it has been observed in some Gram- species.

I constructed a simple mixing model that mixes diatom frustules, planktonic protein, bacterial protein and bacterial peptodoglycan. The model attempts to minimize the differences between the mixture and the average composition of high organic matter sediments. The best model fit results in an average mixture of 50% plankton + bacterial protein, 25% diatom frustule and 25% peptidoglycan (Figure 2). The model output is sensitive to changes in the proportions of diatom frustule and peptidoglycan. A balance between frustules and peptidoglycan is required in order to balance the alanine, serine, and threonine contents of the sediment. Conversely, the model is insensitive to the composition of either planktonic or bacterial protein. For example, increasing bacterial protein at the expense of the planktonic component allows better resolution of lysine and phenylalanine at the expense of the serine, aspartic acid and alanine. Thus, it is not possible to resolve whether the protein-like component is derived exclusively from plankton or bacteria. Even though the mixing model does not take diagenesis into account, it reproduces the composition of an average organic carbon-rich sediment surprisingly well (Figure 2). Only the sedimentary enrichment in arginine cannot be reproduced within the statistical uncertainty of the measurements.

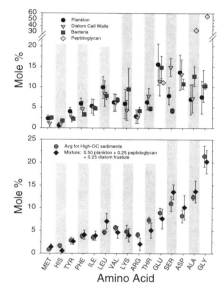

Figure 2. Upper panel illustrates the mole percent composition (± 1 std) of planktonic sources (diatom cell wall-free; n=7), diatom cell walls (n=6), bacteria (with associated peptidoglycan; n=3) and average peptidoglycan (n=1 average value from Koch 1990). Lower panel illustrates the average composition of organic carbon-rich marine sediments (± 1 standard deviation, n=14 stations) and the composition of a mixture that is 50% plankton protein, 25% peptidoglycan and 25% diatom cell wall.

Non-protein amino acids, when expressed as a mole percentage, are inconsequential in most source materials. Non-protein amino acids only become a significant component of the amino acid pool at very low organic matter concentrations, mostly in deep sea sediments (Keil et al. in press). High organic matter sediments are along the continuum between fresh materials and deep sea sediments, but cluster closer to fresh materials, consistent with the protein amino acid mixing model results (data not shown). In order to further investigate the relationship between amino acid compositions and diagenesis, Dauwe & Middelberg (1998) have used principal component analysis to categorize the changes in amino acid compositions. The Dauwe degradation

index ranks changes in individual amino acids in a similar general scheme as has been shown for the overall data compilation; enrichment in threonine, arginine, and glycine and depletion in isoleucine, leucine and histidine. Applying their algorithm to the data compiled here illustrates that the amino acid composition of marine sediments is continually changing toward more degraded materials as a function of sedimentary organic matter content (data not shown; see Keil et al. in press). Thus, the two degradative indexes (non-protein amino acids and the Dauwe index) both support the suggestion that amino acids in high-organic systems are not markedly altered by past degradation.

An interesting result of the data compilation is that there are no known sites of highly degraded amino acids (mole percent non-protein amino acids > 5 or Dauwe degradative index < 0.5) that correspond to organic carbon-rich sediments. In fact, sediments from both the Peru upwelling region (Henrichs et al. 1984) and the Mexican margin (Keil et al. submitted) contain amino acids that better reflect sources rather than highly altered amino acid remains. One possible interpretation of this may be that transit through the water column (even an oxygen-deficient one; Lee & Cronin 1982, Lee & Wakeham 1988) alters the amino acid content of the settling material, but the large-scale changes in composition which happen during diagenesis do not occur in high carbon content sediments.

ACKNOWLEDGEMENTS

I am thankful to Elizabeth Tsamakis for analytical work on the Washington and Mexican margin samples presented as part of this compilation. I also thank Þórarinn Arnarson, John Hedges and Allan Devol for thoughtful discussion. This research was supported by US-NSF grants OCE9618265 and OCE9711792.

REFERENCES

Burdige D. J. & C. S. Martens 1988. Biogeochemical cycling in an organic-rich coastal marine basin:10. The role of amino acids in sedimentary carbon and nitrogen cycling. *Geochimica et Cosmochimica Acta* 52: 1571-1584.

Cowie G. L. & J. I. Hedges 1994. Biochemical indicators of diagenetic alteration in natural organic matter mixtures. *Nature* 369: 304-307.

Cowie G. L. & J. I. Hedges 1996. Digestion and alteration of the biochemical constituents of a diatom (*Thalassiosira weissflogii*) ingested by an herbivorous zooplankton (*Calanus pacificus*). *Limnology and Oceanography* 41: 581-594.

Dauwe B. & J. J. Middelburg 1998. Amino acids and hexosamines as indicators of organic matter degradation state in North Sea sediments. *Limnology and Oceanography* 43: 782-798.

Haugen J.-E. & R. Lichtentaler 1991. Amino acid diagenesis, organic carbon and nitrogen mineralization in surface sediments from the inner Oslofjord, Norway. *Geochimica et Cosmochimica Acta* 55: 1649-1661.

Hecky R. E., K. Mopper, P. Kilham & E. T. Degens 1973. The amino acid and sugar composition of diatom cell walls. *Marine Biology* 10: 323-331.

Hedges J. I. & R. G. Keil 1995. Sedimentary organic matter preservation: an assessment and speculative synthesis. *Marine Chemistry* 49: 81-115.

Henrichs S. M. 1987. Early diagenesis of amino acids and organic matter in two coastal marine sediments. *Geochimica et Cosmochimica Acta* 51: 1-15.

Henrichs S. M. 1993. *Early diagenesis of organic matter: the dynamics (rates) of cycling of organic compounds*. In: M. A. Engel & S. H. Macko (eds), *Organic Geochemistry*: 101-117. New York: Plenum Press.

Henrichs S. M. & J. W. Farrington 1984 Peru upwelling region sediments near 15S. 1. Remineralization and accumulation of organic matter. *Limnology and Oceanography* 29: 1-19.

Henrichs S. M., J. W. Farrington & C. Lee 1984. Peru upwelling region sediments near 15°S. 2. Dissolved free and total hydrolyzable amino acids. *Limnology and Oceanography* 29, 20-34.

Keil R. G., E. Tsamakis & A. H. Devol *submitted*. Amino Acid Compositions and OC:SA ratios indicate enhanced preservation of organic matter in Pacific Mexican Margin sediments.

Keil R. G., E. Tsamakis & J. I. Hedges *in press*. Early diagenesis of particulate amino acids in marine systems. In: S. Macko, G. Goodfriend, & M. Fogel (eds). *Amino Acids in Geological Systems: A tribute to Ed Hare*.

Koch A. L. 1990. Growth and form of the bacterial cell wall. *American Scientist* 78: 327-341.

Lee C. 1988. Amino acid chemistry and amine biogeochemistry in particulate material and sediments. In *Nitrogen cycling in coastal marine environments* (ed. T. H. Blackburn & J. Sorensen), pp. 126-141. John Wiley & Sons, Ltd.

Lee C. & C. Cronin 1982. The vertical flux of particulate nitrogen in the sea: decomposition of amino acids in the Peru upwelling area and the equatorial Pacific. *Journal of Marine Research* 40: 227-251.

Lee C. & S. G. Wakeham 1988. Organic matter in seawater: Biogeochemical processes. In *Chemical Oceanography*, Vol. 9, pp. 49: 1-44. Academic Press.

Nguyen R. T. & H. R. Harvey 1997. Protein and amino acid cycling during phytoplankton decomposition in oxic and anoxic waters. *Organic Geochemistry* 27: 115-128.

Pedersen T. F., G. B. Shimmield & N. B. Price 1992. Lack of enhanced preservation of organic matter in sediments under the oxygen minimim on the Oman Margin. *Geochimica et Cosmochimica Acta* 56: 545-551.

Shemesh, A., S. Macko, C. Charles, & G. Rau 1993. Isotopic evidence for reduced productivity in the glacial southern ocean. *Science* 262: 407-410.

Tyson R. V. 1987. The genesis and palynofacies characteristics of marine petroleum source rocks. *Geological Society Special Publications* 26: 47-67.

Wakeham S. G., C. Lee, J. I. Hedges, P. J. Hernes & M. L. Peterson 1997. Molecular indicators of diagenetic status in marine organic matter. *Geochimica et Cosmochimica Acta* 61: 5363-5369.

Whelan J. K. 1977. Amino acids in a surface sediment core of the Atlantic abyssal plain. *Geochemica et Cosmochimica Acta* 41: 803-810.

Whelan J. K. & K.-C. Emeis 1992. Sedimentation and Preservation of Amino Compounds and Carbohydrates in Marine Sediments. In J. K. Whelan & J. W. Farrington (eds), *Organic matter: productivity, accumulation, and preservation in recent and ancient sediements*: 176-200. New York: Columbia University Press.

Geochemistry of the Earth's Surface, Ármannsson (ed.)© 1999 Balkema, Rotterdam, ISBN 90 5809 073 6

Microbial-silica interactions in modern hot spring sinter

K.O. Konhauser, V.R. Phoenix & S.H. Bottrell
School of Earth Sciences, University of Leeds, UK

D.G. Adams
Department of Microbiology, University of Leeds, UK

I.M. Head
Fossil Fuels and Environmental Geochemistry, Postgraduate Institute, University of Newcastle, Newcastle upon Tyne, UK

ABSTRACT: Examination of siliceous sinter from the Krisuvik hot spring, Iceland has revealed a consortium of silicified cyanobacteria forming discrete layers within the authigenic crusts. All cyanobacteria are mineralised, with epicellular silica ranging in thickness from < 2 μm coatings on individual cells to regions where entire colonies are cemented together in an amorphous silica matrix tens of microns thick. Most cells appear intact and do not show evidence for intracellular silicification, implying that they were still viable when sampled. Those which had their cytoplasm replaced by silica presumably lysed in-situ, retaining intact sheaths and trichomes as the only evidence of their former existence. The silicification of natural cyanobacteria closely resembles experimental results which show epicellular silica precipitates of similar morphology and size.

1. INTRODUCTION

Intimately-layered microcrystalline siliceous sediments are a striking feature of present-day hot spring environments, but the mechanisms of silicification (and crucially the role of microorganisms) is unknown. Studies using Transmission Electron Microscopy (TEM) and Scanning Electron Microscopy (SEM) have shown that microorganisms within the sinter are commonly silicified (Ferris et al. 1986, Schultze-Lam et al. 1995, Konhauser & Ferris 1996, Jones et al. 1997a, b). The mineralisation associated with bacterial cells occurs as spheroidal grains (100s of nm to 2 μm in diameter) both extracellularly, on the sheath (an external polysaccharide that encompasses the cell), and intracellularly, within the cytoplasm after the cells lyse (Schultze-Lam et al. 1995, Konhauser & Ferris 1996, Jones et al. 1997a). Often, the silica particles coalesce, such that individual precipitates are no longer distinguishable; entire colonies becoming cemented together in a siliceous matrix several micrometers thick. Eventually only the sheath and cell wall of the original organic framework, or some remnants of cytoplasm, remain recognisable (Schultze-Lam et al. 1995, Konhauser & Ferris 1996).

The above observations indicate that silicification begins when the microorganisms are living, and continues for some time after their death. The bacteria at hot springs appear to facilitate silicification by providing reactive interfaces for

silica adsorption, thereby reducing the activation energy barriers to nucleation, and allowing surface chemical interactions that sorb more silica from solution to take place. From the ease with which silica forms on bacteria, it appears that the interfacial free energy between the two solid surfaces is smaller than the interfacial free energy between the initial colloidal silica precipitate and the solution. In this way, the bacterium functions as a reactive interface, or template, for heterogeneous nucleation. Because a sufficient supply of silica is generally available in hot spring effluent (in excess of mineral solubility), continued adsorption results in the surface sites becoming saturated, such that mineral growth occurs. After bacteria initiate silica precipitation, continued growth occurs autocatalytically and abiogenically, due to the increased surface area generated by the small silica phases.

Although the presence of silicified microorganisms in sinter is well documented, there is at this point no clear understanding of how important indigenous microflora are to the overall silicification process. In this current study, we have sampled a section of sinter for detailed microscopic analysis to ascertain whether biomineralisation is a key process in sinter formation. We demonstrate here that silica-cyanobacteria interactions account for a significant fraction of the sinter, with some evidence to suggest a seasonal effect. Experimental silicification of *Calothrix* sp., isolated from within the sinter, also shows clear similarities between the morphology, size and cellular nucleation sites with those observed in nature.

2. METHODOLOGY

A section of finely-laminated siliceous sinter was collected from a shallow pool of water (< 1 cm) that surrounded the main vent in Krisuvik, Reykjanes Peninsula, Iceland. Samples were made into polished blocks for SEM, carbon coated and then analysed using a Camscan SEM operating at 20 kV, equipped with a Link analytical EDS detector for qualitative determination of mineral chemistry.

The filamentous cyanobacterium *Calothrix* sp. was isolated from the upper surface of the sinter and cultured in BG11 liquid medium at 30°C. Filaments were then cultured on agar plates using nitrate deficient BG11 nutrients until an even spread across the plates was achieved. A silica solution containing 300 ppm Si was prepared with $Na_2SiO_4.5H_2O$ in deionised water and neutralised to pH 7 with 3M HCl. The agar plate inoculated with *Calothrix* sp. was then placed in a 2L beaker containing 1000 ml of this solution. The cyanobacteria were allowed to mineralise over 20 days, during which time the plates were transferred to a fresh solution every 2 to 3 days to ensure a continual supply of solute, thus replicating to the flowing system of a hot spring. Samples of culture were collected every 2 to 3 days and fixed in 2% gluteraldehyde.

Fixed cells were processed for TEM as described by Konhauser at al. (1994). Samples prepared for TEM analysis were viewed with a Philips CM10 TEM operating at 100kv, equipped with Selected Area Electron Diffraction (SAED) for determination of mineral structure. Qualitative biomineral composition was determined using a Philips CM20 TEM/STEM operating at 200kv, equipped with an Oxford Instruments Energy-Dispersive X-ray Spectrometer (EDS).

3. OBSERVATIONS

3.1 Sinter at Krisuvik

The sample of sinter consists of an approximately 7 mm section with approximately ten laterally extensive laminations, a thick, wedge-shaped feature and numerous poorly-defined horizons (Figure 1). Within the overall profile, there appears to be two very distinct lamina types that repeatedly alternate throughout the sinter, a "microbial" layer in which biomineralised microorganisms are associated with amorphous silica, while a finely-laminated silica layer appears to be completely inorganic in composition (i.e. there is no evidence of

microorganisms). In all cases, the microbial layers have a sharply defined base, overlying the finely-laminated silica, and a gradational upper surface into the finely-laminated silica. The microbial bands, which vary from 50 μm to 150 μm in thickness, are often laterally connected to denser colonies which occur in dips in the laminae. In these particular instances, it appears as though the colonies have filled in the undulating topographical feature that presumably existed at the time of colonisation. In one example, a microbial colony extended approximately 3 mm in thickness.

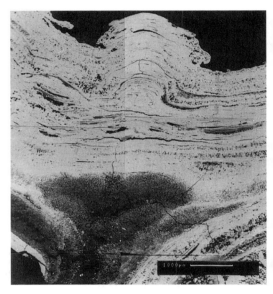

Figure 1. SEM image of a section of siliceous sinter from Krisuvik, Iceland.

Within the microbial layers the dominant cells are commonly filamentous in structure, with some having a clear branching form. Most filaments also display a preferred vertical orientation, with several filaments aligned towards the sediment-water interface (Figure 2). Based on the filamentous and branching morphology, the size range of individual cells (from 5 μm to 8 μm) and the blue-green coloration of the cells under the transmitted light microscope it appears as though the sinter contains the photosynthetic cyanobacterium *Fischerella* sp. We have also isolated and cultured the cyanobacterium *Calothrix* sp., and observed bacterial cells directly attached to the sheaths of the cyanobacteria (this observation relates to the *Fischerella* sp.). Alternating horizons of light (Figure 2) to more heavily mineralised microorganisms (Figure 3) commonly exist within the same layers. These horizons of different intensity biomineralisation grade into one another. It is interesting to note that in the more silicified regions colonies may be encrusted in a silica matrix tens of microns in thickness, while those in the more organic regions have < 2 μm silica coatings on individual cells. However, all cyanobacteria analysed throughout the sinter were consistently encrusted in silica, the silicification predominantly on the external sheath of the microorganisms. Intracellular silicification was infrequently observed, suggesting that most cells were viable at the time of sampling. Those cells that had lysed within the sinter are generally found within relatively non-biomineralised regions dominated by featureless organic material (e.g. exopolymers and lysed cells), and are observed with their sheaths and trichomes intact (Figure 2).

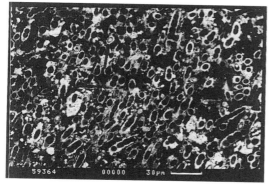

Figure 2. SEM image of a lightly biomineralised horizon within the sinter. Note the preferred filament orientation towards the top of the sample (i.e. sediment-water interface) and evidence of intracellular mineralisation (arrow).

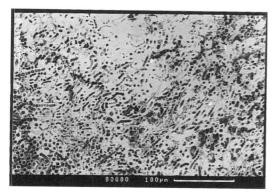

Figure 3. SEM image of highly biomineralised horizon within sinter.

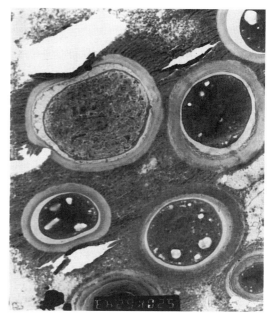

Figure 4. TEM image of experimentally silicified *Calothrix* sp. cells. Note the complete encrustation of the cells in epicellular silica and the presence of intact cytoplasm, indicating the cells were viable upon sampling.

3.2 Experimental silicification

After five days, examination by SEM and TEM revealed amorphous silica precipitates beginning to form on the filaments incubated in the silica solution. Silicification of these filaments steadily progressed and by twelve days much of the colony was heavily encompassed in a silica matrix. Mineralisation around each filament was commonly between 1 μm and 5 μm thick; the extent of mineralisation around each filament was often restricted by the close proximity of other cells (Figure 4). Other areas of the culture exhibited little mineralisation, these were possibly new growths of filaments which had not obviously had sufficient time to extensively silicify. Silicification progressed to the extent that by twenty days, thin crusts of amorphous silica had developed over entire areas of the colony. These crusts were probably the result of silica precipitation exceeding the growth rate of the colony. Close examination by TEM revealed that mineralisation occurred only upon the extracellular sheath. Mineralisation was not observed upon the cell wall or cytoplasm; these components appearing intact and in healthy condition.

4. CONCLUSIONS

Detailed examination of sections of siliceous sinter confirm that biomineralisation of cyanobacteria comprises a major component of the sinter's overall structure, with nearly half the thickness attributed to microbial-silica interactions. All cyanobacteria observed under TEM clearly show the silicification of individual cells, with the thickness of the silica matrix varying in different horizons. In heavily mineralised regions, the silica matrix extends to encompass entire colonies of cells.

The microorganisms examined from the Icelandic sinters are dominated by photosynthetic microorganisms that rely on sunlight for energy generation. The vertical orientation of the cells in the microbial layers highlights their phototactic response, and indicates the cell's attempt to maintain its optimal position in the sinter with regards to the surface. Although only conjecture at this point, the gradational top surface suggests that silica precipitation must have at some point exceeded the cyanobacteria's ability to compensate (i.e. grow upwards); this possibly reflects their natural slow/death phase during the dark winter months of Iceland. In turn, the defined base suggests rapid recolonisation of the inorganic precipitate surface by free-living cyanobacteria in the effluent, once conditions for surface attachment became favourable (i.e. Spring).

Future studies will focus on evaluating the factors which lead to lamination formation, in particular, how microbial activity varies throughout the year. Furthermore, in hot spring environments, where the hydrothermal effluent is commonly saturated with respect to amorphous silica, the primary impediment for phototactic microorganisms would seem to be how to keep pace with continuous accretion. Of obvious interest then is how do completely encrusted cells maintain their optimal position within an ever-thickening sinter?

ACKNOWLEDGEMENTS

This work was supported by a Royal Society Research Grant to KOK and a Natural Environment Research Council studentship to VRP. We further thank the Icelandic Research Council for their permission to undertake scientific research.

REFERENCES

Ferris, F.G., T.J. Beveridge & W.S. Fyfe, W.S. 1986. Iron-silica crystallite nucleation by bacteria in a geothermal sediment. *Nature* 320:609-611.

Jones, B., R.W. Renault & M.R. Rosen 1997a. Biogenicity of silica precipitation around geysers and hot-spring vents, North Island, New Zealand. *J. Sed. Res.* 67:88-104.

Jones, B., R.W. Renault & M.R. Rosen 1997b. Vertical zonation of biota in microstromatolites associated with hot springs, North Island, New Zealand. *Palaios* 12:220-236.

Konhauser, K.O., S. Schultze-Lam, F.G. Ferris, W.S. Fyfe, F.J. Longstaffe, F.J. & T.J. Beveridge 1994. Mineral precipitation by epilithic biofilms in the Speed River, Ontario Canada. *Appl. Environ. Microbiol.* 60:549-553.

Konhauser, K.O. & F.G. Ferris 1996. Diversity of iron and silica precipitation by microbial mats in hydrothermal waters, Iceland: Implications for Precambrian iron formations. *Geology* 24:323-326.

Schultze-Lam, S., F.G. Ferris, K.O. Konhauser & R.G. Wiese 1995. In situ silicification of an Icelandic hot spring microbial mat: Implications for microfossil formation. *Can. J. Earth Sci.* 32:2021-2026.

Geochemistry of the Earth's Surface, Ármannsson (ed.)© 1999 Balkema, Rotterdam, ISBN 90 5809 073 6

The role of charred organic matter in the pedogenesis of Chernozems

I. Kögel-Knabner & M.W.I. Schmidt
Lehrstuhl für Bodenkunde, Technische Universität München, Germany

J.O. Skjemstad
CSIRO, Land and Water, Glen Osmond, S.A., Australia

E. Gehrt
Niedersächsisches Landesamt für Bodenforschung, Hannover, Germany

ABSTRACT: The presence of charcoal from vegetation fires was investigated in Axp and Axh horizons originating from a catena of Chernozemic soils south of Hannover using a suite of complementary methods (high energy ultraviolet photo-oxidation, scanning electron microscopy, solid state ^{13}C nuclear magnetic resonance, lignin analysis by CuO oxidation). This 10 km color-catena of four Chernozemic soils, which were very similar in chemical and physical properties, showed a strong relationship between color and the content of charred organic carbon. Charred organic matter contributed up to 45 % of the bulk soil organic carbon and up to about 8 g/kg of the bulk soil in the color-catena of Chernozemic soils. Finely divided charred organic carbon seems to be a major constituent of Chernozemic soils in Germany and supposedly plays an important role in the pedogenesis of Chernozemic soils. The presence of charred organic matter in Chernozems has major implications for the formation of refractory organic matter in soils.

1 INTRODUCTION

The presence of charcoal from vegetation fires can have a major impact on soil organic matter composition, turnover and formation. Charcoal may increase the amount of aromatic C and contribute a relatively inert type of carbon to the SOM pool (Skjemstad et al. 1996). From charred organic carbon highly aromatic humic acids can be extracted (Haumaier & Zech 1995, Skjemstad et al. 1996).

Using a combination of high energy ultraviolet photo-oxidation, scanning electron microscopy and solid state ^{13}C NMR spectroscopy, it was possible to identify and determine the charcoal content of Australian soils (Skjemstad et al. 1996). Australian grassland soils which were under aboriginal fire management presumably for thousands of years, are characterized by black A horizons with up to 30 % of the soil carbon present as charcoal, whereas adjacent forested soils which were not subjected to regularly burning, display a gray color and small contributions of charcoal (Skjemstad et al. 1997).

A color-catena of four Chernozemic soils was sampled to investigate the potential contribution of charred organic carbon (charcoal) to these soils and a possible relationship between soil color and the chemical structure of soil organic matter (Schmidt et al. 1999).

2 MATERIAL AND METHODS

Soil samples represented a color-sequence changing from black to gray color (Chernozems and Greyzems) which developed on loess in the region south of Hannover in Germany (Table 1). Carbon and nitrogen contents were determined by dry combustion in duplicate with an Elementar Vario EL and for the various fractions investigated during the photo-oxidation procedure, carbon was determined by a modified chromic acid digestion procedure. Scanning electron microscopy was carried out on a Cambridge Stereoscan S250 on samples coated with 20-30 nm of carbon. Elemental characterization was performed using a Link AN1000 EDX analyzer. Alkaline CuO oxidation was carried out as described by Kögel-Knabner (1995). Soil color was determined with dry samples. The degree of darkness was measured with an ASD field spectrometer under sunlight, and is reported as Y-CIE values, which can be converted to Munsell values and correspond closely to human observations.

The content of char in soils was determined by the technique of high energy ultra violet photo-oxidation in combination with ^{13}C nuclear magnetic resonance (NMR) spectroscopy. Previous studies showed that a number of materials found in soils, including wood, lignin, and humic acids, could all be destroyed by the high energy photo-oxidation

Table 1: Chemical and physical properties of the studied soils
        * according to FAO (1994), and ** according to AG-Boden (1994)

	Classification*	Horizon**	Depth	Inorg. C	Org. C	Total N	C/N	pH	Sand	Silt	Clay
			cm	g/kg				CaCl$_2$	(g/100 g)		
1	Haplic Chernozem	Axp	0-20	<0.1	23.6	2.1	11	7.4	3.1	75.2	21.7
		Axh	20-60	<0.1	17.8	1.3	14	7.6	1.6	74.6	23.9
2	Haplic Chernozem	Axp	0-20	<0.1	18.3	1.5	12	7.2	3.0	80.2	16.8
		Axh	20-45	<0.1	11.9	1.4	9	7.2	3.1	79.3	17.6
3	Haplic Greyzem	Axp	0-20	9.0	5.0	0.5	10	7.5	3.3	78.7	18.0
		Axh	20-45	2.0	12.8	1.2	11	7.4	1.6	79.6	18.7
4	Haplic Greyzem	Axp	0-20	<0.1	13.5	1.4	10	7.0	2.4	79.6	18.0
		Axh	20-45	<0.1	13.3	1.1	12	7.3	0.9	77.0	22.1

process, provided they were exposed to ultraviolet radiation in the presence of excess oxygen. Charred organic matter, however, was not destroyed by this treatment. This method was combined with ^{13}C NMR spectroscopy to quantify the content of natural soil organic matter and charred organic carbon in the studied soils (Skjemstad et al. 1996).

3 RESULTS AND DISCUSSION

The studied soils developed from almost identical parent material (loess), display similar chemical and physical properties and are all under agricultural management with similar crop rotations (Table 1). The striking difference was a gradual change from a black to a gray color.

^{13}C CP MAS NMR spectra were obtained before and after UV photo-oxidation. Previous studies showed that all labile organic matter is removed by photo-oxidation for 2 hours and only physically or chemically protected matter remained (Skjemstad et al. 1996). The spectra of soil 1 are shown as an example (Fig. 1). The spectrum before photo-

oxidation shows signals in the alkyl C region (-10 to 45 ppm) representing methylene carbon (33 and 30 ppm). Signals in the O-alkyl C region (45 to 110 ppm) indicate the presence of polysaccharide and alcoholic structures. Signals near 56 ppm are attributed to methoxyl groups and to Cα in proteins. The shoulder at 65 ppm is due to C-6 carbon in polysaccharides. Signals at 74 ppm corresponded with ring carbon C-5, C-3 and C-2, whereas resonances at 104 ppm are due to anomeric C-1 carbon. Resonances in the aryl C region (145 to 110 ppm) are attributable to protonated and C-substituted aryl carbon (116 and 131 ppm). Dominant signals centered around 130 ppm with few additional peaks in the O-aryl C region are typical for the presence of char. The presence of lignin units would have been indicated by signals around 130 ppm in combination with additional signals for O-aryl C structures (165 to 145 ppm), which are of minor importance here. The carbonyl C region (190 to 165 ppm) is dominated by a set of resonances at 173 ppm most likely due to carboxyl and amide carbon. The aldehyde/ketone C region (220 to 190 ppm) revealed only low signal intensities.

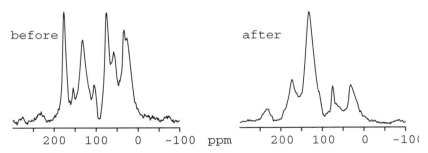

Figure 1: ^{13}C CP MAS NMR spectra of the A horizon of soil 1 obtained before and after UV photo-oxidation.

These results indicated the presence of C-substituted aromatic carbon in the studied horizon. If this signal was due to char, the material would resist photo-oxidation, and the resulting spectrum would show high signal intensities for aryl C. In fact, a higher proportion of soil organic carbon resisted photo-oxidation in soil 1 (Axh: 21 %) compared to soil 4 (1 %). After photo-oxidation, the spectrum was dominated by an aryl C peak at 130 ppm with spinning side bands present at 230 ppm and 30 ppm. At 30 ppm these signals overlapped with resonances of alkyl C, and minor signals indicated the presence of some O-alkyl C (75 ppm) and carboxyl C (170 ppm). Skjemstad et al. (1993) demonstrated that carbonyl bands remaining after photo-oxidation were most likely due to carboxylic acids. The absence of signals in the O-alkyl C region indicated the absence of carbohydrate carbon which is typical for char. Signals from aryl C (Table 2) were highest in the Axh horizon of soil number 1 (37 % of the total signal intensity) and lowest in the Axp horizon of soil 4 (12 %). After photo-oxidation, only small proportions of carbon were left, consistently

decreasing from soil 1 to 4 both for Axp (21 to 10 %) and Axh horizons (45 to 10 %). C-substituted aromatic structures present in char, will be under-estimated by ^{13}C NMR spectroscopy because cross polarization (CP) instead of Bloch decay measurements were used (Skjemstad et al. 1996). Thus, aryl C values determined by CP MAS were corrected (aryl$_{corr}$) and included in Table 2. In the Axh horizon of soil 1, almost all (95 %) organic carbon was present in C-substituted aromatic structures.

To further investigate the nature of this material, its morphology was observed by scanning electron microscopy (results not shown here). Many of the larger particles showed a morphology characteristic of the xylem structure found in woody material. These and many of the fine particles were probed with an EDX system and showed the absence of elements with an atomic weight higher than sodium, with the exception of sulfur. Very few particles could be identified as mineral particles. This suggests that many of the particles are organic in nature and probably are char.

Table 2: Organic carbon in the investigated soils and fractions. C content, C-species as determined by ^{13}C CP MAS NMR spectroscopy, calculated content of char and lignin parameters are compared in the bulk soil and in the <53 μm fraction after high energy ultra violet photo-oxidation (UV).

Hori-zon*	Sample	Organic carbon		C species							Char		Lignin
		g/kg of bulk soil	%	ketone	Carboxyl	O-aryl	aryl	O-alkyl	alkyl	aryl$_{corr}$**	g/kg of bulk soil	% C	g VCS/ kg C
						% of total signal intensity							
1 Axp	bulk soil	23.6	100	1	14	5	21	32	26	-	-	19	6.6
	UV	4.9	21	1	13	7	53	16	11	84	4.1		-
Axh	bulk soil	17.8	100	0	17	6	37	22	18	-	-	45	1.7
	UV	8.0	45	1	13	9	65	6	7	95	7.6		-
2 Axp	bulk soil	18.3	100	2	12	4	18	37	28	-	-	14	11.2
	UV	4.2	23	3	13	8	46	13	17	77	3.2		-
Axh	bulk soil	11.9	100	1	14	3	19	36	28	-	-	23	2.5
	UV	3.9	33	2	14	8	51	12	12	82	3.2		-
3 Axp	bulk soil	5.0	100	3	13	4	14	42	24	-	-	5	17.8
	UV	0.7	14	4	10	7	27	33	19	52	0.4		-
Axh	bulk soil	12.8	100	5	14	5	15	41	21	-	-	3	10.4
	UV	1.3	10	6	12	7	22	34	20	44	0.6		-
4 Axp	bulk soil	13.5	100	2	14	4	12	40	28	-	-	2	14.9
	UV	0.8	6	1	10	8	24	33	23	48	0.4		-
Axh	bulk soil	13.3	100	3	14	4	13	40	27	-	-	3	7.3
	UV	1.3	10	4	9	8	26	33	20	50	0.7		-

-: not determined, n.d.: not detectable, * according to AG-Boden (1994), ** aryl$_{corr}$: aryl C content corrected for Bloch decay

To visualize a potential interdependence between char and lignin content (Table 2) and soil color, data was displayed as a color-sequence from black to gray colored soils in Figure 2. As a trend, progressing from dark to light colored soils in the color-sequence (i) organic matter became less resistant to photo-oxidation, (ii) contributions of char to the bulk soil C decreased, whereas (iii) contributions of lignin to the bulk soil C increased. Contributions from char were always highest in the subsoils compared to the surface horizons, both normalized to organic C content and normalized to bulk soil mass.

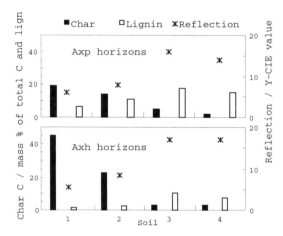

Figure 2: Content of char and lignin compounds detected in the Axp and Axh horizons of soils 1 to 4 displayed as a color-sequence. Data for charred organic matter (mass % of total carbon) and lignin (g VSC kg⁻¹) is displayed on the left axis. Reflection is displayed at the right axis and given as Y-CIE value, which is an equivalent to the Munsell value.

4 CONCLUSIONS

The photo-oxidation method allows the identification of char in German Chernozems. Char contributes up to 45 % to the bulk soil organic carbon, which is equivalent to approximately 8 g/kg of the bulk soil. The color-sequence of soils showed a strong relationship between color and the content of char, suggesting that the presence of char dominates color in these soils. In the Chernozems studied here, char is a major contributor to total organic carbon, but the origin of this fire induced form of organic carbon remained unclear. The char could originate from natural vegetation fires in the postglacial vegetation or from post-mesolithic human use of fire for the clearing of forests and subsequent agricultural management practices. Due

to its recalcitrant nature, char can be preserved in the pedosphere for long periods of time. This has major implications for the processes of pedogenesis in Chernozemic soils and for the sequestration of C in these soils.

ACKNOWLEDGMENTS

The work was financially supported by the Deutsche Forschungsgemeinschaft (Ko 1035/6-1 and 2). H.-J. Altemüller (Braunschweig) provided helpful comments and literature on charred particles in Chernozems.

REFERENCES

AG-Boden 1994. *Bodenkundliche Kartieranleitung*. E. Schweizerbart´sche Verlagsbuchhandlung, Stuttgart.

FAO 1994. *FAO-Unesco. Soil map of the world, revised legend.*, Rome, Italy.

Haumaier, L. & W. Zech 1995. Black carbon-possible source of highly aromatic components of soil humic acids. *Org. Geochem.* 23:191-196.

Kögel-Knabner, I. 1995. Composition of soil organic matter. In K. Alef & P. Nannipieri (eds.), *Methods in applied soil microbiology and biochemistry*. London: Academic Press.

Schmidt M.W.I., J.O. Skjemstad, E. Gehrt & I. Kögel-Knabner (1999): Charred organic matter in Chernozemic soils from Germany? - A study using high energy ultraviolet photo-oxidation and ^{13}C CPMAS NMR spectroscopy. *Eur. J. Soil Sci.* 50, in press.

Skjemstad, J.O., P. Clark, L.J. Head & S.G. MacClure 1993. High energy ultraviolet photo-oxidation: a novel technique for studying physically protected organic matter in clay- and sand-sized aggregates. *J. Soil Sci.* 44:485-499.

Skjemstad, J.O., P. Clarke, J.A. Taylor, J.M. Oades & S.G. McClure 1996. The chemistry and nature of protected carbon in soil. *Aust. J. Soil Res.* 34:251-271.

Skjemstad, J.O., P. Clark, A.Golchin, & J.M. Oades 1997. Characterization of soil organic matter by solid state ^{13}C NMRspectroscopy. In C. Cadish & K. Giller (eds.), *Driven by nature: plant litter quality and decomposition*: 253-271. Cab International.

Geochemistry of the Earth's Surface, Ármannsson (ed.)© 1999 Balkema, Rotterdam, ISBN 90 5809 073 6

Organic matter loss and alteration during black shale weathering

S.T. Petsch & R.A. Berner
Yale University, New Haven, Conn., USA

T.I. Eglinton
Woods Hole Oceanographic Institution, Woods Hole, Mass., USA

ABSTRACT: Weathering of organic matter [OM] in ancient sedimentary rocks provides strong controls on carbon cycling in the earth's surface environments and on the O_2 abundance in earth's atmosphere. The rate and nature of OM remineralization is best revealed by examining active weathering environments rich in organic matter. Samples have been collected from several black shale weathering profiles. Analyses reveal a progressive loss of total organic carbon [TOC] upon weathering (from 60% to near 100% in the most weathered samples). Pyrite loss parallels or precedes OM loss in all profiles. Detailed analyses of OM isolated from several depths within each profile indicate little if any systematic change in the structure or composition of residual OM during weathering. These results suggest that for a variety of organic carbon-rich rocks, weathering results in non-selective loss of OM in spite of nearly a 10-fold loss of OM during weathering.

1 INTRODUCTION

1.1 *Atmospheric oxygen content*

A remarkable feature of the earth's surface environment through time is the approximate constancy of atmospheric O_2 abundance over the past 600 million years. The mass of atmospheric oxygen ($\sim38\times10^{18}$ moles O_2) is relatively small compared with the fluxes that produce and consume O_2, i.e. photosynthesis and respiration (releasing/taking up on the order of 10^{21} moles O_2 my^{-1}, respectively) and oxidation of reduced minerals and volcanic gases and remineralization of organic matter [OM] in ancient sedimentary rocks (each consuming on the order of 10^{18} moles O_2 my^{-1}). Were these fluxes to not balance, the abundance of atmospheric O_2 could be greatly diminished or inflated within a few short million years.

1.2 *Measures and models of paleo-pO$_2$*

No direct means exists to measure the abundance of O_2 in the atmospheres of the geologic past (beyond what is available in the young polar ice sheets). Thus geologic evolution of atmospheric O_2 must be inferred and constrained by other means. Features such as the absence of global-scale ocean anoxia and wildfires suggest that pO_2 has neither fallen nor risen, respectively, by more than a factor of 2 during the past 600my (Watson et al. 1978; Holland 1984; Chaloner, 1989).

Various numerical models have been devised that attempt to describe variations in the sources and

sinks of atmospheric O_2 through geologic time and the resultant effect on pO_2. A common theme through these models is the concept of geochemical cycling. By arguing that on million year timescales O_2 production can be approximated by the burial rate of OM in sediments (i.e. *net* photosynthesis minus respiration rates), authors of these models recognized that cycling of carbon through atmospheric, biologic and sedimentary reservoirs provides the dominant controls on O_2 over geologic time. Some models depend on esimates of sediment abundance through time: periods of enhanced burial of rocks rich in OM represent periods of elevated O_2 production (Berner & Canfield, 1989). Other models employ the $^{13}C/^{12}C$ isotopic discrimination generated during photosynthesis to rearrange the $\delta^{13}C$ record of marine carbonates into a history of OM burial rates and O_2 production (Berner, 1987, Lasaga, 1989). A third set of models does not so much explicitly calculate an O_2 history as develop a dynamic description of the response of atmospheric to various geologic, biologic and oceanographic "forcings" (Holland, 1994; Van Cappellen & Ingall, 1997; Petsch & Berner, 1998).

1.3 *Remineralization of ancient sedimentary OM*

One limitation of the above models is that remineralization and weathering of OM is poorly constrained. A rough estimate of global remineralization is provided by balancing total OM burial to total remineralization so that pO_2 remains constant ($\sim4\times10^{18}$ moles O_2 my^{-1}). However, this average calculation provides no information on the factors that control this rate, limiting understanding of how the rate of OM weathering and O_2 consumption may vary through geologic time.

Various observations indicate that OM weathering does not proceed to completion on the continents and that the efficiency of remineralization is not 100%. Previous OM weathering studies have consistently measured appreciable but not complete TOC loss from shales during weathering (Leythäuser, 1973; Clayton & Swetland, 1978; Lewan, 1980; Littke et al. 1991); these results suggest that some fraction of OM is not completely remineralized in the outcrop and is transported in dissolved or particulate form to downstream sediments. OM in modern sediments can be traced to ancient rocks based on compound identification and isotopic signatures (Sackett et al. 1974; Rowland & Maxwell, 1984; Barrick et al. 1980). Compound-specific coupled ^{14}C and ^{13}C isotope analyses have revealed a significant source of ^{14}C-dead OM entering modern depositional environments (Goñi et al. 1997; Eglinton et al. 1997; Goñi et al. 1999). An intriguing question then remains: How much and what composition of OM escapes weathering and is redeposited in modern sediments? In this paper we describe our efforts to understand the response of the abundance and composition of OM to weathering in black shales.

2 METHODS

2.1 Sample selection and sampling methods

Weathering profiles were collected from the formations listed in Table 1. Formations were chosen to represent a range of OM types. Sampling sites were selected where the full transition from unaltered shale to highly weathered material is exposed, typically in a recent roadcut. The depth of the weathered zone developed on these rocks (corresponding to both color change and TOC loss) is approximately 4 m at all sites. Between 10-30 500 g samples were obtained at intervals of 20-50 cm along the exposed profiles. Samples were obtained from a single stratigraphic horizon at each site to avoid any bed-to-bed heterogeneity.

Table 1. Selected OM-rich formations in this study

Name	Age	Location	TOC	OM type
New Albany	L.Dev.	KY, USA	9-12%	II
Marcellus	M.Dev.	NY, USA	8-10%	II
Woodford	L.Dev.	OK, USA	15-20%	II
Monterey	Mioc.	CA, USA	8-10%	II-S
Green River	Eoc.	UT, USA	15-18%	I

2.2 Analyses

Total carbon content was determined using a Leco CR12 carbon analyzer. Carbonate content was de-

termined on combusted samples (450°C, 24 hrs.); organic carbon content was calculated by difference. Pyrite sulfur abundance was determined by $CrCl_2$ reduction of pyrite, liberation of H_2S, precipitation and iodometric titration of ZnS. Powdered whole rock samples were ultrasonically agitated for 15 minutes in 93:7 v/v CH_2Cl_2/CH_3OH. Kerogen was isolated from extracted rock by demineralizing in PTFE bombs (40°C) under N_2 using standard HCl/HF digestion techniques. Elemental analyses (CNS/O) of demineralized kerogens were performed using a Carlo Erba EA1108 elemental analyzer.

FTIR spectra of demineralized kerogens were obtained using a BioRad FTS175 spectrometer at 4 cm[-1] resolution on KBr pellets containing 10% kerogen. Samples were prepared for Curie point pyrolysis-gas chromatography [Py-GC] by pressing 1-2 μg kerogen with internal standard (poly-t-butylstyrene) onto an Fe-Ni wire. Flash pyrolysis (610°C, 5 s.) was effected using a FOM-3LX Curie point pyrolyzer/Horizons RF generator coupled to an HP 5890-II gas chromatograph. Separation of pyrolysis products was achieved using a Restek Rtx-1 capillary column (60 m × 0.32 mm i.d.; film thickness 0.5 μm; temperature program 0°C 5 min., ramp 3°C min[-1] to 320°C, held at 320°C for 20 min.; flame ionization detector).

3 RESULTS

3.1 TOC and pyrite abundance

In all studied weathering profiles, TOC and pyrite content decreases from least to most weathered samples (Figure 1). Surface samples contain between 0.2% and 5% TOC, corresponding to factor of 2 to factor of 10 organic carbon loss relative to deepest (least weathered) samples. Measured TOC content at depth in all profiles agrees well with pub-

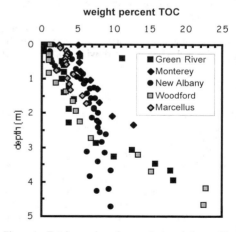

Figure 1. Total organic carbon content variations with depth in 5 selected black shale weathering profiles.

lished values for these formations, indicating that weathering has not occurred at depth. Pyrite loss parallels or precedes TOC loss at all sites, and at all sites pyrite content approaches 0% at or below the top (most weathered) sections of the profiles. This suggests that pyrite weathering has reached completion at these sites.

3.2 *Kerogen N/C and S/C ratios*

Calculated values for the weight percent C_{org}, N_{org}, and S_{org} are expressed as atomic N/C and S/C ratios in Figure 2. The ratios are constant through each profile, suggesting little change in kerogen elemental composition during weathering. The calculated ratios are also consistent with the values expected for these OM types: almost invariant N/C regardless of type, with low, medium and high S/C ratios for the Green River, New Albany/Woodford and Monterey Fms., respectively.

mole/mole ratio

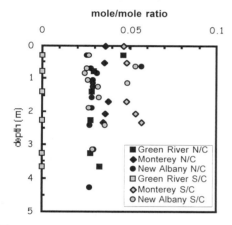

Figure 2. Mole/mole ratio variations in N_{org}/C_{org} and S_{org}/C_{org} in 3 selected black shale weathering profiles.

3.3 *Infrared spectroscopy*

Although not employed in a true quantitative fashion, FTIR spectra from a suite of samples within the same weathering profile was used to reveal trends in the relative composition of kerogen isolates in individual profiles. In general, all samples within a weathering profile generated similar FTIR spectra; however slight trends with progressively weathered samples could be discerned. Among these are a decrease in absorbance of bands centered at 2900 and 2800 cm^{-1} (assigned to alkyl -CH$_2$- and -CH$_3$ bonds) and an increase in absorbance of bands centered near 1700 and 1600 cm^{-1} (assigned to C=O and C=C bonds). These trends are consistent for all studied black shales, with the exception that alkyl absorbance variations are small in the Green River weathering profiles.

3.4 *Flash Pyrolysis*

Pyrograms of all kerogen isolates are dominated by the expected homologous series of C_5 to C_{30+} *n*-alk-1-ene/*n*-alkane doublets, with smaller contributions from alkylbenzenes and branched hydrocarbons. Pyrograms from all samples within an individual weathering profile are very similar, in spite of the significant change in TOC from unaltered to highly weathered shale. This suggests that kerogen in weathered samples, although reduced in abundance, fairly represents the composition of OM at depth.

To reveal trends in pyrolysate composition not apparent upon inspection of pyrograms, values for several composition indices were calculated (Table 2). Variation in these indices within single profiles was minimal. Thiophene abundance, aromaticity, chain length and isoprenoid abundance indices are all reasonably constant through each profile, suggesting little change in pyrolysate composition with weathering. The sole exception is the Green River profile, which shows consistent increase in chain length index values with increased weathering.

Table 2. Pyrolysate composition indices

index	compound peak area ratio
thiophene	*2-methylthiophene* to *toluene+ oct-1-ene ($C_{8:1}$)*
aromaticity	*toluene+ ethylbenzene+ o,m,p-xylene* to $C_{7:1}+C_{8:1}+C_{9:1}$
chain length	$C_{21:1}+C_{22:1}+C_{23:1}+C_{24:1}$ to $C_{6:1}+C_{7:1}+C_{8:1}+C_{9:1}$
isoprenoid	*prist-1-ene+ C_{14} isoprenoid* to $C_{13:1}+C_{15:1}+C_{17:1}$

4 DISCUSSION AND CONCLUSION

4.1 *TOC and pyrite profiles and rate of OM loss*

Weathering was more extensive in the Green River and Woodford outcrops than at the New Albany, Marcellus or Monterey sites. This is unlikely to be related to OM type, as the Woodford, New Albany and Marcellus are of similar age, OM type and maturity. Rather, lithology and physical erosion may play a role in generating these differences. The New Albany, Marcellus and Monterey outcrops expose less cohesive, more friable and permeable shales than do the outcrops of the Woodford (a chert) and Green River (a limestone). Low permeability in a cohesive, well-cemented and erosion-resistant rock may result in less exposure to O$_2$ and lower water-rock contact times.

The Marcellus and New Albany Shales are of similar age and OM type, and are exposed in regions of comparable climate. The close agreement between New Albany and Marcellus TOC profiles suggests that these outcrops have reached steady state between physical erosion and oxidation and leaching of OM. The Marcellus outcrop was

273

scoured by advance and retreat of the Laurentide Ice Sheet, exposing an unweathered surface 15kya. This provides an upper limit on the time required to reach steady state for this lithology and climate (15ky), which corresponds to a minimum of 15gC m^{-2} ky^{-1} of OM lost from this outcrop. Pyrite date suggest that pyrite weathering proceeds more slowly than pyrite loss at all sites; these data could be used to constrain the maximum rate of OM loss.

4.2 *Non-selective OM loss*

Elemental analyses, FTIR spectra and Py-GC data all suggest that OM weathering is predominantly non-selective. There is no observed loss or gain of N_{org} or S_{org} relative to TOC, indicating no preferential reactivity of heteroatomic moieties within the kerogen. Py-GC data (both pyrograms and derived indices) likewise reveal no relative accumulation or loss of alkylaromatic versus *n*-alkyl moities, branched versus *n*-alkyl moities, long-chain versus short chain moieties or thiophene versus alkylaromatic moieties.

FTIR spectra do suggest minor loss of aliphatic C-H bonds and gain of carbonyls and olefins. However, considerably more organic matter has been liberated during weathering than accumulates as C=O or C=C bonds. It is proposed that the gains seen in FITR spectra do not represent a slower weathering rate for carbonyl or olefin moieties, but rather accumulation of these groups as intermediate weathering products. Accumulation of these groups may be masked in pyrolysis experiments by (1) the already high abundance of *n*-alk-1-enes generated during pyrolysis and (2) poor resolution of non-GC-amenable, oxygen-containing pyrolysis products.

4.3 *Summary*

Weathering of black shales results in the loss of 60% to almost 100% of OM. The lack of complete OM weathering in most studies outcrops indicates that appreciable organic carbon transported to downstream environments where it may subsequently remineralize or be redeposited in modern sediments. The rate of weathering is likely dominated by physical erosion and water-rock contact time, and not by any inherent relative reactivities of specific OM components. In the studied outcrops, removal of OM is non-selective, with no significant preferential enrichment or removal of selected moieties or heteroatoms. The mechanism of removal of OM during weathering is unclear, but data are consistent with a model of slow, non-specific oxidation followed by rapid cleavage of oxidized products and release/dissolution of low molecular weight fragments.

REFERENCES

Barrick, R.C., Hedges, J.I. & Peterson, M.L. 1980. Hydrocarbon geochemistry of the Puget Sound region: I, sedimentary acyclic hydrocarbons. *Geochim. Cosmochim. Acta* 44(1349-1362).

Berner, R.A. & Canfield, D.E. 1989. A new model for atmospheric oxygen over Phanerozoic time. *Am. J. Sci.* 289(333-361).

Berner, R.A. 1987. Models for carbon and sulfur cycles and atmospheric oxygen: application to Paleozoic history. *Am. J. Sci.* 287(177-196).

Chaloner, W.G. 1989. Fossil charcoal as an indicator of paleoatmospheric oxygen level. *J. Geol.Soc.*146(171-174).

Clayton, J.L. & Swetland, P.J. 1978) Subaerial weathering of sedimentary organic matter. *Chem. Geol.* 54(149-155).

Eglinton, T.I., Benitez-Nelson, B.C., Pearson, A., McNichol, A.P., Bauer, J.E. & Druffel, E.R.M. 1997. Variability in radiocarbon ages of individual organic compounds from marine sediments. *Science* 277(796-799).

Goñi, M.A., Ruttenberg, K.C. & Eglinton, T.I. 1997. Sources and contribution of terrigenous organic carbon to surface sediments in the Gulf of Mexico. *Nature* 389(275-278).

Goñi, M.A., Ruttenberg, K.C. & Eglinton, T.I. 1998. A reassessment of the sources and importance of land-derived organic matter in surface sediments from the Gulf of Mexico. *Geochim. Cosmochim. Acta* 62(3055-3075).

Holland, H.D. 1984. *The Chemical Evolution of the Atmosphere and Oceans.* Princeton: Princeton Univ. Press.

Holland, H.D. 1994. The phosphate-oxygen connection. *EOS, 1994 Ocean Science Meeting* 75(96).

Lasaga, A.C. 1989. A new approach to isotopic modeling of the variation of atmospheric oxygen through the Phanerozoic. *Am. J. Sci.* 289(411-435).

Lewan, M.D.1980. *Geochemistry of Vanadium and Nickel in Organic Matter in Sedimentary Rocks.* Ph.D. thesis, Univ. Cinncinnati.

Leythäuser, D. 1973. Effects of weathering on organic matter in shales. *Geochim. Cosmochim. Acta* 37(113-120).

Littke, R., Klussmann, U., Krooss, B. & Leythäuser, D. 1991. Quantification of loss of calcite, pyrite and organic matter during weathering of Toarcian black shales and effects on kerogen and bitumen characteristics. *Geochim. Cosmochim. Acta* 55(3369-3378).

Petsch, S.T. & Berner, R.A. 1998. Coupling the geochemical cycles of C, P, Fe, and S: the effect on atmospheric O$_2$ and the isotopic records of carbon and sulfur. *Am. J. Sci.* 298(246-262).

Rowland, S.J. & Maxwell, J.R. 1984. Reworked triterpenoids and steroid hydrocarbons in a recent sediment. *Geochim. Cosmochim. Acta* 48(617-624).

Sackett, W.M., Poag, C.W. & Eadie, B.J. 1984. Kerogen recycling in the Ross Sea, Antarctica. *Science* 185(1045-1047).

Van Cappellen, P. & Ingall, E.D. 1997. Redox stabilization of the atmosphere and oceans by phosphorus-limited marine productivity. *Science* 271(493-496).

Watson, A., Lovelock, J.E. & Margulis, L. 1978. Methanogenesis, fires and regulation of atmospheric oxygen. *Biosystems* 10(293-298).

Geochemistry of the Earth's Surface, Ármannsson (ed.) © 1999 Balkema, Rotterdam, ISBN 90 5809 073 6

Photosynthetic controls on the silicification of cyanobacteria

V.R. Phoenix & K.O. Konhauser
School of Earth Sciences, University of Leeds, UK

D.G. Adams
Department of Microbiology, University of Leeds, UK

ABSTRACT: The silicification of a functioning cyanobacteria (*Calothrix* sp.) is shown to be restricted to outer surface of the sheath. In addition, polycationized ferritin (a probe 11 nm in diameter) labelled the sheath's outer surface but failed to penetrate the sheath matrix, indicating its impermeablity to particles of this size and greater. A model is proposed to explain the restriction of silicification to the outer surface of the sheath. It is suggested that cyanobacterial photosynthesis creates a moderately alkali environment (pH 7-9) adjacent to the sheath, causing silica to form colloids. These colloids are too large to penetrate the sheath, and hence mineralisation is restricted to the outer surface. It is further proposed that the partial diffusion barrier created by the sheath allows very high photosynthetically induced pH levels (> pH 10) to build up inside the sheath matrix. Under these conditions silica is both highly soluble and in its monomeric state, and is thus less able to bind to the sheath matrix.

1 INTRODUCTION

It is believed that bacterial photosynthesis can influence the mineralisation of the cell by mediating the pH of it's surrounding microenvironment. For example, the consumption of CO_2 and/or HCO_3^{2-} during photosynthesis results in the production of hydroxyl ions, creating an alkali environment around the vicinity of the microorganism, (Miller & Coleman 1980, Verrecchia et al. 1995). This biomediated increase in pH has been proposed to induce the precipitation of carbonate biominerals around cyanobacteria such as *Synechococcus* sp.; carbonates being insoluble in high pH conditions (Thompson & Ferris 1990). Furthermore, the high abundance of carbonate sphericules inside the sheath of certain cyanobacteria are probably due to the sheath acting as a partial diffusion barrier, in essence allowing very high pH levels to build up inside the sheath matrix (Verrecchia et al. 1995), thus encouraging carbonate precipitation.

Recent studies of the silicification of the cyanobacteria *Calothrix* sp. have demonstrated that it may continue to function once mineralised, providing silicification is restricted to the extracellular sheath (Phoenix et al. 1998). As these microbes still function during silicification, then photosynthesis may also influence silica precipitation by controlling the pH of their surrounding microenvironment.

We speculate that silica mineralisation would act in an opposite manner to the carbonate precipitation described above, as unlike carbonates, silica is more soluble in high pH conditions. Hence the high pH build up within the sheath would discourage silica precipitation, thus silicification would not be observed inside the sheath of functioning cells. We hypothesise that pH would decrease outside the partial diffusion barrier of the sheath, allowing precipitation of silica on its outer periphery. Once the cell lyses, there would be no photosynthetically induced high pH, and silicification could then occur inside the sheath matrix. Furthermore, silica occurs dominantly as colloids in solutions saturated with respect to amorphous silica at pH's present in the natural environment (pH 6-9; Shimada & Taruntani 1979). It is possible that these colloids may be unable to penetrate the sheath.

In the current study we test the above model by silicifying the cyanobacteria *Calothrix* sp. in a laboratory setting to determine the location of silica biominerals upon the sheath. Furthermore, we investigate the permeability of the sheath to determine if this extracellular organic material may act as a filter to colloidal silica. This is performed using polycationized ferritin (PCF), a protein containing an iron core. This protein is attracted to electronegative sites within the sheath in a similar manner to which metals are bound, however its 11 nm diameter provides an indication of the ability of particles of such size to penetrate the sheath.

2 METHODS

2.1 Silicification of cyanobacteria

Agar plates were firstly inoculated with the cyanobacteria *Calothrix* sp., previously isolated from the Krisuvik hot spring, Iceland, and incubated until an even spread of filaments across the plates was achieved. A silica solution containing 300 ppm Si was then prepared using $Na_2SiO_4.5H_2O$ in deionised water and neutralised to pH 7 with 3M HCl. The agar plates were then placed in 2L beakers containing 1000 mls of this solution. The cyanobacteria were allowed to mineralise over 20 days, during which time the plates were transferred to a fresh solution every 2 to 3 days to ensure continual supply of solute, thus replicating the flowing system of a hot spring. Samples of culture were collected every 2 to 3 days and fixed in 2% gluteraldehyde.

2.2 Labelling of the sheath with Polycationized Ferritin (PCF)

The PCF was prepared as described by Graham & Beveridge (1994) to a final concentration of 250µg/ml. Samples of the cyanobacteria *Calothrix* sp. were cultured in solute enriched liquid media (BG11, +1040 ppm Cl⁻) to induce sheath growth. Filaments were then homogenised, washed once in distilled water, and then resuspended in 0.05M HEPES (pH 6.8) to an optical density of 1.1 at OD_{600}. The suspension was then divided into two equal portions. To ensure that bacterial metabolism, such as proton motive force (Urrutia et al. 1992), did not influence the binding of PCF, both functioning and lysed cells were labelled. Lysed cells were prepared by prefixing one of the cell suspensions in 4% gluteraldehyde to both kill the cells, and prevent disruption of sheath structure due to decay. Thus the permeability of the sheath of the lysed cells would be identical to that of the functioning cells. Both the lysed and functioning cell suspensions were then washed three times in distilled water and resuspended in HEPES buffer (to 1.1 at OD_{600}).

500µL portions of the above cell suspensions were then added to 500µL aliquots of PCF solution and incubated on a rotary stage for 15 minutes at room temperature. These were then pelleted by centrifugation, washed 5 times in 1.5 ml aliquots of HEPES buffer, fixed in 4% glutaraldehyde for 1 hour and processed immediately for transmission electron microscopy (TEM).

3 OBSERVATIONS

After 20 days in the mineralising solution many filaments had become encrusted in a thick silica matrix often several µm thick (Figure 1). These silica crusts occurred exclusively upon the extracellular sheath. Furthermore, mineralisation was restricted to the outer periphery of the sheath as shown in Figure 1.

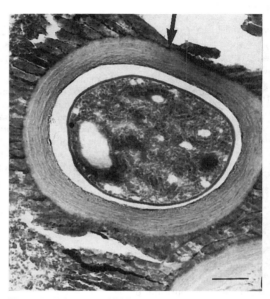

Figure 1. Laboratory silicified cyanobacteria *Calothrix* sp. Note the extensive mineralisation upon the outer surface of the sheath (arrow), yet the sheath matrix contains no mineralisation. Scale bar = 1µm

No mineralisation occurred inside the sheath matrix. The silica crusts surrounding the cells appeared to form from colloids of over 20 nm in diameter. This corroborates data from Konhauser & Ferris (1996) who observed that silicification of bacteria in hot springs appeared to result from the merger of colloids. No intracellular mineralisation was observed, either upon the cell wall, or within the cytoplasm; the cells appearing intact and in viable condition. The sheaths of these cyanobacteria were well developed (often 1 µm thick) and appeared densely packed and fibrous in nature (Figure 1). During silicification the pH of the mineralising solution remained between pH 7-9, this moderate alkalinity was probably maintained by photosynthesis.

In all cases the PCF was bound to the outer surface of the sheath, indicating its strong anionic characteristic and hence high affinity to bind metals.

Figure 2. TEM micrograph of PCF (small arrow) bound to the outer surface of the sheath of a lysed cell. Note the PCF does not penetrate the sheath matrix (large arrow). Scale bar = 1µm

This reinforces the observation that cyananobacteria provide viable nucleation sites for mineral precipitation (Konhauser & Ferris 1996). The PCF failed to penetrate the sheaths of functioning cells, suggesting the sheath had acted as a filter and was impermeable to particles of this size (i.e. 11 nm). Furthermore, the PCF failed to penetrate the sheath of lysed cells, confirming that this characteristic was a result of the sheaths permeability and not a result of cell metabolism (Figure 2).

4 CONCLUSIONS

4.1 Silica solution chemistry

The solution chemistry of silica changes dramatically over the pH range 6 - 10 (e.g. Shimada & Tarutani 1979). Significantly, cyanobacterial photosynthesis can readily alter solution pH between these values and thus they can have a significant effect on the silica solution chemistry of their surrounding environment. Early proposals for the absorption of silica onto organic surfaces presumed monosilicic acid was the primary species bound (Leo & Barghoorn 1976). However, studies have shown that more than 60% dissolved silica is expressed as polymer when silica concentrations exceed the solubility of amorphous silica (Ilher 1965). Furthermore, this polymeric fraction exists as colloids. Shimada & Tarutani (1979) showed that these colloids are dominant, with approximately 80% SiO_2 expressed as polymer (i.e. colloids), in solution between pH 6-9. However, above pH 10

the polymer is unstable and the monosilicic acid is once again dominant. Also, Depasse & Watillon (1970) described the coagulation of silicia colloids by cation bridging. These observations suggest that in hot springs saturated with respect to amorphous silica, where circum-neutral to mildly alkali pH's are common and cations are in abundance, coagulation of silica into large colloids is expected. Significantly, Heaney & Yates (1998) discovered only the colloidal fraction absorbed onto organic surfaces, no monomer was sorbed. It therefore appears that under such conditions silica exists predominantly as colloids between pH 6-9, as a monomer between pH's 1-6 and >9, with only the colloids absorbing onto organic surfaces (Heaney & Yates 1998).

4.2 Silicification model

By combining silica solution chemistry with the filtering effect provided by the sheath, and the photosynthetically induced alkalisation of a cells microenvironment, a model for restriction of mineralisation to the outer surface of the sheath has been developed.

The microenvironment surrounding a cell is divided into two key areas, as shown in figure 4. In the outer zone adjacent to the sheath, pH is elevated to moderately alkaline conditions (pH 7-9) by photosynthesis. At this pH, silica is present in the form of colloids, which are bound by hydrogen bonding, and possibly cation bridging, to the sheath. However, due to the impermeability of the sheath,

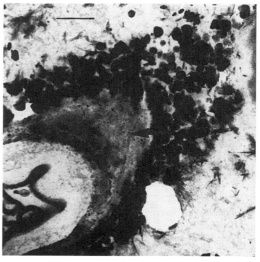

Figure 3. TEM micrograph of a cyanobacteria undergoing lysis. Note the specled nature of the sheath matrix. The speckles are small mineral growths (arrow). Scales bar = 1 µm

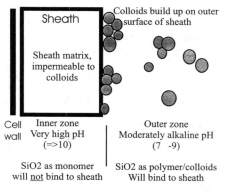

	Colloids build up on outer surface of sheath
Sheath	
Sheath matrix, impermeable to colloids	

Cell wall	Inner zone Very high pH (=>10)	Outer zone Moderately alkaline pH (7 -9)
	SiO2 as monomer will not bind to sheath	SiO2 as polymer/colloids Will bind to sheath

Figure 4. Theoretical model for the silicification of cyanobacteria.

the colloids cannot penetrate into the sheath matrix, and hence silica is only bound to the sheath's outer surface.

Inside the sheath (inner zone), the diffusion barrier created by the sheath creates a much higher alkalinity, with pH values envisaged to be greater than 10. Any silica within this zone would be both highly soluble and in its monomeric state, thus unable to bind to anionic sites on the sheath. Thus, a functioning cell accumulates silica only on the outer surface of the sheath. Upon cell lysis, there is no photosynthetically imposed high pH within the sheath matrix and thus mineralisation occurs here (Figure 3). It is also probable that a loosening of the sheath polymers into a more permeable array during lysis allows colloids to penetrate this zone.

ACKNOWLEDGEMENTS

This work was supported by a Royal Society Research Grant to K.O.K and a Natural Environment Research Council award to V.R.P. We acknowledge the assistance of Slobodan Babic in Microbiology Laboratory, and thank the Icelandic Research Council for their permission to undertake the scientific research at Krisuvik.

REFERENCES

Depasse, J. & A. Watillon. 1970. The stability of amorhous colloidal silica. *Journal of Colloid and Interface Science* 33: 430-438.

Graham, L.L. & T.J. Beveridge 1994. Structural differentiation of the *Bacillus subtilis* 168 cell wall. *Journal of Bacteriology* 176: 1413-1421.

Heaney, P.J. & D.M. Yates 1998. Solution chemistry of wood silicification. *Geological Society of America annual meeting, 1998, abstracts with programs.*

Ilher, R.K. 1965. *Nature* 207: 472-473.

Konhauser, K.O. & F.G. Ferris 1996. Diversity of iron and silica precipitation by microbial mats in hydrothermal waters, Iceland: Implications for Precambrian iron formations. *Geology* 24: 323-326.

Leo, R.F. & E.S. Barghoorn 1976. *Harvard University Botanical Museum Leaflets* 25: 1-46.

Miller, A.G. & B. Coleman 1980. Evidence for HCO_3^- transport by the bluegreen alga (cyanobacterium) *Coccochloris Peniocystis*. *Plant Physiology* 2: 397-402.

Phoenix, V.R., K.O. Konhauser, D.G. Adams & S.H. Bottrell 1998. The role of biomineralisation as an ultraviolet shield: implications for Precambrian life. *Geological Society of America annual meeting, 1998, abstracts with programs.*

Shimada, K. & T. Tarutani 1979. Gel chromatographic study of the polymerization of silicic acid. *Journal of Chromatography* 168: 401-406.

Thompson, J.B. & F.G. Ferris 1990. Cyanobacterial precipitation of gypsum, calcite, and magnesite from natural alkaline lake water. *Geology* 18: 995-998.

Urrutia Mera, M., M. Kemper, R. Doyle & T.J. Beveridge 1992. The membrane-induced proton motive rforce influences the metal binding ability of *Bacillus subtilis* cell walls. *Applied and Environmental Microbiology* 58: 3837-3844.

Verrecchia, E.P., P. Freytet, K.E. Verrecchia & J.L. Dumont 1995. Sperulites in calcrete laminar crusts - Biogenic $CaCO_3$ precipitation as a major contributor to crust formation. *Journal of Sedimentary Research Section A-Sedimentary Petrology and Processes* 65: 690-700.

Geochemistry of the Earth's Surface, Ármannsson (ed.)© 1999 Balkema, Rotterdam, ISBN 90 5809 073 6

Surface area, iron oxide and organic carbon relationships in sediments

S.W. Poulton

School of Earth Sciences, University of Leeds, UK

ABSTRACT: The mode of transport of particulate iron oxides in Huanghe (Yellow River) and Amazon sediments has been evaluated in an attempt to understand possible processes affecting iron oxide deposition in marine sediments. Iron oxides are apparently strongly associated with the edges of sheet silicate minerals, with the result that a strong linear association is evident with respect to surface area. Comparison of iron oxide surface loadings in suspended and deltaic sediments from the Huanghe suggests that iron oxides may be lost from particles of all grain sizes during marine sedimentation. However, different grain size fractions appear to be affected to differing extents, depending on the proximity to the initial contact of fresh and saltwater. The differential flocculation of sheet silicates has been proposed as a possible mechanism for explaining iron oxide losses in marine sediments relative to riverine inputs.

1 INTRODUCTION

The behaviour of particulate iron oxides during deposition in the marine environment may exert an important influence on the formation of pyrite in marine sediments. The availability of iron oxides limits the formation of pyrite in many of the continental margin sediments studied to date (Canfield 1989). By contrast, sediments deposited in certain inner shore environments (e.g. deltas, estuaries) and in pelagic settings are generally not iron-limited with respect to pyrite formation (Berner 1984, Aller et al. 1986). In addition, mass-balance models for the delivery of iron oxides to marine sediments suggest that iron oxides may be disproportionately trapped in inner shore environments (Poulton 1998). Therefore, in order to adequately model the global iron and sulphur cycles it is important to understand the processes that control the sequestering of iron oxides in marine sediments.

Suspended river particulates are by far the most important source of iron minerals to the oceans (Martin & Windom 1992). Thus, a greater understanding of iron oxide deposition in the marine environment may be achieved by consideration of the mode of transport of iron oxides in fluvial systems. The association of iron oxides with particle surfaces in river sediments is well-documented (e.g. Gibbs 1977). However, previous studies have tended to focus on theoretical considerations of surface area and on microscopic observations on coarser (>500 μm) fractions, both of which do not adequately represent the majority of sediments deposited in the marine environment.

The present study is aimed at providing detailed information on the mode of transport of particulate iron oxides in the Huanghe (Yellow River) and the River Amazon, by combining geochemical data with surface area measurements and electron microscope observations. In addition, observations on the iron oxide loading of particles in the Huanghe are compared with surface loadings in Huanghe deltaic sediments, to provide information on the possible processes affecting iron oxide deposition in marine sediments. The surface loadings of organic carbon in sediments have recently been the focus of considerable attention (e.g. Mayer 1994, Keil et al. 1997, Ransom et al. 1998), and relationships between organic carbon and both surface area and iron oxides are also considered here.

2 METHODOLOGY

Suspended sediments were collected for the Huanghe close to the mouth of the river, prior to any saltwater intrusion. Samples were filtered (0.45 μm) in the field, and gently removed from the filters using a scalpel. Three surface sediments were collected from the Huanghe delta (samples HD01-3 represent sediments collected at increasing distances from the river mouth). Bed sediments were collected from the River Amazon at Macapa, Brazil.

All sediments were washed to remove salts and separated into various grain-size fractions; >38 μm by sieving and <38 μm by density settling in water. Discrete organic particles were removed in a sodium metatungstate solution with a density of 1.9 g/cm³.

Sediments were subjected to a buffered (pH = 4.8) sodium dithionite solution for one hour to extract the iron oxides that are highly reactive towards dissolved sulphide (Canfield 1989, Raiswell et al. 1994). Iron analysis was by AAS. Aluminium was measured (by ICP-AES) following a HF-HCLO$_4$-HNO$_3$ total metals extraction. Organic C was measured (using a Carlo Erba 1106 Elemental Analyzer) on samples pre-treated with dilute HCl to remove any carbonate phases. Relative surface area was measured by the single-point BET method.

Table 1. Geochemical and surface area data for the Huanghe and Amazon after removal of discrete organic matter.

Grain size (μm)	Iron oxides (wt%)	Organic C (wt%)	Surface area (m²/g)
Huanghe <2	2.12	1.79	31.7
2-6	1.27	1.24	15.6
6-15	0.72	0.36	8.4
15-25	0.67	0.17	7.3
25-38	0.66	0.39	8.1
38-45	0.94	0.34	15.8
45-63	0.99	0.57	15.2
63-90	1.00	0.54	16.6
90-125	1.01	0.40	15.0
Amazon <2	2.60	3.13	15.4
2-6	1.78	1.37	8.3
6-15	0.89	0.83	5.5
15-25	0.62	0.43	4.1
25-38	0.42	0.14	2.6
38-45	0.35	0.13	2.2
45-63	0.36	0.14	2.6
63-90	0.53	0.16	3.6
90-125	0.92	0.86	4.6

3 RESULTS AND DISCUSSION

3.1 Huanghe and Amazon river particulates

Geochemical and surface area data relating to grain-size fractions in the Huanghe and Amazon are presented in Table 1. The iron oxide and organic carbon concentrations show the expected increase in finer-grained fractions, indicative of sorption to particle surfaces with increased relative surface areas in finer particles. An interesting feature of the data, however, concerns the concomitant increases in iron oxides, organic carbon and surface area in the coarsest fractions (Table 1). Figure 1 demonstrates the direct linear relationships between surface area and both iron oxides and organic carbon (relationships are significant at the 0.1% level, except organic carbon in the Huanghe which is significant at the 1% level). Mayer (1994) notes that organic carbon concentrations in soils and sediments are equivalent to a monolayer of organic matter over all surfaces, and suggests that much of the organic matter is sequestered inside surface roughness features on minerals and bioclasts. Thus, increases in organic matter (and by analogy iron oxides) in coarser fractions may be attributed to a relatively greater degree of surface roughness. However, Ransom et al. (1998) suggest that the concentrations of matrix-linked organic matter in sediments are related to the suite of clay minerals present. From the perspective of the present study, it is obviously important to further evaluate these observations in relation to the transport and deposition of iron oxides.

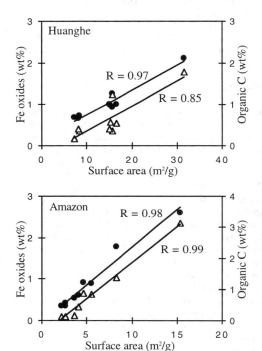

Figure 1. Iron oxide (closed circles) and organic carbon (open triangles) concentrations as a function of surface area for Huanghe and Amazon sediment grain size fractions.

Electron microscope studies (TEM and SEM) of iron oxides in the Huanghe and Amazon (Poulton 1998) suggest that iron oxide associations with mineral surfaces are rarely apparent. Crystalline iron oxides were observed occasionally infilling surface pits in coarser-grained (>100 μm) aluminium silicates. Iron oxides were also observed in association with clay minerals in finer fractions. In these cases, the iron oxides appear to occur as small spheres of 10-20 nm in diameter, associated solely with the edges of clay mineral plates. These iron oxide spheres show some tendency to cluster, forming irregular aggregates, and often appear to be enveloped by organic matter. Thus, the morphology of iron oxides appears to vary between coarser and finer mineral fractions. However, in all cases iron oxides were observed only in association with sheet silicates. The results of X-ray diffraction analyses suggest that sheet silicates may be important constituents of both the fine-grained and coarse-grained fractions of these sediments. However, these data are of little quantitative use due to the difficulties involved in the measurement of relatively small proportions of different clay mineral suites. In order to test the hypothesis that iron oxides in sediments are largely associated with sheet silicates, aluminium has been used as a proxy for the contents of these minerals in different grain size fractions (Figure 2). The combined Huanghe and Amazon data-sets demonstrate a strong relationship between iron oxides and Al, suggesting that sheet silicates do indeed transport the majority of the particulate iron oxides in these river systems. The fact that this relationship is apparent in the combined data-sets (in addition to when the two rivers are considered in isolation) suggests that this is a fundamental relationship, presumably related to weathering processes, which may well be applicable on a global scale. There is some doubt in Figure 2 as to whether the true relationship between iron oxides and aluminium is best expressed linearly or as a curve passing through the origin. Ransom et al. (1998) suggest that organic carbon is associated to differing extents with certain clay mineral suites, and is preferentially sequestered in smectite-rich sediments relative to clay fractions dominated by chlorite (note that the relationship between organic carbon and aluminium in Huanghe and Amazon sediments is strikingly similar to the iron oxide/aluminium relationship in Figure 2; Poulton 1998). At least with regard to Amazon sediments, smectites dominate in the finest grain size fractions, whereas chlorite (and micas) are dominant in the coarser fractions (Gibbs 1977). This suggests that if the relationship in Figure 2 can be considered to be best expressed by a non-linear function, then iron oxides may also be preferentially associated with certain sheet silicate groups relative to others.

3.2 Iron oxide deposition in the Huanghe delta

The strong relationship evident between iron oxides and surface area suggests that surface area may be an ideal conservative property for evaluating the fate of riverborne iron oxides during marine deposition. The Huanghe deltaic system is ideal for such a study since over 90% of the sediments discharged originate from a single source, namely loess (Gong and Xiong 1980). Thus, temporal variations in sediment compositions are minimised, particularly when distinct grain size fractions are considered.

Table 2 details the surface loadings of iron oxides in Huanghe suspended river particulates and in surface muds from the Huanghe deltaic region (note that no iron oxides have been converted to pyrite in these sediments). The data presented in Table 2 suggest that close to the river mouth (HD01) deltaic surface muds are depleted in finer (<2-15 μm) and coarser (38-125 μm) grain size fractions relative to the riverine input. By contrast, the medium grain size fractions are enriched in iron oxides in this deltaic sample. The varying proportions of the different sediment fractions in riverine sediments relative to sample HD01, have the result that iron oxide surface loadings are equivalent in the bulk samples. At more distal sites, all grain size fractions become progressively more depleted relative to Huanghe river particulates (Table 2), with the result that bulk sediments from the most distal site (HD03) are significantly depleted in iron oxides relative to the riverine input.

The data presented in Table 2 suggest that iron oxides may be 'lost' from all particle sizes during marine deposition, and that these losses do not appear to be solely due to physical sorting processes. Whilst there is some doubt as to the actual mechanism of iron oxide loss, a differential flocculation process cannot be discounted. The samples richest in sheet silicate minerals (i.e. the finest and coarsest fractions; see earlier) may undergo coagulation soon after contact with seawater, forming larger particles that may be preferentially deposited further inshore relative to the present sampling sites (note that such flocs are

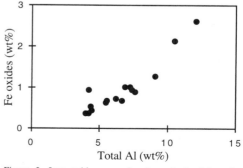

Figure 2. Iron oxides as a function of aluminium for the combined Huanghe and Amazon data-sets.

unstable and unlikely to survive the sampling process). Such a process would be consistent with the commonly observed deposition of large proportions of riverine sediment at the contact of river and saltwater (see Eisma 1993). By contrast, the sheet silicate suite present in the medium grain size fractions may not undergo such rapid flocculation, with the result that further from the river mouth (i.e. sample HD01) these grain size fractions are actually enriched in iron oxides relative to the riverine input (Table 2). Finally, at the most distal site (HD03), all sediment fractions are depleted in iron oxides as a result of the flocculation and preferential deposition of sheet silicate minerals further inshore.

Table 2. Surface area loadings of iron oxides in Huanghe suspended river and delta sediments. Deltaic sediments were collected from progressively more distal sites in the order HD01<HD02<HD03.

Grain size (μm)	Iron oxides/surface area (mg Fe/m^2)			
	River	HD01	HD02	HD03
Bulk	0.66	0.65	0.64	0.51
<2	0.67	0.54	0.49	0.40
2-6	0.81	0.85	0.75	0.54
6-15	0.86	0.74	0.70	0.59
15-25	0.92	1.37	1.12	0.62
25-38	0.82	1.05	0.99	0.66
38-45	0.60	0.57	0.67	0.50
45-63	0.65	0.67	0.60	0.55
63-90	0.60	0.57	0.61	0.50
90-125	0.67	0.64	0.56	0.43

4 CONCLUSIONS

Iron oxides are apparently transported largely in association with sheet silicate minerals in fluvial systems. The surface area of mineral size fractions is strongly influenced by the nature and quantities of the sheet silicate minerals present. This has the result that iron oxides appear to display a strong linear relationship with respect to surface area.

The inferred behaviour of riverine particles during deposition in the Huanghe delta suggests that inner shore environments may operate as a trap for large quantities of iron oxides. The apparent loss of iron oxides from bulk sediments deposited at the most distal deltaic site relative to the riverine input (approximately 25%), equates well with global estimates of particulate iron oxide losses from riverine sediments relative to continental margin and pelagic sediments (Poulton 1998).

The data presented here provide supporting evidence relating to proposed models for the genesis of marine ironstones during the Phanerozoic (e.g.

Young 1989). Inner shore environments appear to be a potential source of iron oxide-rich sediments, which may subsequently be reworked and redeposited on the adjacent continental shelf during changes in sea level, thus increasing the potential for ironstone formation.

REFERENCES

Aller, R.C., J.E. Mackin & R.T. Cox 1986. Diagenesis of Fe and S in Amazon Inner Shelf muds: Apparent dominance of Fe reduction and implications for the genesis of ironstones. *Cont. Shelf Res.* 6:263-289.

Berner, R.A. 1984. Sedimentary pyrite formation: An update. *Geochim. Cosmochim. Acta* 48:605-615.

Canfield, D.E. 1989. Reactive iron in marine sediments. *Geochim. Cosmochim. Acta* 53:619-632.

Eisma, D. 1993. *Suspended matter in the aquatic environment.* Springer-Verlag, 299pp.

Gibbs, R.J. 1977. Transport phases of transition metals in the Amazon and Yukon rivers. *Geol. Soc. Am. Bull.* 88:829-843.

Gong, S.Y. and G.S. Xiong 1980. The origin and transport of sediment of the Yellow River. In *Proc. UNESCO Int. Symp. River Sedimentation, Beijing* 1:43-52.

Keil, R.G., L.M. Mayer, P.D. Quay, J.E. Richey and J.I. Hedges 1997. Loss of organic matter from riverine particles in deltas. *Geochim. Cosmochim. Acta* 61:1507-1511.

Martin, J.M. & H.L. Windom 1991. Present and future roles of ocean margins in regulating marine biogeochemical cycles of trace elements. In R.F.C. Mantoura, J.M. Martin & R. Wollast (eds), *Ocean Margin Processes in Global Change*: 45-67. Chichester: Wiley.

Mayer, L.M. 1994. Relationships between mineral surfaces and organic carbon concentrations in soils and sediments. *Chem. Geol.* 114:347-363.

Poulton, S.W. 1998. *Aspects of the global iron cycle: Weathering, transport, deposition and early diagenesis.* PhD Thesis. Leeds: University of Leeds.

Raiswell, R., D.E. Canfield & R.A. Berner 1994. A comparison of iron extraction methods for the determination of degree of pyritization and the recognition of iron-limited pyrite formation. *Chem. Geol.* 111:101-110.

Ransom, B., D. Kim, M. Kastner & S. Wainwright 1998. Organic matter preservation on continental slopes: Importance of mineralogy and surface area. *Geochim. Cosmochim. Acta* 62:1329-1345.

Young, T.P. 1989. Phanerozoic ironstones: An introduction and review. In T.P. Young & W.E.G. Taylor (eds), *Phanerozoic Ironstones*, Geology Society, London, Special Publication 46:51-64.

Geochemistry of the Earth's Surface, Ármannsson (ed.)© 1999 Balkema, Rotterdam, ISBN 90 5809 073 6

The impact of anthropogenic carbon types on the chemical composition of soil organic matter

C. Rumpel & R. F. Hüttl
Department of Soil Protection and Recultivation, Brandenburg Technical University, Cottbus, Germany

I. Kögel-Knabner
Lehrstuhl für Bodenkunde, Technische Universität München, Germany

ABSTRACT: In the Lusatian mining district, organic matter of recultivated mine soils under forest consists of a mixture of lignite, lignite ash, lignite dust (anthropogenic carbon) and organic matter derived from decomposing plant material. The aim of the study was to assess the impact of anthropogenic carbon types on the chemical composition of the soil organic matter. Bulk soil samples and physical fractions were analysed for structural composition by ^{13}C CPMAS NMR spectroscopy and for lignite content by ^{14}C activity measurements. Both methods yield corresponding results. ^{13}C CPMAS NMR spectrometry shows that anthropogenic carbon types contribute greatly to the chemical composition of the soil organic matter, leading to an increase of alkyl and aromatic carbon species (0-45 and 110-160 ppm). In contrast to natural soils the contribution of carbon species characteristic for plant litter increases with age in the surface soil horizon.

1 INTRODUCTION

In the Lusatian mining district, lignite-rich mine soils under forest contain two organic carbon types: anthropogenic carbon present as lignite, lignite dust or lignite-derived ash and recent carbon present as organic matter derived from decomposing plant litter (Rumpel et al. 1998a). Radiocarbon measurements were used recently to quantify lignite carbon contribution because lignite does not have ^{14}C activity anymore (Rumpel et al. 1998a,b). Measurements of the radiocarbon activity of the mine soils indicate after correction that dead (= anthropogenic) carbon is a major contributor to the organic carbon of the mineral soil as well as the forest floor. In the forest floor, lignite carbon content may be due to airborne lignite-derived carbon input to the rehabilitated mine sites, which are frequently found in the vicinity of coal-fired power plants or briquette factories (Schmidt et al. 1996; Rumpel et al. 1998b). The objective of this study was to assess the impact of anthropogenic carbon on the chemical composition of the organic matter in bulk soil samples as well as physical fractions.

2 MATERIAL AND METHODS

The study was carried out in the Lusatian mining district, in the eastern part of the Federal Republic of Germany. At the study sites, CaO rich ash from a briquette factory and a NPKMg-fertilization were applied to the spoil prior to planting. After afforestation, plant material accumulated and was incorporated into the mineral soil. Samples were taken from the forest floor horizons (L, Oh) and the upper centimetres of the mineral soil, referred to as the Ai horizon from a chronosequence of pine stands (black pine, 11 years old, Scots pine 17 and 32 years old) and a red oak stand (36 years old). The subsoil (Cv horizon) at 1 m depth was also sampled. For chemical analysis an aliquot was ground. Carbon and nitrogen contents have been determined with a Leco CHN 1000 analyser.

Particle-size fractionation of the Ai horizon under red oak was carried out after ultrasonic dispersion (Christensen 1992). The amount of ultrasonic energy has been calibrated following the procedure suggested by Schmidt et al. (1999). 150 J/ml were used to obtain a complete dispersion of aggregates. Sand, silt and a clay fraction were recovered. Density fractionation of the Ai horizon to obtain three density fractions (< 1.6 g/cm³, 1.6-2.0 cm³ and > 2 cm³) was carried out using sodium polytungstate $Na_6(H_2W_{12}O_{40})$ according to Golchin et al. (1994). The ^{13}C CPMAS NMR spectra were recorded with a Bruker MSL-100 NMR spectrometer. The contact time was 1 ms and the pulse delay was 250-350 ms. The solid-state ^{13}C NMR spectra were recorded as free induction decay (FID) and integrated using the integration routine of the spectrometer. The chemical shift regions 0-45 ppm, 45-110 ppm, 110-160 ppm and 160-220 ppm were assigned to alkyl C, O-alkyl C, aromatic C and carboxylic C, respectively (Wilson 1987). The variation of integration data of signals due to the treatment of a well resolved FID (fourier transformation, phasing and baseline correction) is 5% (Knicker 1993). NMR-analyses were carried out after treatment with 10% HF as described by Schmidt et al. (1997).

Radiocarbon (^{14}C) age was obtained using the conventional macro-technique of liquid scintillation as described by Becker-Heidmann et al. (1988). Assuming that the sample contains a mixture of recently formed SOM derived from plant litter and dead carbon (carbon without ^{14}C activity), the amount of lignite C present in the sample can be obtained as a percentage of the total organic carbon:

$$((100 - y) \times k)/ (y + (100\text{-}y)) \qquad (1)$$

In this equation y is the ^{14}C activity and k the correction factor for the elevated ^{14}C content of recent plant material which was chosen 115 % modern carbon (Geyh, pers. comm.).

3 RESULTS AND DISCUSSION

3.1 Chemical parameters

In all soils, the carbon content of the parent substrate is high compared to natural soils (Cv horizon, Table 1). In the Ai horizons, carbon contents of up to 180 g/kg are recorded while the C/N values are slightly lower than those of the Cv horizons. The carbon content of the mineral soil indicates the presence of anthropogenic carbon types. Using ^{14}C activity measurements, Rumpel et al. (1998a) showed that carbon in the subsoil of lignite-containing mine soils is present as lignite. In the Ai horizons, recent carbon contribution of up to 50 % recorded (Rumpel et al. 1998a), indicating that both organic matter types are mixed in this horizon. During mine soil development an increasing accumulation of recent carbon was observed (Rumpel et al 1998a).

3.2 ^{13}C CPMAS NMR spectroscopy

The impact of anthropogenic carbon on the chemical composition of the SOM was observed by ^{13}C CPMAS NMR spectroscopy. ^{13}C CPMAS NMR spectra of the carbon types present in lignite-rich mine soils are presented in Figure 1.

Two distinct peaks in the chemical shift regions 0-50 ppm and 110-160 ppm can be observed in the spectra of both lignite samples, representing

Table 1. Chemical parameters of the forest floor and the mineral soil

Horizon	Sampling depth cm	pH	Carbon g/kg	C/N
11 year old pine				
L	3.0-2.0	4.3	453	50
Of2	1.5-0.0	5.9	234	29
Ai	0.0-2.0	5.6	102	41
Cv	-100	2.5	36	52
17 year old pine				
L	4.0-3.0	4.7	507	64
Of	3.0-0.0	5.0	320	28
Ai	0.0-1.0	5.0	80	33
Ai2	1.0-3.0	5.8	46	42
Cv	-100	3.0	18	36
32 year old pine				
L	7.0-5.0	4.5	514	55
Of	5.0-0.0	5.0	431	28
Ai	0.0-5.0	5.6	180	33
Cv	-100	3.3	36	41
36 year old red oak				
L	3.0-2.0	4.6	462	68
Oh	2.0-0.0	6.7	224	20
Ai	0.0-5.0	6.9	110	20
Cv	-100	3.2	37	93

n.d. = not determined

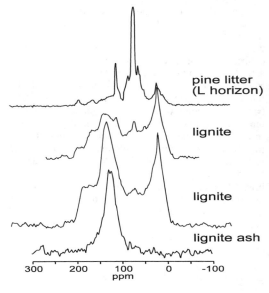

Figure 1: ^{13}C CPMAS NMR spectra of carbon types present in lignite-rich mine soils

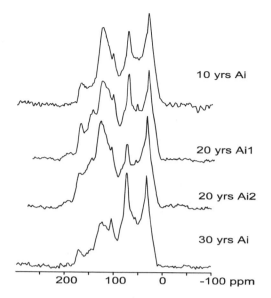

Figure 2: ^{13}C CPMAS NMR spectra of the Ai horizons under the pine chronosequence

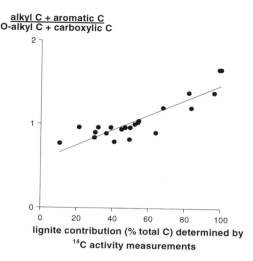

Figure 3: Relationship between ^{13}C CPMAS NMR data and ^{14}C activity measurements

aliphatic and aromatic carbon species (Meiler & Meusinger 1990). Additionally, lignite 1 shows a signal at 72 ppm, which indicates together with the peak at 105 ppm the presence of a small fraction of polysaccharides. Further, the signals at 56 and 150 ppm are characteristic of lignin. A spectrum of ash from a lignite-fired power station shows that the carbonaceous particles of ash are highly aromatic. The differences between the two lignite samples can be explained by the small scale differences of the formation environment. Polysaccharides and lignin have been found to be able to survive the early coalification (Bates & Hatcher 1989). This may have been the case for lignite 1. The spectrum of lignite 2 indicates that geochemical conditions (pressure and temperature) generated a higher degree of coalification. Burning of lignite leads to loss of the aliphatic carbon species as indicated by the ^{13}C CPMAS NMR spectrum of lignite ash. The ^{13}C CPMAS NMR spectra of plant litter are characterised by a dominant signal at 72 ppm with a shoulder at 105 ppm and signals at 56 ppm, 119 ppm and 150 ppm, which are most probably assigned to polysaccharides and lignin. These signals are characteristic for forest soil organic matter (Kögel-Knabner 1993, Baldock & Preston 1995).

It was demonstrated that the peak at 72 ppm decreases during humification and that the peaks in the carboxyl region increase due to oxidation processes (Kögel-Knabner 1993). An additional increase was attributed either to selective

preservation of cutan/suberan (Augris et al. 1998) or the contribution of microbial material (Golchin et al. 1996). In the Ai horizons of lignite-rich mine soils (Figure 2), additional signals in the alkyl and aromatic spectral regions (0-50 and 110-160 ppm) indicate contribution of anthropogenic carbon. The contribution of alkyl and aromatic carbon species decreases in the Ai horizon with increasing age of the forest stands. This observation is the opposite of natural soils, where those carbon species tend to increase (Kögel-Knabner 1993). The ratio of signal intensities A = (alkyl C + aromatic C)/(O-alkyl + carboxylic C) was proposed as an indicator for the presence of lignite in soils (Schmidt et al. 1996, Rumpel et al. 1998a,b). Schmidt et al. (1996), observed A < 1 for an agricultural soil, whereas influence of contamination with brown coal emissions increased the ratio (A > 1). A linear relationship (r^2 = 0,79 **) between the lignite content as determined by ^{14}C activity measurements and data from ^{13}C CPMAS NMR spectroscopy can be obtained for bulk soil samples from different mine sites and physical fractions of an Ai horizon (Figure 3).

These data indicate for a range of different samples that both methods yield corresponding results and that lignite contribution to a bulk soil sample or physical fraction of a soil leads to a higher contribution of alkyl and aromatic carbon species in the ^{13}C CPMAS NMR spectra.

4 CONCLUSION

In the young lignite-containing mine soils of Lusatia (Germany), anthropogenic carbon has a great influence on the chemical composition of the organic matter. Increased anthropogenic carbon contribution as recorded by ^{14}C activity measurements is reflected by an increase of alkyl and aromatic carbon species in the ^{13}C NMR spectra. With site age plant-derived carbon species show an increased contribution to the organic matter of the Ai horizon.

ACKNOWLEDGEMENTS

The Deutsche Forschungsgemeinschaft is acknowledged for financial support. We thank H. Knicker for recording the ^{13}C CPMAS NMR spectra. H.-D. Lüdemann is acknowledged for providing the NMR facilities. P. Becker-Heidmann is thanked for the ^{14}C activity measurements.

REFERENCES

Augris N., J. Balesdent, A. Mariotti, S. Derenne, & C. Largeau 1998. Structure and origin of insoluble and non-hydrolyzable, aliphatic organic matter in a forest soil. *Org. Geochem.* 28: 119-125.

Baldock, J.A. & C. M. Preston 1995. Chemistry of carbon decomposition processes in forests as revealed by solid-state carbon-13 nuclear magnetic resonance. In W.W. McFee & J.M. Kelly (eds) *Carbon forms and functions in forest soils*: 89-118. Madison: SSSA.

Bates, A.L. & P.G. Hatcher 1989. Solid-state ^{13}C NMR studies of a large fossil gymnosperm form the Yallourn Open Cut, Latrobe Valley, Australia. *Org. Geochem.* 14: 609-617.

Becker-Heidmann, P. 1989. Die Tiefenfunktion der natürlichen Kohlenstoffisotopengehalte von vollständig dünnschichtweise beprobten Parabraunerden und ihre Relation zur Dynamik der organischen Substanz in diesen Böden. Hamburger Bodenkundliche Arbeiten, 13: 1-225

Christensen, B.T. 1992. Physical fractionation of soil and organic matter in primary particle size and density separates. In B.A. Stewart (ed) *Advances in Soil Science* 20: 2-76. New York: Springer

Golchin, A., J.M. Oades, J.O. Skjemstad & P. Clarke 1994. Study of free and occluded particulate organic matter in soils by solid state 13C CP/MAS NMR spectroscopy and scanning electron microscopy. *Aust. J. Soil Res.* 32: 385-409.

Golchin, A., P. Clarke & J.M. Oades 1996. The heterogeneous nature of microbial products as shown by solid-state 13C CPMAS NMR spectroscopy. *Biogeochem.* 34: 71-97.

Kögel-Knabner, I. 1993. Biodegradation and humification processes in forest soils. In J.-M. Bollag & G. Stotzky (eds) *Soil Biochemistry* 8: 101-137. New York: Marcel Dekker.

Knicker, H. 1993. Quantitative 15N- und 13C-CPMAS-Festkörper und 15N-Flüssigkeits-NMR-Spektroskopie an Pflanzenkomposten und natürlichen Böden. Dissertation, Regensburg.

Meiler, W. & R. Meusinger 1991. NMR of coals and coal products. *Annual reports on NMR spectroscopy* 23: 376-410.

Rumpel, C., H. Knicker, I. Kögel-Knabner, J.O. Skjemstad. & R.F. Hüttl 1998a. Types and chemical compostion of organic matter of lignite-containing mine soils. *Geoderma* 86: 123-142.

Rumpel, C, H. Knicker, I. Kögel-Knabner & R.F. Hüttl 1998b. Airborne contamination of immature soil (Lusatian mining district) by lignite-derived materials: its detection and contribution to the soil organic matter budget. *Water, Air Soil Poll.* 105: 481-492.

Schmidt, M.W.I., H. Knicker, P.G. Hatcher & I. Kögel-Knabner 1996. Impact of brown coal dust on a soil and its size fractions - chemical and spectroscopic studies. *Org. Geochem.* 25: 29-39.

Schmidt, M. W. I., H. Knicker, P.G. Hatcher, & I. Kögel-Knabner 1997. Improvement of 13C and 15CPMAS NMR spectra of bulk soils, particle size fractions and organic material by treatment with hydrofluoric acid (10%). *Europ. J. Soil Sci.* 48: 319-328.

Schmidt M.W.I., C. Rumpel & I. Kögel-Knabner 1999. Evaluation of an ultrasonic dispersion method to isolate primary organomineral complexes from soils. *European Journal of Soil Science*, submitted.

Wilson, M.A. 1987. *NMR-Techniques and Application in Geochemistry and Soil Chemistry.* Oxford: Pergamon Press.

5 Marine and sedimentary geochemistry

Geochemistry of the Earth's Surface, Ármannsson (ed.)© 1999 Balkema, Rotterdam, ISBN 90 5809 073 6

Invited lecture: Tropical, mobile mud belts as global diagenetic reactors

Robert C.Aller & Panagiotis Michalopoulos
Marine Sciences Research Center, SUNY, Stony Brook, N.Y., USA

ABSTRACT: Sedimentary dynamics rather than net accumulation rate per se is a major factor controlling between-environment differences in diagenetic processing of sedimentary debris and global patterns of elemental storage. Tropical mobile mud belts, such as characterize the Amazon delta and its downdrift coastal regions, are dramatic examples of the effect of massive remobilization and refluxing of sedimentary deposits on reactions in the oxic coastal ocean. Repetitive mixing, reoxidation, and continuous incorporation of reactive organic and inorganic material results in the functional geological equivalent to a suboxic, fluidized bed reactor. These suboxic systems are characterized by efficient decomposition of organic material, generation of nonsulfidic Fe-rich authigenic mineral suites, and rapid reverse weathering.

1 INTRODUCTION

Diagenetic processes and elemental cycling rates in sedimentary deposits underlying oxic or anoxic waters are often related to the local net accumulation rate of sediment. Such correlations are used extensively in recent global diagenetic models. Although net sedimentation is clearly a central factor governing diagenetic reaction balances and material storage, there are a number of major oxic depositional environments in which differences in accumulation rate per se appear not to be a primary control of between-environment diagenetic properties. In addition, an emphasis on local net properties such as vertical net accumulation can obscure processes and products which are best understood as resulting from multi-dimensional properties of a system rather than strictly local phenomena. One such example comes from the net preservation patterns of sedimentary organic carbon. There are broad correlations between net sedimentation rate, the fraction of organic flux to the seafloor that is eventually buried, the reactivity of organic material driving diagenetic reactions, and the dominance of particular diagenetic oxidants (Canfield 1993, Tromp et al. 1995). Major exceptions to these simple correlations are found in deltaic regions where much of the Earth's riverine sediment flux is deposited and diagenetic products stored. Large parts of these deltaic regions store far less organic carbon and of much lower average reactivity than might be expected based on simple

extrapolations from other environments (Mayer 1994, Keil et al. 1997, Aller 1998). Despite their extraordinarily high net accumulation rates, deltaic regions in general act as efficient remineralizers of both marine and terrestrial organic matter. Key factors determining efficient remineralization are cyclic reworking of sediment, reoxidation, and metabolite flushing within the depositional environment as a whole. Residence time of material within an energetic depositional regime is a critical control on reaction type, rates, and extent, and thus diagenetic products. In some tropical deltaic systems, such as the Amazon, frequent physical reworking and reoxidation of tropically-weathered debris is especially effective at promoting the dominance of suboxic diagenetic pathways, in particular Fe cycling, and the diagenetic formation of nonsulfidic authigenic minerals such as Fe-rich clays and carbonates.

2 DISCUSSION

2.1 Tropical, Mobile Mud Belts

High energy, muddy environments characterized by massive deposition and sedimentary recycling are not restricted to the surficial deltaic depocenters proper but can extend along large sections of continental shelves. Dramatic examples again come from the tropics. The Amazon River and delta spawn an extensive mobile mud belt stretching

>1100 km from northern Amapá to Venezuela (Eisma & Van der Marel 1971, Froidefond et al. 1988, Allison et al. 1995, 1999). The recycling flux of sedimentary debris associated with this single shelf system, not including inter mud bank shoreline erosion - redeposition, is likely several times greater than the global annual riverine sediment flux to the oceans. Weathered terrestrial debris rich in metal oxides is mixed with fresh marine organic material, sea water, and biogenic silica in these deposits to form what is essentially a massive, suboxic fluidized bed reactor (Fig. 1). Periodic entrainment of new biogenic debris, reoxidation, flushing of metabolites, and reexposure to photo radiation ensures efficient organic matter remineralization and sustained suboxic conditions. Although marine organic material dominates diagenetic reductants, isotopic composition of pore water ΣCO_2 and sediment C_{org} demonstrate that terrestrial organic matter is also efficiently decomposed. The presence and quantity of natural radiochemical (^{234}Th, ^{210}Pb) and biological tracers in shoreface deposits show that significant cross-shelf transport and three-dimensional interactions occur regularly in these systems. The high mobility and transient nature of the deposits make it difficult to quantify mass fluxes, but measurements in coastal deposits of northern Amapá and along French Guiana indicate high ΣCO_2 production rates typically in the range of 0.1 - 0.5 mM d^{-1} over the upper 0.1 - 1 m, suggesting extensive processing of relatively reactive organic debris. Despite frequent disburbance of sediment, dissolved oxygen penetrates only a few millimeters into these deposits, and decomposition occurs predominantly under suboxic, nonsulfidic conditions. The mobility of sediment inhibits infaunal macrobenthos and bioturbation except in localized regions, and the benthic biomass is dominated by bacteria The distinctive differences between diagenetic reaction balances in these high energy, suboxic tropical mud belts and other more commonly studied depositional regimes are reflected in the accumulated authigenic Fe mineral suites.

2.2 Authigenic minerals and reverse weathering

Whereas the total quantity of diagenetically-reduced Fe found in the Amazon delta and inner shelf deposits is at least comparable to and often greater than found in temperate coastal muds (~ 100 - 360 μmol Fe(II) g^{-1}), a large fraction is present as Fe-rich carbonates and clays rather than sulfides (Figure 2). Sediment C/S ratios are typically >6 in these same deposits, also indicative of nonsulfidic diagenesis. Similar diagenetic conditions appear to characterize portions of other tropical shelf regions such as the Gulf of Papua, Papua New Guinea in Oceania, where a large proportion of the global riverine sediment

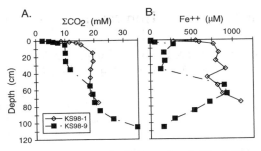

Figure 1. Pore water distributions of total CO_2 (mM) and dissolved Fe^{++} (μM) in upper ~ 1 m at two sites within a migrating mud wave near Sinnemary, French Guiana. Amazon delta and mobile coastal mud wave deposits along the NE coast of S. America are characterized by suboxic diagenetic reactions (Fe, Mn cycling) and intense remineralization of reactive organic material (net C/N ~ 4.6 – 6, from stoichiometric models of pore water composition).

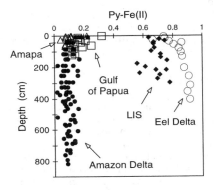

Figure 2: The proportion of diagenetically reduced Fe(II) present as pyrite (Py-Fe(II)) is distinctly different in mobile mud deposits of the Amazon delta and Amapá coast of New Guinea) compared to either typical temperate estuarine (Long Island Sound , USA) or open shelf muds (Eel River delta,USA). The mobile tropical mud deposits are suboxic and dominated by nonsulfidic authigenic mineral suites. The most reactive initial Fe pools (e.g. 0.2 M oxalate or 1 N HCL 2 – 4 hr leach)also tend to be ~ 2 – 3 X higher in the tropical deposits than the two temperate cases shown. Diagenetically S. America or Gulf of Papua deltaic deposits (Papua reduced Fe is inferred from total S (as pyrite) and 6N HCl leachable Fe(II) (15 minutes, 25° C), and ranges from ~ 100 to 360 μmol g^{-1} in these deposits (Data after Aller & Blair 1996, Sommerfield et al. 1999, Aller & Michalopoulos 1996, unpubl.).

flux enters the ocean (Alongi 1995, Aller, unpubl.). There are, however, also definite differences between mobile tropical deposits, (for example concentrations of pore water Fe^{++} are typically lower and spatially less extensive in Gulf of Papua compared to S. American deposits), that may reflect

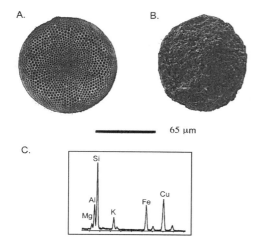

A.

B.

65 μm

C.

Si

Al

Mg

K

Fe

Cu

Figure 3: Biogenic silica is rapidly converted into aluminous, authigenic clays within tropical mobile mud deposits. A). Centrate diatom frustule with partially altered surface layer. B). Diatom-derived authigenic clay aggregate characteristic of complete alteration. Matrix material is also sometimes agglutinated with authigenic clay. C). TEM-EDS spectrum of an authigenic clay crystallite from frustule microstructure. The Cu peak represents grid background. TEMs of completely converted frustules often demonstrate pseudomorphic retention of original porous diatom frustule microstructure (Michalopoulos et al. 1999). (Examples after: Michalopoulos et al. 1999, Michalopoulos & Aller, unpubl.).

differences in the type of weathered material delivered from the drainage basin or other peculiarities of the depositional environment.

Carbon isotopic analyses, microprobe analyses, and operational leaches of the sedimentary carbonate pool in the case of the Amapá coastal and Amazon delta topset deposits indicate that most of the disseminated sedimentary carbonate debris is diagenetic (~0.1 - 0.2 % C), and that 30 - 70% is likely siderite or ferroan calcite (Blair & Aller 1995, Michalopoulos et al. unpubl, Zhu & Aller, unpubl.). A portion of the biogenic silica incorporated into these suboxic deposits is rapidly converted to authigenic clays, including Fe- and K-rich forms which can be found in some cases as diatom pseudomorphs (Figure 3). Both laboratory experiments and field observations demonstrate that conversion of biogenic silica to clay can be extremely rapid (months - years). Despite rapid formation, the intense sediment reworking and dissemination of authigenic minerals limit discernable vertical gradients in sediment composition and complicate simple calculation of fluxes from distributions. Operational leaches and diagenetic models of nonsteady - state pore water

K^+ and F^- profiles indicate, however, that a significant fraction of the Amazon River silica flux to shelf waters (>20%) is converted to authigenic clays in Amazon shelf sediments and has not been accounted for in the traditional methods used to estimate biogenic SiO_2 burial (DeMaster et al. 1983, Rude & Aller 1994, Michalopoulos et al. 1999). Reactive Al-oxides delivered from intensely weathered regions of tropical drainage basins presumably favor rapid authigenic clay formation, directly tying oceanic reaction patterns in suboxic shelf muds to terrestrial drainage basin processes(Porrenga 1967, Odin 1988, Aplin 1993). The occurence of rapid formation and significant quantities (several weight %) of authigenic clay in major depocenters requires a reevaluation of factors controlling cycles of elements such as K and Si. The relative importance of energetic mud belts as diagenetic environments must also vary as a function of sea stand, further complicating inferences of glacial-interglacial differences in oceanic processes.

3 CONCLUSIONS

Repetitive mobilization of muddy sediments in energetic oxic shelf and deltaic systems is a critical factor governing the eventual diagenetic modification of terrestrial and marine debris. Dramatic examples of the role of sedimentary dynamics rather than net accumulation rate per se are particularly evident in tropical regions where highly weathered fine-grained material is supplied and reacts under repetitive oxic - suboxic conditions. The relative roles of tropical mobile mud deposits in diagenetic processing and elemental storage on a global scale remain to be determined in detail. It is clear, however, that the elemental recycling fluxes are huge, and that the reaction pathways and resulting storage products differ from those normally emphasized in considerations of global cycles. All evidence to date indicates that the highly mobile, suboxic deposits often found in tropical shelf and deltaic regions are sites of both the efficient remineralization of organic material, and of rapid and extensive reverse weathering reactions.

ACKNOWLEDGEMENTS

Research supported by USA NSF Chemical Oceanography Program.

REFERENCES

Aller, R. C. 1998, Mobile deltaic and continental shelf muds as suboxic, fluidized bed reactors. *Mar. Chem.* 1: 143-155.

Aller, R.C. & N.E.Blair 1996. Sulfur diagenesis and burial on the Amazon shelf: Major control by physical sedimentation processes. *Geo.-Mar. Lett.* 16: 3-10.

Aller, R. C. & P. Michalopoulos 1996. Controls on Fe diagenesis and authigenic mineral formation in terrigenous, nearshore environments. In S.H. Bottrell (ed), *Fourth International Symposium on the Geochemistry of the Earth's Surface*: 15-18. Leeds: Univ. of Leeds.

Allison, M.A., C.A. Nittrouer & G.C. Kineke 1995. Seasonal sediment storage on mudflats adjacent to the Amazon River. *Mar. Geol.* 125: 303-328.

Allison, M.A., M.T. Lee, A.S. Ogston, & R.C. Aller 1999. Origin of Amazon mudbanks along the northeastern coast of South America. *Mar Geol.* (submitted).

Alongi, D.M. 1995. Decomposition and recycling of organic matter in muds of the Gulf of Papua, northern Coral Sea. *Cont. Shelf. Res.* 15: 1319-1337.

Aplin, A.C. 1993. The composition of authigenic clay minerals in recent sediments: links to the supply of unstable reactants. In D.A.C. Manning, P.L. Hall & C.R. Hughes (eds), *Geochemistry of clay-pore fluid interactions*: 81-106. London: Chapman and Hall.

Blair, N.E. & R.C. Aller 1995. Anaerobic methane oxidation on the Amazon shelf. *Geochim. Cosmochim. Acta* 59: 3707-3715.

Canfield, D.E. 1993. Organic matter oxidation in marine sediment. In R. Wollast, L Chou, & F. Mackenzie (eds), NATO-ARW *Interactions of C, N, P and S Biogeochemical Cycles and Gobal Change*: 333-365. New York: Springer.

DeMaster, D.J., G.B. Knapp & C.A. Nittrouer 1983. Biological uptake and accumulation of silica on the Amazon continental shelf. *Geochim. Cosmochim. Acta* 47: 1713-1723.

Eisma, D. & H.W. Van der Marel 1971. Marine muds along the Guyana coast and their origin from the Amazon basin. *Contr. Mineral. Petrol.* 31: 321-334.

Froidefond, J.M., M. Pujos, & Andre X. 1988. Migration of mud banks and changing coastline in French Guiana. *Mar. Geol.* 84: 19-30.

Keil, R.G., L.M. Mayer, P.D. Quay, J.E., Richey, & J.I. Hedges 1997. Loss of organic matter from riverine particles in deltas. *Geochim. Cosmochim. Acta* 61: 1507-1512.

Mayer, L.M. 1994. Surface area control of organic carbon accumulation in continental shelf sediments. *Geochim. Cosmochim. Acta* 58: 1271-1284.

Michalopoulos, P., R.C. Aller, & R.J., Reeder 1999. Conversion of diatoms to clay minerals during early diagenesis in tropical, continental shelf muds. *Geology* (submitted).

Odin, G.S. (ed.) 1988. *Green Marine Clays*. Amsterdam: Elsevier.

Porrenga, D.H. 1967. Clay mineralogy and geochemistry of Recent marine sediments in tropical areas. PhD. Dissertation. Amsterdam: Univ. Amsterdam.

Rude, P.D. & R.C., Aller 1994. Fluorine uptake by Amazon continental shelf sediment and its impact on the global fluorine cycle. *Cont. Shelf Res.* 14: 883 – 907.

Sommerfield, C.K., R.C. Aller, & C.A Nittrouer 1999. Modern and ancient diagenetic environments of the Eel River Basin (USA): Role of bioturbation on C-S-Fe signatures. *J. Sed. Res.* (in press).

Tromp, T.K., P. Van Cappellen, & R.M. Key 1995. A global model for the early diagenesis of organic carbon and organic phorsphorus in marine sediments. *Geochim. Cosmochim. Acta* 59: 1259-1284.

Geochemistry of the Earth's Surface, Ármannsson (ed.)© 1999 Balkema, Rotterdam, ISBN 90 5809 073 6

Invited lecture: Clay mineral diagenesis at Great Salt Lake, Utah, USA

Blair F. Jones
US Geological Survey, Reston, Va., USA

Ronald J. Spencer
Department of Geology, University of Calgary, Alb., Canada

ABSTRACT: X-ray diffraction and major element chemical analysis of the fine clay fraction in lacustrine sediment deposited since ~25KA from the Great Salt Lake and adjacent Bonneville salt flats are compared with colloidal sediment from inflow streams. Results indicate that diagenesis in such environments involves authigenesis of magnesium silicate from saline pore fluids through interstratification with detrital smectite.

1 INTRODUCTION AND BACKGROUND

Interest in global mass budgets for the elements and related geochemical processes has long fueled consideration of the alteration or diagenesis of sediments at the earth's surface. Especially difficult is the delineation of the role of silicate authigenesis, particularly reactions involving the 2:1 layer clay minerals, which are ubiquitous and appear to exert major influences on the distribution of magnesium and silica in natural waters.

The clay minerals of Great Salt Lake (GSL) sediments were studied as part of larger interdisciplinary investigations, particularly focused on the sedimentary record obtained in cores collected from the southern arm of the main basin in 1979-80 and 1995-6. The lake elevation during both coring efforts was close to a long-term historical average near 1280 m (4200 ft.) above sea level. Construction of a railroad causeway across the central lake in 1959 restricted circulation between the north and south arms (Figure 1). Because of halite precipitation in the north arm, all cores were obtained in the south arm of the lake. Samples selected for clay mineral analysis were taken largely from core sites close to the middle, or in the south-central part of the south arm. Fine clays in shallow sediments from the lowest segments of the Bonneville salt flats west of the GSL were also examined; these two basins were connected when the overall lake level elevation exceeded 1285 m.

Spencer et al. (1984) divided the GSL strata into five major units based on their overall sedimentologic and biostratigraphic characteristics. This and later stratigraphic evidence (Thompson et al. 1990; Oviatt et al. 1992) indicates that prior to about 25-27 KA the basin was occupied by a shallow lake or ephemeral playa (units IV AND V). A perennial lake, which has existed since, rose in roughly two stages, first to the Stansbury shoreline (1372 m), and then after a 50m recession, more

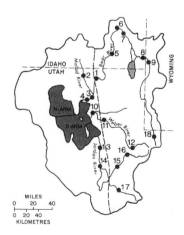

Figure 1. Outline map of Great Salt Lake basin. The numbers show locations sampled for inflow-stream fine clays. The line across the main body of the lake indicates the location of the causeway separating the north and south arms. The drainage divide for the Bonneville basin is directly west of the northern arm. Cores of lake sediments were obtained from the middle and south-central parts of the southern arm, and the lowest portions of the Bonneville salt flats.

rapidly to the Bonneville level, over 300 meters above historic stands (unit III). At about 15 KA the lake overflowed to the north. The freshest water conditions have been correlated with the highest lake levels. After overflow, the lake level paused about 100 meters lower (Provo level) and then fell to much lower levels that are comparable to present-day elevations (unit II). The lake was subject to only relatively minor level changes (<15m) through the Holocene (unit I). Spencer et al. (1984) provide striking mineralogic and biologic evidence from cores for significant increase in salinity accompanying the fall of lake level.

The bulk non-clay mineralogy of the sediments (Spencer et al., 1984) is dominated by quartz in the lower part of the cores. Quartz abundance declines sharply to about 15% in most of the upper sections. The deepest core intervals are low in $CaCO_3$. Carbonate content increases moderately with lake level rise until an abrupt increase to nearly 50% by weight occurs corresponding to the major drop from Bonneville levels. The marked increase in carbonate is accompanied by a progressive increase in the magnesium content of calcite to about 11 mole percent. Thereafter, the dominant carbonate is aragonite. Minor amounts of dolomite persist throughout the core, but only reach substantial quantities just above the calcite maximum. The carbonate mineral assemblage on the Bonneville salt flats is similar, but with some material containing more dolomite (Bissell & Chilingarian 1962), and possibly magnesite (Graf et al. 1961).

2 RESULTS AND DISCUSSION

Assuming the fine clay is the most homogenous and reflective of diagenetic change, the ultrafine fractions of more than 50 samples from cores, plus fine clay from the Bonneville salt flats, were examined. The material was washed to reduce salt and separated by high-speed centrifugation. Clay fractions, typically <0.1 micron in equivalent spherical dimension, were oriented on glass slides and examined by x-ray diffraction after air drying, glycol solvation, or heating to 500 degrees C; additional material was chemically analyzed. Representative diffractograms from main lake sediments are given in Figure 2. A very small amount of carbonate, and probably iron oxyhydroxide, persists into some ultrafine fractions, but other non-layer silicates are not detected. Smectite, with lesser amounts of illite and kaolinite, is the principal clay mineral in all samples. Although this is the same assemblage identified by Grim et al. (1960) in <2 μm fractions of lacustrine sediments throughout the GSL area, the proportion of smectite is significantly greater in the fine clays.

There are no unambiguous differences in the diffraction patterns with respect to a definite stratigraphic position or overall depth in the cores. At most, there is a vague trend upward toward poorer resolution of peaks slightly greater expansion of the largest 2:1 basal reflection, reduction in the principal kaolin peak, and an increase in low-angle background characteristic of interstratification.

In contrast to the x-ray diffraction results, chemical analyses (Table 1) show that the fine clays from the saline GSL sediment of the last 10 KA, and fine clay from the Bonneville salt flats, contain significantly more magnesia and less alumina than the fine clay of fresh lake periods or the average <0.1 μm fraction of sediment from the major inflow stream systems. This difference increases upward in the core to near the sediment-water interface. The basic chemical differences seen in the fine clays of the original lake cores are corroborated by samples from later cores, and by comparison to fine clays from a section of high-stand Bonneville lake sediments described by Oviatt (1986) from outcrop along the Bear River.

The chemical analyses of Table 1 are recast into 2:1 layer phyllosilicate (smectite - type) formulae in Table 2. Such calculations further emphasize the role of magnesium in the diagenetic alteration of the clays, especially in the increase in sum of the octahedral ions. The assignment of Mg to permanent octahedral sites, rather than to interlayer exchange positions, is supported by the x-ray diffraction evidence for increased interstratification.

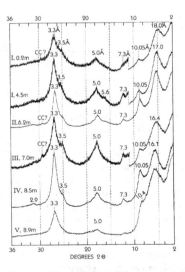

Figure 2—X-ray (CuKα) diffractograms of oriented Great Salt Lake fine clay (<0.1um). Basal spacings are marked in angstrom units. Note the gradual upward increase in (001) spacings, accompanied by decrease in relative intensity.

Table 1. Representative major oxide chemical analysis of Great Salt Lake area clays, <0.1-μm fraction

Core	Depth (m)	Unit	----------------Anhydrous weight percent----------------							
			SiO_2	Al_2O_3	Fe_2O_3	MgO	CaO	Na_2O	K_2O
G	0.5	I	57.93	12.59	6.43	19.45	1.49	0.07	2.04
G	2.4	I	57.73	12.89	7.35	16.27	3.42	0.17	1.16
G	5.3	I	56.95	15.68	8.67	12.48	3.39	0.29	2.54
C	6.3	II	58.01	16.32	10.03	10.40	2.60	0.15	2.49
C	6.9	III	56.71	23.90	8.89	4.66	1.52	0.35	4.01
C	7.8	IV	58.71	19.96	9.95	6.20	1.32	0.34	3.51
C	8.9	V	59.03	20.94	8.30	5.99	1.02	1.43	3.30
Inflow Ave.		----	56.42	23.90	10.85	3.93	1.47	0.34	3.10
Bear River+.			55.20	24.21	11.05	4.13	1.52	0.40	3.49
Bonneville^			57.25	12.21	5.50	22.20	*	0.24	2.30

*Analyzed calcium omitted and assumed to be all carbonate; + Bonneville Fm. Section, along Bear River, Honeyville quadrangle, UT; ^lowest elevations, Bonneville Salt Flats.

Table 2. Chemical formulae for representative analyses Great Salt Lake area clays, <0.1-μm fraction

Core	Depth (m)	Unit	----------------Cations per 22 negative charges---------------							
			Si	Al	Fe^{3+}	Mg	Ca	Na	K	Σ oct.
G	0.5	I	3.59	0.92	0.30	1.80	0.10	0.01	0.16	2.61
G	2.4	I	3.61	0.95	0.35	1.52	0.23	0.02	0.17	2.43
G	5.3	I	3.57	1.16	0.41	1.17	0.23	0.04	0.20	2.31
C	6.3	II	3.62	1.20	0.47	0.97	0.17	0.02	0.20	2.26
C	6.9	III	3.52	1.75	0.42	0.43	0.10	0.04	0.32	2.12
C	7.8	IV	3.65	1.46	0.47	0.57	0.09	0.04	0.28	2.15
C	8.9	V	3.65	1.53	0.39	0.55	0.07	0.17	0.26	2.12
Inflow Ave.			3.50	1.75	0.51	0.36	0.10	0.04	0.25	2.12
Bear River +			3.45	1.78	0.52	0.38	0.10	0.05	0.28	2.13
Bonneville ^			3.55	0.44	0.26	2.05	*	0.03	0.18	2.75

*Analyzed calcium omitted and assumed to be all carbonate. + Bonneville Fm. Section, along Bear River, Honeyville quadrangle, UT; ^lowest elevations, Bonneville Salt Flats.

Earlier studies of the saline lake environment demonstrate the neoformation of authigenic trioctahedral stevensite or kerolite interstratified with detrital dioctahedral smectite (Jones 1986).

More general chemical evidence of the reaction of waters with sediment, relatively independent of the inhomogeneities of the clay fractions, can be illustrated in comparison of major element ratios for lacustrine fine clays with those from inflow stream sediments. No substantial difference in mineralogy between river and high-stand lacustrine fine clays are seen by x-ray diffraction. Plots are presented (Figures 3-5) for potash, iron, silica and magnesia versus alumina, which is assumed to be conserved in the solid phase. Values for lacustrine fine clays are plotted according to depth in core. Data for the river sediments are plotted in proportion to the distance from the lake edge. No consistent trends exist in the potash and iron ratios; the lake clay values are only slightly higher than for the river clays. This

contrasts with a plot for the silica/alumina ratio, Figure 3, and is markedly different from a similar plot for the magnesia/alumina ratios, Figure 5. Despite the erratic variations in SiO and MgO/Al ratios for the saline strata, none of these values are as low as those for the stream inflow or Lake Bonneville fine clays. This clearly indicates the authigenic development of magnesium silicate in the saline lacustrine environment.

The strong similarity in chemistry and mineralogy between the fine clay fractions of the stream sediments and high-stand freshwater lake sediments suggests that the essential process of magnesium silicate authigenesis is the uptake of magnesium and silica from saline pore fluids. This is confirmed for magnesium by examination of pore fluid profiles (Spencer et al. 1985), whereby the large decrease in Mg^{2+} in pore fluid at the top of the sediment column contrasts with the simple diffusion profile for K

Pore-fluid silica concentration profiles, on the other hand, show more scatter and actually increase

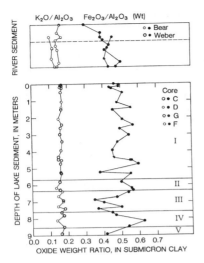

Figure 3—Ratios of potash and ferric oxide to alumina for fine clays from Great Salt Lake and stream inflow sediments. Core stratigraphic intervals from Spencer et al. 1984, are indicated by Roman numerals as shown in Figure 2; all depths are approximate. Points representing samples from the streams are plotted according to distance from river mouth; the short line connects points for the Weber River system and an average from the Jordan River drainage. The dashed line approximates the position of the mountain front.

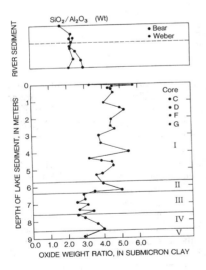

Figure 4—Ratio of silica to alumina for fine clays from Great Salt Lake and stream inflow sediments. Symbols and explanations are the same as in Figure 3.

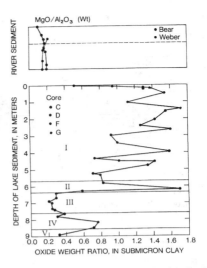

Figure 5—Ratio of magnesia to alumina for fine clays from Great Salt Lake and stream inflow sediments. Symbols and explanations are the same as in Figure 3.

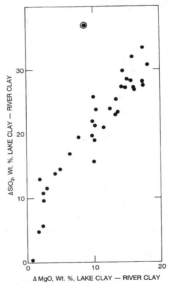

Figure 6—Plot of differences of silica and magnesia content between fine clay, normalized to a constant alumina content, from Great Salt Lake sediments and the average from stream inflow. Results for material sampled at the sediment-water interface are indicated by a circled point. Note the generally linear relation achieved above a lake-stream sediment-silica content difference of ten percent.

through the bulk of the saline strata, before decreasing irregularly downward. The silica results are most readily explained by the non-uniform dissolution of diatom frustules within the lake sediment as a source of pore-fluid silica and the formation of clay as a sink.

The approximate Mg/Si stoichiometry of the authigenic magnesium silicate formed in saline lacustrine sediments is delineated by the subtraction of the average magnesia and silica content of fluvial fine clays from those of the <0.1 μm lacustrine clay, normalized against the alumina content (Fig 6). The plot of the normalized differences illustrates an increasing straight-line relationship beyond a differential silica concentration of ten percent, indicating a Mg/Si ratio of nearly one. This suggests a Mg/Si stoichiometry that is even in excess of that required for kerolite (hydrous talc), previously suggested (Eberl et al. 1982; Jones & Weir 1983; Jones 1986) to form in these situations, and is probably accounted for by an added exchange reaction .

3 CONCLUSIONS

These investigations of the Great Salt Lake area fine clays further support the suggestions of Jones (1986) that a principal mechanism for incorporation of magnesium and silica in sediments is the formation of a Mg-silicate interstratification in a pre-existing detrital 2:1 phyllosilicate. Like carbonate or salt precipitation, this reaction is apparently driven by increased solute concentration, but takes place within the sediment by interaction of pore fluid with closely associated solids. As a result, variable surface kinetics interfere with attempts to describe the process solely in mass action terms, and further complicate the evaluation of Mg partition between silicate and carbonate phases in the diagenetic environment.

ACKNOWLEDGEMENTS

We wish to thank O.P. Bricker, R.S. Thompson and E. Callender for review, C.G. Oviatt, R.M. Forester, and J.L. Mason for discussion and field guidance, and S.L. Rettig, D.M. Webster, B.L. Coleman, K. M. Conko, and M. M. Shapira for technical support.

REFERENCES

Bissell, H.J. & G.V. Chilingar 1962. Evaporite-type dolomite in salt flats of western Utah. *Sedimentology*: 1:200-210.

Eberl, D.D., B.F. Jones & H.N. Khoury 1982. Mixed-layer kerolite/stevensite from the Amargosa Desert, Nevada. *Clays and Clay Minerals*, 30:321-326.

Graf, D.L., A.J. Eardley, & N.F. Shimp 1961. A preliminary report on magnesium carbonate formation in Glacial Lake Bonneville. *J. Geology* 69: 219-223.

Jones, B.F. 1986. Clay mineral diagenesis in lacustrine sediments. In *Studies in Diagenesis*: U.S. Geological Survey Bulletin 1578: 291-300.

Jones, B.F. & A.H. Weir 1983. Clay minerals of Lake Abert, an alkaline saline lake. *Clays and Clay Minerals*, 31:161-172.

Oviatt, C.G. 1986. Geologic map of the Honeyville quadrangle, Box Elder and Cache counties, Utah. *Utah Geological and Mineral Survey Map 89 (1:24,000)*.

Oviatt, C.G., D.R. Currey & D. Sack. 1992. Radiocarbon chronology of Lake Bonneville, Eastern Great Basin, USA. *Paleogeography, Paleoclimatology, Paleoecology* 99:225-241.

Spencer, R.J., M.J. Baedecker, H.P. Eugster, R.M. Forester, M.B. Goldhaber, B.F. Jones, K. Kelts, J. McKenzie, D.B. Madsen, S.L. Rettig, M. Rubin & C.J. Bowser 1984. Great Salt Lake, and precursors, Utah: the last 30,000 years. *Contributions to Mineralogy and Petrology* 86:321-334.

Spencer, R.J., H.P. Eugster, B.F. Jones & S.L. Rettig 1985. Geochemistry of Great Salt Lake, Utah. I: Hydrochemistry since 1850. *Geochimica et Cosmochimica Acta* 49: 727-737.

Thompson, R.S., L.J. Toolin, R.M. Forester, & R.J. Spencer 1990. Accelerator-mass spectrometer (AMS) radiocarbon dating of Pleistocene lake sediments in the Great Basin. *Paleogeography, Paleoclimatology, Paleoecology* 78: 301-313.

Geochemistry of the Earth's Surface, Ármannsson (ed.)© 1999 Balkema, Rotterdam, ISBN 90 5809 073 6

Invited lecture: Pore water dissolved sulfide profiles: Chemical controls of thickness and magnitude

R. Raiswell

Department of Earth Sciences, Leeds University, UK

ABSTRACT: Pore water dissolved sulphide profiles can be quantified by a steady state diagenetic model which describes the rate of microbial sulphide production and the rate of reaction with iron minerals. The model predicts that Log ω/x = 1.94 log ω - 1.55 where ω is the sedimentation rate and x is the depth of the dissolved sulphide profile. Data from 17 modern marine sediments show a close fit to the model prediction, indicating that the persistence of dissolved sulphide is principally controlled by organic matter reactivity.

1 INTRODUCTION

Studies of sulfate reduction in modern sediments have concentrated on near-surface regions, where organic matter is most reactive, rates of sulfate reduction are highest and most pyrite is formed (e.g. Berner 1970, 1984, Goldhaber & Kaplan 1974, Canfield 1989). By contrast comparatively little attention has been paid to the deeper regions of sulfate reduction, where dissolved sulfide accumulates. However a significant number of studies have been made over depth intervals which are long enough to show complete (or near-complete) porewater dissolved sulfide profiles, which increase from negligible concentrations to a maximum before decreasing back to negligible levels. Goldhaber & Kaplan (1974) recognised that these depth variations in dissolved sulfide arose from variations in the rate of sulfide production by microbial sulfate reduction, and the rate of sulfide removal by reaction with iron minerals. Dissolved sulfide only accumulates once iron oxides have been consumed (Canfield & Raiswell 1991) and thereafter silicate iron is the most abundant reactant available in the sediment (although the conversion of FeS to FeS_2 may be important in some cases; Rickard 1997). This observation suggests these dissolved sulfide profiles could provide important information on the rates at which silicate iron minerals react with sulfide in different depositional environments. This paper will analyse the literature profiles of dissolved sulfide from seventeen different depositional environments, (spanning a wide range of sedimentation rates from 0.003 to 10 cm yr^{-1}) and examine the kinetic controls on the depth ranges over which dissolved sulfide persists.

2 RESULTS

Table 1 (based on Raiswell 1997) reports data from sediments deposited under oxygenated or partially oxygenated bottom waters which show complete (or near-complete) porewater profiles of dissolved sulfide. Complete profiles describe the passage from negligible concentrations of dissolved sulfide, through a maximum and back again to negligible levels. Near-complete profiles pass through a maximum and show a sufficiently well-defined decreasing limb to allow extrapolation of the profile to define the depth at which zero concentrations of dissolved sulfide would be reached. In most cases there is sufficient data to provide an adequate constraint on the maximum dissolved sulfide concentrations and the thickness of the zone of sulfidic porewaters. However at four localities (Black Sea Station 5, Bornholm Deep φ7, FOAM and the Japan Trench) the deepest data point records a dissolved sulfide concentration which has not yet decreased to less than 20% of the maximum dissolved sulfide concentration. Extrapolation of the data to find the depth at which sulfide concentrations reach zero may be in error at these localities.

Best-fit curves through the profiles were generated by eye, where no curves were presented by the authors. These curves were used to define the depth range over which dissolved sulfide persists (x; see Table 1) and the maximum dissolved sulfide concentration. The Peru Margin has the highest dissolved sulfide concentrations (12 mM; Mossmann et al. 1991) but these are still below the levels which are toxic to sulphate reducing bacteria. Reis et al. (1992) have shown that undissociated

299

Table 1. Sulfide profiles in modern sediments (for data sources see Raiswell 1997)

Location	Sed. Rate (cm/yr)	H$_2$S Depth Range (cm)	Exp.Time (yr)
Cape Look Out	10.3	74	7.2
Black Sea St 15	0.90	41	45.5
Sachem	1.50	75	50
Saanich	1.00	173	173
Black Sea St 5	0.70	38	54.3
Loch Etive	0.90	70	77.8
Bornholm φ12	0.29	71	245
Guaymas	0.273	200	733
Sal Si Puedas	0.316	390	1230
Carmen	0.187	260	1390
Pescadero	0.046	100	2170
Bornholm φ7	0.150	454	3030
FOAM	0.100	460	4600
Gotland φ31	0.023	323	14040
Japan Tr 6158	0.003	450	1.5×10^4
Peru Marg 680	0.003	8200	2.7×10^6
Peru Marg 686	0.017	4700	2.7×10^5

H$_2$S is the toxic species (HS$^-$ predominates in marine waters) and that concentrations of 16 mM H$_2$S are needed to inhibit sulphate reduction.

Almost all the dissolved sulfide profiles are assymmetrical with depth, such that decreasing sulfide concentrations occupy a greater thickness of sediment than do increasing concentrations. Only two localities appear to show the opposite behaviour (Black Sea Station 15 and Bornholm Deep φ7), and extrapolation errors may be present in both curves (see above). Table 1 also estimates a sediment exposure time to dissolved sulfide as x/ω, where ω is the sedimentation rate (cm/yr). Values of exposure time range over six orders of magnitude, suggesting that there are very large differences in the reactivity of sediments towards dissolved sulfide.

3 APPROACH

Estimates for the rate constants of iron mineral sulfidation (Canfield et al. 1992) indicate that silicate iron and magnetite are the most likely phases to be removing sulfide over the depth ranges being considered, for two reasons. Firstly, Canfield (1989) and Canfield & Raiswell (1991) have shown that reactions between iron oxides (lepidocrocite, ferrihydrite, haematite and goethite) and dissolved sulfide are rapid enough to remove sulfide as fast as it is produced by sulfate reduction, hence dissolved

sulfide only accumulates in porewaters once all the iron oxides have reacted to form pyrite. Secondly, organic matter is unlikely to react with dissolved sulfide to any significant extent. There is a growing body of evidence to suggest that sulfur is incorporated into organic matter close to the sediment surface (Mossmann et al. 1991, Aplin & Macquaker 1993, Raiswell et al. 1993), probably via partially oxidised sulfur species. Indeed the Peru Margin sediments studied by Mossmann et al. (1991) and Raiswell & Canfield (1996) contain an unusually thick zone of sulfidic porewaters (nearly 3 million years sediment exposure time to dissolved sulfide; Table 1), yet show no evidence of sulfur removal by organic matter. Consistent with this Raiswell & Canfield (1996) showed that reactions between silicate iron and dissolved sulfide accounted for the removal of all the dissolved sulfide from the Peru Margin porewaters. These arguments indicate that silicate iron and magnetite (which are usually more abundant than FeS) are most likely to be the major sinks for dissolved sulfide for all the sediments in Table 1.

On this basis the model which Raiswell & Canfield (1996) used to describe porewater sulfide profiles in Peru Margin sediments can be generally applied. This model assumes that sulfide removal is first order with respect to Fe mineral concentration, and that there is no dependancy on sulfide concentration. These kinetics are supported by the analysis of Fe distributions in the Peru Margin sediments. If sulfide is removed as pyrite;

$$\text{Sulfide removal} = - 2 \, k_{Fe} \, C_{Fe} \, (1 - \varphi)/\varphi \qquad (1)$$

where k_{Fe} is the reaction rate constant (yr^{-1}) for Fe minerals, C_{Fe} is the concentration of unreacted iron per cm^3 solids, and φ is the porosity. Sulfide removal is in units of micromoles per cm^3 of porewater per year and the factor of two represents the stoichiometric coefficient linking sulfide removal to pyrite formation. Sulfide is produced by sulfate reduction using the simple one G model (Berner 1980)

$$\text{Sulfide production} = 0.5 \, k_{ORG} \, G \, (1 - \varphi)/\varphi \qquad (2)$$

where k_{ORG} is the rate constant (yr^{-1}) for organic matter decay by sulfate reduction, G is the concentration of organic carbon in micromoles per cm^3 solid, and 0.5 is the stoichiometric coefficient linking the production of sulfide to the oxidation of organic carbon. Equations 1 and 2 can be substituted into the general diagenetic equation (Berner 1980) to derive the following expression;

$$D_S \, d^2C_S/dx^2 - \omega \, dC_S/dx$$
$$+ 0.5 \, k_{ORG} \, G_o \, (1 - \varphi)/\varphi \, \exp(k_{ORG} \, x/\omega)$$
$$- 2 \, k_{Fe} \, C_{Fe,o} \, (1 - \varphi)/\varphi \, \exp(k_{Fe} \, x/\omega) = 0 \qquad (3)$$

where D_S is the sediment diffusion coefficient for sulfide (cm^2 yr^{-1}), C_S is the concentration of dissolved sulfide (micromoles per cm^3 porewaters), and G_o and $C_{Fe,O}$ are the initial concentrations of organic matter and reactive iron respectively. This equation is solved with the boundary conditions that C_S is zero at $x = 0$ and dC_S/dx is zero as x tends to infinity. Raiswell & Canfield (1996) give the solution to equation 3 as;

$$C_S = \frac{[0.5\, k_{ORG}\, G_o\, (1 - \varphi)/\varphi]\, [\, 1 - \exp(-k_{ORG}\, x/\omega)]}{D_S\, (k_{ORG}/\omega)^2 + k_{ORG}}$$

$$- \frac{[2\, k_{Fe}\, C_{Fe,O}\, (1 - \varphi)/\varphi]\, [\, 1 - \exp(-k_{Fe}\, x/\omega)\,]}{D_S\, (k_{Fe}/\omega)^2 + k_{Fe}} \quad (4)$$

Putting $C_S = 0$ makes x the lower depth at which dissolved sulfide becomes negligible (in effect the thickness of the sulfidic zone) and cancelling D_S and $(1 - \varphi)/\varphi$ and rearranging produces;

$$\frac{1 - \exp(-k_{ORG}\, x/\omega)}{1 - \exp(-k_{Fe}\, x/\omega)} = \frac{4\, C_{Fe,O}\, (k_{ORG} + \omega^2)}{G_o\, (k_{Fe} + \omega^2)} \quad (5)$$

To a first approximation $\exp(k_{ORG}\, x/\omega) = 1 - k_{ORG}\, x/\omega + (k_{ORG}\, x/\omega)^2/2$ and $\exp(k_{Fe}x/\omega) = 1 - k_{Fe}\, x/\omega + (k_{Fe}\, x/\omega)^2/2$, which allows equation 5 to be simplified to;

$$\frac{2\, k_{ORG} - k^2_{ORG}\, x/\omega}{2\, k_{Fe} - k^2_{Fe}\, x/\omega} = \frac{4\, C_{Fe,O}\, (k_{ORG} + \omega^2)}{G_o\, (k_{Fe} + \omega^2)}$$

which can be re-arranged to show that

$$\omega/x = \frac{4\, C_{Fe,O}\, k^2_{Fe}\, (k_{ORG} + \omega^2)\, - k^2_{ORG}\, G_o\, (k_{Fe} + \omega^2)}{8\, C_{Fe,O}\, k_{Fe}\, (k_{ORG} + \omega^2)\, - 2\, k_{ORG}\, G_o\, (k_{Fe} + \omega^2)}$$

This complex expression can be reorganised to express ω/x as the sum of two terms;

$$\frac{k_{Fe}}{2 - \{k_{ORG}\, G_o\, (k_{Fe} + \omega^2)\}/\, 2\, C_{Fe,O}\, k_{Fe}\, (k_{ORG} + \omega^2)} +$$

$$\frac{k_{ORG}}{2 - \{8\, C_{Fe,O}\, k_{Fe}\, (k_{ORG} + \omega^2)\}/\, k_{ORG}\, G_o\, (k_{Fe} + \omega^2)} \quad (6)$$

which shows that ω/x varies mainly with either k_{Fe} or k_{ORG} depending critically on the term $\{8\, C_{Fe,O}\, k_{Fe}\, (k_{ORG} + \omega^2)\}/\{k_{ORG}\, G_o\, (k_{Fe} + \omega^2)\}$.

Thus, when this expression is << 2, it follows that the second term in equation 6 approaches $k_{ORG}/2$. Furthermore the first term in equation 6 becomes small and negative, because
$\{8C_{Fe,O}\, k_{Fe}\, (k_{ORG} + \omega^2)\}/\{k_{ORG}G_o\, (k_{Fe} + \omega^2)\} << 2$
is the same as;
$\{k_{ORG}G_o(k_{Fe} + \omega^2)\}/\{8C_{Fe,O}\, k_{Fe}(k_{ORG} + \omega^2)\} >> 0.5$
or $\{k_{ORG}G_o\, (k_{Fe} + \omega^2)\}/\{2C_{Fe,O}\, k_{Fe}(k_{ORG} + \omega^2)\} >> 2$
The converse applies when
$\{8C_{Fe,O}\, k_{Fe}(k_{ORG} + \omega^2)\}/\{k_{ORG}\, G_o\, (k_{Fe} + \omega^2)\} >> 2$
and the first term in equation 6 approaches $k_{Fe}/2$, whereas the second term becomes small and negative. Thus the model predicts that ω/x (the sediment exposure time to dissolved sulfide) will tend to be controlled either by the rate constant for organic matter decay by sulfate reduction ($k_{ORG}/2$) or by the rate constant for the sulfidation of silicate iron ($k_{Fe}/2$).

The data in Table 1 produce a best fit line (solid line in Figure 1) given by;

$$\log \omega/x = -2.03 + 1.49 \log \omega \qquad r = 0.96$$

$$\text{or} \quad x = 0.01\omega^{1.49+/-0.18} \quad (7)$$

One solution to equation 6 is that $\omega/x = k_{ORG}/2$, and the rate constant of organic matter decay by sulphate reduction is given by Toth & Lerman (1977, up-dated by Tromp et al.1995) as;

$$k_{ORG} = 0.057\, \omega^{1.94+/-0.22}$$

Figure 1 therefore shows the envelope (dashed lines) defined by;

$$k_{ORG}/2 = 0.028\, \omega^{1.72} \text{ and } 0.028\, \omega^{2.16} \quad (8)$$

The data are reasonably well-described by this line, even though this is a rather less good fit to the data than the best fit line of equation 7. This important result indicates that the sediment exposure time to dissolved sulfide (ω/x) can be considered to be a function of $k_{ORG}/2$ and that equation 8 can be used to predict the thickness of sulfidic porewaters (x).

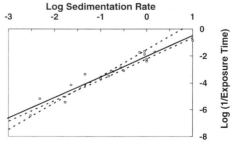

Figure 1. Data from Table 1 for sulfide profiles

The fit of the data to the line $k_{ORG}/2$ indicates that the first term in equation 6 must be insignificant and the Peru Margin data from Raiswell & Canfield (1996) have $C_{Fe,O} = 300$ micromoles cm_s^{-3}, $G_o = 4755$ micromoles cm_s^{-3}, $k_{Fe} = 1$ to 4×10^{-7} yr^{-1}, $k_{ORG} = 5 \times 10^{-6}$ yr^{-1} and $\omega = 0.005$ cm yr^{-1} so that $\{ 8\, C_{Fe,O}\, k_{Fe}\, (k_{ORG} + \omega^2)\}/\{ k_{ORG}\, G_o\, (k_{Fe} + \omega^2)\}$ is approx. 0.01 to 0.03. Thus the alternative solution to equation 6 is ($k_{Fe}/2$) which will operate in sediments that contain more iron, and lower concentrations of organic C, than the Peru Margin.

4 CONCLUSIONS

1. The steady state diagenetic model of Raiswell & Canfield (1996) can be used to predict pore water dissolved sulfide profiles in modern marine sediments deposited in oxygenated bottom waters.

2. The model predicts that ω/x (the reciprocal of sediment exposure time to dissolved sulfide) = $0.028\omega^{1.94+/-0.22}$. This relationship shows a close fit to measurements of dissolved sulfide profiles at 17 modern sediment localities.

3. Even the maximum dissolved sulfide concentrations observed (12mM) are probably below levels at which sulphate reduction is inhibited.

REFERENCES

Aplin, A. & J.S. Macquaker 1993. C-S-Fe geochemistry of some modern and ancient marine muds and mudstones. *Phil. Trans. Roy. Soc. London* A344:89-100.

Berner, R.A. 1970. Sedimentary pyrite formation. *Amer. J. Sci.* 268:1-23.

Berner, R.A. 1980. *Early diagenesis: A theoretical approach.* Princetown University Press.

Berner, R.A. 1984 Sedimentary pyrite formation: An update. *Geochim. Cosmochim. Acta* 48:605-615.

Canfield, D.E.1989. Reactive iron in marine sediments. *Geochim. Cosmochim. Acta* 53:619-632.

Canfield, D.E. & R. Raiswell 1991. Pyrite formation and fossil preservation. In P.A. Allison & D.E.G. Briggs (eds), *Taphonomy: Releasing the data locked in the fossil record*: 337-387. Plenum Press: New York.

Canfield, D.E., R. Raiswell & S.H. Bottrell 1992. The reactivity of sedimentary iron minerals toward sulfide. *Amer. J. Sci.* 292:659-683.

Goldhaber, M.B. & I.R. Kaplan 1974. The sulfur cycle. In E.D.Goldberg (ed), *The Sea* 5:569-655.

Mossmann, J., A.C. Aplin, C.D. Curtis & M.L. Coleman 1991. Geochemistry of inorganic and organic sulphur in organic-rich marine sediments from the Peru Margin. *Geochim. Cosmochim. Acta* 55:3581-3595.

Raiswell, R., S.H. Bottrell, H.J. Al-Biatty & M. Md.Tan 1993. The influence of bottom water oxygenation and reactive iron content on sulfur incorporation into bitumens from Jurassic marine shales. *Amer. J. Sci.* 293:569-596.

Raiswell, R. & D.E. Canfield 1996. Rates of reaction between silicate iron and dissolved sulfide in Peru Margin sediments. *Geochim. Cosmochim. Acta* 60:2777-2787.

Raiswell, R.1997. A geochemical framework for the application of stable sulphur isotopes to fossil pyrtization. *J. Geol. Soc. London* 154:343-356.

Reis, M.M., J.S. Almeida, P.C. Lemos & M.J.T. Carrondo. 1992. Effect of hydrogen sulfide on growth of sulphate reducting bacteria. *Biotech. Bioeng.* 40:593-600.

Rickard, D.T. 1997. Kinetics of pyrite formation by the H_2S oxidation of iron(II) monsulfide in aqueous solution beteen $25^{\circ}C$ and $125^{\circ}C$: the rate equation. *Geochim. Cosmochim. Acta* 55:1505-1514.

Toth, D.J. & A. Lerman 1977. Organic matter reactivity and sedimentation rates in the ocean. *Amer. J. Sci.* 277:265-285.

Tromp, T.K., P. Van Cappellen & R.M. Ker 1995. A global model for early diagenesis of organic carbon and organic phosphorus in marine sediments. *Geochim. Cosmochim. Acta* 59:12159-1284. 1995.

Geochemistry of the Earth's Surface, Ármannsson (ed.) © 1999 Balkema, Rotterdam, ISBN 90 5809 073 6

Sulfur isotope fractionation during bacterial sulfate reduction

V. Brüchert, C. Knoblauch & B. B. Jørgensen
Max Planck Institute for Marine Microbiology, Bremen, Germany

ABSTRACT: The isotopic fractionation during bacterial sulfate reduction was determined in batch culture experiments with new strains of sulfate-reducing bacteria that grow at temperatures between -1.8 and 20°C. The observed isotopic fractionations for the two isolates are significantly smaller than commonly measured isotopic differences between sulfate and sulfide in temperate and cold marine sediments. For all cultures, sulfate reduction rates decrease with decreasing temperature. However, in one strain the isotopic fractionation was independent of the sulfate reduction rates and the temperature at which the bacteria were grown whereas another culture showed an increasing isotopic fractionation with decreasing temperature. Phylogenetic differences between the strains, differences in the cellular electron transport pathways during substrate oxidation, and the formation of sulfur intermediates at very low rates of sulfate reduction are inferred to be the dominating factors controlling the observed isotopic fractionations.

1 INTRODUCTION

Bacterial sulfate reduction is the dominant metabolic process in anoxic marine sediments (Jørgensen 1982). It has long been known that the reduction of sulfate to sulfide is coupled to a pronounced enrichment of sulfide in ^{32}S, yet the factors controlling the isotopic enrichment are still insufficiently understood. Rates of bacterial sulfate reduction and temperature have been suggested to exert the dominating control on variations in isotopic fractionation. Here we report experimental results on sulfur isotopic fractionations by 4 strains of sulfate-reducing bacteria to quantify the effects of temperature and sulfate reduction rates and to determine if additional processes can affect the isotopic fractionation. Sulfate reducers were isolated from arctic marine habitats off the coast of Svalbard, Spitsbergen. This is the first report on isotopic fractionations by sulfate-reducing bacteria that are capable of growing at temperatures of very cold marine habitats and at temperatures that are common to most temperate marine environments worldwide.

2 METHODOLOGY

Pure cultures of the sulfate-reducing strains LSv54, ASv26, LSv514, and ASv20 were grown in batch culture at temperatures of -1.8, 2, 4, 9, 16, and 20°C using the anaerobic medium described in Widdel and Bak (1992). Organic substrates were lactate (20 mM final concentration) for strains LSv54 and LSv514 and acetate (20 mM final concentration) for strains ASv26 and ASv20. Most cultures were grown to the stationary phase of growth. Samples for the isotopic determination of sulfate and sulfide were taken in the lag phase, the early exponential, the late exponential, and the stationary phase. At each time point, concentrations of sulfate and sulfide were determined. Sulfate was determined by unsuppressed anion chromatography, and sulfide was determined by the methylene blue method (Cline 1969). At each of the above time points, 10 ml aliquots of the cultures were incubated for 6 hours with $^{35}SO_4$ (100 kBq) radiotracer to determine specific sulfate reduction rates for each growth phase. An additional aliquot of the culture was used for cell counting. Sulfate reduction rates could thus be reported as moles of sulfate reduced per bacterial cell rather than on a volumetric basis. To determine the stable isotopic composition 20 % zinc acetate was added to the remaining aliquot to terminate bacterial activity and to precipitate dissolved sulfide. Dissolved sulfate and precipitated zinc sulfide were separated by acid distillation and trapping of H_2S in $AgNO_3$. Dissolved sulfate was recovered as $BaSO_4$. The isotopic composition of dissolved sulfate and sulfide were determined by continuous flow isotope ratio monitoring gas chromatography mass spetrometry according to methods described in Giesemann et al. (1994).

3 RESULTS AND DISCUSSION

All analyzed strains show a characteristic decrease in sulfate reduction rates with temperature. In general, sulfate reduction rates decrease from the exponential to the stationary phase of growth. Strain LSv54, which oxidizes lactate to acetate and is not capable of complete substrate oxidation showed a small isotopic fractionation of 4.6 ‰ at 9°C (Figure 1). To our knowledge this is the smallest isotopic fractionation ever reported for sulfate-reducing bacteria. Strain ASv26, which oxidizes acetate to CO_2 and is thus capable of complete substrate oxidation showed a fractionation of 22 ‰ (Figure 2). Phylogenetic differences based on 16S rRNA sequences indicate that these two strains belong to two different new genera of sulfate-reducing bacteria (Knoblauch et al. in press). The isotopic fractionation in strain ASv26 was constant over the viable temperature span for this bacterial strain. Furthermore, the isotopic fractionation is constant for the different growth stages. In strain LSv54 sulfide had a lower isotopic composition at -1.8°C than at 9°C while the evolution of the isotopic composition of sulfate was consistent with a Rayleigh-type fractionation during reduction of sulfate by 4.6 ‰ (Figure 1). These differences can be reconciled with the formation of intermediate sulfur species that are enriched in ^{34}S relative to sulfide. One of the proposed pathways during dissimilatory sulfate reduction includes the formation of the intermediates thiosulfate and trithionate during the reduction of sulfite (Widdel and Hansen 1992). The turnover time of these intermediates at low rates of sulfate reduction may be significantly longer creating an additional sulfur pool that was not accounted for in the Rayleigh model. Thus the observed isotopic differences at different temperatures of strain LSv54 may reflect the formation of a larger pool of sulfur intermediates in the cellular sulfate reduction process at -1.8°C. The continous trend in the isotopic composition of sulfate indicates that the intermediate sulfur species are not reoxidized to sulfate.

The observed sulfur isotopic fractionations in these pure culture experiments are significantly smaller than the isotopic differences observed between dissolved sulfide and sulfate in coexisting sediments and in most anoxic sediments from temperate environments. In these sediments the

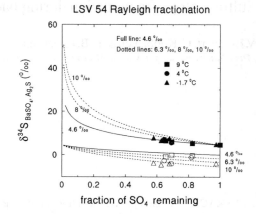

Figure 1. Isotopic fractionation of sulfate and sulfide in batch culture experiments with strain LSv54. Closed symbols represent sulfate and open symbols represent sulfide. Regression lines indicate the different isotopic fractionations.

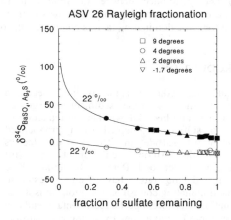

Figure 2. Isotopic fractionation of sulfate and sulfide in batch culture experiments with strain ASv26. Closed symbols represent sulfate and open symbols represent sulfide. Regression lines indicate the isotopic fractionation.

observed isotopic differences between dissolved sulfide and sulfate are generally between 36 and 58 ‰ and imply additional fractionations during the oxidative sulfur cycle.

4 CONCLUSIONS

Our results suggest that naturally observed isotopic differences between sulfate and sulfide may not only be caused by variations in the rates of bacterial sulfate reduction. Genetic and physiological differences between different genera of sulfate-reducing bacteria may also exert a strong influence on the isotopic fractionation. The electron transport pathways for the oxidation of organic substrates differ significantly between complete- and incomplete-oxidizing sulfate reducers (Widdel and Hansen 1992; Cypionka 1994). Also, two different electron transport pathways have been proposed for the reduction of sulfite to sulfide in sulfate-reducing bacteria. These physiological differences can be reflected in phylogenetic differences. In marine sediments, different genetic populations and populations with different adaptations to temperature will result in different gross isotopic fractionations during sulfate reduction. However, comparison of the experimental results with field data confirms previous findings that additional fractionations must exist in the oxidative part of the sedimentary sulfur cycle. The observed fractionations are too small to explain the naturally occurring isotopic differences of porewater sulfate and sulfide. Therefore additional fractionations must occur either during disproportionation or sulfide oxidation.

REFERENCES

Cline J.D. 1969. Spectrophotometric determination of hydrogen sulfide in natural waters. *Limnology and Oceanography* 14: 454-459.

Cypionka H. 1994. Sulfate transport. In *Methods in Enzymology 243*, 243, 3-14.

Giesemann A., H.J. Jäger, A.L. Norman, H.R. Krouse, & W.A. Brand 1994. On-line sulfur isotope determination using an elemental analyzer coupled to a mass spectrometer. *Analytical Chemistry* 66: 2816-2819.

Jørgensen B.B. 1982. Mineralization of organic matter in the sea bed - the role of sulphate reduction. *Nature* 296: 643-645.

Knoblauch C., K. Sahm, & B.B. Jørgensen (in press) Psychrophilic sulphate reducing bacteria isolated from permanently cold arctic marine sediments: description of *Desulfofrigus oceanense* gen. nov., sp. nov., *Desulfofrigus fragile* sp. nov., *Desulfofaba gelida* gen. nov., *Desulfotalea psychrophila* gen. nov., sp. nov., and *Desulfotalea arctica*, sp. nov. International Journal of Systematic Bacteriology.

Widdel F. & T.A. Hansen 1992. The dissimilatory sulfate- and sulfur-reducing bacteria. In H. Balows, H.G. Trüper, M. Dworkin, W. Harder, K.-L. Schleifer (eds), *The Procaryotes* 1:583-624. Springer.

Widdel F. & F. Bak 1992. Gram-negative mesophilic sulfate-reducing bacteria. In H. Balows, H.G. Trüper, M. Dworkin, W. Harder, K.-L. Schleifer (eds), *The Procaryotes* 4:3352-3378. Springer.

Geochemistry of the Earth's Surface, Ármannsson (ed.)© 1999 Balkema, Rotterdam, ISBN 90 5809 073 6

Mineral exploration under sedimentary cover in Australia: CRC LEME basins program

P. de Caritat, D. Gibson, N. Lavitt, E. Papp & E. Tonui – *Cooperative Research Centre for Landscape Evolution and Mineral Exploration, c/o Australian Geological Survey Organisation, Canberra, A.C.T., Australia*

S. Hill – *Cooperative Research Centre for Landscape Evolution and Mineral Exploration, c/o University of Canberra, A.C.T., Australia*

M. Killick – *Cooperative Research Centre for Landscape Evolution and Mineral Exploration, c/o CSIRO Exploration and Mining, Perth, W.A., Australia*

ABSTRACT: The Australian continent has a unique geological and weathering history, presenting particular problems for Australian mineral exploration. CRC LEME was established in 1995 to aid the discovery of concealed world-class ore deposits through knowledge of the Australian landscape evolution. The Basins Program of CRC LEME focuses on understanding the processes of landscape evolution, weathering of, and dispersion from, ore systems existing beneath and within sedimentary basin sequences. It is concerned with the study of the interaction of water and rock at various scales, and in diverse geological settings. Through mapping, remote sensing, geochemical (including hydrogeochemical) investigations and geophysical surveys, the Basins Program aims to generate models of erosion, transport and sedimentation, of sediment diagenesis and weathering, and of (physical and hydromorphic) dispersion of geochemical and/or isotopic anomalies. Together, these are expected to yield improved exploration techniques in areas of shallow basin cover and extensive weathering. Various ongoing and proposed projects focus on a range of Australian basins.

1 INTRODUCTION

Some regions of Australia have been emergent for a considerable length of time and its landscapes are thus partly very ancient. The lack of recent large-scale glaciations or major tectonic uplift has resulted in a relatively stable continent through much of the Phanerozoic. This has in turn translated into relatively long time periods available for weathering of the rocks making up the continental surface. The history of continental drift experienced by Australia has, however, ensured that a variety of climatic zones have affected this weathering history through time. This unique combination of tectonic stability and shifting paleolatitudes has resulted in a complex history of landscape evolution in Australia. It is this history that the Cooperative Research Centre for Landscape Evolution and Mineral Exploration is attempting to unravel in order to improve exploration methods for ore deposits in this weathered landscape.

2 CRC LEME

The Cooperative Research Centre for Landscape Evolution and Mineral Exploration (CRC LEME) is an unincorporated joint venture between the Australian National University (ANU), the University of Canberra (UC), the Australian Geological Survey Organisation (AGSO) and the Commonwealth Scientific and Industrial Research Organisation's Division of Exploration and Mining (CSIRO DEM). It was established in July 1995 with a seven-year support grant from the Australian Commonwealth Government's Cooperative Research Centres Program. CRC LEME's purpose is to aid the discovery of concealed world-class ore deposits through knowledge of the Australian landscape evolution. Australia has been endowed with rich mineral resources, and the mineral industry provides about one third of the nation's export income.

CRC LEME has four broad objectives:
- To establish a framework to promote greater understanding of the three-dimensional evolution of the Australian landscape by integrating geomorphic, geological and geochemical processes and concepts;
- To translate this knowledge into a greatly improved ability to recognise major mineral deposits in provinces of strategic importance to the Australian mineral industry;
- To integrate industry-supported research with education, particularly postgraduate and post-doctoral opportunities, in order to ensure a suitable skill base for the exploration, mining and general earth-science communities, and for fur-

ther research;

- To inform and guide decision-makers of the relevance and contribution of this field of research to Australia's future.

CRC LEME now has a staff of about 53 (full-time equivalents). Personnel is located within the Core Parties in Perth (Headquarters, CSIRO DEM) and Canberra (ANU, UC, AGSO), as well as within three regional nodes (Adelaide, Brisbane and Sydney). The budget for the functioning of LEME is provided by the CRC Grant of AU$16 million over seven years (about 27% of the total resources), in-kind contributions form the four Core Parties (56%), and external funding mainly from the exploration and mining industry and State Government Departments (17%). The amount of industry support is substantial, and has increased steadily during the Centre's first 3 years. CRC LEME has four 'regional' programs: Program 1 - Regolith and Landscapes of the Australian Shields; Program 2 - Regolith and Landscapes of the Tasman Fold Belt; Program 3 - Basins and Landscape Evolution; and Program 4 - Synthesis of Regolith-Landscape Evolution. In addition, an Education Program and an Applications

Program exist to ensure proper training and technology transfer. Strategic Themes have been identified to be: Mapping the regolith in 3D; Expressions of ore systems in the regolith; Exploration in areas of basin cover; Regolith evolution; Terminology and classification; Synthesis GIS and expert system; Education and training.

3 THE BASINS PROGRAM

3.1 *Objectives*

Program 3, Basins and Landscape Evolution, has for objective to develop exploration procedures for mineral deposits concealed within and beneath sedimentary rocks in shallow basins and at basin margins, through a comprehensive understanding of the nature and evolution of the sedimentary sequences (with contained waters), regolith and landforms. Figure 1 shows the location of major sedimentary basins in Australia, and illustrates that a significant portion of the land is covered by basin sequences. In the first instance, we are concerned with areas of

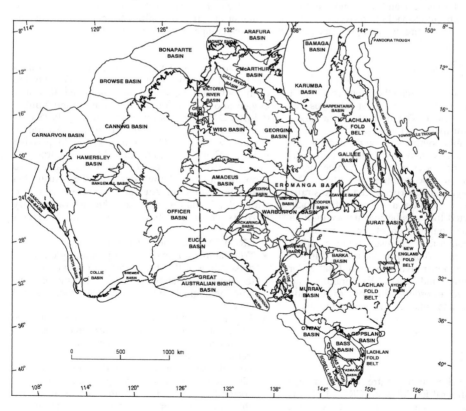

Figure 1. Location of major sedimentary basins in Australia

relatively thin sediment cover (<300 m), i.e. basin margins.

Specific objectives of the Basins Program are:
- To select sedimentary basins of prime exploration potential;
- To liaise with industry to gain access to a variety of ore deposit types where orientation studies can be carried out;
- To establish the sedimentary, landform, regolith and ore deposit settings and to unravel the history of landscape evolution of the selected areas by mapping at appropriate scales, collating existing information and interpreting observations;
- To characterise the chemical and mineralogical evolution of the sediments resulting from diagenesis and weathering, focussing on processes by which concealed mineralisation can give rise to geochemical and isotopic anomalies in the near surface regolith/sediment or within the formation or ground waters.

3.2 Strategies

Since we are dealing with both physical (sediment provenance, paleodrainage, etc.) and chemical (water-rock interaction, geochemical dispersion, etc.) processes (Figure 2), several strategies must be adopted to tackle the above objectives.

Firstly, we unravel the history of erosion in the highlands, and transport and deposition of the resulting sediments within basins, particularly as it relates to paleoclimatic, hydrologic and tectonic conditions, through a combination of regolith/landscape mapping and stratigraphic analysis.

Secondly, we investigate whether sedimentological tools such as the study of paleocurrents and paleochannels, basin analysis or basin fill modelling, can be useful in locating detrital ore deposits within the sedimentary sequence.

Thirdly, we characterise the chemical and mineralogical evolution of the sediments in response to diagenesis and weathering, with specific attention to processes by which concealed mineralisations (within and beneath the sedimentary cover) can give rise to geochemical anomalies in the near-surface regolith/sediments or within the formation or ground waters.

All this is carried out in close collaboration with Programs 1 and 2 in basins that cover shield and Tasman Fold Belt rocks with the greatest exploration potential, and/or that are of greatest scientific interest.

3.3 Activities

A significant proportion of the Program's activities concentrates on the sedimentary margins of the Broken Hill/Olary block (New South Wales, South Australia and Victoria). There, outputs currently in preparation include:

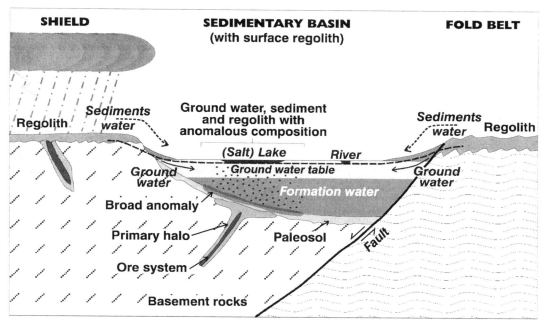

Figure 2. Schematic cross-section of processes affecting mineral exploration in sedimentary basins

- 1:500,000 scale regolith-landform maps of Broken Hill and Curnamona (hardcopy, CD ROM and explanatory notes);
- 1:100,000 scale regolith-landform map of Broken Hill and environs (hardcopy, CD ROM and explanatory notes);
- 1:25,000 scale regolith-landform maps of Balaclava, Redan, Thackaringa and Triple Chance (hardcopy, CD ROM and explanatory notes);
- Airborne hyperspectral mapping (HYMAP^{TM}) in the Thackaringa area;
- Geochemistry and mineralogy of the sediments/regolith, and geochemistry of groundwater in the Mundi Mundi plains;
- Geochemistry and mineralogy of the sediments/regolith on the Benagerie ridge;
- Hydrogeology and hydrogeochemistry in the Curnamona province;
- Diagenesis of sediments in the Murray basin.

Activities in progress in other areas include: regolith mapping and hydrogeochemistry around the Mt Isa block (Queensland); hydrogeochemical and geophysical investigations of paleodrainage systems at the headwaters of the Darling basin (New South Wales); hydrology and climate evolution of the Lake Lewis basin (Northern Territory); authigenic gold precipitation in the Carnarvon basin (Western Australia). In addition to this, short courses/workshops (hydrogeochemistry in exploration; denudation history of the Yilgarn craton) and various technical booklets and publications are being prepared.

A key issue over the next few years will be to improve our understanding of how geochemical dispersion occurs in sedimentary basins over extended periods of time and under varying tectonic and climatic conditions.

3.4 *Research outcomes*

The main products of the Program will include:
- Basin-scale models of erosion, transport and deposition of sediments from surrounding areas (provenance and paleocurrent/paleodrainage studies);
- Similar models of secondary mineralisation within basins;
- Geochemical maps of ground/formation water geochemistry in selected basins;
- Landscape evolution models for selected regions, and relationship to element distribution;
- Improved geochemical exploration models for basin-covered areas;
- Regolith-landform maps of selected districts;
- Refined landscape mapping techniques and protocols;
- Cost- and technology-effective exploration techniques.

4 CONCLUSIONS

CRC LEME is an unincorporated joint venture between ANU, UC, AGSO and CSIRO DEM established in 1995 to aid the discovery of concealed world-class ore deposits through knowledge of the Australian landscape evolution. It provides a unique opportunity to bring together industry, government and academia in formulating and addressing the principal challenges facing the exploration and mining industry in its search for further hidden mineral wealth in a weathered land, and thereby maintaining this sector's lead in the export income of Australia.

The Basins Program of CRC LEME focuses on understanding the processes leading to landscape evolution, weathering and dispersion of ore systems existing beneath and within sedimentary basin sequences. As such, it is devoted to the study of water-rock interaction at scales ranging from microscopic to regional, and in settings ranging from surficial to deep within rock formations. Through mapping of the regolith and the landscape (including using remotely sensed techniques), and through sampling and analysing regolith, sediment and formation and ground waters, the Basins Program aims to generate models of erosion, transport and sedimentation, of sediment diagenesis and weathering, and of dispersion of geochemical anomalies that overall are expected to yield improved exploration techniques in areas of shallow basin cover and extensive weathering. Ongoing and future projects focus mainly on and around the Broken Hill/Olary block, the and Mt Isa block.

ACKNOWLEDGEMENTS

We wish to thank CRC LEME colleagues and students for their contributions to the Basins Program. This paper presents part of the work program of the Cooperative Research Centre for Landscape Evolution and Mineral Exploration, and is published with the permission of the Director of the CRC, and with the permission of the Executive Director, AGSO. More information about CRC LEME is posted on the World Wide Web at *http://leme.anu.edu.au*

Geochemistry of the Earth's Surface, Ármannsson (ed.)© 1999 Balkema, Rotterdam, ISBN 90 5809 073 6

Reactivity of biogenic silica from surface ocean to the sediments

S. Dixit & P. Van Cappellen
Georgia Institute of Technology, Atlanta, Ga., USA

ABSTRACT: Flow-through reactor experiments were performed to quantify the solubility and dissolution kinetics of biogenic silica under physico-chemical conditions relevant to marine environments. Results show that processes occurring in the water column and in the sediments lead to significant reductions in silica solubility and reactivity. The dissolution kinetics of fresh diatomaceous plankton cultures are higher than those of biogenic silica found in surficial deep-sea sediments by a factor of about 30. These high dissolution rates explain the efficient recycling of silica observed in the upper ocean. At the sea floor, incorporation of Al into the silica structure reduces the apparent solubility of biogenic silica. The results further indicate that pore waters in sediments with high detrital (lithogenic) contents remain undersaturated with respect to the apparent silica solubility, due to formation of authigenic alumino-silicates. Aluminosilicate precipitation was simulated in the laboratory by supplying Al- and silicic acid-containing solutions to reactors in which biosiliceous sediment was suspended. As expected, the precipitation rate increases with the concentrations of dissolved silicic acid and dissolved Al(III).

1 INTRODUCTION

Biogenic silica produced in the surface ocean undergoes dissolution while settling to the ocean floor. Dissolution continues in the sediments during early diagenesis. However, the physico-chemical conditions in the water column are significantly different from those in the sediments. Furthermore, the reactivity of silica may be different due to aging processes. Therefore, it is important to quantify the changes in solubility and reactivity of biogenic silica before the opal record can be used to estimate paleoproductivity, paleoenvironmental conditions, or to reconstruct the oceanic silica cycle over geologic time.

We have used stirred flow-through reactors to quantify the solubility and dissolution kinetics of biogenic silica under simulated marine conditions. The results from the laboratory studies are used to explain field observations about the dissolution of biogenic silica in the water column and the build-up of pore water silicic acid in deep sea sediments.

2 SAMPLES AND METHODS

Diatom cultures and biosiliceous sediments from the Indian sector of the Southern Ocean were used in this study. Diatom cultures of *Chaetoceros mulleri* were grown in coastal sea water medium. The cultures were concentrated by centrifugation and then washed with a 10% H_2O_2 solution.

A detailed description of the Southern Ocean sediment sample locations and compositions can be found elsewhere (Van Cappellen & Qiu 1997a). Briefly, the biogenic silica content in the sediments varies from 15 to 86 wt.%. The non-siliceous fraction of the sediments is composed of variable amounts of organic matter, $CaCO_3$, and detrital (lithogenic) material.

The dissolution kinetics and solubility of biogenic silica were measured using flow-through reactors (Van Cappellen & Qiu 1997a). By changing the silicic acid concentration of the input solution from undersaturation to supersaturation, both precipitation and dissolution kinetics were measured. Solubility was then estimated, by interpolation, as the concentration of silicic acid where there was no net production or consumption of silicic acid.

Precipitation of alumino-silicates was induced inside the reactor by flowing input solutions containing both silicic acid and dissolved aluminum. Traditional batch experiments were also performed in order to simulate the build-up of pore water silicic acid in sediments having variable detrital to opal contents. A wide range of detrital to opal ratios was tested by mixing variable amounts of biosiliceous ooze with kaolinite or basalt.

3 RESULTS

3.1 *Solubility*

Dissolution of lithogenic detritus deposited at the sea floor releases dissolved Al(III) to the pore waters which can then be incorporated into the outer layers of biogenic silica fragments. This is evident from the positive correlations observed in the Southern Ocean sediments between pore water Al(III), wt.% detrital, and Al/Si ratio of the surface layers of diatom fragments (Al/Si ratios were measured by Van Bennekom, NIOZ, using microprobe technique).

The results of this study show that incorporation of Al into the biogenic silica matrix reduces silica solubility. There is about a 15% decrease in the apparent solubility with a seven fold increase in the Al/Si ratio of the diatom surface layers (Figure 1). This, in part, explains the reported solubility differences between fresh plankton (lower Al/Si) and biogenic silica recovered in sediments (higher Al/Si).

In addition, pressure (water depth) and temperature effects on the solubility should be taken into account when comparing silica solubilities (Figures 1 & 2). The *in situ* silica solubility at 4000 m water depth is about 18% higher than that at atmospheric pressure (Wiley 1974). This estimate assumes that the pressure effect on the solubility of biogenic silica is identical to that of amorphous silica. Typical pore water silicic acid profiles show a downward increase in the concentration until it levels off at a quasi-constant value (the asymptotic concentration) at depths of around 10-30 cm below the water-sediment interface (McManus et al. 1995 and references therein). The traditional interpretation of the asymptotic silicic acid profile is that it corresponds to the equilibrium solubility of biogenic silica. The silica solubilities of Southern Ocean sediments measured experimentally in this study were corrected for *in situ* temperature and pressure, and compared with the asymptotic silicic acid concentrations measured in the pore waters of the sediments (Figure 2). The asymptotic silicic acid concentration decreased with increase in the detrital to opal ratios in the sediments. In sediments containing more than 36 wt.% SiO$_2$ the experimental solubilities agree very well with the asymptotic concentrations, suggesting that the pore waters reached equilibrium with the biogenic silica. This result also implies that the flow-through reactor technique produces meaningful silica solubilities that can be extrapolated to the natural environment. For sediments with detrital to biogenic opal ratios greater than 0.75, the pore waters remained undersaturated with respect to the experimental

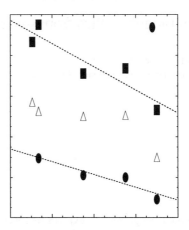

Figure 1. Silica solubility of surficial Southern Ocean sediments, at different temperatures. The solubility decreases with increasing Al/Si ratios measured in the surface layers of diatom fragments of the sediments.

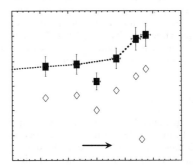

Figure 2. Silica solubilities measured in flow-through reactor experiments are compared to asymptotic silicic acid concentrations measured in Southern Ocean sediments. Pore water undersaturation with respect to the *in situ* silica solubility increases with increasing detrital to opal ratio of the sediment.

solubilities. Also, the difference between silica solubility and asymptotic silicic acid concentration increased with increasing detrital to opal ratios in the sediments (Figure 2). This suggests that additional kinetic factors such as aging and/or authigenic clay formation affect the build-up of pore water silicic acid concentrations (Van Cappellen & Qiu 1997a).

The role of lithogenic material in controlling the asymptotic silicic acid concentration was further studied in batch reactors. Detrital to biogenic opal ratios were varied by mixing variable amounts of biosiliceous oozes with kaolinite or basalt. The steady state concentration of silicic acid in these experiments mimics the relationship found between the asymptotic silicic acid concentration in Southern Ocean sediments and the ratio of detrital content and opal concentration (Figure 3).The differences observed between kaolinite and basalt show that the asymptotic silicic acid concentration is also a function of the type of lithogenic material in the sediments.

The results from this study suggest that the wide range of asymptotic silicic acid concentrations (200 - 900 µM) found in marine sediments can be partly explained by differences of *in situ* silica solubilities and by the interaction between deposited biogenic silica and lithogenic material.

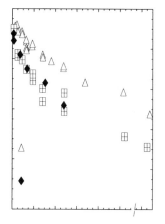

Figure 3. Asymptotic pore water silica concentration and experimental steady state silicic acid concentration plotted against the ratio of %detrital and %opal. The experiments were conducted at 50• C. Note the similar trends between experimental and field data.

3.2 *Kinetics*

At low undersaturation, the dissolution rate of biogenic silica depends linearly on the degree of undersaturation; at higher undersaturation, the dependence switches to an exponential one (Van Cappellen & Qiu 1997b). This is important because the degree of undersaturation in the water column falls in the exponential regime of the rate law, whereas the undersaturation found in most marine sediments falls in the linear part. Therefore, the use of a linear rate law over the entire range of undersaturation, as done in most previous studies, would significantly underpredict the amount of biogenic silica dissolving in the water column. Dissolution rate measurements further show that the reactivity of biogenic silica decreases substantially with depth in the Southern Ocean sediments. The decrease in reactivity is explained by a progressive reduction of the defect density of the silica surfaces (aging) through dissolution and reprecipitation of silica (Van Cappellen & Qiu 1997b). The Al/Si ratio of the diatom surface layers does not affect the dissolution kinetics, other than through the effect on solubility discussed previously.

Oceanographic field measurements have shown that a large fraction (18 - 100%) of biogenic silica produced in the photic zone redissolves within the upper 100 m of the water column (Nelson et al. 1995 and references therein). In order to explain this efficient recycling of silica, the dissolution rates measured on recently deposited biosiliceous oozes were extrapolated to surface water conditions of the Southern Ocean. The experimental kinetics, however, underpredicted the observed extent of dissolution in the surface ocean by factors of up to 10.

This was true, even after correcting for differences in temperature, pressure, solubility (Al content), and specific surface area. Therefore, we investigated the dissolution of fresh plankton cultures in the laboratory. The dissolution kinetics of fresh biogenic silica followed the same non-linear rate law

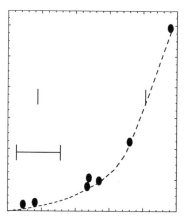

Figure 4. Alumino-silicate precipitation kinetics onto biosiliceous oozes are a function of the dissolved silicic acid and Al concentrations. Rates are expressed as mass dissolved Al(III) precipitated per unit mass of biogenic silica, per unit time. Bars indicate ranges of $[H_4SiO_4] \times [Al]$ generally found in pore waters of coastal and deep-sea environments.

observed for the sediments, but the dissolution rates were about 30 times faster (rates normalized to the mass of biogenic silica). The much higher reactivity of fresh biogenic silica therefore explain the rapid recycling of silica observed in the surface waters. The results further indicate that the reactivity of biogenic silica continuously decreases, from the time the siliceous organism dies until the ultimate preservation of a silica fragment in the sedimentary column.

Precipitation of authigenic alumino-silicate minerals was induced inside the reactor by flowing solutions containing dissolved silicic acid and aluminum. The precipitation rate increased with increases in either the dissolved silicic acid or aluminum concentration (Figure 4). This suggests that the silicic acid build-up in the batch experiments (Figure 3) may be limited by the formation of alumino-silicate phases. The relatively high rates of precipitation imply that the precipitation of authigenic alumino-silicates can play a major role in the silica cycle. Efforts are ongoing to characterize the precipitates.

4 CONCLUSIONS

In this study some of the key parameters and processes controlling the solubility and dissolution kinetics of biogenic silica have been identified and quantified. The following conclusions can be drawn:

1) The detrital fraction of sediments acts as a source of soluble Al(III) which can be incorporated into the surface lattice of biosiliceous fragments causing a decrease of the silica solubility.

2) Below a threshold availability of soluble Al(III), pore waters reach equilibrium with the *in situ* solubility of biogenic silica. At high detrital to biogenic silica ratios, the pore waters remain undersaturated with respect to the *in situ* solubility of biogenic silica. Additional kinetic factors, including authigenic clay formation, control the build-up of pore water silicic acid.

3) The dissolution kinetics of biogenic silica follow a distinctly non-linear rate law. At low undersaturation, the dissolution rate follows a linear rate law but it switches to an exponential one at higher undersaturation.

4) The reactivity of fresh biogenic silica is much higher than that of biogenic silica found in sediments. This high reactivity explains why a large fraction of marine biogenic silica production is recycled within the upper ocean.

5) The precipitation rate of authigenic alumino-silicates is a function of the dissolved Al(III) and silicic acid concentrations. The relatively fast kinetics observed in this study suggests that authigenic clay formation can not be ignored in the overall silica budget.

REFERENCES

McManus, J., D.E. Hammond, W.M. Berelson, T.E. Kilgore, D.J. Demaster, O.G. Ragueneau & R.W. Collier 1995. Early diagenesis of biogenic opal: Dissolution rates, kinetics, and paleoceanographic implication. *Deep-Sea Research II* 42: 871-903.

Nelson, D.M., P. Tréguer, M.A. Brzezinski, A. Leynaert & B. Quéguiner 1995. Production and dissolution of biogenic silica in the ocean: Revised global estimates, comparison with regional data and relationship to biogenic sedimentation. *Global Biogeochemical Cycles* 9: 359-372.

Van Bennekom, A. J., A.G.J. Buma & R.F. Nolting 1991. Dissolved aluminum in the Weddell-Scotia confluence and effect of Al on the dissolution kinetics of biogenic silica. *Marine Chemistry* 35: 423-434.

Van Cappellen, P & L. Qiu 1997a. Biogenic silica dissolution in sediments of the Southern Ocean: I. Solubility. *Deep-Sea Research II* 44: 1109-1128.

Van Cappellen, P & L. Qiu 1997b. Biogenic silica dissolution in sediments of the Southern Ocean: II. Kinetics. *Deep-Sea Research II* 44: 1129-1149.

Wiley, J.D. 1974. The effect of pressure on the solubility of amorphous silica in seawater at 0• C. *Marine Chemistry* 2: 239-250.

Geochemistry of the Earth's Surface, Ármannsson (ed.) © 1999 Balkema, Rotterdam, ISBN 90 5809 073 6

Evolution of Mn contents of Neogene carbonates along the Bahamas transect (ODP leg 166). Relationship of geochemical data to sea-level changes

L. Emmanuel & B. Vincent
Centre des Sciences de la Terre, U.M.R. C.N.R.S. 5561, Biogéosciences, Université de Bourgogne, France

M. Renard
Département de Géologie Sédimentaire, E.S.A. C.N.R.S. 7073, Université P. et M.Curie, France

E. H. De Carlo
Department of Oceanography, SOEST, University of Hawaii, USA

ABSTRACT : The evolution of trace element contents (e.g. Mn) in Oligocene to Upper Pliocene carbonate sediments at ODP Sites 1005, 1003, 1007 (from proximal to distal with respect to platform margin) was used to study the geochemical response to sea level changes along the western flank of the Great Bahamas Bank.The geochemical data reflect the patterns of sedimentation that result from repetitive fluctuations in relative sea level. Based on interpretation of Mn and sequence-stratigraphic interpretations, we identify geochemical cycles that correspond to second-order changes of sea level. Each cycle contains numerous distinct subordinate geochemical sequences (labeled G1 to G19) that have been interpreted as third-order changes of sea level. Correlation of seismic sequences with geochemical sequences has shown that most seismic reflectors are sequence boundaries. Some of the seismic reflectors, however, correspond to maximum flooding events. Geochemical cycles show a good agreement with the Cenozoic chronostratigraphic framework of European basins, except for the major maximum flooding that occurs in the early Miocene.

1 INTRODUCTION

One of the primary objectives of the Ocean Drilling Program (ODP) Leg 166 Bahamas Transect was to recognize the effects of eustatic sea-level fluctuations on the margin of a carbonate platform (Eberli et al. 1997). To achieve a thorough understanding of the sedimentary response to sea-level changes, a transect from shallow-water to deep-water settings is necessary. Four sites (1005, 1003, 1007 and 1006, from proximal to distal) were drilled on the western flank of the Great Bahamas Bank (GBB) through prograding carbonate sequences, ranging in age from late Oligocene to Holocene (Figure 1).

Fluctuations (e.g. of Mn) may have been controlled by variations in submarine hydrothermal activity, that modifies chemical element production or consumption in mid-oceanic ridge environments. Sea level fluctuations also influence the chemical element input/output to the ocean via the erosion/sedimentation balance. These variations are considered to be global, and bulk carbonate geochemistry data could be potentially useful in both stratigraphic and paleoceanographic reconstructions. In conclusion, chemical stratigraphy can be used to characterize variations in the amounts of numerous elements in marine carbonates with time.

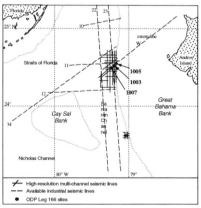

Figure 1. Sampling sites

The aims of this study are (1) to show a geochemical subdivision within Miocene to Pliocene sedimentary deposits from four drilling sites (1005, 1003, 1007 and 1006) using Mn contents, and (2) to compare the deposits' geochemical subdivisions with sequential interpretation from sedimentological and seismic data. Reference is made to similar comparisons at various scales in ancient basin deposits (Emmanuel 1993, Emmanuel & Renard 1993, Corbin 1994) and shallow-water environments (Vincent et al. 1997).

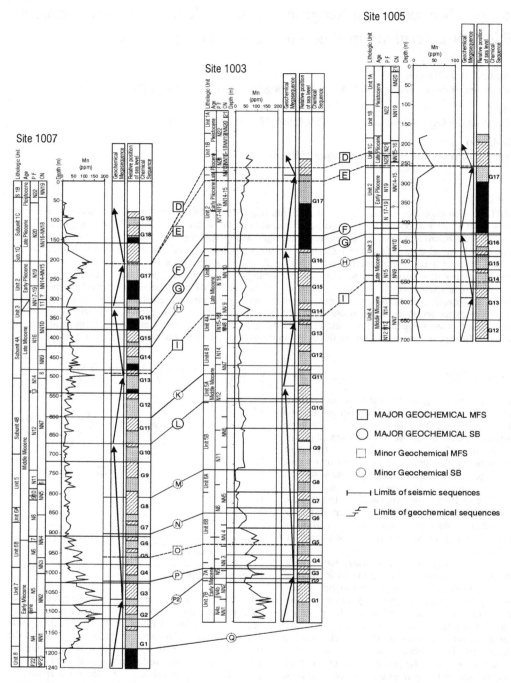

Figure 2. Depth profiles of Mn content in carbonate fraction of cores from sites 1003, 1005 and 1007.

Six hundred and ninety samples for geochemical analysis have been collected from three sites in Miocene and Pliocene deposits. The samples come from differents holes within the sites :
- Site 1005—> 50 samples
- Site 1003—> 135 samples
- Site 1007—> 305 samples

Geochemical measurements have been done on the carbonate fraction of the bulk samples using an atomic absorption spectrometer (AAS) Perkin Elmer 3300. Chemical analyses have been performed on raw powders. After crushing, samples were dissolved in 1N acetic acid, and the solutions were used for AAS measurements.

2 RESULTS

Depth profiles for Mn content, in the carbonate fraction of the sediments, at sites 1003, 1005, and 1007 are shown in Figure 2. In the studied interval of the different sites, the percentages of aragonite and dolomite are low and do not interfere with Mn amplitude variations. The last part is characterized by relatively low Mn content (< 40ppm) with low recorded signals. Site 1003 has been chosen as reference site for this geochemical study because it shows the most complete sedimentary succession, ranging in age from late Oligocene to Holocene. Therefore sampling intervals are shorter in order to improve the geochemical signal for the Neogene.

2.1 Site 1005

The Mn content shows small amplitude variations between 7 and 51 ppm. The Mn content decreases from 19 ppm (695 mbsf) to 10 ppm (650 mbsf). The values increase between 639.04 mbsf and 565.96 mbsf, including two shifts, with 18 ppm and 19 ppm Mn content, respectively. Between 565 mbsf and 300 mbsf, the Mn content is constant and low (between 7 and 10 ppm). Above, the Mn content increases sharply to reach a maximum value of 51 ppm at 260 mbsf. This maximum corresponds to the the early late Pliocene boundary. Above this limit, the Mn content decreases to 15 ppm.

2.2 Site 1003

The evolution of the Mn curve is characterized by three main parts. The lower part (between samples 166-1003C-87R-7 and 166-1003C-60R-2) shows a decreasing trend. The central part (between samples 166-1003C-59R-1 and 166-1003C-12R-1) shows a gradual but slight increase from 8 to 58 ppm between samples 166-1003C-10R-1 and 166-1003A-14X-1.

2.3 Site 1007

The evolution of Mn at Site 1007 is very similar to that at Site 1003, with more well-marked variations. The same trends are observed but show a superimposed succession of high frequency fluctuations. Mn contents are relatively low (between 3 and 216 ppm) along the whole record but higher than at Site 1003.

3 DISCUSSION

3.1 Interpretation of Mn variations

The variations in Mn content are related to sea level fluctuations, i.e. relative paleodepth. Consequently, positive shifts in Mn along the curves characterize successive maximum depths. The record of sea level rise is almost instantaneous on the geological time scale because Mn residence time in sea water is evaluated to be around 1300 years (G. Blanc, pers. comm. 1998). In addition the detrital input due to continental erosion could be an important source of Mn. Nevertheless, this input into the GBB environment is probably very low if not negligible.

3.2 Geochemical correlations

3.2.1 Interpretation of sea level changes from Mn contents

Manganese is the most interesting element for sequential analysis beacause of its relationship with the paleodepth variations. The determination of geochemical sequences is based on the same method as the one used for basin sedimentary deposits (Emmanuel & Renard 1993, Corbin 1994).

3.2.2 Characterization of low frequency geochemical cycles

Before proposing a sequence interpretation, major high amplitude trends are identified on the geochemical curves. These are linked to low frequency events. The time lag between two major Mn content shifts are between ca. 2 and 6-7 My (calculated from sedimentation rates), indicating second order cycles (Vail et al. 1991). The Mn content curves for the three sites show three (1007 and 1003) and two (1005) periods of high contents separated by low, relatively constant gaps.

3.2.3 Characterization of high frequency geochemical sequences

These high frequency geochemical sequences are described using relative variations of Mn contents inside major trends (low frequency cycles). The results are given in Figure 2. Nineteen geochemical sequences (labelled G1 to G19) have been identified. All are present at Site 1007. Four are missing at Site 1003. For instance, sequence G2 is not present because of an important hiatus at this level. Sequences G18 and G19 have not been identified in the 1003 Site because samples were missing. Only six sequences (G12 to G17) have been observed at Site 1005 due to the depth of drilling. The succession of the various units is also used for third order correlations between sites and to propose a sequence stratigraphy interpretation of seismic data.

3.2.4 *Correlation between geochemical sequences and seismic sections*

For the three sites (1005, 1003 and 1007), fourteen Seismic Sequence Boundaries (SSB) have been identified (labelled D to Q, Eberli et al. 1997). All the SSB can be correlated with notable geochemical variations within the resolution of the seismic reflection. In addition, many of these variations can be used to indicate either a maximum of regression or a maximum of transgression i.e., SB or MFS of third order sequences. Therefore, the SSB F, G, H, K, L, M, N, P, P2 and possibly Q correspond to geochemical SB (of sequences G17, G16, G15, G12, G11, G9, G7, G4, G2 and G1, respectively) as defined above. The SSB D, E, I and O correspond rather to geochemical MFS (of sequences G17 -for SSB D and E-, G13 and G5, respectively). In addition, it is possible to rank the SSB sequentially using geochemical interpretation. SSB L and F or G are interpreted as a major SB of second order of sea level change. SSB I, D and E are interpreted as major second order MFS . The major MFS identified in the early Miocene (N5 zone, sites 1003 and 1007) does not correspond to a SSB.

A comparison with Cenozoic chronostratigraphic framework of the European basins has been made. The second order megasequences, characterized by geochemical interpretation, coincide with the T/R facies cycles of the Cenozoic chronostratigraphic framework (Hardenbol et al. 1999), except for the early Miocene maximum transgression.

4 CONCLUSION

Variations of the geochemistry of the ocean are well recorded in pelagic sediments. Variations in Mn contents are found to be particularly useful to correlate sea level fluctuations. In this study, a sequential geochemical interpretation of Mn variations were made at three sites (1005, 1003 and 1007) in the Neogene carbonates of the Bahamas. The high Mn contents seem to be related to major hydrothermal events associated either with a strike-slip fault zone (active during the Miocene in SW Cuba) or to the mid-Atlantic ridge. The maximum Mn content is considered to correspond to maximum depth, i.e., the geochemical MFS. The lowest values of Mn content are related to lowstand units. Three major geochemical megasequences have been identified during the early Miocene - late Pliocene interval. The first megasequence has a major MFS located in the early Miocene (N5 zone), the second at the middle Miocene-late Miocene boundary (N14 zone), and the third at the early Piocene - late Pliocene boundary (N19 zone). The maximum regression between megasequences 1 and 2 occurred during middle Miocene (N12 zone). The maximum regression between megasequences 2 and 3 can be located during the early Pliocene (N17 zone).

Sequential divisions of the sedimentary successions with third order units allow the geochemical signal to be compared with seismic sequence boundaries. A correlation between boundaries of geochemical and seismic sequences has been proposed and shows a good agreement with the Cenozoic chronostratigraphic framework of European basins.

REFERENCES

Corbin, J.C. 1994. Evolution géochimique du Sud-Est de la France : influence des variations du niveau marin et de la tectonique. *Mém. Sci. Terre* 94-12. , Paris : UPMC.

Eberli, G.P., P.K. Swart, M.J. Malone et al. 1997. *Proc. ODP, Init. Repts.* 166: College Station, TX (Ocean Drilling Program).

Emmanuel, L. 1993. Apport de la géochimie des carbonates à la stratigraphie séquentielle. Application au Crétacé inférieur du domaine vocontien. *Mém. Sci. Terre* 93-5. Paris : UPMC.

Emmanuel, L. & M. Renard 1993. Carbonate geochemistry (Mn,δ^{13}C, δ^{18}O) of the late Tithonian-Berriasian pelagic limestones of the Vocontian though (SE France). *Bull. Centres Rech. Explor.-Prod. Elf Aquitaine* 17(1): 205-221.

Hardenbol, J., J.Thierry, M.B. Farley, T Jacquin, P.C. De Graciansky & P.R Vail. 1999. Cenozoic chronostratigraphy. In P.C De Graciansky., J Hardenbol., T.Jacquin, P.R. Vail. & M.B Farley. (eds.), *The Mesozoic-Cenozoic chronstratigraphic framework, Sequence Stratigraphy of European Basins.* Spec. Publ. Soc. Econ. Paleontol. Mineral (in press).

Vincent, B., L.Emmanuel, J.P. Loreau & J. Thierry 1997. Caractérisation et interprétation de cycles géochimiques sur la plate-forme bourguignonne (France) au Bajocien-Bathonien. *C. R. Acad. Sci. Paris* 325 : 783-789.

Geochemistry of the Earth's Surface, Ármannsson (ed.) © 1999 Balkema, Rotterdam, ISBN 90 5809 073 6

A stochastic approach to bioirrigation modeling in sedimentary environments

C. M. Koretsky, P. Van Cappellen & C. Meile
Georgia Institute of Technology, Atlanta, Ga., USA

ABSTRACT: The shapes, sizes and distribution of infaunal burrows exert a strong control on solute transport in sedimentary environments due to irrigation of burrows with overlying water. Because bioirrigation is spatially and temporally heterogeneous in shallow-water, organic-rich sediments, it may be difficult to represent using strictly deterministic models. In order to account explicitly for the spatial and temporal variability of burrow distributions encountered in shallow-water sediments, a 3-dimensional stochastic model has been developed to relate the density and morphology of burrows to bioirrigation coefficients from a nonlocal transport model. The basic assumption is that the number of burrows intersected at any depth is a proxy for the surface area of exchange between burrows and the porous sediment matrix. The results of this model, while qualitatively in agreement with those from a nonlocal transport model, suggest that the concentrations of reactive solutes inside burrows are depth-dependent.

1 INTRODUCTION

The activities of burrowing macrofauna may exert a strong control on sediment and porewater biogeochemistry. This is because the organisms periodically flush their burrows with overlying water, for example to rid their habitats of toxic metabolic products, a process termed bioirrigation (e.g., Boudreau 1996). In so doing, overlying water may come directly into contact with porewaters at depth within the sediment and diffusion of solutes may occur laterally, as well as vertically, into the porous sediment maxtrix.

Bioirrigation has been modeled previously using deterministic 1-, 2-, and 3-dimensional transport models. For example, Aller (1980) developed a "tube-model" which depicts burrows as evenly spaced vertical cylinders of fixed radii and lengths. It has been shown that this model is equivalent to a 1-dimensional nonlocal transport model with an irrigation coefficient (α), representing the mass transfer intensity, through

$$\frac{dC}{dt} = D\frac{\partial^2 C}{\partial x^2} - \alpha(x)(C - C_b) + R = 0 \qquad (1)$$

where t is time, x is depth with respect to the sediment-water interface, C is the horizontally averaged concentration of a given solute at depth x, C_b is the concentration of the solute inside the burrow, D is the sediment diffusion coefficient, and R is the net rate of production of the solute as a function of depth (Boudreau 1996). Typically, C_b has been assumed to be depth-independent and equal to C_o, the concentration of the solute at the sediment-water interface. In this model, α is proportional to the amount of surface area through which solute exchange occurs.

Although Equation 1 may be very useful for extracting irrigation coefficients as a function of depth for specific environments, information regarding burrow distributions is typically not taken into account in deriving irrigation profiles using this approach (e.g., Matisoff and Wang 1998). In this study, a stochastic model is used to describe the dependence of $\alpha(x)$ on burrow densities and morphologies, assuming that the number of burrows encountered at a given depth in the sediment serves as a proxy for irrigation intensity. This allows information regarding the seasonal or spatial variation of burrow densities, shapes and sizes in a given environment to be used to assess the effects of temporal and spatial variation of bioirrigation.

2 MODEL

In the stochastic model presented in this paper, the sediment column is represented as a 3-dimensional grid of points over which burrows are distributed. Burrows are represented using ten end-member shapes (Figure 1). The mean total burrow density, the probability of occurrence for each burrow shape and the mean and standard deviations of the lengths of the segments of these burrow shapes are specified in an input file.

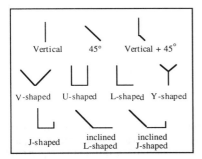

Figure 1. Burrow shapes used in this study.

Using the specified criteria, burrows are distributed over the grid by employing the following procedure. First, the user-specified mean total burrow density (BD_m) is combined with a Rayleigh cumulative distribution function (cdf),

$$BD = \sqrt{-2 * BD_m^2 * \log(1 - R_{num})} \qquad (2)$$

where R_{num} is a random number between 0 and 1, to calculate the total burrow density (BD). A Rayleigh cdf was used because it cannot result in negative burrow densities. Next, the total burrow density is distributed over the possible burrow shapes by combining randomly generated numbers with the probability of occurrence for each shape. For each burrow, a Gaussian distribution is used with the given segment length means and standard deviations to calculate the size of each burrow segment. For horizontal or angled segments, the direction of burrowing is assigned to north, south, east or west, with equal probability for each.

Next, each burrow is assigned coordinates in the grid by choosing two random numbers to locate the burrow exit at the grid surface. If another burrow has already been assigned to either the surface coordinates, or any of the coordinates at depth (that is, if the burrow "collides" with another), then a new set of surface coordinates is chosen to locate the burrow. In this way, the burrows are distributed so that they never intersect one another. If a given burrow cannot be placed in the grid after a large number of attempts (for example, 500), then the lengths and directions of the segments of that burrow are determined again. To eliminate edge effects, burrows that exit the grid are

continued at the opposite edge (i.e., a periodic boundary condition is imposed).

3 RESULTS

The stochastic approach described above has been used in the present study to model bioirrigation at three sites in a saltmarsh at Sapelo Island, GA. These three sites include an unvegetated, heavily bioturbated creek bank, a levee site vegetated by tall-form *Spartina alterniflora*, and a ponded marsh site vegetated by short-form *S. alterniflora*. The macrofauna responsible for the burrowing activity at these sites are polychaete worms, mud crabs and fiddler crabs (Basan and Frey 1977). Fiddler crabs are far more abundant than mud crabs at all three sites, and create much larger and deeper burrows than do polychaete worms. Thus, in this study it is assumed that the majority of solute exchange by bioirrigation occurs via fiddler crab burrows. Fiddler crab populations and approximate shapes and lengths of burrows that have been measured for these types of sites (Teal 1958; Basan and Frey 1977) were used in this study to estimate mean burrow densities, shapes and lengths at the three Sapelo Island saltmarsh sites (Table 1).

Table 1. Probabilities of occurrence and means and standard deviations of burrow lengths (cm) used to calculate profiles shown in Figure 2. Data on burrow shapes and sizes from Basan and Frey (1977). Mean burrow densities of 27, 61 and 65 (per m[2]) for the ponded marsh (all *Uca pugnax*), levee (all *U. pugnax*) and creek bank sites (*U. pugnax* and *U. pugilator*), respectively, were used in the model calculations.

Burrow Type		Ponded Marsh	Levee	Creek Bank
45°	Probability	.25	.25	.13
	Depth ±Std	12 ± 2	20±2	20±3
Straight	Probability	.25	.25	.13
	Depth ± Std	12 ± 2	20±2	20±3
U-shaped	Probability	.50	.50	.10
	Depth ± Std	8 ± 2	12±1	12±1
	Length ± Std	5 ± 1	5±1	5±1
L-shaped	Probability	-	-	.16
	Depth ± Std	-	-	15±1
	Length ± Std	-	-	5±1
Inclined L-shaped	Probability	-	-	.16
	Depth ± Std	-	-	15±1
	Length ± Std	-	-	5±1
J-shaped	Probability	-	-	.16
	Depth ± Std	-	-	15±1
	Length ± Std	-	-	5±1
	Hook ± Std	-	-	3±1
Inclined J-shaped	Probability	-	-	.16
	Depth ± Std	-	-	15±1
	Length ± Std	-	-	5±1
	Hook ± Std	-	-	3±1

The data in Table 1 were used with the stochastic model described above to calculate the number of burrows intersected at any given depth at the three sites. These results are depicted in Figure 2, which also shows ± one standard deviation in the number of burrows encountered with depth for 10000 iterations at the creek bank. Although the standard deviation is fairly large, on the order of 50% of the number of burrows encountered at a given depth, this mostly reflects variations in the total burrow density. For any given simulation, the shapes of the curves shown in Figure 2, reflecting the shapes of the corresponding bioirrigation coefficient profiles, remain essentially unchanged.

It can also be seen in Figure 2 that the concentration of burrows at any given depth, and therefore the magnitude of the irrigation coefficient (α), is expected to be least at the ponded marsh site, and that the depth of irrigation is predicted to be considerably greater at the creek bank and levee sites than at the ponded marsh site.

An inverse modeling approach has recently been used to quantify irrigation coefficients and irrigation intensity as a function of season at the three Sapelo Island saltmarsh sites (Koretsky et al. 1998; Meile et al. 1999). Using Equation 1, with C_b for all depths set equal to the concentration of sulfate at the sediment-water interface (C_o), and using measured sulfate concentration and reduction rate profiles with sulfate diffusion coefficients corrected for temperature, salinity and tortuosity, Meile et al. (1999) applied a simplex optimization algorithm to extract irrigation coefficients as a function of depth. Resulting irrigation coefficient profiles from August, 1997 for the ponded marsh and creek bank sites are depicted in Figure 3.

In general, from Figures 2 and 3, it can be seen that both the stochastic and the inverse models indicate that $\alpha(x)$ is greatest at the creek bank site and least at the ponded marsh site. In addition, a comparison of Figures 2 and 3 reveals similar trends in the depth of irrigation. The nonlocal transport model results shown in Figure 3 indicate that irrigation in the ponded marsh sediment persists to 26 cm depth and irrigation in the creek bank occurs at depths as great as 40 cm. This is consistent with the maximum depths of burrow penetration (approximately 23 cm at the ponded marsh versus 37 cm at the creek bank) predicted by the stochastic model.

Qualitatively, the shapes of the curves in Figures 2 and 3 are similar; both show a rapid decrease in the irrigation coefficient with depth, as has been observed in previous studies (see references in Boudreau 1996). However, it is clear that the irrigation coefficients regressed using Equation 1 decrease much more rapidly as a function of depth than do the curves shown in Figure 2. This discrepancy may be due to some of the assumptions regarding burrow

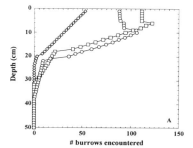

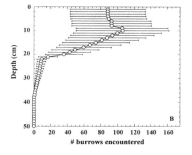

Figure 2. Average number of burrows encountered as a function of depth; calculated using the average of 10000 iterations with the data given in Table 1. (A) Open circles correspond to creek bank data, open squares indicate levee and open diamonds indicate ponded marsh. (B) Error bars represent one standard deviation for 10000 simulations of the creek bank data.

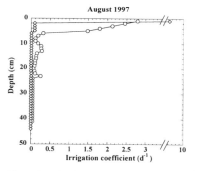

Figure 3. Irrigation coefficients as a function of depth from the inverse model described by Meile et al. (1999). Open circles indicate creek bank; open diamonds indicate ponded marsh. Parallel lines indicate break in the x-axis scale.

shapes used in the stochastic model. For example, burrows in the stochastic model were assumed to have a constant diameter. However, in reality, they often exhibit widening near the sediment-water interface (e.g., Basan and Frey 1977).

A more likely source of the discrepancy in the shapes of the curves in Figures 2 and 3 comes from

321

the assumption used in the inverse model that C_b is constant and equal to C_0. Especially in the summer months, sulfate concentrations decrease rapidly as a function of depth, and sulfate reduction rates are extremely high (up to 3.5 mM·d^{-1}). Thus, the concentration of sulfate within the burrows may not be constant; more likely, the sulfate concentration in the burrows decreases with depth. Therefore, the bioirrigation coefficients calculated using the inverse model with a constant C_0 will be systematically underestimated, with the error increasing as a function of depth. Thus, we propose, as a working hypothesis, that C_0 does not accurately reflect the concentration of water in the burrows, particularly during the summer months.

4 CONCLUSIONS

A stochastic model of burrow distributions has been developed to function as a link between ecological data and bioirrigation mass transfer coefficients. Using burrow shapes, sizes, densities and probabilities of occurrence, it is possible to create profiles of the number of burrows encountered as a function of depth under different ecological scenarios. These profiles can be used as a proxy for the available surface area for solute exchange as a function of depth.

As an example, the stochastic model was used with literature data regarding fiddler crab populations at three sites along a transect of a saltmarsh on Sapelo Island, GA to calculate burrow density profiles as a function of depth. The general trends with respect to intensity and depth of irrigation predicted by this model are in agreement with the results of inverse model calculations of irrigation coefficients at these same sites.

However, bioirrigation coefficients calculated using the inverse nonlocal transport model, with the assumption that C_b is constant and equal to C_0, decrease much more rapidly as a function of depth than would be expected from the stochastic model results. This may indicate that the use of a constant C_b, equal to the concentration at the sediment-water interface, is inappropriate in this shallow-water, organic-rich setting because the concentration of sulfate in the burrows most likely decreases as a function of depth. This results in an underestimation of the bioirrigation coefficients, with an increasing error with depth.

Porewater chemistry may be influenced profoundly by bioirrigation, particularly in sedimentary environments with large populations of burrowing macrofauna. The model presented in this study has the potential to exploit ecological data regarding changes in macrofauna populations due to seasonal or other environmental factors in order to elucidate the resultant effects on bioirrigation intensity. In addition, it highlights the limitations of assuming a constant flushing concentration, C_0, as is often done in the application of nonlocal transport models of bioirrigation.

5 REFERENCES

Aller R.C. 1980. Quantifying solute distribution in the bioturbated zone of marine sediments by defining an average microenvironment. *Geochimica et Cosmochimica Acta* 44: 1955-1965.

Basan P.B. & Frey R.W. 1977. Actual paleontology and neoichnology of salt marshes near Sapelo Island, Georgia. In Crimes T.P and Harper J.C. (eds.), *Trace Fossils 2*: 41-70. London: Steel House Press.

Boudreau B. 1996. *Diagenetic Models and Their Implementation*. New York: Springer-Verlag.

Koretsky C.M., Van Cappellen P. and Kostka J. 1998. Seasonal dependence of porewater irrigation in a saltmarsh at Sapelo Island, GA: Implications for sediment biogeochemistry. *Geological Society of America, Abstracts with Programs*.

Matisoff G. & Wang X. 1998. Solute transport in sediments irrigated by freshwater infaunal bioirrigators. *Limnology and Oceanography* 43: 1487-1499.

Meile C., Koretsky C., Van Cappellen P. and Kostka J. 1999. Spatial and temporal variation in bioirrigation: An inverse approach. *Fifth International Symposium on the Geochemistry of the Earth's Surface*, August 12-20, Reykjavik, Iceland (this volume).

Teal J.M. 1958. Distribution of fiddler crabs in Georgia salt marshes. *Ecology* 39: 185-193.

Geochemistry of the Earth's Surface, Ármannsson (ed.) © 1999 Balkema, Rotterdam, ISBN 90 5809 073 6

Spatial and temporal variations in bioirrigation: An inverse approach

C. Meile, C. Koretsky & P. Van Cappellen
Georgia Institute of Technology, Atlanta, Ga., USA

J. Kostka
Skidaway Institute of Oceanography, Savannah, Ga., USA

ABSTRACT: An inverse model has been developed to quantify bioirrigation in the surface sediments of a salt marsh located on Sapelo Island, GA. The results show that irrigation is higher along the unvegetated creek bank than in the interior of the marsh. They also reveal intense and deep irrigation activity during the winter and late summer months. The seasonal and spatial variations are in qualitative agreement with reported behavior of the macrofauna at the study site.

1 INTRODUCTION

Macrofaunal feeding and burrowing activities redistribute chemical constituents and may have pronounced effects on pore water composition, solid sediment geochemistry and microbial activity in aquatic sediments. An appropriate description of these processes is critical to the understanding of the spatial and temporal patterns of early diagenesis in surface sediments with active macrofauna. The intensity of bioirrigation is typically represented by a nonlocal exchange function (Emerson et al. 1984). To assess irrigation intensity with depth, an inverse model has been developed and applied to measured profiles of sulfate concentration and sulfate reduction rate, at two sites, and for three different times during the year, in a salt marsh located on Sapelo Island, GA.

2 MODEL

An iterative inverse approach is presented, which is used to solve for the irrigation coefficient (α) with depth by minimizing the difference between measured and calculated sulfate concentrations. Conservation of mass is described in one dimension and at steady state, with terms accounting for diffusion, reaction and irrigation:

$$\frac{\partial C}{\partial t} = \frac{\partial}{\partial z}\left(D\frac{\partial C}{\partial z}\right) - R + \alpha(C_0 - C) = 0 \qquad (1)$$

where C is concentration of sulfate (C_0 is the concentration at the sediment-water interface), t is time, z is depth below the sediment-water interface; α is the depth-dependent irrigation coefficient, R is the sulfate reduction rate and D is the molecular diffusion coefficient corrected for tortuosity (Ullman & Aller 1982), temperature and salinity.

Directly solving for $\alpha(z)$ using Equation 1 requires much less computational effort than an inverse approach but has the following disadvantages:
- direct forward solutions tend to oscillate,
- direct forward solutions are unconstrained by basic geochemical principles, and
- uncertainties in measured quantities are not taken into account.

To solve Equation 1 numerically, the depth range of interest (0 to approx. 40 cm) is divided into a finite number of boxes of 1 cm thickness, which corresponds to the approximate spatial resolution of the measurements. Equation 1 is then recast into a matrix equation using a central difference scheme:

$$\underline{A} * \bar{C} = \bar{d} \qquad (2)$$

where the vector $\bar{C}$ contains the concentrations of each depth segment, $\bar{d}$ includes the sulfate reduction rates and the irrigation coefficients, and the elements of matrix $\underline{A}$ depend on the diffusion and irrigation coefficients.

The procedure for calculating the irrigation coefficient α as a function of depth is as follows:

1 An initial guess of the irrigation coefficient profile is made that is consistent with specified constraints (e.g. $C \geq 0$, $\alpha \geq 0$). To make this initial guess, the measured concentrations are assigned an uncertainty and the maximum and minimum contribution of diffusion to the measured values is computed with Equation 1. From this calculation and the measured sulfate reduction rates, the

initial guess for the irrigation coefficient profile may be narrowed down considerably.

2 Based on this irrigation coefficient profile, a concentration profile is calculated using Equation 2.
3 The deviation between the calculated and measured concentrations in each box is computed. If the deviation exceeds the specified uncertainty, the irrigation coefficient profile is modified using the simplex algorithm, a standard minimization technique (Press et al. 1989), and the procedure returns to step 2.
4 When the difference between measured and calculated concentration in each box is smaller than the specified limit, the two boxes with the most similar α are combined and the procedure begins again at step 2 with the new depth discretization (one box less).

Step 4 gradually reduces the number of adjustable parameters. Generally, a smaller deviation between measured and calculated concentrations is achieved if more free parameters are used. However, application of a statistical F-test reveals the significance (or lack thereof) of the improved fit of a model with more free parameters, and provides an objective method for choosing the irrigation coefficient profile that best describes the measurements with the least number of fitted parameters.

Once the final irrigation coefficient profile has been obtained, we can vary the value of α, at any given depth, while keeping the rest of the depth profile unchanged. In this manner, it is possible within each depth interval to constrain the range of α that is consistent with the convergence criteria (allowable range).

To address the uniqueness of a given solution, a set of Monte Carlo simulations are carried out, where the initial guess is randomly chosen from the possible solution range (step 1 above). Furthermore, the simplex algorithm is modified to prevent the solution from being trapped in a local minimum that does not meet the specified convergence criteria.

The work presented here differs from that of Berg et al. (1998) in two important aspects:

- Uncertainties of the measurements are incorporated into the model calculations during the selection of the initial guess, and when deciding whether the deviation between measurement and calculation is significant or not. This avoids overinterpretation of the data.
- Constraints on the irrigation coefficient and the calculated concentrations can be imposed. This allows us to incorporate geochemical knowledge and avoids physically meaningless results (e.g., negative concentrations or irrigation coefficients), even if they give a very good fit to the data.

3 DATA AND FIELD SITE

Sapelo Island is a barrier island approximately 5 miles off the Georgia mainland. Its highly productive salt marshes are dominated by *Spartina alterniflora* grass and are densely inhabited by fiddler crabs, causing high irrigation intensities. Data from two sites on a transect from the main island towards a tidal creek are used in this study. The ponded marsh site is located in the interior of the marsh, vegetated by a shorter form of *S. alterniflora*; the creek bank site is located along a tidal creek and is unvegetated.

Concentration profiles were measured with diffusion equilibrators ("peepers") and represent an average over horizontal and temporal scales of several centimeters and weeks, respectively. Sulfate was analyzed by a turbimetric method as described in Tabatabai (1974). Sulfate reduction rates were determined in sediment slurries by adding radiolabeled sulfate (Jørgensen 1978).

4 RESULTS

A typical irrigation coefficient profile obtained by the inverse model is presented in Figure 1. The number of boxes giving the best representation of the measurement was iteratively reduced from an initial value of 42 to only 4 boxes.

As can be seen, the irrigation coefficient, which reflects burrow density and flushing, decreases rapidly with depth. Comparison of irrigation coefficients derived in this study with those from the literature (e.g., Emerson et al. 1984, Wang & Van Cappellen 1996, Martin & Banta 1992, Schlüter et al. in review) shows that bioirrigation coefficients in the salt

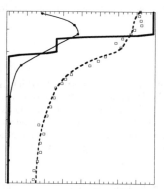

Figure 1. Ponded marsh site, January 1998. The solid thick line represents the calculated irrigation coefficient profile, filled circles represent measured sulfate reduction rates (SSR), open squares correspond to measured SO_4^{2-} concentrations. The dashed line is the SO_4^{2-} profile corresponding to the final irrigation coefficient profile.

marsh tend to be higher than in other estuarine and coastal marine sediments.

4.1 Variation of irrigation with site and season

To analyze the changing influence of macrofaunal activity on pore water exchanges, with respect to site and season, α is integrated over the entire depth range (Figure 2). This integrated quantity can be interpreted as the irrigation transfer velocity across the sediment-water interface. However, care must be taken in depth zones where the measured sulfate concentrations are within the uncertainty of the flushing concentration C_0, because very high irrigation coefficients may yield a mathematically acceptable, but meaningless, solution. Hence, for these depth zones, α values are calculated by interpolation of values from adjacent depth zones.

The higher values of the integrated irrigation coefficient at the creek bank, as compared to the ponded marsh site, are consistent with higher crab population densities at the creek bank (Teal 1958) which are reflected by the greater number of burrows found at the creek bank than at the ponded marsh site (1040 and 415 apertures per m^2, respectively, Basan & Frey 1977). Model results (not shown) also indicate that irrigation is deeper in January than in August. This agrees with the increased depth of burrowing of fiddler crabs during winter reported by Wolfrath (1992).

According to our results, irrigation is the dominant solute transport process at all seasons and sites investigated. Total irrigation fluxes of sulfate into the sediment are calculated by integrating $\alpha(C_0-C)$ over the top 39 cm of sediment (Figure 3). The fluxes are found to be significantly higher in the summer than in the winter. At first sight, this seems to disagree with the results presented in Figure 2, which show that the integrated irrigation coefficients are similar in summer and winter. However, the sulfate reduction rates in summer are about twice as high as in winter. These high rates lead to a more pronounced sulfate depletion of the pore waters and, hence, to higher values of (C_0-C). Furthermore, the sulfate reduction rates are highest in the top-most sediment, causing a significant drop in sulfate concentration near the sediment-water interface, that is, in the region of maximum irrigation coefficients. In the winter, because of lower sulfate reduction rates, values of (C_0-C) are smaller and low irrigation fluxes are observed even though irrigation coefficients are relatively high. High sulfate reduction rates also result in a high irrigation flux at the creek bank site in June. However, due to a low irrigation intensity, the irrigation flux at the ponded marsh site in June is small despite a rapidly decreasing sulfate concentration with depth.

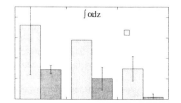

Figure 2. Irrigation transfer velocities (integrated irrigation coefficient over the top 39 cm) at the creek bank and the ponded marsh site for Aug'97, Jan'98 and June'98, respectively; error bars indicate ± 1 standard deviation. No error bar is shown for the creek bank in winter. This is because the flushing concentration C_0 lies within the uncertainty interval of the measured sulfate concentrations in the top 20 cm and extrapolation from this depth to the surface is questionable; therefore only the median is plotted.

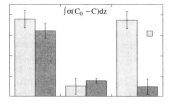

Figure 3. Flux of sulfate into the top 39 cm of sediment due to irrigation at the creek bank and ponded marsh site. Error bars indicate ± 1 standard deviation based on the 3 best profiles from 4 sets of Monte Carlo simulations. The simulations differ slightly in the method used to combine boxes (step 4 above), eliminating biases when reducing the number of free parameters.

4.2 Limitations

A number of limitations to the preliminary results presented here should be mentioned:
- Advection by tidal pumping is neglected in Equation 1. Due to the low permeability of the sediments at the study sites this is likely to have a minor effect. This assumption is in agreement with measured temperature profiles at the study sites.
- The sulfate profiles are assumed to be at steady state. Calculation of the temporal derivative in Equation 1 by linear interpolation between seasons does indicate that this assumption is justified, even though the salt marsh ecosystem is clearly not in a steady state over the whole year. In addition, the residence time of sulfate, calculated by dividing the sulfate stock by the depth-integrated sulfate reduction rate, varies between 9 and 40 days, which indicates that treating the sulfate profiles as a set of consecutive (quasi-)steady states may be a reasonable approximation.
- Production of sulfate through reoxidation of sul-

fide is not included in the reaction term. Although oxygen penetration depths are only on the order of millimeters, oxygen can be flushed deeper into the sediments through the burrows. At the ponded marsh site, oxygen may also be brought to greater depth through plant roots. Therefore, the reported values of α represent upper limits in the uppermost centimeters of the sediment.

5 CONCLUSIONS

The irrigation coefficient profile shape reflects burrow density, size and morphology (Koretsky et al. 1999). The temporal and spatial variability in the irrigation coefficient profiles obtained here agree, at least qualitatively, with ecological data on macrofaunal activity.

Modeling results suggest generally more intense and deeper macrofaunal activity at the creek bank than at the ponded marsh site and lower irrigation coefficients in early summer, compared to summer or winter. However, the flux of a dissolved species across the sediment-water interface due to irrigation does not only depend on the intensity of burrow flushing, but also on the differences between flushing concentration (C_0) and the pore water concentrations (C). For instance, high sulfate reduction rates near the sediment-water interface are observed in summer, causing steep gradients in the sulfate concentration profile. Low pore water concentrations, combined with a relatively intense macrofaunal activity, are responsible for the high irrigation fluxes of sulfate in summer.

As pointed out by Matisoff and Wang (1998) a mathematical formulation of irrigation using a non-local exchange function describes most of the solute transport processes very well but depends strongly on how α values are parameterized. The approach presented here allows α values to be retrieved objectively. The irrigation coefficients can then be used to model early diagenesis of other reactive species by applying existing reaction-transport codes, such as STEADYSED (Wang & Van Cappellen 1996) or CANDI (Boudreau 1996).

REFERENCES

Basan, P. B. & R. W. Frey 1977. Actual-palaeontology and neoichnology of salt marshes near Sapelo Island, Georgia. In T. P. Crimes and J. C. Harper (eds) *Trace Fossils* 2: 41-70. Steel House Press.

Berg, P., N. Risgaard-Petersen & S. Rysgaard 1998. Interpretation of measured concentration profiles in sediment pore water. *Limnol. Oceanogr.* 43(7): 1500-1510.

Boudreau, B. P. 1996. A method-of-lines code for carbon and nutrient diagenesis in aquatic sediments. *Computers Geosci.* 22(5): 479-496.

Emerson, S., R. Jahnke & D. Heggie 1984. Sediment-water exchange in shallow water estuarine sediments. *J. Mar. Res.* 42: 709-730.

Jørgensen, B. B. 1978. A comparison of methods for the quantification of bacterial sulfate reduction in coastal marine sediments. I. Measurement with radiotracer methods. *Geomicrobiol.* 1(1): 11-27.

Koretsky, C. M., P. Van Cappellen & C. Meile 1999. A stochastic approach to bioirrigation modeling in sedimentary environments. *Fifth International Symposium on the Geochemistry of the Earth's Surface.*

Martin, W. R. & G. T. Banta 1992. The measurement of sediment irrigation rates: A comparison of the Br- tracer and ^{222}Rn/^{226}Ra disequilibrium techniques. *J. Mar. Res.* 50: 125-154.

Matisoff, G. & X. Wang 1998. Solute transport in sediments by freshwater infaunal bioirrigators. *Limnol. Oceanogr.* 43(7): 1487-1499.

Press, W. H., B. P. Flannery, S. A. Teukolsky & W. T. Vetterling 1989. *Numerical Recipes (Fortran Version).* Cambridge University Press.

Schlüter, M., E. Sauter, H.-P. Hansen, & E. Suess, in review. Modelling seasonal variations of bioirrigation in coastal sediments. *Geochim. Cosmochim. Acta.*

Tabatabai, M. A. 1974. A rapid method for determination of sulfate in water samples. *Env. Lett.* 7(3): 237-243.

Teal, J. M. 1958. Distribution of fiddler crabs in Georgia salt marshes. *Ecol.* 39(2): 185-193.

Ullman, W. J. & R. C. Aller 1982. Diffusion coefficients in nearshore marine sediments. *Limnol. Oceanogr.* 27(3): 552-556.

Wang, Y. & P. Van Cappellen 1996. A multicomponent reactive transport model of early diagenesis: Application to redox cycling in coastal marine sediments. *Geochim. Cosmochim. Acta* 60(16): 2993-3014.

Wolfrath, B. 1992. Burrowing of the fiddler crab Uca tangeri in the Ria Formosa in Portugal and its influence on sediment structure. *Mar. Ecol. Progr. Ser.* 85: 237-243.

Geochemistry of the Earth's Surface, Ármannsson (ed.)© 1999 Balkema, Rotterdam, ISBN 90 5809 073 6

Input of terrestrial organic matter to the Baltic Sea – The Pomeranian Bight as a case study

Anja Miltner & Key-Christian Emeis
Section of Marine Geology, Baltic Sea Research Institute, Rostock, Germany

ABSTRACT: Terrestrial organic matter may play an important role in the carbon cycle of land-enclosed marginal seas like the Baltic Sea. We used the lignin signature of surface sediments to investigate the input and transport of terrestrial organic matter from the river Oder to the Arkona Basin. Approximately 40% of the sedimentary organic matter in Pomeranian Bight surface sediments was of terrestrial origin. The main lignin source in the study area was nonwoody angiosperm tissue. This corresponds to the dominant vegetation of the drainage area. Peat-like material found at one station consisted of gymnosperm wood. This material did not contribute significantly to bulk lignin. We therefore conclude that terrestrial organic material is transported together with fine particles.

1 INTRODUCTION

More than 75% of the global active carbon are assimilated in terrestrial ecosystems (Louchouarn et al. 1997). A large portion of this carbon is remineralized after a few months, another part is transformed to relatively stable soil organic matter, and a certain amount is transported to marine sediments. Once buried in these sediments, terrestrial carbon is no longer an active pool of the biogeochemical carbon cycle. The accumulation of terrestrial organic carbon in marine sediments is especially pronounced in coastal zones and, as a consequence, in marginal seas enclosed by land, such as the Baltic Sea. These areas may therefore play an important role in the carbon cycle and contribute to the long-term fixation of atmospheric carbon.

The river Oder is the fifth largest river discharging into the Baltic Sea. The material provided by the Oder accumulates mainly in the Arkona Basin, which receives sedimentary material mainly from this particular river (Leipe et al. 1995). The distance between the river mouth and the sedimentary basin is short. Therefore we can assume only minor or negligible degradation and diagenetic alteration of the terrestrial organic matter during transport. The region off the Oder mouth is therefore an excellent model area for the investigation of organic matter transport from terrestrial systems to marine sediments.

Lignin is a macromolecule composed of phenylpropene units. It is synthesized exclusively by vascular plants. As these are restricted to terrestrial ecosystems, lignin may be used as a tracer for terrestrial organic matter in marine sediments. Lignin from different tissue types is of variable composition. Therefore some conclusions about the origin of terrestrial organic matter may be drawn from lignin compositional data.

We studied the transport of terrestrial material from the Oder lagoon to the Arkona Basin. The aims of the study were to estimate the contribution of terrestrial carbon and carbon from eroded strata at the seafloor to total organic carbon in Pomeranian Bight sediments and to elucidate the transport mechanisms of terrestrial organic matter.

2 MATERIALS AND METHODS

We sampled surface sediments at four stations on a transect from the Oder mouth to the Arkona Basin and at one station in the Oder lagoon (Fig. 1). The exact positions of the stations as well as the TOC contents of the surface sediments are given in Table 1. The sediments in the Pomeranian Bight were sampled on a cruise with r/v Professor Albrecht Penck in June 1998. The Oder lagoon sediment was taken on a cruise with r/v Professor Albrecht Penck in March 1998.

The samples were freeze-dried and analyzed for total carbon and total inorganic carbon with an ELTRA Megalyt CS 1000/S C/S analyzer. Total organic carbon (TOC) was calculated by difference.

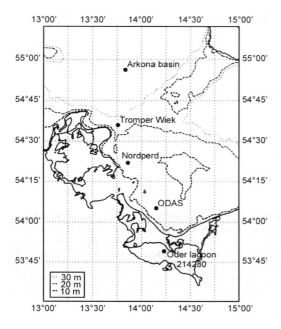

Figure 1. Map of the sampling locations

Table 1. Short description of the stations

station	Position	Water depth m	TOC content %
Oder lagoon	53°49.20N 14°09.96E	5.3	8.2
ODAS	54°04.85N 14°09.52E	16	0.29
Nordperd	54°21.96N 13°51.73E	21.5	1.10
Tromper Wiek	54°36.06N 13°45.64E	27	1.56
Arkona Basin	54°56.14N 13°49.95E	41	5.58
Nordperd peat	same as Nordperd		47.35

n.a.: not applicable

Lignin signatures were determined by CuO oxidation according to Hedges & Ertel (1982).

The contents of the individual phenolic compounds released by CuO oxidation are given in mg phenolic C/g TOC. The lignin contents of the sediments are proportional to the summed concentrations of vanillic and syringic acids and aldehydes and p-coumaric and ferulic acid (V+S+C). A simple mixing model is then used to estimate the contribution of terrestrial organic matter to marine sediments by the V+S+C contents of the sediments.

During diagenesis, the abundance of the acids in-

creases relative to that of the aldehydes. The acid-to-aldehyde ratios of the vanillyl and the syringyl moieties, $(ac/al)_V$ and $(ac/al)_S$, respectively, are therefore a measure of diagenetic alteration of lignin.

The relative abundance of different moieties depends on the origin of the lignin. In general, lignin derived from gymnosperm tissue does not contain syringyl phenols, whereas angiosperm tissue is composed of approximately equal amounts of syringyl and vanillyl phenols. The cinnamic acids (p-coumaric and ferulic acids) are typical for nonwoody tissue (Hedges et al. 1988). The syringyl-to-vanillyl and the cinnamyl-to-vanillyl ratios (S/V and C/V, respectively) are thus indicative of the origin of the lignin in marine sediments.

3 RESULTS AND DISCUSSION

3.1 Contribution of terrestrial organic matter to Pomeranian Bight sediments

The contents of lignin-derived phenols in the surface sediments from the Pomeranian Bight are approximately half those in the sediments from the Oder lagoon (Fig. 2). As we can assume that marine organic matter does not contain any lignin, we can estimate that about 40% of the sedimentary organic matter in the surface sediments of the Oder lagoon is terrestrially derived. The reason for the relatively low value at the ODAS station might be that material from the Oder lagoon does not remain in this region, but is transported further offshore during storms.

The lignin contents are similar at all Pomeranian Bight stations, and thus the contribution of terrestrial organic matter appears to be constant throughout the transect. This is in contrast to the expected decrease

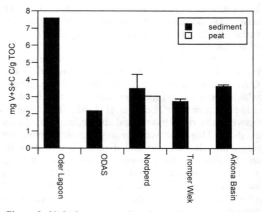

Figure 2. V+S+C contents of surface sediments in the Oder lagoon and the Pomeranian Bight

of terrestrial organic matter with increasing distance from the coast (Hedges et al. 1997; Voß & Struck 1997). The reason might be the short distance of the investigated stations from the Oder mouth and the relative stability of lignin compared to other organic compounds. A decrease of lignin contents may only be observable on a larger scale. Biomarkers which represent bulk organic C, such as $\delta^{13}C$, can be expected to yield lower results for the contribution of terrestrial organic matter to marine sediments.

3.2 Diagenetic alteration of lignin in Pomeranian Bight sediments

It is important to estimate the degree of diagenetic alteration before discussing the contribution of individual tissue types, because the ratios indicative of lignin origin may change during diagenesis (Goñi et al. 1993; Miltner & Zech 1998).
The $(ac/al)_V$ ratio of the investigated sediments range from 0.29 to 0.37, $(ac/al)_S$ from 0.30 to 0.37 (Fig. 3). There is no significant difference between the $(ac/al)_{V,S}$ ratios in sample 214280 from the Oder lagoon and the Pomeranian Bight sediments.
A similar range of $(ac/al)_{V,S}$ as in our sediments is frequently observed in marine sediments (Hedges et al. 1988; Prahl et al. 1994). They are at the low range of soil material (Louchouarn et al. 1997). The diagenetic alteration of lignin in the Baltic Sea sediments is therefore only moderate. S/V and C/V are apparently not or only slightly influenced by diagenesis and can thus be used as lignin source indicators. Even if there was a small shift in lignin composition due to diagenesis, we can compare our sediments, because they are all altered to a similar degree.

3.3 The origin of the lignin in surface sediments of the Pomeranian Bight

The S/V ratios of the sedimentary lignin are in the range from 0.42 to 0.65, C/V varied from 0.22 to 0.30 (Fig. 4). This lignin composition corresponds to a mixture of woody and nonwoody as well as gymnosperm and angiosperm tissue as the lignin source. Ertel & Hedges (1985) give average values for the lignin signature of plant material of different tissue types. Based on these average values and a formula given by the authors, we assessed the approximate contribution of each tissue type to total lignin. We estimate that nonwoody angiosperm tissue contributes most to total lignin. The next most abundant component is gymnosperm tissue. Woody angiosperm tissue is only a minor component of the lignin in Pomeranian Bight sediments. This corresponds well to the vegetation of the drainage area of river Oder, which consists mainly of deciduous temperate forest and agricultural plants.
The compositional similarity of lignin in all samples including the one from the Oder lagoon indicates that the river Oder is the common source for lignin in the surface sediments in the Pomeranian Bight. The lignin composition at the Arkona Basin station, however, differs slightly, but significantly from that at the other stations. This points to an additional source of lignin for that region, probably from the north. In addition, hydrodynamic sorting may cause shifts in lignin composition, because lignin from certain sources may be attached to particles of different size fractions (Goñi et al. 1998). At present, we can not estimate how much this process contributes to the observed change in lignin composition.

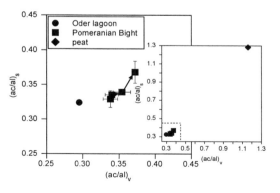

Figure 3. Acid-to-aldehyde ratios of surface sediments from the Oder lagoon and the Pomeranian Bight. The arrows point from the shallow to the deep stations.

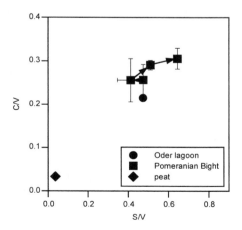

Figure 4. S/V vs. C/V of surface sediments from the Oder lagoon and from the Pomeranian Bight. The arrows point from the shallow to the deep stations.

3.4 The peat-like material at the Nordperd station

At one of the stations, large (ca. 5 mm) organic, peat-like particles were found. This material was analyzed separately in order to elucidate the origin of this material and to estimate how much it contributed to total lignin. The material yields about the same concentration of lignin-derived phenols as the bulk sediment, if referred to TOC (Fig. 2).

The $(ac/al)_V$ and $(ac/al)_S$ ratios are much higher than in the bulk sediment (Fig. 3). In contrast, S/V and C/V are close to zero (Fig. 4). This lignin signature indicates that the lignin was highly oxidized and contained only traces of syringyl and cinnamyl phenols. We therefore conclude that the peat-like material consists nearly completely of highly degraded gymnosperm wood. The degradation must have occurred in an aerobic environment. It is therefore not probable that the material originates from a bog.

The compositional differences to the bulk sediment are pronounced. The peat-like material obviously does not contribute significantly to total organic carbon in the surface sediments of the Pomeranian Bight. This indicates that coarse particulate material eroded at the seafloor is not an important source of terrestrial organic carbon to these sediments. Terrestrial organic matter must therefore be transported in a different form, probably sorbed to fine particles consisting mainly of clay minerals with organic matter sorbed to them.

4 CONCLUSIONS

According to their lignin contents, about half of the organic matter in Pomeranian Bight sediments is land-derived. Due to the short transport distance and the relative stability of lignin, we cannot find an offshore decrease in the contribution of terrestrial organic matter. The terrestrial organic matter in the Pomeranian Bight sediments is only slightly modified by diagenesis. Most of the sedimentary lignin originates from nonwoody angiosperm tissue. This corresponds to the vegetation in the drainage area of the Oder river, which is the common source for terrestrial organic matter in the sediments. Only in the Arkona Basin, an additional source appears to contribute to the terrestrial organic matter. The observed change in lignin composition may, however, also be due to hydrodynamic sorting of particles carrying organic matter with differing lignin composition. The transport of organic matter from the Oder lagoon to the Pomeranian Bight sediments probably takes place in fine particulate form. Coarse particles of morphologically recognizable plant residues and erosion of peat at the seafloor do not contribute significantly to terrestrial organic matter in Pomeranian Bight sediments.

ACKNOWLEGDEMENTS

We thank the crew of the r/v Professor Albrecht Penck for their assistance during sampling. The sample from the Oder lagoon was supplied by Dr. Thomas Leipe (Baltic Sea Research Institute, Warnemünde). Mrs. Dagmar Benesch determined the TC and TIC contents of the samples. The study was financially supported by the European Union (BASYS 3a)

REFERENCES

Ertel, J.R. & J.I. Hedges 1985. Sources of sedimentary humic substances: vascular plant debris. *Geochimica et Cosmochimica Acta* 49: 2097 - 2107.

Goñi, M., A., B. Nelson, R.A. Blanchette & J.I. Hedges 1993. Fungal degradation of wood lignins: Geochemical perspectives from CuO-derived phenolic dimers and monomers. *Geochimica et Cosmochimica Acta* 57: 3985 - 4002.

Goñi, M.A., K.C. Ruttenberg & T.I. Eglinton 1998. A reassessment of the sources and importance of land-derived organic matter in surface sediments from the Gulf of Mexico. *Geochimica et Cosmochimica Acta* 62: 3055 - 3075.

Hedges, J.I. & J.R. Ertel 1982. Characterization of lignin by capillary chromatography of cupric oxide oxidation products. *Analytical Chemistry* 54: 174 - 178.

Hedges, J.I., R.G. Keil & R. Benner 1997. What happens to terrestrial organic matter in the ocean? *Organic Geochemistry* 27: 195 - 212.

Hedges, J.I., C.A. Wayne & G.L. Cowie 1988. Organic matter sources to the water column and surficial sediments of a marine bay. *Limnology and Oceanography* 33: 1116 - 1136.

Leipe, T., T. Neumann & K.-C. Emeis 1995. Schwermetallverteilung in holozänen Ostseesedimenten - Untersuchungen im Einflußbereich der Oder. *Geowissenschaften* 13: 470 - 478.

Louchouarn, P., M. Lucotte, R. Canuel, J.-P. Gagné & L.-F. Richard 1997. Sources and early diagenesis of lignin and bulk organic matter in the sediments of the Lower St. Lawrence Estuary and the Saguenay Fjord. *Marine Chemistry* 58: 3 - 26.

Miltner, A. & W. Zech 1998. Beech leaf litter lignin degradation and transformation as influenced by mineral phases. *Organic Geochemistry* 28: 457 - 463.

Prahl, F.G., J.R. Ertel, M.A. Goñi, M.A. Sparrow & B. Eversmeyer 1994. Terrestrial organic carbon contribution to sediments on the Washington margin. *Geochimica et Cosmochimica Acta* 58: 3035 - 3048.

Voß, M. & U. Struck 1997. Stable nitrogen and carbon isotopes as indicator of eutrophication of the Oder river (Baltic Sea). *Marine Chemistry* 59: 35 - 49.

Geochemistry of the Earth's Surface, Ármannsson (ed.)© 1999 Balkema, Rotterdam, ISBN 90 5809 073 6

Pyritization at the Holocene/Late Pleistocene transition in the Black Sea sediments: Sulfur species and their isotopic composition

L. Neretin, M. E. Böttcher & B. B. Jørgensen
Biogeochemistry Department, Max-Planck Institute for Marine Microbiology, Bremen, Germany

I. I. Volkov
P.P.Shirshov Institute of Oceanology of Russian Academy of Sciences, Russia

H. Lüschen
Institute of Chemistry and Biology of the Marine Environment, Germany

ABSTRACT: This paper presents the high-resolution data on sulfur compounds distribution and their isotopic composition for a gravity core from the western part of the Black Sea. Pyritization processes in the upper part of lake beds is considered in detail and potential sources of hydrogen sulfide are discussed. The pyrite sulfur isotope composition is used as an indicator of paleoconditions in the basin during Late-Quaternary time.

1 INTRODUCTION

At the present time the Black Sea is the world's largest anoxic basin with a total inventory of hydrogen sulfide of about 4.6×10^9 t. History of the Black Sea during Quaternary time is the succession of several transgressions and regressions influenced by climatic changes (glacial and interglacial periods). The last interglacial period is still continuing and started ~ 10.5 ky B.P. when sea level rise led to the connection between the Mediterranean and the Black Sea and density stratification in the water column formed. Formation of strong pycnocline restricted exchange between surface oxygenated waters and lower layers, and anoxic conditions in the deep appeared. First study of sulfur compounds isotopic composition in the Black Sea sediments by Vinogradov et al. (1962) dated this event by about 7.5-8 ky B.P. Recent detailed studies of varve chronology and ^{14}C dating for the sediments confirmed these conclusions and showed that the onset of anoxic conditions in the water column was synchronous across the basin over a depth range of 200-2200 m at 7.5 ky B.P. and an organic-rich sapropel was produced at the same time. Before the first Mediterranean water invasion and anoxia formation a time span of about 1.5-3.5 kyr is assumed (Jones & Gagnon 1994; Arthur & Dean 1998).

A change from a low salinity lake to a brackish-marine basin initiated the formation of a sulfidization front and subsequent post-depositional pyritization of the upper lake beds (Strakhov 1963). Pyrite formation in these anaerobic conditions is regulated by H_2S flux and proceeds according to the scheme: $FeS \rightarrow FeS_x \rightarrow FeS_2$. Time dependent pyritization leads to the preservation of acid volatile sulfides, which are seen as a black band(s) – hydrotroilite layer in HI_1 sequence. Elemental sulfur (or reduced sulfur intermediates in pore waters) was considered as the intermediate species in the subsequent transformation of AVS into pyrite (Volkov 1984). In this paper we present data on sulfur species distribution and their isotopic composition for a gravity core recovered in the western Black Sea in 1997. Mechanism of pyritization and distribution of isotopic composition among sulfur species are discussed.

2 METHODS

Core was recovered at St.7 (43°31.'32 N, 30°13.'84 E, H=1201 m) in the western part of the Black Sea during the 1997 cruise of the R/V "Petr Kotzov" of the Max-Planck-Institute for Marine Microbiology using a 10-cm diameter plastic-barrel gravity corer. Total organic carbon content was determined coulometrically applying a Stroehlein Coulomat 702-LI after removal of carbonate. Carbonate content was calculated based on Ca content determined by XRF. Reduced solid sulfur species (acid volatile sulfide, elemental sulfur before and after AVS extraction and pyrite) were determined according to the conventional extraction scheme. At some horizons elemental sulfur before and after acid volatile sulfide (AVS) extraction was measured. All sulfur samples were converted to Ag_2S and $^{34}S/^{32}S$ ratios were measured by means of combustion isotope-ratio-monitoring mass spectrometry (C-irmMS) using Carlo Erba elemental analyser connected to a Finnigan MAT 252 mass spectrometer via a Finnigan MAT Conflo II split interface. $^{34}S/^{32}S$ ratios are given in $\delta^{34}S$ notation with respect to the V-CTD standard.

3 RESULTS AND DISCUSSION

The hydrotroilite layer is located in the upper part of Hl_I sediments and characterised by accumulation of AVS and elemental sulfur with concentrations up to 0.658% d.w. and 0.152% d.w., respectively. Results confirmed time-dependent transformation of iron sulfides into pyrite, but did not reveal any significant redistribution of iron in the vicinity of the black band as assumed earlier (Volkov 1984). Pyrite forms from the existing iron pool and the process seems to include sulfidization of iron-bound silicates. Our data on distribution and isotopic composition of elemental sulfur showed its essential role in pyrite formation. Iron sulfides in the hydrotroilite layer are characterized by a non-stoichiometric composition - $FeS_{1.10-1.33}$. Their transformation into pyrite starts when S^0/AVS ratio exceeds 0.2-0.3 and is regulated by the availability of elemental sulfur (or sulfur intermediates). Mass-balance calculation in the uppermost part of the hydrotroilite layer supposes additional contribution of H_2S or sulfur intermediates upon pyrite formation. Elemental sulfur formation from the H_2S reaction with iron hydroxides proceeds after sulfidization of reactive iron.

Interpretation of pyrite sulfur isotopic composition in the Late Quartenary sediments of the Black Sea is complicated by the superimposition of several factors that influence on $\delta^{34}S$ of its initial source – hydrogen sulfide: kinetic isotope effect upon sulfate reduction, changing of salinity (sulfate availability) and post-depositional transformation. Based on $\delta^{34}S$-FeS_2 the core is divided into three zones: Zone I – from the upper part below to Hl_{II}/Hl_I boundary with the average value –26.2±7.3 (-37.4 - -11.0)‰, the upper part of Hl_I – Zone II – with continuous enrichment of pyrite S by ^{32}S from +15.0 to –0.9‰ and the lower part of Hl_I sediments (Zone III) with $\delta^{34}S$-FeS_2 values in the range –13.0 - -0.1‰. $\delta^{34}S$-FeS_2 in the sapropel layer does not vary significantly and is in the range –37.4—22.7‰, large scattering is observed below in Hl_{IIb} sediments from –31.2 to –11.0‰. An abrupt shift in isotopic composition of pyrite in the lower part of this unit seems to be the indication for the first Mediterranean water invasion in the basin. Taking into account that this sedimentary sequence was deposited during the time of critical changes in the Black Sea thermohaline structure, we suppose that the salinity regime at least in the western part of the Black Sea after the first saline water invasion was non-stable.

Sporadic invasions of Mediterranean water or (and) time dependent changing of river discharge at the beginning of interglacial period could be the reasons for such instability and could influence isotopic composition of sulfate and consequently, $\delta^{34}S$ of pyrite sulfur.

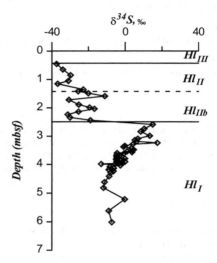

Figure 1. Isotopic composition of pyrite sulfur at St.7

From the lower part of the sapropel to the surface $\delta^{34}S$-FeS_2 is relatively stable, that means already established conditions of deposition environment. The main factor sustaining anoxic conditions in water column is density stratification. Thus if one would expect perturbations of salinity regime during the beginning of the Holocene, establishment of permanent anoxic conditions is improbable at that time.

Biostratigraphy data show that at the end of the sapropel deposition (Hl_{Ia}) (Nevesskaya 1965) salinity in the surface water reached the contemporary level of about 18‰. Thus, if we extrapolate the present conditions to the residence time of H_2S of about 100-200 yr (Bezborodov & Eremeev 1993) and take into account sedimentation rates within the sapropel 3.8-8.9 cm/kyr (Calvert & Karlin 1998), that the formation of the anoxic zone in the Black Sea would be "written" in a sedimentary sequence of about 2 cm. Therefore our data show that we can not accept a time span of about 3-4 kyr for the anoxic water column development in the Black Sea as hitherto been assumed.

Literature and our data on $\delta^{34}S$-FeS_2 in Hl_I sediments are presented in Fig. 2. In earlier studies sulfidization in the upper part of Hl_I sediments was attributed to H₂S diffusion from the above lying sapropel (Strakhov 1963; Volkov 1984). The isotopic composition of pyrite sulfur contradicts with the "diffusion theory". δ^{34}S-FeS$_2$ in the sapropel is significantly lighter than in the upper Hl$_I$ sediments.

"Reservoir effect" on sulfate sulfur and consequently pyrite S isotopic compositions could explain ^{34}S enrichment with depth in the upper Hl$_I$ sequence observed at St. 55 (Figure 2a) and St.BS4-14GC (Figure 2b). At the same time, ^{32}S enrichment with depth observed at our station (Figure 2c) and St.55 can not be explained by the "reservoir effect" and requires an additional source of hydrogen sulfide with ^{32}S enriched isotopic composition to the mentioned above. Under such conditions, anaerobic methane oxidation can be considered as a potential process (Böttcher et al., in prep.; Jørgensen et al., in prep.).

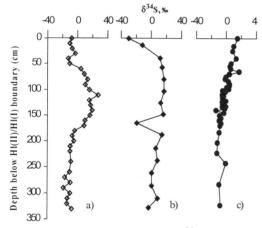

Figure 2. Vertical distribution of δ^{34}S-FeS$_2$ in the upper Hl$_I$ sediments: at St.55 (42°53.'8 N, 31°07.'3 E, H=2150 m) (Nikolaev 1995) (a), St.BS4-14GC (43°4.'7 N, 34°1.'6 E, H=2218 m) (Calvert et al., 1996) (b) and St. 7 – our data (c).

In the lower part of the core we recovered typical Euxinic sediments. δ^{34}S-FeS$_2$ in these sedimentary beds is –7.0±4.4 (-18.6 - -0.1)‰ (n=42) (Vinogradov et al. 1962; Nikolaev 1995; our data). Salinity in the New Euxinic basin water was about 7‰ (Nevesskaya 1965). Based on emperical relationship derived by Migdisov et al. (1983) the isotopic fractionation factor in these sediments seemed not to exceed 20‰ and contemporaneous sulfate isotopic composition was below 13‰. This is in accordance with the present δ^{34}S-SO$_4$ values in the brackish zones of the Azov Sea with the same salinity range that was contemporaneous for the Euxinic Black Sea.

4 CONCLUSIONS

1. Sulfidization of the upper Euxinic sediments of the Black Sea proceeds according to the scheme FeS FeS$_x$ FeS$_2$, where x=1.10-1.33. Elemental sulfur (or (and) sulfur intermediates) have an important role in pyrite precipitation from AVS. The H₂S-diffusion controlled process apparently involves sulfidization of iron-bound silicates.

2. δ^{34}S-FeS$_2$ data suggest that after the first invasion of Mediterranean water (beginning of the sapropel deposition – sapropelic layer - Unit Hl$_{IIb}$) salinity in the basin was quite variable and excluded the possibility of anoxia formation in the water column. Anoxic conditions in water seem to be established during 100-200 yr after density stratification was formed and coincided with the deposition of the sapropel layer.

3. Pyritization in the upper Hl$_I$ lake beds cannot be driven by diffusion of hydrogen sulfide from the sapropel, but is superimposed by additional contribution of H₂S possibly from anaerobic methane oxidation.

4. Our data suppose that the models of stationary diagenesis are not applicable for the situation observed at the contrast geochemical boundary in the Black Sea sediments and have a wider geochemical significance for the interpretation of paleorecords in sedimentary basins.

REFERENCES

Arthur, M.A. & W. E . Dean 1998. Organic matter production and preservation and evolution of anoxia in the Holocene Black Sea. *Paleoceanography* 13(4): 395-411.

Bezborodov, A.A. & V.N. Eremeev 1993. *The Black Sea: oxic/anoxic interface*. Sevastopol: MGI AN Ukraine.

Calvert, S.E. & R.E. Karlin 1998. Holocene sapropel of the Black Sea. *Geology* 26(2): 107-110.

Calvert, S.E., H.G. Thode, D.Yeung & R.E. Karlin 1996. A stable isotope study of pyrite formation in the Late Pleistocene and Holocene sediments of the Black Sea. *Geochim. Cosmochim. Acta* 60(7): 1261-1270.

Jones, G. A. & A. R. Gagnon 1994. Radiocarbon chronology of Black Sea sediments. *Deep-Sea Res.* 41(3): 531-557.

Migdisov, A.A., A. B. Ronov & V. A . Grinenko 1983. The sulphur cycle in the lithosphere. Part I. Reservoirs in the global biogeochemical sulphur cycle. In M.V. Ivanov & J.F. Freney (eds.), *The global biogeochemical sulfur cycle*, SCOPE Report 19: 25-94. New York: Wiley.

Nevesskaya, L.A. 1965. *Late – Quaternary Bivalvia of the Black Sea, systematics and ecology.*

Moscow: Nauka.

Nikolaev, S.D. 1995. *Isotope paleogeography of the continental seas.* Moscow: VNIRO.

Strakhov, N.M. 1963. About some features of the Black Sea sediments diagenesis. *Lithology and mining resources* 1: 7-27.

Vinogradov, A. P., V.A. Grinenko & V.I. Ustinov 1962. Isotopic composition of sulfur species in the Black Sea. *Geochemistry* 10: 851-873.

Volkov, I.I. 1984. *Sulfur geochemistry in ocean sediments.* Moscow: Nauka.

Geochemistry of the Earth's Surface, Ármannsson (ed.)© 1999 Balkema, Rotterdam, ISBN 90 5809 073 6

Morphology and sulfur isotope composition of early diagenetic pyrite

J.K. Nielsen
Geological Institute, University of Copenhagen, Denmark

Y. Shen & D.E. Canfield
Biological Institute, University of Odense, Denmark

ABSTRACT: Sulfur isotope analysis on pyrite in fine-grained sediments of the Upper Permian Ravnefjeld Formation, East Greenland, reveals that the large fractionation (up to 52.2 ‰) between pyrite and the contemporaneous seawater sulfate, which indicate that additional fractionation accompanying disproportionation of sulfide intermediates must have been put into effect. Various pyrite morphotypes in both the laminated and the bioturbated units has shown to be an useful indicator of ocean chemistry.

1 GEOLOGICAL SETTING

Calcareous, organic-rich to organic-lean mudshales and siltstones of the Upper Permian Ravnefjeld Formation, Foldvik Creek Group, were deposited in the East Greenland Basin under a sea level rise, probably under a highstand. Fluctuations in sea level (Figure 1) are implied by growth of time-equivalent platform and build-up carbonates of the Wegener Halvø Formation. For example, a relative sea-level rise was recognized as vertical and lateral growth of bryozoan-submarine cement mounds contemporaneously with a transition of bioturbated to laminated units (Stemmerik 1995, Piasecki & Stemmerik 1991).

Absence of an in- and epifauna for two laminated units indicates a non- to low-oxygen regime, while three interbedded bioturbated units with minor residual sedimentary structures indicate well oxidized bottom water conditions (Piasecki & Stemmerik 1991).

2 MATERIAL

Pyrite was examined in fresh drill core material from Triaselv in Jameson Land, East Greenland (GGU core 303102), where the mudshales and siltstones are thermally immature and no signs of secondary metal or sulfur enrichment exists. 50 bulk samples were extracted for pyrite sulfur and analyzed for sulfur isotope relative to CDT with an accurate better than 0.5 ‰.

3 STABLE ISOTOPES

Sulfur isotope results on pyrite (-41.2 to -28.2 ‰) indicates bacterial sulfate reduction (Figure 1). The fractionation of sulfur isotope is in order of max. 52.2 ‰ when compared to the sulfur isotope composition of contemporaneously seawater sulfate (ca. 11 ‰) (Scholle 1995). It has been shown that the pure cultures and natural populations of sulfate reducing bacteria can only produce sulfide depleted in ^{34}S by 4 to 46 ‰ compared with the initial sulfate (Habicht & Canfield 1997, Canfield & Teske 1996). Therefore we conclude that the large fractionations of pyrite in bioturbated units are caused by the re-oxidation of the sulfide followed by additional fractionation accompanying disproportionation of sulfide intermediates (Canfield & Thamdrup 1994).

4 MORPHOLOGY

The pyrite consists of finely disseminated micrometer-sized grains in the laminated units, while the bioturbated units show a concentration of grains inside and around small-sized burrows (Nielsen & Pedersen 1998). Morphology of pyrite grains is here used for the characterization of the pore water in the two sedimentary unit types.

Different morphotypes such as framboids and pyritized fossils and Fe-rich silicates are presence in the laminated units. These morphotypes pro-bably indicate a non-consolidated substrate marked by a transition from iron to sulfide dominated pore

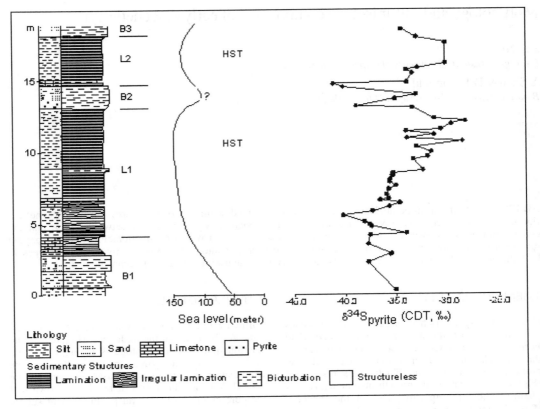

Figure 1. Lithology, sea level change and sulfur isotope composition of pyrite of the Ravnefjeld Formation, Foldvik Creek Group, East Greenland. The fivefold division of the Ravnefjeld Formation is shown by the labels L1 and L2 for laminated units and B1 to B3 for bioturbated units. Unit boundaries are gradual between L1 to B2 and L2 to B3 and are otherwise sharp (Piasecki, pers. comm. 1999, Piasecki & Stemmerik 1991).

The sea level curve and sedimentary log are from Piasecki & Stemmerik (1991).

water zones (Canfield & Raiswell 1991).

Agglomerates of euhedral morphotypes and framboids in tubes, which partially turned into euhedral forms, are found in the bioturbated units (Nielsen & Pedersen 1998, Piasecki & Stemmerik 1991). Pyrites in the burrows suggest a locally low saturation level of sulfide.

5 CONCLUSION

The $\delta^{34}S$ data has demonstrated that the sulfide re-oxidation with subsequent disproportionation is responsible for the large fractionation between pyrite and seawater sulfate. We suggest that the occurrence of various pyrite morphotypes in the sedimentary units of Ravnefjeld Formation were correlated to saturation levels of sulfide.

REFERENCES

Canfield, D.E. & R.Raiswell 1991. Pyrite Formation and Fossil Preservation. In P.A.Allison & D.E.G.Briggs (eds), *Taphonomy: Releasing the Data Locked in the Fossil Record*: 337-387. New York: Plenum Press.

Canfield, D.E. & A.Teske 1996. Late Proterozoic rise in atmospheric oxygen concentration inferred from phylogenetic and sulphur-isotope studies. *Nature* 382: 127-132.

Canfield, D.E. & B.Thamdrup 1994. The Production of ^{34}S-Depleted Sulfide During Bacterial Disproportionation of Elemental Sulfur. *Science* 266: 1973-1975.

Habicht, K.S. & D.E.Canfield 1997. Sulfur isotope fractionation during bacterial sulfate reduction in organic-rich sediments. *Geochim. Cosmochim. Acta* 61: 5351-5361.

Nielsen, J.K. & M.Pedersen 1998. Hydrothermal activity in the Upper Permian Ravnefjeld Formation of central East Greenland - a study of sulphide morphotypes. *Geol. Greenland Survey Bull.* 180: 81-87.

Piasecki, S. & L.Stemmerik 1991. Late Permian anoxia in central East Greenland. In R.V.Tyson & T.H.Pearson (eds), *Modern and Ancient Continental Shelf Anoxia:* 275-290. London: Geological Society Special Publication 58.

Scholle, P.A. 1995. Carbon and Sulfur Isotope Stratigraphy of the Permian and Adjacent Intervals. In P.A.Scholle, T.M.Peryt & D.S.Ulmer-Scholle (eds), *The Permian of Northern Pangea:* 133-149. Berlin Heidelberg: Springer-Verlag.

Stemmerik, L. 1995. Permian History of the Norwegian-Greenland Sea Area. In P.A.Scholle, T.M.Peryt and D.S.Ulmer-Scholle (eds), *The Permian of Northern Pangea:* 98-118. Berlin Heidelberg: Springer-Verlag.

Geochemistry of the Earth's Surface, Ármannsson (ed.)© 1999 Balkema, Rotterdam, ISBN 90 5809 073 6

Sulfate reduction rates and organic matter composition in sediments off Chile

C.J.Schubert, T.G.Ferdelman & B.Strotmann
Department of Biogeochemistry, Max Planck Institute for Marine Microbiology, Bremen, Germany

ABSTRACT: Various organic geochemical parameters were determined in four sediments cores off the coast of central Chile. Marine and terrestrial contributions to the organic matter fraction were estimated using C/N ratios, carbon isotopic composition, protein and chlorin concentrations. Additionally, sulfate reduction rates (SRR) were determined. A comparison between the cores showed one station (14) on the shelf in front off the Bay of Concepcion to be dominated by terrestrial organic carbon, whereas two other stations in the bay (4,7), and another stations offshore (18) include mainly marine organic matter. The distribution pattern of SRR of Station 14 off the bay differed considerably from both stations 4 and 7 in the bay and bay entrance respectively and Station 18 on the shelf. SRR of Station 14 did not exhibit such a striking quasi-exponential drop-off as at the other stations, but instead stay at a relatively high level throughout the core. We attributed this pattern by the difference in organic carbon composition between Station 14 and the other stations.

1 INTRODUCTION

The region off central Chile (Concepción) is characterized by intense upwelling during summer (December to April) that leads to one of the highest primary productivity rates of the world oceans (9.6 gC m^2 d^{-1}, Fossing et al. 1995). Even during non-upwelling times primary productivity rates are still high compared to other ocean areas. Due to this high rate of primary production, the water column is oxygen depleted between 30 and 300 m water depth.

One feature distinguishing this area from other parts of the oceans is the massive occurence of the sulphur bacterium *Thioploca* spp. in sediments accumulating below oxygen depleted bottom waters (Gallardo 1977, Fossing et al. 1995). This bacterium is able to concentrate nitrate in a vacuole and using it to oxidize hydrogen sulphide. Calculations showed that 16 to 34% of the produced hydrogen sulphide in the sediments are oxidized by *Thioploca* spp., although sulfate reduction rates are extremely high in this region (up to 4670 nmol cm^{-3} d^{-1}, Ferdelman et al. 1997).

We compared sulfate reduction rate measurements with the organic carbon composition of sediment cores. We investigated the organic carbon composition in respect to their terrestrial and marine source. Bulk characterization of the organic matter using C/N values and the carbon isotopic composition were performed, as well as total proteins and chlorins, the degradation products of chlorophyll were measured. α-amyrin that is restricted to higher plants has been used as a biomarker for terrestrial input. Additionally, we have measured sulfate reduction rates to investigate their relation to the composition of organic matter.

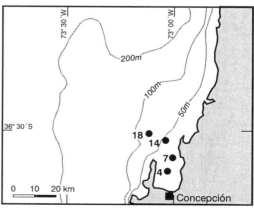

Figure 1. Sampling stations

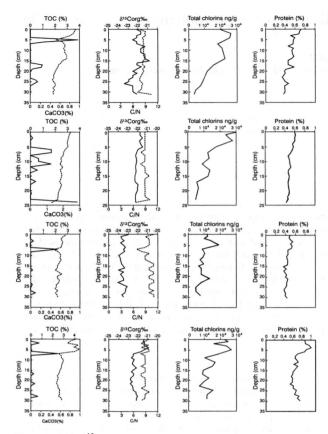

Figure 2. TOC, δ^{13}C, total chlorine and protein in cores 4, 7,14 and 18

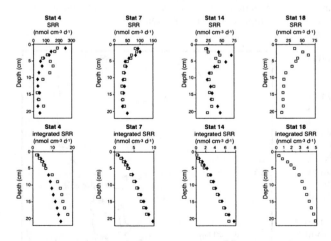

Figure 3. Sulfate reduction rate at stations 4, 7, 14 and 18

Four multicorer cores from the shelf off Chile were collected in March 1998 in a period which was strongly influenced by an El Niño event. Two sediment cores (Station 4 and 7) were taken in the Bay of Concepción in 29 and 37 m water depth, two cores (Station 14 and 18) off the Bay on the shelf in 57 and 87 m water depth (Figure 1).

Total (TC) and organic carbon (OC) was determined by combustion/gas chromatography (Carlo Erba NA-1500 CNS analyzer) with a precision of ±1.2 % on the bulk and carbonate-free subsamples, respectively.

For the determination of $\delta^{13}C$ values, decarbonated samples were combusted in an online Heraeus element analyser and the evolved CO_2 was passed to a Finnigan Delta isotope-ratio mass spectrometer in a continous flow of helium. Results are reported in the δ notation, $\delta^{13}C=\{(^{13}C/^{12}C)_{sample}\ /(^{13}C/^{12}C)_{standard}-1\}$ per mil, relative to VPDB and the measurement precision is better than ±0.2‰.

For the determination of lipids, freeze-dried and gently ground sub-samples (~ 2 g) were extracted by successive sonication and centrifugation in methanol, methanol/methylene chloride (1:1) and methylene chloride, respectively. The extracts were saponified (6% KOH), further extracted with hexane, and derivatized with BSTFA (Sigma) prior to injection onto a HP1 chromatographic column (50 m length, 0.32 mm I.D., 0.17 μm).

For the determination of chlorins, which include a whole suite of degradation products of chlorophyll, freeze dried sub-samples (~ 0.6 g) were extracted by threefold sonication and centrifugation in acetone. The samples were cooled with ice under low light conditions during extraction to prevent decomposition of the chlorins. Sediment extracts were measured fluorimetrically (Hitachi F-2000 fluorometer) immediately after extraction. Chlorophyll a (Sigma) which was acidified with a few drops of hydrochloric acid was used as a standard. The precision of the method was ±10 %.

Total proteins were measured after hydrolysation of the ground sediment with 0.5 N sodium hydroxide (Bradford, 1976). The hydrolysate was measured photometrically after complexation using a Shimadzu UV-160A spectrophotometer.

Sulfate reduction rates were determined using a whole-core $^{35}SO_4^{-2}$ incubation method (Jørgensen, 1978). Reduced ^{35}S was determined using the one-step Cr-II destillation (Fossing & Jørgensen, 1989). Sulfate was measured using nonsuppressed ion chromatography with a Waters 510 HPLC pump, Waters WISP 712 autosampler, Waters IC-Pak anion exchange column (50 x 4.6 mm), and a Waters 430 conductivity detector (Ferdelman et al., 1997).

2 RESULTS

At Station 4, organic carbon concentrations varied from 3.7 % at the top to 1.9 % at 31 cm (Figure 2). After a sharp decrease in the uppermost 5 cm to 2 %, values increased again to 3 % at 15 cm and declined rapidly to the core end. C/N values vary from 4 to 10 (Ø = 7) which basically could be interpreted as a mainly marine organic matter source with small amounts of terrestrial organic matter being mixed in. This is supported by the carbon isotopic composition which varies from -20.6 to - 22.0 ‰ VPDB (Ø = -21.5 ‰ VPDB).

At Station 7, OC values declined linearly from 3.3 to 1.7 % from the core top to 24 cm (Figure 2). C/N values are higher and have a narrower range from 7 to 10 (Ø = 9) indicating a moderate increase in terrestrial organic matter. This is also demonstrated in the $\delta^{13}C$ values which are slightly lighter and range from - 21.5 to - 22.3 ‰ VPDB (Ø = -22.1 ‰ VPDB).

At Station 14, OC values were relatively stable downcore and varied between 1.9 and 2.8 % (Figure 2). C/N values range from 7 to 11 (Ø = 10) indicating higher terrestrial organic material being mixed in, which is supported by light $\delta^{13}C$ values of - 23.0 to - 24.0 ‰ VPDB (Ø = -23.6 ‰ VPDB).

At Station 18, organic carbon concentrations at the top 6 cm varied around 4 % (Figure 2). Values declined rapidly at 7 cm and varied the entire core downwards at values around 2.5 %. C/N values decrease to an average of 9 (8 to 10) demonstrating again decreasing amounts of terrestrial organic matter. This is also supported by slightly heavier carbon isotope values averaging at -22.2 ‰ VPDB (- 20.9 to - 22.8 ‰ VPDB).

At Station 4 SRR peaked at the surface at 185 - 250 nmol cm^{-3} d^{-1} and decreased quasi-exponentially with depth to rates of 40 - 82 nmol cm^{-3} d^{-1} at depths of 19 cm (Figure 3).

Station 7 exhibited a sub-surface SRR peak of 86 - 100 nmol cm^{-3} d^{1} at 2 cm and decreased to rates of 31 - 38 nmol cm^{-3} d^{-1} at 19 cm (Figure 3).

Station 14 differed considerably from both stations 4 and 7 in the bay and bay entrance respectively and Station 18 on the shelf (Figure 3). SRR peaked at 4 to 5 cm depth at Station 14 did not exhibit such a striking quasi-exponential drop-off as at the other stations. Although peak rates of sulfate reduction were low, between 47 and 69 nmol cm^{-3} d^{-1}, SRR remained at a relatively higher rate at depth of 23 to 48 nmol cm^{-3} d^{-1} at a depth of 19 cm. Cumulative sulfate reduction rates (mmole m^{-2} d^{-1}) at Station 14 increased nearly linearly with depth.

Station 18 exhibited SRR profiles similar to that for Station 7, with a peak of 62 nmol cm^{-3} d^{-1} at 2 to 3 cm and then rapidly decreasing to rates of 12 to 14 nmol cm^{-3} d^{-1} at depth (Figure 3).

3 DISCUSSION

Most interesting is the obvious difference between Station 14 and the other 3 sites, both with respect to the pattern of sulfate reduction as a function of depth and the biomarker indices for terrestrial versus marine organic matter input.

Based upon the C/N ratios, $\delta^{13}Corg$ and the concentrations of chlorins, and proteins, Stations 4, 7, and 18 appear to be similarly dominated by marine input, whereas Station 14 exhibits a clear terrestrial signal. Although Station 14 is in the the middle of the transect from enclosed bay to open shelf, it is clear that the organic carbon deposited there is significantly different in origin than that of the other stations.

The differing inputs of terrestrial versus marine organic matter are also manifested in the SRR profiles. Where marine inputs dominate, whether that be in the bay (Stations 4 and 7) or on the open shelf (Station 18), volumetric rates of sulfate reduction exhibit relatively high rates near the surface with rapidly, nearly exponentially decreasing rates with depth. The depth at which 1/2 of the cumulative sulfate reduction occurs is 5 to 6 cm for Stations 4, 7, and 18, while it falls between 9 and 11 cm for Station 14. This appears to be independent of the overall depth integrated rate. The overall depth integrated rates showed a gentle decrease going from the bay to the shelf and is independent of core top organic carbon contents. Thus, although the peak volumetric rates appear to be lower at the terrestrially influenced Station 14, the overall depth integrated rates are actually in line with those for the other stations.

What causes the observed difference between the SRR at the site influenced by terrestrial input and the other marine influenced stations? One possibility is that the kinetics of degradation of terrestrial material is slower than that for protein-rich marine material. The terrestrial material is degraded more slowly, but is eventually degraded. A second possibility is that the marine organic matter being degraded is simply diluted by a non-reactive components, including terrestrially derived organic matter. The lower Corg concentrations at Station 14 would support the latter scenario. ^{210}Pb fluxes to Station 14 do not appear anomalously low or high, nor do they indicate a significantly higher sedimentation rate. However, there is no clear answer to this problem at the present.

Nevertheless, the combination of bulk organic carbon parameters and biomarkers with a direct measure of organic carbon reactivity, promises to provide us with important insights into the relationship between the type of organic matter input and it's subsequent rate of degradation under real conditions.

4 CONCLUSION

Whereas, three cores (Stations 4 and 7 in the Bay of Concepción and Station 18 on the shelf) appear to be similarly dominated by marine organic matter input, the organic fraction of the core from Station 14 is strongly influenced by terrestrial organic matter. This is demostrated by higher C/N ratios, lighter $^{13}Corg$ values, lower protein and lower chlorine concentrations. Additionally, α-amyrin, a biomarker restricted to higher plants could only be so far observed in the sediments of Station 14. SRR of Station 14 differed considerably from both stations 4 and 7 in the bay and bay entrance respectively and Station 18 on the shelf. Cumulative sulfate reduction rates (mmole m^{-2} d^{-1}) at Station 14 increased nearly linearly with depth. The striking difference in the SRR as well as in the cummulative sulfate reduction rates of Station 14 compared to Station 4, 7, and 18 can be explained by the difference in the organic carbon composition. It may be that the marine organic matter being degraded is simply diluted by non-reactive components, including terrestrially derived organic matter or that the kinetics of terrestrially derived organic matter degradation is slower. This work shows that a combination of bulk organic carbon parameters and biomarkers with a direct measure of organic carbon reactivity, promises to provide important insights into the relationship between the type of organic matter input and it's subsequent rate of degradation under real conditions.

REFERENCES

Ferdelman, T.G., C. Lee, S. Pantoja, J. Harder, B.M. Bebout, & H. Fossing, 1997. Sulfate reduction and methanogenesis in a Thioploca-dominated sediment off the coast of Chile. Geochim. et Cosmochim. Acta 61(15): 3065-3079.

Fossing, H., V.A. Gallardo, B.B. Jørgensen, M. Hüttel, L.P. Nielsen, H. Schulz, D.E. Canfield, S. Sorster, R.N. Glud, J.K. Gundersen, J. Küver, N.B. Ramsing, A. Teske, B. Thamdrup, and O. Ulloa, 1995. Concentration and transport of nitrate by the mat-forming sulphur bacterium Thioploca. Nature 374: 713-715.

V.A. Gallardo 1977. Large benthic microbial communities in sulphide biota under Peru-Chile subsurface countercurrent. Nature 286: 331-332.

Geochemistry of the Earth's Surface, Ármannsson (ed.)© 1999 Balkema, Rotterdam, ISBN 90 5809 073 6

Long-term external forcing of fluvial dynamics and sediment composition

L. A. Tebbens & A. Veldkamp
Laboratory of Soil Science and Geology, Wageningen Agricultural University, Netherlands

ABSTRACT: Source rock weathering, erosion, transport and sedimentation determine fluvial sediment composition. The long-term (10^3-10^5 years) interaction of spatial complex response and temporal non-linear behaviour in the fluvial system requires a thorough quantification of fluvial dynamics to assess the climatic impact on sediment composition. Semi 3-D forward modelling was performed to quantify the 250-0 ka BP sediment fluxes and bulk geochemical sediment composition of the River Meuse. A scenario of minimal discharges and maximum hillslope process dynamics during cold glacial periods alternating with maximum discharge and minimal hillslope processes during prolonged interstadials or interglacials largely succeeded to reproduce measured bulk geochemical trends registered in the 15-0 ka BP fine-grained sedimentary record of the Meuse lower reach.

1 INTRODUCTION

Most fine-grained fluvial sediments are ultimately derived from the erosion of pre-weathered saprolites and/or of clay-rich horizons of soil profiles. Consequently, fluvial sediment composition will vary due to 1) source rock diversity in the drainage basin (provenance), 2) physical and chemical weathering of these source rocks under varying climatic conditions and 3) subsequent erosion and transport pathways within the fluvial system (Johnsson 1993).

The weathering rate and the residence time of weathering material in the sediment production areas determine how the weathering processes modify fluvial sediment composition. Climate (temperature and precipitation) directly determines the weathering rate. It also directly influences the residence time of weathering material in fluvial drainage basins via vegetation cover and indirectly via long-term forcing of the intensity of denudation processes and fluvial erosion or deposition.

Fluvial sediment supply can be weathering-limited or transport-limited. In a weathering-limited system, the residence time of the weathering products in the upstream catchments is too short for weathering to leave clear climatic imprints on sediment composition. In a transport-limited system, residence time is long enough for climate to overprint provenance characteristics. Bulk geochemical analysis has proven a powerful tool to rapidly quantify the natural compositional variability of fluvial sediments (Tebbens et al. 1998).

Many fluvial drainage basins experienced the Quaternary alternation of cold and dry glacial stadials and interstadials and warm and humid interglacials. On a time-scale of 10^3-10^6 years and at the spatial scale of a fluvial drainage basin, these climatic changes interacted with tectonics and sea-level changes to influence the occurrence, timing, and magnitude of erosional and depositional events. Since large relative changes in the quantity of sediment fluxes will also affect their basic mineralogy and bulk geochemistry, it is interesting to study the long-term (10^3-10^5 years) effects of climate-controlled fluvial dynamics and hillslope processes on fluvial sediment composition.

2 RIVER MEUSE CASE STUDY

2.1 *Geological setting*

This contribution presents a 250-0 ka BP case study of the entirely rain-fed and climatically sensitive 874-km River Meuse, draining an area of 33,000 km². The upper reach lithology comprises Jurassic marls in northern France, and the middle reach exhibits Palaeozoic low-grade metapelites in the low mountain range (Belgian Ardennes), giving way to loess-covered Cretaceous limestones in the southern

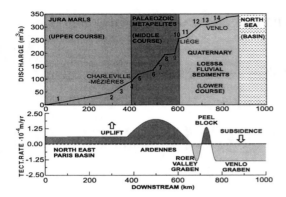

Figure 1. River Meuse drainage area characteristics (top). The successive downstream tributaries are numbered. Major ones are 2: Chiers, 4: Semois, 6: Lesse, 7: Sambre, 9: Ourthe, Vesdre, Amblève and 11: Geul, Roer. The bottom part outlines the (modelled) tectonic setting.

Netherlands (Figure 1). All these rocks are found in a sloping landscape typical for an actively uplifting tectonic setting. Downstream of the major Feldbiss fault, the Meuse starts to flow over its own sandy to silty alluvium in the low-relief lower reach as it enters the complex Roer Valley Graben of the subsiding southern North Sea basin. Sea level influenced the lowermost reaches (>800-km) of the Dutch combined Rhine-Meuse delta only during interglacials. Clay mineral supply from the upstream areas has changed in time to cause relatively low K_2O-contents and high Al_2O_3-values in Holocene lower reach sediments versus high K_2O-contents and low Al_2O_3-contents in Late-glacial sediments (Tebbens et al. 1998).

2.2 Modelling sediment flux and composition

A thorough analysis of fluvial dynamics must incorporate external forcing by tectonics, climate and sea-level change to allow for complex-response and non-linear fluvial behaviour acting on large spatio-temporal scales. Semi 3-D forward-modelling of the development of the River Meuse longitudinal profile in response to external forcing enabled the authors to quantify the sediment fluxes of the main river for the period 250 to 0-ka BP and for the entire drainage basin (Tebbens 1999). The FLUVER2 model (cf. Veldkamp & Van Dijke 1998) calculated sediment fluxes for successive 1-km profile segments, assuming mass continuity and dynamic equilibrium in the fluvial system.

The presented scenario included frequent climatic perturbations of the balance between the sediment supply to and sediment transport capacity (via the discharge) of the fluvial system. Well-dated GRIP

ice-core normalised $\delta^{18}O$-data formed the basis of this climatic input (Dansgaard et al. 1993). A climate and relief feedback mechanism allowed to vary hillslope supply in time and along the river profile. The semi 3-D longitudinal profile model calculates highest hillslope supply during the coldest periods (glacials) to simulate short residence times of weathering material in sloping sediment production areas. Lowest hillslope supply is calculated in low-relief areas and during the warmest stages (interglacials, prolonged interstadials) to represent periods characterised by long residence times.

The varying contributions of modelled sediment fluxes resulting have been linked to the bulk geochemical composition of the tributary source catchments (see references in Tebbens 1999). The ratio of K_2O to Al_2O_3 (K/Al-ratio) was used to quantify downstream and long-term compositional changes of the main river sediment flux. Like weathering duration, the weathering rates in the tributary catchments might have varied too. However, the input K/Al-ratios were kept constant in time for all tributary catchments during the whole simulation. This enables us to assess the pure effects of climate-controlled hillslope processes and internal river dynamics (provenance and complex-response) on the evolution of sediment composition.

3. RESULTS

3.1 Downstream and temporal variation

The compositional variability along the longitudinal profile is largest owing to mixing of the sediment supply of several tributary catchments, each having their own local catchment compositions.

Figure 2 visualises the modelled K/Al-ratios versus the downstream position of every 1-km segment along the entire longitudinal profile and for two climatically contrasting time-slices, namely 21.9-ka BP and 6-ka BP. The 21.9-ka BP time-slice represents the Late Pleniglacial coldest part in the GRIP-core, characterised by low discharges and high hillslope supply inputs. The 6-ka BP time-slice represents the mid-Holocene climatic optimum, with high discharges and low hillslope supply inputs.

The 21.9-ka cold climate generated high hillslope contributions in the upper and middle reach sloping areas (0-661-km). Simultaneously, the combination of these high hillslope contributions with low discharges and low local K/Al-ratios (385-505-km) led to a depositional phase in the fluvial system and effectively diluted the constant K/Al-ratio of the main river sediment flux coming in from the lithologically homogeneous Jurassic area (0-385 km). The same high hillslope contributions combined with high tributary K/Al- ratios increase the

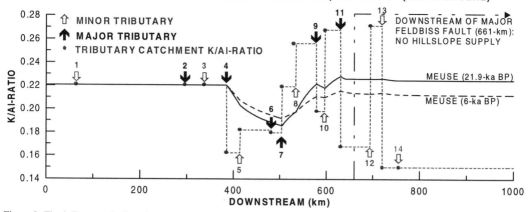

TIME SLICES: 21.9-ka BP (WEICHSELIAN PLENIGLACIAL) AND 6-ka BP (MID HOLOCENE)

Figure 2. The influence of tributaries on the modelled sediment composition of the River Meuse for two climatically contrasting time-slices (Late Pleniglacial, 21.9-ka BP) and Mid Holocene (6-ka BP). Tributary K/Al-ratios are derived from Hindel et al. (1996) and Van der Sluys et al. (1997): see references in Tebbens (1999). Legend for major tributaries indicated in caption of Figure 1.

K/Al-ratio again from 505-630-km.Hillslope processes hardly contribute in the low-relief lower reach (>661-km). Moreover, river discharge and sediment load have grown so large that downstream tributaries have progressively less influence on the magnitude of the sediment flux K/Al-ratio.

The 6-ka BP K/Al-ratio curve (Figure 2) represents a situation, with climatically induced lower hillslope supply. The curve amplitude is smaller, suggesting that a decreasing hillslope supply from the tributary catchments will decrease the relevance of local K/Al-ratios for the composition of the main river sediment flux too. Accordingly, the influence of local hillslope supply in sloping areas on the K/Al-ratio of the main river sediment flux is dampened during interglacial periods, while the influence of local supply is strengthened during glacial periods.

These results indicate that external forcing of changing hillslope contributions and internal river dynamics can cause changes in long-term sediment composition, without having incorporated direct effects of increased climatic weathering intensity. Now the main downstream and temporal compositional trends have been established, it is interesting to check whether the modelled K/Al-ratios reflect actual trends in measured data.

3.2 Measured and modelled data

Figure 3 demonstrates the results of measured and modelled K/Al-ratios for the 15-0 ka BP time-frame (including the ~12-ka cold Younger Dryas event) for one section along the longitudinal profile, namely the lower reach in the subsiding Venlo Graben (Figure 1, ~762-km). Holocene fine-grained residual

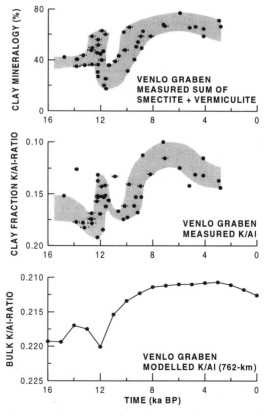

Figure 3. Measured and modelled clay mineralogical and geochemical data for the Venlo Graben section. Note reverse y-axes for middle and bottom pictures and similar temporal trends.

345

infil-lings contained more smectite and vermiculite than Late-glacial sediments (Figure 3, top). Consequently, measured Holocene K/Al-ratios in separated clay fractions were lower than Late-glacial K/Al-ratios (Figure 3, middle; note reverse y-axis) and so were measured bulk K/Al-ratios (Tebbens et al. 1998; not shown). Although compared size fractions were not fully similar, the temporal evolution of modelled bulk K/Al-ratios strikingly resembles the measured clay and bulk trends (Figure 3, bottom; reverse y-axis).

4 DISCUSSION

The qualitatively good temporal correspondence suggests that the modelled climatic forcing of hillslope supply and internal fluvial dynamics is able to mimic the measured evolution of K/Al-ratios in fine-grained sediments in the Meuse fluvial system to a great extent. Therefore, the FLUVER2 model simulation corroborates the hypothesis that the interaction of long-term (10^3-10^5 years) external forcing variables - more specifically the indirect climatic control of hillslope processes and internal river dynamics - effectively changes the composition of fluvial sediments in the right direction. The absolute values and magnitude of change do not completely match, however. These imperfections can originate from tributary catchment model input K/Al-ratios that were based on slightly different grain size fractions than the ones Tebbens et al. (1998) calculated from their analyses performed on bulk geochemical sampling. Furthermore, rising mean annual temperature and precipitation during the Late-glacial may have augmented weathering rates in the sediment production areas, but this was not incorporated in the model. Finally, large influxes of aeolian loess originating from outside the river drainage area are likely to have influenced Late Weichselian sediment composition. However, these fluxes have not been quantified yet and therefore they could not be included in the model calculations.

5 CONCLUSIONS

1. By modelling Late Quaternary fluvial dynamics on a basin-wide scale and linking the calculated sediment flux to fluvial overbank bulk geochemical data of tributary catchments, it is possible to simulate sediment compositional changes along the longitudinal profile of the main river. This opens new perspectives to reconstruct and interpret long-term (10^3-10^5) provenance- and climate-controlled bulk sediment compositional changes in Quaternary fluvial systems.

2. A scenario in which sediment supply was simulated to shift from glacial weathering-limited to interglacial transport-limited conditions, performed well to predict the temporal evolution of bulk composition as well as the direction of bulk geochemical changes for sediments in the Meuse lower reach. This suggests that climatically controlled shifts from glacial weathering-limited sediment supply to interstadial or interglacial transport-limited sediment supply can explain the greater part of natural compositional variation in River Meuse fine-grained sediments.

3. Additional research is needed to quantify the impact of extra-basinal aeolian influxes on fluvial sediment supply and bulk geochemical composition.

REFERENCES

Dansgaard, W., S.J. Johnsen, H.B. Clausen, D. Dahl-Jensen, N.S. Gundestrup, C.U. Hammer, C.S. Hvidberg, J.P. Steffensen, A.E. Sveinbjörnsdottir, J. Jouzel & G. Bond, 1993. Evidence for general climatic instability of past climate from a 250-kyr ice-core record. *Nature* 364: 218-220.

Johnsson, M.J. 1993. The system controlling the composition of clastic sediments, In M.J. Johnsson & A. Basu (eds), *Processes controlling the composition of clastic sediments*: 1-19, Geological Society of America Special Paper 284, Boulder, Colorado.

Tebbens, L.A., A. Veldkamp & S.B. Kroonenberg 1998. The impact of climate change on the bulk and clay geochemistry of fluvial residual channel infillings: the Late Weichselian and Early Holocene River Meuse sediments, The Netherlands. *Journal of Quaternary Science* 13: 345-356.

Tebbens, L.A. 1999. *Late Quaternary evolution of the Meuse fluvial system and its sediment composition. A reconstruction based on bulk sample geochemistry and forward modelling.* PhD Thesis. Wageningen: Wageningen Agricultural University.

Veldkamp, A. & J. J. Van Dijke 1998. Modelling long-term erosion and sedimentation processes in fluvial systems: A case study for the Allier/Loire system. In G. Benito, V.R. Baker, & K.J Gregory (eds), *Palaeo-hydrology and environmental change*: 53-66. New York: Wiley.

Geochemistry of the Earth's Surface, Ármannsson (ed.) © 1999 Balkema, Rotterdam, ISBN 90 5809 073 6

Rare earth element chemistry of rain water

J. Zhang, T. Ishii & S. Yoshida
National Institute of Radiological Sciences, Ibaraki, Japan

C.-Q. Liu
Institute of Geochemistry, Chinese Academy of Sciences, Guiyang, People's Republic of China

ABSTRACT: Rain water collected from Tokyo, Japan, and surroundings and from the East China Sea have been measured for their contents of rare earth elements (REEs) and major ions. The chemical compositions of these samples suggest the influence of both aerosol and seawater, permitting also evaluation of their contribution from these two sources have variable REE compositions and show different shale-normalized REE abundances. Particle-rain water interaction, the pH of the rain water or speciation of REEs are probably responsible for the variable abundance and fractionation of REEs in these waters.

The yearly calculated rainfall over the oceans (3.910^{17} kg/yr.) results in fluxes for individual REEs ranging from about 2 times for heavy REEs, about 6 for middle REEs and about 28 times for light REEs of the river fluxes. The errors of the estimates may be large because of the large variation of the REE concentrations in rain water. However, it is concluded that the atmospheric inputs of individual REEs, especially of the wet deposition fluxes, is much more significant relative to those of dry deposition and riverine flux to the ocean.

1 INTRODUCTION

There are several important processes affecting the oceanic geochemistry of rare earth elements (REEs): riverine, atmospheric and hydrothermal input, adsorption and scavenging onto particles, coprecipitation, ion pairing and redox reactions. Among these the source of the REEs in the oceans play the most important role in control the oceanic geochemistry of the REEs. A recent study (Greaves et al. 1994), which has put an emphasis only on dry deposition and release of REEs by reaction with seawater, indicates the importance of the atmospheric inputs in controlling the oceanic REE geochemistry, as compared with riverine and hydrothermal inputs. Since an input through dissolution of the REEs from particles by rain water and introduction in the dissolved phase to seawater could be important but has been studied much less, we initiated a research on the REE chemistry of rain water, with the purpose of understanding the behavior of the REEs during rain-formation processes and the significance of this input to the ocean.

2 SAMPLES AND ANALYTICAL PROCEDURE

The rain water samples were collected from five locations, two over the East China Sea and three in the eastern Japan. The samples on land were collected at the same dates or in the same rain events, and are located 18km, 9km and 40m from the coast respectively.

After collection, all rain water samples were filtered immediately through 0.22 μm membrane filters, and the samples used for REE determination were acidified to pH=1.6. Major cations were determined with an ICP-AES, anions with an ion chromatography. The method of preconcentration and determination for REE in the rain waters were the same as those of Shabani et al. (1990) and Zhang et al. (1994). The accuracy and precision of REE measurements were better than ±3%.

3 RESULTS AND DISCUSSION

3.1 Chemistry and Origin of Major Ions

The pH values of the rain water samples range from 5.5 to 6.1, which is slightly higher than the annual mean values of 4.5 - 5.8 at each site in Japan from 1989 to 1993 (Hara et al. 1995). Since alkaline dust, (a large portion of soil dust contains calcium carbonate) often dominates precipitation chemistry, the relatively high pH values in these rain waters may be due to relatively high concentration of alkaline dust.

Seven major ions in the rain waters were measured. The total dissolved solids (TDS) are most

347

abundant in those samples collected near the coastline and over the sea. These rain waters, unlike rain water in continental areas (Sanusi et al. 1996, Avila 1996, Mamane and Gottlieb 1995, Tuncel and Ungor 1996), are characterized by high concentrations of Na^+ and K^+, with remarkably $Na^+ > Ca^{2+}$ and $K^+ > Ca^{2+}$. For anions, Cl^- and NO_3^- dominate in most samples.

There are, in general, three main origins for the dissolved solids in rain water, which are marine, anthropogenic and terrigenic (mineral dust). The presence of nitrate in rain water can be attributed to an input of gaseous nitric acid, and used as a tracer of anthropogenic pollution (Ezcurra et al. 1988, Colin et al. 1989, Sanusi et al. 1996). The sulphate chemistry is slightly complicated, but this component is of marine origin. Na is generally considered as a tracer of the marine source, although there may be a small contribution of terrigenic origin from crustal aerosols. Consequently, these tracer elements are used to identify the sources of the dissolved solids in the rain water samples.

All rain water samples have higher SO_4^{2-}/Na^+ ratios than average sea water, whereas some samples have higher Cl^-/Na^+ ratios and some lower. The lower Cl^-/Na^+ ratios can partially be explained by the presence of Na of terrigenic origin, and those $Cl^-/Na+$ ratio larger than 1.17, suggest that Cl^- may not only be marine but there is a contribution from anthropogenic sources. It may result from the stack gases of refuse incineration, combustion and decomposition of organochlorine compounds, such polyvinyl chloride. High SO_4^{2-}/Na^+ ratios are also indicative of anthropogenic origin of Cl^- in these samples. K^+ does not vary in association with $Na+$ in these samples, suggesting their different origins.

3.2 REE Chemistry

The REE concentrations show significant differences between rain waters from different locations but also among rain events at the same location. As compared with the surface seawater of the Pacific and Indian Oceans, these rain waters have similar overall ranges of heavy REE concentrations but clearly higher light REE concentrations. For samples from the same locations, all rain waters except those of Tokyo show larger variations in the concentrations of heavy REEs than those of light REEs. However, for rain waters from different locations, the light REE concentration shows about 10 fold variations, larger than the variation of heavy REE concentration. The REE compositions of these rain waters, as seen from the shale-normalized REE patterns (Figure 1), are largely variable, ranging from a shale-like to LREE-enriched compositions. The rain waters collected at Tokyo have flat patterns, indicating a shale-like composition, while the rain waters from Chofu, a medium industry city, pose the largest en-

richment of light REEs, with variable heavy REE abundances and relatively constant light REE abundances among these samples. It is noteworthy that the rain waters of Chofu and Nakaminato, which were collected on the same dates or same rain events, demonstrate similar REE patterns, although there is fractionation between light and heavy REEs. Several factors, such as source, pH of precipitation, solid state speciation of REEs and duration of aqueous-solid phase interaction, are considered to affect the REE composition of rain water.

3.3.1 Sources

The REEs in rain water are most likely derived from the washout of various types of aerosols. The REEs in aerosols like Cu, Zn and Pb (Chester et al. 1994) is the case for trace metals probably have three main types of globally important sources. They include 1) low temperature crustal weathering (crustal source), 2) a variety of mainly high temperature anthropogenic processes (anthropogenic source) and 3) sea-salt or sea-spray generation (oceanic source). For the crustal source, the REE composition of average crustal composition or shale may be taken as a reference, and those of bulk surface sea water as a refer-

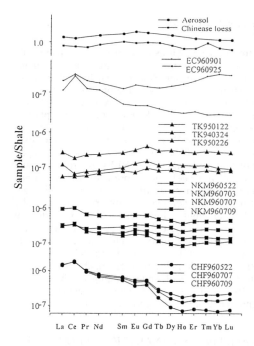

Figure 1. Shale-normalized REE patterns of the rain waters and the aerosol (data from Greaves et al. 1994) and the Chinese loess (data from Liu et al. 1993)

ence for the oceanic source. However, very little is known of the REE composition of the anthropogenic source. It is, thus, difficult to assess the principal sources of the REEs in aerosols. From a simple comparison of the shale-normalized REE patterns between aerosol, loess and the rain waters, the REEs in the rain waters of Tokyo might have been mainly

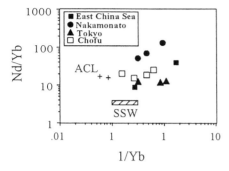

Figure 2. Variation of Nd/Yb vs. 1/Yb for the rain waters and also the average Chinese loess (data from Liu et al. 1993) and the bulk surface sea water of the Pacific Ocean)

derived from a crustal source, since their REE patterns are loess- or aerosol-like. To identify the sources or components of the REEs in the rain waters, the Nd/Yb fractionation is illustrated versus the change of 1/Yb value in Figure 3. As seen from this figure, a simple mixing between three components, i. e., crustal, oceanic and anthropogenic source, can account for the REE compositional variations for most of the rain waters. The anthropogenic source is probably characterized by different degrees of light REE-enrichment and low heavy REE concentration. Mixing mainly of such anthropogenic source with crustal and oceanic sources may explain the REE composition signature of Chofu , and of East China Sea and Nakaminato rain waters. The large percentages of sea spray (PSS) in the rain waters of Naka-minato and East China Sea support the important contribution of sea spray to their REE compositions.

3.3.2 Solid state speciation of REEs in aerosol

The REE composition of the rain water is further affected by the extent to which they are leached from aerosols. The rain water dissolution of the REEs from an aerosol is related to their solid state speciation, the manner in which they are partitioned between the host components of both the crustal and anthropogenic material in the parent material, which in turn is dependent on the original source of the REEs. Chester et al. (1989, 1994) considered that the most important binding fractions of trace metals such as Cu, Zn and Pb in aerosols are 1) an ex-

changeable association, which refers to those fractions which are associated with the surfaces of both inorganic and organic aerosol components in the most loosely held forms, 2) an oxide and carbonate association, which refers to all associations with carbonates and oxides, and includes metals which are in surface positions on phases such as hydrous iron and manganese oxides, clays and organic matter but which are not part of the most loosely held exchangeable association; 3) a refractory association, which refers to trace metals associated with the crystalline matrix of minerals and with refractory organic matter. The leaching experiments carried out by Chester et al. (1989, 1994) show that the trace metals such as Cu, Zn and Pb have higher proportions of their total concentrations in exchangeable associations in anthropogenic-rich than crust-rich samples. It is these exchangeable trace metals which are the most mobile in the aqueous environment. No study has been carried out to determine the solid state speciation of the REEs in aerosols. However, the experimental results of these authors could be referred to assess the solid state speciation of the REEs in aerosol.

In the rain waters of Chofu, which were less influenced by oceanic source than those of Nakaminato, the light REEs have been much enriched over the heavy REEs. The crustal source can not account for such a REE composition. The REE composition of this rain water is probably produced by dissolution of an adsorption phase or exchangeable fraction, since the REEs adsorbed onto particulate oxides in equilibration with sea water show light REE enrichment (Koeppendastrop and De Carb 1992), and rain water more acidic than sea water can easily wash out of the REEs from the surface of aerosols. Since anthropogenic-rich aerosols have higher proportions of exchangeable associations or adsorbed trace metals (Chester et al. 1994), the dissolved solids in the rain waters of Chofu are most likely of anthropogenic origin. This is in accordance with the enrichment of pollution components such as NO_3^- and SO_4^{2-} in these rain waters. Although the rain waters of Tokyo are also enriched in such anthropogenic components, their REE compositions do not show enrichment of light REEs, and are similar to that of crustal source. This decoupling between major ion and the REE composition could be the cause of different degrees of particulate-water interaction. This is because more intensive interaction between rain water and aerosol, which depends on several factors such as pH of rain water, temperature and interaction time, will wash out more REEs from the crystalline matrix of particulate minerals.

3.3.3 Acidity of rain water

The scavenging rate of the REEs from aerosols by rain water is largely dependent on the pH of this water, because cation adsorption on various soil components shows strong pH dependence (Davis and Kent 1990). According to the relationship identified by several authors (e.g. Chester et al. 1990) between the pH of rain water and the extent to which trace metals are dissolved from aerosols, Zn and Pb, for example, have been shown to be more soluble in acidic than in basic rain waters. It is considered that the REEs in aerosols have similar behavior as Pb, i.e. their mobiling increases with decreasing pH of the rain water. However, the pH values of these rain waters do not show significant variations. Moreover, the pH value in rain water changes with time during a rain event, and is often higher at the start of the rain event. Further, knowledge is required about the pH dependence of fractionation between REEs. At present, it is not possible to assess the effect of pH on the REE composition of rain water. For this reason, a survey of the REE fractionation in rain water is needed.

3.4 Atmospheric Fluxes of REEs Into the Ocean

Greaves et al. (1994) have estimated atmospheric fluxes of REEs through dry deposition and release of REEs by reaction with seawater. Their results show that the dry flux of the REEs are comparable with river fluxes, and larger fluxes of light REE, compared with heavy REE fluxes. Based on a yearly rainfall amount of 3.910^{17} kg/yr. over the oceans, our calculation results in fluxes of individual REEs ranging from about 3 times for heavy REEs to about 15 times for light REEs of dry deposition fluxes. The cerium has the highest flux in wet deposition. These results indicate that the atmospheric inputs of the REEs, especially of the wet deposition fluxes are much more significant relative to those of dry deposition and riverine flux to the oceans.

4 CONCLUSIONS

1. Rare earth element abundance in rain water, although very low, can be determined precisely with solvent extraction method followed by ICP-MS analysis, and have the significance for understanding the rare earth element chemistry of rain water.

2. Fractionation between rare earth elements depends on chemical composition and the physicochemical features of rain water.

3. Atmospheric input of REEs of wet deposition is more significant relative to dry deposition and riverine flux.

REFERENCES

Avila, A. 1996. Time trends in the precipitation chemistry at a mountain site in northeastern Spain for the period 1983-1994. *Atmo. Environ.* 30: 1363-1373.

Chester, R., G.F. Bradshaw & P.A. Corcoran 1994. Trace metal chemistry of the North Sea particulate aerosol; concentrations, sources and sea water fates. *Atmo. Environ.* 28: 2873-2883.

Chester, R., K.J.T. Murphy & F.J. Lin 1989. A three stage sequential leaching scheme for the characterization of the sources and environmental mobility of trace metals in the marine aerosol. *Environ. Technol. Lett.* 10: 887-900.

Chester, R., Nimmo M., Murphy K.J.T. & Nicholas E. 1990. Atmospheric trace metals transported to the Western Mediterranean: data from a station on Cap Ferrat. *Water Pollut. Reps.* 20: 597-612.

Colin, J.L., D. Renard, V. Lexsoat, J.L. Jaffrezo, J.M. Gros & B. Srauss 1989. Relationship between rain and snow acidity and air mass trajectory in eastern France. *Atmo. Environ.* 23: 1487-1498.

Davis, J.A. & D.B. Kent 1990. Surface energy and adsorption at mineral/water interfaces: An introduction. In:. M.F. Hochella, Jr. & A.F. White (eds), Mineral-Water Interface Geochemistry *Reviews in Mineralogy* 23. Mineralogical Society of America.

Ezcurra, A., H. Casado, J.P. Lacaus & C. Garcia 1988. Relationships between meteorological situations and acid rain in Spanish Basque country. *Atmo. Environ.* 12: 2779-2789.

Greaves, M. J., P.J. Statham & H. Elderfield 1994. Rare earth element mobilization from marine atmospheric dust into seawater. *Mar. Chem.* 46: 255-260.

Hara, H., M. Kitamura, A. Mori, I. Noguchi, T. Ohizumi, S. Seto, T. Takeuchi & T. Deguchi 1995. Precipitation chemistry in Japan 1989-1993. *Water, Air, and Soil Pollution* 85: 2307-2312.

Koeppenkastrop, D. & E.H. De Carlo 1992. Sorption of rare-earth elements from seawater onto synthetic mineral particles: An experimental approach. *Chem. Geol.* 95: 251-263.

Liu, C.-Q., A. Masuda, A. Okada, S. Yabuki, J. Zhang & Z.-L. Fan 1993. A geochemical study of loess and desert sand in Northern China: Implications for continental crust weathering and composition. *Chemical Geology* 106: 359-374.

Mamane, Y. and J. Gottlieb 1995. Then rears of precipitation chemistry in Haifa, Israel. *Water, Air, and Pollution* 82: 549-558.

Sanusi, A., H. Wortham, M. Millet & P. Mirabel 1996: Chemical composition of rain water in Eastern France. *Atmo. Environ.* 30: 59-71.

Shabani, M. B., T. Akagi, H. Shimizu & A. Masuda, 1990. Determination of trace lanthanides and yttrium in seawater by inductively coupled plasma mass spectrometry after preconcentration with solvent extraction and back extraction. *Anal. Chem.* 62: 2709-2714.

Tuncel, S.G. & S. Ungor 1996. Rain water chemistry in Ankara, Turkey. *Atmo. Environ.* 30: 2721-2722.

Zhang, J., H. Amakawa & Y. Nozaki 1994. The comparative behaviors of yttrium and lanthanides in the seawater of the North Pacific. *Geophys. Res. Lett.* 21: 2677-2680.

6 Mineralogy, microbes and chemistry of weathering

Geochemistry of the Earth's Surface, Ármannsson (ed.)© 1999 Balkema, Rotterdam, ISBN 90 5809 073 6

Invited lecture: Exploring microbial controls on mineral weathering

J. F. Banfield, S. A. Welch, A. E. Taunton, C. M. Santelli & W. W. Barker
Department of Geology and Geophysics, University of Wisconsin-Madison, Wis., USA

ABSTRACT: We have measured the effects of specific organic polymers, small organic acids, and metabolic byproducts on mineral dissolution rates and reaction stoichiometries to quantify field-based observations. Microbial-mineral interactions can be driven by metabolic need for elements such as phosphorus. Geochemical and mineralogical data suggest that in organo-phosphate-limited regions of weathering profiles, microbial activity correlates with enhanced apatite dissolution, suppression of secondary phosphate precipitation, and ultimately, dissolution of all phosphates. Microbial colonization is strongly localized by phosphate mineral surfaces. Enhanced apatite dissolution was demonstrated experimentally by comparing reaction rates in controls with rates in phosphate-limited heterotrophic cultures and abiotic experiments containing either acetate or oxalate. For silicates, microbial interactions can enhance or suppress weathering rates. We have shown that organisms can derive energy by catalysis of oxidation of Fe^{2+} liberated by olivine dissolution. However, the resulting increased Fe^{3+} in solution decreases the olivine dissolution rate compared to abiotic controls.

1 INTRODUCTION

A wide diversity of microorganisms colonize the Earth's near sub-surface region. These organisms can potentially effect reactions involving minerals. Microbial populations are limited by available pore space, ambient physical conditions, and availability of suitable energy sources and electron acceptors. Where sufficient porosity exists, cells frequently attach to mineral surfaces. In this case, interactions may occur between mineral surface atoms and organic molecules (e.g., extracellular polymers). Effects of high molecular weight extracellular polymers on silicate reactivity have been quantified by measuring plagioclase feldspar dissolution rates and stoichiometries in solutions containing various polysaccharides (including alginates) at several pHs (Welch et al. 1999). In acidic solutions, rate increase is due to protons associated with carboxylic acid functional groups. Metal binding, which changes solution saturation state and leads to biomineralization, also enhances the rate of dissolution. However, under near neutral pH conditions, the dissolution rate is inhibited by these polymers.

In regions where direct cell colonization does not occur, microorganisms can indirectly affect mineral dissolution and precipitation through production of soluble organic molecules, especially acids (e.g., Berthelin & Belgy 1979, Stillings et al. 1996).

It is well known that microbes can modify dissolution rates and solution speciation by catalysis of redox reactions involving sulfides and oxides (e.g., Nordstrom & Southam 1997). For example, some bacteria and archaea can generate metabolic energy by catalyzing the oxidation of ferrous iron (which is inorganically inhibited at low pH). Under anoxic conditions organisms can utilize ferric iron (and other inorganic compounds) as terminal electron acceptors for growth (e.g., using organic carbon as the energy source). Such processes lead to solubilization and precipitation of a variety of minerals (Lovely & Phillips 1988).

The extent to which organisms can generate energy from oxidation of silicate minerals, and thereby affect mineral dissolution, is unknown. A variety of potential energy sources are available, but reduced iron in biotite, amphiboles, pyroxenes, and olivine, is potentially the most abundant and important. Consequently, it is appropriate to explore the possibility of microbial growth by oxidation of reduced iron released by silicate dissolution.

Here we present examples from our recent work that illustrate the dramatic role that both heterotrophic and lithotrophic microorganisms may play in determining the behavior of some biologically and geochemically important elements. Experimental results contribute to our understanding of processes documented in field-based investigations and begin to define the range of

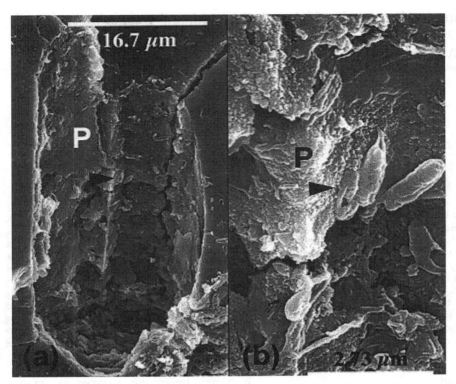

Figure 1 Secondary phosphates after apatite. (b): Higher magnification image of area in (a). Arrow indicates one of a number of cells on the secondary phosphate mineral surface (b) visible in both images

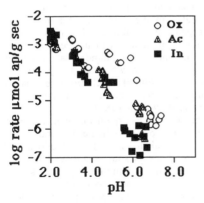

Figure 2: Apatite dissolution: in oxalate ○, acetate △, and inorganic ■

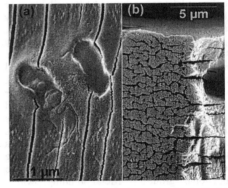

Figure 3: SEM images fayalite reacted biotically (a) and abiotically (b)

conditions under which minerals may sustain microbial populations in the subsurface.

2 FIELD STUDIES: PHOSPHATES

Prior investigations have documented the dissolution of apatite and the formation of typically insoluble secondary phosphate phases during early weathering of a granite from Southern NSW, Australia (Banfield & Eggleton 1989). Very high lanthanide abundances in the lower weathering profile clearly indicate lanthanide addition, implying secondary phosphate solubilization elsewhere. Using geochemical data, mineralogy, and microbe distribution patterns (Taunton et al. in prep.), we have shown a strong correlation between location in the weathering profile, abundance of microorganisms, and lack of persistence of secondary phosphates. We infer that, due to the scarcity of organic matter, microbial communities in the weathering profiles are phosphate limited. Thus, microorganisms populate secondary phosphate surfaces (Figure. 1) and promote their dissolution by indirect or direct means. This result has important implications for understanding microbial controls on soil fertility, and also for research that involves use of "insoluble" secondary phosphates to immobilize heavy metal and radionuclides at contaminated sites.

3 EXPERIMENTAL: PHOSPHATES

We have initiated a series of experiments in order to evaluate the effects of microorganisms and their metabolic byproducts on phosphate mineral reactivity (Figure 2). Heterotrophic organisms from the field site were cultured under phosphate-limited conditions in the presence of apatite and apatite plus biotite. In both cases, several hundred micromoles phosphate was released to solution. In the biotite plus apatite experiments, pH decreased from 7 to ~3-4, but in the apatite only experiments the pH remained near neutral. The dissolved phosphate concentrations in biological experiments are about two orders of magnitude greater than control values. Results suggest that equivalent solubilization is achieved by different types of organic ligands.

Specific organic ligands such as acetate and oxalate were tested for their ability to effect rates of phosphate mineral dissolution. Results show pH-dependent rate enhancement, followed by saturation and precipitation of secondary minerals. Primary mineral surfaces become highly etched. Morphological differences exist between surfaces reacted abiotically versus biologically, suggesting organic molecules specifically interact with surface sites.

4 LITHOTROPHS AND FE-SILICATES

Experiments were conducted to determine whether silicate dissolution can support growth of lithotrophic microorganisms and to test the effect of metabolism on dissolution rates. Fayalite was reacted at pH 2 and 3 with cultures of *Thiobacillus*

ferrooxidans. Microbial growth occurred. Solutions were non-stoichiometric relative to fayalite. Biomineralization of iron oxyhydroxides, often associated with cell surface polymers (Figure 3), occurred (Santelli et al. in prep.).

Rates for abiotic controls (agitated media) were comparable to fayalite dissolution rates under anoxic (N_2 purged) conditions and to anoxic rates reported previously (Casey et al. 1993). Rates of accumulation of Fe and Si in solution in abiotic experiments were much faster than rates in biotic experiments (an order of magnitude faster at pH 2). Characterization of reacted minerals verifies that this is due to more extensive dissolution of fayalite and not to different amounts of secondary mineral precipitation. As expected, biotic solutions contain more ferric iron. Rate suppression due to iron oxidation was simulated by addition of ferric iron to abiotic controls. We attribute rate suppression to interactions between ferric ions and the fayalite surface.

Fayalite (001) surfaces are highly unreactive. In abiotic experiments, the extent of etching of (010) and (100) surfaces is very different. This is not true for biotic experiments. The change in relative reactivity of (100) and (010) suggests that rate suppression in biotic experiments may involve binding of ferric ions, or ferric iron oxyhydroxide or oxide precipitation, on (010).

In biotic experiments using natural fayalite, regular channels (tens of nanometer wide, spaced on the sub-micron-scale) develop parallel to (001). These channels are very similar to those previously reported from naturally weathered olivine (Smith et al. 1987). Channels may be localized by subtle ordering of ferric iron and vacancies.

5 ACKNOWLEDGMENTS

This research was supported by grants US Department of Energy and DE-FG02-93ER14328 and National Science Foundation EAR-9706382

REFERENCES:

Banfield JF & RA Eggleton 1989 Apatite replacement and rare earth mobilization, fractionation and fixation during weathering. *Clays and Clay Minerals* 37: 113-127

Berthelin J & G Belgy 1979 Microbial degradation of phyllosilicates during simulated podzolization. *Geoderma* 21:297

Casey WH JF Banfield, HR Westrich & L McLauglin 1993 What do dissolution experiments tell us about natural weathering? *Chemical Geol.ogy* 105: 1

Lovley DR & EJP Phillips 1988 *Appl. Environ Microbiol.* 51: 683-689.

Nordstrom DK, G Southam 1997 MSA Rev.Min 35: 361

Smith K, AR Milnes & RA Eggleton 1987 Weathering of basalt: Formation of iddingsite. *Clays and Clay Mins* 35:418.

Stillings LL JI Drever, SL Brantely, Y Sun & R Oxburgh 1996 Rates of feldspar dissolution at pH 3-7 with 0-8 mM oxalic acid. *Chem. Geol.* 132:79-89

Welch SA WW Barker & J.F. Banfield 1999 Microbial extracellular polysaccharides and plagioclase dissolution. *Geochim. Cosmochim. Acta*, in press.

Geochemistry of the Earth's Surface, Ármannsson (ed.) © 1999 Balkema, Rotterdam, ISBN 90 5809 073 6

Invited lecture: Abiotic vs. biotic dissolution of hornblende

S. L. Brantley, L. Liermann, B. Kalinowski & S. Givens
Department of Geosciences, Pennsylvania State University, University Park, Pa., USA

C. G. Pantano & A. Barnes
Department of Material Science and Engineering, Pennsylvania State University, University Park, Pa., USA

ABSTRACT. In this work, the rates and mechanisms of weathering and Fe release from hornblende were investigated under abiotic and biotic conditions. Under abiotic conditions, the dissolution rate of hornblende increased from pH 6 to 1, and stoichiometry of Fe and Si release rates varied with pH. For biotic experiments, bacteria from an Adirondack soil were isolated that could extract Fe from the hornblende in minimal media. We showed that these bacteria enhanced the release rate of Fe from hornblende over rates in sterile culture, produced low molecular weight organic acids and acidic biofilms, enhanced etching of hornblende underneath biofilms, produced catecholate siderophores, changed the stoichiometry of hornblende dissolution compared to abiotic dissolution, and (perhaps) recycled siderophores for Fe transport. Enhanced Fe release rates from hornblende are also reported for experiments without bacteria but with a commercially available derivative of a natural hydroxamate siderophore added to the solution.

1 HORNBLENDE DISSOLUTION AND METAL RELEASE

Although the average crustal abundances and concentrations in surface waters of Fe, Mn, Zn, Ni, Cu, Co, and Mo are extremely low, each of these metals is used in bacterial enzymes, coenzymes, and cofactors by common anaerobic and aerobic soil bacteria. While it is well known that bacteria excrete siderophores to extract Fe from their environment, it is not understood how these siderophores attack minerals to provide the Fe(III), nor is it understood how bacteria extract micronutrients other than Fe. One primary silicate mineral that contains these micronutrients and is the principal source of Fe and Mg to natural waters from soils developed on granite is the mineral hornblende. In this investigation we investigated the dissolution of hornblende under biotic and abiotic conditions.

2 ABIOTIC HORNBLENDE DISSOLUTION

2.1 *Dissolution rates*

Flow-through reactors were used for inorganic aerobic dissolution experiments for measuring hornblende dissolution at pH 1, 2, 3 and 6. Reactant solutions (HCl-H_2O) were pumped through reactors containing powders and polished planchets of hornblende. Reactors were continuously agitated in a bath at ambient temperature and solution was collected at the outlet.

BET (Brunauer-Emmett-Teller) surface area of powder (A, 0.33 m^2/g) measured before dissolution, was used in the following equation to determine the rate of dissolution of hornblende during the experiment:

$$R = \frac{Q(C_{out} - C_{in})}{Am} \qquad (1)$$

where R is the rate of element release (mol/m^2 s), Q is the pump rate (l/s), C_x is the concentration in the effluent (mol/l, $x = in$ or out for in- or outlet), and m is the mass (g) of hornblende. The rate of Si release from hornblende (6.2 x 10^{-11} mol m^{-2} s^{-1}) is comparable to published data at low pH, considering that differences in silicate dissolution rate as large as an order of magnitude are not uncommon. We observed that dissolution decreases over ~11 months and is generally nonstoichiometric. Fitting our data to the customary rate equation yields an exponent that is typical of silicates, $R = k a_{H+}^{0.5}$ (where k = rate constant and a_{H+} = activity of proton).

3 EFFECT OF BACTERIA ON HORNBLENDE DISSOLUTION

3.1 Isolated bacteria

Laboratory strains of bacteria are routinely cultured in media with high nutrient concentrations. Such strains may have lost the ability to obtain metals from sources other than solution. Thus, for all our experiments, we isolated fresh bacteria from soil. In these isolations, we used inocula from a soil in the Adirondack mountains (NY, USA) containing hornblende in enriched medium. After collection, these isolates were tested to see if they could extract nutrients from hornblende. Partial sequencing of the 16s rRNA gene (Nucleic Acid Facility, Life Sciences Consortium, Penn State) of two aerobes capable of mobilizing Fe from hornblende showed that neither of the isolates is found in the ribosomal database project (Department of Microbiology, Univ. of Illinois). However, based upon sequencing similarities, the two isolates are closely related to the genera *Streptomyces* and either *Arthrobacter* or *Micrococcus*. SEM (scanning electron microscopic) analysis of the latter species shows a rod-coccus growth cycle, typical for arthrobacter but not for micrococcus. We refer to it here as an arthrobacter.

We identified two strategies for mobilizing Fe from hornblende used by both species (Liermann et al. 1999a,b): 1) the bacteria secreted low molecular weight organic acids (LMWOA); 2) the bacteria secreted siderophores. These two strategies are described below.

3.2 Acid production

To investigate the acid production, pH within microbial biofilms was measured with microelectrodes (Liermann et al. 1999b). We used both hornblende crystal and synthesized hornblende glass which exhibited a homogeneous composition (i.e. the crystal has secondary phases while the glass is homogeneous). We also used a low-Fe glass for comparison. The values of the pH difference, $pH_{bulk\ medium}$ - $pH_{mineral-water\ interface}$, were larger in unbuffered ($\Delta pH \leq 0.6$) than buffered medium ($\Delta pH \approx 0.04$). Growth of the *Arthrobacter* sp. also resulted in a larger pH difference between bulk and biofilm than the *Streptomyces* sp. by up to a factor of 10 (Figure 1).

SEM analysis of hornblende after biofilm removal showed enhanced weathering along polishing scratches, although interpretation was difficult due to secondary phases. Low-Fe glass surfaces were also more pitted after bacterial incubation. SEM analysis of high-Fe glass surfaces revealed no differences between abiotic and bacteria-incubated samples The growth rate and production of

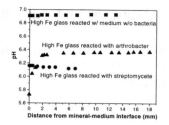

Figure 1. Gradients in pH as measured using microelectrodes *in situ* in biofilm growth experiments in unbuffered medium. The zero point on the x axis refers to the hornblende glass-water interface. Changes in pH are presumably due to LMWOA, which are produced in greater quantities by the arthrobacter than the streptomycete. The hornblende (high Fe) glass was synthesized to have the same bulk chemical composition as the natural hornblende crystal, but without the added complexity of secondary phases such as chlorite.

LMWOA by the arthrobacter exceeds that of streptomyces, consistent with greater ΔpH in arthrobacter biofilms (Figure 1). LMWOA commonly produced by microbes include 2-ketogluconic, lactic, acetic, citric, oxalic, pyruvic, and succinic. Our preliminary HPLC (high performance liquid chromatography) data indicates production of benzoic and butanoic acids by the streptomyces, but we have no data yet for the arthrobacter. Some acids may have been produced by both..

3.3 Siderophore production

In buffered media, the streptomyces and arthrobacter increased the Fe release rate from hornblende over the first week up to 5x to 10x over abiotic controls respectively (Liermann et al., 1999a; Kalinowski et al., 1999). Two different catecholate siderophores produced by the isolates and characterized by HPLC and mass spectrometry (MS) are presumed to cause this Fe release enhancement. Without *Streptomyces* sp. present, Fe/Si and Al/Si release rate ratios were lower than hornblende starting material. In comparison, with *Streptomyces* sp., the Fe/Si release ratio was increased while the Al/Si ratio remained relatively unchanged. Therefore, stoichiometry of dissolution depends upon biotic vs abiotic conditions.

4 EFFECT OF DFAM ON HORNBLENDE DISSOLUTION

4.1 DFAM

Hornblende dissolution was analyzed in the presence of deferoxamine mesylate (DFAM), a hydroxamate siderophore derivative. Deferoxamnine is one of many siderophores produced by streptomycetes. Addition of DFAM increased the dissolution rate non-linearly according to the rate equation, $R = (8.8 \times 10^{-13}) C^{0.5}$ where R is the release rate of Fe (mol $m^{-2} s^{-1}$) and C is the concentration (mol l^{-1}) of DFAM. The DFAM also increased release of Al and Si from hornblende. Differences in stoichiometry of dissolution in the presence of DFAM vs. streptomycete may be related to selectivity of catecholates *vs* hydroxamates.

A second set of observations also emphasizes the difference in stoichiometry (and implicitly, mechanism of dissolution) in comparisons of siderophore-promoted (biotic) and proton-promoted (abiotic) dissolution of hornblende. XPS (X ray photoelectron spectroscopy) of hornblende planchets after dissolution in the presence of *Arthrobacter* sp. revealed a substantial drop in the Fe/Si ratio of the hornblende surface after removal of the bacteria (ratio ≈ 0.1) as compared to control samples treated in media without bacteria (ratio ≈ 0.2) (Kalinowski et al., 1999). In contrast, XPS measurements of polished planchets of hornblende buried in a Gore Mountain soil for several weeks have shown Fe/Si ratios that are elevated above blank values (> 0.20). Understanding how soil microbiota affect the stoichiometry of dissolution and precipitation at the mineral surface is of particular importance in understanding the rate and mechanism of weathering of hornblende.

Of additional interest, when we added the streptomycete in the presence of DFAM (a foreign siderophore for this bacterium) at three concentrations, we consistently observed enhanced Fe release approximately 2 to 3 times over the rate with siderophore alone. Using the rate equation, a smaller increase would have been predicted, suggesting recycling of the siderophore or enhanced production of another siderophore by the streptomycete in the presence of DFAM.

5 SUMMARY

We measured the rates of abiotic hornblende dissolution and observed 1) that dissolution increased as pH decreased from pH 6 to 1, and 2) that stoichiometry of dissolution varied with pH. We also isolated bacteria from an Adirondack soil that could extract Fe from the hornblende surface in minimal medium. We showed that these bacteria could 1) enhance Fe release over sterile control experiments 2) create acidic microenvironments in biofilms 3) produce catecholates 4) change the stoichiometry of hornblende dissolution and 5)

(perhaps) recycle siderophores. In addition, we used surface spectroscopic and microscopic techniques to identify effects of bacteria on the mineral surface.

REFERENCES

Kalinowski B.E., Liermann, L.J. Brantley, S.L., & Pantano, C.G. 1999, Microbial surface reactions on hornblende as studied by x-ray photoelectron spectrometry and scanning electron microscopy. *Subm.Geochim. Cosmochim. Acta.*

Liermann, L. J., Kalinowski, B. E., Brantley, S. L. & Ferry, J. G. 1999a, Role of bacterial siderophores in dissolution of hornblende . *Subm. Geochim. Cosmochim. Acta.*

Liermann, L.J., Kalinowski, B.E., Barnes, A.S. Zhou, X., & Brantley, S.L. 1999b, Measurements of pH by microelectrodes in biofilms grown on silicate surfaces. *Subm. Chem. Geo.*

Geochemistry of the Earth's Surface, Ármannsson (ed.) © 1999 Balkema, Rotterdam, ISBN 90 5809 073 6

Processes controlling Ca and Mg mobility of natural waters in Skagafjördur, N-Iceland

A. Andrésdóttir & S. Arnórsson
Science Institute, University of Iceland, Reykjavík, Iceland

ABSTRACT: In the Valley of Skagafjördur, N-Iceland, geothermal activity is widespread. Natural waters from the area and the highlands south of the valley have been divided into four groups on the basis of their composition and geology as 1) surface waters 2) soil waters 3) groundwaters in the lowlands and 4) groundwaters in the highlands. Soil waters show the highest CO_2 partial pressure and Ca and Mg contents but the lowest pH, around 7. The pH of groundwaters is around 10 and their CO_2 partial pressure and Ca and Mg contents are low. The Ca content of groundwaters is controlled by the solubility of calcite and that of Mg by the solubility of amorphous magnesium-silicate of the composition $Mg_4SiO_3O_{10-x}(OH)_{2x}$. All the other water types are undersaturated with these minerals. Results indicate that vegetative soil cover enhances the release of Ca and Mg into rivers and streams from rock derived material.

1 INTRODUCTION

Geothermal activity is widespread in the Valley of Skagafjördur, N-Iceland. Natural waters in the area and in the highlands south of the valley have been sampled for chemical analysis of major and trace elements, along with measurements of selected stable isotopes. This contribution only deals with the main chemical characteristics of the waters which form the basis for their classification into different groups, as well as the processes controlling the Ca and Mg contents of these waters. The bedrock in the study area consists mainly of flood basalts of Miocene age. The oldest rocks are exposed in the northeastern part (Saemundsson et al. 1980). A description on the bedrock geology and hydrology of the southern tributary valleys of Skagafjördur and their surroundings has been given by Hjartarson et al. (1998). In the southernmost part of the study area formations are of Plio-Pleistocene age and the formations in the highlands are located near the active volcanic zone (Jakobsson 1979). Permeability of the bedrock is low except where fractures occur. Younger formations are more permeable than the older ones. Occurrence of warm and hot springs and drillholes in the area has been described by Kristmannsdóttir et al. (1984) and Karlsdóttir et al. (1991).

2 CHEMICAL CARACTERISTICS

2.1 Types of water

Waters from the have been divided into four groups on the basis of their chemical composition and geological setting. They are: (1) surface waters; rivers, streams and lakes (2) soil waters, emerging from soil that is high in organic matter (3) groundwaters in the lowlands ranging from non-thermal to thermal (4) groundwaters in the highlands south of the Skagafjördur valley system.

The surface waters in the Skagafjördur area contain 10-100 ppm dissolved solids but most values lie between 40 and 80 mg/l. Soil waters are higher in dissolved solids, 30-260 mg/l, largely due to higher Ca, Mg and HCO_3 contents. In both these water types the main anions are Cl and HCO_3 and the main cation Na, but also Ca and Mg in the case of soil waters. The dissolved solids content of the thermal groundwaters, which have temperatures ranging from ambient to 90°C, is 60-310 mg/l and higher than that of most non-thermal groundwaters, which contain 30-100 mg/l. The dissolved solids content increases rather smoothly with rising temperature. Silica is the predominant constituent in these waters and Na the most abundant cation. The main anions are HCO_3, Cl, SO_4 and ionized silica, their relative concentrations being quite variable. Waters from springs in the highlands possess some chemical and isotopic differences from groundwaters in the low-

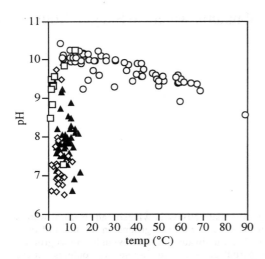

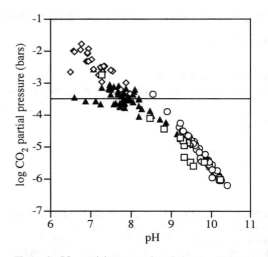

Figure 1. pH of natural waters in Skagafjördur. Triangles: surface waters, diamonds: soil waters, circles: groundwaters in the lowlands, boxes: groundwaters in the highlands.

Figure 2. CO_2 partial pressure in relation to pH in natural waters in Skagafjördur. Symbols have the same notation as in Figure 1. The reference line is the atmospheric CO_2 pressure.

lands that can be related to permeability and, for that reason, they have been considered a separate group. Most groundwaters contain more SiO_2 than surface and soil waters. Rising temperature and high pH leads to silica dissolution and ionization. The water from the springs in the highlands is relatively low in SiO_2. A probable explanation is short residence time in the younger and more permeable bedrock in this part of the area causing less rock dissolution.

2.2 pH, Ca, Mg and CO_2 partial pressure

pH, CO_2 partial pressure and the state of calcite and magnesium-silicate saturation were calculated for water samples at the in situ temperature with the aid of the aqueous speciation program WATCH (Arnórsson et al. 1982, Bjarnason 1994). In surface waters pH is most often in the range of 7-8 (Figure 1). Soil waters tend to have lower pH. Groundwaters, both thermal and non-thermal have considerably higher pH. It reaches maximum of around 10 at 25°C but decreases with increasing temperature above 25°C. Waters emerging in organic soil have the highest CO_2 partial pressure (Figure 2) and the highest Ca and Mg contents (Figures 3 and 4), whereas groundwaters are lowest in these components, and the CO_2 partial pressure is much lower than that of the atmosphere. Surface waters are generally higher in Ca and Mg than groundwater and have CO_2 partial pressures similar to that of the atmosphere.

3 CONTROLLING PROCESSES

3.1 Water-rock interaction

According to Gíslason & Eugster (1987a,b) the pH of surface waters in Iceland is controlled by two counteracting processes, i.e. consumption of protons by dissolution of rock-forming minerals which act as bases, and generation of protons by supply of CO_2 to the water from atmospheric and organic sources. When surface waters seep into the ground and become isolated from the atmospheric and organic sources of CO_2, progressive interaction with the rock leads to increased water pH until a balance is reached between proton consumption by mafic mineral dissolution and hyroxide ion consumption by precipitation of secondary OH-bearing minerals (Arnórsson et al. 1995).

The process controlling the compositional characteristics of the soil waters is little doubt CO_2 flux to the water from decaying organic matter in the soil. This causes decrease in pH and increase in CO_2 partial pressure, resulting in rise of aqueous Ca and Mg concentrations due to increased degree of undersaturation of Ca-and Mg-bearing minerals in the soil (Figures 3 and 4).

In groundwaters rock dissolution quickly leads to a rise in pH, to as much as 10. The main proton donor to groundwaters isolated from the atmosphere, when the pH has risen sufficiently, is ionization of silica dissolved from the rock. Silica ionization along with possible precipitation of secondary OH-bearing minerals with increasing temperature will tend to lower the pH of the water (Andrésdóttir et al. 1998).

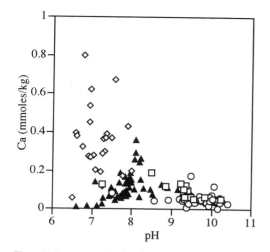

Figure 3. Ca concentration in relation to pH in natural waters in Skagafjördur.

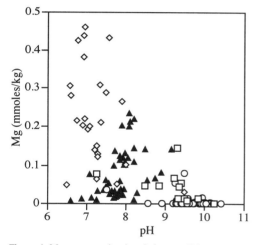

Figure 4. Mg concentration in relation to pH in natural waters in Skagafjördur. Same symbol notation as in Figure 1

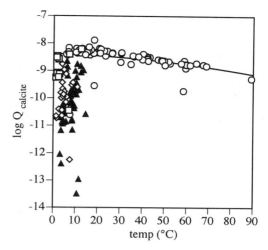

Figure 5. The state of calcite saturation in natural waters in Skagafjördur. The curve indicates calcite solubility according to Plummer & Busenberg (1982). Same symbol notation as in Figure 1.

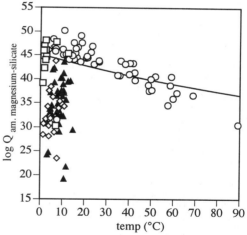

Figure 6. The state of amorphous magnesium-silicate saturation in natural waters in Skagafjördur. The curve indicates the solubility of amorphous magnesium-silicate ($Mg_4SiO_3O_{10-x}(OH)_{2x}$) according to Gunnarsson (1999) Symbols have the same notation as in Figure 1

Moulton & Berner (1998) studied the effect of plants on weathering in western Iceland. According to them the rate of release of Ca and Mg to streams and vegetation is two to five times higher in areas of new forest development than the release of these components to streams in barren areas.

3.2 Mineral saturation

If one looks at the state of calcite and magnesium-silicate saturation of natural waters in Skagafjördur in Figures 5 and 6, it is evident that the low Ca and Mg content of the groundwaters is controlled by the solubility of calcite, on one hand, and amorphous magnesium-silicate, on the other. All other water types are undersaturated with these minerals. This will lead to leaching of Ca and Mg from the rock and soil to surface waters and especially soil waters. Most of the groundwaters have closely approached calcite and amorphous magnesium-silicate saturation.

4 CONCLUSIONS

It is concluded that vegetative soil cover must have a marked influence on Ca and Mg mobility and therefore, its transport from sites of weathering to the ocean. Accordingly organic soil cover can be expected to influence the global cycle of CO_2 through enhanced Ca and Mg transport to the ocean and their subsequent precipitation as calcium- and magnesium-carbonates.

REFERENCES

Andrésdóttir, A., S. Arnórsson & Á.E Sveinbjörnsdóttir 1998: Geochemistry of natural waters in Skagafjördur, N-Iceland. In Arehart, G.B & J.R. Hulston (eds.): *Water-Rock Interaction*: 477-481. Rotterdam: Balkema.

Arnórsson, S., S. Sigurðsson & H. Svavarsson 1982. The chemistry of geothermal waters in Iceland. I. Calculation of aqueous speciation from 0° to 370°C. *Geochim. Cosmochim. Acta* 46:1513-1532.

Arnórsson, S., S.R. Gíslason & A. Andrésdóttir 1995. Processes controlling the pH of geothermal waters. *Proc. World Geothermal Congress, Florence, Italy* 2: 957-962.

Bjarnason, J.Ö.1994. *The speciation program WATCH version 2.1*. The National Energy Authority, Reykjavík: 7p.

Gíslason, S.R. & H.P. Eugster 1987a. Meteoric water-basalt interactions. I. A laboratory study. *Geochim. Cosmochim. Acta* 51: 2827-2840.

Gíslason, S.R. & H.P. Eugster 1987b. Meteoric water-basalt interactions. II. A field study in northern Iceland. *Geochim. Cosmochim. Acta* 51:2841-2855

Gunnarsson, I. 1999. *Experiments on the solubility of amorphous silica and magnesium-silicate under hydrothermal conditions*. Unpublished. M.S.-thesis, Reykjavík: University of Iceland: 73 p.

Hjartarson, Á., G.Ó. Fridleifsson & Th.H. Hafstad 1998. *Bedrock in the valleys of Skagfjördur and tunneling routes* (in Icelandic). Unpublished. report of the National Energy Authority, OS-97020, Reykjavík: 55p.

Jakobsson, S.P. 1979. Outline of the petrology of Iceland. *Jökull* 29: 57-73.

Karlsdóttir, R., G.I. Haraldsson, A. Ingimarsdóttir, Á. Gudmundsson & Th.H. Hafstad 1991. *Skagafjördur: Geology, geothermal activity, fresh water and exploration drillings* (in Icelandic). Unpublished. report of the National Energy Authority, OS-91047/JHD-08, Reykjavík: 96 p.

Kristmannsdóttir, H., M.J. Gunnarsdóttir, R. Karlsdóttir, G.I. Haraldsson & H. Jóhannesson 1984. *Geothermal activity in Skagafjördur* (in Icelandic). Unpublished. report of the National Energy Authority, OS-84050/JHD-09, Reykjavík: 107 p.

Moulton, K.l.& R.A. Berner 1998. Quantification of the effect of plants on weathering: Studies in Iceland. *Geology* 26: 895-898.

Plummer, L.N & Busenberg, E. 1982. The solubilities of calcite, aragonite and vaterite in CO_2-H_2O solutions between 0 and 90°C, and an evaluation of the aqueous model for the system $CaCO_3$-CO_2-H_2O. *Geochim. Cosmochim.* Acta 46:1011-1040.

Saemundsson, K., L. Kristjánsson, I. McDougall & N.D. Watkins 1980. K-Ar dating, geological and paleomagnetic study of a 5-km lava succession in Northern Iceland. *Jour. Geoph. Res.* 85: 3628-3646.

Geochemistry of the Earth's Surface, Ármannsson (ed.)© 1999 Balkema, Rotterdam, ISBN 90 5809 073 6

The weathering of Mount Cameroon: 2. Water geochemistry

M. F. Benedetti & J. Boulegue – *UPRESA CNRS 7047, Lab. Géochemie et Métallogénie, Univ. Paris, France*

A. Dia, M. Bulourde & C. Chauvel – *CNRS UPR 4661, Géosciences Rennes, Campus Beaulieu, France*

F. Chabaux, J. Riotte & B. Fritz – *Centre de Géochimie de la Surface, E.O.S.T., Strasbourg, France*

M. Gérard & J. Bertaux – *Laboratoire des Formations Superficielles, IRD, Bondy, France*

J. Etame – *Université de Douala, Département des Sciences de la Terre, Cameroun*

B. Deruelle – *UPRESA, CNRS 7047, Laboratoire de l' Université Paris, France*

Ph. Ildefonse – *Laboratoire de Minéralogie-Cristallographie, Université Paris, UMR 7590 et IPGP, France*

ABSTRACT: In the present study we have focussed our effort on the early stage of the weathering of dated basaltic flows of the Mount Cameroon. We report results on the chemical composition of water samples collected around the volcano. We were able to characterize the end members implicated in the weathering processes: the glassy matrix, rain water and some weathering products. The water rock ratio were calculated with the help of the Strontium isotopic compositions of the water samples

1 INTRODUCTION

Meteoric solution percolate and react with the minerals of the rocks they permeate, generating weathering minerals. When they are formed, these minerals also react with the percolating solution from season to season. In this manner, microscopic and macroscopic structures of soils and alterites are generated (Nahon 1991). The study of the solutions flowing through thick weathering profiles has established that their chemistry depends not only of the dissolution reactions of the primary minerals but also of the equilibrium with the mineral phases at the top of the profiles (Benedetti at al. 1994). Moreover, the few studies integrating water chemistry and mineralogy have shown remarkable discrepancy between minerals/solutions equilibria and recognized mineralogical associations (Sarazin et al. 1982). In most case , at the weathering front very disordered mineral phases are observed while more ordered minerals are present in the upper part of the profiles. It has been shown that the amorphous material could remain in soils up to 250 000 years and that their chemical composition depends of the chemistry of the solutions percolating at the weathering front (Parfitt & Wilson, 1984). It therefore very important to study the early stages of the weathering process to understand the nature of the different chemical reactions controlling the formation of these amorphous materials. In tropical humid conditions the weathering is maximum and favor such studies. Dilute meteoric waters respond readily to dissolution and precipitation reactions and then can be used to define the geochemical nature of such reactions. The dated lava flow of the Mount Cameroon provide a unique opportunity to study solute acquisition by meteoric

waters in basaltic terranes, to quantify and estimate the rate of the weathering. Here, we report the results of the investigation of the chemical composition of natural waters percolating the lava flows of different age and at different weathering stages.

2 SAMPLING

The Mount Cameroon (4095m) is the most accessible active volcano in inter-tropical Africa. The area has dated volcanic lava flow from 1815 to 1982 (Déruelle et al. 1987). The mount Cameroon is facing the sea this geographical location creates a remarkable climatic regime especially on the SW side of the volcano. The rainfall is one of the highest in the world (10 to 12m/year) at Debundcha on the west side at the foot of the Mount Cameroon. The mean temperature at the sea level is ranging from 26 to 29°C with a relative humidity ranging from 75 to 80%. On the south east side of the volcano the rainfall is much less 1800 mm/year at the city of Ekona. Such climatic conditions associated to recent volcanic activity create unique conditions to study early stages of the weathering processes on basic rocks. In December 96 and May 1997, we have sampled in this area 36 river and spring samples all around the volcano. Spring samples were taken at high altitude seeping from unweathered and fractured lava flow. Other samples were taken at lower altitudes on older lava flows as well as in rivers flowing around the volcano. For each sample, t°, pH and alkalinity were determined in the field. major and trace elements were determined in the laboratory on different fractions: fraction 1, pore size 0.2μ HCl acidification for major elements determinations, fraction 2 pore size filtration 0.01µ HNO₃ acidification for trace elements determinations and isotopic composition determinations (Sr, U).

3 RESULTS AND DISCUSSION

Major ion chemistry show a strong correlation with the altitude of the sample. This is shown in FIGURE 1 where the Chloride concentrations are reported as function of the altitude of the sample. The higher the altitude the lower the Cl concentration. This reflects both the effect of the sea water contribution in the rainfall for the low altitude samples and the dilution of the rain water as it reaches higher altitudes.

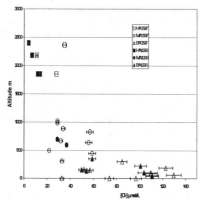

Figure 1: Cl- concentrations as function of the altitude of sampling. HA high altitude, MA medium altitude, BA low altitude;

A similar trend is observed for alkalinity which increases as altitude decreases. This trend could be the result of a longer time of interaction between the percolating water and the fresh lava for the water samples having a lower altitude.

The good correlation between alkalinity and dissolved silica shown in Figure 2 suggest that major elements concentrations are only controlled by the weathering processes and that secondary reactions with organic matter or vegetation are not important. This is also supported by Figure 3 showing the sum of cations as function of alkalinity. in this figure a perfect correlation is obtained indicating that secondary reactions are not controlling the major ion chemistry. Their concentrations in solution is the result of the release of the ion from the basaltic lava as shown in the companion paper investigating the weathered mineral phases (Gérard et al. this volume).

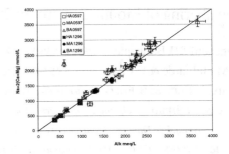

Figure 2: Alkalinity as function of the dissolved silica concentration. HA high altitude, MA medium altitude, BA low altitude

In Figure 4 two mixing hyperboles are observed and corresponding to two types of waters. One trend for the high altitude spring samples and the other trend for the low altitude water samples. Both trends however are related to the same mixing poles which are identified in this figure: the rain water and the basaltic rocks.

Spring waters collected during the dry season have almost uniform Sr isotopic compositions. ^{87}Sr/^{86}Sr ratios vary only between 0.70346 and 0.70355. These values are slightly higher than those of the basaltic lavas (0.70335) showing the influence of a source with more radiogenic Sr isotopes. We suspect that the radiogenic Sr end-member could have its origin in aerosols from seawater. Measurements of ^{87}Sr/^{86}Sr ratios on spring waters collected during the wet season on various sites around the volcano should help confirming this interpretation.

At the scale of the Mount Cameroon volcano, the data obtained on the 0.1mm-filtered waters outline important variations of the (^{234}U/^{238}U) activity ratios from 1.04 to 1.4. The high altitude springs (2400-1000m high) have low ^{234}U-^{238}U disequilibrium, between 1.04 and 1.08, whereas water samples of intermediate altitude (900-200m high) have higher (^{234}U/^{238}U) activity ratios, from 1.1 to 1.4. Spring waters of high and intermediate altitude define good positive correlation in both the (^{234}U/^{238}U) vs. U/Sr diagram and the (^{234}U/^{238}U) vs. Cl diagram, which can be interpreted in terms of mixings between different water sources, i.e. aquifer, with different Cl, U/Sr and ^{234}U/^{238}U signatures. The water samples from lower altitude (<100m high) define a very scattered domain when plotted in the two former diagrams. The U/Sr ratios as well as the Cl concentrations of these waters reach higher value s than in waters of the two other groups. A contribution of waters from surficial horizons of the soils and the weathered profiles could account for the geochemical characteristics of the low altitude samples.

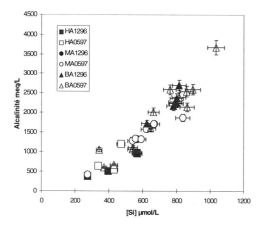

Figure 3: Sum of cations as function of alkalinity. HA high altitude, MA medium altitude, BA low altitude.

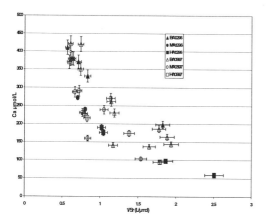

Figure 4: Total calcium concentrations as function of 1/ total Sr concentrations.

Activity diagrams are used to discuss the chemistry of the water with respect to the weathering products identified in the companion paper (Gérard et al., this volume). However, such diagrams can only give general equilibrium trends towards which the geochemical system could or will tend. The major minerals observed on the high altitude weathered lava are amorphous Fe-hydroxides, Al-rich allophanes, and only scarce 7 Å-halloysite and gibbsite (about 1%) may be detected by XRD and FTIR (Gérard et al., this volume). For the older lava at lower altitude halloysite and iron oxides are the major mineral phases. Saturation index were tested and all samples are supersaturated with respect to gibbiste, kaolinite,

quartz, and Fe-oxides and hydroxides. Supersaturation with respect to Al bearing minerals could be partly due to organic complexes that are not taken into account during the calculations. Activity diagrams similar trends and therefore it can be –concluded that Si and Al concentrations in the different water samples are controlled by the above mentioned amorphous mineral phases. The high iron concentrations measured in most samples could also be due to the presence of the hydrous-ferric oxides which may be smaller than the cut-off size used in this study (i.e. 0.01μ).

With the help of the Sr isotopic data it is possible to estimate the water rock ration that can account for the measured chemical composition of the different water samples analyzed in this study. The isotopic composition of the local rain water was estimated to be 0.709 since the volcano is located close to the sea water. These assumptions made it is possible to calculate the water rock ratio. Calculated values are ranging from 6500 to 30000, according to the origin of the water sample. With this mixing model we calculate the rain water Sr concentration. The concentrations is found to be equal to 23 nanoM. this values is in good agreement with expected values for the area. Calculated concentrations for major ions are compared to measured values it appears that the major elements would be more soluble than Sr. This could be due to preferential weathering of the glassy matrix with a lower Sr content or to preferential trapping of Sr in secondary minerals these hypothesis are currently under investigation.

4 ACKNOWLEDGEMENTS

This work was supported by the PEGI/CNRS/ORSTOM program (Programme d'Etude de la Géosphère Intertropicale). We thank IRD and especially Michel Molinier for their support during our field work.

REFERENCES

Benedetti et al. 1994 Water rock interactions in tropical catchments: field rates of weathering and biomass impact *Chem Geol.* 118: 203-220.

Déruelle et al 1987 J. *African Earth Sci.* 6,:197-214

Gérard et al this volume

Nahon, D. 1991. *Introduction to the petrology of soils and Chemical Weathering.*, NewYork: John Wiley

Parfitt & Wilson 1984 *Catena* 7,:1-8

Sarazin, G. Ph. Ildefonse, & J.P. Mueller 1982 *Geochim Cosmochim Acta* 46: 1267-1279

Geochemistry of the Earth's Surface, Ármannsson (ed.) © 1999 Balkema, Rotterdam, ISBN 90 5809 073 6

A new approach for determining the $^{87}Sr/^{86}Sr$ ratio of the 'granitoid weathering component'

T.D. Bullen & A.F. White
US Geological Survey, Menlo Park, Calif., USA

T.G. Huntington & N.E. Peters
US Geological Survey, Antlanta, Ga., USA

ABSTRACT: Sr isotopes can be used in granitoid catchment studies to partition base cations to mineral weathering and atmospheric sources if the $^{87}Sr/^{86}Sr$ ratio provided by each source can be quantified. Although the $^{87}Sr/^{86}Sr$ ratio of the atmospheric component is relatively easy to quantify, the $^{87}Sr/^{86}Sr$ ratio of the weathering component is more difficult to quantify due mainly to incongruent weathering. To determine the $^{87}Sr/^{86}Sr$ ratio of the weathering component, we propose a graphical method in which the array of Nb/Sr and $^{87}Sr/^{86}Sr$ ratios of weathering residues in a rock or soil profile is regressed to the Nb-free component. We apply the technique to weathering profiles at a catchment in the Georgia (USA) Piedmont, and show that the calculated $^{87}Sr/^{86}Sr$ ratio of the weathering component is that of plagioclase in the bedrock weathering environment, and on average that of stream baseflow in the saprolite and colluvium weathering environments.

1 INTRODUCTION

Sr isotopes are increasingly being used to constrain base cation sources and mineral dissolution rates in studies of granitoid weathering. Sr isotopes have been particularly useful for distinguishing mineral weathering and atmospheric sources of base cation pools in catchments nested in granitoid materials (e.g., Miller et al. 1993, Bailey et al. 1996), for determining the relative weathering rates of biotite and plagioclase in granitoid weathering systems (e.g., Blum and Erel 1995, 1997, Bullen et al. 1998), and for constraining the evolution of the cation exchange pool in soil chronosequences developed on granitoid alluvium (Bullen et al. 1997).

For the Sr isotope mass balance approach to be successful, the $^{87}Sr/^{86}Sr$ ratio of Sr derived from all sources must be known. For example, $^{87}Sr/^{86}Sr$ of the "atmospheric component" can be determined through analysis of precipitation and dry fall. On the other hand, $^{87}Sr/^{86}Sr$ of the "weathering component" is far more problematic to determine due to the multi-mineralic character of granitoids and the chemical complexity of the constituent minerals and their weathering products. Regardless, an important persistent result of the above studies is that $^{87}Sr/^{86}Sr$ of the weathering component generally differs significantly from that of the whole-rock or whole-soil material undergoing weathering.

Thus the question arises of how to best determine $^{87}Sr/^{86}Sr$ of the weathering component in a complex natural system. Previous approaches have included using $^{87}Sr/^{86}Sr$ obtained in weak acid leaches of the weathering material (e.g., Miller et al. 1993), using $^{87}Sr/^{86}Sr$ derived from Sr mass balance comparisons between adjacent horizons within a weathering profile (e.g., Bailey et al. 1996), and using $^{87}Sr/^{86}Sr$ of stream baseflow to approximate $^{87}Sr/^{86}Sr$ of the weathering component (Clow et al. 1997). Each of these approaches is probably valid for the individual studies in which they were employed, but their general application becomes problematic due to differences in physical, hydrological and weathering characteristics among catchments.

2 A NEW APPROACH

We propose that a more general approach is to consider variations of $^{87}Sr/^{86}Sr$ in a weathering profile to be correlated with losses of Sr relative to an "immobile" trace element in the weathering material, such as the high-field strength element Nb. We prefer to use a true trace element rather than other immobile elements such as Ti and Zr which are stoichiometric constituents of certain heavy minerals in the granitoids. Thus, Ti and Zr concentrations may be controlled to a greater degree by sedimentary processes that tend to accumulate the heavy minerals.

Relative to the starting granitoid material, extraction of Sr as a weathering component will result in a residue having complementary Sr isotope composition determined by simple mass balance. This approach is ideally suited to graphical interpretation, as shown in Figure 1. On this diagram, in which $^{87}Sr/^{86}Sr$ is plotted against the inverse of Sr concentration, a conservative mixing array (derived by indexing Sr to Nb or some other conservative element) plots as a straight line. As shown in Figure 1, the weathering component, which contains essentially no Nb, is determined simply by extending the line connecting the starting and residue compositions to the $^{87}Sr/^{86}Sr$ axis.

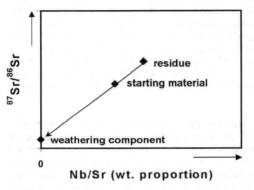

Figure 1. Cartoon showing linear mixing relationship among components of a hypothetical granitoid weathering system

3 APPLICATION TO A COMPLEX NATURAL WEATHERING SYSTEM

The granitoid weathering system at the Panola Mountain Research Watershed, located in the Piedmont province of Georgia, USA, provides an excellent natural example to test this technique. This site has been the focus of an extensive watershed research program since the mid-1980's (see references in Huntington et al. 1993), and there is a considerable record of stream and ground water, biomass, and mineral phase chemical and isotopic analysis with which to compare results.

The watershed is underlain by granodiorite emplaced ~320 Ma ago in a host rock that contains amphibolite and muscovite schist. The average mineralogy of the fresh granodiorite (determined from point counts) consists of plagioclase (31%), quartz (27%), K-feldspar (20%), biotite (12%), muscovite (8%), and minor hornblende, magnetite, apatite, zircon and calcite. Sr concentrations and

$^{87}Sr/^{86}Sr$ ratios of the aluminosilicate minerals are given in Bullen et al. (1998).

3.1 Field and analytical techniques

The samples discussed here were collected from the upper portion of the watershed, which is underlain solely by the granodiorite, from profiles developed both in residuum at a ridge top and in colluvium below the ridge near the stream draining the catchment. The residuum profile consists of 1.5m of soil overlying 3m of saprolite, which in turn overlies weathered to fresh bedrock. The colluvium profile is approximately 4m deep. Samples from the saprolite portion of the residuum and from the colluvium were collected using a hand auger to the point of refusal. Samples from the weathered bedrock portion of the residuum were collected as a continuous core using a diamond drill rig.

The <2mm fraction of saprolite and colluvium samples were first stripped of exchangeable cations using 0.1M ammonium acetate buffered to pH=5.5 with acetic acid. The dried residues were ground to a fine powder in an agate mortar. Crushed samples of weathered bedrock were ground to a fine powder using an alumina shatter box. We point out that a tungsten carbide shatter box is inappropriate for this treatment due to the use of Nb as a binder in the casing materials that could cause contamination.

Powders were dissolved in Teflon beakers in a mixture of HF, HNO_3, and HCl at 40°C to the point at which no visible residue was evident. Accessory zircons were probably not dissolved using this technique, however Nb is not a significant constituent of this recalcitrant mineral. An aliquot of the rock dissolution was analyzed for major and trace element concentrations using a Perkin-Elmer Elan 6000 inductively-coupled plasma mass spectrometer. Precision for all elements discussed here is better than 2%. Sr in the remainder of the dissolved material was separated by cation exchange, and the purified Sr was analyzed for $^{87}Sr/^{86}Sr$ using a Finnigan MAT 261 thermal ionization mass spectrometer. Details of these techniques are given in Bullen et al. (1997).

3.2 Relative Sr loss in the profiles

The Sr/Nb ratio of samples from the residuum and colluvium profiles are plotted vs. depth in Figure 2. In this case, we plot the Sr/Nb ratio to provide an index of the amount of Sr that has been lost to weathering relative to the immobile element Nb. In the weathered bedrock portion of residuum profile, the uppermost sample at approximately 5m depth has lost essentially no Sr relative to that in the fresh rock at the profile base, while at intermediate depths

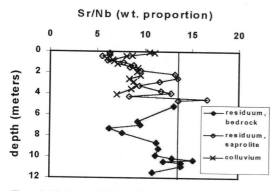

Sr/Nb (wt. proportion)

Figure 2. Variation of Sr/Nb with depth in the profiles through granodiorite residuum and colluvium. The vertical line corresponds to the Sr/Nb ratio of the fresh Panola granodiorite.

approximately half the original Sr has been lost to weathering. The persistence of Sr at the upper margin of the weathered bedrock profile is remarkable considering that our chemical data shows that greater than 90% of Ca and Na, and thus of plagioclase have been lost from the bedrock at depths less than 9m. The mechanism for separating the behavior of Sr from that of Ca, as well as the mineral repository for the Sr derived from plagioclase weathering are as yet unclear.

Relative Sr concentrations in the saprolite portion of the residuum profile follow a complex pattern at depth, but tend to decline regularly from about 3m to the surface. Zones represented by samples having high Sr/Nb may have been relatively impervious to weathering fluids due to a variety of geological factors. Relative Sr concentrations in the colluvium profile likewise follow a complex pattern, but intriguingly tend to increase from 1m depth to the surface. We interpret this increase to indicate that the colluvium has been deposited episodically, and that the surface samples are relatively young compared to those at depth. Due to this structural complexity, it is probably impractical to interpret the colluvium samples as a single weathering profile, although the weathering reactions may be consistent throughout its development.

3.3 Sr isotope variations in the bedrock weathering environment

In Figure 3, $^{87}Sr/^{86}Sr$ ratios of samples from the weathered bedrock profile are plotted vs. Nb/Sr to allow determination of the Sr isotope component derived from weathering. The data regress to a $^{87}Sr/^{86}Sr$ ratio of 0.7085; this is the isotopic composition of the Sr that has been variably removed from these samples by weathering. This Sr

isotopic composition is nearly identical to that of plagioclase in the fresh granodiorite ($^{87}Sr/^{86}Sr=0.7081$; Bullen et al. 1998), suggesting that plagioclase weathering is the dominant control on Sr isotope variability in this portion of the weathering profile. This is consistent with the Ca and Na loss calculations mentioned above. Losses of other major elements in the weathered bedrock are generally less than 10%, indicating only negligible weathering of reactive phases such as biotite and K-feldspar. The slightly greater $^{87}Sr/^{86}Sr$ value of the calculated weathering component relative to that of plagioclase is consistent with a minor contribution of Sr from either biotite or K-feldspar, as well as from the inter-granular calcite.

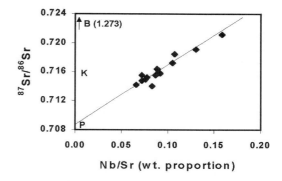

Nb/Sr (wt. proportion)

Figure 3. Variation of $^{87}Sr/^{86}Sr$ with Nb/Sr ratio of weathered bedrock samples. Line is least squares regression through data; the line intersects the $^{87}Sr/^{86}Sr$ axis at a value of 0.7085. $^{87}Sr/^{86}Sr$ values of minerals in Panola granodiorite are given by symbols: P=plagioclase, K=K-feldspar; B=biotite.

3.4 Sr isotope variations in the saprolite and colluvium environments

The $^{87}Sr/^{86}Sr$ ratio of samples from the residuum saprolite and colluvium profiles are plotted vs. Nb/Sr in Figure 4. Because there is considerably more scatter in the data than for the weathered bedrock samples, simple average compositions of the saprolite and colluvium are plotted as well. The saprolite and colluvium data all plot below the regression defined by the weathered bedrock samples (line "a" in Figure 4), implying a significant Sr contribution due to weathering of either K-feldspar or biotite or both.

The regression through the saprolite and colluvium samples (line "b" in Figure 4) intersects the $^{87}Sr/^{86}Sr$ axis at 0.7133. Although this value does not correspond to the $^{87}Sr/^{86}Sr$ of any single

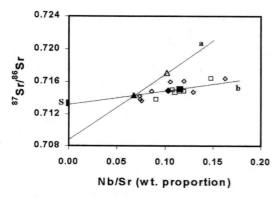

Figure 4. Variation of $^{87}Sr/^{86}Sr$ with Nb/Sr ratio of residuum saprolite and colluvium samples. Open diamonds = saprolite; open squares = colluvium; solid diamond and square are average saprolite and colluvium, respectively. Solid triangle = average fresh bedrock; open triangle = average weathered bedrock. Line "a" is regression from Figure 3; line "b" is regression through colluvium and saprolite samples, and intersects the $^{87}Sr/^{86}Sr$ axis at 0.7133. Band labeled "S" on y-axis corresponds to range of $^{87}Sr/^{86}Sr$ of stream baseflow.

mineral in the granodiorite, it does lie within the small range of $^{87}Sr/^{86}Sr$ ratios measured thus far for stream baseflow from the upper portion of the catchment. This result, although somewhat surprising, does suggest that $^{87}Sr/^{86}Sr$ of stream baseflow may provide an excellent approximation of the weathering component in the hydrologically-dynamic portion of a catchment. Moreover, because the saprolite and colluvium samples represent the result of long-term weathering, it is likely that the $^{87}Sr/^{86}Sr$ ratio of the weathering component has not changed significantly over their weathering history. Obviously there may be significant differences in $^{87}Sr/^{86}Sr$ of the weathering component at the fine scale, but these heterogeneities are presumably averaged out at the catchment scale.

4 CONCLUSIONS

The results of this study demonstrate that weathering in the bedrock and saprolite/colluvium environments provides strikingly different $^{87}Sr/^{86}Sr$ to catchment waters. We suggest that this combined trace element and Sr isotope approach may provide a novel and powerful means for determining the $^{87}Sr/^{86}Sr$ of the weathering component in a variety of catchment and weathering settings. As more trace-element and Sr isotope data sets on weathered catchment materials become available, we should be able to rigorously test the validity of using stream baseflow as the ideal proxy for the weathering component.

REFERENCES

Bailey, S.W., J.W. Hornbeck, C.T. Driscoll & H.E. Gaudette 1996. Calcium inputs and transport in a base-poor forest ecosystem as interpreted by strontium isotopes. *Water Resour. Res.* 32: 707-719.

Blum, J.D. & Y. Erel 1995. A silicate weathering mechanism linking increases in marine $^{87}Sr/^{86}Sr$ with global glaciation. *Nature* 373: 415-418.

Blum, J.D. & Y. Erel 1997. Rb-Sr isotope systematics of a granitic soil chronosequence: the importance of biotite weathering. *Geochim. Cosmochim. Acta* 61, 15: 3193-3204.

Bullen, T.D., A.F. White, A.E. Blum, J.W. Harden & M.S. Schulz 1997. Chemical weathering of a soil chronosequence on granitoid alluvium: II. mineralogic and isotopic constraints on the behavior of strontium. *Geochim. Cosmochim. Acta* 61, 2: 291-306.

Bullen, T.D., A.F. White, D.V. Vivit & M.S. Schulz 1998. Granitoid weathering in the laboratory: chemical and Sr isotope perspectives on mineral dissolution rates. In G.B. Arehart & J.R. Hulston (eds), *Proceedings of the 9th International Symposium on Water-Rock Interaction*: 383-386. Rotterdam: Balkema.

Clow, D.W., M.A. Mast, T.D. Bullen, & J.T. Turk 1997. Strontium 87 / strontium 86 as a tracer of mineral weathering reactions and calcium sources in an alpine/subalpine watershed, Loch Vale, Colorado. *Water Resour. Res.* 33(6): 1335-1351.

Huntington, T.G., R.P. Hooper, N.E. Peters, T.D. Bullen, & C. Kendall 1993. Water, energy, and biogeochemical budgets at Panola Mountain Research Watershed, Stockbridge, Georgia--a research plan. *U.S.G.S. Open-file Report*: 93-55.

Miller, E.K., J.D. Blum, & A.J. Friedland 1993. Determination of soil exchangeable-cation loss and weathering rates using strontium isotopes. *Nature* 362: 438-441.

Geochemistry of the Earth's Surface, Ármannsson (ed.)© 1999 Balkema, Rotterdam, ISBN 90 5809 073 6

Weathering of granitic rocks at sub-tropical climate: An example from Harsit granitoid, NE Turkey

S.Ceryan
Department of Geology, GMF, Karadeniz Technical University, Trabzon, Turkey

C.Sen
Department of Geology, MMF, Karadeniz Technical University, Trabzon, Turkey

ABSTRACT: Secondary mineral developments and changing of fabric features due to chemical weathering and hydrothermal alteration are investigated in detail from a set of samples which were collected from weathering profiles developed on Harsit granitic rocks, NE Turkey. With the increasing degree of chemical weathering, the amount of unaltered mineral and micro petrographical indexes decrease and the amount of micro fissures and voids increase. The most resistance mineral for the chemical alteration is quartz. It can even be found as a fragments in completely weathered rocks. Feldspars start to alter to serisite, epidote and beidellite in hydrothermal stage and, disintegrates, non-stoichiometrically dissolve and decompose to clay minerals during weathering. Vermiculite, chlorite, epidote and Fe-Ti oxides are hydrothermal stage products of biotite and amphiboles and these alteration products are not stable under surface weathering and decompose to clay and ironeous compounds.

1 INTRODUCTION

Economically, it is important to determined that weathering state and stability of rocks. Close relationship between mass movement and weathered materials especially in mountainous areas (e.g. Eastern Pontides, NE Turkey) strongly effects human life sites. Although many parameters may affect the weathering processes, besides to rock properties, climate (and related vegetation) is one of the key parameter itself. Values of average annual temperature and precipitation indicate that the study area lies in sub-tropical climate and according to Peltier (1950) in area at moderate chemical weathering. In this study, weathering of Upper Cretaceous granite batholith at the sub-tropical climate where evapo-transpiration reaches up to 480 mm, is studied.

2 SAMPLE PREPARATION AND ANALYTICAL METHODS

Samples were collected from weathering zones where rock framework still prevailed. After obtaining sections from weathered portion of the rocks, the samples were impregnated with polymerized methyl methacrylate. Micro petrographical properties were examined with transmitted light microscope and scanning electron microscope (SEM) equipped with an energy dispersive X-ray spectrometer (at the University of New Brunswick, Canada). Using the same sections, chemistry of individual minerals was determined using JEOL Super probe 700 (at the University of New Brunswick). Prepared clay-size fractions from the weathered rocks blocks were examined by X-ray diffractometer using Ni-fillered CuKα radiation (at the Hacettepe University, Turkey). The whole samples of each weathered zones were analyzed by X-ray fluorescense (XRF) at the Karadeniz Teknik University, Turkey.

3 MINERALOGICAL AND FABRIC CHANGES

This section includes the observed sequence of events that occurred during the weathering of each of the major minerals in the granite and of resulting fabric changes.

3.1 Unaltered granite

The fresh granites examined are fine-medium (0.5-10 mm) grained, hypidiomorph to xenomorph textured, and modally consist of 22±3 % quartz, 29±5 % K-feldspar, 35±5 % plagioclase, 9±1 % hornblende, 3±1 % biotite and 2±0.5 % opaque minerals. Their micro petrographic indexes (sound constituents / unsound constituents, Dearman and Irfan 1978) are higher than 5.6; amount of micro crack plus voids 0.15 to 1.25 vol. percent and unaltered mineral contents are above 85 vol. percent. Although no visible signs of weathering, hydrothermal alteration cause some changes on minerals. While less than 10 percent of plagioclases have been altered to serisite and epidote, orthoclases are general unaltered and rarely show acid effects. Less than 5 percent hornblende and biotite are altered to chlorite, and small amount of epidote and Fe-oxides.

Grain boundaries are tight but tectonic and regional stresses have caused some openings. Regular and systematic cracks have developed. Crack surfaces are clean, unfilled and not discolored.

3.2 Slightly weathered granite

These samples can be distinguished from fresh granite by the effects of hydrothermal alteration. Micro petrographic indexes range from 5.6 to 2.1; amount of micro cracks plus voids are 0.85 to 3.2 vol. percent and unaltered mineral contents are between 67-90 vol. percent. Slight color changes occurs due to increasing weathering of biotite. Structurally controlled feldspar dissolution has begun. Serisite and epidote have developed on 5 to 30 percent of the plagioclases. Less than 25 percent orthoclase show dissolution effects. Less than 60 percent biotite and 40 percent hornblende have altered to epidote, chlorite and oxide minerals.

Some micro cracks developed on biotite due to secondary mineral development related to expansion on the lamely structure. Fe-oxide discolor appears in some micro cracks and discontinuities but grain boundaries are still tight.

3.3 Moderately weathered granite

Micro petrographic indexes range from 2.5 to 1.0; amounts of micro cracks plus voids are between 3 to 9 vol. percent and unaltered minerals make 50 to 72 percent of samples. Less than 60 percent og the plagioclase have decomposed to serisite, epidote and clay minerals. Due to dissolution, semi-opaque amorphous aluminosilicates and itches can be seen on plagioclases by microscope. Honey comb structure have started to develop on orthoclase surfaces. Chlorite, epidote, Fe-oxides and clay minerals growths make 20 to 60 percent of biotite and hornblende (Fig 1).

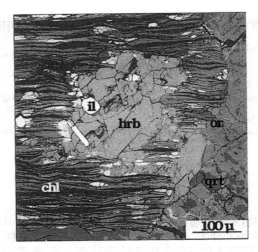

Figure 1. Replacement of the hornblende (hrb) by the chlorite (chl). Ilmenite (il) and other Fe-Ti oxides occur during this decomposition reactions.

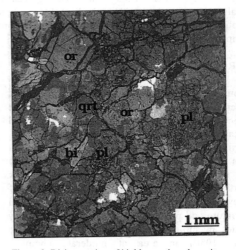

Figure 2. Disintegration of highly weathered granite.

Compared to slightly weathered rocks, amounts of micro cracks increase rapidly. Quartz and plagioclases show fine developed irregular micro cracks. Due to increasing dissolution rates, porosity and permeability have increased. Most of the grain surfaces are discolored by iron oxides.

3.4 Highly weatered granite

Micro petrographic indexes range from 1.2 to 0.7; micro cracks plus voids amount reaches 11.0 vol.

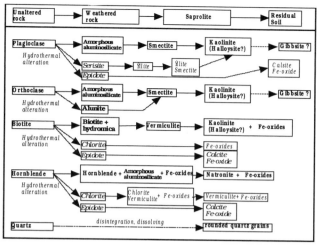

Figure 3. Possible weathering and alteration paths of minerals in the Harsit granitoid

percent. In this stage the amount of clay minerals has significantly increased. More than 65 percent plagioclase turns to epidote and clay minerals. Due to serisite to clay mineral transition, amount of serisite decreases. 25 to 50 percent orthoclase show dissolution effects and on their surfaces clay minerals has started to develop. Amounts of clay minerals and opaque minerals increase on biotite and hornblende.

Micro cracks and grain boundaries are well open (Fig. 2). Quartz and significant amount of plagioclases were disintegrated. Fe-oxides discolor the openings, some are filled by clay minerals.

3.5 Completely weathered granite

Micro petrographical indexes are less than 0.85; amount of micro crack plus voids are up to 25.0 percent. Contains 25 to 45 percent unaltered minerals. Most of the plagioclases have disintegrated and decomposed to clay minerals. Less than 35 percent orthoclase survives and itches, dissolution canals covers their surfaces. Epidote which is a hydrothermal alteration product, has decomposed to chlorite, calcite and small amount of opaque minerals. Most of the biotite and hornblende has turned to chlorite, Fe-oxides and clay ninerals.
Most of the plagioclases (pl) have decomposite to clay minerals. Some orthoclases (or) are relatively unaltered. Quartz (qrt) is unaltered but disintigrated along the grain boundaries.

Most micro cracks and void space are connected, the rock has weakened and it can be broken by hand but rock texture are still preseved.

3.6 Residual soil

Original texture of granite is completely destroyed. Almost all feldspars have decomposed to clay minerals. Due to dissolution, edges of some quartz grains are rounded. Ferromagnesian minerals have decomposed to clay minerals and Fe-oxides.

4 CONCLUSIONS

Possible decomposition and hydrothermal alteration paths of minerals of the Harsit Granite are summarized in Figure 3. Decomposition paths of the minerals suggest that chemical washing reaches significant amounts but the washing rate allows the step by step development of intermediate products.

REFERENCES

Dearman, W. R.& T. Y. Irfan 1978. Assessment of the degree of weathering in granite using petrographic and physical index tests. *International symposium on deterioration and protection of stone monuments* 2-3, UNESCO, Paris.
Peltier L. C.1950. The geographical cycle in periglacial regions as it is related to climatic geomorphology. *Annuals of Assoc. of American Geographers* 40:214-236.

Geochemistry of the Earth's Surface, Ármannsson (ed.) © 1999 Balkema, Rotterdam, ISBN 90 5809 073 6

Clays and zeolites record alteration history at Teigarhorn, eastern Iceland

Th. Fridriksson, P.S. Neuhoff & D. K. Bird
Department of Geological and Environmental Sciences, Stanford University, Calif., USA

S. Arnórsson
Science Institute, University of Iceland, Reykjavík, Iceland

ABSTRACT: Meteoric fluid-basalt interaction at Teigarhorn, Iceland, resulted in three stages of mineral paragenesis: near-surface alteration (I), burial metamorphism (II), and hydrothermal alteration (III). Celadonite and silica pore linings precipitated during Stage I. Stage II is characterized by mixed-layer chlorite/smectite (C/S) (10-85 % chlorite) pore rim followed by regional zeolite zone assemblages (scolecite and heulandite + stilbite + mordenite). Chlorite contents of Stage II C/S increased with time in response to increasing temperature during burial. Stage II zeolites formed after burial but before rift-related dike intrusion ceased. Stage III involved precipitation of quartz, heulandite, stilbite, and other zeolites in fractures and breccias around the dike intrusions. Temperatures at the ends of Stages II and III were ~90 and ~120 °C, respectively.

1 INTRODUCTION

Depth- or temperature-controlled mineral zones in regionally metamorphosed lavas and geothermal systems are sensitive indicators of thermal gradients and fluid composition in the crust. For example, zones characterized by trioctahedral smectite, corrensite and/or mixed layer chlorite/smectite (C/S), and chlorite occur in low grade metabasalts and active geothermal systems (Schiffman and Fridleifsson, 1991; Kristmannsdóttir, 1976; Bevins, et al. 1991). Similarly, zeolite assemblages are often depth controlled, with as many as five separate "zeolite zones" developed with increasing grade (Walker, 1960; Neuhoff et al., 1997).

Some of the most accessible and best exposed examples of zeolite facies metabasalts occur in Eastern Iceland. Coastal and glacial erosion in fjords created near vertical outcrops that have been known as sources for zeolite minerals for over 200 years. Among the most famous of these localities is Teigarhorn, a small farm along Berufjörður (Fig. 1). Tertiary basalt outcrops at this locality offer three-dimensional views of the geologic, chemical, and hydrologic evolution of basaltic crust. The mineralogical evolution of the lavas at Teigarhorn are described below, and presented in more detail in a manuscript submitted to *American Journal of Science*.

Figure 1. Location of study area.

2 MINERAL PARAGENESIS

Alteration of the lavas at Teigarhorn occurred during at three paragenetic stages that are genetically distinct and distinguished by textures, mineral assemblages, and geologic relations:

2.1 *Stage I*

Paragenetic stage I, the earliest alteration at Teigarhorn, is characterized by lining of primary pore space with botryoidal or layered celadonite and silica minerals (Fig. 2). All other mineral assemblages are either precipitated on top of or cross cut Stage I alteration. Celadonite typically formed before silica. Representative compositions of celadonite from Teigarhorn are shown in Table 1 Figure 3.

2.2 *Stage II*

Paragenetic stage II is characterized by continued infilling of primary pore space and partial replacement of groundmass and phenocryst minerals (including albitization of plagioclase). Pore and grain-scale textures are preserved except for infilling of vesicles. The first phase to precipitate in primary pore space during Stage II is C/S (Fig. 2). Brown to brownish- green radiating bundles of C/S line primary pore space. Bundles nucleated on pore walls or on Stage I minerals and grew towards the center of the pores. Stage II C/S compositions (Table 1; Fig. 3) lie along a binary compositional join between ferro-magnesian chlorite and ferro-magnesian trioctahedral smectite. The percentage of chlorite interlayers in stage II clays (calculated from electron microprobe analysis [EMPA]; Schiffman and Fridleifsson, 1991) is between 12 and 85% (Table 1).

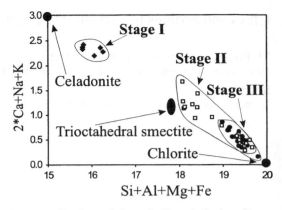

Figure 3. Chemistry of Stage I, II, and II clays from Teigarhorn (normalized to 28 O charge equivalents) as a function of total exchangeable ion content ($2*Ca + Na + K$) and the total Si, Al, Mg, and Fe content.. Ideal, end-member celadonite, chlorite, and trioctahedral smectite are plotted as large black symbols. Most Stage II and III clays are mixtures of chlorite and trioctahedral smectite.

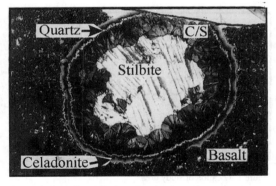

Figure 2. Stage I and II alteration developed in a vesicular top of a lava flow. The 4mm wide vesicle is rimmed by early celadonite followed by silica (recrystallized to quartz). The inner rim is Stage II chlorite/smectite interlayer clay; the last filling phase is stilbite (plane polarized light).

Stage II C/S minerals became more chlorite-rich with time. In general, the earliest clays precipitated near vesicle walls. Figure 4 shows the results of a linear EPMA traverse from the wall of a vesicle through a Stage II clay pore lining (expressed as percent chlorite in each analysis as a function of distance from vesicle wall). It can be seen in Figure 4 that the chlorite content of the clay increases from the vesicle wall towards the center of the vesicle.

Table 1. Representative clay analysis from Teigarhorn, eastern Iceland determined by EMPA

Sample	ISMP2 celad.	94-53 C/S	94-92 C/S	94-94 C/S
Stage	I	II	II	III
SiO_2[†]	45.55	29.96	34.25	28.58
Al_2O_3	6.70	13.49	9.83	13.58
FeO*	14.06	30.62	24.09	33.81
MnO	0.00	0.35	0.00	0.30
MgO	3.83	11.52	10.68	11.02
CaO	0.19	0.56	1.85	0.32
Na_2O	0.11	0.12	0.20	0.00
K_2O	7.89	0.24	0.64	0.20
total	78.32	86.86	81.54	87.80
% Chlorite	-	77.2	27.3	90.9

[†] weight percent of oxide component
* all iron assumed to be Fe^{2+}

The last phases to crystallize during Stage II are quartz and chalcedony along with abundant scolecite, heulandite, and stilbite that are indicator minerals for the two regional zeolite zones that outcrop at Teigarhorn: the mesolite-scolecite zone described by Walker (1960), underlain by a zone characterized by stilbite and heulandite. This boundary suggests a temperature during Stage II zeolite formation of ~90 °C (Kristmannsdóttir and Tómasson, 1978). In the mesolite-scolecite zone, radiating bundles of scolecite fibers generally fill all primary porosity remaining after clay precipitation. Typical vesicle fillings in the heulandite-stilbite zone consist of stilbite, heulandite, and mordenite intergrowths with sporadic epistilbite.

The systematic zoning observed in mixed C/S minerals precipitated in vesicles during stage II (Fig. 4), is interpreted, by analogy to C/S trends observed in prograde low-temperature metamorphism and in active geothermal systems (Bevins et al. 1991; Kristmannsdóttir, 1979), to reflect the increase in temperature during burial of the lava pile. High chlorite contents (85%) of the latest stage II clays and the absence of low-temperature zeolite precursors in stage II zeolite assemblages suggest that precipitation of mixed C/S minerals continued until the lavas reached maximum temperature. High chlorite contents in late Stage II clays contrast with the absence of chlorite below 200°C in Icelandic geothermal systems (Kristmannsdóttir, 1979). However, chlorite-rich clays are commonly observed in basalts metamorphosed under conditions similar to those inferred at Teigarhorn (e.g. Bevins et al., 1991). The regional zeolite zone assemblages (Walker, 1960) formed after the lavas reached maximum burial depth and temperature.

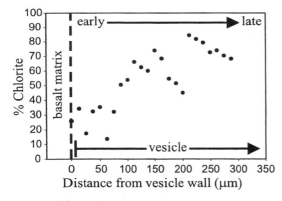

Figure 4. Temporal changes in the compositions of Stage II mafic phyllosilicates depicted as a function percent chlorite. The compositions of the phyllosilicates trend from early smectitic clays (near the vesicle wall) to relatively late chloritic clays (towards the center of the vesicle).

2.3 *Stage III*

The latest alteration at Teigarhorn (Stage III) is a local feature developed in secondary pore space (fractures, breccias, etc.) formed during dike intrusion. Brecciation during Stage III entrains Stage I and II minerals; rafts of vesicle walls coated with celadonite and Stage II heulandite are common in outcrop within Stage III breccias. Lavas within a few

meters of the dikes are extensively fractured with little post-stage II mineral precipitation. At distances greater than three meters from the dikes brittle deformation hosts Stage III mineral assemblages (Fig. 5). Most Stage III alteration is developed in irregular fractures and breccias. The apertures of these features are variable, ranging from microscopic veins to open breccias up to 0.8 m across.

The earliest phases during Stage III, clinoptilolite, quartz, and C/S, line fractures that are filled with later zeolites and calcite. The zeolites occur as vein fillings (Fig. 5), often forming the large, museum grade euhedral crystals for which Teigarhorn is famous. Zeolite-filled veins most often contain euhedral stilbite, heulandite, or stilbite + heulandite + mordenite. Scolecite, laumontite, and calcite typically occur in large, open cavities within breccias.

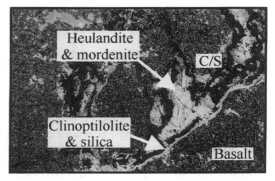

Figure 5. Stage III alteration in a dike-related fracture. (horizontal field of view 7.2 mm). Fracture walls are lined with dark C/S; the last filling phase is light colored heulandite with mordenite intergrowths.

Laumontite and the high chlorite content of C/S (55-90% chlorite; Table 1; Fig. 2) in Stage III assemblages indicates temperatures slightly higher than the maximum temperatures during Stage II. Laumontite occurs with stilbite, heulandite, and scolecite in Icelandic geothermal systems at ~120°C (Kristmannsdóttir and Tómasson, 1978), similar to fluid inclusion homogenization temperatures in Stage III calcite (115±8 °C).

3 PETROGENETIC HISTORY

The three paragenetic stages observed in Teigarhorn basalts represent phases in the metamorphic and tectonic evolution of Icelandic crust. A petrogenetic history of Teigarhorn rocks, summarized in Figure 6, began with precipitation of celadonite and opal at low temperatures during early stages of burial. This

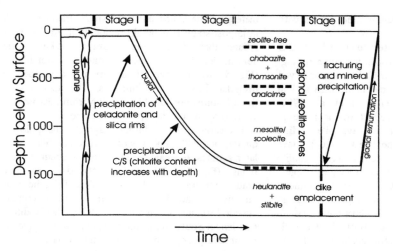

Figure 6. Spatial and temporal development of mineral assemblages observed at Teigarhorn.

was followed by formation of stage II smectites during burial and postburial precipitation of stage II zeolites. Finally, diking and associated fracturing lead to increased fluid flow resulting in precipitation of stage III mineral assemblages at temperatures comparable to the maximum stage II temperatures.

Cross cutting relationships between stage II and III assemblages demonstrate that stage II zeolite precipitation was complete before dike intrusion initiated stage III alteration. The dikes at Teigarhorn are probably related to the Alftafjordur central volcano (Blake, 1970) but likely did not feed the volcanic pile (e.g., Gudmundsson, 1983). Stage III alteration was a local phenomenon associated with late dikes after burial. Stage II zeolitization was thus complete while still within the volcanic zone.

The change in Stage II mineralogy from C/S to zeolites results from an increase in the concentration of Ca^{2+} relative to Mg^{2+} and Fe^{2+} in the fluids. Iron and magnesium necessary for precipitation of C/S were provided by rapid hydrolysis of olivine and glass. Albitization of plagioclase released Ca^{2+} to the fluids, leading to saturation with Ca zeolites such as scolecite, heulandite, and stilbite. The absence of calcite indicates low f_{CO_2} during Stage II.

4 CONCLUSIONS

1. The mineralogical alteration observed at Teigarhorn results from near-surface weathering, burial metamorphism, and hydrothermal alteration.

2. Clay and zeolite paragenesis at Teigarhorn reflects the tectonic history of eastern Iceland.

3. Regional zeolite zones formed after the lavas were buried but before passage of the rift zone.

REFERENCES

Bevins, R.E. Robinson, D. & Rowbotham G. 1991. Compositional variations in mafic phyllosilicates from regional low-grade metabasites and application of the chlorite geothermometer. *J. Met. Geol.* 9:711-721.

Blake, D.H. 1970. Geology of the Alftafjordur volcano, a Tertiary volcanic center in south-eastern Iceland. *Sci. Icel.* 2: 43-63.

Gudmundsson, A. 1983. Form and dimensions of dykes in eastern Iceland. *Tectonophysics* 95: 295-307.

Kristmannsdóttir, H. 1979. Alteration of basaltic rocks by hydrothermal activity at 100-300 °C. In M. Mortland and V. Farmer (eds.) *Developments in Sedimentology*: 359-367. Amsterdam: Elsevier.

Kristmannsdóttir, H. & Tómasson, J. 1978. Zeolite zones in geothermal areas in Iceland. In L.B. Sand and F.A. Mumpton (eds.) *Natural Zeolites*: 227-284. Oxford: Pergamon.

Neuhoff, P.S. Watt, W.S. Bird, D.K. & Pedersen, A.K. 1997. Timing and Structural Relations of Regional Zeolite Zones in Basalts of the East Greenland Continental Margin. *Geology* 25:803-806.

Schiffman, P. & Fridleifsson, G.Ó. 1991. The smectite-chlorite transition in drillhole NJ-15, Nesjavellir geothermal field, Iceland: XRD, BSE and electron microprobe investigations. *J. Met. Geol.* 9:679-696.

Walker, G.P.L. 1960. Zeolite zones and dike distribution in relation to the structure of the basalts of eastern Iceland. *J. Geol.* 68515-528.

Geochemistry of the Earth's Surface, Ármannsson (ed.)© 1999 Balkema, Rotterdam, ISBN 90 5809 073 6

Weathering of Mount Cameroon: 1. Mineralogy and geochemistry

M. Gérard & J. Bertaux – *Laboratoire des Formations Superficielles, IRD, Bondy, France*

Ph. Ildefonse – *Laboratoire de Minéralogie-Cristallographie, UPMC, UMR 7590 et IPGP, Paris, France*

M. Bulourde, C. Chauvel & A. Dia – *Géosciences Rennes, CNRS UPR 4661, Campus Beaulieu, France*

M. Benedetti & J. Boulègue – *Lab. de Géochimie et Métallogénie, UPMC, UPRESA CNRS 7047, Paris, France*

F. Chabaux & B. Fritz – *Centre de Géochimie de la Surface, EOST, Strasbourg, France*

J. Etame – *Université de Douala, Département des Sciences de la Terre, Cameroun*

I. Ngounouno – *Université de Ngoundéré, Département des Sciences de la Terre, Cameroun*

B. Deruelle – *Université Paris, UPRESA CNRS 7047, France*

ABSTRACT: Chemical weathering of Mount Cameroon was investigated through the study of three soil profiles developed on young tephra (810 - 4530 years). Mineralogical data indicate that short ordered mineral constituents (allophane and ferrihydrite) are the prevailing secondary phases. These phases are related to the weathering of glass which represents a significant proportion of tephra materials. Geochemical data obtained on major elements evidenced that the weathering of glass is characterized by a depletion in Si, Mg, Ca, Na, K and an enrichment in Ti, Al, Fe. Major and trace element concentrations of bulk samples highlight a severe depletion of alkaline elements (Na, K, Sr, Rb)

1 INTRODUCTION

A geochemical and mineralogical study of the earliest steps of weathering and pedogenetic processes on volcanic rocks subject to humid equatorial climatic conditions has been performed. The chosen target, Mount Cameroon, is an active volcano located on the Cameroon volcanic line. The oldest lavas are upper Miocene. The recent activity is widespread on the volcano flanks and the ages of the eruptions are precisely known (1815, 1909, 1922, 1925, 1959, 1982). Eruptions at the summit have been reported in 1954 and 1982. Mount Cameroon is 4095 m high and has a massive volcanic horst structure. Due to its shape, altitude and location near the sea, extreme and variable conditions of humidity and temperature are observed between the seaward and the landward flanks of the volcano. Rainfalls reach 10-12 m/year on the southwest flank along the seacoast, while they are only 2 m/year on the southeast flank. Temperature variations range from 26-29°C at sea level to 0° C at the summit.

Altered and fresh massive lavas, together with soils developed on pyroclastic deposits were collected at different altitudes and flank locations. Thirty-six river and spring waters were also collected during the dry season and the small rainy season in order to determine the present chemical erosion (Benedetti & al., this volume).

In this paper, we focus on young soils developed on pyroclastic deposits because they represent a material reactive to weathering solutions. In contrast, historical massive lavas. These massive lavas do not show significant weathering features and chemical transformations as evidenced by major and trace element concentrations, and strontium, neodymium and lead isotopes (Bulourde & al. 1999).

2 MATERIALS AND METHODS

Three soil profiles are presented here. They are representative of the landscape of the upper part of Mt Cameroun above the tropical forest (>1800 m). Two, located at the same elevation 2540 m, are associated with well shaped scoria cones and differ by their vegetation. The first one, JC, is covered by savanna grass and a C^{14} age of 810 ± 60 years was obtained on coals sampled at 60 cm depth. The second one, VC, covered by scarce trees and ferns, gave a C^{14} age of 1395 ± 100 years on coals sampled at 30 cm depth. The last soil profile, HU, was sampled in savanna at 2190 m. It consists of 110 cm of loose weathered material covering a weakly weathered scoriacious lava flow. A C^{14} age of 4530 ± 80 years was obtained on coals burried at 75 cm depth. Due to the periodicity of volcanic activity, the three soil profiles are characterized by the alternation of burried paleosoils and less weathered tephra deposits. This situation is particularly well suited to constrain alteration kinetics.

Undistributed soil samples were collected in order to perform petrographic observations combined with in-situ microprobe analyses. Major and trace element data were acquired by ICP-AES and ICP-MS, respectively, on bulk samples. Mineralogical data were performed both by XRD and FTIR on bulk and fine fractions (0.2-0.45 μm). Organically bound Al_p, Fe_p and Si_p were measured using 0.1 M $Na_4P_2O_7$ reagent. Al_o, Fe_o and Si_o, extracted by 0.15 M ammonium oxalate at pH 2, are associated to organic matter and short range ordered minerals (alumino-silicates and ferrihydrites). XAS data were obtained at Fe-K edge at LURE (Orsay, France) on fine fractions in order to characterize short range ordered hydrated iron oxides.

3 RESULTS AND DISCUSSION

3.1 *Mineralogical data*

Total carbon analyses (3-15 % C) show that topsoils are characterized by variable organic matter contents. The highest values are obtained for HU topsoils (0-25/30 cm), whereas the lowest value is found in the first two centimeters of VC profile where three tephra layers have been distinguished. Al, Si and Fe extracted by ammonium-oxalate and Na-pyrophosphate indicate that these young soils contain significant amounts of organically bound metals, alumino-silicates (allophanes) and ferrihydrite. Al_p and Fe_p associated to organic matter in organic-rich horizons range between 0.7-1.8 % and 0.3-0.7 % respectively.

Allophane and ferrihydrite, estimated by the Si_o-Si_p (Parfitt & Henmi 1982) and Fe_o-Fe_p (Childs 1992), range between 3 and 18 % allophane and 0.5 to 5 % ferrihydrite.

Mineralogical results obtained by XRD and FTIR on bulk and separated fractions (between 0.2 μm and 0.45 μm) confirm the dominant occurrence of short range ordered mineral components such as allophane and ferrihydrite over crystalline weathering minerals (Figure 1). Only scarce 7 Å-halloysite and gibbsite (about 1%) may be detected by XRD and FTIR.

Intimate association of organic matter (C=C, C=O, C-N, CH_2 and CH_3 groups) with mineral components is highlighted by FTIR spectroscopy (Figure 2).

FTIR features are consistent with Al-rich allophanes with imogolite-like local order (344 cm-1 feature). Fe-XAS data performed on the fine fraction (profile HU, sample 4) indicate that the molecular environment of iron is consistent with hydrated ferric oxide (Fe-O = 1.99 Å ± 0.1 Å) with prevailing edge-sharing octahedra (Fe-Fe = 3.06 Å ± 0.1 Å) (Figure 3).

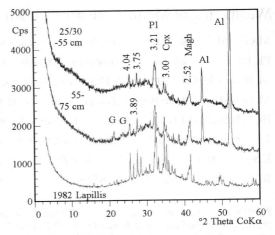

Figure 1: Examples of XRD patterns of fine fractions (<0.2 μm) from 25/30-55 and 55-75 cm horizons in HU profile compared to fresh lapillis (1982 eruption). G: gibbsite, Cpx: clinopyroxene, Pl: plagioclase, Magh: maghemite, Al: sample holder.

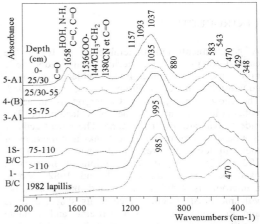

Figure 2: FTIR spectra of fine fractions (0.2 -0.45 μm) from HU profile compared to fresh lapillis (1982 eruption)

3.2 *Microprobe data*

We concentrated our study on the chemical evolution of lapilli weathering at the microscopic scale.

Various stages of lapilli weathering were recognized and chemically characterized: unweathered brown parent glass in the center of some lapilli grains, hydrated yellow glass, and yellowish-brown cortex. Depending on the efficiency of weathering, some lapilli may be fully hydrated and oxidized and may contribute by

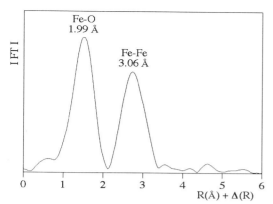

Figure 3: Radial distribution function of the fine fraction from sample 4 in HU profile (Fe-K edge at 10 K). Calculated Fe-O and Fe-Fe distances are indicated.

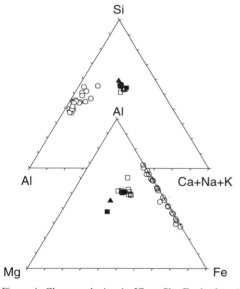

Figure 4. Glass weathering in JC profile. Fresh glass (open squares); hydrated glass (open circles). Tholeitic (filled triangle) and alkalin basaltic (filled circles) glass composition from Noak (1981) and Furnes (1975). Filled square is bulk composition of unweathered lapillis sampled at the summit of Mt Cameroon.

fragmentation to an aggregated brown-orange soil matrix. Small microlites of plagioclase, clinopyroxene, and titanomagnetite are always present and unweathered.

The chemical transformations associated to glass weathering include a depletion in Ca, Na, K, Mg and Si and an enrichment in Ti, Al, Fe in the weathered lapillis (Figure 4). The soil matrix is noticeably enriched in Al and rather depleted in Fe by comparison to the weathered lapillis. Al/Si ratios characterize Al-rich allophanes.

3.3 Bulk chemical data

The three studied profiles display similar major element trends. They are all characterized by a depletion in SiO_2, CaO, MgO, Na_2O and K_2O. The amplitude of the depletion increases from the JC profile to the VC profile and reaches about 15-20 % of the original concentrations in the HU profile (Figure 5). This relationship suggests that the degree of depletion is related to the age of the soil profile.

The same observations apply to the trace elements. All reputably immobile elements are slightly enriched relative to the chosen reference. This is probably related to the departure of a significant proportion of major elements such as MgO, CaO and SiO_2. However, few trace element do not follow this trend: severe depletion of Rb and Sr are observed in the three profiles while Ba is depleted only in the most altered HU profile. Surprisingly, Cs does not behave as Rb and is not depleted in any of the three profiles. As a consequence, the Rb/Cs ratio is highly variable and is always lower than that of the fresh lavas.

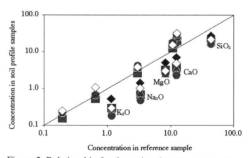

Figure 5: Relationship for the major element concentrations between the reference sample (on the x axis) and the various samples along the HU profile (on the y axis).

4 CONCLUSION

Chemical weathering of Mount Cameroon was investigated through the study of three soil profiles developed on young tephra (810-4530 years). Mineralogical data indicate that short ordered mineral coomponents (allophane and ferrihydrite) are the prevailing secondary phases. These phases are related to the weathering of glass which represents a sigificant proportion of tephra materials. Geochemical data obtained on major elements evidenced that the weathering of glass is characterized by a depletion in Si, Mg, Ca, Na, K and an enrichment in Ti, Al, Fe. Major and trace element concentrations of bulk samples highlight a severe depletion of alkaline elements (Na, K, Sr, Rb). The degree of depletion is related to the age of the soil profile.

ACKNOWLEDGEMENTS

This work was supported by the PEGI/CNRS/IRD (ORSTOM) program (Programme d'Etude de la Géosphère Intertropicale). We thank IRD and especially Michel Molinier for their support during our field work. This is an IPGP contribution and INSU contribution.

REFERENCES

Bulourde M., Chauvel C., Dia A., Déruelle B., Ildefonse P., Gérard M., Chabaux F. & Ngounouno I. 1999. No incipient chemical weathering in present-day basalts exposed to humid tropical conditions: evidence from Mount Cameroon lavas. *EUG* X.

Parfitt R.L. & Henmi T. 1982. Comparison of an oxalate-extraction method and an infrared spectroscopic method for determining allophane in soil clays. *Soil Science and Plant Nutrition* 28: 183-190.

Childs C. W. 1992. Ferrihydrite: A review of structure, properties and occurence in relation to soils. *Z. Pflanzenernähr. Bodenk.* 155: 441-448.

Furnes H. 1975. Experimental palagonitisation of basaltic glasses of varied composition. *Contrib. Miner. Petrol.* 50: 105-113.

Noak Y. 1981. La palagonite: caractéristiques, facteurs d'évolution et mode de formation. *Bull. Minéral.* 104: 36-46.

Geochemistry of the Earth's Surface, Ármannsson (ed.)© 1999 Balkema, Rotterdam, ISBN 90 5809 073 6

Weathering of a Pb-mineralized sandstone, Ardèche, France: Lead speciation

Ph. Ildefonse, G. Morin, F. Juillot & G. Calas – *Laboratoire de Minéralogie Cristallographie, Université Paris, France*

J.C. Samama – *E.N.S.G., Vandoeuvre les Nancy, France*

G.E. Brown Jr – *Department of Geological and Environmental Sciences, Stanford University, Calif., USA*

P. Chevallier & P. Populus – *LURE, Université Paris, Orsay, France*

ABSTRACT: Chemical forms of lead were determined and quantified in a naturally enriched soil (1330 - 2055 mg/kg Pb) developed on a geochemical anomaly. XRD and EXAFS data evidence that plumbogummite $(PbAl_3(PO_4)_2(OH)_5.H_2O)$, a low solubility mineral $(K_s=10^{-29.4})$, is the main lead phase along the soil profile investigated. Lead is also present as inner-sphere Pb(II) complex sorbed onto manganese (hydr)oxides and onto soil organic matter. Variations of the relative proportions of these forms of lead along the profile studied indicate that i) the proportion of plumbogummite is almost constant ii) Pb(II)-manganese (hydr)oxides complexes are gradually replaced by Pb(II)-organic matter complexes upward the soil profile. Besides, geochemical balance based on Zr invariant suggests a global loss of lead during soil formation.

1 INTRODUCTION

Lead is a widespread pollutant which generally accumulates in topsoil horizons as a result of surface/water reactions with soil components. Recent in-situ studies of Pb-contaminated soils reported that Fe/Mn-(hydr)oxides and soil organic matter (Morin et al. 1999) as well as Pb-bearing phosphates (Cotter-Howells et al. 1994) often act as scavengers for lead in soils.

In the present study, chemical forms of lead have been investigated in a naturally enriched soil with lead concentration in the same order of magnitude as those encountered in polluted soils. This natural system can be viewed as an analogue for Pb-contaminated soils in terms of lead levels, although the contamination proceeded through the weathering of the bedrock instead of originating from anthropogenic sources. Determining chemical forms of lead in such a soil allowed identification of Pb-immobilization processes that are efficient at the geological time-scale.

2 MATERIALS AND METHODS

2.1 Geological setting

In the area of Largentière (Ardèche, France), lead-zinc mineralization occurs as stratabound anomalies in stratigraphic levels from the lower Triassic to middle Lias. Among them, a relatively high grade anomaly (locally over 5000 mg/kg Pb) is hosted by upper Triassic arkose. Systematic analyses of fresh rocks from these levels, sampled from drill cores, revealed that lead, zinc and barium occur as galena, sphalerite and barite respectively.

2.2 Samples studied and analytical methods

One soil profile, developed on mineralized arkose from upper Triassic level, was sampled. Three distinct horizons, referred to as A-, B- and C-horizons (Table 1), were recognized. Lead concentrations in bulk samples were determined by ICP-MS.

Table 1. Chemical characterization of the samples studied.

Horizon	A	B	C
Depth (cm)	0 - 8	8 - 20	20 - 30
pH (H₂O)	5.5	6.0	6.5
TOC (wt. %)	5.6	1.6	0.1
Pb (mg/kg)	1330	2055	1874

Undistributed soil samples were embedded in an epoxy resin and prepared as thin sections mounted on pure silica SUPRASIL® slides. Elemental maps were collected on these thin sections in step scan mode on 1 mm² areas, with 20 μm steps and counting 20 seconds per step using x-ray micro-Fluorescence (μ-SXRF) on the FLUO-D15A beamline at LURE/DCI radiation facility (Orsay, France).

Bulk samples (< 2 mm) and isolated fine fractions (< 2 μm) were studied by powder XRD and by EXAFS spectroscopy at the Pb L_{III}-edge. Quantitative mineralogy was determined by Rietveld analysis (XND code, Bérar, 1990) using 5 wt.% fluorite as an internal standard.

Pb L$_{III}$ EXAFS data were recorded, at 10-20 K, in fluorescence detection mode on the wiggler beam-line BL IV-3 at SSRL (Stanford, USA).

Model compounds, including plumbogummite (PbAl$_3$(PO$_4$)$_2$(OH)$_5$.H$_2$O) (San Martin, Spain) and adsorbed Pb(II) species onto birnessite (MnO$_2$) and onto humic substances, were also studied for comparison with natural samples.

3 RESULTS

3.1 Mineralogy

XRD analysis indicated that orthoclase and quartz are the main mineral components of the bulk samples which also contain minor amounts of smectite, kaolinite, illite and barite (BaSO$_4$).

Plumbogummite (PbAl$_3$(PO$_4$)$_2$(OH)$_5$.H$_2$O) was unambiguously identified by XRD in the bulk samples of all three soil horizons. This phase was also recognized by SEM-EDS with an average P/Pb molar ratio of 1.94±0.10, as measured by EMPA, in agreement with the ideal plumbogummite stoechiometry Pb/P = 2. The low amount of Ba ($\Box$ 0.01 mol%) and S ($\Box$ 0.02 mol%) attested for the pure end-member character of this phase in the soil studied.

The <2µm fractions were significantly enriched in plumbogummite (Figure 1a). Indeed, this phase accounted for about 65% of the total lead content in these fractions, as derived from quantitative Rietveld analysis.

3.2 Spatial distribution of lead

Elemental maps obtained by µ-SXRF are displayed on Figure 2, together with the corresponding optical photograph. These maps allow to distinguish two main pools for Pb : i) sandy and silty black grains enriched in Pb, Fe and Mn, and ii) clayey matrix with diluted Mn, Fe and Pb. A strong Pb-Mn correlation is evidenced on these maps although this correlation is much poorer in the clay matrix where quite no manganese is present. Similar features were observed on the three horizons studied, although the best Pb-Mn correlation was observed in the C-horizon.

Under SEM-EDS observations, lead was found in association with Mn-(hydr)oxides coatings (few microns thick) and with silty aggregates of Mn- and Fe-(hydr)oxides. These observations strongly suggest that a fraction of lead is not incorporated in plumbogummite and could occur as sorbed species onto manganese and/or iron (hydr)oxides surfaces.

3.3 EXAFS data

Pb L$_{III}$-edge EXAFS data of the bulk C-horizon are displayed on Figure 3, together with the EXAFS functions and RDFs of plumbogummite and of Pb(II) sorbed onto birnessite. The spectrum of the C-horizon (Figure 3a) exhibits strong similarities with that of plumbogummite (Figure 3b). However, characteristic features observed at k values higher than 6 Å^{-1} could be attributed to the contribution of lead sorbed onto manganese (hydr)oxides (Figure 3c).

EXAFS spectrum and RDF of the bulk C-horizon A spectrum are both well reproduced by summing the contributions of 50% plumbogummite and 50% Pb(II) sorbed onto birnessite (Figure 3a). This result confirms that lead is present under two distinct forms in this C-horizon.

EXAFS data from bulk samples of both A- and B-horizons indicate that plumbogummite is the main form for lead in these horizons. EXAFS analyses allow to show that lead enrichment in the < 2 µm fractions is directly related to increase of the plum-

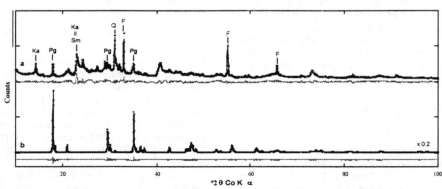

Figure 1. Experimental and calculated XRD patterns of (a) <2µm fraction of the B-horizon with fluorite as an internal standard and (b) reference plumbogummite sample. Pg: plumbogummite; Q: quartz ; Sm : smectite; Il: illite ; Ka: kaolinite; F: fluorite.

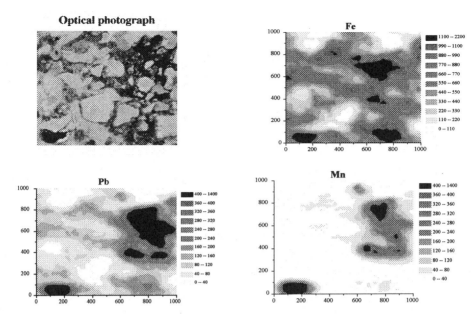

Figure 2. Elemental maps obtained by μ-SXRF for Fe, Pb and Mn of a 1mm x 1mm area on a thin section of the B-horizon and corresponding optical photograph (upper left). Concentrations are relative (counts) on elemental maps obtained by μ-SXRF.

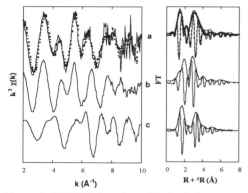

Figure 3. EXAFS data from (a) the bulk C-horizon; plain line: experimental; dotted line : calculated as linear combination of (b) plumbogummite and (c) Pb(II) sorbed onto birnessite.

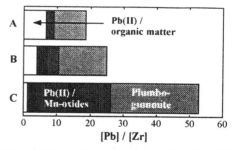

Figure 4. Geochemical balance of lead species obtained from least-squares fitting of EXAFS functions. In each A, B and C horizon, the sum of the contributions of individual lead species is equal to the total lead content of the horizon and is divided by the Zr concentration in the corresponding horizon.

bogummite fraction, i.e. 60±5% of the total lead in the < 2 μm fraction of the B-horizon.

Besides, linear decomposition of EXAFS data indicate that the part of lead adsorbed onto manganese (hydr)oxides decreases upward the soil profile (from 50% to 22% and 11% in the C-, B- and A- horizons, respectively), as shown in Figure 4. On the contrary, an increase in the proportion of Pb(II)-organic complexes is observed (from <5% to 18% and 35% in the C-, B- and A- horizons, respectively).

4 DISCUSSION AND CONCLUSION

Normalizing the amounts of the different lead species to the concentration of a geochemical invariant, i.e. zirconium, gives additional constraints on lead transfers related to weathering in the soil studied. As shown in Figure 4, the $[Pb_{total}]/[Zr]$ ratio significantly decreases in the B- and A-horizons with respect to the C-horizon. That could be interpreted as an absolute loss of lead during soil genesis. Weathering of plumbogummite and of Pb(II)-manganese oxides complexes could be explained by the chelating role of soil organic matter, since the lead released in topsoil horizons appears to be partly sequestered by organic matter.

Although the [plumbogummite]/[Zr] ratio significantly decreases upward the soil profile, this mineral appears to be more resistant than Pb(II)-manganese (hydr)oxides complexes which almost disappear in the organic rich topsoil horizon. Plumbogummite is a rare mineral known to occur as a secondary phase after lead sulfide alteration in the oxidized zones of orebodies (Nriagu, 1984). The relative stability of plumbogummite in the soil studied is consistent with its low solubility product ($K_S=10^{-29.4}$) under Earth surface conditions (Eighmy et al. 1997, Nriagu 1984). In addition, the precipitation of plumbogummite during soil formation processes can not be excluded, since this mineral is expected to be thermodynamically more stable than pyromorphite at nearneutral pH and room temperature, provided that the phosphate activity is low enough (Nriagu 1984).

The present study shows that lead speciation significantly varies along the soil profile and that lead is strongly bound to organic matter in topsoil horizons where T.O.C. is high. Such an affinity of Pb(II) for organic matter was also observed by EXAFS in topsoils horizons of Pb-contaminated soils (Morin et al. 1999). Low solubility minerals like Pb-phosphates were proposed by many authors (Ma et al. 1993, Ruby et al. 1994, Cotter-Howells & Caporn 1996, Laperche et al. 1996) for in-situ lead immobilization. Our results suggest that plumbogummite should be more efficient than surface complexes for sequestering lead in soils over geological time-scales.

ACKNOWLEDGMENTS

The authors wish to acknowledge John Ostergren for providing EXAFS spectra of various model compounds, M.C. Boisset for kindly providing the Pb-humate sample and the Pb-birnessite sample and L. Touret (École des Mines de Paris) for the plumbogummite sample. The authors are indebted to the SSRL staff, especially the SSRL Biotechnology group, for their technical assistance and efficiency during our beam time. This work was supported by the cooperative CNRS-NSF research program under grants CNRS-INT-5914 and NSF-INT-9726528, and by the PROSE98 CNRS/INSU Program and by IPGP.

REFERENCES

Bérar J.F. 1990. Reduction of the number of parameters in real time Rietveld refinement. *IUCr. Sat. Meeting Powder diffraction*, Toulouse.

Cotter-Howells J. D. & S. Caporn 1996. Remediation of contaminated land by formation of heavy metal phosphates. *Applied Geochem.* 11: 335-342.

Cotter-Howells J. D., P. E. Champness, J. M. Charnock & R.

A. D. Pattrick 1994. Identification of pyromorphite in mine-waste contaminated soils by ATEM and EXAFS. *J. Soil Sci.* 45: 393-402.

Eighmy T. T., B. S. Crannell, L. G. Butler, F. K. Cartledge, E. F. Emery, D. Oblas, J. E. Krzanowski, J. D. Eusden, E. L. Shaw & C. A. Francis 1997. Heavy metal stabilization in municipal solid waste combustion dry scrubber residue using soluble phosphate. *Environ. Sci. Technol.* 31: 3330-3338.

Laperche V., S. J. Traina, P. Gaddam & T. J. Logan 1996. Chemical and mineralogical characterizations of Pb in a contaminated soil: Reactions with synthetic apatite. *Environ. Sci. Technol.* 30: 3321-3326.

Ma Q. Y., S. J Traina. & T. J. Logan 1993. In situ lead immobilization by apatite. *Environ. Sci. Technol.* 27: 1803-1810.

Morin G., J. Ostergren, J. J. Rytuba, F. Juillot, Ph. Ildefonse, G. Calas & G. E. Brown Jr. 1999. XAFS Determination of the chemical form of lead in smelter-contaminated soils and mine tailings: Importance of sorption processes. *American Mineralogist* 84/3: 420-434

Nriagu J. O. 1984. Phosphate minerals: their properties and general modes of occurrence. In J. O. Nriagu & P. B. Moore (eds) *Phosphate Mineral:* 1-136. London: Springer.

Ruby V. M., A. Davis & A. Nicholson 1994. *In situ* formation of lead phosphates in soils as a method to immobilize lead. *Environ. Sci. Technol.* 28: 646-654.

Samama, J.C. 1968. Contrôle et modèle génétique de minéralisation en galène de type "Red-Beds", gisement de Largentière, Ardèche, France. *Miner. Deposita* 3: 261-271.

Geochemistry of the Earth's Surface, Ármannsson (ed.) © 1999 Balkema, Rotterdam, ISBN 90 5809 073 6

Petrography and geochemistry of the Gümüsler (Nigde) area gneisses and amphibolites, Central Turkey

H. Kurt & H. Bas
Geology Department, Selçuk University, Konya, Turkey

M. Arslan
Geology Department, Karadeniz Technical University, Trabzon, Turkey

ABSTRACT: The objective of this study is to investigate the petrography and geochemistry of the Gümüsler (Nigde) area gneisses and amphibolites, Central Turkey. Amphibolites and gneisses are commonly very difficult to determine in areas whether sedimentary or igneous origin. The Gümüsler rocks are interpreted to be of igneous origin based up on Niggli trends and trace element characteristics. The gneisses were mainly calc-alkaline whereas amphibolites were mainly tholeiitic in character, and both of them reflect volcanic arc setting precursor. Field relationships and chemical data suggest that the amphibolites formed as metamorhism of basic tuff wheras gneisses likely from acidic tuff and related rocks.

1 INTRODUCTION AND GEOLOGICAL SETTING

The study area is located at the east of Gümüsler, about 5 km to Nigde (Figure 1). The area is a part of the Nigde massif which contains varied metamorphic basement rocks (Nigde Group) of almandine-amphibolite facies grade (Ozgur et al. 1984, Göncüoglu 1981, 1986). The lower most unit of the group is made up of gneiss with intercalations of amphibolite, marble and quartzite, but in the upper levels marble is dominant (e.g., Göncüoglu 1981, 1986, Isler and Büyükgidik 1994, Akçay 1995). Pre-Upper Cretaceous age metamorphics, which mainly comprises amphibolite, gneisses and marble, form the basement of the study area (Figure 1). The metamorphosed units are cut by aplitic, micropegmatitic and pegmatitic dykes which are spatially related to the Sb-Hg-W deposits (Akçay et al. 1995). All these rocks are overlain unconformably by Pliocene aged tuffs (Isler & Büyükgidik 1994).

Fresh gneisses are beige, white and varying tones of grey in colour with the altered surfaces being pale brownish. Amphibolites with dark green and black colours occur as small lenses or thin layers between gneisses. The most common lithology is a medium-grained, massive to thinly banded and foliated amphibolite interlayered and infolded with gneiss and quartzite. These amphibolites occur as discontinuous, highly deformed lenticular pods aligned parallel to the NE-SW regional strike of the host rocks.

2 PETROGRAPHY

The gneisses consist mainly of quartz (30-60%), plagioclase (20-45%), K-feldspar (10-30%), up to 5% biotite, and minor sericite, tourmaline, zircon, apatite. The rocks have heteroblastic texture. The quartz grains are generally small (0.4-0.6 mm in diameter) with slightly curved grain boundaries and have weak undulating extinction and preferred orientation. Some quartz porphyroblast are aligned parallel to the schistosity. The plagioclase twin

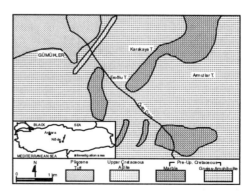

Figure 1. Location and simplified geological map of the study area (modified from Ýsler & Büyükgýdýk 1994; Akçay et al. 1995).

lamellae are frequently warped and distorted, and it often shows sericitic and kaolinitic alteration. The K-feldspar (0.5 to 1 mm) generally shows perthitic texture and is commonly altered to sericite and kaolinite.

The amphibolites are fine-grained, conspicuously well foliated. Subhedral amphibole grains form 65 to 75 volume present of the rocks. The rocks contain mainly hornblende, plagioclase, quartz, pyroxene in a nematoblastic texture in which both plagioclase and hornblende have smooth grain boundaries. The average hornblende grain size is 0.4 mm, but some grains reach a length of 1mm. These large grains contain inclusions of epidote and quartz. The hornblendes are replaced by actinolite in rims. The green to light green hornblende shows strong pleochroism from dark to light brown. The grains often display crystallographic preferred orientation. The plagioclase is 15 to 20 volume percent in the rocks, and contains abundant epidote and quartz inclusions. Albite twinning is characteristic and occasionally a few microperthite grains have been observed. Quartz is up to 10 volume percent. The quartz grains have undulating extinction and some deformation lamellae. Pyroxene porphyroblast about

5% occur in some rocks. The pyroxene is grey and small pyroxene flakes crystallised parallel to the foliation Accessory calcite, sphene and apatite are also present in the rocks.

3 GEOCHEMISTRY

Samples from gneisses and amphibolites were analysed for major and trace elements composition by XRF method (Table 1). Zr/TiO_2 versus Ni and Niggli al-alk versus c plots have been used to test for sedimentary and igneous origin of the rocks (Figure 2). All rocks plot within the igneous fields.

On the K versus Rb diagram (Figure 3), all samples closely follow an igneous trend. SiO_2 versus FeO*/MgO plot shows that gneisses fall mainly in the calc-alkaline field but amphibolites in the tholeiitic field (Figure 3).

Rb in gneisses is generally less than 138 ppm and in amphibolite less than 13 ppm, which are more like those in acidic and basic igneous compositions, respectively. Furthermore, Th content of the rocks is again similar to that of acid and basic compositions of igneous rocks.

Table 1. Representative major (wt.%) and trace (ppm) element analyses of the Gümüsler gneiss and amphibolites.

Samp.	Gneiss					Amphibolite		
	1	2	3	4	5	6	7	8
SiO_2	67.4	71.0	69.2	69.6	69.4	49.8	47.2	48.8
TiO_2	0.2	0.4	0.1	0.2	0.3	0.3	0.3	0.3
Al_2O_3	15.3	13.9	16.2	14.9	14.7	11.5	16.3	15.2
FeO^*	1.8	2.9	0.9	1.4	1.8	7.7	10.2	10.4
MnO	0.0	0.1	0.1	0.0	0.0	0.2	0.2	0.2
MgO	1.1	0.7	0.3	0.9	0.4	11.3	9.1	9.4
CaO	2.4	4.9	3.9	4.4	2.3	14.1	12.9	13.5
Na_2O	2.8	2.5	2.4	3.5	2.0	0.6	1.3	0.8
K_2O	4.8	0.1	2.4	0.6	4.3	0.3	0.2	0.2
P_2O_5	0.2	0.1	0.1	0.1	0.2	0.0	0.1	0.1
LOI	2.9	3.3	4.1	3.5	3.4	4.2	2.1	1.1
Total	98.9	99.9	99.7	99.1	98.8	100	99.9	100
Ba	894	160	717	149	848	362	225	289
La	13	0	7	11	13	bdl	bdl	bdl
Ce	25	0	10	32	26	bdl	bdl	bdl
Zr	129	114	94	103	136	19	15	10
Y	10	4	6	8	8	10	7	24
Sr	186	168	213	144	190	72	99	76
U	4	2	2	4	8	1	5	7
Rb	126	4	65	32	138	13	9	6
Th	bdl	bdl	2	6	5	bdl	bdl	1
Pb	51	3	41	34	50	5	6	13
Ga	15	13	16	17	16	8	12	13
Zn	32	23	9	16	38	44	60	61
Cu	bdl	11	bdl	7	bdl	19	74	18
Ni	15	bdl	7	30	15	111	28	61
Co	5	4	bdl	6	3	39	44	50
Cr	21	73	76	115	87	925	67	129
Nb	11	3	3	bdl	5	7	4	5

bdl= below the detection limit*: all Fe is calculated as FeO..

4 CONCLUSIONS

Field relations indicate that the intercalated amphibolite and marbles infolded in gneisses in the Gümüsler (Nigde) area. Thus, precursors of the amphibolites were deposited together with limestones in the same basin. Geochemical data indicate that both amphibolites and gneisses are igneous in origin. Amphibolites formed as

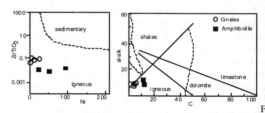

Figure 2. Zr/TiO_2 versus Ni (Winchester et al. 1980) and Niggli c versus al-alk (Van de Kamp 1969) plots of the Gümüsler gneisses and amphibolites.

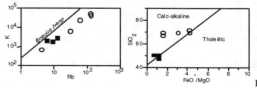

Figure 3. K versus Rb (Shaw 1968) and SiO_2 vs. FeO*/MgO (Miyashiro 1974) plots.

metamorphism of basic volcanics, most probably basaltic tuff, whereas gneisses as metamorphism of acidic volcanics. Conclusively, the precursor of the rocks were produced by different products of magmatism.

REFERENCES

Akçay, M. 1995. Gümüsler (Nigde) yöresi Sb+Hg+W cevherlesmelerinin jeolojik, mineralojik ve altýn potansiyeli yönünden incelenmesi. *Türkiye Jeoloji Bülteni* 38(2): 23-34.

Akçay, M., C.J. Moon & B.C Scott. 1995. Fluid inclusions and chemistry of tourmalines from the Gümüsler Sb-Hg-W deposits of the Nigde massif (Central Turkey). *Chem. Erde.* 55: 225-236.

Göncüoglu, M.C. 1981. Nigde Masifinde viridin gnaysýn kökeni. *Türkiye Jeoloji Kurultay• Bülteni,* 24 (1): 45-51.

Göncüoglu, M.C. 1981. Nigde Masifinin jeolojisi: Ýç Anadolu'nun Jeolojisi Sempozyumu. *35. Türkiye Jeoloji Kurultay:* 16-19.

Göncüoglu, M.C. 1986. Geochronological data from the southern part (Nigde area) of the Central Anatolian Massif: *Bull. Mineral Research and Exploration Institute of Turkey* 105-106: 83-96.

Ísler, F. & H. Büyükgídík1994. Gümüsler (Nigde) yöresinin jeolojisi ve petrografisi. *Ç.Ü Müh-Mim Fak. Dergisi,* 9 (1-2): 207-216.

Miyashiro, A. 1974. Volcanic rock series in island arcs and active continental margins. *Am. J. Sci.* 274: 321-355.

Shaw, D.M. 1968. A review of K-Rb fractionation trends by covariance analysis. *Geochim. Cosmochim. Acta* 32: 573-602.

Oygür, V., H.E. Erkale, N. Erkan & N. Karabalýk1978. Nigde Masifi demir cevherlesmeleri maden jeolojisi raporu. *M.T.A. Report No.* 6851.

Van de Kamp, P.C. 1969. Origin of amphibolites in the Beartooth Mountains-new data and interpretation. *Bull. Geol. Soc. Am.* 80: 1127-1136.

Winchester, J.A., R.G. Park & J.G. Holland 1980. The geochemistry of Levisian semipelitic schists from the gairloch district, Wester Ross. *Scott. J. Geol.* 16:165-179.

Geochemistry of the Earth's Surface, Ármannsson (ed.)© 1999 Balkema, Rotterdam, ISBN 90 5809 073 6

Replacement and self-organization in the weathering of Mn-rich shales at Moanda, Gabon

Enrique Merino
Geological Sciences, Indiana University, Bloomington, Ind., USA

Daniel Nahon
Laboratoire Géochimie de l'Environnement, Université d'Aix-Marseille III, Les Milles, France

Yifeng Wang
Sandia National Laboratories, Carlsbad, N.Mex., USA

ABSTRACT: Micronodular rhodocrositic shales at Moanda have weathered into several thick, self-organizational layers of Mn oxides. Each layer formed by two successive pseudomorphic replacements, rhodocrosite by manganite and manganite by pyrolusite. Writing the observed constant volume into the replacement redox reactions leads to uncovering the mineral reactions that were coupled to, and that drove, those replacements. The *repeated* replacement layers were produced by reaction-transport probably involving the catalytic action of manganese oxides on their own growth, in a framework of reactants (downward-moving oxygenated water and upward-moving parent shale) "interdiffusing" through the weathering profile.

1 INTRODUCTION

Tropical weathering at Moanda has produced a striking profile consisting of a few thick horizons, each 0.1 to 0.5 m thick, of manganese oxides that constitute a rich Mn ore. Petrographic descriptions (by Weber et al. 1979, Nahon et al. 1983, and us) indicate that the lowermost $-1/5$ of each layer consists of manganite (MnOOH) pseudomorphs of the original rhodocrosite micronodules of the parent rock and of manganite overgrowths. The uppermost $-4/5$ of each layer consists of pyrolusite (MnO$_2$), which pseudomorphically replaced the manganite, since it still contains tiny bits of manganite left unreplaced. The preservation of the original micronodular morphology through the two replacements indicates that both replacements conserved bulk volume. Each layer cuts across parent-rock bedding (Figure 1). The manganitic lowermost portion of each layer is concretionary, suggesting downward growth into the rhodocrositic parent rock.

The objective here is to construct a set of mineral reactions that simultaneously account for the main textural features observed, namely the *repetitive* layers, the *volume-preserving* redox replacements, and the manganite overgrowths. These reactions are the basis for quantitative dynamic modeling in progress; numerical results of the modeling should validate or invalidate the feasibility of the mechanisms incorporated in it.

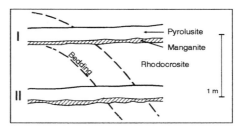

Figure 1. Replacement layers of Mn oxides cut across bedding. Manganite replaces rhodocrosite in lower fifth of layer. Pyrolusite replaces manganite in upper four-fifths of each layer.

2 REPLACEMENT REACTIONS

The replacement of rhodocrosite by manganite entails both a change of manganese oxidation number from 2 to 3 *and* constant volume. The oxidant is taken to be molecular oxygen dissolved in the percolating water. The conventionally-adjusted redox reaction would be

$$2MnCO_{3(rhod)} + 0.5O_{2(aq)} + 3H_2O =$$
$$= 2MnOOH_{(mang)} + 2CO_3^{--} + 4H^+, \qquad (1)$$

but this reaction does not preserve solid volume, since two rhodocrosite formulas have greater volume than two manganite formulas. Taking the numbers of oxygens to proxy for formula volumes, we need to reduce the number of oxygens in the 2 rhodocrosites on the left to a total of 4 (to equal the 4 in the two manganites on the right), and this requires multiplying the 2 rhodocrosites by 2/3, adding 2/3 Mn^{++}, and adjusting the rest. The redox reaction that *also* preserves solid volume turns out to be:

$$4MnCO_{3(rhod)} + 2Mn^{++} + 1.5O_{2(aq)} + 9H_2O =$$
$$= 6MnOOH_{(mang)} + 4CO_3^{--} + 12H^+, \qquad (2)$$

where the equal number of oxygens in the four rhodocrosites and the six manganites roughly represents the equality in volume required by the pseudomorphic replacement. (This reaction can be balanced more accurately with coefficients calculated from the formula volumes of rhodocrosite and manganite, see Merino et al.(1993), but the stoichiometric coefficients so derived would be clumsy.)

3 COUPLED REACTIONS

The significant geochemical consequence of the constant-volume redox reaction (3) is that it requires an input of dissolved Mn^{++} on the left side (which was not true of the conventionally-adjusted reaction (2)). The constant-volume adjustment thus indicates that the local replacement system must be open. The incoming dissolved Mn^{++} undoubtedly percolates from above, after being released by some other reaction overhead – probably the congruent dissolution of rhodocrosite,

$$MnCO_{3(rhod)} = Mn^{++} + CO_3^{--} . \qquad (3)$$

Thus *two* coupled mineral reactions, rhodocrosite dissolution above and replacement of rhodocrosite below, must occur simultaneously at two points in the profile. The coupling element is the aqueous manganese that travels from the dissolution site to the replacement site. The situation is similar to the weathering of plagioclase (Merino et al. 1993, reactions A,B and Figure 3) and other minerals, which also proceeds by dissolution above and linked replacement below.

The supply of aqueous manganese also drives the growth of manganite overgrowths on the replaced micronodules, according to

$$2Mn^{++} + 0.5O_2 + 3H_2O = 2MnOOH_{(mang)} + 4H^+. \qquad (4)$$

The replacement of manganite by pyrolusite in the upper 4/5 of each layer also involves both an increase in the Mn oxidation number (from 3 to 4) and volume conservation. It can be written as

$$2MnOOH_{(mang)} + 0.5O_{2(aq)} = 2MnO_{2(pyrol)} + H_2O. \qquad (5)$$

In this case the conventionally-adjusted redox reaction already comes out approximately preserving mineral volume (itself for simplicity proxied as before by the number of oxygens in the mineral formulas). That is, in this case, by chance, the supply of dissolved oxygen alone drives a constant-volume replacement.

4 FEEDBACK

The repeated layers of manganese oxides at Moanda must be self-organizational, because they cannot be explained by calling on features of the parent rock that might have guided the growth of the authigenic layers – since the Mn-rich layers cut across the parent rock bedding. (Self-organization refers generally to a system's autonomous structuring, one driven by its own dynamics, not forced externally.) Necessary conditions for self-organization are disequilibrium and positive feedback (Nicolis & Prigogine 1977, Ortoleva et al. 1987). Many cases of geochemical self-organization were reviewed by Merino (1984, 1987) and have been modelled by Wang & Merino (1995 and references therein).

We suppose the feedback at Moanda, and at the basis of the quantitative model in progress, is the known autocatalytic effect that Mn oxide has on its own production according to reactions (2,4) above, in the framework of the "counterdiffusion" of the two general reactants -- the original rhodocrositic shale, which moves *up* through the profile, and oxygenated rainwater, which moves *down* through the profile. The profile is simply the reaction zone where reactions (2-5) take place. (Repeated crystal layers produced by interdiffusion of reactants through a gel can be made easily with commercially available kits.)

5 DYNAMICS

Downgoing oxygenated acid water dissolves some rhodocrosite near or at the top of the profile. The aqueous Mn^{++} released leads to supersaturation with respect to manganite at some level further down. Manganite then replaces rhodocrosite at that level according to reaction (2), starting to form the first manganite layer. The manganite grown accelerates its own further growth, which locally depletes the local aqueous Mn^{++} concentration. Upon continued rhodocrosite dissolution, a new supersaturation zone (and thus a second, deeper, manganite replacement layer) can now be produced, but only sufficiently further down from the first replacement layer to escape the Mn^{++} depletion its growth produces. Eventually, a third layer nucleates and grows sufficiently deeper than the second, and so on. The process runs out when no more rhodocrosite is left at the top of the system to dissolve and supply Mn^{++} (though it might continue by now starting to redissolve the top oxide layer).

This dynamics is what the modeling in progress is expected to show quantitatively. The model consists of continuity differential equations, one for each relevant species, containing dissolution, growth, diffusion, advection, and the supposed catalytic feedback exerted by manganese oxides on their own production. This feedback enters into the rate laws used. Stoichiometric coefficients of the *volume*-adjusted reactions such as (2,5) above enter the model equa-tions as multipliers in appropriate source terms. One test of the quantitative modeling will be whether the predicted spacing between layers coincides with the observed spacing, which is on the order of one meter.

REFERENCES

Merino, E. l984. Survey of geochemical self patterning phenomena. In G. Nicolis & F. Baras (eds), *Chemical Instabilities: Applications in Chemistry, Engineering, Geology, and Materials Science*, NATO Adv. Sci. Series C 120: 305-328. Rotterdam: Reidel.

Merino, E. 1987. Textures of low-temperature self-organization. In R. Rodriguez-Clemente & Y. Tardy (eds) *Geochemistry of the Earth's Surface*: 597-610. Madrid, Consejo Superior de Investigaciones Cientificas & Centre National de la Recherche Scientifique.

Merino, E., D. Nahon & Y. Wang 1993. Kinetics and mass transfer of replacement: application to replacement of parent minerals and kaolinite by Al, Fe and Mn oxides during weathering. *Amer. J. Science* 293: 135-155.

Nahon, D., A. Beauvais, J.L. Boeglin, J. Ducloux & P. Nziengui-Mapangou 1983. Manganite formation in the first stage of the lateritic manganese ores in Africa: *Chemical Geology* 40: 25-42.

Nicolis, G. & I. Prigogine 1977. *Self-Organization in Non-Equilibrium Systems*. New York: John Wiley.

Ortoleva, P., E. Merino, C. Moore & J. Chadam 1987. Geochemical self-organization, I. Feedbacks and quantitative modeling. *Amer. J. Science* 287: 979-1007.

Wang, Yifeng & E. Merino 1995. Origin of fibrosity and banding in agates from flood basalts. *Amer. J. Science* 295: 49-77.

Weber et al, 1979, Epigénies manganesifères successives dans le gisement de Moanda, Gabon. *Sciences Geologiques*. Bull 32: 147-164.

Geochemistry of the Earth's Surface, Ármannsson (ed.)© 1999 Balkema, Rotterdam, ISBN 90 5809 073 6

Effects of different forest types on soil leachates in eastern Iceland

Ragnhildur Sigurdardóttir
Icelandic Forest Research, Mógilsá, Mosfellsbær, Iceland & School of Forestry and Environmental Studies, Yale University, New Haven, Conn., USA

Kristiina A. Vogt & Daniel J. Vogt
School of Forestry and Environmental Studies, Yale University, New Haven, Conn., USA

ABSTRACT: Soil leachate chemistry was studied in monocultures of *Larix sibirica* Lebed., *Pinus contorta* Dougl., and *Betula pubescens* Ehrh. in Hallormsstaður Forest, Eastern Iceland. Soil leachates were analyzed for NO_3-N, NH_4-N, PO_4-P, Mg, Ca, Zn, K, Fe, Na, Cl, SO_4, inorganic-C (IC) and organic-C (TOC) concentrations. The *Larix* stands impacted the greatest number of elements studied in the ecosystem, with significantly higher soil leachate concentrations of NO_3, K, Fe, Zn and TOC than in either *B. pubescens* or *P. contorta*. *Betula* had significantly higher leachate concentrations of Ca and IC than the other two species, and the Na, Cl and SO_4 concentrations were significantly higher in the *Pinus* systems. Seasonality of element leaching was also observed, with high spring fluxes before active growth in the deciduous systems, and a reduction of certain nutritional elements during the growing season. These results suggest a strong influence of biological factors on element cycles. Elements studied with no observed significant species differences in chemical concentrations of soil leachates were PO_4-P, NH_4-N and Mg.

1 INTRODUCTION

Forests characterized by either evergreen conifers or broadleaved deciduous tree species show differences in tolerance to environmental constraints and chemical characteristics of tissues related to leaf structure and longevity.

Evergreen conifers are commonly considered to have an ecological advantage in harsh environments (cold and nutrient poor) over deciduous species, due to an improved carbon balance (Mooney 1972, Schulze et al. 1977), improved resource-use efficiency (Mooney and Gulmon 1982), and by avoiding the additional carbon cost of producing a full complement of new foliage every year (Sprugel 1989). The deciduous conifer, genus *Larix*, is an important exception with a widespread distribution in some of the harshest environmental conditions of forest ecosystems (Ostenfeld and Larsen 1930).

Evergreen and deciduous tree species have been found to differ in foliar nutrient concentrations (Bormann & Likens 1970, Alban 1982). Evergreen leaves tend to have lower concentrations of Ca, Mg, K, P, N, total ash content and pH. At senescence, leaf litter from deciduous species tends to have higher nitrogen levels and lower levels of lignin, waxes and other more recalcitrant or resistant compounds than evergreen conifers (Carlyle & Malcolm 1986, Taylor et al. 1989, Son & Gower

1991). The chemical and recalcitrant nature of evergreen litter reduces rates of mineralization and decay, resulting in slower decomposition rates. Since leaf litterfall quantity is similar in deciduous and evergreen forests (Chabot & Hicks 1982), the total amount of nutrients in plant residues deposited annually, as well as the rate of their release through decomposition, varies greatly depending on their leaf structure and longevity. Thus, leaf structure and longevity may influence the chemical composition of the soil solution.

By examining global carbon budgets, Vogt et al. (1995) found that deciduous broadleaved forests in the boreal zone accumulated more organic matter overall than boreal evergreen coniferous forests, although evergreen forests are far more common under these climatic conditions. Boreal, broad-leaved deciduous forests accumulate more carbon in soil organic matter, but coniferous forests tend to have greater standing biomass and accumulations of carbon on the forest floor (Vogt et al. 1995, Vogt et al. 1997). This differential allocation of carbon within the different ecosystems suggests that trends in leaf structure and longevity may affect the concentrations and characteristics of soil solution C, which further may affect weathering rates of soil minerals and the chelation of trace metals.

The purpose of this paper is to compare element concentrations of the soil leachates from three different forest types, cultivated under common environmental conditions, in order to test hypotheses on whether different trends in leaf structure and longevity have an effect on soil leachate chemistry.

2 METHODS

2.1 Study site and sampling design

Soil solution chemistry was studied in single species plantations of *Larix sibirica* (Siberian larch) and *Pinus contorta* (lodgepole pine), and in naturally regenerated stands of *Betula pubescens* (downy birch) in Hallormsstaður Forest, eastern Iceland. The forest is situated in a valley, at 65°05´N and 14°75´W, about 70 km from the ocean. Hallormsstaður Forest, like other woodland areas in the country, is dominated by birch, the only native tree species that forms natural forests (Sigurðsson 1977).

The research plots were selected at sites where the soil environment, vegetative dominance and land-use history were similar prior to the establishment of each stand. All selected stands had reached an optimal canopy closure, had comparable slopes, aspects, site qualities, and deep soils (> 50 cm deep) with no surface bedrock present.

The bedrock in the area is mostly Tertiary basalt, which is covered by a 0.5-2-m thick layer of aeolian andisol (Cryands) (Arnalds 1990, Blöndal 1995). A preliminary examination has shown that between the aeolian deposits and the bedrock is a layer of glacial deposited gravel (C-horizon). Soil pits were dug within each research plot, which revealed homogenous layers of rock-free mineral soil with intermittent ash-layers, which qualitatively supports assumptions of original homogeneous soil origins and conditions.

The research plots consist of three replicate stands of each species. On these sites, a total of nine 20×20 m permanent plots have been established. The conifers were approximately 40-years-old, but the *Betula* sites were somewhat older, representing the area of the birch woodland into which the plantations were established.

Soil water leachates were collected by using three tension porous-cup lysimeters in each plot - a total of 9 lysimeters for each of the three species. The lysimeters were installed at 35-cm depths to sample leachate directly below the main zone of fine roots within the mineral soil. Soil water was collected monthly at equal sampling intervals, from 28-30 May 1998 to 29-31 October 1998, and a time series covering the growing season, until after senescence in the late fall.

2.2 Chemical analyses of soil water

NO_3, NH_4, Cl and PO_4 ions were analyzed on a Perstorp Analytical Enviroflow 3500 auto-analyzer, Ca, Mg, Na, Fe and Zn on a Perkin Elmer Optima 3000 ICP, SO_4 on a DIONEX DX 500 ion chromatograph, K on a Perkin Elmer Atomic Absorption Spectrophotometer, and IC, and TOC on a Shimadzu TOC-5000A (Total Organic Carbon) analyzer.

Data variability within replicate sites, between forest types, and sampling dates, was tested with a series of two way ANOVAs using the SAS statistical analysis package (SAS institute 1988).

3 RESULTS

The Cl concentrations of the *Pinus* systems were significantly higher than those of the *Larix* and *Betula* forests at all sampling dates (Table 1), with the exception of late October. During the growing season, there were no significant differences in Cl concentrations between the *Larix* and the *Betula* forests, with the exception of the first (May) and last (October) sampling dates (Figure 1). The *Betula* forests showed no seasonal variation in Cl concentrations. However, the *Pinus* stands showed a steady, but slow, increase in concentrations throughout the time series. Na concentrations were significantly higher in *Pinus* than in *Larix* and *Betula* forests for all sampling dates, with the exception of late August. No significant seasonal variations were found for SO_4.

Table 1. Average chemical composition of soil leachates in different forest systems (mg l^{-1}). Means with the same letter are not significantly different ($P < 0.05$) among the different ecosystems.

Species	Cond mS	pH	Na	Cl	SO₄	NH₃	NO₃	PO₄	Ca	Mg	K	Fe	Zn	TOC	IC
Betula	0.08b	7.3 a	5.78 b	3.68 c	1.97 c	0.017 a	0.028 b	0.027 a	6.23 a	3.49 a	0.17 b	0.007 b	0.014 b	7.19 b	5.52 a
Larix	0.07b	7.3 a	5.74 b	6.62 b	3.09 b	0.019 a	0.064 a	0.028 a	5.44 b	3.00 a	0.35 a	0.024 a	0.040 a	8.53 a	3.70 b
Pinus	0.10a	7.2 a	8.19 a	12.10 a	4.52 a	0.015 a	0.005 b	0.023 a	5.28 b	3.20 a	0.17 b	0.005 b	0.019 ab	7.51 ab	3.42 b

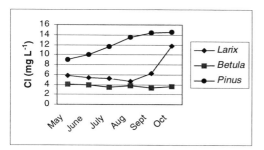

Figure 1 Species-specific and seasonal variation of Cl concentration of soil leachates in Hallormsstadur Forest (1998).

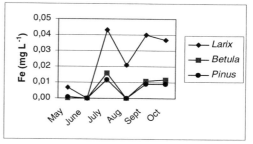

Figure 2 Species-specific and seasonal variation of Fe concentration of soil leachates in Hallormsstadur Forest (1998).

Most sampling dates showed significantly lower SO_4 concentrations in *Betula* forest leachates than in *Pinus*, but neither significantly differed from the *Larix* leachates.

A significant seasonal variation in NH_4 concentration was observed, with higher concentrations throughout the peak-growing season, but no species-specific patterns were observed. NO_3 concentrations were significantly higher in the *Larix* forests than in the *Betula* and *Pinus* forests (Table 1). Some seasonal variations were present in the NO_3 concentrations of the soil leachates, with peak values in late July for both coniferous species. The trend of NO_3 concentrations for the *Betula* forests contrasted with the other species by having a continuous drop in NO_3 throughout the time series.

The average PO_4 concentrations of soil leachates were not significantly different among the different forest types (Table 1). There were, however, dramatic patterns in the seasonal availability of PO_4 in the *Pinus* systems.

Betula forests had significantly higher concentrations of Ca than the coniferous systems (Table 1). There were significant seasonal effects for both deciduous systems, with higher concentrations of Ca during the growing season. higher concentrations of Ca than *Larix* in July There were no significant species-specific differences in the leachate concentrations of Mg. There was, however, a seasonal effect, with reduced concentrations for all species during the growing season.

The concentrations of K in soil leachates were significantly higher in the *Larix* forests than in the *Pinus* and *Betula* forests (Table 1). Seasonal trends in K concentration also varied among forest types. *Larix* systems showed no significant variation in K between sampling dates, while *Betula* forests had peak K concentrations in May and July, and *Pinus* forests had peak values in July.

Fe and Zn showed a clear species-specific pattern, with *Larix* forests having significantly higher concentrations than *Pinus* and *Betula* forests (Table 1). The release of Fe also showed some seasonal trends, with the highest concentrations occurring in late July and in the fall. Interestingly, there was a negligible availability of Fe in May and June, with no significant differences between the different systems. Also there was a dramatic rise in Fe availability in the *Larix* systems compared with the other species (Figure 2), starting in late July and continuing throughout the time series. Seasonality was observed for Zn concentrations in the soil leachates, with a flux of Zn in the early spring (May) and a dramatic drop at the time of leaf expansion.

Betula forests had higher concentrations of IC than the *Larix* and *Pinus* forests (Table 1). These values contrasted with the amount of TOC in those systems, where the *Larix* systems had significantly higher concentrations than *Betula*. *Larix* forests also had higher concentrations of TOC on all sampling dates between late July and October. *Larix* systems were, furthermore, the only systems that did not show a steady decrease in the concentration of TOC throughout the time series. All forests types showed similar seasonal patterns in the concentrations of IC, with significantly reduced concentrations in late June, but otherwise fairly constant values (except for a high IC flux for *Larix* and *Betula* in May).

4 DISCUSSION

A comparison of the effects of *Larix*, *Betula* and *Pinus* on nutrient leaching from the Hallormsstaður forest ecosystems shows that larch had the greatest impact on ecosystem elemental cycling for the elements studied. Biologically mediated elements were extensively leached from the *Larix* forests relative to the other species studied. The higher levels of N, K, Fe, Zn, and organic-C leaching from the *Larix* systems suggest an increase in the availability of these elements at the ecosystem level. Organic-C and Fe have been found to be important for soil forming processes (Boudot et al. 1986), suggesting that *Larix* is having a greater impact on soil formation than *Betula* or *Pinus*.

Past studies have shown the importance of trees in controlling soil development by their maintenance of certain elements in the biological cycles (e.g., Vogt et al. 1987). Part of this effect occurs through the decomposition of litter produced by the different tree species (Swift et al. 1979, Vogt et al. 1986). The amount of organic carbon leaching from the *Larix* stands facilitates the flux of nutrients required at the ecosystem level. Organic carbon has been found to be an important component of the ecosystem when the soils are either andisols, the predominant soil order of Iceland (Arnalds 1990), or spodosols, the predominant soil order elsewhere in subpolar Boreal Forest ecosystems (Antweiler and Drewer 1988, Dahlgren and Marrett 1991).

The fact that P leaching was not changed under *Larix*, despite the increased leaching of NO_3, is important because of the limitation of P as a nutrient and the high inherent P retention of andisols (Wada 1985). The leachate concentrations of PO_4 in Hallormsstaður Forest were about 1/10 of the lower range of soil solution concentrations in agricultural systems, or similar to the world average concentration for natural river water (0.025 mg L^{-1}) (Berner and Berner 1996).

The ability of *Larix* to increase the mobility and availability of both macro- and micronutrients may partly explain the success of this exotic species in colonizing degraded sites in Iceland. For example, *Larix* as a species has been shown to have superior establishment abilities in land reclamation projects in Iceland as compared to other forest species (S. Blöndal, personal communication).

Betula has been found to accumulate approximately 10 times more Zn in its tissues than any other co-occurring tree species in northeastern United States (Thomas Siccama, personal communication). The significantly lower Zn leachate concentrations in the *Betula* systems compared with *Larix*, and the dramatic drop in concentration during the growing season suggest this trend exists also for *B. pubescens* in Iceland, as well as suggesting an existence of Zn deficiency in the native birch systems.

Betula has been implicated as an effective Ca cycler and is able to increase Ca levels where it grows. In New England, USA, high Ca levels have been attributed to *Betula's* deep rooting and concomitant access to soil Ca. Some co-occurring species (e.g., *Picea* spp.) have been shown to be less efficient in acquiring Ca and the mortality of this species has been linked to its lack of a co-associated species that supplies or enhances Ca availability (Lawrence et al. 1999). The Ca cycled by *Betula* could be important in maintaining the health of certain species that are less efficient in acquiring Ca. This may help explain the higher productivities achieved in mixed species stands compared to monocultures (Vogt et al. 1995).

The concentrations of K in the soil leachates from the *Larix*, *Pinus* and *Betula* forests (0.351, 0.170 and 0.167 mg L^{-1}, respectively) were very low compared to published values for soil solutions and soil leachates (e.g., Barber et al, 1962, Reisenauer 1964, Graustein 1981). These low values may be explained by the adsorption complex above the 35-cm lysimeter depth and the relatively low K content of Icelandic basalt (Chorley 1959, Jakobsson 1984). The low concentrations K in of soil leachates in this study may also represent K deficiencies, especially in the *Betula* and *Pinus* systems.

Soil solution Ca and Mg are usually present in fairly high concentrations and considerably higher than the concentrations of K (Mengel & Kirkby 1987). Generally the Ca concentration of the soil solution is about 10 times higher than that of K (Mengel & Kirkby 1987). In the Hallormsstaður Forest, Ca concentrations were 15, 37, 31 times higher than K concentrations in the soil leachates for *Larix*, *Betula* and *Pinus* forests, respectively. The *Larix* forests are, therefore, the only forest type in Hallormsstaður Forest which has a relatively typical K:Ca ratio. Thus, the high Ca values compared to the low K values in the *Pinus* and *Betula* forests may further induce K deficiency in those systems. It has been found that when the K/(Ca+Mg) ratio decreases, there is an decrease in the uptake potential for K by plant roots because of the competition between these cations on the exchange sites (e.g., Malavolta 1985).

Application of KCl fertilizer on volcanic soils tends to increase the Cl and Ca concentrations in leaves, but decreases K concentrations (cf. von Uexküll, 1985). It is thought that this pattern is due to the effects of Cl in the promotion of Ca absorption to the detriment of K. This could explain the patterns that exist in the volcanic soils of the Hallormsstaður Forest. Thus, Cl concentrations found in the soil leachates in the *Pinus* forests might be inducing a K deficiency.

The high concentrations of Cl and Na may reflect a large influence of marine aerosols in the precipitation. The ratio of Cl:Na (on weight basis) in the soil leachates of Hallormsstaður Forest is 1.15, 0.64, and 1.5 for *Larix*, *Betula* and *Pinus* forests respectively. The Cl:Na ratio of seawater is 1.8 (Millero & Sohn 1992). The ratio of Cl:Na is thus close to the seawater ratio in the *Pinus* forest, but Na seems to be in excess in the *Larix* and especially the *Betula* forests. A possible reason is a higher uptake rate of Cl than Na in the deciduous forests, and a greater uptake rate of both Na and Cl in the deciduous forests than the *Pinus* forests, which may not utilize these elements to the same extent biologically.

While determining the origin of groundwater composition, Na released from mineral weathering is deduced by subtracting an amount equivalent to the measured Cl concentration from the total measured Na concentration (Berner & Berner 1996). This method assumes that Cl is non-biological (no uptake by trees) and the accumulation of Cl in soil water is the result of pure water being lost through stomata by transpiration (e.g., Likens et. al. 1977, Berner & Berner 1996). This view is refuted by the results of the Hallormsstaður study, where the differences in Cl and Na concentrations cannot be explained by differences in transpiration rates of the different forest systems alone. If Cl is only dependent upon inputs from precipitation, there should be no significant differences between the *Larix* and *Betula* stands when there is no foliage transpiration occuring (before leaf expansion and after senescence). Also the concentration of Cl should be smaller in the *Pinus* systems, since evergreen coniferous forests are assumed to transpire less water than deciduous forests (Kramer & Boyer 1995). Furthermore, if the Cl concentrations of these systems were only dependent upon inputs from precipitation and outputs of pure water through the process of transpiration, there should be a peak in Cl concentrations during the growing season. This is not apparent in any of the systems studied, since the Cl concentration is constant throughout the growing season for *Larix* and *Betula*, but increases steadily throughout for the *Pinus* systems, a pattern which does not correlate with the pattern of precipitation. Lastly, peak values in Cl concentrations should not occur after the end of growing season and senescence (for *Larix*) or after the growing season (for *Pinus*), when transpiration is minimal.

Cl is recognized as less independent from biological activity than hitherto thought (e.g., Likens et al. 1977) is indeed an essential mineral nutrient in higher plants (e.g., Walker & Leigh 1981, Stout & Cleland 1982, von Uexküll 1985). There are, furthermore, species-specific differences in the uptake of Cl. K and Cl have been found to be the two nutrients taken up and removed in the largest amounts in a study of coconut plantations in the Ivory Coast (von Uexküll 1985). Some of the nutritional effects previously attributed to K are now attributed to Cl, but its physiological role is not clear. (von Uexküll 1985, Mengel & Kirkby 1987).

The seasonality of element leaching also suggests a strong control of the biology on the element cycles of these forests. Many of the elements were leaching at some of the highest rates just prior to the active growth of *Betula* (e.g., NO_3, Mg, Ca, Zn, K, C), *Pinus* (e.g., Zn, C), and *Larix* (e.g., Mg, Ca, Zn, C). Decomposers are active during this time but the trees are not in a state during which they take up nutrients or control their flux. The reduction of some of the nutritional elements during the growing season is assumed to be influenced by their biologic utilization. High spring fluxes of Mg, Ca, NO_3, PO_4, Zn, K, inorganic- and organic-C of the deciduous species may represent leaching losses before the beginning of leaf expansion. These patterns in the early spring are not well defined for the evergreen conifer, except for a high flux of organic-C, supporting the hypothesis of Monk (1966), that nutrient cycles in evergreen stands should be "tighter," with less leakage from the system by leaching than in deciduous stands.

5 CONCLUSIONS

The species chosen for this experiment, characterized by differences in leaf structure and longevity, have significant effects on nutrient cycling and availability when grown under similar environmental conditions. Several different processes may explain the differences in chemical concentrations of the soil leachates. For example, these processes may be affected by different rates of decomposition, different chemical concentrations of leaf litter and living tissues, and differences in nutrient-use-efficiencies. Soil acidification due to the decomposition of coniferous litter vs. deciduous litter (Chabot & Hicks 1982) was not sufficient to affect the soil leachate pHs as all the stands had a pH of approximately 7. Evergreen coniferous forests have larger leaf areas, which enhance precipitation interception (Sprugel 1989), and tree form has been found to affect the interception of wind-blown dust (Graustein 1981). Graustein (1981) found that spruce and fir trees, in a mountainous area of New Mexico, trapped considerable quantities of wind-blown dust, which was washed off by precipitation and became part of the throughfall. These factors could certainly explain some of the differences found between the forest types.

The species-effects can thus be both biological and abiotic in character. However further studies in these systems at the ecosystem level are necessary to determine what the driving variables are within the biological component that affect nutrient cycling. From a management point of view, it may, however, be concluded that since the species studied during this study had such different effects, it may be relevant to consider mixed species plantations instead of single species plantations to create the most healthy ecosystems. At times, certain understory plants may provide functions similar to a tree species and, in conjunction with a tree species, provide similar results as a 'mixed species' system. Therefore, managing for the impact of a species on nutrient cycles should incorporate both tree species and other species that influence ecosystem functions.

REFERENCES

Alban, D.H. 1982. Effects of nutrient accumulation by aspen, spruce, and pine on soil properties. *Soil. Sci. Soc. Am. J.* 46:853-861.

Antweiler, R.C. & J.I. Drever 1988. The weathering of a late Tertiary volcanic ash: Implication of organic solutes, *Geochim. Cosmochim. Acta.* 47:623-629.

Arnalds, O. 1990. Characterization and erosion of andisols in Iceland. Dissertation. Texas A&M University, Texas.

Barber, S.A., J.M. Walker, & E.H. Vasey 1962. Principles of ion movement through the soil to the plant root. In G.J. Neale (ed.), *Trans. J. Meet. Comm. 4, 5 Int. Soc. Soil. Sci Soil Bureau, P.B.*, Lower Hutt: New Zealand.

Berner, E.K. & R.A. Berner 1996. *Global Environment: Water, Air, and Geochemical Cycles.* New Jersey: Prentice Hall.

Blöndal, S. 1995. Innfluttar trjátegundir í Hallormsstaðaskógi (Exotic tree species in Hallormsstadur Forest). Iceland Forest Service.

Bormann, F.H. & G.E. Likens 1970. Chemical analyses of plant tissues from the Hubbart Brook ecosystems in New Hampshire. *Yale Univ. Sch. For. Bull.* 79.

Boudot, J.P., B.A. Bel Hadj, & T. Chone 1986. Carbon mineralization in andosols and aluminum-rich highland soils. *Soil Biol. Biochem.* 18:457-461.

Carlyle, J.C., & D.C. Malcolm 1986b. Larch litter and nitrogen availability in mixed larch-spruce stands. II. A comparison of larch and spruce litters as a nitrogen source for Sitka spruce seedlings. *Can. J. For. Res.* 16: 327-329.

Chabot,B.F. & D.J. Hicks 1982. The ecology of leaf life spans. *Ann. Rev. Ecol. Syst.* 13:229-259.

Chorley, R.J. 1959. *Water, Earth and Man.* Methven and Co.Ltd.

Dahlgren, R.A. & F.C. Ugolini 1989. Formation and stability of imogolite in a tephritic Spodosol, Cascade Range, Washington, USA. *Geochim. Cosmochim. Acta* 53:1897-1904.

Dahlgren, R.A. & D.J. Marrett 1991. Organic carbon sorption in arctic and subalpine Spodosol B horizons. *soil Sci Soc. Amer. J.* 55:1382-1390.

Dahlgren, R.A. & F.C. Ugolini 1991. Distribution and characterization of short-range-order minerals in Spodosols from the Washington Cascades. *Geoderma* 48:391-413.

Graustein, W.C. 1981. The effect of forest vegetation on solute acquisition and chemical weathering: A study of the Tesuque watersheds near Santa Fe, New Mexico. Ph.D. Dissertation, Yale University, New Haven, CT.

Jakopsson, S. 1984. Íslenskar bergtegundir III. Þóleiít (Icelandic rock types. III. Tholeit). *Náttúrufræðingurinn* 53:53-59.

Kramer, P.J. and J.S. Boyer 1995. *Water Relations of Plants and Soils.* San Diego: Academic Press.

Lawrence G., K.A. Vogt, M. David, D. Vogt, J. Tilley, P. Wargo & M. Tyrrell Expected 1999. Chapter 5. Atmospheric deposition effects on surface waters, soils, and forest productivity in northeastern United States— Advances since NAPAP. In: *Responses of Northern Forests to Environmental Change.* New York: Springer-Verlag.

Likens, G.E., F.H. Bormann, R.S. Pierce, J.S. Eaton, & N.M. Johnson 1977. *Biogeochemistry of a Forested Ecosystem.* New York: Springer-Verlag.

Malavolta, E. 1985. Potassium status of tropical and subtropical region soils. In R. Munson (ed.), *Potassium in Agriculture*: 164-200. Madison:ASA-CSSA-SSSA.

Mengel, K. & E.A. Kirkby 1987. *Principles of Plant Nutrition.* Bern: International Potash Institute.

Millero & Sohn 1992. *Chemical Oceanography.* Boca Raton: CRC Press.

Monk, C.D. 1966. An ecological significance of evergreenness. *Ecology* 47:504-505

Mooney, H.A. 1972. The role of carbon balance of plants. *Annu. Rev. Ecol. Syst.* 3: 315-346.

Mooney, H.A. & S.L. Gulmon 1982. Constraints on leaf structure and function in reference to herbivory. *BioScience* 32: 198-206.

Ostenfeld, C.H. & C.S. Larsen 1930. The species of the genus *Larix* and their geographical distribution. *Biol. Medd. K. Dan. Vidensk. Selsk.* 9: 1-106.

Reisenauer, H. 1964. Mineral nutrients in soil solution. In P.L. Altman & D.S. Dittmer (ed.) *Environmental Biology*: 507-508.Bethseda: Federation of the American Societies for Experimental Biology.

Sas Institute 1988. *SAS/STAT User's Guide.* Cary: SAS Institute.

Schulze, E.-D., M. Fuchs, & M.I. Fuchs 1977. Spatial distribution of photosynthetic capacity and performance in mountain spruce forest of northern Germany. III. The significance of the evergreen habit. *Oecologia* 30:239-248.

Sigurðsson, S. 1977. Birki á Íslandi, útbreiðsla og ástand (Birch woodland in Iceland, its distribution and condition). p. 146-172. In: *Skógarmál.* Reykjavík: Prentsmiðjan Oddi hf.

Son, Y. & S.T. Gower 1991. Aboveground nitrogen and phosphorous use by five plantation-grown trees with different leaf longevities. *Biogeochemistry* 14:167-191.

Sprugel, D.G. 1989.The relationship of the evergreenness, crown architecture, and leaf size. *Am. Nat.* 133:465-479.

Stout, R.G. and R.E. Cleland 1982. Evidence for a Cl⁻ -stimulated Mg ATPase proton pump in oat toot membranes. *Plant Physiol.* 32:267-289.

Swift, M.J., O.W. Heal & J.M. Anderson 1979. *Decomposition in Terrestrial Ecosystems.* Berkley: University of California Press.

Taylor, B.R., D. Parkinson & W.F.J. Parsons 1989. Nitrogen and lignin content as predictors of litter decay rates: a microcosms test. *Ecology* 70:97-104.

Vogt, K.A., R.A. Dahlgren, F. Ugolini, D. Zabowski, E.E. Moore, & R.J. Zasoski 1987. Above- and belowground: II. Pools and circulation of Al, Fe, Ca, Mg, K, Mn, Cu, Zn and P in a subalpine Abies amabilis stand. *Biogeochemistry* 4: 295-311.

Vogt, K.A., C.C. Grier & D.J. Vogt 1986. Production, turnover, and nutrient dynamics of above- and belowground detritus of world forests. *Adv. Ecol. Res.* 15:303-377

Vogt K., D. Vogt, S. Brown, J. Tilley, R. Edmonds, W. Silver & T. Siccama 1995. Forest floor and soil organic matter contents and factors controlling their accumulation in boreal, temperate and tropical forests. *Advances in Soil Science*, 159-178.

Vogt, K.A., D.J. Vogt, P. Palmiotto, P. Boon, J. O'Hara & H. Asbjornsen 1997. Review of root dynamics in forest ecosystems grouped by climate, climatic forest type and species. *Plant and Soil* 187:159-219.

Von Uexküll, H.R. 1985. Potassium nutrition of some tropical plantation crops. In R. Munson (ed.), *Potassium in Agriculture*: 929-954. Madison:ASA-CSSA-SSSA.

Wada, K. 1985. The distinctive properties of Andosols. *Adv. Soil Sci.* 2:173-229.

Walker, R.R. & R.A. Leigh 1981. Characterization of a salt-stimulated ATPase activity associated with vacuoles isolated from storage roots of red beet (*Beta vulgaris* L.). *Planta.* 153:140-149.

Geochemistry of the Earth's Surface, Ármannsson (ed.)© 1999 Balkema, Rotterdam, ISBN 90 5809 073 6

Weathering of basic magmatic rocks under Mediterranean conditions

A. Singer

The Seagram Center for Soil and Water Sciences, Faculty of Agricultural, Food and Environmental Quality Sciences, Hebrew University of Jerusalem, Rehovot, Israel

ABSTRACT: Three major weathering forms of Pleistocene basalts under moderately humid Mediterranean conditions are recognized: *a weathering crust*, several millimeters thick, *a weathering zone*, several centimeters thick, and thoroughly altered basalt *saprolite*. The clay fractions in the weathering crust and zone consist of smectite and kaolinite with halloysite, while in the saprolite, smectite is the only clay mineral formed. SiO_2/Al_2O_3 ratios change from 6-7 in the rock to 2-3 in the crust clay, and 3-5 in the soil clay and this decrease represents maximum desilication obtained. With soil development, a resilication of soil clays relative to crust clay takes place due to aeolian accumulation of quartz. The mobility of minor elements, assessed according to their depletion rate from the weathered rocks, is Sr > Mn, Cu > Co, Ni > Zn > Cr.

Three major weathering forms of Pleistocene basalts under moderately humid Mediterranean conditions are recognized: *a weathering crust*, several millimeters thick, at the interface with the massive basalt; *a weathering zone*, several centimeters thick, associated with vesicular basalt; and thoroughly altered basalt *saprolite*, mostly obtained under prolonged water-saturated conditions. The clay fractions in the weathering crust and zone consist of smectite and kaolinite with halloysite, while in the saprolite, smectite is the only clay mineral formed. SiO_2/Al_2O_3 ratios change from 6-7 in the rock to 2-3 in the crust clay, and 3-5 in the soil clay and this decrease represents maximum desilication obtained.

The thin weathering crust is characterized by a sharp concentration change of some of the chemical constituents across the rock-soil interface (Fig. 1). Large losses of Mg, Ca, Na, K and smaller ones of Si have occurred in the clay fraction of the crust and are accompanied by the nearly complete disappearance of mafic minerals. Plagioclase persists in part in the clay fraction of the crust indicating a greater stability of this mineral. Dissolution of the basalt evidently is not stoichiometric. Gislason and Eugster (1987) have experimentally shown that while basaltic glass dissolves stoichiometrically at 25°C, this is not so for the crystalline basalt. Clay minerals formed in the course of the weathering include smectite, halloysite and kaolinite. The chemical and mineralogical transitions across the sharply defined

weathering front are associated with changes in the pH environment of the unaltered rock, of the crust and of the soil. This weathering crust can be compared to the "weathering rind" of an alkali basalt pebble from France examined by Jongmans et al. (1993). The depletion of Na and K from that rind was stronger, that of Mg, Ca and Si somewhat weaker than in the weathering crust on the basalt from Israel. Altogether the trends are similar. The lower removal rate of K is probably related to the uptake of K by the neoformed 2:1 clay minerals (Fig. 2). Saprolites are usually associated with vesicular basalt. Argillation of the primary minerals into kaolinite, halloysite, smectite and iron oxides is accompanied by significant depletions in Mg, Ca, Na, K, Ti and Si, which, however, are smaller than those occurring during the rock-soil interface type of weathering described previously. This decrease in the depletion of chemical constituents is explained by a decrease in the intensity of leaching due to the presence of the vesicles. The depletion range of major elements in the course of basalt weathering is similar to that observed by Eggleton et al. (1987) during the weathering of eastern Australian basalts. Saprolitic argillation of some other basalt flows resulted in the formation of a single clay mineral, smectite, from all primary alumosilicates, and was accompanied by a relatively small depletion in silica, with SiO_2/Al_2O_3 5 or higher. This form of weathering is attributed to the influence of water saturation conditions. Plagioclase pseudomorphs consisted of

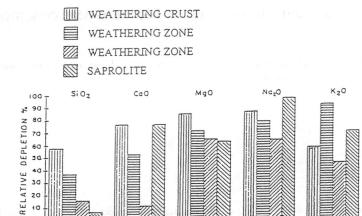

Figure 1. Relative depletion of five major chemical constituents in the weathering crust, weathering zone and saprolite from Pleistocene basalt in the Galilee, Israel.

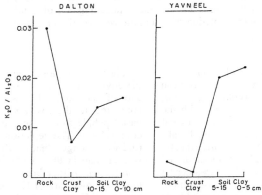

Figure 2. K_2O/Al_2O_3 ratios in the basalt rock, the basalt weathering crust and soil clay from two soil horizons in two vertisolic soils formed on basalt in a humid Mediterranean (Dalton) and semi-arid Mediterranean (Yavneel) climate.

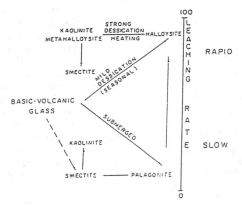

Figure 3. Alteration and transformation sequence of basic pyroclastic material under humid Mediterranean conditions.

halloysite in the vesicular and of smectite in the saprolitic type of weathering.

With soil development, a resilication of soil clays relative to crust clay may take place in the deeper, stable profiles. This development is due to aeolian accumulation of quartz in the coarser size fractions, and Fe and K enrichment in the clay fraction. Vertisols represent the mature, stable end products of basalt weathering on uneroded surfaces in the semi-arid and sub-humid parts of the eastern Mediterranean. In the more humid parts, Red and Brown Mediterranean soils (Xeralfs) become dominant.

Under more humid Mediterranean conditions of the upper Golan Heights (precipitation ~900 mm yr^{-1}), the weathering of basalt, scoria and tuff-lapilli resulted in kaolinitic/halloysitic soil clays (Fig. 3).

The weathered rocks were considerably depleted. From the clays, alkali and alkaline earth metals were nearly completely eliminated except for K, whose depletion was less complete. Molar SiO_2/Al_2O_3 ratios decreased from 4.7-7.7 in rocks to 3.2-4.6 in the clays. The mobility of minor elements, assessed according to their depletion rate from the weathered rocks, was Sr > Mn, Cu > Co, Ni > Zn > Cr. The relative capacities of the minor elements to be retained in the clays were: Zn, Mn > Ni, Co, Cu > Cr, Sr.

From the clay fractions, only Sr and Cr had been severely depleted. For Ni, Co, Cu, Zn and Mn, the depletion had been slight to moderate. The only moderate depletion of Fe from the clays indicates that part of the Fe-bearing primary minerals had undergone decomposition and released that element which was subsequently incorporated into the clay fraction, mainly in the form of "free iron" (the rubification process) but also in the lattice of phyllosilicates. On scoria-derived red soil clays, which contain relatively large amounts of hematite, most of the iron had been preserved in the clay fraction. The depletion of Ti from the clay fraction is more marked than that of iron, because most of that ion is contained in very stable primary minerals that are not likely to undergo pronounced weathering under Mediterranean conditions. In clays produced by tropical weathering, in contrast, Ti accumulates mainly in the form of anatase (Mohr et al., 1972). Apatite is fairly resistant towards weathering under alkaline conditions, as shown by the persistence of P in the weathered rocks. Under the mildly acid conditions of the soil environment, the stability of apatite is decreased and the clays are considerably depleted in P.

Weathering of basic pyroclastic rocks under humid Mediterranean conditions differs from that of basalt inasmuch as the proportion of secondary 1:1 clays is larger. Poor diffraction lines, as well as electron microscopy, suggest that this mineral is meta-halloysite. According to Silber et al. (1994), this mineral, with a Si/(Al+Fe) molar ratio of one, can be identified as "embryonic" halloysite. Allophane, a common alteration product of pyroclastics under humid conditions, could not be definitely identified. It is postulated that the long, dry season characteristic for the Mediterranean climate, interferes with allophane formation.

REFERENCES

Eggleton, R.A., Foudoulis, C. & D. Varkevisser 1987. Weathering of basalt: changes in rock chemistry and mineralogy. *Clays and Clay Minerals* 35:161-169.

Gislason, S.R. & H. Eugster 1987. Meteoric water – basalt interactions. I. A laboratory study. II. A field study in N.E. Iceland. *Geochim. Cosmochim. Acta* 51:2827-2855.

Jongmans, A.G., Veldkamp, E., Van Breemen, N. & I. Staritsky 1993. Micromorphological characterization and microchemical quantification of weathering in an alkali basalt pebble. *Soil Sci. Soc. Am. J.* 57:128-134.

Mohr, E.C., Van Baren, F.A. & J. Van Schuylenborgh 1972. *Tropical soils.* Third Edition. Mouton-lehtiar Baru – Van Hoeve. The Hague. 481 pp.

Silber, A., Bar-Yosef, B., Singer A. & Y. Chen 1994. Mineralogical and chemical composition of three tuffs from northern Israel. *Geoderma* 63:123-144.

Geochemistry of the Earth's Surface, Ármannsson (ed.) © 1999 Balkema, Rotterdam, ISBN 90 5809 073 6

Dissolution rate and weathering of porous rhyolites from Izu-Kozu Island, Japan

T. Yokoyama
Mineralogical Institute, University of Tokyo, Japan

J. F. Banfield
Department of Geology and Geophysics, University of Wisconsin-Madison, Wis., USA

ABSTRACT. The rate of dissolution of rhyolite has been determined through analysis of physical and chemical data from four compositionally similar lavas ranging in age from ~1,000 to ~40,000 years. The lavas, which consist primarily of aluminosilicate glass, are extremely porous and permeable, and this causes fairly homogeneous weathering. After a small correction of volume-based chemical abundance data (assuming Fe and Ti are approximately conserved), rates of release of elements from samples can be calculated from linear trends in element abundance vs. time plots. Glass dissolution rates can be inferred by accounting for elements retained in secondary aluminosilicates. Results suggest that glass dissolution rates are about three orders of magnitude slower than published rhyolite dissolution rates at comparable temperatures. The discrepancy cannot be attributed to restricted access of fluids or intermittent water supply. We propose that the rates are suppressed because solutions are frequently close to saturation.

1. INTRODUCTION

The dissolution of volcanic glass and formation of secondary products during weathering influences the geochemistry of solutions, soils, sediments, and changes of strength of rocks. The glass dissolution rate determines fluxes of elements such as Si, Na, Ca, Mg, K into sea water, with direct implications for biosphere and atmosphere processes. The magnitude of weathering rates, as determined by laboratory and field-based analyses, have been the topic of many research and discussion. Well-constrained field-based values are essential to test and validate laboratory results, but they are difficult to obtain. Several studies based on watershed mass-balance have been published (e.g. Velbel 1985, Paces 1983). These data revealed a significant discrepancy (several orders of magnitude) between field and laboratory rate measurements. Often this has been attributed to difficulties in determination of specific reactive surface area in the field (Velbel 1993).

The goal of the current study is to directly determine rhyolite weathering rates through analysis of weathering progress in lavas of known age and to test the hypothesis that, if reactive surface areas are well constrained, field rates should closely match laboratory rates. Ideally, sites used for this type of analysis should have experienced constant climatic conditions for the duration of weathering. Our work

has involved samples from Kozu Island, where prior studies indicate climate buffering by the surrounding ocean, with temperature variations of less than 6 °C over 40,000 years (Sawada & Handa 1998). Generally, lavas with different ages have different chemical and mineral compositions, partly due to crystallization differentiation. However, previous studies of the Kozu island rhyolites indicate that the chemical and mineral compositions of the lavas have been surprisingly constant over the last several tens of thousands of years (Taniguchi 1977, Isshiki 1982). Porous rhyolites were chosen because of their simple mineralogy (> 86% glass), high porosity and permeability (Oguchi et al. 1994), and limited change in reactive surface area over time due to the weathering mechanism. The very high porosity and abundant cooling joints in the lava domes allows deep water penetration. Consequently, the rhyolitic rocks investigated here have characteristics of homogeneous-deep-weathering (Oguchi et al. 1994, Taniguchi 1980). This allows determination of element loss directly from changes in measured chemical compositions as a function of time.

2. SAMPLES

Samples from four glassy rhyolitic domes were studied. The ages of domes have been variously constrained: Mt. Tenjo at ~1,100y B.P. by historic records, Mt.Kobe by hydration layer analysis at ~2,600y B.P., Mt.Osawa by carbon dating at 20,000y B.P., and Mt.Awano-mikoto at ~40,000y B.P. by hydration layer analysis (Oguchi et al. 1994). All consist of > 86% glass. Glass and bulk densities, surface areas, and porosity of samples from each lava dome were reported by Oguchi et al. (1994).

3. WEATHERING CHARACTERIZATION

To verify that the initial compositions of the four lavas were similar, fresh part of glass was analyzed using an electron microprobe (EPMA). The results (Table 1) indicate that initial chemical compositions of the four lavas are approximately equal.

Scanning electron microscope (SEM) image show that the glass contains abundant, elongate pores that are arranged so as to confer high permeability (Figure 1). These provide easy access of fluids to glass surfaces during weathering. Little alteration is apparent on the smooth glass surfaces of 1,100y B.P. and 2,600y B.P. rocks. In contrast, alteration products are evident on the glass surfaces of the ~20,000 and ~40,000y B.P. rocks (Figures 2,3). The ~20,000y B.P. glass is covered with a ~300-500 nm thick layer composed of spheres of two different sizes (Figure 3). The first are about 20 nm in diameter and the second are about 100-200 nm in diameter. The ~40,000y B.P. glass surface is covered with a thick alteration layer composed mostly of the larger particles. The 100-200 nm (see Figure 3) particles are identified as halloysite, based on their size, morphology, and chemical composition as determined by transmission electron microscopy and energy-dispersive X-ray analysis (EDX). The smaller particles (see Figure 3) are close to allophane in composition, as determined by EDX.

4. DISSOLUTION RATE AND DISCUSSION

Bulk compositions were determined by XRF in order to estimate rates of loss of elements from the glass. It is necessary to either convert weight-based abundances to volume-based abundances using measured sample densities or to assume an element is immobile to obtain volume-based abundances by inference. Calculations using published density values (Oguchi et al. 1994) indicate some gain of normally very insoluble components (Al_2O_3, FeO, TiO_2) with weathering. This may be due to slight compaction or colloidal transport. Given that EPMA maps reveal that Fe, and possibly Ti, are condensed in altered products, we assumed Fe (and Ti) conservation. The effect of this correction on the

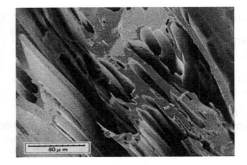

Figure 1. SEM image illustrating high porosity of the slightly weathered glass

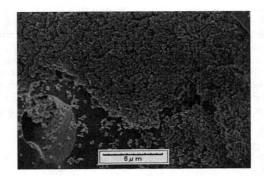

Figure 2: SEM image of alteration on the glass of ~ 40,000 B.P.

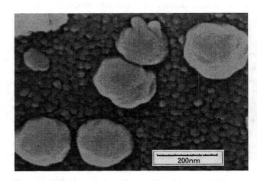

Figure 3: SEM image showing small and large particles within the altereation layer.

determined rate is small (undetectable if rates are rounded to the nearest order of magnitude).

A rate of Si release of ~10^{-18} moles/cm^2/sec was calculated from the specific surface area, age, and slopes of the linear trends revealed when the change

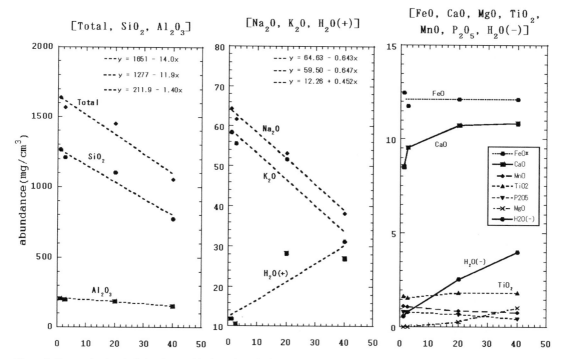

Figure 4 Changes in chemical abundance with elapsed weathering time [corrected for Fe conservation

Table 1. EMPA data for the composition of fresh glass (*1 below detection limit; *FeO = FeO + Fe₂O₃)

	MtTenjo 1.1ka	MtKobe 2.6ka	MtOsawa 20ka	MtAwano-Mikoto 40ka
SiO2	79.0	78.2	77.5	77.9
Al2O3	12.4	12.4	12.2	12.1
TiO2 *1	0.07	0.06	0.09	0.09
FeO *	0.60	0.61	0.57	0.53
CaO	0.35	0.35	0.51	0.54
MnO *1	0.08	0.08	0.07	0.07
K2O	3.84	3.90	3.81	3.67
P2O5	0	0	0	0
Na2O	3.92	3.82	3.71	3.78
MgO *1	0.06	0.06	0.08	0.08
H2O				
Total	100.4	99.5	98.5	98.7
No of data	59	57	60	68

in chemical composition (corrected) is plotted against time (Figure 4). Both model calculations, based on change of chemical composition, and SEM observations indicate 0.3 cm³ of altered products per cm³ of ~40,000y B.P. rock. Based on this result, the precipitation rate was estimated and true dissolution rate was calculated. The resulting rate is ~2 x 10⁻¹⁸ moles/cm²/sec. This value is ~10³ lower than

experimental dissolution rate of rhyolitic glass, measured at 25 °C and pH 6.2 in closed system (White 1983).

The historic mean temperatures around Kozu island, reported by Sawada & Handa (1998), range between 23–27 °C over the past 25,000 years. Consequently, the discrepancy between field and laboratory rates is probably not attributable to temperature. Rainfall is high, ranging from 138–328 mm per month (Japan Meteorological Agency 1990). Given the high porosity and permeability and fairly uniform distribution of reaction products, restricted access of fluids to glass surfaces is probably not a major factor.

One explanation for the comparatively slow rate of natural rhyolite glass dissolution may be that solutions are close to saturation relative to the experimental solutions. When it rains, water penetrates into lava dome along cooling joints and then into the finer scale porosity where it is retained, reacts with glass surfaces, reaches saturation, and precipitates secondary aluminosilicate phases. The dependence of rates on saturation state is well known. Given that we can rule out most other factors, we attribute the significant difference between laboratory and field rates to this effect.

409

REFERENCES

Isshiki, N. 1982. *Geology of the Kozushima district. Geological survey of Japan*

Japan Meteorological Agency 1990. *Climatic table of Japan* 1: 478

Oguchi, T. & T. Hatta & Y. Matsukura 1994. *Changes in rock properties of porous rhyolite through 40,000 years in Kozushima Island, Japan. Geographical Review of Japan* 67A: 775-793• • •

Paces, T. 1983. *Rate constants of dissolution derived from the measurements of mass balance in hydrological catchments. Geochimica et Cosmochimica Acta.* 47: 1855-1863

Sawada, K. & N. Handa 1998. *Variability of the path of the Kuroshio ocean current over the past 25,000 years. Nature* 392: 592-595

Taniguchi, H. 1977. *Volcanic Geology of Kozu-shima, Japan. Kasan, Bulletin of the Volcanological Society of Japan* 22: 133-147

Taniguchi, H. 1980. *Some volcano-geological significances of the hydration layer observed in the glassy groundmass of Kozu-shima Rhyolite. Kasan, Bulletin of the Volcanological Society of Japan* 25: 217-229

Velbel, MA. 1985. *Geochemical mass balances and weathering rates in forested watersheds of the southern Blue Ridge. Am. J. Sci.* 285: 904-930

Velbel, MA. 1993. *Constancy of silicate-mineral weathering-rate ratios between natural and experimental weathering. Chem.Geol.*105: 89-99

White A.F. 1983. *Surface chemistry and dissolution kinetics of glassy rocks at 25°C. Geochimica et Cosmochimica Acta* 47 :805-815

7 Geochemical thermodynamics and kinetics

Geochemistry of the Earth's Surface, Ármannsson (ed.) © 1999 Balkema, Rotterdam, ISBN 90 5809 073 6

Invited lecture: A general mechanism for multi-oxide solid dissolution and its application to basaltic glass

E. H. Oelkers & J. Schott
Laboratoire de Géochimie, Université Paul-Sabatier, Toulouse, France

S. R. Gíslason
Science Institute, University of Iceland, Reykjavík, Iceland

ABSTRACT: A general model describing the dissolution rates and mechanisms of multi-oxide minerals and glasses is summarized. Dissolution proceeds via the sequential equilibration of metal-proton exchange reactions until no further viable structure remains. The relative rates of these exchange reactions are deduced from the dissolution rates of corresponding single oxide minerals. The final step in many cases is the liberation of partially detached Si-tetrehedrals, which is likely H_2O rather than proton catalyzed. Application of this formalism to basaltic glass dissolution yields a rate equation similar to that for the alkali-feldspars; logarithms of laboratory measured 'far from equilibrium' rates are proportional to $\log (a^3_{H^+} / a_{Al^{3+}})$ where a_l refers to the activity of the subscripted aqueous species.

1 MULTI-OXIDE SOLID DISSOLUTION

The bulk of minerals and glasses found in nature are comprised of more than one type of metal-oxygen bond. The dissolution of these solids may proceed via the breaking of either one or more than one type of metal-oxygen bond depending on the structure of the solid, and the relative rates at which each of these bonds break. The relative rates at which the metal-oxygen bonds break can be deduced from two basic assumptions: 1) the relative rates of breaking any metal-oxygen bond is similar regardless of the identity of the mineral and 2) bonds of partially detached metals react faster than those that are fully attached to the structure. Consideration of the dissolution rates of single oxide and hydroxide minerals suggests that the relative rate at which metal-oxygen bond break follows the order Na ≈ K > Ca > Mg > tetrehedrally coordinated Al > Si > octahedrally coordinated Al (see also Casey 1991). The breaking of each metal-oxygen bond of a multi-oxide solid apparently occurs via proton exchange reactions that liberate the metal (M_i) according to

$$M_i< + z_iH^+ \longleftrightarrow M_i^{z_i+} + H_{z_i}< \qquad (1)$$

where the symbol < indicates that the species is attached to the surface and z_i represents the charge of

the ith aqueous metal ion. Dissolution proceeds via the sequential equilibration of the metal-proton exchange reactions in the order of their relative reaction rates until no further viable structure remains. As the last of these sequential exchange reactions destroys the structure it cannot equilibrate, and is thus irreversible. By analogy with the dissolution of single oxides it seems likely that each exchange reaction consists of several steps (e.g. the sorption of one or more protons or other aqueous species at the surface followed by the breaking of the metal-oxygen bonds), however, with the exception of the final exchange reaction that destroys the structure such sorption reactions are transparent. It is possible that other ions may exchange for $M_i<$, in particular when there are abundant ions of similar charge and size, for example the exchange of Na for K in K-feldspar during albitization. In such cases provision must be made to account for this competing reaction when computing dissolution rates. It seems likely, however, in the absence of such similar ions, that hydrogen is the dominant ion involved in the exchange reaction owing to its small size; a proton radius is $\sim 10^{-13}$ cm, whereas the typical metal ion radius is $\sim 10^{-8}$ cm.

Note that in some cases not all metal-oxygen bonds present in the structure need be broken to completely destroy a mineral. For example, the forsterite structure which contains both Mg-O and Si-O bonds, can be destroyed by breaking only Mg-O bonds (see Oelkers 1999), and the anorthite structure, which contains Ca-O bonds, and tetrahedral Al-O and Si-O bonds is completely destroyed by breaking only the Ca-O and Al-O bonds (see Oelkers & Schott 1995). In both cases Si-O bonds do not need to be broken to dissolve the mineral. A summary of various mineral dissolution mechanisms is presented in Table 1.

The extent to which each metal-proton exchange reaction completely liberates the metal from the near surface can be deduced from the law of mass action of reaction (1) which can be written

$$K_i = \frac{a_{M_i z+} X_{H_{z_i}<}}{a_{H^+}^{z_i} X_{M_i<}} \qquad (2)$$

where a_i stands for the activity of the subscripted aqueous species, X_i represents the mole fraction of

the indicated species at the mineral surface, and K_i designates the equilibrium constant for the ith metal exchange reaction. Assuming that metal sites contain either the original metal or proton(s) requires

$$1 = X_{H_{z_i}<} + X_{M_i<} \qquad (3)$$

which can be combined with Equation (3) to yield

$$X_{H_{z_i}<} = K_i\left(\frac{a_{H^+}^{z_i}}{a_{M_i z_i+}}\right) \Bigg/ \left(1 + K_i\left(\frac{a_{H^+}^{z_i}}{a_{M_i z_i+}}\right)\right) \qquad (4)$$

The degree to which the denominator in Equation (4) exceeds unity controls the extent to which the metal M_i is removed from the near surface potentially leading to leached layer formation. For a metal that is dominated by free aqueous ions in solution Equation (4) suggests that leached layer formation is favored by low pH and low M_i

Table 1. Summary of dissolution mechanisms of some minerals and basaltic glass.

	Reaction	Alkali-Feldspar	Anorthite	Muscovite	Enstatite	Wollas-tonite	Forsterite	Basaltic Glass
	Alkali metal -H exchange	Step 1	↓	Step 1		↓		Step 1
Decreasing Equilibration Rate	Ca-H exchange reaction		Step 1			Step 1	↓	Step 2
	Mg-H exchange reaction		↓		Step 1		Mineral Destroyed	Step 3
	Tetrahedral Al-H exchange reaction	Step 2	Mineral Destroyed	Step 2				Step 4
	Breaking Si-O bonds[1]	Mineral Destroyed		Mineral Destroyed	Mineral Destroyed	Mineral Destroyed		Solid Destroyed

1) The breaking of Si-O bonds most likely involves H_2O absorption rather than an Si for H exchange reaction (see Dove & Crerar 1990). In the reactions described in this table, however, the breaking of Si-O bonds is the final step of the dissolution process destroying the mineral structure.

concentration. The size of the leached layer is, however, related to the rate of M_i and H^+ diffusion into and out of the mineral structure relative to the leached layer dissolution rate Equations describing dissolution rates consistent with these mechanisms can be deduced considering transition state theory. In accord with this theory, the forward dissolution rate (r_+) is proportional to the concentration of the final surface species formed prior to the final irreversible step; this surface species is commonly referred to as the precursor complex. This precursor complex consists of the final partially detached metal oxide of the reaction sequence, and in some cases sorbed catalytic species. The concentration of this species can be deduced from the law of mass action for the formation of this precursor complex from the original solid, comprising exchange reactions (1) and the possible catalytic sorption reactions. The overall dissolution rate (r) contains in addition, provision for the reverse reaction(s) ($r = r_+ - r_-$, where r_- refers to the rate of the reverse reaction), which in this case is the reprecipitation of either the original solid or a partially depleted leached surface layer. The degree to which 1) the overall long term dissolution rate depends on the

$$\left(\frac{a_{H^+}^z}{a_{M_i^{z+}}} \right), \text{ and } 2)$$

aqueous activity ratio

equilibrium can be attained with a partially depleted leached surface layer that is distinct from the equilibrium of the overall solid is directly related to the degree to which the denominator in Equation (4) exceeds unity, which itself depends on 1) the pH, 2) the aqueous concentration of the metal M_i, 3) the aqueous speciation of this metal, and 4) the value of the equilibrium constant K_i.

2 APPLICATION TO BASALTIC GLASS

Because of its widespread occurrence on the ocean floor and in volcanic terrains, its emission during volcanic eruptions, and its relatively rapid dissolution rate, basaltic glass plays a major role in the global flux and cycling of numerous metals and nutrients. Moreover, basaltic glass is the major host rock of the Mururoa Atoll French nuclear test site, and it has been used as an analog for various radioactive waste forms. Consequently, significant efforts have been focused on understanding the dissolution rates and mechanisms of this solid.

A representative chemical formula for basaltic glass is $Si_3AlFe_{0.5}Ca_{0.7}Mg_{0.77}Na_{0.33}K_{0.03}O_{10}$. The structure of this glass is such that the removal of all

cations other than Si leads to a viable, though partially detached structure of Si-O bonds. It seems reasonable, therefore to assume that its dissolution consists of the sequential equilibration of metal-proton exchange reactions leading to the formation of partially liberated Si atoms, which dissolve irreversibly to complete the dissolution process (see Table 1). This mechanism is consistent with hydrogen depth profiling and XPS analyses reported by Schott (1990) (see also Berger et al. 1987) which show that dissolving basaltic glass surfaces are depleted in network modifying cations such as Na, Ca, and Mg, which are rapidly and reversibly exchanged with protons. Several other observations which reveal the extent to which the exchanged cations participate in the dissolution rate of the partially leached layer include

—The pH dependence of its dissolution rate mimics that of aluminum oxide mineral solubility: it exhibits a sharp increase with decreasing pH at acidic conditions and a more gentle increase with increasing pH at neutral to basic conditions. The minimum rate is found at ~pH=6 at low temperatures, but this minimum moves to lower pH with increasing temperature (Guy & Schott 1989).

—At far from equilibrium conditions and at constant pH, basaltic glass dissolution rates are 1) independent of aqueous silica activity, 2) decrease substantially with increasing aqueous aluminum activity, and 3) increase with increasing oxalic acid concentration in mildly acidic solution (Gíslason et al. 1999).

—Its dissolution rates approach zero as the chemical affinity of the overall Al, Fe, Si-oxide structure network hydrolysis reaction approaches zero (Daux et al. 1997).

Taking account of these observations, it seems likely that at the solution compositions considered in the laboratory, Na, K, Ca, and Mg depletion from the basalt surface are sufficient (the denominators in Equation (4) for their exchange reactions is sufficiently greater than unity) for the dissolution rate of the outer leached layer to be independent of their aqueous concentration and a unique equilibrium can be obtained with this K, Na, Ca, and Mg depleted leached layer. In contrast, it appears that sufficient Al and perhaps Fe are present at the surface to affect the concentration of the partially detached Si-O precursor complex.

Data regression indicates that all experimentally measured basaltic glass dissolution rates reported in the literature are consistent with

$$r = k_+ s \left(\left(\frac{a_{H^+}^3}{a_{Al^{3+}}} \right) \middle/ 1 + K_{Al} \left(\frac{a_{H^+}^3}{a_{Al^{3+}}} \right) \right)^n$$

$$\times \left(1 - exp(-A / RT)\right) \qquad (5)$$

where, k_+ designates a rate constant, s refers to the glass-solution interfacial surface area, a_i stands for the activity of the subscripted aqueous species, K_{Al} signifies the equilibrium constant for the Al/proton exchange reaction, A refers to the chemical affinity of the Na, Ca, Mg depleted layer dissolution reaction, R designates the gas constant, T represents absolute temperature, and n corresponds to a stoichiometric coefficient of ~ 1/3. Note the pre-exponential part of this equation is identical to that

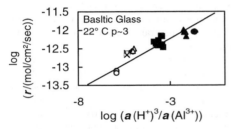

Figure 1. Logarithm of measured basaltic glass steady state dissolution rates as a function of log $(a_{H^+}^3/a_{Al^{3+}})$. Filled symbols correspond to data obtained in Al-free inlet solutions; rates represented by the filled squares, triangles, and circles were obtained in inlet solutions that contained 0, 5×10^{-4}, and 1×10^{-3} mol/kg oxalate respectively. Open symbols were generated from oxalate free inlet solutions containing 5, 10, and 50 ppm Al, respectively for the triangles, crosses and circles. Symbols represent data reported by Gíslason et al. (1999) and the linear curve represents a fit of these data to Equation (5).

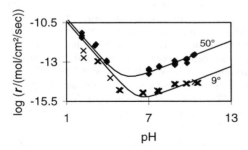

Figure 2. Variation of measured basaltic glass steady state dissolution rates as a function of pH at 9 and 50° C. Symbols represent data reported by Gíslason et al. (1999) and the curves corresponds to a preliminary fit of these data to Equation (5).

proposed to describe alkali feldspar dissolution rates (see Oelkers et al. 1994). Use of Equation (5) together with an Arrhenius relationship yields dissolution rates as a function of temperature, and taking account of solution speciation calculations permits description of rates as a function of pH, and the aqueous activities of Si, Al, and organic acids. Several examples are provided in Figures 1 and 2. The effect of aqueous Fe on these rates cannot be evaluated with the present data set.

REFERENCES

Berger, G., J. Schott & M. Loubet 1987. Fundamental processes controlling the first stage of alteration of a basalt glass by seawater: an experimental study between 200 and 320° C. *Earth Planet. Sci. Let.* 84:431-445.

Casey, W.H.. 1991. On the relative dissolution rates of some oxide and orthosilicate minerals. *J. Coll. Interf. Sci.* 146:586-589.

Daux, V., C. Guy, T. Advocat, J-L. Crovisier & Stille 1997. Kinetic Aspects of basaltic glass dissolution at 90° C: Role of silicon and aluminum. *Chem. Geol.* 142:109-128.

Dove, P. M. & D. A. Crerar 1990. Kinetics of quartz dissolution in electrolyte solutions using a hydrothermal mixed flow reactor. *Geochim. Cosmochim. Acta*, 54:955-970.

Gíslason S.R., E. H. Oelkers & J. Schott 1999. In preparation.

Guy, C. & J. Schott 1989. Multisite surface reaction versus transport control during the hydrolysis of a complex oxide. *Chem. Geol.*, 78:181-204.

Oelkers, E.H. 1999. A comparison of enstatite and forsterite dissolution rates and mechanisms. In B. Jamtveit and P. Meakin (eds) *Growth, Dissolution and Pattern Formation in Geosystems* Rotterdam: Kluwer Academic Publishers (in press).

Oelkers, E. H. & J. Schott 1995. Experimental study of anorthite dissolution and the relative mechanism of feldspar hydrolysis. *Geochim. Cosmochim. Acta* 59:5039-5053.

Oelkers, E.H., J. Schott & J.-L. Devidal 1994. The effect of aluminum, pH, and chemical affinity on the rates of aluminosilicate dissolution reactions. *Geochim. Cosmochim. Acta* 58:2011-2024.

Schott, J. 1990. Modeling of the dissolution of strained and unstrained multiple oxides: The surface speciation approach. In W. Stumm (ed) *Aquatic Chemical Kinetics:* 337-366 New York: John Wiley and Sons

Geochemistry of the Earth's Surface, Ármannsson (ed.) © 1999 Balkema, Rotterdam, ISBN 90 5809 073 6

Metal speciation in surficial geological fluids

J. Schott
Laboratoire de Géochimie, CNRS, Université Paul-Sabatier, Toulouse, France

G. S. Pokrovski
Centre de Recherche sur la Synthèse des Minéraux, CNRS, Orléans, France

J. D. Kubicki
Department of Geosciences, Pennsylvania State University, University Park, Pa., USA

ABSTRACT: The first part of this paper gives a short overview of metal speciation in aqueous solutions and its trends in the Periodic Table. Recent results of spectroscopic measurements, molecular orbital studies, and the kinetics of ligand exchange reactions are reported and their consequences for metal speciation and reaction kinetics in surficial environments are discussed. In the second part of this presentation, we report on-going studies of Al, Ga, and Fe hydrolysis in the presence of aqueous silica in which potentiometric titrations, spectroscopic (XAFS, NMR) measurements and molecular orbital calculations are combined. Results show considerable changes in metal hydrolysis with the formation of metal-silica complexes that hamper the formation of polymeric hydroxide species. This study allows for the first time determination of the structure and thermodynamic properties of some of the metal-silicate complexes formed.

1 OVERVIEW OF METAL SPECIATION

Metal ion hydrolysis and complexation with organic and inorganic ligands plays a major role in the control of many physical, chemical, and biological processes occurring in aquatic natural systems, from weathering and redox reactions to water quality, biogeochemical cycling and isotope distribution. The first goal of this presentation is to give a short overview of metal speciation in aqueous solutions and its trends in the Periodic Table. Recent spectroscopic and molecular orbital studies greatly enhance our understanding of metal coordination, the mechanisms and kinetics of ligand exchange reactions, the trends in metal hydrolysis and complexation, the electronic and steric constraints on complex formation, and the usefulness of linear free energy relationships in both speciation and kinetics. These results provide new tools to understand and model the processes that control the behavior of metals both at and out of equilibrium. In particular, it is shown that knowledge of metal speciation is crucial to model the kinetics of many superficial processes (redox and adsorption reactions, mineral dissolution/precipitation).

2 Al AND Ga HYDROLYSIS IN Si-BEARING SOLUTION AND THE FORMATION OF METAL-SILICA COMPLEXES.

This study takes account of potentiometric and spectroscopic (XAFS, NMR) measurements and molecular orbital calculations. XAFS spectroscopy was used to quantify at the atomic scale the interactions of gallium (10^{-4}-10^{-3} m) with aqueous silica (0-0.05 m) over a wide range of pH (1-11). Results show that in Ga-bearing alkaline solutions (pH > 8), Ga is coordinated with 4 oxygens in the first shell with Ga-O average distances of 1.85 Å. This is consistent with the tetrahedral $Ga(OH)_4^-$ species being dominant at pH > 7 in silica-free aqueous solutions (Diakonov et al. 1997, Bénézeth et al. 1997). In the presence of silica, Si is detected in the second Ga shell with Ga-Si distances of 2.2 and 2.8 Å, suggesting the formation of Ga-O-Si chemical bonds analogous to those observed in Al-Si aqueous complexes (Pokrovski et al. 1998). In strongly acidic (pH < 2) gallium nitrate solutions (0.001 m), Ga is coordinated to 6 oxygens with a mean Ge-O distance of 1.97 Å; this corresponds to the hydrated Ga^{3+} cation, $Ga(H_2O)_6^{3+}$ (Munoz-Paez et al. 1995). At higher pH (pH > 3), an important second shell is observed that corresponds to Ga-Ga pairs with a mean distance of 3.3 ± 0.2 Å, indicating the presence of polymerized Ga hydroxide complexes having Ga-O-Ga bonds. However, in the presence of aqueous silica at the same pH this

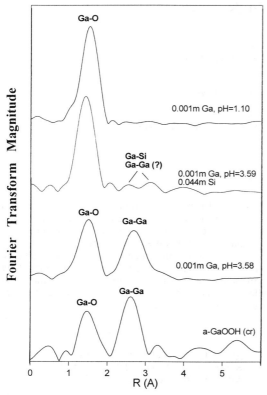

Figure 1. Fourier transforms of gallium-bearing aqueous solutions EXAFS spectra at 25°c vs. the distance R to nearest neighbors from the absorbing atom (gallium).

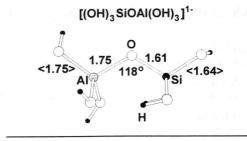

$[(OH)_3SiOAl(OH)_3]^{1-}$

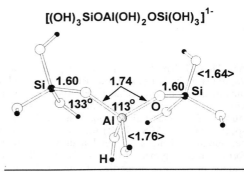

$[(OH)_3SiOAl(OH)_2OSi(OH)_3]^{1-}$

**Distances in Angstroms
"< >" indicates an average of T-(OH) bonds**

Figure 2. Optimized geometries (HF/6-32G*) of aluminum silica aqueous complexes.

feature disappears while Ga coordination number is reduced from 6 to 4 as shown by XANES spectra. This demonstrates that silicic acid hampers the formation of polymeric hydroxide species via complexing with gallium. The stability of $AlH_3SiO_4^{2+}$, the aluminum-silica complex formed in acid solutions, has been recently quantified in the temperature range 25-300°C at P_{sat} (Pokrovski et al. 1996, Salvi et al. 1998). It was found that the stability of $AlH_3SiO_4^{2+}$ strongly increases with temperature and this complex can dominate Al speciation in strongly acidic hydrothermal fluids. The formation of aluminum-silica complexes in alkaline conditions (8 < pH <11) was investigated via potentiometric titrations and ^{27}Al NMR studies of dilute Al- and Si-bearing solutions (10^{-4} <(Al, Si) < 10^{-3} m). The results demonstrate the formation of $[(OH)_3Si-O-Al(H_2O)_5]^{2+}$ in acid solutions, and of the single Al-substituted Q^1_{Al} silica dimer, Q^2_{Al} silica trimer, Q^3_{Al} silica tetramer, and Q^4_{Al} silica pentamer in neutral to alkaline solutions . It is

interesting to note that in the presence of aqueous sodium, the silica tetramer, Q^3_{Al}, can be viewed as the aqueous precursor of albite (Pokrovski et al. 1998, Diakonov et al. 1996). However, the silica dimer Q^2_{Al} dominates aqueous aluminum speciation at silicic acid concentrations typical of most superficial natural fluids.

This study allows for the first time determination of the thermodynamic properties of the aluminum-silica dimer ([$(OH)_3Al$-O-Si $(OH)_3]^-$ and trimer $([(OH)_3Si$-O-$Al(OH)_2$-O-$Si(OH)_3]^-)$ Molecular orbital calculations using the HF/3-21G** and HF/6-31G* basis sets (HF = Hartree-Fock approximation) have been performed on aluminum-silica clusters, including $Al(H_2O)_6^{3+}$, $[Al(OH)_4]^-$ .2(H_2O), $[(OH)_3Al$-O-$Si(OH)_3]^-$, $[(OH)_3Si$-O-$Al(OH)_2$-O-$Si(OH)_3]^-$ (see calculation details in Sykes et al. 1997). The model results yield the structures, energetics, and vibrational spectra of these species. The optimized geometries for the above aluminosilicate complexes are shown in Fig2. Calculated ^{27}Al NMR chemical shifts are very close to the experimental values, confirming experimental shift assignments (Table 1). Moreover, self-consistent Isodensity polarized continuum

Table 1. ^{27}Al NMR chemical shifts (in ppm)

	HF/6-31G*	Experimental value
Al^{3+}	0	0
Al(OH)$_4^-$	78	80.5
Q^{1}Al	74	74
Q^{2}Al	68	69.5

model (SCIPCM) calculations with the HF/6-311+G** basis set predict a ΔE value (-10 kJ/mol) for $[(OH)_3Al-O-Si(OH)_3]^-$ formation reaction which is in close agreement with the experimentally derived enthalpy (-17 kJ/mol).

The results reported in this study provide new insights on metal transport and control in natural systems. For example, it is shown that in most natural waters from basaltic or granitic areas (Iceland, Pyrénées, etc.), aluminum speciation is dominated by aqueous aluminosilicate complexes. It follows that accurate calculation of mineral dissolution/ crystallization rates, solubility, and mass transport during water-aluminosilicate interaction requires explicit provision for Al-Si complexing.

REFERENCES

Bénézeth, P., I.I. Diakonov, G S. Pokrovski, J.-L. Dandurand, J. Schott & I. L. Khodakovsky 1997. Gallium speciation in aqueous solution. Experimental study and modelling: Part 2. Solubility of α-GaOOH in acidic solutions from 150 to 250°C and hydrolysis constants of gallium (III) to 300°C. Geochim. Cosmochim. Acta 61: 1345-1357.

Diakonov, I., G. Pokrovski, J. Schott, S. Castet & R. Gout 1996. An experimental study of sodium-aluminum complexing in crustal fluids. Geochim. Cosmochim. Acta 60: 197-211.

Diakonov, I.I., G. S. Pokrovski, P. Bénézeth, J. Schott, J.-L. Dandurand & J. Escalier 1997. Gallium speciation in aqueous solution. Experimental study and modelling: Part 1. Thermodynamic properties of Ga(OH)$_4^-$ to 300°C. Geochim. Cosmochim. Acta 61: 1333-1343.

Munoz-Paez, A. et al. 1995. EXAFS study of the hydration structure of Ga^{3+} aqueous solutions. Comparison of data from two laboratories. J. Phys. IV France , Colloque C2: 647-648.

Pokrovski, G. S., J. Schott, J.-C. Harrichoury, & A. S. Sergeyev 1996. The stability of aluminum silicate complexes in acidic solutions from 25 to 150°C. Geochim. Cosmochim. Acta 60: 2495-2501.

Pokrovski, G.S., J. Schott, S. Salvi, R. Gout & J. D. Kubicki 1998. Structure and stability of alumino-silicate complexes in neutral to basic solutions. Experimental study and molecular orbital calculations. Min. Mag. 62A: 1194-1196.

Salvi, S., G. S. Pokrovski & J. Schott 1998. Experimental investigation of aluminum-silica aqueous complexing at 300°C. Chem. Geol. 151: 51-67.

Sykes, D., J. D. Kubicki & T. C. Farrar 1997. Molecular orbital calculation of ^{27}Al and ^{29}Si NMR parameters in Q$_3$ and Q$_4$ aluminosilicate molecules and implications for the interpretation of hydrous aluminosilicate glass NMR spectra. J. Phys. Chem. A 101: 2715-2722.

Geochemistry of the Earth's Surface, Ármannsson (ed.)© 1999 Balkema, Rotterdam, ISBN 90 5809 073 6

The relative abundance of Al-species in natural waters in Iceland

S. Arnórsson

Science Institute, University of Iceland, Reykjavík, Iceland

ABSTRACT: The concentrations of total dissolved aluminium in equilibrated geothermal waters tend to increase with rising temperature but to decrease with increasing water salinity. The most abundant aqueous aluminium species is the Al-Si dimer. The aluminate ion generally accounts for the rest of the dissolved Al. Only in moderately saline to saline waters at temperatures in excess of 250°C do the Na- and K-aluminate ion pairs and the neutral Al-hydroxy species constitute a significant fraction of the total dissolved aluminium. The concentrations of less hydrolysed Al-species and of Al-fluoride and Al-sulphate complexes are trivial in ordinary geothermal waters.

1 INTRODUCTION

Much progress has been made in recent years in characterizing the thermodynamic properies of aqueous aluminium species and of Al-silicates, thus improving understanding of mineral saturation in natural waters, both surface- and groundwaters. Aluminium ion hydrolyses strongly in aqueous solution and complexes with various ligands such as sulphate and fluoride. Additionally, the aluminate ion complexes with Na^+ and K^+ (Diakonov et al. 1996, Pokrovskii & Helgeson, 1997) and with aqueous silica (Pokrovski et al. 1998). The purpose of this study is to examine the relative abundance of aqueous Al-species in waters discharged from drillholes in Iceland ranging in temperature from ambient to over 300°C. Earlier studies have shown that these waters, when above about 50°C, have closely approached equilibria with minerals for all major components except chloride (Arnórsson et al. 1983).

2 ALUMINIUM SPECIATION

The WATCH chemical speciation program (Arnórsson et al. 1982, Bjarnason 1994) was used to calculate the speciation distribution in the drillhole waters including those containing Al. Figure 1 shows the relative distribution of hydrolysed Al-species and pairs with SO_4, Na, K and Si. Al-F species are not shown. The SO_4 and F ion pairs occur in trivial concentrations relative to the aluminate ion ($Al(OH)_4^-$), as well as Al^{+3}, $AlOH^+$ and $Al(OH)_2^+$ (Figure 1). The neutral species, $Al(OH)_3^0$, is 10-1000 times less abundant than the aluminate ion. At low temperatures the concentrations of the Na- and K-aluminate ion pairs

are very low but their relative abundance increases with temperature. Yet, they are always considerably less abundant in the dilute Icelandic geothermal waters than the aluminate ion, even in the hottest waters. By contrast, the Al-Si dimer is more abundant than the aluminate ion. According to the experimental data on the thermodynamic stability of this complex, it is the most abundant Al-bearing species in the dilute Icelandic geothermal waters, accounting for 60-80% of the total dissolved aluminium, the $Al(OH)_4^-$ species constituting 97-99% of the rest.

3 CONSTRAINTS IMPOSED BY MINERAL-SOLUTION EQUILIBRIA

It has been demonstrated that Icelandic geothermal waters above about 50°C closely approach equilibrium with low-albite, microcline and silica minerals (chalcedony at <180°C; quartz at >180°C) (Arnórsson 1975, Stefánsson & Arnórsson 1999). This also appears to be so for most geothermal systems in the world, at least, above 150-200°C. From the following reaction:

$$NaAlSi_3O_8 + 2H_2O = 3SiO_2 + NaAl(OH)_4^0 \quad (1)$$

it is seen that simultaneous equilibrium between low-albite and quartz (or chalcedony) fixes the activity (concentration) of the Na-aluminate ion pair. Similarly, equilbrium between microcline and either of the silica minerals controls the activity of the K-aluminate species. The activities of these species take the value of the respective equilibrium constant. Accordingly, these activities vary uniquely with temperature and pressure. Pressure, in the range

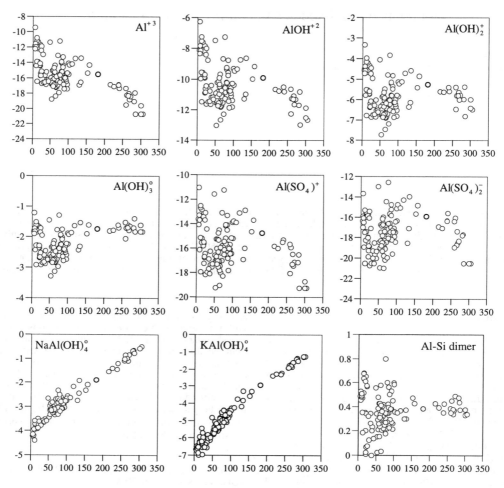

Figure 1. The abundance of aqueous aluminium species in drillhole waters from Iceland relative to the aluminate ion. The vertical axes show relative abundance on logarithmic scale and the horizontal axes the temperature of the water in °C.

occurring in groundwater systems has little effect of the values taken by mineral-solution equilibrium constants compared to temperature. This fact and the prevalence of the alkali-feldspar/silica mineral equilibria explains the fine relationship between temperature and the calculated activities of the Na- and K-aluminate complexes for the Icelandic well waters (Figure 1).

Albite/silica mineral reactions may be written as

$$NaAlSi_3O_8 + 2H_2O \rightleftharpoons Na^+ + Al(OH)_4^- + 3SiO_2 \quad (2)$$

From this reaction it is seen that the activities of Na^+ and $Al(OH)_4^-$ are inversely related at equilibrium.

Increasing water salinity (= increased Na concentrations) involve decreasing activity of the aluminate ion. From the relationship between temperature and the Na-aluminate ion pair activitiy and that involving Na^+ and $Al(OH)_4^-$, it is possible to relate the relative abundance of $Al(OH)_4^-$ and $NaAl(OH)_4^o$ to temperature and salininty (Na concentrations) in equilibrated geothermal waters (Figure 2). The same applies, of course, to $Al(OH)_4^-$ and $KAl(OH)_4^o$. Figure 2 shows that the aluminate ion dominates over the corresponding Na and K ion pairs at temperatures below 300°C for very dilute waters (<10 and <1 mmoles/kg of Na and K, respectively) but in waters of moderate salinity and above 250°C these ion pairs are

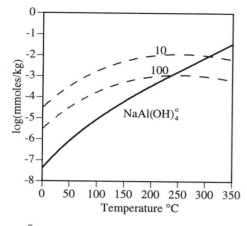

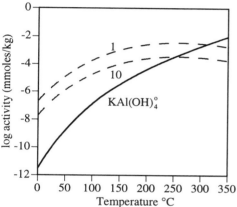

Figure 2. Calculated activities of the Na- and K-aluminate ion pairs in water in equilibrium with low-albite/microcline and quartz (solid curves). The activity of the aluminate ion in equilibrium with the same minerals and at selected aqueous sodium and potassium ion activities are also shown (dashed curves).

more abundant than $Al(OH)_4^-$.

On the basis of cation/proton relationships in geothermal waters at overall equilibrium (Arnórsson et al. 1983) it is possible to estimate for such water how the activity ratio of $Al(OH)_3^0/Al(OH)_4^-$ varies with the water salinity and temperature. Figure 3 shows the relationship between Na^+/H^+ and $Al(OH)_4^-/OH^-$ activity ratios and temperature. It has been chosen here to express the aluminium relationship in this way rather than as a ratio between Al^{+3} and H^+. It can be shown that

$$\frac{K_{H2O}^3}{K_{1-4}} \cdot \frac{[Al^{+3}]}{[H^+]^3} = \frac{[Al(OH)_4^-]}{[OH^-]} \tag{3}$$

where square brackets denote activities, K_{H2O} the dissocia-tion constant of water and K_{1-4} the dissociation constant for the reaction:

$$Al(OH)_4^- = Al^{+3} + 4OH^- \tag{4}$$

From the constants K_{Na-H} and K_{Al-OH} (Table 1) and the following dissociation reaction

$$Al(OH)_4^- = Al(OH)_3^0 + OH^- \tag{5}$$

it can be shown that

$$\frac{[Al(OH)_3^0]}{[Al(OH)_4^-]} = \frac{K_4}{K_{Na-H} \cdot K_{H2O}}[Na^+] \tag{6}$$

or the activity ratio of $Al(OH)_3^0/Al(OH)_4^-$ is equal to temperature dependent constants times the sodium ion activity. K_4 is the dissociation constant for equation (5) above, K_{H2O} the dissociation constant for water and K_{Na-H} the Na^+/H^+ activity ratio in equilibrated geothermal waters (Table 1). Figure 4 shows how the relative abundance of $Al(OH)_3^0$ and $Al(OH)_4^-$ change with temperature at two preselected Na^+ activity values.

Table 1. The temperature dependence of the logarithm of the dissociation constants for some aqueous Al-complexes and of the logarithm of activity ratios in equilibrated geothermal waters. T in Kelvin.

(1) $-60.857 + 2213.16/T - 23.022 \cdot 10^{-6} \cdot T^2 + 22.722 \cdot logT$
(2) $-96.222 + 4814.11/T - 21.812 \cdot 10^{-6} \cdot T^2 + 33.686 \cdot logT$
(3) $901.84/T + 0.756$
(4) $75.79 - 158.90/T + 18.628 \cdot 10^{-6} \cdot T^2 - 28.056 \cdot logT$
(5) $-190.16 + 7016.73/T - 24.812 \cdot 10^{-6} \cdot T^2 + 67.472 \cdot logT$

(1): $NaAl(OH)_4^0$, (2): $KAl(OH)_4^0$, (3): Al-Si dimer, (4): Na^+/H^+, (5): $Al(OH)_4^-/OH^-$

4 CONCLUSIONS

The Al-Si dimer is the dominant Al-bearing aqueous species in natural waters. $Al(OH)_4^-$ is second in abundance and accounts for 97-99% of the remaining dissolved Al. As geothermal waters generally equilibrate with low-albite, microcline and quartz (or chalcedony), it follows that the concentrations of the Na- and K-aluminate ion pairs are a function of the water temperature only. It is only in waters of moderate salininty (ionic strength >0.1) with tempera-

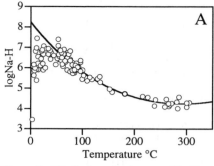
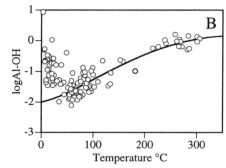

Figure 3. (A) Sodium ion/proton (Na-H) and (B) aluminate ion/hydroxide (Al-OH) ion activity ratios in equilibrated (>50°C) geothermal drillhole waters from Iceland. Also shown are <50°C non-equibrated drillhole waters, recognized by their lower and higher Na-H and Al-OH values, respectively. The curves were obtained by multiple linear regression of the data points for waters above 50°C. Their equation is shown in Table 1.

tures in excess of about 250°C that these ion pairs are about equal to, or more abundant, than $Al(OH)_4^-$. At higher salinities these ion pairs become relatively more abundant at lower temperatures. The $Al(OH)_3^0$ species is less abundant than $Al(OH)_4^-$ in all equilibrated geothermal waters except in those above 250°C and with salinity equal to that of seawater, or higher. Although the relative abundance of the aqueous aluminium species is predictable in equilibrated geothermal waters hosted in rocks from basaltic to silicic in composition, this is, of course not so for non-equilibrated waters. When incorporating the Al-Si dimer into speciation calculations, waters below 200°C are close to equilibrium with low-albite and micro-cline, or slightly supersaturated, whereas waters

above 200°C are undersaturated with these minerals. This is not consistent with field data which show that low-albite and microcline are typical as secondary minerals in active geothermal systems (Stefánsson & Arnórsson 1999). It may be that the Al-Si dimer is less stable at high temperatures than indicated by the experimental results presented by Pokrovski et al. (1998) although it is also considered possible that error in other thermodynamic data may be involved.

REFERENCES

Arnórsson, S. 1975. Application of the silica geothermometer in low-temperature hydrothermal areas in Iceland. *Am. J. Sci.* 275: 763-784.

Arnórsson, S., S. Sigurdsson & H. Svavarsson 1982. The chemistry of geothermal waters in Iceland.I. Calculation of aqueous speciation from 0° to 370 °C. *Geochim. Cosmo-chim. Acta* 46: 1513-1532.

Arnórsson, S., E. Gunnlaugsson & H. Svavarsson 1983. The chemistry of geothermal waters in Iceland. II. Mineral equilibria and independent variables controlling water compositions. *Geochim. Cosmochim. Acta* 47: 547-566.

Bjarnason, J. Ö. 1994. The speciation program WATCH version 2.1. *The National Energy Authority, Reykjavík* 7 p.

Diakonov, I., G. Pokrovski, J. Schott, S. Castet & R. Gout 1996. An experimental and computational study of sodium aluminium complexing in crustal fluids. *Geochim. Cosmo-chim. Acta* 60: 197-211.

Pokrovskii, V. A. & Helgeson, H. C. 1997. Thermodynamic properties of aqueous species and the solubilities of minerals at high pressures and temperatures: the system, Al_2O_3-H_2O-KOH. *Chem. Geol.* 137: 221-242.

Pokrovski, G.S., J. Schott, S. Salvi, R. Gout & J.D. Kubicki 1998. Structure and stability of aluminium-silica complexes in neutral to basic solution, Experimental study and molecular orbital calculations. *Min. Mag.*, 62A: 1194-1195.

Stefánsson, A. & S. Arnórsson 1999. Feldspar saturation state in natural waters. *Geochim. Cosmochim. Acta* submitted.

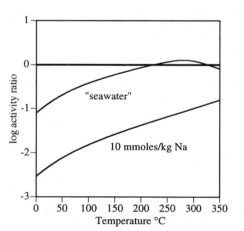

Figure 4. The aqueous $Al(OH)_3^0/Al(OH)_4^-$ activity ratio in equilibraed waters with sodium concentration equal to that of seawater and 10 mmoles/kg, as indicated.

Geochemistry of the Earth's Surface, Ármannsson (ed.)© 1999 Balkema, Rotterdam, ISBN 90 5809 073 6

The dissociation constants of Al-hydroxy complexes at 0-350°C and P_{sat}

S. Arnórsson & A. Andrésdóttir
Science Institute, University of Iceland, Reykjavík, Iceland

ABSTRACT: The temperature dependence of the dissociation of Al-hydroxy complexes in aqueous solution is described by the following equations:

$$logK_1 = -39.183 - 62.84/T - 29.515 \cdot 10^{-6} \cdot T^2 + 13.339 \cdot logT$$

$$logK_2 = 43.638 - 3272.37/T - 9.745 \cdot 10^{-6} \cdot T^2 - 16.354 \cdot logT$$

$$logK_3 = 16.962 - 2096.03/T - 10.262 \cdot 10^{-6} \cdot T^2 - 6.818 \cdot logT$$

$$logK_4 = 97.711 - 5237.63/T + 9.840 \cdot 10^{-6} \cdot T^2 - 35.498 \cdot logT$$

where K_1 to K_4 represent equilibrium constants for stepwise dissociation of $Al(OH)_n^{3-n}$ according to the reactins shown in Table 1. The equations are valid in the range 0-350°C at 1 bar below 100°C and at vapour saturation pressures at higher temperatures. They are based on assessment of published experimental data on Al-hydroxy complex dissociation.

1 INTRODUCTION

Knowledge of the thermodynamic properties of the aqueous aluminium hydroxy complexes and other aluminium species is critical for evaluating the state of saturation of aluminium bearing minerals in natural waters, both under weathering and geothermal conditions. Much experimental work has been carried out in recent years to determine the dissociation constants of the Al-hydroxy complexes over a range of temperatures (see Figure 1). At elevated temperatures solubility experiments involving boehmite have proved to be the most useful.

2 Al-HYDROLYSIS CONSTANTS

The majority of available experimental results on Al-hydrolysis constants are summarized in Figure 1. On the basis of these results temperature equations have been obtained which describe the temperature dependence of Al-hydrolysis constants in the temperature range 0-350°C at 1 bar below 100°C and at vapour saturation pressures (P_{sat}) at higher temperatures (Table 1).

In general results for the first hydrolysis constant compare well (Figure 1). There is, however, considerable discrepancy between results for the second to the fourth hydrolysis constants. Values derived for these constants based on variation in boehmite solubility as a function of pH are very sensitive to minor measurement deviations and the

modelling of pH of the experimental solutions at high temperature. As discussed by Arnórsson (1999) the dominiant Al-hydroxy species in very many natural waters is $Al(OH)_4^-$. However, the $Al(OH)_3^0$ species is most abundant in neutral pH and weakly acidic waters under weathering conditions. It may also be important in high-temperature geothermal waters (>250°C) of relatively high salinity. Thus, reliable data on the dissociation of the aluminate ion is important for those engaged in studies of Al-mineral-solution reactions both in the weathering environment and in geothermal systems.

3 THE THERMODYNAMIC PROPERTIES OF Al-HYDROXY COMPLEXES

The derivation of the equations describing the temperature dependence of the Al-hydrolysis constants was based on selected values from experimental results for logK and the entropy of reaction at 25°C and 1 bar and subsequent multiple linear regression of the experimental data at other temperatures. ΔG_r^o and ΔH_r^o for the hydrolysis reactions at 25°C and 1 bar were derived from the selected logK and ΔS_r^o data (Table 2).

From Wesolowski's (1992) thorough evaluation of gibbsite solubility in alkaline solution and the thermodynamic properties of gibbsite (Table 3) the standard Gibbs energy and enthalpy of formation from the elements of $Al(OH)_4^-$ at 25°C and 1 bar have

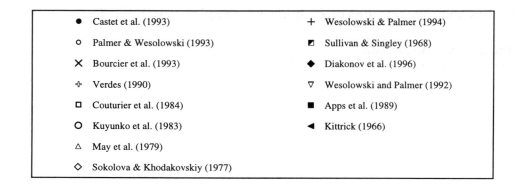

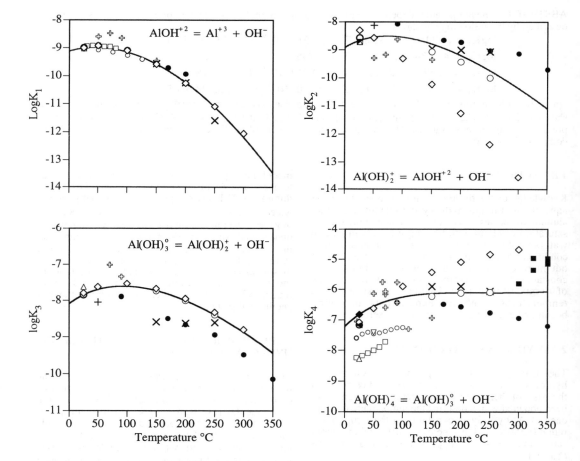

Figure 1. Summary of experimental data on Al-hydrolysis constants. The source of the data is shown at the top of the figure.

Table 1. The temperature dependence of Al-hydrolysis constants and of the boehmite solubilty constant in alkaline solution. The equations are valid in the range 0-350°C at 1 bar below 100°C and at vapour saturation pressures at higher temperatures.

Diss. const.	Reaction	Temperature equation. T in Kelvin
K_1	$AlOH^{+2} \rightleftharpoons Al^{+3} + OH^-$	$-39.183 - 62.84/T - 29.515 \cdot 10^{-6} \cdot T^2 + 13.339 \cdot \log T$
K_2	$Al(OH)_2^+ \rightleftharpoons AlOH^{+2} + OH^-$	$43.638 - 3272.37/T - 9.745 \cdot 10^{-6} \cdot T^2 - 16.354 \cdot \log T$
K_3	$Al(OH)_3^\circ \rightleftharpoons Al(OH)_2^+ + OH^-$	$16.962 - 2096.03/T - 10.262 \cdot 10^{-6} \cdot T^2 - 6.818 \cdot \log T$
K_4	$Al(OH)_4^- \rightleftharpoons Al(OH)_3^\circ + OH^-$	$97.711 - 5237.63/T + 9.840 \cdot 10^{-6} \cdot T^2 - 35.498 \cdot \log T$
	$AlOOH + OH^- + H_2O \rightleftharpoons Al(OH)_4^-$	$-37.335 + 1,069.97/T - 1.092 \cdot 10^{-6} \cdot T^2 + 13.202 \cdot \log T$

Table 2. Standard Gibbs energy, enthalpy and absolute entropy data at 25°C and 1 bar for Al-hydroxy complex dissociation selected in this study. The reported standard enthalpies of reaction were calculated from the Gibbs energy and entropy data.

Reaction	ΔG_r° J/mole	ΔH_r° J/mole	ΔS_r° J/mole/K
$AlOH^{+2} \rightleftharpoons Al^{+3} + OH^-$	51429[a]	4318	-158.01[b]
$Al(OH)_2^+ \rightleftharpoons AlOH^{+2} + OH^-$	49488[c]	12219	-125.0[d]
$Al(OH)_3^\circ \rightleftharpoons Al(OH)_2^+ + OH^-$	44808[b]	12814	-107.31[b]
$Al(OH)_4^- \rightleftharpoons Al(OH)_3^\circ + OH^-$	38929[e]	22259	-55.91[b]

[a]Average of experimental data. [b]Kuyunko et al. (1983). [c]Average of experimental values given by May et al. (1983) and Kuyunko et al. (1983). [d]Estimated in the present study. [e]Diakonov et al. (1996).

Table 3. Standard thermodynamic properties of Al^{+3}, Al-hydroxy complexes, gibbsite and boehmite at 25°C and 1 bar and temperature equation for standard apparent Gibbs energies (ΔG°) valid in the range 0-350°C at 1 bar below 100°C and at vapour saturation pressures at higher temperatures.

Species	ΔG_f° J/mole	ΔH_f° J/mole	S° J/mole/K	$\Delta G_T^\circ = k_1^* \cdot T + k_3^* + k_5^* \cdot T^3 + k_7^* \cdot T \cdot \log T$			
				k_1^*	k_3^*	$k_5^* \cdot 10^6$	k_7^*
Al^{+3}	-492,121	-529,040	-291.5	-1913.94	478,782	50.384	753.595
$AlOH^{+2}$	-700,750	-763,368	-144.4	-2297.70	-646,390	-274.693	864.761
$Al(OH)_2^+$	-907,438	-1,005,597	-30.3	-1095.91	-875,440	-221.285	407.470
$Al(OH)_3^\circ$	-1,109,446	-1,248,421	66.1	-404.80	-1083,973	-177.771	132.742
$Al(OH)_4^-$	-1,305,575	-1,500,690	111.1	1832.22	-1348,651	250.592	-691.073
OH^-	-157,200[a]	-230,010[a]	10.90[a]	366.37	-166.402	239.979	-144.209
H_2O_{liq}	-237,140[a]	-285,830[a]	69.95[a]	436.13	-238,788	0.596	-174.041
$Al(OH)_{3,gib}$	-1,154,900[b]	-1,293,180[c]	68.44[b]				
$AlOOH_{boeh}$	-918,115	-996,390[d]	37.2[d]	314.96	-923,148	-10.889	-120.071

[a]Cox et al. (1989). [b]Hemingway et al. (1977). [c]Hemingway & Robie (1977). [d]Hemingway et al. (1991). The data on ΔG_T° for OH^- and H_2O are consistent with those given by Shock & Helgeson (1988) and Helgeson & Kirkham (1974), respectively.

been estimated as -1,305,575 J mol[-1] and 1,500,690 J mol[-1], respectively. The standard absolute entropy of this species at the same conditions is 111.1 J mol[-1] K[-1]. From data on the thermodynamic properties of boehmite and its solubility in alkaline solution (see Table 3) an equation has been derived which describes the temperature dependence of the apparent standard partial molal Gibbs energy for $Al(OH)_4^-$ (Table 3). Subsequently, temperature equations for the apparent standard partial molal Gibbs energies of the other Al-hydroxy complexes were obtained from those of $Al(OH)_4^-$ and the respective equilibrium constants for the Al-hydrolysis reactions (Table 2).

REFERENCES

Apps, J.A., J.M. Neil & C.H. Jun 1989. Thermo-chemical properties of gibbsite, bayerite, boehmite, diaspore, and the aluminate ion between 0 and 350°C. *Div. Waste Mgt., Office of Nucl. Material Safety and Safeguards, US NRC NUREC/CR-5271-LBL 21482.*

Arnórsson, S. 1999. The relative abundance of Al-species in natural waters in Iceland. *This volume.*

Bourcier, W.L., K.G. Knauss & K.J. Jackson 1993. Aluminium hydrolysis constants to 250°C from boehmite solubility measurements. *Geochim. Cosmochim. Acta* 57: 747-762.

Castet, S., J-L. Dandurand, J. Schott & R. Gout 1993. Boehmite solubility and aqueous aluminium speciation in hydrothermal solutions (90-350°C): Experimental study and modeling. *Geochim. Cosmochim. Acta* 57: 4869-4884.

Couturier, Y., G. Michard & G. Sarazin 1984. Constantes de formation des complexes hydroxydés de l'aluminium en solution aqueuse de 20 a 70°C. *Geochim. Cosmochim. Acta* 48: 649-659.

Cox, J.D., D.D. Wagman & V.A. Medvedev 1989. *CODATA key values for thermodynamics: New Yorkm Hemisphere Publishing Corp.*, 271 pp.

Diakonov, I., G. Pokrovski, J. Schott, S. Castet & R. Gout 1996. An experimental and computational study of sodium aluminium complexing in crustal fluids. *Geochim. Cosmochim. Acta* 60: 197-211.

Helgeson, H.C. & D.H. Kirkham 1974. Theoretical prediction of the thermodynamic behavior of aqueous electrolytes at high pressures and temperatures. I. Summary of the thermodynamic/electrostatic properties of the solvent. *Am. J. Sci.* 274: 1089-1198.

Hemingway, B.S. & R.A. Robie 1977. Enthalpies of formation of low albite ($NaAlSi_3O_8$), gibbsite ($Al(OH)_3$), and $NaAlO_2$, revised values for $\Delta H_{f,298}^o$ and $\Delta G_{f,298}^o$ of aluminium silicates. *U. S. Geol. Surv. Jour. Res.*5: 413-429.

Hemingway, B.S., R.A. Robie, J.R. Fisher & W.H. Wilson 1977. The heat capacities of gibbsite, $Al(OH)_3$, between 13 and 480 K and magnesite, $MgCO_3$, between 13 and 380 K and their standard entropies at 298.15 K, and the heat capacites of Calorimetry Conference benzoic acid between 12 and 316 K. *U. S. Geol. Surv. Jour. Res.* 5: 797-806.

Hemingway, B.S., R.A. Robie & J.A. Apps 1991. Revised values for the thermodynamic properties of boehmite, AlOOH, and related species in the system Al-H-O. *Amer. Mineral.* 76: 445-457.

Kittrick, J.A., 1966. The free energy of formation of gibbsite and $Al(OH)_4^-$ from solubility measure-ments. *Soil Science Soc. Amer. Proc..* 30: 595-598.

Kuyunko, N.S., S.D. Malinin & I.L. Khodakovskiy 1983. An experimental study of aluminium ion hydrolysis at 150, 200, and 250°C: *Geochem. Intl.* 20: 76-86.

May, H.M., P.A. Helmke & M.L. Jackson 1979. Gibbsite solubility and thermodynamic properties of hydroxy-aluminium ions in aqueous solution at 25°C. *Geochim. Cosmochim. Acta* 43: 861-868.

Robie, R.A., B.S. Hemingway & J.R. Fisher 1979. Thermo-dynamic properties of minerals and related substances at 298.15 K and 1 bar (10^5 Pascals) pressure and at higher temperatures. *U. S. Geol. Surv. Bulletin 1452* 436 pp.

Shock, E.L. & H.C. Helgeson 1988. Calculation of the thermodynamic and transport properties of aqueous species at high pressures and temperatures: Correlation algorithms for ionic species and equation of state predictions to 5kb and 1000°C. *Geochim. Cosmochim. Acta* 52: 2009-2036.

Sokolova, N.T. & I.L. Khodakovskiy 1977. The mobility of aluminium in hydrothermal systems: *Geochem. Intl.* 14: 105-112.

Sullivan, J.H. & J.E. Singley 1968. Reactions of metal ions in dilute aqueous solution; hydrolysis of aluminium. *J. Amer. Water Works. Assoc.* 60: 1280-1287.

Verdes, G., 1990. *Solubilité des hydroxides d'aluminium entre 20 et 300°C. Propriétés thermodynamiques des principales espèces naturelles du systèmes Al_2O_3-H_2O.* Toulouse: Université Paul-Sabatier: Unpublished thesis.

Verdes, G., R. Gout & S. Castet 1992. Thermodynamic properties of the aluminate ion and of bayertite, boehmite, diaspore and gibbsite. *Eur. J. Mineral.* 4: 767-792.

Wesolowski, D.J., 1992. Aluminium speciation and equilibria in aqueous solution: I. The solubility of gibbsite in the system Na-K-Cl-OH-$Al(OH)_4$ from 0 to 100°C. *Geochim. Cosmochim. Acta* 56: 1065-1091.

Wesolowski, D.J. & D.A. Palmer 1994. Aluminium speciation and equilibria in aqueous solutions: V. Gibbsite solubility at 50°C and pH 3-9 in 0.1 molal NaCl solutions (a general model for aluminium speciation; analytical methods). *Geochim. Cosmochim. Acta* 58: 2947-2969.

Iron monosulphide stability: Experiments with sulphate reducing bacteria

Liane G. Benning
Department of Geosciences, Pennsylvania State University, University Park, Pa., USA (Presently: School of Earth Sciences, University of Leeds, UK]

Rick T. Wilkin
Astrobiology Research Center, Pennsylvania State University, University Park, Pa., USA

Kurt O. Konhauser
School of Earth Sciences, University of Leeds, UK

ABSTRACT: The formation of Fe-S compounds was investigated in systems containing sulphate-reducing bacteria (*Desulfovibrio desulfuricans*) and inorganic sources of iron using a new geochemical-biological approach. The experiments were conducted under strictly anaerobic conditions in nutrient-depleted synthetic seawater. The iron monosulphide mackinawite was the only stable phase even after more than 2500 hours. Pyrite formation did not occur under the conditions of the experiments. These results reaffirm previous findings that the principal role of sulphate-reducing bacteria in forming iron sulphides is in the production of bisulphide ions.

1 INTRODUCTION

Iron sulphides are important components of the biogeochemistry of diagenetic environments and are widely used as environmental indicators in both modern sediments and ancient sedimentary rocks. It is well documented that the formation of sedimentary iron sulphides is controlled by a complex interplay between incompletely understood biotic and abiotic processes (e.g., Berner 1970; Goldhaber and Kaplan 1974). Issatschenko (1929) observed a conspicuous relationships between bacteria and iron sulphide precipitates, and Thiessen (1920) noted an obvious association between pyrite and organic matter preserved in coals. However, only a few studies have attempted to unravel the importance of inorganic-organic interactions to iron sulphide formation (Rickard 1969). This contrasts the considerable attention given to the inorganic processes and reaction pathways leading to the formation of mackinawite, greigite, and pyrite (e.g., Berner 1970; Schoonen and Barnes 1991; Wilkin and Barnes 1996; Rickard 1997; Benning et al. 1999). It appears that sulphate-reducing, sulphide-oxidising, and iron-reducing bacteria all play crucial roles in the sedimentary geochemical cycles of Fe, S, and C; in particular, producing redox chemistry favourable for pyrite formation. Therefore, in this study, we explore the formation of iron sulphides in experimental systems with thriving populations of the sulphate-reducing bacteria, *Desulfovibrio desulfuricans*.

It is generally believed that the principal role of sulphate-reducing bacteria in iron sulphide formation (neglecting magnetotactic bacteria; see Bazylinski and Moskowitz 1997; Pósfai et al. 1998) is the production of bisulfide ions, i.e., the bacteria are indirectly related to mineral precipitation. This conclusion is based largely upon comparisons of the crystallographic and morphologic properties of mineral products (e.g., sulphides, oxides) formed in biogenic and non-biogenic systems (Rickard 1969; Bazylinski and Moskowitz 1997). However, an indirect control has recently been challenged for iron sulphides, by new evidence suggesting that the activity of sulphate-reducing bacteria can dramatically increase the rates of mineral formation processes in the Fe-S system (e.g., Ravin and Southam 1997).

Consequently, to compare rates of inorganic and biogenic precipitation of iron monosulphides and pyrite, experiments were conducted with various inorganic iron sources and with viable cultures of *D. desulfuricans*, an ubiquitous sulphate reducing bacteria as the only source of reduced sulphur. However, to make a valid geochemical comparison, the composition of the traditional growth media was modified, and excluded were components that potentially complicate the direct comparison of rates. It is expected that these experiments will resolve whether the activity of sulphur metabolising bacteria induces changes in the stabilities of the various iron sulphide phases.

2 METHODS

2.1 *Microorganisms and Anaerobic Medium*

A pure culture of the sulphate-reducing bacteria *Desulfovibrio desulfuricans* (ATCC # 29578) was anaerobically grown in a modified Baars's medium. Oxygen-free conditions were ensured in all experiments by using anaerobic tubes sealed with butyl rubber stoppers and aluminium crimp seals. The sealed tubes were flushed prior to use for 15 minutes with O_2-free nitrogen. A seawater-type medium was supplemented with Na-lactate (80 mM) as the sole carbon source. The anaerobic media was prepared with doubly distilled water that was boiled for 25 minutes and cooled under constant O_2-free nitrogen. The salts per litre of medium were (in grams): NaCl, 27.47; KCl, 0.75; $CaSO_4\cdot2H_2O$, 1.72; $MgSO_4$, 2.41; $MgCl_2\cdot6H_2O$, 6.10; NH_4Cl, 0.32; and, $NaPO_4\cdot12H_2O$, 0.38. The pH of the medium was adjusted with O_2-free NaOH to a value of about 7.5.

In the attempt to keep the media (i.e., the system) as simple as possible, several components of the original growth medium (ATCC medium #1250), which do not serve metabolic purposes and are usually added as indicators, were omitted. In the original medium, with increasing pH, a white, amorphous Ca-Mg-phosphate salt precipitated. Consequently, to prevent this precipitation, the phosphate concentration was decreased to 1/6 of the original value. Omitted were also sodium citrate, which serves to hold traces of iron in solution; resazurin, which is a colour indicator for anaerobic conditions; and, ferrous ammonium sulphate, used as an indicator of sulphate reduction due to the precipitation of iron sulphides. In addition, yeast extract was not added as it clouded the medium. This new, simplified medium was injected under a constant stream of O_2-free nitrogen gas into the sealed anaerobic tubes. In separate anaerobic tubes, ferric and ferrous iron stock solutions (0.05 and 0.005 M) were prepared in a similar O_2-free environment. The media and the iron stock solutions were sterilised at 121 °C for 20 minutes. The tubes with the minimal media were subsequently inoculated with the bacterial culture and equilibrated at 34 °C in an oven.

Due to the nutrient-deficient media, the growth of the *D. desulfuricans* was slowed down, and instead of reaching its maximum growth after 48 h (Postgate 1984), the production of H_2S (i.e., sulphate reduction) was detected only after 3-5 days (see also Figure 1). After 10 days, tubes with new, sterile minimal media were autoclaved and inoculated with an aliquot of the stock culture and with 1 mL of ferric or ferrous iron solution. The addition of the iron stock solution promoted instantaneous formation of a black iron sulphide precipitate. At time intervals of 1, 2, 6 and 15 weeks a tube from each set (ferric and ferrous) was sampled with sterile syringes and the fluid and solid compositions were monitored.

2.2 *Analysis*

The pH of the filtered solutions was measured at 25 °C using a sulphide-tolerant glass electrode. Total dissolved iron ($\bullet$ $Fe_{(aq)}$) was determined using inductively-coupled plasma spectroscopy (ICPS), and total sulphate ($\bullet$ SO_4^{2-}) was determined with an ion chromatograph (conductivity mode, AS4-A column). Total reduced aqueous sulphur ($\bullet$ $S_{red,\ tot}$) was determined with a S-coulometer (UIC Coulometer System 140) by reacting an aliquot of solution filtered through a 0.2 μm polycarbonate filter with 6 N HCl and titrating the produced H_2S gas. The filtered solids were dried under a vacuum and characterised using X-ray diffraction (XRD). A fraction of each sample was prepared for scanning electron microscopy (SEM-EDS) by filtering onto sterile 0.2 μm filters, followed by fixation in 2% glutaraldehyde, repeated washes in 0.1 m phosphate buffer (pH=7.4), and a series of dehydration steps in ethanol. This was followed by critical point drying in bone dry liquid CO_2 at 31.5 °C, 1100 psi, and finally gold coating. This procedure ensured the preservation of the bacterial morphology and avoided lysing. The number of cells per mL of solution and their viability was determined by flow cytometry and fluorescence microscopy after treating with a BacLight DNA dye. A total count of 1-2 x 10^8 cells per mL was determined in all cultures at maximum growth. The ratio between iron monosulphides (Acid-Volatile Sulphides, AVS; mackinawite and greigite) and pyrite (Chromium-Reducible Sulphur), was determined via a sequential extraction method by using a S-coulometer. Freshly filtered samples were reacted first with boiling 6 N HCl producing H_2S gas corresponding to the AVS fraction, while in a second step, $CrCl_2$ was used as the reactant yielding H_2S gas from the dissolution of pyrite and/or elemental sulphur. In addition, the suspended particles in all sample tubes, as well as the dried solids, were monitored for their magnetic character.

3 RESULTS AND DISCUSSION

The XRD analysis revealed the persistence of iron monosulphide phases in all samples. The principal reaction product was mackinawite, displaying different degrees of ordering depending on reaction time. Pyrite and elemental sulphur were not observed even after 105 days (i.e., 2500 h) of ageing and reaction between the initial precipitate (amorphous mackinawite) and the H_2S produced during sulphate reduction. In some sample tubes from the week 15 period greigite was observed and magnetisation could be detected. However, concomitant experiments did not show the formation of greigite and, therefore, its occurrence is believed to be a result of partial oxidation due to failure of the tube seals and not due to bacterial involvement.

Bacterial activity was monitored via the time-dependent decrease in sulphate concentrations and increase in hydrogen sulphide concentrations. For example, in an experiment where Fe^{2+} was added, a plot of time vs. sulphate and reduced sulphur (Figure 1) shows that a maximum H_2S concentration is reached after ~ 2 weeks. The production rate for H_2S generally corresponds to the sulphate consumption rate.

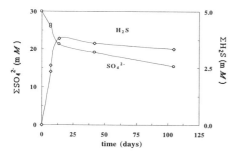

Figure 1. Sulphate reduction vs. H_2S production in an experiment spiked with ferrous iron. Note the different scales for the two y-axes. The decrease in H_2S concentration after about 2 weeks can be interpreted as either partial escape through the rubber stoppers or due to consumption/ageing.

Following this maximum, the concentration of sulphide slightly decreases although sulphate reduction continues, albeit at a slower rate.

The injection of the iron stock solutions induced immediate precipitation of mackinawite and the concentration of iron decreased to a stable value of 0.2 - 0.5 ppm. In experiments with ferrous iron this precipitation occurs following:

$$Fe^{2+}(aq) + H_2S(aq) \Leftrightarrow FeS(s) + 2H^+ \qquad (1)$$

The pH of all solutions initially dropped due to this precipitation reaction. However, subsequently, due to the buffering of the initial solution and the H_2S produced during sulphate reduction, the pH of all solutions stabilised within one log unit of the initial value.

SEM observations show a mature culture of D. desulfuricans together with iron monosulphide particles (Figure 2). Whether the nucleation of the iron monosulphides occurs on bacterial cell walls is unclear. Transmission electron microscopy (TEM) studies are currently underway to address this issue.

The instantaneous precipitation of iron monosulphide demonstrates that their formation follows a purely inorganic pathway. The persistence of mackinawite in anaerobic systems with H_2S as the

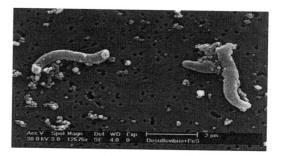

Figure 2. Scanning Electron microphotograph of D. desulfuricans and iron monosulphide precipitates, sampled after 7 days of reaction. SEM-EDX patterns indicate an approximate 1:1 ratio between iron and sulphur intensities.

only reduced sulphur source, is in accordance with the results of most inorganic experimental studies (e.g., Berner 1970; Rickard, 1969; Schoonen and Barnes 1991; Wilkin and Barnes 1996; Benning et al. 1999). Similar to purely inorganic systems, in these experiments, the transformation to greigite and/or to the stable iron sulphide, pyrite, occurs only when an oxidising agent is available.

Previous studies exploring iron sulphide formation processes in biogenic systems are inconclusive. The direct involvement of bacteria in producing sedimentary sulphides was advocated by Issatschenko (1929), ZoBell (1961), Kramarenko (1962), and Ravin and Southam (1997). On the other hand, Freke and Tate (1961), Baas Becking and Moore (1961), Rickard (1969), Hallberg (1972), and Herbert et al. (1998) favour indirect mechanisms whereby sulphate-reducing bacteria only produce hydrogen sulphide, a conclusion in agreement with the results of the present study.

4 SUMMARY AND IMPLICATIONS

Low temperature experiments with the sulphate-reducing bacteria D. desulfuricans show that the formation processes of Fe-S compounds follow inorganic reaction pathways. D. desulfuricans do not catalyse the formation of pyrite. The iron monosulphide mackinawite was the only stable phase even after more than 2500 hours. These results reaffirm previous findings that the principal role of sulphate-reducing bacteria in forming iron sulphides is in the production of bisulphide ions. On-going experiments in similar biogenic systems explore reaction processes with mixed reactive iron sources and intermediate sulphur species.

REFERENCES

Baas Becking, L.G.M. & Moore, D. 1961. Biogenic sulfides. *Econ. Geol.* 56:259-272.

Bazylinski, D.A. & Moskowitz, B.M. 1997. Microbial biomineralization of magnetic iron minerals: microbiology, magnetism, and environmental significance. In J.F. Banfield & K.H. Nealson (eds), *Geomicrobiology: Interactions between microbes and minerals, Rev.Min.* 35:181-223.

Benning, L.G., Wilkin, R.T., & Barnes, H.L. 1999. In situ determination of reaction pathways in the Fe-S system below 100°C. *Chem. Geol.*, in review.

Berner, R.A. 1970. Sedimentary pyrite formation. *Amer. J. Sci.* 268:1-23.

Freke, A.M. & Tate D. 1961. The formation of magnetic iron sulphide by bacterial reduction of iron solutions. *J. Biochem. Microbiol. Tech. Eng.* 3: 29-39.

Goldhaber, M.B. & Kaplan, I.R. 1974. The sulfur cycle. In E.D. Goldberg (ed) *The Sea* 5:569-655, Wiley:New York.

Hallberg, R.O. 1972. Iron and zinc sulfides formed in a continuous culture of sulfate-reducing bacteria. *N. Jb. Miner. Mh.* 11:481-500.

Herbert, R.B., Benner, S.G., Pratt, A.R., & Blowes, D.W. 1998. Surface chemistry and morphology of poorly crystalline iron sulfides precipitated in media containing sulfate-reducing bacteria. *Chem. Geol.* 144: 87-97.

Issatschenko, B.L. 1929. Zur Frage der biogenischen Bildung des Pyrits. *Int. Rev. Gesamten Hydrobio. Hydrogeog.* 22:99-101.

Kramarenko, L.E. 1962. Bacterial biogenesis in underground waters of some mineral fields and their geologic importance. *Mikrobiologiya* 31: 694-701.

Pósfai, M., Buseck, P.R., Bazylinski D.A., & Frankel, R.B. 1998. Reaction sequence of iron sulfide minerals in bacteria and their use as biomarkers. *Science* 280: 880-883.

Postgate, J.R. 1984. *The sulphate-reducing bacteria, 2nd edition.* Cambridge University Press: New York.

Ravin, D. & Southam, G. 1997. Bacterially mediated, rapid, low temperature diagenesis of ferrous sulfide under reducing conditions. *GSA Abs. Prog.* 29: A363.

Rickard D.T. 1969. The microbiological formation of iron sulphides. *Stock. Contrib. Geol.* 20:49-66.

Rickard, D.T. 1997. Kinetics of pyrite formation by the H_2S oxidation of iron (II) monosulfide in aqueous solutions between 25 and 125°C: The rate equation. *Geochim. Cosmochim. Acta* 61:115-134.

Schoonen, M.A.A. & Barnes, H.L. 1991. Reactions forming pyrite and marcasite from solution: I. Nucleation of FeS_2 below 100°C. *Geochim. Cosmochim. Acta* 55:1495-1504.

Thiessen, R. 1920. Occurrence and origin of finely disseminated sulfur compounds in coal: *Trans. Amer. Inst. Min. Metal.* 63:913-931.

Wilkin R.T. & Barnes H.L. 1996. Pyrite formation by reaction of iron monosulfides with dissolved inorganic and organic sulfur species. *Geochim. Cosmochim. Acta* 60:4167-4179.

ZoBell, C.E. 1961. Importance of microorganisms in the sea. *Proc. Low Temp. Microbiol. Symp.*:107-132, Camden: New Jersey.

Uranium speciation in groundwaters from Paraná basin, Brazil

D. M. Bonotto

Instituto de Geociências e Ciências Exatas, UNESP, Rio Claro, São Paulo, Brazil

ABSTRACT: This paper describes the results of a regional study involving the sampling of 60 pumped tubular wells drilled at the Paraná sedimentary basin, Brazil, which was carried out with the purpose of evaluating the U speciation in the Botucatu-Pirambóia aquifer. Uranium proved to be intensively dissolved even under the enhanced reducing conditions occurring at the most confined zones of the aquifer, and Eh-pH diagrams were utilized to evaluate the influence of temperature and pressure on the migration of the U-species within the aquifer.

1 INTRODUCTION

Uranium in the hydrologic environment is of special interest because of its economic importance and its chemical and radiotoxicity and that of some of its daughter nuclides. It is a good natural chemical analogue for some transuranic elements like plutonium or neptunium, which are present in high-level radioactive wastes, and, therefore, the geochemical behaviour of uranium under oxidizing and reducing conditions must be very well known, since the potential reservoirs for some transuranic elements should be located within reduced zones for the safe disposal for a time scale of at least tens of thousands of years (Brookins 1984). In general terms, uranium has been considered very insoluble under reducing conditions, and the most active etch solution of U occurs in the recharge zone of an aquifer, where oxidizing conditions generally prevail (Osmond & Cowart 1976, Langmuir 1978).

Uranium that goes into solution can migrate over long distances, mainly because of its ability to form complexes with bicarbonate and carbonate ions, which are most stable in solutions whose pH is greater than 7.5 (Langmuir 1978). However, in order to evaluate properly the dissolution of U it is important to take into account several parameters, among them, the temperature and pressure, that can affect significantly the kinetics and thermodynamic properties of the mineral phases, with implications for their dissolution or precipitation. For instance, a saturated solution of calcite in equilibrium with CO_2 at 3×10^{-3} atm contains about 75 mg/l of Ca^{2+} at 5°C, but only 40 mg/l at 30°C (Faure 1991), showing, as a result, that dissolution of this mineral

is favored at lower temperatures, whereas precipitation is expected at higher temperatures. Therefore, as a consequence of such behaviour, the migration of uranium may be affected by the well recognized complexation of the uranyl ion (UO_2^{2+}) with bicarbonate/carbonate ions (Langmuir 1978). The purpose of this paper is to characterize the pH and Eh in the waters of the huge aquifer Botucatu-Pirambóia (Tacuarembó, Misiones, Guarany), located in the South American continent, for evaluating the U speciation in the aquifer, as well the influence of temperature and pressure on the transport of uranyl species.

2 SAMPLING AND RESULTS

The Botucatu-Pirambóia aquifer of Triassic-Jurassic age has continental dimensions, an average thickness of 300-400 m, and is composed of silty and shaly sandstones of fluvial-lacustrine origin (the Pirambóia formation), and of variegated quartzitic sandstones accumulated by eolian processes under desertic conditions (the Botucatu formation) (Gilboa et al. 1976). Situated within an intercratonic basin, the potentiometric surface of the water shows that about 70% of its total area can be ascribed to artesian conditions, and recharge occurs by direct infiltration of rainwater in the outcrop area, which is about 98,000 km^2 (Rebouças 1988). The percolating water moves from the phreatic exposed areas that surround the entire basin towards its central part (Silva 1983). The sampling of the Botucatu-Pirambóia aquifer was performed at 51 localities in São Paulo, Mato Grosso do Sul and Paraná States, where the 60 groundwater samples reported by Bonotto (in press) were also collected for pH and Eh determinations. The available data describing the wells allowed the geostatic pressure, P, to be estimated from the equation (Castany 1982):

$$P = Pa + \rho gh \qquad (1)$$

where Pa is the atmospheric pressure, ρ is the average density of the terrain, g is the gravity acceleration, and h is the depth of the top of the aquifer. Table 1 summarizes the range of the obtained values for the parameters considered in this paper.

Table 1. Results of measurements for 60 groundwater samples from Botucatu-Pirambóia aquifer.

| PARAMETER | UNIT | VALUE | |
		MINIMUM	MAXIMUM
Depth of the well	m	86	4582
Geostatic pressure	bar	0.9	430.4
Temperature	°C	20	70
pH	-	4.0	9.9
Eh	mV	-88	+632
Uranium	μgL^{-1}	0.01	4.82

3 pH AND Eh RELATIONSHIPS

The chemical data obtained for the studied groundwaters were plotted in an Eh-pH diagram (Fig. 1), which shows that most of the data fall into

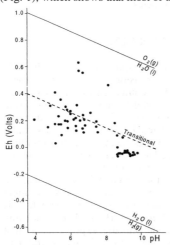

Figure 1. Data for groundwaters from the studied area plotted on an Eh-pH diagram.

the transitional to reducing field, even under the more acidic and basic conditions. A linear relationship ($r = 0.77$) between the pH and the geostatic pressure was observed (Fig. 2b), showing that the ground water becomes more basic as it flows from the border of the basin towards its central part, in the direction of the dip of the geological units. Such an increase of basicity is attributed to the increased dissolution of sodium, in agreement with the same trend (Bonotto, in press, Silva 1983). An inverse significant correlation ($r = -0.50$) was also observed between the redox potential and the geostatic pressure (Fig. 2a), suggesting that the reducing conditions in the aquifer are favored when the more confined zones are reached, a situation very commonly referred to in the literature (Brookins 1988).

4 URANIUM SPECIATION

Uranium is actively dissolving in groundwaters from the studied part of the aquifer, as demonstrated by measured values of up to about 5 μgl^{-1} for dissolved U. The complexation of the uranyl ion (UO_2^{2+}) with bicarbonate/carbonate ions in the studied aquifer was pointed out by Bonotto (in press) on the basis of a significant linear relationship that was found between these parameters. As expected on the basis of the great depths the aquifer reaches (almost 2 km) and the thick confining basaltic cover, the occurrence of water at high temperatures (above 50°C) was demonstrated. It is important to evaluate how this aspect can affect the transport of dissolved uranium.

Thus, the obtained data for the groundwaters were initially plotted in the Eh-pH diagram for uranium species at 25°C and 1 bar pressure, i.e. the system U-C-O-H often utilized to investigate the low temperature solution-mineral equilibria of U in sedimentary basins (Langmuir 1978). Figure 3a shows that most of the data fall into the fields of carbonate, dicarbonate and tricarbonate uranyl species, with some of them falling into the field of well-crystallized uraninite, $UO_{2(c)}$. At low temperatures, the U(IV) amorphous oxyhydroxides generally precipitate instead of the crystalline UO_2. Since no occurrence of crystalline UO_2 has been reported in the studied aquifer, then, it is likely that any deposition of U nuclei would take place initially in a highly disordered state, approximating "amorphous" UO_2, as observed by Andrews & Kay (1983) for the East Midlands groundwaters in England. Therefore, the Eh-pH field in Figure 3a shows that carbonate complexes are stable in the aqueous phase for practically all the groundwaters and that no UO_2 deposition should occur anywhere in the studied part of the aquifer.

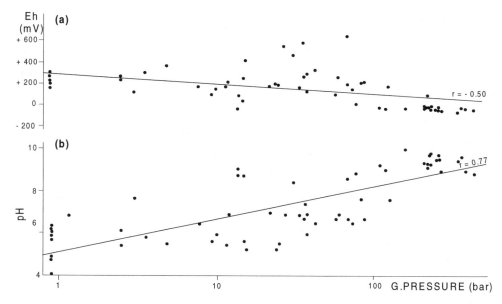

Figure 2. The (a) redox potential Eh and (b) pH of groundwaters from the studied area plotted against the geostatic pressure in the aquifer.

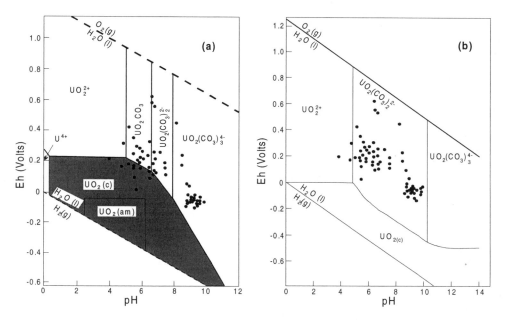

Figure 3. Data for groundwaters from the studied area plotted on Eh-pH diagrams of uranyl carbonate complexes at (a) 25°C, 1 bar pressure (Langmuir 1978) and (b) 100°C, 1 bar pressure (Brookins 1984).

The data obtained for the groundwaters are plotted in a 100°C, 1 bar Eh-pH diagram diagram in order to evaluate the effect of temperature (Fig. 3b). Stability-field boundaries for "amorphous" UO_2 are not represented in Figure 3b because this phase is only formed at low temperatures. The range of Eh and pH in which primary U mineralization occurs is much less well known at 100°C than at 25°C. At 100°C the stability field of uranyl dicarbonate, into which probably all the groundwaters fall, expands at the expense of the uranyl carbonate field, which disappears. Thus, the relative proportions of uranyl species transported as carbonate, dicarbonate or tricarbonate ions may be affected by temperature. Because increasing pressure has only a slight effect on equilibrium constants at temperatures below 300°C, the Eh-pH boundaries calculated for 1 bar conditions can be used for pressures up to ~500 bars (Helgeson et al. 1969), which are higher than the highest estimated value of 430 bar for the studied aquifer.

5 CONCLUSIONS

Significant relationships were observed between the geostatic pressure, pH and Eh in the studied part of the aquifer, indicating the occurrence of basic and reducing conditions at the most confined zones of the aquifer. Uranyl carbonate complexes are stable in the aqueous phase for practically all the analysed groundwaters, and the relative proportions of uranyl carbonate, dicarbonate or tricarbonate ions are affected by temperature.

ACKNOWLEDGEMENTS

The author thanks the International Atomic Energy Agency (IAEA Research Contract No. 9723/Regular Budget Fund) and CNPq (Conselho Nacional de Desenvolvimento Científico e Tecnológico)-Brasil for financial support of this investigation.

REFERENCES

Andrews, J.N. & R.L.F. Kay 1983. The U contents and $^{234}U/^{238}U$ activity ratios of dissolved uranium in groundwater from some Triassic sandstones in England. *Isotope Geoscience* 1: 101-117.

Brookins, D.G. 1984. *Geochemical aspects of radioactive waste disposal*. New York: Springer-Verlag.

Brookins, D.G. 1988. *Eh-pH diagrams for geochemistry*. Berlin: Springer-Verlag.

Castany, G. 1982. *Principes et méthodes de l'hydrogéologie*. Paris: Dunod.

Faure, G. 1991. *Principles and applications of inorganic geochemistry*. New York: MacMillan Publishing Co.

Gilboa, Y., F. Mero & I.B. Mariano 1976. The Botucatu aquifer of South America, model of an untapped continental aquifer. *J.Hydrol.* 29: 165-179.

Helgeson, H.C., T.H. Brown & R.H. Leeper 1969. *Handbook of theoretical activity diagrams depicting chemical equilibria in geologic systems involving an aqueous phase at one atm and 0° to 300°C*. San Francisco: Cooper & Company.

Langmuir, D. 1978. Uranium solution-mineral equilibria at low temperatures with applications to sedimentary ore deposits. *Geochim. Cosmochim. Acta* 42: 547-569.

Osmond, J.K. & J.B. Cowart 1976. The theory and uses of natural uranium isotopic variations in hydrology. *At. Energy Rev.* 14: 621-679.

Rebouças, A.C. 1988. Groundwater in Brazil. *Episodes* 11: 209-214.

Silva, R.B.G. 1983. *Estudo hidroquímico e isotópico das águas subterrâneas do aqüífero Botucatu no Estado de São Paulo*. São Paulo: USP-São Paulo University (PhD Thesis).

Geochemistry of the Earth's Surface, Ármannsson (ed.)© 1999 Balkema, Rotterdam, ISBN 90 5809 073 6

Recent advances in trace element geochemical modelling

J. Bruno, D. Arcos, E. Cera, L. Duro, S. Jordana & C. Rollin
QuantiSci SL, Avgda. Universidad Autònoma, Cerdanyola del Vallès, Spain

M. J. Gimeno, L. Pérez del Villar & J. Peña
CIEMAT, Spain

ABSTRACT: The general processes governing trace element behaviour in 2 different sites are presented and discussed. The measured trace element groundwater concentrations in selected samples from these sites have been compared with the results from kinetic and thermodynamic modelling. The results are discussed in the light of the mineralogical data on the occurrence of trace elements in the sites. In general, trace element mobility and retardation are controlled by their co-precipitation and co-dissolution with major mineral phases, particularly Fe(III) oxy-hydroxides and calcite.

1 INTRODUCTION

Trace element mobility in the geosphere is controlled by their aqueous, surface and solid state interactions with the main geochemical components. In previous publications we have introduced the co-precipitation and co-dissolution concepts which quantitatively link trace metal behaviour to the cycling of major components, particularly Fe and C and their application to trace element behaviour in the Poços de Caldas (Brazil) and El Berrocal (Spain) systems (Bruno et al. 1992 and Bruno et al. 1998).

In this work we present a further extension of this approach by reporting on the main findings from the geochemical model testing undertaken at two Natural Analogue sites: the Oklo fossil reactor system in Gabon and the Palmottu uranium deposit in Finland.

2 THE OKLO AND PALMOTTU GEOCHEMICAL SETTINGS

2.1 *The Oklo natural analogue system*

The fossil natural nuclear reactor systems of Oklo are located in two different uranium deposits. The most important is the Okélobondo deposit, which contains 16 nuclear reactors. This deposit is located at the Northwestern edge of the Franceville basin, along the basement uplift of Mounana. An additional reactor has been located in the Bagombé uranium deposit in the southern part of the Franceville basin (Naudet 1991). The location of these unique fossil reactors prompted the study of these systems as analogues to the disposal of spent nuclear fuel.

In Figure 1 we include an schematic cross section with the main geological units of the Okélobondo system.

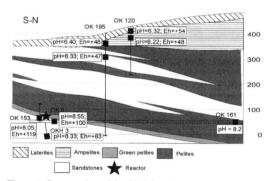

Figure 1. Cross section of the Okélobondo system.

2.2 *The Palmottu uranium deposit*

The Palmottu uranium deposit occurs within a zone of metamorphosed supracrustal and sedimentary rocks that extends from SW Finland into central Sweden. The Palmottu U-Th mineralisation is related to the latest stages of the orogenic events in Finland about 1.8-1.7 Ga. ago. The Palmottu region is characterised by granites and highly metamorphosed migmatitic rocks (Blomqvist et al. 1995).

A schematic view of the system studied is presented in Figure 2.

Figure 3 summarises in a Pipper diagram the main geochemical data of the Oklo and Palmottu samples used in the trace element geochemical modelling of these systems.

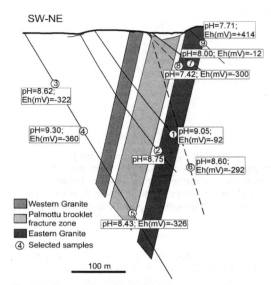

Figure 2. Cross section of the Palmottu system.

2.3 Groundwater chemistry

The groundwater samples selected from Oklo are all of the Na-Ca-HCO$_3$-type, although they can be classified in two groups; one more acidic corresponding to surficial waters, and a more alkaline and more oxidised corresponding to the waters found close to the reactor zone.

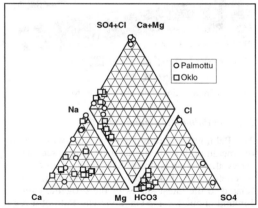

Figure 3. Piper diagram integrating the chemical water data of the Oklo and Palmottu samples used in the modelling exercise.

The Palmottu groundwater samples are classified in three different groups. The Ca-Na-HCO$_3$-type which

are diluted, slightly acidic and oxidant surface waters. The Na-Cl-SO$_4$ type is characteristic of deeper waters which are less diluted and less oxidant than the first type. And the Na-Cl type which is very reducing and located at great depth.

The selected waters in Palmottu are in general very close to equilibrium with calcite and slightly oversaturated with dolomite. Concerning the hydroxides, gibbsite is over-saturated in all the samples while the iron hydroxides are close to equilibrium for the most oxidant waters. These observations seem to indicate that carbonates and iron hydroxides could control the chemistry of some trace elements.

3 MODELLING RESULTS AND DISCUSSION

3.1 Trace element modelling in Okélobondo

In a first stage, a geochemical modelling exercise was performed to test the ability of the modellers to predict the concentrations of trace metals in the groundwaters sampled in Bagombé (Duro & Bruno 1998). In a second stage the work was focused on the prediction of the aqueous concentration of uranium in groundwater samples from Okélobondo focusing the exercise on the surroundings of the reactor zone OK84. Based on the hydrochemistry and the mineralogy of the site the first step in the modelling work was the selection of the pure uranium solid phases more likely to exert a solubility control. The calculated uranium concentrations based on this selection were compared to the actual measurements at the different boreholes. In addition and based on the fact that the analyses of the redox state of Okélobondo groundwaters indicates that most of the samples are controlled by the $Fe^{2+}/Fe(OH)_3$(am) redox couple, it was worthy to apply a co-precipitation approach (Bruno et al. 1995) to the association between iron and uranium. The results obtained are presented in Figure 4.

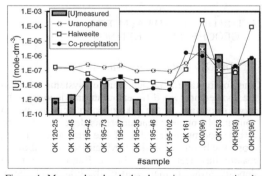

Figure 4. Measured and calculated uranium concentration by assuming equilibrium with pure uranium-minerals and a co-precipitation approach with iron(III) hydroxide.

3.2 *Trace element modelling in Palmottu*

The behaviour of 13 trace metals has been studied by using nine selected groundwater samples as well as the mineralogical information.

The pure solid phases more likely to occur in these waters were firstly identified and the corresponding solubilities calculated for the different trace elements studied. However in most of the cases the calculated solubilities overestimated the measured concentrations of the trace metals in groundwaters. In such cases co-dissolution and co-precipitation processes of mixed solid phases were checked to explain the behaviour of these trace elements in this natural system. As an example of the work performed the modelling results obtained for strontium are given in Figure 5.

As can be observed in Figure 4, strontianite solubi-lity overestimates strontium concentrations in all water samples while considering a dissolution pro-cess of a mixed carbonate solid phase with calcium the calculated strontium concentrations agree satisfactorily with the measured ones.

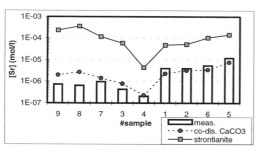

Figure 5. Measured and calculated strontium concentrations by assuming equilibrium with a pure solid phase and a co-dissolution process with calcite.

4 CONCLUSIONS

The experimental observations on trace element associations together with the geochemical modelling work performed in Oklo and Palmottu give additional evidence of the linkage between trace element and major component cycling, particularly C, Fe and S.

Fe(III) oxy-hydroxide and calcite precipitation and dissolution equilibria appear to be critical processes to control trace element mobility in groundwater environments.

The co-precipitation/co-dissolution modelling approaches previously proposed appear to give a simple but consistent representation of the fate and mobility of trace metals in natural water systems.

5 ACKNOWLEDGEMENTS

The financial support of Enresa and SKB to this work is gratefully acknowledged.

6 REFERENCES

Blomqvist R., J. Suksi, T. Ruskeeniemi, L. Ahonen, H. Niini, U. Vuorinen & K. Jakobsson, 1995. The Palmottu natural analogue project. Summary Report 1992-1994. *Report YST-88.*

Bruno, J., J. de Pablo, L. Duro & E. Figuerola, 1995 Experimental and modelling study of U(VI)-Fe(OH)$_3$ coprecipitation equilibria: The application of the solubility conditional constant. *Geochim. Cosmochim Acta* 59, 20: 4113-4123

Bruno, J., J.E. Cross, J. Eikenberg, I.G. McKinley, D. Read, A. Sandino & P. Sellin, 1992. Testing models of trace elements geochemistry at Poços de Caldas in The Poços de Caldas Project: Natural Analogues of Processes in Radioactive Waste Repository. *Journal of Geochemical Exploration* 45: 451-470.

Bruno, J., L. Duro, J. de Pablo, I. Casas, C. Ayora, J. Delgado, M.J. Gimeno, J. Peña, C. Linklater, L. Pérez del Villar & P. Gómez, 1998. Estimation of the concentrations of trace metals in Natural Systems. The application of co-precipitation and co-dissolution approaches to trace element mobility in El Berrocal (Spain) and Poços de Caldas (Brazil). *Chem. Geol.* 151, 1/4: 277-291.

Duro, L. & J. Bruno, 1998. Blind prediction modelling (BPM) exercise in Oklo. 1st. stage: Bagombé. in: D. Louvat & C. Davis (eds.) Proceedings of the first joint EC-CEA workshop on the Oklo-natural analogue. Phase II project held in Sitges, Spain. *EUR 18314 EN.*: 261-275.

Naudet R., 1991. Oklo: *Des reacteurs nucléaires fossiles. Étude physique.* Collection CEA. Eyrolles.

Geochemistry of the Earth's Surface, Ármannsson (ed.) © 1999 Balkema, Rotterdam, ISBN 90 5809 073 6

Redox buffering effect on iron and uranium minerals in Palmottu, Finland

E. Cera, J. Bruno & L. Duro
QuantiSci SL., Avgda. Universidad Autònoma, Cerdanyola del Vallès, Spain

ABSTRACT: The capacity of a deep geological system to buffer the effect of an oxic intrusion is investigated in Palmottu (Finland). Oxidation of iron(II) and uranium(IV) minerals are the main processes contributing to the restablishment of anoxic conditions after an oxidant intrusion in the system. The main parameters identified as affecting the ReDucing Capacity (RDC) of the system are, among others: Wetted mineral surface area, porosity of the host rock, mineral abundance and magnitude of the oxic disturbance.

1 INTRODUCTION

Due to the difficulty associated with the measurement of stable in situ redox potentials, Scott & Morgan (1990) proposed the use of extensive magnitudes to define the redox state of geological systems. They used the concepts of OXidising and ReDucing Capacities (OXC and RDC) in analogy to magnitudes such as acidity or alkalinity in acid/base systems.

The RDC of a system gives an estimation of its capacity to accept oxidants, i.e., it is a chemical sum of the maximum amount of oxidants that the system is able to buffer. The generic definition of RDC is as follows:

$$RDC = \sum_i n_i \times [\mathrm{Re}\,d] - \sum_j n_j \times [Ox] \qquad (1)$$

where $[\mathrm{Re}d]$ and $[Ox]$ are, respectively, the concentration of reducing and oxidising species present in the system, and n_i and n_j are the number of electrons involved in the redox reactions.

The main contributors to OXC and RDC in crystalline bedrock are iron minerals (oxides, sulphides and iron bearing silicates). In most deep ground waters the redox state is governed by electron transfer between Fe(II) and Fe(III) species. Therefore, the largest reductive capacity of such systems will be given by their content of Fe(II).

At pH > 4, the oxidation of Fe(II) is fast and leads to the precipitation of amorphous Fe(III) solid phases, $Fe(OH)_3 \cdot nH_2O$. In fact, the redox state of most natural waters can be explained in terms of the equilibrium $Fe^{2+} \Leftrightarrow Fe(OH)_3(s)$.

In a reducing environment, the sulphur/sulphate system might be considered as an important redox couple. However, the chemical inactivity of sulphate in the absence of bacterial catalysis probably restricts its role as active redox species.

The presence of other redox sensitive components such as uranium, may act as oxygen scavenger and can contribute to the redox buffering capacity of the system.

The RDC concept has been previously developed for a hypothetical high level nuclear waste (HLNW) repository system (Bruno et al. 1996) where UO_2 was the main contributor to the RDC of the system. Bruno & Duro (1996) have recently applied this concept to a hypothetical natural system, with Fe(II) as the main contributor to the RDC of the system.

In this work we will focus on the contribution of both uranium dioxide and Fe(II) to the RDC of the natural system of Palmottu. The role of both redox systems on buffering the redox state of deep ground waters after the intrusion of an oxic disturbance in the system will be assessed.

2 OBJECTIVES

The objectives of this work are:
1. To study the redox buffering capacity of the $UO_2/UO_{2.33}$ and $Fe(II)/Fe(OH)_3$ couples.
2. To quantify the relative contributions of the oxidation processes involved to the oxidant uptake.
3. To calculate the evolution of the ReDuctive Capacity (RDC) of the Palmottu system with time

after the occurrence of an oxidant intrusion and to identify the critical parameters affecting it.

3 DEFINITION OF THE SYSTEM

We will simulate the evolution of the system after an initial pulse of oxygen.

According to the mineralogy and the hydrogeochemistry of the area under study the following processes have been included in the model as possible oxidant sinks:

a) Uraninite oxidation and dissolution

$$UO_2(s) + 0.166\ O_2 + 0.667\ HCO_3^- \rightarrow$$
$$0.33\ UO_2(CO_3)_2^{2-} + 0.33\ H_2O \quad (2)$$

with a rate in mole·m^{-2}s^{-1}:

$$R_{ox} = 2.28 \cdot 10^{-7} \times \left[O_2(aq)\right]^{0.5} \quad \text{(Torrero 1995)} \quad (3)$$

$$R_{diss} = 1.1 \cdot 10^{-12} \times \left[H^+\right]^{-0.3} \quad \text{(Bruno et al. 1991)} \quad (4)$$

b) Oxidation of aqueous Fe(II)

$$Fe^{2+} + 2.5\ H_2O + 0.25\ O_2(aq) \rightarrow Fe(OH)_3(s) + 2\ H^+ \quad (5)$$

with a rate in mole·dm^{-3}·s^{-1} (Stumm & Morgan 1981)

$$R_{Fe} = 5 \cdot 10^{-14} \times \left[O_2(aq)\right] \times \left[H^+\right]^{-2} \times \left[Fe(II)\right] \quad (6)$$

c) Pyrite oxidation by oxygen

$$FeS_2(s) + 15/4\ O_2 + 7/2\ H_2O \rightarrow Fe(OH)_3(s) + 2\ SO_4^{2-} + 4\ H^+ \quad (7)$$

with a rate in mole·m^{-2}·s^{-1} (Nicholson et al. 1990):

$$R_{py} = -\frac{d\left[FeS_2\right]}{dt} = 1.4 \cdot 10^{-6} \times \left[O_2(aq)\right] \quad (8)$$

We have also included dissolution of biotite as an additional source of Fe(II), Biotite → Fe(II), occurring at a rate (Malmström et al. 1995):

$$R_{biot} = 2.68 \cdot 10^{-9} \times \left[H^+\right]^{0.51} + 2.04 \cdot 10^{-18} \times \left[H^+\right]^{-0.65}$$
$$\text{mole Fe·m}^{-2}\text{·s}^{-1} \quad (9)$$

3.1 Definition of the Electron Reference Level (ERL)

A redox ladder (Scott & Morgan 1990) has been built in order to select correctly the ERL and identify the oxidant and the reductant species of our system (Figure 1). By selecting the $UO_{2.33}$ (U_3O_7) as the ERL, the species above this compound on the left side of the redox ladder will act as oxidants and the species below the $UO_{2.33}$ on the right side of the redox ladder will act as reductants. Therefore, in the system under study, the oxidant will be oxygen, and the reductants will be uraninite, pyrite and Fe^{2+}.

In summary, the reductive capacity of our geochemical system will be given by:

$$RDC_t = 0.667 \times \left[UO_2\right]_t + \left[Fe(II)\right]_t +$$
$$+15 \times \left[FeS_2\right]_t - 4 \times \left[O_2(aq)\right]_t \quad (10)$$

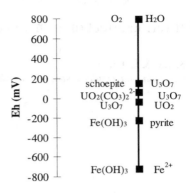

Figure 1. Redox ladder as defined by Scott & Morgan (1990).

3.2 Parameters of the model

The main parameters that we need to perform the calculations are:

a) Specific surface area of pyrite, biotite and uraninite; SA_{pyrite}=13.6 m^2·kg^{-1}; $SA_{biotite}$=1810 m^2·kg^{-1} $SA_{uraninite}$=3.6 m^2·kg^{-1}.

b) Weight percentage of pyrite, biotite and uraninite in the rock; W_{pyrite}=$W_{biotite}$=1%; $W_{uraninite}$=0.1% (Blomqvist et al. 1995).

c) Density of the rock; ρ = 2700 kg·m^{-3}.

d) Porosity of the media; ϕ = 1%.

e) Wetted Surface of the mineral; WS_{pyrite}=$WS_{biotite}$=$WS_{uraninite}$=1%. This parameter gives the surface of the mineral that is able to enter into contact with ground water.

f) Number of surface sites available for oxidation on the surface of pyrite and uraninite; n_{pyrite}=1.575·10^{-2} mole·m^{-2}, (Kornicker & Morse 1991); $n_{uraninite}$=2.86·10^{-1} mole·m^{-2} (Casas 1989).

4 CALCULATION OF THE RDC IN THE PALMOTTU SYSTEM

We have considered the intrusion of oxygenated water equilibrated with a content of oxygen equal to 0.1 times the one given by equilibration with the atmospheric $O_2(g)$ partial pressure.

In order to calculate the RDC of this geochemical system according to the previous equation, the evolution of uraninite and pyrite co-ordination sites with time, and the variation in the Fe(II) and oxygen concentrations with time were studied.

Evolution of the uraninite surface sites with time:

$$\frac{d\left[> UO_2\right]}{dt} = -R_{ox} \times \frac{SA \cdot W \cdot \rho \cdot WS}{\phi} +$$
$$+ R_{diss} \times \frac{SA \cdot W \cdot \rho \cdot WS}{\phi} \quad (11)$$

Evolution of the aqueous ferrous iron with time:

$$\frac{d[Fe(II)]}{dt} = R_{biot} \times \frac{SA \cdot W \cdot \rho \cdot WS}{\phi} - R_{Fe} \quad (12)$$

Evolution of the pyrite co-ordination sites with time:

$$\frac{d[>FeS_2]}{dt} = -R_{py} \times \frac{SA \cdot W \cdot \rho \cdot WS}{\phi} \quad (13)$$

Evolution of the oxygen concentration with time:

$$\frac{d[O_2]}{dt} = -0.166 \times R_{Ox} \times \frac{SA \cdot W \cdot \rho \cdot WS}{\phi} -$$
$$-0.25 \times R_{Fe} - \tfrac{15}{4} \times R_{py} \times \frac{SA \cdot W \cdot \rho \cdot WS}{\phi} \quad (14)$$

the following boundary conditions apply at t=0:
$[>UO_2]_0 = n_{uraninite} = 2.86 \cdot 10^{-3}$ mole·dm^{-3}; $[Fe(II)]_0 = 0$;
$[>FeS_2]_0 = n_{pyrite} = 1.575 \cdot 10^{-4}$ mole·dm^{-3}.
$[O_2]_0 = 2.5 \cdot 10^{-5}$ mole·dm^{-3}.

The uptake of oxygen due to each one of the processes is shown in Figure 2.a. By assuming a lower initial oxygen content in the intruding groundwater ($3.12 \cdot 10^{-6}$ mole·dm^{-3}), we obtain the results shown in Figure 2.b.

The evolution of the reductive capacity of the system in both cases studied is shown in Figure 3.

As can be inferred from figure 2.a and 2.b, the process of oxidation of uraninite is the main contributor to the uptake of oxygen. The relative importance of the other two processes of oxygen uptake (FeS$_2$ or aqueous Fe(II) oxidation) depends on the initial pulse of oxygen. When the initial oxygen content increases, the uptake of oxidant due to pyrite oxidation is more important than the one caused by the oxidation of aqueous iron. This is the direct consequence of two factors: a) The rate of pyrite oxidation is a function of the oxygen concentration. Therefore, the lower the oxygen, the slower the oxidation rates. b) Fe(II) is released due to biotite dissolution, which is, in principle, independent of the oxygen content. This causes a continuous supply of iron able to scavenge oxygen from the system.

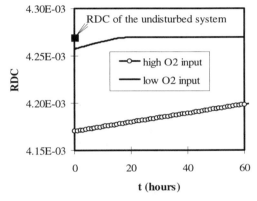

Figure 3. Evolution of RDC with time.

We can observe (Figure 3) an initial sharp decrease in the RDC of the system due to the intrusion of oxidants and a further smooth reestablishment of the RDC, i.e., of the anoxic conditions due to the uptake of oxidants by the reducing species present in the system.

4.1 Influence of the wetted mineral surface area

One of the geochemical parameters that directly affects the influence of each process involved in the oxygen uptake in a system is the surface of the mineral exposed that can enter into contact with the groundwater (Wetted Surface).

The time required consuming 50% of the oxygen (t_{50}) of the system as a function of the WS considered is shown in Figure 4.

When the WS increases, the oxidation processes speed up and, therefore, the time needed to re-establish reducing conditions is reduced.

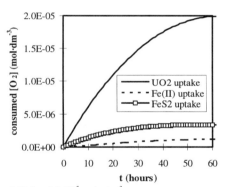

a) $[O_2]_0 = 2.5 \cdot 10^{-5}$ mole·dm^{-3}.

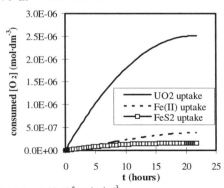

b) $[O_2]_0 = 3.12 \cdot 10^{-6}$ mole·dm^{-3}.

Figure 2. Uptake of oxygen due to each oxidation process.

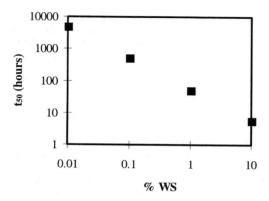

Figure 4. Time needed for the consumption of 50% of the oxygen intrusion as a function of %WS.

5 CONCLUSIONS

The main process affecting the oxygen uptake in the system studied is the oxidation of uraninite.

The most important parameters affecting the RDC of the system are:

- Porosity and density. If ϕ increases $\Rightarrow$ R_{py} and R_{ox} decrease.
- Percentage of minerals in the bedrock. If weight percentage increases $\Rightarrow$ major oxygen uptake due to the oxidation of the minerals involved.
- Percentage of wetted surface area (WS). If WS increases $\Rightarrow$ major oxygen uptake due to the oxidation of the minerals involved.
- Number of sites available for oxidation on the surface of the mineral.

REFERENCES

Blomqvist R., J. Suksi, T. Ruskeeniemi, L. Ahonen, H. Niini, U. Vuorinen & K. Jakobsson 1995. *The Palmottu Natural Analogue. Summary Report 1992-1994.* Geological Survey of Finland, Nuclear Waste Disposal Research, *Report YST-88.*

Bruno J., I. Casas & I. Puigdomènech 1991. The kinetics of dissolution of UO_2 under reducing conditions and the influence of an oxidized surface layer (UO_{2+x}): Application of a continuous flow through reactor. *Geochim. Cosmochim. Acta* 55: 647-658.

Bruno J. & L. Duro 1996. Estimation of the influence of the Fe(II) bearing clays on the RDC of deep groundwater systems. Application of the results on biotite weathering obtained at KTH. *QuantiSci Internal Report.*

Bruno J., E. Cera, L. Duro, T.E. Eriksen & L.O. Werme 1996. A kinetic model for the stability of spent fuel matrix under oxic conditions. *J. Nucl. Mater.* 238: 110-120.

Casas I. 1989. Estudios físico-químicos de la disolución del UO_2. *Ph.D. Thesis*, Univ. Autònoma de Barcelona.

Kornicker W.A. & J.W. Morse 1991. Interactions of divalent cations with the surface of pyrite. *Geochim. Cosmochim. Acta* 55, 2159-2172.

Malmström M., S. Banwart, L. Duro, P. Wersin & J. Bruno 1995. Biotite and chlorite weathering at 25°C. *SKB TR 95-01.*

Nicholson R.V., R.W. Gillham & E.J. Reardon 1990. Pyrite oxidation in carbonate-buffered solution: 2. Rate control by oxide coatings. *Geochim. Cosmochim. Acta* 54: 395-402.

Scott M.J. & J.J. Morgan 1990 Energetics and conservative properties of redox systems in: *Chemical Modeling of Aqueous Systems*, Chapter 29 (American Chemical Society, 1990) p. 368.

Stumm, W. & J.J. Morgan 1981. *Aquatic Chemistry.* John Wiley and Sons.

Torrero M.E. 1995. Estudio de la disolución del UO_2 como análogo químico de la matriz del combustible nuclear gastado. Influencia de los principales parámetros fisicoquímicos que definen los repositorios en medio salino y granítico. *PhD Thesis.* Univ. de Barcelona.

Geochemistry of the Earth's Surface, Ármannsson (ed.)© 1999 Balkema, Rotterdam, ISBN 90 5809 073 6

Reactive solute transport modeling of thermally-perturbed and irradiated Boom Clay

J. Delgado, L. Montenegro, J. Samper, A. Vázquez & R. Juncosa
E.T.S. Ingenieros de Caminos, University of A Coruña, Spain

ABSTRACT: Heat and radiation-triggered processes influence the geochemistry of the Boom Clay porewaters. The CERBERUS experiment (Control Experiment with Radiation of the Belgian Repository for Underground Storage) was aimed at studying these processes by taking clay samples and collecting porewater throughout the five-year time extent of the experiment. Reactive solute transport modeling was pursued using available data. Results obtained indicate that currently available codes can successfully reproduce the experimental trends while they give detailed insights into the geochemical processes leading to changes of the Boom Clay system.

1 INTRODUCTION

Deep storage in selected geological formations is being considered in many countries as a highly reliable solution for high level radioactive waste disposal. However, a key point for regulatory authorities, and of the major public concern, is the assessment of the long-term safety of disposal sites. Among the European Commission financed projects, CERBERUS is one of the firsts where in situ geochemical determinations have been performed in addition to measurements of dose rates and thermal and hydraulic parameter.

2 EXPERIMENT DESCRIPTION

A detailed description of the experiment can be found in Bonne et al. (1992) and Noynaert et al. (1998). Only a brief summary of the main aspects of the test will be given here. The CERBERUS test was aimed at simulating the near-field effects in an argillaceous environment (the Boom Clay Formation) of a Cogema HLW-canister after 50 years cooling time. This canister can be characterized by the gamma activity of ^{137}Cs and its daughter ^{137}Ba (1.53 PBq per canister for each isotope) and by a thermal power of 580 W. The test was performed during 5 years (1989-1994) using a ^{60}Co source of 400 TBq and two electric heaters delivering each 363 W.

The test was designed to collect experimental in situ data about thermo-hydro-mechanical processes

in the Boom Clay when submitted simultaneously to heat and radiation. In situ experimental work included thermal and hydraulic tests, pH/E_H measurements, collection of gases produced by radiolysis and corrosion (H_2 and CH_4) and water sampling. At the end of the heating phase of the test, samples were taken to study the behavior of the engineered barrier materials (backfill material, waste capsule, overpack, canister and host rock).

3 CODE DESCRIPTION

The code used for the thermo-hydro-geochemical modeling (THG) is CORE-LE (Xu 1996, Samper et al. 1998). The code deals with transient or steady state thermodynamic systems under local chemical equilibrium conditions. It accounts for a wide range of geochemical processes including homogeneous and heterogeneous reactions (aqueous speciation, redox, solid dissolution/precipitation, gas equilibria, ion exchange, surface complexation, and decay series). The chemical formulation is based on the ion association theory and uses an extended version of the Debye-Hückel equation (B-dot) for the activity coefficient of aqueous species.

Flow and transport equations are solved using triangle finite element methods for spatial discretization while implicit, explicit, or any other finite difference schemes are used for time discretization. CORE-LE can cope with problems having heterogeneous physicochemical properties and irregular boundaries. It can be used with heterogeneous and anisotropic two-dimensional media.

Transport processes included in the code are advection, molecular diffusion, and mechanical dispersion.

CORE-LE can cope with any number of aqueous, exchanged and adsorbed species, minerals, and gases. Heat transport is solved at each time step. Computed nodal temperatures are used to update diffusion coefficients, equilibrium constants and other parameters needed for calculating activity coefficients.

Thermodynamic data used by CORE-LE is a reworked version of the DATA0.COM.V7.R22a from the EQ3/6 software package (Wolery 1979).

Code development has been funded by ENRESA, the spanish agency in charge of the management of HLW disposal, within the framework of EU R&D program.

4 CONCEPTUAL MODEL

The main features of the conceptual model developed are: (1) Boom Clay remains always fully saturated. According to previous TH modeling, water evaporation is not likely. Therefore, a liquid phase plus a number of solid phases are considered. (2) Due to the excavation of the Hades main drift (the underground research laboratory of SCK•CEN at Mol, Belgium) and the CERBERUS well, water flows while the experiment is being performed. The ^{60}Co source and the two heaters induce a thermal gradient across the near field of the Boom Clay. The temperature variation affects flow parameters such as hydraulic conductivity. (3) Transport processes considered are advection, mechanical dispersion, and molecular diffusion. (4) The major heat transport process is conduction along both the solid and liquid phases. The combined thermal conductivity of the solid + liquid is treated by CORE-LE as a function of water saturation. For the present modeling, heat production due to nuclear decay is not taken into account. (5) Radiolysis of water is considered the only radiation-triggered relevant process. Radiolysis produces both oxidizing and reducing agents, H_2O_2 and H_2 being the most relevant (Choppin et al. 1996; Beaufays et al. 1994). H_2O_2 readily decomposes into O_2 in the presence of water what allows modeling the oxidation of the Boom Clay through a constant mass flux of molecular oxygen. (6) For all the chemical processes the assumption of chemical equilibrium is granted. (7) The set of chemical processes included in the model is mineral dissolution/precipitation; aqueous chemical speciation; electron transfer; ion exchange; and gas equilibria.

5 CRITICAL EVALUATION OF DATA

Geochemical data have been taken from the sources given next. Boom Clay mineralogy from Griffault et al. (1996); Undisturbed and disturbed Boom Clay pore water compositions are taken from references cited in Noynaert et al. (1998).

In order to check the reliability of measurements, a pre-modeling critical assessment was done. It was acknowledged that most of the analyses were performed under laboratory conditions, which are not representative of the true in situ conditions. Therefore, they had to be back converted into the conditions found at the sampling point. These corrections have proven to be of the major help during the modeling. However, some other problems were identified when dealing with the experimental database of CERBERUS: (1) Data concerning chemical composition of the undisturbed pore water of the Boom Clay are scarce. Therefore, the initial chemical conditions have to rely on very few analyses. (2) Some of the analyses corresponding to the first sampling surveys are affected by contamination (electrode glass bulb dissolution, KCl leak from the salt bridge of the electrode). This affects measured concentrations of Cl, K, B, and Si. Total Fe content could also be affected due to potential corrosion of the piezometer and/or steel tubing used to collect water at the sampling point. (3) Measured aluminum content (~ 10^{-5} m) is much greater than the expected values in equilibrium with silicates (clay minerals and feldspars). The reason for that is unclear and awaits further research. (4) pH/E_H measurements in the first sampling surveys are not very reliable due to fast aging of the electrodes and the necessity of collecting large water samples prior to the measurement.

5.1 *Initial chemical conditions*

Initial chemical conditions were pursued trying to ensure consistency between the best available hydrochemical data and known rock mineralogy. The residence time of water in the Boom Clay is not well known but surely longer than 30.000 years (Beaufays et al. 1994). Taking into accounts its low permeability and large-scale mineralogical homogeneity, it is reasonable to assume chemical equilibrium between the rock and its pore water. The consistency check was performed with the aid of the EQ3NR code. In general, modeling results agree with available data for undisturbed-clay samples.

6 MODELING

Model complexity was increased in a stepwise manner. The main steps followed were: (a) 1-D thermo-hydraulic modeling up to a distance of 30 m of the source; (b) 1-D THG modeling considering only the effect of heating up to a distance of 1 m of the axis of the experiment; (c) Oxidation of the Boom Clay by a prescribed mass flux of molecular $O_{2(aq)}$.

The results of a previous thermo-hydrodynamic numerical model indicate that water flow around the experiment was nearly negligible, with velocity vectors pointing towards the axis of the CERBERUS well and slightly upwards. Therefore, it was decided to carry out the THG modeling without water flux.

6.1 Heating of the Boom Clay

Thermal steady state is reached in about one year at a distance of 1 m, the farthest point of the modeling domain. Therefore, all the chemical variations observed later than this time are due to solute transport.

The increase in temperature induces a drop in E_H that is more pronounced in the hottest area (left part of Fig. 1). On the other hand, pH shows a similar evolution becoming more acidic than the initial values. According to observations made in the Boom Clay, its buffering capacity is due to the presence of carbonates and iron-bearing minerals. Final pH and E_H values reflect this equilibrium condition at any considered temperature. However, dissolution and precipitation of certain amounts of minerals reset the chemical composition of the pore water, and by extension that of the exchange complex.

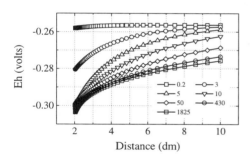

Figure 1. Evolution of E_H as a function of distance in the near field of the CERBERUS experiment. Numbers account for days.

6.2 Heating and oxidation of the Boom Clay

The mass flux of molecular oxygen has been calibrated in order to obtain the best results when compared to measured porewater composition of the Boom Clay. The flux was imposed at the nodes closest to the ^{60}Co source (3.0×10^{-5} mol/day). The

calibrated mass flux compares well with the theoretical value of 5.0×10^{-5} mol/day calculated from the source activity and radiation chemical yield for $O_{2(g)}$ (0.1 molecules/100 eV, Beaufays et al. 1994).

In addition to the features observed in the only-heating phase E_H evolves according to the reactions triggered by the oxygen flux (Fig. 2). The oxidation front propagates with time. It is not noticeable at distances larger than half a meter after the 5-year duration of the experiment. Computed values agree fairly well with the measurement, at the observation point at the final time.

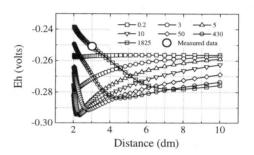

Figure 2. Evolution of E_H as a function of distance in the near field of the CERBERUS experiment. Numbers account for days. Big dot represent measured data.

Increase in silica concentration reflects the temperature dependence of silica-bearing phases in the Boom Clay. Calculated value is in good agreement with measured value.

Conservative elements (F, B, Cl) give identical results to those of the previous stage, showing no changes in their composition as a function of time.

pH drops to values much lower than in the only-heating step. Measured value is well matched by the calculated value. In the zone where oxidation is taking place pyrite dissolves, releasing iron, sulfur, and lowering the pH of the pore water. Siderite tends also to dissolve due to the coupled effect of pH, temperature, and oxidation. Increasing iron concentration induces goethite precipitation. However, since calcite dissolves due to the pH drop, the amount of carbonate released induces at some time an inversion in the dissolving tendency of siderite. When siderite precipitates, goethite starts to dissolve. These coupled processes do not translate into a massive sequence of dissolution/precipitation of minerals. In fact, the amounts of dissolved or precipitated minerals are rather small at the end of the experiment. Additional runs were performed with the same system but for longer times (up to

5000 years) Their results indicate that more than 2000 years are needed in order to exhaust the amount of calcite present in the Boom Clay.

Significant differences are found in the results obtained for the total solute concentrations of some elements when compared with the only-heating step. Iron tends to develop a concentration front due to the coupled effect of oxidation plus equilibria with goethite, siderite, and pyrite. Calculated concentrations are about one half order of magnitude lower than the measured values. Explanation for that can be found in the fact that measured iron concentrations are subjected to some uncertainty. Total dissolved sulfur increases steadily due to pyrite dissolution. Its calculated concentration matches quite nicely measured values.

Total carbonate in solution is the result of both calcite and siderite dissolution and precipitation. Near the boundary where oxidation takes place TIC is governed by calcite dissolution up to the point when siderite starts to precipitate. At the other boundary, where oxidation has not yet arrived, TIC contents are much lower and similar to those obtained in the heating step. Measured value is in fair agreement with model results. Computed total dissolved Ca, Mg and Na agree fairly well with measured values but K is underestimated. The reason for underestimation might be found in the low potassium content of the pore water (and its associated uncertainty) and the fact that some relevant mineral phase such as K-feldspar are not considered. Ca in solution increases since the beginning, increasing also the amount of Ca in the exchange complex. This is due to calcite dissolution and the necessity of keeping simultaneous equilibrium between the solution and the exchanger according to the Mass Action Law. To enter into the exchange complex it must displace K, Na, and Mg, which concomitantly increase into the pore water. Changes in the solid due to ion exchange are rather small and, perhaps, analytically undetectable. The same can be said about dissolution and precipitation of minerals. In summary, observed variations in porewater chemistry can be fairly well matched with the present THG model.

7 CONCLUSIONS

The most important geochemical effects induced over the Boom Clay by the CERBERUS experiment are a slight oxidation and changes in the porewater composition. These changes can be successfully modeled with the aid of CORE-LE. These encouraging results show that reactive transport codes are amenable of application to predict the long-term behavior of complex systems such as those to be found in HLW repositories.

ACKNOWLEDGEMENTS. This research has been performed with funds from the European Commission (FI4W-CT95-0008) and the Xunta de Galicia (XUGA11802B98). CORE-LE development has been favored with R&D funds from ENRESA.

REFERENCES

Beuafays, R., W. Blommaert, J. Bronders, P. de Cannière, P. del Marmol, P. Henrion, M. Monsecour, J. Patyn, & M. Put 1994. *Characterization of the Boom clay and its multilayered hydrogeological environment.* Nuclear Science & Technology Report. Final Report. Contract FI1W/0055. EUR 14961 EN.

Bonne A., H. Beckers, R. Beaufays, M. Buyens, J. Coursier, D. de Bruyn, A. Fonteyne, J. Genicot, D. Lamy, P. Meynendonckx, M. Monsecour, B. Neerdael , L. Noynaert, M. Voet, & G. Volckaert 1992. *The HADES demonstration and pilot project on radioactive waste disposal in a clay formation.* SCK•CEN. Final Report. Contract FI1W/0004 B. EUR 13851 EN.

Choppin, G., J. Liljenzin, & J. Rydberg 1996. *Radiochemistry and Nuclear Chemistry.* London: Butterworths & Heinmann.

Griffault, L., T. Merceron, J. Mossmann, B. Neerdael, P. de Cannière, C. Beaucaire, S. Daumas, A. Bianchi, & R. Christen 1996. *Projet Archimede-Argile: Acquisition et régulation de la chimie des eaux en milieu argileux.* Rapport Final. EUR 17454.

Noynaert L., G. Volckaert, P. de Cannière, P. Meynendonckx, S. Labat, R. Beaufays, M. Put, M. Aertsens, A. Fonteyne, & F. Vandervoort 1998. *The CERBERUS Project: A demonstration test to study the near field effects of a HLW-canister in an argillaceous formation.* Final Report (July 1, 1990 - December 31, 1996). Contract FI 2W/0003B. EUR 18151 EN.

Samper, J., R. Juncosa, J. Delgado & L. Montenegro 1998. *CORE-LE-2D: A code for water flow and reactive solute transport.* User's manual – Draft version. University of A Coruña.

Wolery, T. 1979. *Calculation of chemical equilibrium between aqueous solution and minerals: The EQ3/6 software package.* Lawrence Livermore National Laboratory Report UCRL-52658.

Xu, T. 1996. *Modeling non-isothermal multi-component reactive solute transport through variably saturated porous media.* Ph.D. dissertation. University of A Coruña.

Geochemistry of the Earth's Surface, Ármannsson (ed.) © 1999 Balkema, Rotterdam, ISBN 90 5809 073 6

New data on the standard Gibbs energy of H_4SiO_4 and its effect on silicate solubility

I. Gunnarsson & S. Arnórsson
Science Institute, University of Iceland, Reykjavík, Iceland

ABSTRACT: Equations describing the temperature dependence of quartz and amorphous silica solubilities in pure water in the range 0-350°C have been obtained by multiple linear regression of experimental solubility data on these minerals. New experimental data indicate that quartz is more soluble below 50°C than has generally been accepted to date. From the quartz solubility constant at 25°C and the standard Gibbs energies of formation from the elements (ΔG_f^o) of quartz and liquid water, $\Delta \overline{G}_f^o$ of $H_4SiO_4^o$ was calculated to be -1,309,181 J/mole. From the ΔG_r^o-temperature slope for quartz solubility at 25°C and the absolute entropies (S^o) of quartz and liquid water a value of 178.85 J/mole/K was obtained for $\overline{S}^o$ of $H_4SiO_4^o$. The standard apparent partial molal Gibbs energy of $H_4SiO_4^o$ in the temperature range 0-350°C, as calculated from quartz solubility, on the one hand, and amorphous silica solubility, on the other, are within 500 J/mole. This is considered to support the validity of the new data on quartz solubility at low temperatures. The new standard apparent partial molal Gibbs energy data on $H_4SiO_4^o$ indicate that all silicate minerals are more soluble under Earth's surface conditions than has generally been accepted to date.

1 INTRODUCTION

Knowledge of the thermodynamic properties of aqueous silica ($H_4SiO_4^o$) is very important for interpretation of interaction between silicate minerals and water. Silicate mineral solubility is generally retrieved from the thermodynamic properties of the mineral itself and the aqueous species formed upon its dissolution, including $H_4SiO_4^o$. Therefore, any changes in the selected standard Gibbs energy of $H_4SiO_4^o$ have an effect on the calculated solubility of silicate minerals.

The thermodynamic properties of $H_4SiO_4^o$ are generally calculated from the solubility of quartz. New data for quartz solubility below 100°C (Rimstidt 1997) indicate that quartz is more soluble at low temperature than has generally been accepted to date. These new data have large effect on the calculated values for the standard apparent partial molal Gibbs energy of $H_4SiO_4^o$ below 100°C.

In this contribution temperature equations for quartz and amorphous silica solubilities were obtained in the temperature range 0-350°C by multiple linear regression of experimental solubility data. These experiments were carried out at 1 bar below 100°C and P_{sat} at higher temperatures. From the equation for quartz solubility, on the one hand, and amorphous silica solubility, on the other, the

standard apparent partial molal Gibbs energy of $H_4SiO_4^o$ has been calculated in the range 0-350°C. The maximum difference in the standard Gibbs energy values so calculated is 470 J/mole.

2 QUARTZ AND AMORPHOUS SILICA SOLUBILITY.

Equations describing quartz and amorphous silica solubilities in pure water have been obtained in the temperature range 0-350°C by multiple linear regression of experimental solubility data. The solubility data for quartz cover the range 21-350°C and those for amorphous silica 25-350°C. The quartz solubility data selected below 100 °C to obtain the temperature equation are those of Rimstidt (1997) and van Lier (1960). The following equations describe the temperature dependence of the quartz and amorphous silica solubilities :

$$\log K_{quartz} = -34.188 + 197.47 \cdot T^{-1} \\ -5.851 \cdot 10^{-6} \cdot T^2 + 12.245 \cdot \log T \quad (1)$$

$$\log K_{am.silica} = -15.433 - 151.60 \cdot T^{-1} \\ - 2.977 \cdot 10^{-6} \cdot T^2 + 5.464 \cdot \log T \quad (2)$$

where T is in K. The equations are valid in the range 0-350°C at 1 bar below 100°C and at P_{sat} at higher temperatures. In Figure 1 a comparison is made between the quartz solubility curve obtained in this study and curves from various researchers. Below 100°C the curves can be divided into two groups, those of Walther & Helgeson (1977) and Fournier & Potter (1982) which indicate lower solubility, and those proposed by Rimstidt (1997), von Damm (1991) and us (Figure 1b) which indicate higher solubility in this temperature range.

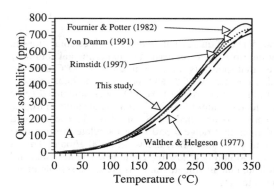

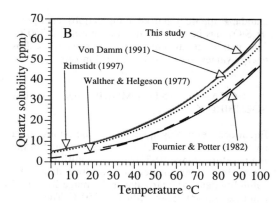

Figure 1. Quartz solubility curves presented by various researchers.

The amorphous silica solubility curve obtained in the present study is compared with solubility curves presented by others in Figure 2. The curves are in excellent agreement up to 200°C, but at higher temperature the curve proposed by Fournier (1977) indicates higher solubility and above 250 °C the curve presented by Rimstidt & Barnes (1980) indicates higher solubility than the curve obtained in this study.

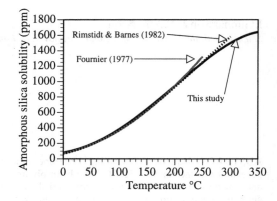

Figure 2. Amorphous silica solubility curves presented by various researchers.

3 THERMODYNAMIC PROPERTIES OF $H_4SiO_4^o$

The dissolution of quartz and amorphous silica in water can be described by the equation:

$$SiO_{2,(s)} + 2H_2O_{(l)} = H_4SiO_4^o \qquad (3)$$

where s represents the solid phase. From this reaction it is seen that knowledge of ΔG_f^o (the standard molal Gibbs energy of formation from the elements), ΔH_f^o (the standard molal enthalpy of formation from the elements) and S^o (the standard absolute entropy) of the minerals and liquid water, as well as the corresponding thermodynamic properties for the dissolution reaction (ΔG_r^o, ΔH_r^o, ΔS_r^o), enables calculation of $\Delta \overline{G}_f^o$, $\Delta \overline{H}_f^o$ and $\overline{S}^o$ for the $H_4SiO_4^o$ species. Multiplication of equation (1) by $-RT_r\ln(10)$ and differentiation with respect to T yields ΔG_r^o and ΔS_r^o values of 21380 J/mole and -2.49 J/mole/K, respectively at 25°C. From $\Delta H_r^o = \Delta G_r^o + T_r\Delta S_r^o$ a ΔH_r^o of 20638 J/mole is obtained at 25°C. To obtain data on $\Delta \overline{G}_f^o$, $\Delta \overline{H}_f^o$ and $\overline{S}^o$ for $H_4SiO_4^o$ at 25°C and 1 bar, we have selected the corresponding data on quartz and liquid water as given by Richet et al. (1982) and Cox et al. (1989), respectively. The results are as follows: $\Delta \overline{G}_f^o$ = -1,391,181 J/mole, $\Delta \overline{H}_f^o$ = -1,461,722 J/mole and $\overline{S}^o$ = 178.85 J/mole/K. The value obtained here for the standard partial molal Gibbs energy at 25 °C and 1 bar is the same for all practical purposes as that obtained by Rimstidt (1997) but it is considerably more negative than the values reported by Shock et al. (1989) and Robie et al. (1979). The difference is 1492 and 1181 J/mole, respectively.

An equation of the form:

$$\Delta G^o_{i,T} = k^*_1 \cdot T + k^*_2 + k^*_3 \cdot T^3 + k^*_4 \cdot T \cdot \log T \qquad (4)$$

describes accurately the temperature variation of the standard apparent Gibbs energy (ΔG^o) of quartz, amorphous silica and liquid water in the range 0-350°C at 1 bar below 100°C and at P_{sat} at higher temperatures. i represents the compounds and k^*_1 to k^*_4 are constants independent of temperature and pressure, and T is in K. Multiplication of equations (1) and (2) with $-RT\ln(10)$ yields equations of the same form as equation (4) for the temperature dependence of ΔG^o_r for the quartz and amorphous silica dissolution reactions. By summing up in appropriate proportions the Gibbs energy-temperature equations for the quartz dissolution reaction, quartz and liquid water an equation is obtained which describes the temperature dependence of the standard apparent partial molal Gibbs energy of $H_4SiO^o_4$. The values of the constants are: $k^*_1=1,790.03$, $k^*_2=-1,338,984$, $k^*_3=1.0988 \cdot 10^{-4}$ and $k^*_4=-686.96$. They are valid in the same P-T conditions as the equation for quartz solubility. Figure 3 shows the difference between calculated standard apparent partial molal Gibbs energy of $H_4SiO^o_4$ obtained in this study and calculated using quartz solubility curves from other researchers. The difference is within 500 J/mole using our quartz solubility curve and those of Rimstidt (1997) and von Damm (1991) but it is larger between our values and those calculated from the solubility curve presented Fournier & Potter (1982) or about 1000 J/mole in the range 25-100°C but decreases with rising temperature above 100°C.

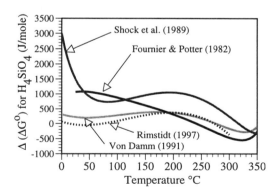

Figure 3. Difference in the standard apparent partial molal Gibbs energy ($\Delta \overline{G}^o$) of aqueous silica, calculated from quartz solubility curves from this study and those indicated in the graph. For Shock et al. (1989) we used their data on aqueous silica ($SiO_{2,aq}$) and added to it data for water to obtain values for $H_4SiO^o_4$.

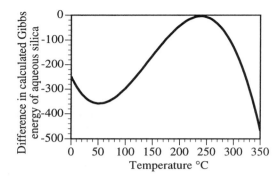

Figure 4. Difference in calculated values for the standard apparent partial molal Gibbs energy ($\Delta \overline{G}^o$) of aqueous silica, as based on quartz solubility, on the one hand, and amorphous silica on the other.

The difference between our results and those of Shock et al. (1989) is around 1000 J/mole in the range 50-250°C but below 50°C it becomes larger and is 3000 J/mole at 0°C. This difference is quite large and has a significant effect on calculated silicate mineral solubilities at low temperature. Figure 4 shows how the standard apparent partial molal Gibbs energy of $H_4SiO^o_4$ compares when calculated from the solubility of quartz, on one the hand, and the solubility of amorphous silica, on the other. Thermodynamic properties of amorphous silica were taken from Robie et al. (1979). The largest difference is 460 J/mole at 350°C. This small difference supports the contention that the quartz and amorphous silica solubility curves obtained here are reliable and indicates that Rimstidt's (1997) results for quartz solubility at low temperature are valid.

4 CONCLUSIONS

Experimentally it is difficult to demonstrate mineral-solution equilibrium for minerals of relatively complex composition, e.g. as most Al-silicates, by approaching equilibrium from under- and supersaturation due to slow kinetics and possibly also because of the formation of unwanted phases. For this reason the solubility of such minerals needs to be calculated from their measured thermodynamic properties and the thermodynamic properties of the aqueous species which form upon their dissolution, including $H_4SiO^o_4$. The new standard apparent partial molal Gibbs energy data on this species indicate that the solubility of all silicate minerals are considerably more soluble at low temperatures than has generally been accepted to date. To demonstrate this in more detail consider albite as an example. The dissolution of albite can be expressed as:

$$NaAlSi_3O_8 + 8H_2O =$$
$$Na^+ + Al(OH)_4^- + 3H_4SiO_4^0 \qquad (5)$$

$H_4SiO_4^0$ appears 3 times in the reaction. Accordingly, changes in the standard apparent partial molal Gibbs energy of $H_4SiO_4^0$, due to higher

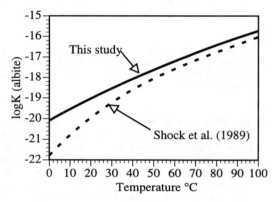

Figure 5. Solubility of albite in water. The broken curve is calculated using data on aqueous silica from Shock et al. (1989) and the solid curve using data on aqueous silica obtained in this study.

solubility of quartz, leads to three times larger change in the calculated solubility of albite. Figure 5 compares solubility curves for low albite using, on the one hand, the data for $H_4SiO_4^0$ obtained here, and on the other, data reported by Shock et al. (1989). Using the data obtained here gives higher solubility for albite at temperatures below 100°C, and in particular below 30°C. The difference at 25°C and 0°C, is 0.8 and 1.7 logK units, respectively.

REFERENCES

Cox J. D., D. D. Wagman & V. A. Medvedev 1989. *CODATA key values for thermodynamics.* Hemisphere Publishing Corp.

Fournier R. O. & J. J. Rowe 1977. The solubility of amorphous silica in water at high temperatures and pressures. *Amer. Mineral* 62: 1052 - 1056.

Fournier R.O. & R. W. Potter 1982. An equation correlating the solubility of quartz in water from 25°C to 900°C at pressures up to 10,000 bars. *Geochim. Cosmochim. Acta* 46: 1969 - 1973.

Richet P., Y. Bottinga, L. Denielou, J. Petitet & C. Tequi 1982. Thermodynamic properties of quartz, cristobalite and amorphous SiO₂: drop calorimetry measurements between 1000 and 1800 K and a review from 0 - 2000 K. *Geochim. Cosmochim. Acta* 46: 2639 - 2658.

Rimstidt J. D. & H. L Barnes 1980. The kinetics of silica-water reactions. *Geochim. Cosmochim. Acta* 44: 1683 - 1699.

Rimstidt J. D. 1997. Quartz solubility at low temperatures. *Geochim. Cosmochim. Acta* 61: 2553 - 2558.

Robie R. A, B. S. Hemingway, & J. R. Fisher 1979. Thermodynamic properties of minerals and related substances at 298.15 K and 1 bar (10⁵ Pascals) pressure and at higher temperatures. *U. S. Geol. Surv. Bull.* 1452: 456 p.

Shock E. L., H. C. Helgeson & D. A. Sverjensky 1989. Calculation of the thermodynamic and transport properties of aqueous species at high pressures and temperatures: Standard partial molal properties of inorganic neutral species. *Geochim. Cosmochim. Acta* 53: 2157 - 2183.

van Lier J. A., P. L. de Bruyn & J. T. G. Overbeek 1960. The solubility of quartz. *J. Phys. Chem.* 64: 1675 - 1682.

von Damm K. L., J. L. Bischoff & R. J. Rosenbauer 1991. Quartz solubility in hydrothermal seawater: An experimental study and equation describing quartz solubility for up to 0.5 M NaCl solutions. *Am. J. Sci.* 291: 977 - 1007.

Walther J. W. & H. C. Helgeson 1977. Calculation of the thermodynamic properties of aqueous silica and the solubility of quartz and its polymorphs at high pressures and temperatures. *Am. J. Sci.* 277: 1315 -1351.

Geochemistry of the Earth's Surface, Ármannsson (ed.) © 1999 Balkema, Rotterdam, ISBN 90 5809 073 6

Estimates of thermodynamic properties for aqueous haloethanes

J. R. Haas

University of North Carolina, Charlotte, N.C., USA

ABSTRACT: Standard partial molal Gibbs free energies and enthalpies of formation and third law entropies for aqueous haloethane species have been estimated using experimental gas-solubility data and correlations involving thermodynamic properties of haloethane gases. Correlations used to estimate aqueous species properties include ethane and thirteen haloethane compounds containing fluorine, chlorine, iodine, bromine, as well as compounds containing more than one type of halogen. These correlations may be used to estimate the properties of a wide range of aqueous haloethane species of environmental interest, provided that gas-phase properties are available. Standard partial molal thermodynamic properties of fourteen additional haloethane species are provided, that have been estimated using these correlations. Estimated values may be used, together with values for minerals, gases, and aqueous species, to calculate the stability of aqueous haloethanes in settings near Earth's surface, and may be used to predict conditions under which haloethane degradation in natural waters will be favored or unfavored energetically.

1 INTRODUCTION

The release and environmental dispersal of haloethane compounds represents a major source of ground and surface water contamination in the industrialized world. Haloethane compounds comprise all chemical species in which one or more hydrogens in the ethane structure (H_3C-CH_3) are replaced by halogens, including species such as hexachloroethane, pentachloroethane, fluoro-, chloro-, bromo- and iodoethane, isomers of difluoro-, dichloro-, diiodo- and dibromoethane, a range of isomers containing 3 or 4 isohalogens (i.e. 1,1,1-trichloroethane), and a wide range of mixed-halogen species. Some haloethane compounds are produced naturally in small quantities by microbes (e.g. Gribble 1994) or through volcanism (e.g. Stoiber et al. 1971), but the predominant input of these species to the biosphere is anthropogenic. Haloethane species are manufactured and used primarily as industrial degreasing agents, adhesives, aerosols, and pesticides, and many of these compounds are regularly released to the environment in large quantities. For example, industrial releases of 1,1,1-trichloroethane in the USA alone in 1996 amounted to approximately 4.0×10^6 kg, while total production-related waste mass exceeded 2.4×10^7 kg (Toxics Release Inventory 1996).

Pervasive contamination of near-surface waters by haloethane compounds has promoted widespread interest in the biogeochemical properties of these chemical species, specifically with respect to their persistence in natural aquatic settings. Microbially-mediated degradation of aqueous halocarbon contaminants has attracted particular interest as a potential mechanism by which halocarbon species may be transformed to less harmful compounds by natural microbial populations. The ability of some aerobic and anaerobic microbes to actively degrade dissolved halocarbon species has been demonstrated experimentaly (e.g. deBruin et al. 1992; Komatsu et al. 1994; Sonier et al. 1994; Chang and Alvarez-Cohen 1996), and in field settings (e.g. Tanhua et al. 1996). These and similar studies demonstrate that biotically-mediated degradation of halocarbons in solution is energetically favorable under some conditions, but such studies do not allow for estimation of the equilibrium stability of a wide range of halocarbon species under conditions significantly different from those in the laboratory.

This paper presents estimates of standard partial molal thermodynamic properties of twenty-seven haloethane compounds of environmental interest, obtained using correlations of gas-phase thermodynamic values and experimentally-derived data describing gas-solubility reactions involving haloethane species. Values for specific haloethane species are presented, but the correlations provided in this work may be used to estimate the standard partial molal Gibbs free energies of formation, enthalpies of formation, and third law entropies of any haloethane species for which corresponding gas-phase data are available.

2 ESTIMATION OF THERMODYNAMIC PROPERTIES OF AQUEOUS SPECIES

2.1 Gibbs free energies of formation

Horvath (1982) provides a review and compilation of experimental values for the solubilities of a wide range of halocarbon species in water under conditions of known halocarbon gas partial pressure. These values with respect to haloethanes are reported for a range of temperatures ranging from approximately 273 K to 373 K. These values may be used to calculate dimensionless Henry's Law partition coefficients (H) at temperature, according to the relation

$$H = \frac{C_{aq}}{C_g} \tag{1}$$

which relates H to the concentration of a dissolved species (C_{aq}) and to the partial pressure of its corresponding gas species (C_g). Values of H can also be expressed in terms of the equilibrium constant (K_{sol}) for the gas-solubility reaction, by recalculation of C_{aq} in terms of molality and C_g in terms of fugacity, yielding the relation

$$K_{sol} = \frac{f}{m} \tag{2}$$

(Stumm and Morgan 1996). Using relation (2), values for standard partial molal Gibbs free energies of haloethane gas-solubility reactions (ΔG^o_r) can be calculated using the expression

$$\Delta G^o_r = -2.303RT \log K_{sol} \tag{3}$$

which relates ΔG^o_r to K_{sol}, temperature in Kelvins (K), and the gas constant (R). Standard partial molal Gibbs free energies of formation of aqueous species ($\Delta G^o_{f,aq}$) were calculated for haloethanes using the corresponding values for the standard partial molal Gibbs free energies of formation of gases ($\Delta G^o_{f,g}$) and ΔG^o_r according to the expression

$$\Delta G^o_{f,aq} = \Delta G^o_r + \Delta G^o_{f,g}. \tag{4}$$

Values of $\Delta G^o_{f,g}$ used in this study were taken from the compilations of Wagman et al. (1982), Reid et al. (1987), and Daubert and Danner (1989).

2.2 Enthalpies of formation and third law entropies

Free energy data derived from experimental gas solubility values were used to regress values of standard partial molal enthalpies of formation for aqueous haloethane species. Using a Van't Hoff assumption of constant enthalpy and entropy as a function of temperature near the reference condition of 25° C, the expression relating standard partial molal Gibbs free energy of a reaction to the standard partial molal enthalpy (ΔH^o_r) and entropy (ΔS^o_r) of reaction

$$\Delta G^o_r = \Delta H^o_r - T\Delta S^o_r \tag{5}$$

gives a straight line in ΔG^o_r and temperature having a slope equal to $-\Delta S^o_r$ and an intercept equal to ΔH^o_r. At temperatures below approximately 300 K, values of ΔG^o_r for haloethane gas-solubility reactions as a function of temperature approximate a straight line, according to the Van't Hoff assumption. For example, Figure 1 illustrates ΔG^o_r values for 1,1,2,2-tetrabromoethane as a function of temperature, along with the least-squares linear regression used to estimate values of ΔH^o_r and ΔS^o_r.

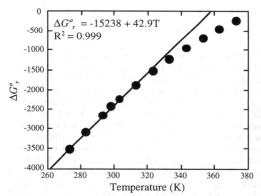

Figure 1. Filled circles indicate ΔG^o_r for the gas-solubility reaction of 1,1,2,2-tetrabromoethane as a function of temperature (K), from data in Horvath (1982); solid line indicates a linear least-squares regression of values below ~320 K.

Values of standard partial molal enthalpies and entropies of formation for aqueous haloethanes were calculated using regressed ΔH^o_r values and properties of the gaseous species, using expressions analogous to that shown in (4). Third law entropy values (S^o_{aq}) were calculated using estimated values of standard partial molal entropies of formation for aqueous haloethanes ($\Delta S^o_{f,aq}$) and values of third law entropies for the elements ($S^o_{element}$) in their reference states (Wagman et al. 1982), according to the expression

$$S^o_{aq} = \Delta S^o_{f,aq} + \Sigma S^o_{element} \quad (6)$$

Values of standard partial molal Gibbs free energies, enthalpies, and third law entropies of aqueous haloethane species for which experimental data at elevated temperatures are available in Horvath (1982) are provided in Table 1 together with values for ethane(aq) from Wagman et al. (1982).

Table 1. Estimated standard partial molal thermodynamic properties for aqueous haloethane species.

Species(aq)	$\Delta G^o_{f,aq}$ [a]	$\Delta H^o_{f,aq}$ [a]	S^o_{aq} [b]
1,1-dichloroethane	-16980	-46970	17.8
1,2-dichloroethane	-18860	-47960	20.8
1,1,1-trichloroethane	-16620	-43490	39.4
1,1,2-trichloroethane	-21500	-53560	22.0
1,1,1,2-tetrachloroethane	-18670	-46390	47.6
1,1,2,2-tetrachloroethane	-20050	-47270	49.3
pentachloroethane	-15380	-44240	54.8
iodoethane	5734	-9955	42.0
bromoethane	-5130	-23200	38.3
1,2-dibromoethane	-2679	-18800	47.5
1,1,2,2-tetrabromoethane	-8355	-41240	-3.5
1,1-difluoroethane	-98700	-118700	46.5
chloropentafluoroethane	-244600	-275800	45.9
ethane	-3886	-24710	26.6

a. cal mol^{-1}. b. cal mol^{-1}K^{-1}

3 PREDICTIVE CORRELATIONS

The values shown in Table 1 may be used to calculate reaction equilibria involving any of the listed aqueous species at 25° C and 1 bar, provided that corresponding values for necessary minerals, gases and aqueous species are available. The species listed in Table 1 include several compounds of acute environmental and geochemical interest, but lack many other single- and mixed-halogen compounds that commonly occur as contaminants in the biosphere.

However, it is possible to make use of strong linear correlations between the thermodynamic properties of gaseous and aqueous haloethane species to estimate the properties of aqueous species for which experimentally derived thermodynamic parameters are unavailable, but for which the corresponding properties of gaseous species are accessible. Figures 2 and 3 depict free energy and enthalpy correlations, respectively, involving the haloethane species listed in Table 1. Both are strong, having R^2 values of ~0.999 or better. In each figure the inset diagram expands the cluster of data points seen in the upper right hand corner of the respective plot. The correlations depicted in Figures 2 and 3 include single-halogen species containing F, Cl, Br, I, and mixed

halogen species containing F and Cl. Consequently, these correlations may be used to predict the properties of a wide variety of aqueous haloethane species, including natural and synthetic mixed-halogen species.

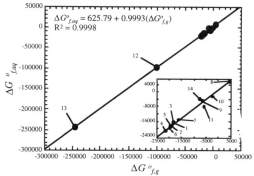

Figure 2. Free energy correlation for haloethane gases and aqueous species. Enumerations refer to (1) 1,1-dichloro-(ethane), (2) 1,2-dichlor0-, (3) 1,1,1-trichloro-, (4) 1,1,2-trichloro-, (5) 1,1,2-tetrachloro-, (6) 1,1,2,2-tetrachloro-, (7) pentachloro-, (8) iodo-, (9) bromo-, (10) 1,2-dibromo-, (11) 1,1,2,2-tetrabromo-, (12) 1,1-difluoro-, (13) chloropentafluoro-, (14) ethane.

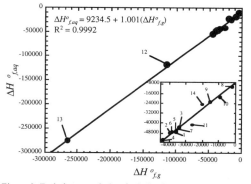

Figure 3. Enthalpy correlation for haloethane gases and aqueous species. Numerical designations are the same as in Figure 2.

Slopes for both correlations are approximately 1.000, indicating that the properties of aqueous haloethane species do not depart significantly from those of gaseous species, apart from small energetic contributions to free energy and enthalpy, expressed in the intercepts of the linear correlations in Figures 2 and 3, that are likely attributable to solvation. The solvation contribution to free energy is less than one kcal/mol, but the enthalpic contribution is more significant (~9.2 kcal/mol), demonstrating the need for explicit formulation of aqueous species properties, as opposed to using gaseous species properties, in thermodynamic calculations involving dissolved haloethanes.

Estimates of the standard partial molal Gibbs free energies and enthalpies of formation and third law entropies of fourteen additional haloethane aqueous species, calculated using the correlations shown in Figures 2 and 3 and gas species data provided in Wagman et al. (1982), Reid et al. (1987), and Daubert and Danner (1989), are provided in Table 2.

Table 2. Estimated standard partial molal thermodynamic properties for aqueous haloethanes. Based on correlations shown in Figures 2 and 3.

Species(aq)	$\Delta G^o_{f,aq}$ [a]	$\Delta H^o_{f,aq}$ [a]	S^o_{aq} [b]
hexafluoroethane	-285600	-335500	-19.34
1,1,1-trifluoroethane	-157200	-193900	-0.976
fluoroethane	-49500	-73990	22.82
1,2-diiodoethane	-18720	-24750	72.72
hexachloroethane	-13570	-43670	61.70
chloroethane	-14160	-42430	12.56
bromotrichloroethane	-103200	-122900	66.30
1,2-dibromotetrafluoroethane	-204600	-241000	13.85
1,2-dichlorotetrafluoroethane	-189700	-232100	10.81
1,1,2-trichlorotrifluoroethane	-145400	-183500	27.70
1,1,2,2-tetrachlorodifluoroethane	-145700	-184600	27.32
halothane ($C_2HBrClF_3$)	-151900	-186000	94.11
1-chloro-1,1-difluoroethane	-99710	-130900	19.93
1,2-difluoroethane	-102800	-132600	13.63

a. cal mol^{-1}. b. cal mol^{-1}K^{-1}

The fourteen species listed in Table 2 represent only a fraction of the potential haloethane compounds for which the correlations shown in Figures 2 and 3 may be used to estimate the properties of their aqueous species, provided that gas-phase data are available. Estimated aqueous species values, including the values listed in Tables 1 and 2, may be used in conjunction with thermodynamic values for minerals, other aqueous species and gases to calculate the thermodynamic properties of reactions involving haloethane species in solution. The environmental and biogeochemical utility of this approach is enormous, and can allow for a quantitative assessment of the energetic favorability or unfavorability of haloethane degradation reactions, including microbially-mediated overall metabolic reactions. Although kinetic limitations may inhibit reactions that are energetically favored, an assessment of the influence of geochemically pertinent factors such as pH, activity and fugacity of reactant and product species on equilibrium haloethane stability will enable a more quantitative evaluation of the potential for degradation in highly variable natural settings. This ap-

proach could also be profitably applied to the optimization of engineered bioremediation strategies, by manipulating natural or artificial parameters to maximize the energetic favorability of reactions that are not kinetically limited.

REFERENCES

Chang H.-L. and Alvarez-Cohen L. (1996) Biodegradation of individual and multiple chlorinated aliphatic hydrocarbons by methane-oxidizing cultures. *Applied and Environmental Microbiology* 62, 3371-3377.

Daubert T. E. and Danner R. P. (1989) *Physical and thermodynamic properties of pure chemicals: data compilation.* Hemisphere Publishing Corp., New York.

deBruin W. P., Kotterman M. J. J., Posthumus M. A., Schraa G. and Zehender A. J. B. (1992) Complete biological reductive transformation of tetrachloroethene to ethane. *Applied and Environmental Microbiology* 58, 1996-2000.

Gribble G. W. (1994) The natural production of chlorinated compounds. *Environmental Science and Technology* 28, 310A-319A.

Horvath A. L. (1982) *Halogenated Hydrocarbons: Solubility-Miscibility with Water.* Marcel-Dekker.

Komatsu T., Momonoi K., Matsuo T. and Hanaki K. (1994) Biotransformation of cis-1,2-dichloroethylene to ethylene and ethane under anaerobic conditions. *Wat. Sci. Tech.* 30, 75-84.

Reid R. C., Prausnitz J. M. and Poling B. E. (1987) *The Properties of Gases and Liquids.* McGraw-Hill.

Sonier D. N., Duran N. L. and Smith G. B. (1994) Dechlorination of trichlorofluoromethane (CFC-11) by sulfate-reducing bacteria from an aquifer contaminated with halogenated aliphatic compounds. *Applied and Environmental Microbiology* 60, 4567-4572.

Stoiber R. E., Leggett D. C., Jenkins T. F., Murrmann R. P. and Rose W. I. (1971) Organic compounds in volcanic gas from Santiaguito volcano, Guatemala. *Geological Society of America Bulletin* 82, 2299-2302.

Stumm W. (1992) *Chemistry of the Solid-Water Interface.* Wiley-Interscience, New York.

Tanhua T., Fogelqvist E. and Basturk O. (1996) Reduction of volatile halocarbons in anoxic seawater, results from a study in the Black Sea. *Marine Chemistry* 54, 159-170.

Toxics Release Inventory (1996) United States Environmental Protection Agency.

Wagman D. D., Evans W. H., Parker V. B., Schumm R. H., Halow I., Bailey S. M., Churney K. L. and Nuttall R. L. (1982) *The NBS Tables of Chemical Thermodynamic Properties; Selected Values for Inorganic and C1 and C2 Organic Substances in SI Units.* American Chemical Society.

Geochemistry of the Earth's Surface, Ármannsson (ed.)© 1999 Balkema, Rotterdam, ISBN 90 5809 073 6

Brucite dissolution at pH 1.74 to 4.7 and at 25°C to 55°C

C-M. Mörth & H. Strandh
Department of Geology and Geochemistry, Stockholm University, Sweden

ABSTRACT: This paper shows that the mechanism of Brucite ($Mg(OH)_2$) dissolution is a complicated pathway of surface reactions, diffusion of ions and reaction between ions in the diffuse double layer created at the surface – solution interface. The mixed dissolution kinetics of Brucite, is shown using rotating disc experimental data and mathematical modelling of rate laws of diffusion and surface reactions.

1 INTRODUCTION

Understanding the complexity of mineral-solution interaction during dissolution is of importance in any attempt to interpret or predict natural weathering of minerals. Surface dissolution kinetics can be complicated by mixed kinetics, i.e. influence of diffusion, by the rate at which the ions are transported from and to the surface, and by surface reactions, by the rate at which the breaking of structural bonds occurs.

Brucite formation has been shown to be a precursor in dissolution processes of the magnesium oxide, periclase (Wogelius et al. 1995). Wogelius et al. (1995) performed dissolution of periclase and observed the first-step changes of the periclase surface into a hydroxylated structure close to that of brucite, predicting this reaction mechanism to be active for most metal oxide minerals. In humid environments all oxide surfaces are covered by hydroxyl groups, which indicates the importance of understanding metal hydroxide dissolution processes. The knowledge of brucite dissolution mechanisms is far from complete, but very complex dissolution behaviour has been observed. Vermileya (1969) determined brucite dissolution to be transport controlled at pH>5, but surface controlled at pH=5, while Casey (1987) calculated brucite dissolution to be a transport controlled reaction.

2 METHODS

The rotating disk was used to run the brucite dissolution experiments, in an apparatus setting described by (Sjöberg & Rickard 1983). The reactive surface of the disc was 0.7 cm² and all the experimental runs were performed under normal air pressure and composition. To keep the pH constant a Mettler automatic titrator was used. All solutions were made of distilled water and the background electrolyte was made of special analytical grade of KCl. The concentration of the KCl was 0.7 M. The pH-electrode was calibrated against NBS-buffer solutions, pH 4.01 and 6.86. The HCl, 0.1 to 0.001 M, used for titrating the reaction solution was of a special analytical grade.

The rotating discs were made of brucite from Gabbs (Nye County, Nevada). This brucite contains a few percent dolomite. Since dolomite has much slower reaction kinetics than brucite, and since the surface area for dolomite is small compared to the surface area of brucite this impurity can be neglected.

The discs were polished with 600 mesh carborundum powder and later washed with 1 M HCl and distilled water. If much of the brucite is dissolved from the disc's surface the surface roughness will increase. This will be observed as an increase in reaction rate with time and will be more pronounced at lower pH. However, all experiments were done in such a way as to assure that the measured rates were unaffected by this phenomenon.

2.1 Finding the reaction rate and diffusion coefficients

The rotating disc method is used to distinguish transport/diffusion and chemically controlled reactions. Sometimes the transport rate and chemical rate are of the same order. The total observed rate is then both transport/diffusion and chemically controlled this is referred to as mixed kinetics. When brucite is dissolving, H^+ ions are transported to the brucite surface or to a reaction zone near the surface where the reaction occurs. The mass transfer reaction can be written:

Table 1. Apparent and calculated diffusion coefficient and the apparent chemical rate constant.

pH_B	D 10^{-4} cm^2 s^{-1}	Corrected D 10^{-4} cm^2 s^{-1}	K_c 10^{-2} cm s^{-1}
1.74	0.64	N/A	1.81
2.00	0.75	N/A	1.92
2.30	1.22	1.19	1.83
2.56	1.35	1.26	1.98
3.00	1.20	0.91	3.40
3.25	1.07	0.60	3.18
4.00	1.33	N/A	6.25
4.30	1.47	N/A	8.31
4.70	1.42	N/A	8.79

$$J_T = D_{H+} / \delta_N (c_b{}^{H+} - c_s{}^{H+}) \tag{1}$$

where J_T is the mass flux per unit area and time, D is the diffusion coefficient, c_b and c_s the bulk and surface concentration, respectively, and δ_N the diffusion boundary layer (DBL) thickness defined as:

$$\delta_N = 1.61 \; D^{1/3} \; \nu^{1/6} \; \omega^{-1/2} \tag{2}$$

where ν is the kinematic viscosity and ω is the rotational velocity.

Combining equations (1) and (2) and assuming a first order chemical reaction rate law, it is possible to write (Sjöberg & Rickard 1983);

$$J_{H+}{}^{-1} = 1.61 \; \nu^{1/2} \; (D^{2/3} c_b{}^{H+})^{-1} \; \omega^{-1/2} + (k_c c_b{}^{H+})^{-1} \tag{3}$$

The measured reported rates are initial rates, which means that the Mg^{2+} concentration in bulk solution is

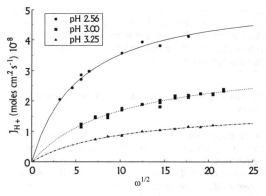

Figure 1. Flux of H$^+$ ions vs w$^{0.5}$. A linear relationship will show a purely transport controlled mechanism, while the graph shows that brucite dissolution is controlled by mixed kinetics.

very low, in practice almost zero. The precision for the calculated rates was well within ±10% and in most runs much better than that. Temperature control was within ±0.2°C at all temperatures.

3 RESULTS AND DISCUSSION

The theory of the rotating disc method states that rates of diffusion controlled reactions will be proportional to the rotational velocity, ω, while a purely chemically controlled reaction will be independent of ω (Figure 1). Furthermore, if the reaction is first order then a plot of 1/rate versus $\omega^{-0.5}$ will give a straight line. In Figure 2 the relation between 1/rate and $\omega^{-0.5}$ is plotted according to equation (3). The chemical rate constants and the calculated diffusion coefficient for H$^+$ diffusion are presented in Table 1.

Figure 1 clearly suggests that the dissolution is controlled by both chemical and transport reactions. It can also be seen in Figure 2 and Table 1 that the chemical rate constant, k_c, increases with increasing pH. The diffusion coefficient, D, for H$^+$ diffusion, is much higher than that reported for H$^+$ diffusion of $0.71 \cdot 10^{-4}$ cm^2/s in 0.7 M KCl (Landsberg et al. 1961), except for the experiments made at pH 1.74 and 2. Equation (3) assumes that the distance for H$^+$ diffusion is the total thickness of the DBL. If the reaction occurs at some distance from the surface this calculated D_{H+} will be too high.

It has been shown that a reaction zone between dissolving and bulk solution species can be established near a dissolving mineral surface (Litt & Serad 1964). Depending on the nature of the reaction the reaction zone can look very different. In the case of brucite dissolution, where H$^+$ ions react with OH$^-$ ions at the surface or near the surface (at a reaction zone), the reaction can be assumed to be very quick and irreversible. The reaction zone will then be very distinct and possible to define (Litt & Serad 1964).

To calculate the diffusion profiles of the reacting species (H$^+$ and OH$^-$ ions) near the zone where the reaction occurs, knowledge is required about the flow of the bulk solution fluid around the rotating disk. It has been shown that the fluid motion around the disk can be described by three second order differential equations (Levich 1962). The axial distance is normalized according to:

$$\xi = (\omega/\nu)^{0.5} \gamma \tag{4}$$

where γ is the actual distance.

The important fluid motion for finding the diffusion profiles is the axial fluid motion, H. Litt & Serad (1964) showed that if species A is diffusing from the bulk towards the surface, and species B from the surface then:

$$\Phi''_A - Sc_A H \Phi'_A = 0 \tag{5}$$

where Φ is the concentration profile of species A and Sc is the Schmidt number

The concentration profile of species B is then written in the same way as for species A:

$$\Phi''_B - Sc_B H \Phi'_B = 0 \qquad (6)$$

The boundary conditions for the equations are different, however. For equation (5) in the interval $0 \leq \xi \leq \xi_R$: at $\xi=0$, $\Phi_A=1$ and at $\xi=\xi_R$ $\Phi_A=0$ and for equation (6) in the interval $\xi_R \leq \xi < \infty$: at $\xi=\xi_R$, $\Phi_B=0$ and when $\xi \to \infty$ then $\Phi_B \to 1$.

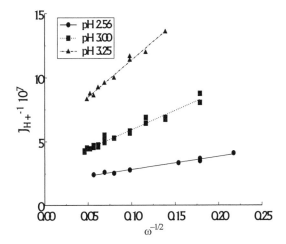

Figure 2. Reciprocal rate vs $w^{-0.5}$ shows straight lines describing mixed kinetics with a first order chemical reaction.

At the reaction zone, $\xi=\xi_R$, then $\Phi_A=0$, $\Phi_B=0$ and $\Phi'_B = \Phi'_A (b\, Sc_B\, C_{A0})/(a\, Sc_A\, C_{B\infty})$, where a and b are stoichiometric coefficients for the reaction, Sc is the Schmidt number for species A and B, C_{A0} is the concentration of species A at the surface and $C_{B\infty}$ is the concentration of species B in the bulk solution.

In the complete pH range the dissolution was found to be controlled by mixed kinetics. At pH 1.74 and 2.00 the chemical rate constants and H^+ diffusion constants are independent of pH (Table 1). The rate of chemical reaction at the surface is directly proportional to the surface concentration of H^+. At higher pH, 2.30, 2.56, 3.00 and 3.25, the higher calculated D_{H+} given by equation (3) suggests a reaction between OH^- and H^+ in the DBL at some distance from the surface. By calculating the position of the reaction zone, as described above (Figure 3), the distance of diffusion is reduced and the calculated value from equation (3) can be adjusted. This gives the corrected D_{H+} values presented in Table 1. As these values are corrected at pH 3.00 and 3.25 are near the measured value in 0.7 M KCl solutions, it is concluded that the correction is valid. At pH 2.30 and 2.56 the corrected D val-

ues are still too high since the reaction zone is very close to the surface. At higher pH, from 4.00, the predicted reaction zone is placed outside the DBL and therefore the reaction zone model is not valid.

The interpretation of the chemical reaction constants given by equation (3) is not obvious considering where the reaction takes place. At low pH, 1.74 to 2.56, both the placement of the reaction zone and the similar low chemical rate constants indicate that the reaction follows a simple first order reaction mechanism dependent on H^+ activity near the surface. At pH 3.00 and 3.25 the chemical rate constants are slightly higher and the reaction zone is further out from the surface, which suggests that the reaction constant must be corrected for the distance of diffusion. Above this pH the chemical rate constant is increasing and the reaction zone is determined to be outside the DBL. The reaction is then to a large degree controlled by transport of ion species and the reaction rate law may be expressed in terms of saturation of $Mg(OH)_2$ in the solution near the surface.

The temperature dependence of the apparent chemical rate constant, k_c, and the apparent diffusion coefficient, D_{H+}, were measured at pH 4. By using the Arrhenius equation, the activation energies of reaction and diffusion were found to be 38 and 14 kJ mole^{-1}. These are of the same order as activation energies determined for MgO of 54 and 15 kJ mole^{-1} respectively (Macdonald & Owen 1971).

4 CONCLUSIONS

The rotating disc can be used to separate the transport rate and the chemical rate. At all investigated pH values the observed data was evaluated assuming a first order chemical reaction. However, the interpretation of the diffusion constant and the chemical rate constant is not obvious. At low pH, 1.74 and 2.00, the diffusion constants for H^+ ions is very near measured values in 0.7 M KCl solutions and the reaction probably takes place at the mineral surface. Above pH 2 the diffusion constant is much higher than expected. By assuming that the reaction takes place in the diffusive boundary layer (DBL) and using a reaction zone model the diffusion constant at pH 3.00 and 3.25 are reduced to measured values. The reaction zone model fails above pH 4, when the predicted zone is outside the DBL. The constant reaction constants in the pH range 1.74 to 2.56 show that the first order reaction rate law is valid and that the rate depends on the H^+ activity near the surface. Above this pH range the rate law may be expressed in terms of brucite saturation near the surface.

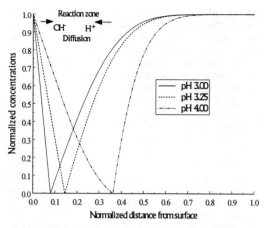

Figure 3. The reaction zone as a function of pH. At pH 4 the zone is outside the diffusive boundary layer at normalized distance of 0.32.

ACKNOWLEDGEMENT

The experimental work and initiation of the project was carried out by Lennart Sjöberg, who tragically passed away in 1996.

REFERENCES

Casey, W.H. 1987. Heterogenous kinetics and diffusion boundary layers: The example of reaction in a fracture. *J. Geophys. Res. Solid Earth Planets* 92:8007-13.

Landsberg, R., W. Geissler & S. Müller 1961. Über einige anwendung der rotierenden scheibenelektroden. *Z. Chem.* 1:169-74.

Levich, V.G. 1962. *Physicochemical hydrodynamics*. New Jersey: Prentice Hall.

Litt, M. & G. Serad 1964. Chemical reactions on a rotating disk. *Chem. Eng. Sci.* 19:867-84.

Macdonald, D.D. & D. Owen 1971. The dissolution of magnesium oxide in dilute sulfuric acid. *Can. J. of Chem.* 49: 3375-3380

Sjöberg, L. & D.T. Richard 1983. *Studies of calcite dissolution reactions: A report of work between 1979-1982.* Stockholm: Meddelanden från Stockholms Universitets Geologiska Institution 254.

Vermileya, D.A. 1969. The dissolution of MgO and Mg(OH)$_2$ in aqueous solutions. *J. Electrochem. Soc.* 116:1179-83.

Wogelius, R.A, K. Refson, D.G. Fraser, G.W. Grime & J.P. Goff 1995. Periclase surface hydroxylation during dissolution. *Geochim. Cosmochim. Acta* 59:1875-81.

Geochemistry of the Earth's Surface, Ármannsson (ed.) © 1999 Balkema, Rotterdam, ISBN 90 5809 073 6

Olivine surface speciation and reactivity in aquatic systems

O.S. Pokrovsky
Laboratory of Environmental Hydrochemistry, Department of Geography, Moscow State University, Russia
& Laboratoire de Géochimie, CNRS (UMR 5563)-OMP, Université Paul-Sabatier, Toulouse, France

J. Schott
Laboratoire de Géochimie, CNRS (UMR 5563)-OMP, Université Paul-Sabatier, Toulouse, France

ABSTRACT: The surface of forsterite (Fo_{92}) in aqueous solutions was investigated using surface titrations, electrokinetic measurements, and X-ray Photoelectron Spectroscopy. A Mg-depleted layer and a Mg-rich brucite-like layer have been shown to form on olivine surface in acid to neutral and alkaline solutions, respectively. The dissolution rates of Fo_{92} were measured at 25°C in a mixed-flow reactor as a function of pH and ΣCO_2. The mechanism of forsterite dissolution includes 1) $Mg^{2+} \leftrightarrow H^+$ exchange reaction on the surface followed by H^+ adsorption in the leached layer in acid solutions and 2) hydration of surface Mg sites with formation of $>MgOH_2^+$ species in alkaline solutions. Carbonate ion was recognized to be a strong inhibitor of forsterite dissolution in alkaline solutions when $pCO_3^{2-}<4$. The influence of dissolved CO_2 on forsterite dissolution rate at pH $\geq$ 9 should be taken into account when modeling the effect of CO_2 on mafic minerals weathering.

1 INTRODUCTION

The chemical weathering of rock-forming minerals is one of the major processes controlling the global cycles of the elements in the hydrosphere. Its modeling requires the accurate description of the structure, energetics and chemistry of the various solid-solution interfaces where it occurs. The surface coordination theory developed by Stumm and coworkers is presently one of the most efficient tool to rationalize this information and derive general rate laws for mineral dissolution. In this approach, the dissolution rate depends on the relative concentration of adsorbed species which promote or retard metal detachment from the surface. The stoichiometry and surface concentration of these precursor species can be obtained by combining surface titrations, adsorption experiments, and electrokinetic measurements with *in situ* spectroscopic studies (via XPS, XAS, DRIFT, AFM, etc.) of the solid/water interface. Such an approach has been taken to characterize forsterite surface speciation and reactivity in aqueous solutions.

2 RESULTS AND DISCUSSION

Surface potentiometric titrations and electrokinetic measurements (streaming potential and electrophoresis techniques) have been combined with an XPS and DRIFT study of hydrated forsterite surfaces and dissolution kinetics experiments performed in a mixed-flow reactor. These results show that at pH<9, a Mg-depleted surface layer (<80 Å in thickness) is formed due to an exchange reaction between a magnesium and 2 protons as also confirmed by the preferential release of Mg over Si during the initial stage of dissolution. The formation of a Si-rich leached layer on the forsterite surface in acid solutions is consistent with previous results of XPS (Fujimoto & Velde 1990) and SEM (Rosso & Rimstidt 1997) observations. At pH>9, the surface exhibits a slight enrichment in Mg as also confirmed by preferential Si release at the initial stage of dissolution. Electrokinetic measurements yield an isoelectric point or $pH_{i.e.p.}$ of 4.5 which is consistent with the dominance of silica sites at olivine surface for pH<9. In contrast, surface titrations result in a point of zero charge or $pH_{p.z.c.}$ of 10 and the development of a large positive charge (up to 10^{-4} mol/m^2) in acid conditions. This may be explained by penetration of H^+ into the first 3-5 molecular layers of forsterite as confirmed by our XPS data that show a distinct increase in the half-peak-height width of the O_{1s} peak. Such penetration of hydrogen into the structure of forsterite under acidic conditions, was also detected using X-ray Absorption Spectroscopy (XAS) and Elastic Recoil Detection Analysis (ERDA) observations (Wogelius & Fraser 1996) as well as various surface analytical techniques based on energetic ion beams (Petit et al. 1990a, b, Fujimoto et al. 1992).

The dissolution rates of Fo_{92} (Figure 1) were measured at 25°C in a mixed-flow reactor as a function of pH (3 to 12), ΣCO_2 (0 to 0.05 M) and aqueous silica concentration (0 to 0.001 M). In CO_2-free solutions, the rates decrease with pH at $3 \leq pH \leq 8$ with a slope close to 0.5 in accord with the results of Blum & Lasaga (1988), Wogelius & Walther (1991) and Oelkers (1999). At pH>8, in contrast to the results of Blum & Lasaga (1988) and Wogelius and Walther (1991), the rates continue to decrease with a smaller slope of ~0.2 and become pH independent for pH$\geq$10. Olivine dissolution rates are inversely proportional to aqueous carbonate and silica activities at $pCO_3^{2-}<4$ and pH>8.5, respectively. These results allow to elucidate the mechanism of forsterite dissolution. In acid to weakly alkaline solutions, H^+-promoted dissolution is controlled by the decomposition of a silica-rich, Mg-free, protonated precursor complex formed by a fast exchange reaction of $2H^+$ for one Mg^{2+}, followed by the rate limiting sorption of $0.5H^+$ on each exchanged site forming $>SiO-H_{0.5}^+$ species linked to the OH groups in the altered layer. This is consistent with the development of a leached layer where residual hydrated silicate groups are either linked to Mg ions deeper in the mineral structure or which polymerize and link to one another (Schott & Berner 1985, Petit et al. 1989, Fujimoto & Velde 1990, Casey et al. 1993). At pH>9, it is the hydration of surface Mg sites with formation of $>MgOH_2^+$ species which controls dissolution. Preferential Si release, observed at the initial stage of dissolution in alkaline solutions, together with our XPS results, is consistent with the formation of a brucite-like layer on the surface. The breaking of Mg-O-Mg bonds in this partially altered surface layer is thus the critical step for forsterite dissolution at these conditions. Far from equilibrium, olivine dissolution is thus controlled by two parallel reactions occurring at silica-rich ($\equiv SiO-H_{0.5}^+$) and hydrated Mg ($>MgOH_2^+$) sites in accord with:

$$R_+ = k_{Si} \cdot \{>SiO-H_{0.5}^+\} + k_{Mg}\{>MgOH_2^+\} \qquad (1)$$

where R_+ is the forward dissolution rate; k_{Si} and k_{Mg} are the rate constants for these two reactions, respectively, and $\{i\}$ is the concentration of the species on the surface or in the altered layer.

It can be seen in Figure 1 that Equation 1 provides an accurate description of forsterite dissolution rate in carbonate-free solutions at 2<pH<12.

The effect of carbonate ions on the forsterite dissolution rates at different pH is illustrated in Figure 2. The rates decrease linearly with $a_{CO_3^{2-}}$ with a slope close to 1. If the hydrolysis of Mg surface

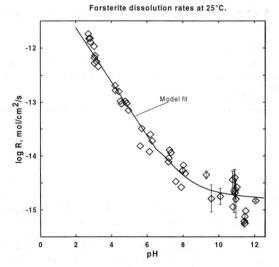

Forsterite dissolution rates at 25°C.

Figure 1. Steady-state forsterite dissolution rates as a function of pH in CO_2-free solutions. The rates are calculated based on Mg and Si release. The solid line was derived from the model as described in the text (Equation 1).

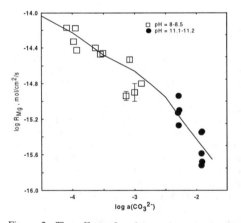

Figure 2. The effect of carbonate ions on the forsterite dissolution rates. The solid line is generated using the model described in the text (Equations 1-2). Note that all solutions are undersaturated with respect to $MgCO_3$.

groups controls forsterite dissolution at pH>9, then the inhibition of dissolution by dissolved carbonate may be explained by the decrease of $\{>MgOH_2^+\}$ according to the following reaction:

$$>MgOH_2^+ + CO_3^{2-} = >MgCO_3^- + H_2O \qquad (2)$$

The solid line in Figure 2, where the dissolution

rates in carbonate-bearing solutions are depicted, was calculated using Equation 1 taking into account reaction (2) with log $K_2 = 2.7$. Similar mechanism of dissolution inhibition by dissolved CO_2 was suggested by Wogelius & Walther (1991), but with the adsorption of one aqueous carbonate group on two magnesium surface centers.

Based on the results obtained in this study and available literature information, one may estimate the effect of CO_2 on the dissolution rates of Ca and Mg silicates at neutral to alkaline pH conditions. For both forsterite (this study) and diopside (Knauss et al. 1993), the removal of atmospheric CO_2 from solution does not appear to have any significant effect on their dissolution rates over a wide range of pH (6 to 12). This is also in agreement with the data for basalt glass and anorthite and augite dissolution (Brady and Carrol 1994, Brady and Gislason 1997) at higher pCO_2 values.

However, addition of carbonate ions to the solution at pH > 8 leads to an important decrease of forsterite dissolution rate when $u_{CO_3^{2-}} > 10^{-4}$ M. Such conditions may occur in the course of basalt or ultramafic rocks weathering by carbonate rich solutions. Note, for example, that at pH 10, $u_{CO_3^{2-}}$ in equilibrium with atmospheric CO_2 is equal to $10^{-1.66}$ corresponding to $\Sigma CO_2 = 0.14$ M(!). Springs from partially serpentinized peridotites and dunites or fresh ultramafic rocks contain up to 0.024 M ΣCO_2 at pHs close to 9 (Barnes and O'Neil 1969). At these conditions, the effect of CO_2 on the reaction rates is of crucial importance and must be taken into account. Another important consequence is that the decrease of olivine dissolution rate with increasing carbonate concentration results in a decrease of atmospheric CO_2 consumption, i.e., unlike for feldspars, there is a negative feedback between pCO_2 and olivine weathering rate.

REFERENCES

Barnes, I. & J.R. O'Neil 1969. The relationship between fluids in some fresh alpine-type ultramafics and possible modern serpentinization, Western United States. *Geol. Soc. Amer. Bull.* 80: 1947-1960.

Blum, A.E. & A.C. Lasaga 1988. Role of surface speciation in the low temperature dissolution of minerals. *Nature* 331: 431-433.

Brady, P.V. & S.A. Carroll 1994. Direct effects of CO_2 and temperature on silicate weathering: Possible implications for climate control. *Geochim. Cosmochim. Acta* 58: 1853-1856.

Brady, P.V. & S.R. Gíslason 1997. Seafloor weathering controls on atmospheric CO_2 and global climate. *Geochim. Cosmochim. Acta* 61: 965-973.

Casey, W. H., M.F. Hochella Jr., & H.R. Westrich 1993. The surface chemistry of manganoferous silicate minerals as inferred from experiments on tephroite (Mn_2SiO_4). *Geochim. Cosmochim. Acta* 57: 785-793.

Fujimoto, K., K. Fukutani, M. Tsunoda, H. Yamashita, & K. Kobayashi 1992. Characterization of olivine-water interface using $^1H(^{15}N, \alpha\gamma)$ ^{12}C resonant nuclear reaction. *Water-Rock Interaction*, Kharaka & Maest (eds), Balkema, Rotterdam, 145-148.

Fujimoto, K. & B. Velde 1990. Dissolution and hydration of olivine under hydrothermal conditions. *EOS* 71: 962.

Knauss, K.G., S.N. Nguyen, & H.C. Weed 1993. Diopside dissolution kinetics as a function of pH, CO_2, temperature, and time. *Geochim. Cosmochim. Acta* 57: 285-294.

Oelkers, E.H. 1999. A comparison of enstatite and forsterite dissolution rates and mechanisms. In *Growth and Dissolution in Geosystems* (B. Jamveit and P. Meakin Eds), 253-257, Kluwer.

Petit, J.-C., J.-C. Dran, A. Paccagnella, & G. Della Mea 1989. Structural dependence of crystalline silicate hudration during aqueous dissolution. *Earth Planet. Sci. Lett.* 93: 292-293.

Petit, J.C., Della Mea, G., Dran, J.C., Magonthier, M.C., Mando, P.A., & Paccagnella, A. 1990a. Hydrated-layer formation during dissolution of complex silicate glasses and minerals. *Geochim. Cosmochim. Acta* 54: 1941-1955.

Petit, J.C., G. Della Mea, & J.C. Dran 1990b. Energetic ion beam analysis in the earth sciences. *Nature* 344: 621-626.

Rosso, J.J. & J.D. Rimstidt 1997. Linked dissolution/ precipitation reactions at the silicate surfaces. *Abstracts of the VIIth Goldschmidt Conference, Tuscon, Arizona*. p.180.

Schott, J. & R.A. Berner 1985. Dissolution mechanism of pyroxenes and olivines during weathering. In *The Chemistry of Weathering* (J.I. Drever Ed.), 35-53, Reidel.

Wogelius, R.A. & D.G. Fraser 1996. Kinetics and mechanisms of surface reactions at the olivine-aqueous fluid interface. *Abstracts of European Research Conference Geochemistry of Crystal Fluids*, Seefeld, Austria, p.20.

Wogelius, R.A. & J.V. Walther 1991. Olivine dissolution at 25°C: Effects of pH, CO_2, and organic acids. *Geochim. Cosmochim. Acta* 55: 943-954.

Geochemistry of the Earth's Surface, Ármannsson (ed.)© 1999 Balkema, Rotterdam, ISBN 90 5809 073 6

Structure and stability of the solvated $(UO_2)^{2+}$ uranyl ion: An in situ XAS study

P.F. Schofield
Department of Mineralogy, Natural History Museum, London, UK

E.H. Bailey
Division of Environmental Science, University of Nottingham, Loughborough, UK

J.F.W. Mosselmans
CLRC Daresbury Laboratory, Warrington, UK

ABSTRACT: Speciation of the U(VI) ion is complex, with many possible species, including polynuclear complexes, within the system $(UO_2)^{2+}$ - $(NO_3)^-$ - H_2O. We have utilised the advantages of X-ray absorption spectroscopy in an *in situ* study of the structure, speciation and stability of the solvated $(UO_2)^{2+}$ uranyl ion in the temperature range 25°C to 250°C and at equilibrium saturated vapour pressures. At all temperatures the $(UO_2)^{2+}$ uranyl ion itself remains linear with U=O distances of 1.77 Å. At low temperatures the $(UO_2)^{2+}$ uranyl ion is coordinated to 5 equatorial H_2O ligands with a U-OH$_2$ distance of 2.42 Å. This species is stable up to 160°C, however, above this temperature there is a decrease in the number of equatorial H_2O ligands. There is no apparent variation of the U-OH$_2$ distance. At 250°C and a 1.0 M concentration of $(NO_3)^-$ there are demonstrable interactions between the $(UO_2)^{2+}$ uranyl ion and the nitrate groups. Within the limits of this study there is no evidence for the existence of polynuclear complexes.

1 INTRODUCTION

1.1 *Background*

The potential for uranium pollution of natural groundwaters is currently a major environmental concern. Consequently, the role of toxic uranium within the environment is the subject of major experimental and field based research programmes focusing upon the solubility, transportation and precipitation of uranium in natural groundwaters. Indeed, uranium concentrations in natural waters can be as high as hundreds of μg l^{-1} in the vicinity of uranium ore deposits such as Koongarra, Australia (Murakami et al. 1997; Payne et al. 1992) or even tens of mg l^{-1} as reported for the evaporation ponds of the San Joaquin Valley (Duff et al. 1997; Bradford et al. 1990). Consequently, a knowledge of the interaction of uranium ions in aqueous media is important for understanding and predicting uranium mobility in the natural environment. The solubility of uranium is redox dependent, with uranium being much more soluble under oxic conditions. As such the uranyl ion, $(UO_2)^{2+}$, with hexavalent uranium, is likely to be the dominant dissolved form of uranium, particularly in acidic, oxidized groundwaters.

The speciation of the $(UO_2)^{2+}$ uranyl ion is complex and thought to be very sensitive to the pH of the solution (Langmuir 1978). A large number of potential uranyl-oxy-hydroxide species has been

proposed based primarily upon potentiometric measurements (Grenthe et al. 1992; Bailey 1994). As the pH of the solution is increased the uranyl ion, present in very acidic solutions, undergoes hydrolysis and forms hydroxy-polynuclear complexes, a process common to many polyvalent cations in solution. As the pH approaches neutrality, amorphous uranyl-hydroxy precipitates form and as such the $(UO_2)^{2+}$ is removed from solution.

In this paper we describe the results of an *in situ* X-ray absorption spectroscopy (XAS) study of the temperature dependence of the structure of the solvated $(UO_2)^{2+}$ uranyl ion. This work is part of a larger study, incorporating effects of pH and $(UO_2)^{2+}$ concentration, although in this report we will focus upon solutions of 0.1 M $(UO_2)^{2+}$ uranyl ion, containing of 0.2 M and 1.0 M concentrations of $(NO_3)^-$, within the temperature range 25°C to 250°C at equilibrium saturated vapour pressures.

1.2 *X-ray Absorption Spectroscopy*

X-ray absorption spectroscopy (XAS) is an element specific probe which can be used *in situ* to provide direct structural information including the type, number, distances and distribution of ligands, about the specific element of interest, in this case uranium. The efficacy of XAS in the study of metal speciation in solutions at elevated temperatures has recently been successfully demonstrated for a range of metal cations in a variety of solution chemistries. Pfund et al. (1994) measured for example the hydration spheres of Sr^{2+} and Kr in supercritical water; Mosselmans et al. (1996) studied speciation changes of $(MoO_4)^{2-}$, Cd^{2+} and Co^{2+} in acidic and chloride bearing solutions; Seward et al. (1996) monitored aquated Ag^+ in acidic solutions; Ragnarsdottir et al. (1998) studied the speciation of Y in nitrate and chloride solutions; Oelkers et al. (1998) looked at Sb in chloride solutions.

2 DATA COLLECTION AND ANALYSIS

A solution of 0.1 M $(UO_2)^{2+}$ and 0.2 M $(NO_3)^-$ was prepared by dissolving uranyl nitrate hexahydrate in de-ionised water, and a second solution was made by adding potassium nitrate to produce a 1.0 M concentration of $(NO_3)^-$. These solutions were sealed in quartz tubes of wall thickness 1 mm and internal diameter of 5 mm and loaded into the heating cell of Seward et al. (1996). At each temperature the solutions were left for 15 minutes to equilibrate prior to the collection of any XAS data, and the first and last spectra were compared to ensure no changes to the solutions had occurred during data collection.

Uranium L_{III}-edge X-ray absorption spectra were collected on station 9.2 of the CLRC Synchrotron Radiation Source, Daresbury, U.K., with the storage ring operating at 2 GeV and a beam current between 100 and 250 mA. Data were recorded in transmission mode employing the rapid scanning QuEXAFS method (Murphy et al. 1995) using a double crystal Si(220) monochromator detuned to give a rejection of 50% of the incident beam. The monochromator was calibrated against the K-edge of an yttrium metal foil, defined at an energy of 17040 eV. Six scans to k = 15 Å^{-1} were collected per temperature.

The L_{III}-edge spectra were summed, calibrated and background subtracted using the SRS Daresbury programs MOTPLOT, EXCALIB and EXBACK. The EXAFS data analysis was performed using the program EXCURV98 (Binsted 1998) with phase shifts calculated *ab initio* using Hedin-Lundqvist exchange potentials and von Barth ground state potentials. Theoretical fits to the experimental EXAFS were produced by refining the parameters Ef (an energy offset), CN (coordination numbers), R (uranium-scatterer interatomic distances) and $2\sigma^2$ (Debye-Waller factor). The coordination number of the first shell of oxygen atoms (i.e. those of the $(UO_2)^{2+}$ uranyl ion) was fixed at 2. Multiple scattering calculations were performed for each spectrum, but no account has yet been taken of possible temperature dependent anharmonic effects.

The structural data gleaned from the analysis of the EXAFS spectra are given in Table 1. Data quality and theoretical fits are shown in Figure 1. Previous work on uranyl solutions shows errors of

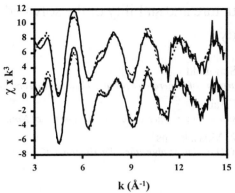

Figure 1 Experimental (solid) and theoretical (dashed lines) k^3-weighted, background subtracted EXAFS spectra for a solution of 0.1 M $(UO_2)^{2+}$, 0.2 M $(NO_3)^-$, at 25°C (bottom) and 250°C (top). Fits include multiple scattering calculations.

Table 1 Structural parameters derived from the non-linear least squares fitting of the EXAFS data.

T (°C)	atom	CN	R (Å)	$2\sigma^2$ (Å^{-2})	R-factor
0.1 M $(UO_2)^{2+}$, 0.2 M $(NO_3)^-$					
25	O	2	1.77	0.004	29
	O	4.5	2.41	0.012	
70	O	· 2	1.77	0.004	34
	O	4.5	2.40	0.013	
115	O	2	1.77	0.005	27
	O	4.9	2.41	0.018	
160	O	2	1.78	0.004	26
	O	4.7	2.41	0.018	
205	O	2	1.77	0.004	31
	O	3.7	2.41	0.016	
250	O	2	1.78	0.004	29
	O	3.1	2.43	0.015	
0.1 M $(UO_2)^{2+}$, 1.0 M $(NO_3)^-$					
25	O	2	1.77	0.004	26
	O	4.2	2.42	0.012	
70	O	2	1.78	0.005	29
	O	4.8	2.42	0.015	
115	O	2	1.78	0.004	21
	O	4.4	2.42	0.015	
160	O	2	1.78	0.004	28
	O	4.4	2.42	0.019	
205	O	2	1.79	0.004	36
	O	3.0	2.43	0.013	
250	O	2	1.77	0.004	34
	O	2	2.32	0.008	
	O	3	2.48	0.013	
	N	1	3.02	0.025	
	O	1	4.17	0.013	

±10% for CN and $(2\sigma^2)$ and ±1% for R (Docrat et al. 1999).

3 DISCUSSION

The XANES part of the XAS spectrum represents multiple scattering events which may involve several shells of back-scattering atoms. As such, this region of the spectrum can be a good indicator of changes in geometry around the central ion. The XANES for the two solutions are dominated by the scattering of the $(UO_2)^{2+}$ uranyl ion, but subtle changes are evident as a function of temperature (Figures 2 and 3). In particular, the shoulder, at about 17180 eV, on the high energy side of the first peak, and the trough at about 17190 eV are dampened as the temperature increases. While this may partially be a result of increased thermal disorder, it also implies temperature related changes in the structure or coordination number of the solvated $(UO_2)^{2+}$ uranyl ion.

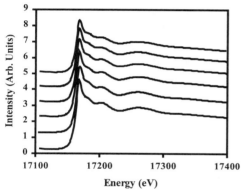

Figure 2 The XANES spectra for a solution of 0.1 M $(UO_2)^{2+}$, 0.2 M $(NO_3)^-$, as a function of temperature, with 25°C at the bottom and 250°C at the top.

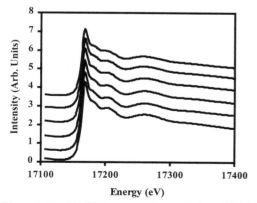

Figure 3 The XANES spectra for a solution of 0.1 M $(UO_2)^{2+}$, 1.0 M $(NO_3)^-$, as a function of temperature, with 25°C at the bottom and 250°C at the top.

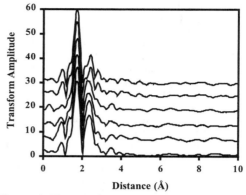

Figure 4 The Fourier transforms of the EXAFS spectra for a solution of 0.1 M $(UO_2)^{2+}$, 0.2 M $(NO_3)^-$, as a function of temperature, with 25°C at the bottom and 250°C at the top.

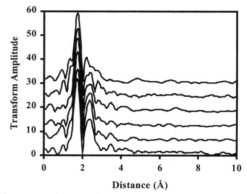

Figure 5 The Fourier transforms of the EXAFS spectra for a solution of 0.1 M $(UO_2)^{2+}$, 1.0 M $(NO_3)^-$, as a function of temperature, with 25°C at the bottom and 250°C at the top.

Stronger evidence of changes in coordination number around the $(UO_2)^{2+}$ uranyl ion can be seen in the Fourier transforms of the EXAFS spectra, shown as a function of temperature in Figures 4 and 5 for solutions containing 0.1 M $(UO_2)^{2+}$ and either 0.2 or 1.0 M $(NO_3)^-$ respectively.

The main peak, at a distance of about 1.8 Å, in the Fourier transforms of Figures 4 and 5 represents the two oxygen atoms that, along with uranium, comprise the linear $(UO_2)^{2+}$ uranyl ion. There is no change in either the position or the intensity of this peak, which implies that the basic structure of the $(UO_2)^{2+}$ uranyl ion remains unaffected by temperature. This is confirmed by the modelling of the EXAFS which consistently shows that the two

U=O bonds of the $(UO_2)^{2+}$ uranyl ion remain at 1.77 Å throughout the entire temperature range. Further evidence for this interpretation can be taken from the presence of a peak in the Fourier transforms at about 3.6 Å. Occurring at twice the distance of the first peak in the Fourier transforms, this peak results from multiple scattering events involving the two oxygen atoms of the $(UO_2)^{2+}$ uranyl ion and reflects the linear nature of this molecule.

Although the first peak in the Fourier transforms does not appear to alter as a function of temperature, there are significant changes in the shape and intensity of the second main peak in the Fourier transforms, which occurs at about 2.4 Å. This peak represents the oxygen atoms of the equatorial H_2O ligands, and a decrease in the intensity of this peak as the temperature is increased, as seen in Figures 4 and 5, implies a decrease in the number of equatorial H_2O ligands about the $(UO_2)^{2+}$ uranyl ion on increasing temperature. This interpretation is confirmed by the modelling of the EXAFS (Table 2).

At 25°C, the modelling of the EXAFS spectra shows that there are 5 equatorial H_2O ligands around the $(UO_2)^{2+}$ uranyl ion with a U-OH_2 of 2.42 Å, a distance which agrees excellently with that found from previous EXAFS studies of the $(UO_2)^{2+}$ uranyl ion at room temperature and pressure (Thompson et al. 1997; Allen et al. 1997). This complex appears to remain stable up to a temperature of 160°C for both solutions, however, at 205°C the number of equatorial H_2O ligands around the $(UO_2)^{2+}$ uranyl ion appears to decrease, with the solution containing 0.2 M (NO_3)⁻, apparently stabilising the species $[UO_2(H_2O)_3]^{2+}$ at 250°C. Within this temperature range, there appears to be no changes in the U-OH_2 distance. The solution containing 1.0 M (NO_3)⁻, however, shows strong evidence in both the Fourier transform (Figure 5) and the modelling of the EXAFS (Table 2) for ion-pairing between the $(UO_2)^{2+}$ uranyl ion and the (NO_3)⁻ nitrate ion, with a second oxygen shell coordination number of 5 and an equatorial U-N distance of 3.02 Å.

REFERENCES

Allen, P.G., J.J. Bucher, D.K. Shuh, N.M. Edelstein & T. Reich 1997. Investigation of aquo and chloro complexes of UO_2^{2+}, NpO_2^+, Np^{4+} and Pu^{3+} by X-ray absorption fine structure spectroscopy. *Inorg. Chem.* 36: 4676-4683.

Bailey, E.H. 1994. The solubility of uranium and thorium in hydrothermal fluids and subduction zones. *Unpublished PhD, University of Bristol.*

Binsted, N. 1998, EXCURV98. CCLRC Daresbury Laboratory Computer Program.

Bradford, G.R., D. Bakhtar & D. Westcot 1990. Uranium, vanadium and molybdenum in saline waters. *J. Environ. Quality* 19: 105-108.

Docrat, T.I., J.F.W. Mosselmans, J.M. Charnock, M.W. Whitele, D. Collison, F.R. Livens, C. Jones & M.J. Edmiston 1999. XAS of tricarbonatouranate(V), $[UO_2(CO_3)_3]^{5-}$, in aqueous solution. *Inorg. Chem.* In Press.

Duff, M.C., C. Amrhein, P.M. Bertsch & D.B. Hunter 1997. The chemistry of uranium in evaporation pond sediment in the San Joaquin Valley, California, USA, using X-ray fluorescence and XANES techniques. *Geochim. Cosmochim. Acta* 61: 73-81.

Grenthe, I., J. Fuger, R.J. Lemire, A.B. Muller, C. Nguyen-Trung, & H. Wanner 1992. *Chemical Thermodynamics Volume 1: The Chemical Thermodynamics of Uranium.* North-Holland.

Langmuir, D. 1978. Uranium solution-mineral equilibria at low temperatures with applications to sedimentary ore deposits. *Geochim. Cosmochim. Acta* 42: 547-569.

Mosselmans, J.F.W, P.F. Schofield, J.M. Charnock, C.D. Garner, R.A.D. Pattrick & D.J. Vaughan 1996. X-ray absorption studies of metal complexes in aqueous solution at elevated temperatures. *Chem. Geol.* 127: 339-350.

Murakami, T., T. Ohnuki, H. Isobe & T Sato 1997. Mobility of uranium during weathering. *Amer. Mineral.* 82: 888-899.

Murphy, L.M., B.R. Dobson, M. Neu, C.A Ramsdale, R.A. Strange & S.S. Hasnain 1995. Quick fluorescence EXAFS - an improved method for collection of conventional XAFS data and for studying reaction intermediates in dilute systems. *J. Synch. Rad.* 2: 64-69.

Oelkers, E.H., D.M. Sherman, K.V. Ragnarsdottir & C.R. Collins 1998. An EXAFS spectroscopic study of aqueous antimony(III)-chloride complexation at temperatures from 25 to 250°C. *Chem. Geol.* 151:21-28.

Payne , T.E., R. Edis, A.L. Herczeg, K. Sekine, Y. Seo, T.D. Waite & N. Yanase 1992. *Groundwater chemistry. Alligator Rivers Analogue Project Final Report.* 7: Sydney, Australian Nuclear Science and Technology Organisation.

Pfund, D.M., J.G. Darab & J.L. Fulton 1994. An XAFS study of strontium ions and krypton in supercritical water. *J. Phys. Chem.* 98: 13102-13207.

Ragnarsdottir, K.V., E.H. Oelkers, D.M. Sherman & C.R. Collins 1998. Aqueous speciation of yttrium at temperatures from 25 to 340°C at P_{sat} : an in situ EXAFS study. *Chem. Geol.* 151:29-40.

Seward, T.M., C.M.H. Henderson, J.M. Charnock & B.R. Dobson 1996. An X-ray absorption (EXAFS) spectroscopic study of aquated Ag+ in hydrothermal solutions to 350C. *Geochim. Cosmochim. Acta* 60: 2273-2282.

Thompson, H.A., G.E. Brown Jr. & G.A Parks 1997. XAFS spectroscopic study of uranyl coordination in solids and aqueous solution. *Amer. Mineral.* 82: 483-496.

8 Geochemistry of crustal fluids and of catastrophic events

Geochemistry of the Earth's Surface, Ármannsson (ed.) © 1999 Balkema, Rotterdam, ISBN 90 5809 073 6

Invited lecture: Progressive water-rock interaction and mineral-solution equilibria in groundwater systems

S. Arnórsson
Science Institute, University of Iceland, Reykjavík, Iceland

ABSTRACT: The concentrations of many components in surface waters are dominated by the extent of mineral dissolution. By contrast, the composition of groundwaters, particularly when temperatures exceed ~ 150°C, are controlled by close approach to equilibrium with secondary minerals for all major components except Cl and B, and sometimes SO_4. Apart from temperature, the most important factors favouring attainment of equilibrium are the reactivity of the rock-forming minerals and the supply of acids to the water, equilibrium being favoured when the minerals are very reactive and the supply of acids limited. Other factors that also contribute include the concentrations of soluble salts in the rock, the surface area between water and mineral, the age of the water and sometimes magmatic gas input. In extreme cases overall equilibrium is closely approached at temperatures as low as 10-20°C.

1 INTRODUCTION

The concentrations of many components in surface waters are too low to saturate the waters with many common rock forming minerals. Even if saturation, or supersaturation, is attained with such minerals sluggish kinetics may prevent their precipitation from solution. Accordingly, the composition of surface waters is strongly influenced by the extent of mineral dissolution. When such waters seep into the ground and react progressively with the rock they evolve towards mineral-solution equilibrium. Geothermal waters with temperatures in excess of 150°C have apparently closely approached equilibrium with secondary minerals for all major components except Cl and B and sometimes SO_4 (Giggenbach, 1981, Arnórsson et al. 1983). How closely equilibrium is approached depends on (1) the temperature of the water, (2) the rate of mass transfer between minerals and solution, (3) the rate of fluxes of matter through the system and (4) the quantity of mass transfer required to reach equilibrium.

2 LOCAL EQUILIBRIUM

All large systems on Earth, such as groundwater systems, are not at overall equilbrium as temperature, pressure and most often also chemical composition vary across these systems. Studies of mineral-solution equilibria/dis-equilibria in geological systems are based on the concept of local equilibrium (Nordstrom & Munoz 1994). This concept assumes that one can isolate a small volume or part of a larger

system so that equilibrium is maintained, or rather, closely approached within this small volume and that all the thermodynamic variables and relationships needed to describe the state of equilibrium can be applied to this sub-system.

It is generally not possible to sample groundwater in contact with specific minerals with the purpose of assessing the state of particular local mineral-solution equilibria. The type of data generally available include the composition of water from drillhole and spring discharges and general information on bedrock mineralogy. The use of such data to demonstrate specific mineral-solution equilibrium/dis-equilibrium conditions in groundwater systems is an approximation at its best.

3 PERMEABILITY AND MIXING

Most common rock types have anisotropic permeability. When this is the case the fluid discharged from a spring or a well is a mixture of fluids that have travelled different distances through the rock from different sources depending on the three dimensional distribution of permeabiilty in the formation. The different source areas within the bedrock may possess different temperatures and pressures. In such instances the composition of the discharged water does not really meet the conditions required to demonstrate local equilibrium/dis-equilibrium because such a demonstration requires the selection of a specific temperature and pressure. Further, when the bedrock composition is variable within the volume of rock from which a spring or drillhole discharge receives its water the conditions implicit to the definition of local equilibrium are not met.

4 PRESSURE AND TEMPERATURE DEPENDENCE OF MINERAL-SOLUTION EQUILIBRIA

Temperatures and pressures lie within about 0-350°C and 1-300 bars in most groundwater systems. Within these ranges of temperature and pressure the standard Gibbs energy of aqueous species and minerals vary very little with pressure compared to temperature. The same applies to mineral solubility constants. For this reason the effect of pressure on the composition of water at equilibrium with specific minerals is not important in most groundwater systems and can be ignored.

Hydraulic head and/or density difference between hot and cold groundwater sustain convection in geothermal systems. When permeability anomalies are associated with vertical or near-vertical fractures, such convection may cause temperature to be quite constant over large depth intervals, within 10-20°C over 2000 m in some low-temperature systems in Iceland where the undisturbed geothermal gradient is 60-100°C/km. Systems with near constant temperature within a large volume of rock provide the best opportunity to study the state of mineral-solution equilbria by calculation of saturation indices for minerals in the rock for water discharged at the surface from springs or drillholes. This is particularly so, if the chemical and mineralogical composition of the rock is the same. As pressure does not significantly affect the composition of the water at equilibrium, it does not matter from what depth the water is derived and reactions as the water ascends to emerge in springs are minimal, as long as temperature is constant.

5 PROGRESSIVE WATER-ROCK INTERACTION

Rainwater, which is low in dissolved solids and slightly acid, is undersaturated with most, if not all common minerals. This water, when it seeps into the soil and bedrock, will have a tendency to dissolve the rock minerals. In the process its dissolved solids content increases and as it does the water appraoches saturation with the minerals of the rock. These minerals may indeed dissolve until the water becomes saturated with them. However, if the rock is igneous or high-grade metamorphic, it is likely that the water will attain saturation with some other minerals that are less soluble than the primary ones before reaching saturation with the primary rock-forming minerals. Primary rock minerals that have formed in a high-energy environment, like minerals of igneous and high-grade metamorphic rocks, are likely to be more soluble than hydrothermal minerals which form in a lower energy environment.

If secondary minerals are less soluble than the primary ones, the water will never reach saturation with the latter. Removal of dissolved constituents by precipitation of secondary minerals from the water ensures that the water remains undersaturated with the primary minerals. As a result they will continue to dissolve and the secondary minerals will continue to precipitate in their place. Given enough time the rock may be completely transformed mineralogically. During this process some material may be added to the rock by the incoming water or extracted from it by the water leaving the system. However, for major components alteration by groundwater generally appears to be close to isochemical. Trace elements may, on the other hand, be added to, or subtracted from, the rock in such amounts during the alteration process that their concentrations in the altered rock differs very much from those in the fresh rock.

The overall water-rock interaction process in groundwater systems, just desribed, is a good example of an irreversible change. Yet, within parts of this system local equilibrium may be closely approached, at least for some components, i.e. partial local equilibrium.

6 RESULTS FOR SELECTED NATURAL WATERS

Iceland provides a unique opportunity to study mineral-solution equilibrium conditions in groundwater systems because of relatively homogeneous bedrock geology and a large number of drillholes that discharge water from ambient temperatures to more than 300°C. Additionally, many of the geothermal systems are associated with near vertical fractures and, due to groundwater convection, temperatures are relatively constant over a large depth interval. M a n y areas in Iceland are sparsely vegetated and, as a result, water seeping into the ground contains minimal dissolved CO_2 from organic sources. For that reason relatively limited reaction (mass transfer) is required with the reactive basaltic minerals to saturate the water with secondary minerals that subsequently precipitate from the water. Studies indicate that equilibrium is closely approached with these minerals at temperatures as low as 40°C (Arnórsson et al. 1983, Arnórsson & Andrésdóttir 1992) and in extreme cases at 10-20°C. As long as the minerals are precipitating, the water must be somewhat supersaturated. The reason for the close approach to equilibrium between solution and secondary minerals is probably governed by the formation of a relatively large surface area between these phases as the secondary minerals nucleate and form many small crystals.

Space does not permit detailed discussion of the state of saturation of natural waters with respect to specific primary and secondary minerals. Only a few examples will be mentioned here that relate to the observed temperature stability of some secondary minerals and problems encountered in deducing the state of equilibrium from calculation of saturation indices.

Groundwaters are generally close to being calcite √

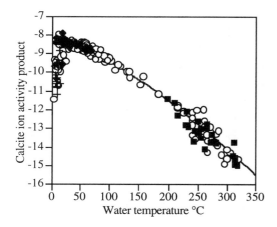

Figure 1. The state of calcite saturation in selected natural waters. The curve represents calcite solubility. Crosses indicate surface waters, circles and filled squares drillhole discharges from Iceland and various other countries, respectively, and filled diamonds spring discharges from Iceland.

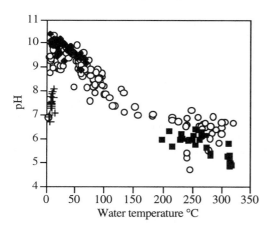

Figure 2. pH of selected natural waters. Symbols have the same signature as in Figure 1.

saturated whereas surface waters are undersaturated (Figure 1). Calcite is an abundant secondary mineral in bedrock at all temmperatures. Evolution towards equilibrium in groundwater systems, particularly in systems hosting rocks of basaltic composition, involves an increase in water pH (Figure 2) that causes an increase in the activity of the carbonate ion. Increase in the concentration of total carbonate carbon also contributes at high temperatures, as well as the retrograde solubility of calcite.

Analcime is a common secondary mineral in basalts at low temperatures (<50°C,), but it is also observed in some high-temperature geothermal systems. Indeed the selected natural water data indicate its stability at low and high temperatures, respectively, but at

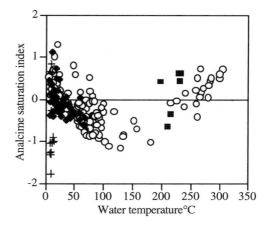

Figure 3. Saturation index for analcime in selected natural waters. Symbols have the same signature as in Figure 1.

intermediate temperatures (50-200°C) this mineral appears unstable (Figure 3). Decreasing supersaturation with increasing temperature in the range 0-50°C appears logical. Lower degree of supersaturation is required to precipitate this mineral from solution as temperature increases. The pattern at high temperatures, on the other hand, does not appear logical and further studies on the thermodynamic data of the solubility of this mineral is required as well as of the aqueous speciation calculations.

Epidote is a very abundant secondary mineral in hydrothermal systems with temperatures in excess of about 230°C and ranging in composition from basaltic to silicic. The composition of this mineral depends on the composition of the host rock. High temperature waters seem to be relatively close to equilibrium with the clinozoisite component of epidote (Figure 4), an observation that matches field data and indicates that high-temperature waters closely approach equilibrium with this mineral. However, some waters at lower temperatures also seem to be saturated although most are undersaturated (Figure 4). Observations of this kind are made for many minerals. The cause is probably small difference between the standard Gibbs energies of many mineral-solution reactions. The error involved in calculating saturation indices is apparently great compared to the actual difference in the standard Gibbs energies (equilibrium constants) for specific mineral-solution reactions.

7 DISCUSSION

Interaction between water and many common rock forming minerals may be viewed as a titration process where the minerals play the role of the base reacting with acids dissolved in the water. In surface waters the most important acid is dissolved CO_2 derived from the atmosphere and decaying organic matter.

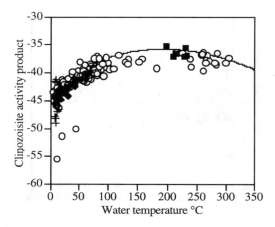

Figure 4. The state of clinozoisite (czo) saturation in selected natural waters. A czo activity of 0.3 is assumed which corresponds well with average epidote composition in geothermal systems hosted in basaltic rocks in Iceland. Symbols have the same signature as in Figure 1.

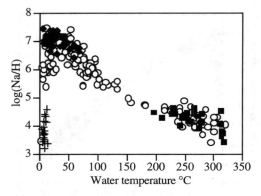

Figure 5. Sodium ion/proton activity ratios (Na/H) in selected natural waters. Symbols have the same signature as in Figure 1.

This acid is also the most important one in many volcanic geothermal systems. Other important acids include boric acid, hydrogen sulphide, aqueous silica and occasionally bisulphate. Progressive water-rock interaction towards equilibrium involves simultaneous increase in the aqueous concentration of cations from the base (minerals) and decrease in hydrogen ion concentrations. For this reason cation/proton activity ratios are a convenient measure of how far water in a particular geological environment has evolved towards equilibrium (Figure 5).

If acids are supplied to the water in relatively large quantities more reaction with the rock forming minerals is needed to bring the system to equilibrium than would be the case when acid supply is limited. If the minerals of the rock constitute strong bases, as e.g. olivine and pyroxene in basalt, it is to be expected that equilibrium is attained faster than in the absence of strong bases, other things being equal, such as temperature, surface area between the reacting phases etc.

Common rock types contain a small quantity of soluble salts (see e.g. Ellis and Mahon 1964, 1967). The amount of these salts in the rock is expected to affect how rapidly mineral-solution equilibria can be attained, at least, if the relative abundance of the cations released into solution as these salts dissolve differ from their relative abundance in solution at equilibrium.

All major components in ground-water systems in the basaltic terrain of Iceland, except for Cl, B, and sometimes SO_4, are observed to be controlled by equilibria with secondary minerals when temperatures are <40°C and in extreme cases at temperatures as low as 10-20°C. Isotopic data indicate that such ground-waters are thousands of years in age. The close approach to mineral-solution equilibria in the Icelandic groundwater systems at low temperatures, as compared to many other such systems in other parts of the world, is considered to be largely a consequence of the high reactivity of the mafic minerals of the basalt, its low content of soluble salts and generally limited supply of CO_2 to water from decaying organic soil due to sparse vegetational cover in many parts of Iceland. At high temperatures (>150°C), on the other hand, equilibrium seems generally to be closely approached in groundwater systems in the world for all major components except Cl, B, and sometimes SO_4, most probably because temperature controls reaction rates when it has reached so high values. High temperature and compositional gradients within groundwater systems may obliterate prevalence of local mineral-solution equilibria due to mixing of water of different temperatures derived from different rock types within the system.

REFERENCES

Arnórsson, S. & Andrésdóttir, A. 1995. Processes controlling the chemical composition of natural waters in the Heppar-Land area in southern Iceland. *International Atomic Energy Agency, IAEA TEC-DOC-788*, 21-43.

Arnórsson, S., Gunnlaugsson, E. & Svavarsson, H. 1983. The chemistry of geothermal waters in Iceland. II. Mineral equilibria and independent variables controlling water compositions. *Geochim. Cosmochim. Acta* 47: 547-566.

Ellis, A. J. & Mahon, W. A. J. 1964. Natural systems and experimental hot water/rock interactions. *Geochim. Cosmochim. Acta* 28: 1323-1357.

Ellis, A. J. & Mahon, W. A. J. 1967. Natural hydrothermal systems and experimental hot water/rock interactions. Part II. *Geochim. Cosmochim. Acta* 31: 519-538.

Giggenbach, W. F. 1981. Geothermal mineral equilibria. *Geochim. Cosmochim. Acta* 45: 393-410.

Nordstrom, D. K. & Munoz, J. L. (1994) *Geochemical thermodynamics*. Cambridge MA, Blackwell Scientific Publications: 493 p.

Geochemistry of the Earth's Surface, Ármannsson (ed.) © 1999 Balkema, Rotterdam, ISBN 90 5809 073 6

Invited lecture: Geochemistry of catastrophic events

G. E. Sigvaldason
Nordic Volcanological Institute, Reykjavík, Iceland

ABSTRACT: Catastrophic events in nature are usually associated with slow accumulation of material and/or energy, followed by a rapid surge between different material reservoirs. If the surge exceeds the buffering capacity of the receiving reservoir adverse environmental conditions may result. Volcanism is a prime example, where accumulated matter and energy escapes suddenly from a crustal or mantle reservoir to be added to the surface reservoirs. Subsurface processes may take long time to reach maturity, the final breakthrough, however, is generally short, from a few minutes to a few years. Selected examples of volcanic events that had serious, sometimes catastrophic, concequences will be addressed.

1 THE LAKI ERUPTION IN ICELAND

The plate tectonic position of Iceland is expressed in its location above a strong mantle plume that is transected by a mid-ocean ridge. Vigorous volcanic activity associated with the North Atlantic mantle plume has formed submarine, aseismic ridges between Greenland and Scotland, with the subaerial Icelandic land mass marking the present position of the mantle plume. The subaerial part of Iceland spans a period of 16 million years, while the submarine plume trails to the east and west record the time since the opening of the North Atlantic ocean some 60 million years ago.

Since the settlement of the country in the 9th century, volcanic activity has adversely affected the population by flowing lava and volcanic ash (tephra) from explosive eruptions. Since the settlement the local effects of these eruptions have been serious, often amplified by deteriorating climate and a series of plagues. One volcano in particular, Hekla in South Iceland, is noteworthy for its fluorine pollution that invariably kills a large number of grazing animals, even if its tephra cover is less than a millimeter in thickness. Due to sparse population and uneven distribution of settlement within the country most eruptions occurr outside populated areas. The high frequency of eruptions, one every fifth year on the average, is therefore less threatening than it might seem. Most of these eruptions are effusive, producing volumes of the order of 0.01 to 1 km^3, but rarely up to 2 km^3. A few explosive eruptions producing silicic tephra have occurred since the settlement, and some basaltic

eruptions occurring beneath water or ice have distributed substantial tephra layers over extended areas.

In the past 1120 years two eruptions stand out as major events in the volcanic history of the country. Both occurred within the same general area in South Iceland but with origin in two separate volcanic systems. Each of these eruptions produced about 14-15 km^3 of lava and both have left significant acid layers in the Greenland ice-core. The first of these eruptions, the Eldgjá eruption, occurred in the year 934 and little is known about its consequences. A detailed contemporary account of the second eruption, with descriptions of the consequences of the volcanic pollution for the human population, is available in english translation (Steingrímsson 1784). This eruption, known as the 1783 Laki eruption, was an effusive lava producing event that lasted for 8 months. With a total production of 15.1 km^3 it represents the highest emission of lava on Earth in historic times. The catastrophic effects of the eruption have initiated quantitative studies on the amount of volcanic volatiles released to the atmosphere. Volatile degassing of erupting lava can be divided into steps based on the solubility of individual volatile components in the liquid lava. During ascent water and carbon dioxide are the first components to reach saturation followed by sulphur components, while hydrocloric- and hydrofluoric acids are the most soluble volatiles. Crystalline phases, formed in the magmatic liquid before ascent, often contain glass inclusions that will show the initial volatile content of the liquid. The difference between the volatile content of these glass inclusions and the volatiles remaining in different volcanic products provides an estimate of the total amount of volatiles escaping to the atmosphere.

The total amount of volatiles released to the atmosphere by the Laki eruption was 235 Megatons H_2O, 122 Mt SO_2, 349 Mt CO_2, 6.8 Mt HCl and 15.1 Mt HF respectively. During the first 40 days the eruption produced about 60% of the total erupted volume. At this eruption rate about 1.7 Mt of SO_2 were released to the atmosphere per day. With decreasing eruption rate the daily release dropped to 0.6 Mt SO_2 in the following 60 days and to 0.1 Mt SO_2 during the waning stages (Thordarson et al. 1996). Introduction of sulphuric acid aerosols into the stratosphere absorbed solar radiation, causing cooling at the Earth´s surface. The climatic effect of the Laki eruption was felt over large parts of the northern hemisphere for at least one year. This is expressed in more than 4°C drop in temperature below the 225 year average in the eastern United States (Sigurdsson 1982) and severe crop failures occurred in Central Europe. In Iceland the pollution by the eruption caused the death of a large percentage of the live stock and that in turn resulted in famine, wich along with direct poisionous effects of the pollution, killed 20% of the human population of the country.

2 THE LAKE NYOS EVENT

While the Laki eruption in Iceland is an event where very large amounts of aggressive volatiles exceed the buffering capacity of the atmosphere on a local and regional scale, we also find examples of volatile escape that, given very special conditions, can lead to events of catastrophic dimensions. The surface expression of deep seated alkaline volcanism is frequently in the form of carbon dioxide springs, at or just above ambient temperature. With progressive crystallisation of solidifying magmatic bodies at depth the volatile components will be expelled and the more aggressive components are consumed in reactions with wall rocks. Carbon dioxide is the only component that survives the passage through the Earth´s crust to appear in carbonated springs at the surface.

The 1600 km long fracture going through Cameroun and Nigeria in East Africa is marked by a chain of alkaline volcanoes. With the exception of Mt Cameroun on the Atlantic coast the fracture has no recorded historic eruptions, but is seismically active. The fracture is lined with a number of small edifices, larger lava flows and a number of maars or tuff cones formed by explosive activity (Barberi et al. 1989).

On August 21, 1986 an enormous carbon dioxide gas cloud arose from Lake Nyos in Northwestern Cameroun. It spread in the surroundng valleys and killed more than 1700 people in the nearby villages (Sigvaldason 1992). Lake Nyos lies at an elevation of 1100 meters and has a surface area of 1.4 km^2 and a volume of 0.17 km^3. It is a maar crater with a central 220 meter deep funnel shaped hollow. Isotopic analysis indicate that the carbon dioxide escaping from the lake is of magmatic origin (Kusakabe et al. 1989). Thermal and chemical stratification in the lake caused accumulation of cabonate from the influx of carbonated ground water. Some unknown triggering mechanism caused partial inversion of the lake water and the release of carbon dioxide that spread downhill with catastrophic consequences (Kling et al. 1989). The distribution of animal and human casualties shows that a lethal concentration of gas reached a height of 120 m above the lake surface and spread subsequently to a distance of 25 km from the lake, following the downhill topography. A 10-20% mixture of carbon dioxide with air is potentially lethal. The gas cloud responsible for death and injury is estimated to have been about 1 km^3, and the amount of pure carbon dioxide released from the lake was about 0.15 km^3.

3 VOLCANISM AND CLIMATE

The above examples of sudden escape of accumulated magmatic volatiles from the Earth´s interior can be compared to other, more dramatic events in recent Earth history. The climatic record obtained from drilling through the Greenland ice shield shows that the past 10,000 years are the longest warm period in the past 250,000 years. This period follows a 100,000 year glaciation when water from the world oceans accumulated to form thick ice masses at higher latitudes. In Iceland, for example, the ice thickness may have reached maximally 2000 meters in the central part of the country (Sigmundsson 1991). This mass of ice is equivalent to that of about 500 meters thick solid rock and causes isostatic depression of the crust. The pressure increase appears to affect the melting behaviour in the mantle towards lower degrees of partial melting and reduced volcanic productivity. (Jull & McKenzie 1996).

The change in climate from glaciation to present day condition occurred within a relatively short time of 1000-2000 years (Dansgaard et al. 1993). The sudden pressure release correlates with very high volcanic production rate that is expressed both in tick ash layers in oceanic sediment cores in the North Atlantic (Lacasse et al. 1995) and an estimated 30-40 times higher volcanic production rate in Iceland than at present (Sigvaldason et al. 1992). If the present rate of volcanic production in Iceland is taken as approximately 0.05 km^3 per year (Thorarinsson 1965; Jakobsson 1972) an amount of up to 2 km^3 per year might have been produced at the Pleistocene-Holocene boundary, with corresponding increase in volatile release to the atmosphere.

REFERENCES

Barberi, F., W. Chelini, G. Marinelli & M. Martini 1989. The gas cloud of Lake Nyos (Cameroun, 1986): Results of the Italian technical mission. *J. Volcanol. and Geothermal Res.* 39:125-134.

Dansgaard, W., S.J. Johnsen, A.B. Clausen, D. Dahl-Jensen, N.S. Gundestrup, C.U. Hammer, C.S. Avidberg, J.P.Steffensen & A.E. Sveinbjörnsdóttir 1993. Evidence for general instability of past climate from a 250-kyr ice-core record. *Nature* 364:218-220.

Jakobsson, S. P. 1972. Chemistry and distribution pattern of Recent basaltic rocks in Iceland. Lithos 5:365-386.

Jull, M. & D.McKenzie 1996. The effect of deglaciation on mantle melting beneath Iceland. *J. Geophys. Res.* 101:21,815-21,828.

Kling, G.W., M.L. Tuttle & W.G. Evans 1989. The evlution of thermal structure and water cemistry in Lake Nyos. *J. Volcanol. Geothermal Res.* 39:151-165.

Kusakabe, M., T. Ohsumi & S. Aramaki 1989. The Lake Nyos gas disaster: chemical and isotopic evidence in waters and dissolved gases from three Cameroonian crater lakes, Nyos, Monoun and Wum. *J. Volcanol. and Geothermal Res.* 39:167-185.

Lacasse, C., H. Sigurdsson, H. Jóhannesson, M. Paterne & S. Carey 1995. The source of Ash Zone 1 in the North Atlantic. *Bull. Volc.* 57:18-32.

Sigmundsson, F. 1991. Post-glacial rebound and astenosphere viscosity in Iceland. *Geophys. Res. Letters* 18:1131-1134.

Sigurdsson, H. 1982. Volcanic pollution and climate: the 1783 Laki eruption. *EOS* 63:601-602.

Sigvaldason, G.E. 1989. International conference on Lake Nyos disaster, Yaoundé, Cameroun 16-20 March, 1987: Conclusions and recommendations. *J. Volcanol. Geothermal Res.* 39:97-107.

Sigvaldason, G.E., K. Annertz & M. Nilsson 1992. Effect of glacier loading/deloading on volcanism: postglacial volcanic production rate of the Dyngjufjöll area, central Iceland. *Bull. Volcanol.* 54:385-392.

Steingrímsson, J. 1784. *Fullkomið rit um Síðueld.* English translation: *Fires of the Earth, The Laki eruption 1783-1784* by K. Kunz. Univ. of Iceland Press, 1998.

Thorarinsson, S. 1965. The Median Zone of Iceland. *In The World Rift System. Geol.Survey of Canada, Paper 66-14:187-211.*

Thordarson, Th., S. Self, N. Óskarsson & T. Hulsebosch 1996. Sulphur, chlorine, and fluorine degassing and atmospheric loading by the 1783-1784 AD Laki (Skaftár Fires) eruption in Iceland. *Bull Volcanol.* 58:205-225.

Geochemistry of the Earth's Surface, Ármannsson (ed.) © 1999 Balkema, Rotterdam, ISBN 90 5809 073 6

Statistical interpretation and modeling of waters in Snæfellsnes, W-Iceland

E. Bedbur, U. Wollschläger & M. Petersen
Institute of Geoscience, University of Kiel, Germany

ABSTRACT: This paper presents the results of geostatistical analysis and hydrochemical inverse modeling of 152 water samples from Snæfellsnes Peninsula (W-Iceland). For each sample 21 parameters were measured. This data set could only be evaluated in its complexity by using advanced statistical and geochemical modeling methods. The factor analysis showed the main factors determining the water composition. The cluster analysis formed groups relating to the origin of the sample. The inverse modeling with the computer code PHREEQC allowed a comparison of two main flow systems. Water-rock interaction was quantified as mass transfer of minerals dissolved or precipitated.

1 INTRODUCTION

Groundwater on the western part of the Snæfellsnes Peninsula (W-Iceland) has its origin in rain water and melt water from the glacier of the volcano Snæfellsj kull. The main aquifers are built by tertiary and quaternary basaltic lavas which show high hydraulic conductivities (k_f) of 10^0-10^{-3} m/s. Less permeable aquifers are built by quarternary subglacially formed hyaloclastites (k_f=10^{-2}-10^{-6} m/s) (Sigurðsson 1990). In July and August 1997 152 water samples from the different aquifers were taken. During field sampling the parameters pH, redox potential (E_H), electric conductivity, temperature, and O_2-content were measured. In the laboratory the contents of Cl^-, SO^-, NO, F^-, Br^-, HCO, Na^+, K^+, Mg^{2+}, Ca^{2+}, Al^{3+}, Si_{tot}, Mn_{tot}, Fe_{tot}, Sr^{2+}, and B_{tot} were determined. Because of short flow paths in silicic rocks at the steep flanks of the volcano only low water-rock interactions could be expected. For most samples TDS was less than 100 mg/l. This set of data for all sample points was analysed with statistical methods as factor and cluster analysis. A forward and an inverse geochemical modeling with the computer code PHREEQC (Parkhurst 1995) allowed the calculation of mass transfers of water-rock interactions within the aquifer, taking into account the analytical error.

2 GEOSTATISTICS

A factor and cluster analysis including the whole data set for each sampling point was carried out (Petersen, unpubl.).

2.1 Factor analysis

The factor analysis yields two main factors. Factor 1 shows high loading (> 0.5) of the parameters el. conductivity, Cl^-, SO^-, Na^+, K^+, Mg^{2+}, Ca^{2+}, Sr^{2+}.

This factor gives the marine influence on the groundwater chemistry. Factor 2 has high loading (> 0.5) in the parameters el. conductivity, HCO, K^+, Mg^{2+}, Ca^{2+}, Si_{tot}, Sr^{2+} reflecting the process of silicate weathering (Table 1).

2.2 Cluster analysis

The cluster analysis allows to differentiate clearly the aquifer material by building two clusters with water samples taken from basaltic and from hyaloclastitic rocks (Figure 1). Three more clusters were built grouping samples with low TDS (melt and rain water), samples with high TDS (marine intrusions and cold carbon dioxide springs) and samples taken near the coast or near CO_2-rich springs.

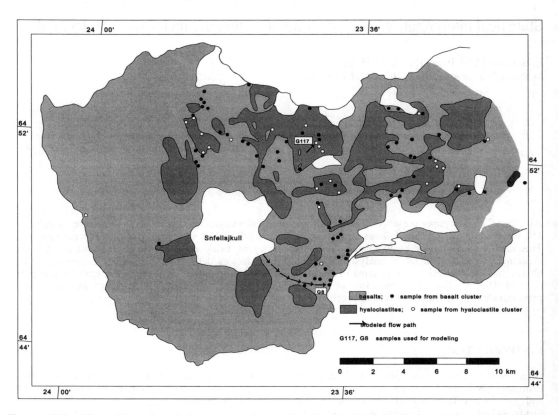

Figure 1. W-Snæfellsnes: Schematic geological map. Sampling locations, basalt and hyaloclastite cluster. Arrows: modeled flow paths

Table 1. Loading of factors (Petersen, unpubl.)

Parameter	Factor 1	Factor 2
temperature	0.24	0.08
electric conductivity	0.78	0.56
pH	0.19	0.20
HCO	0.33	0.91
Cl⁻	0.90	0.17
SO⁻	0.92	0.20
NO	0.41	0.43
Na⁺	0.86	0.46
K⁺	0.79	0.52
Mg²⁺	0.70	0.63
Ca²⁺	0.57	0.73
Al³⁺	-0.09	-0.02
Si_tot	0.18	0.65
Sr²⁺	0.70	0.60

Samples from hyaloclastitic aquifers have a higher pH (8.2-9.3) than samples from the basaltic aquifers (pH 6.4-8.3). These relative high pH values reflect their hydrochemical evolution in a closed system (Gíslason & Eugster 1987a). Basalt and hyaloclastite have the same geochemical composition, but hyaloclastite is rich in basaltic glasses in contrast to crystalline basalt (Gíslason & Eugster 1987a). This results in a higher solubility of the hyaloclastite which is more than ten times higher than the solubility of the crystalline basalt (Gíslason 1985, Gíslason & Arnórsson 1993). The higher solubility results in a higher proton transfer rate which leads to an increase in pH of the solution in a closed aquifer system. The higher solubility should lead to a higher TDS in the water samples taken from hyaloclastites which was not observed in the present study.

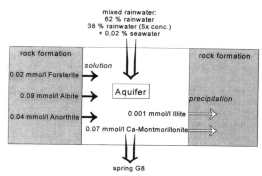

Figure 2. Mass transfers along a flow path in a basaltic aquifer.

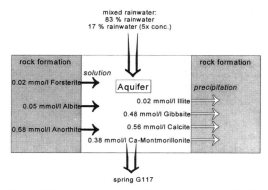

Figure 3. Mass transfers along a flow path in a hyaloclastitic aquifer.

3 THERMODYNAMICAL MODELING

A geochemical modeling with the computer code PHREEQC (Parkurst 1995) gave saturation indices (SI) for the different samples. Also an inverse modeling of selected flow paths in basaltic and hyaloclastitic aquifers (Figure 1) with PHREEQC was carried out (Wollschläger, unpubl.). The results of the modeling are consistent with the results of the cluster analysis. The mineral phases used in the inverse modeling were chosen by data from literature (Bistry 1986, Jakobsson 1972, Gíslason & Eugster 1987b, Gíslason et al. 1993 and Nesbitt & Young 1984) (Table 2). The mass transfer was calculated by using only the secondary phases which have shown supersaturation in the forward modeling.

Depending on the flow path either a melt water or a rain water was used as the initial solution. There is a high correlation of the marine influence on the water with the topographic height (marine born aerosols). This was taken into account by giving the model either a concentrated rain water and/or an input of sea water which could be mixed with the initial water to produce a starting water for the inverse model.

3.1 Flow path in a basaltic aquifer

The sample from the spring G8 (Figures 1, 2) representing water from the basalt cluster was inversely modeled with in input water containing 62% melt water, 38% rainwater (5x concentrated) and 0.02% seawater. In the model the primary minerals forsterite, diopside, albite, anorthite, adularia, and the secondary phases kaolinite, gibbsite, illite and Ca-montmorillonite were used. During the underground passage 0.02 mmol/l

Table 2. Mineral phases used for the inverse modeling

Mineral	Composition
Forsterite	Mg_2SiO_4
Diopside	$CaMgSi_2O_6$
Albite	$NaAlSi_3O_8$
Anorthite	$CaAl_2Si_2O_8$
Adularia	$KAlSi_3O_8$
Illite	$K_{0.6}Mg_{0.25}Al_{2.3}Si_{3.5}O_{10}(OH)_2$
Kaolinite	$Al_2Si_2O_5(OH)_4$
Ca-Montmorillonite	$Ca_{0.16}Al_{2.33}Si_{3.67}O_{10}(OH)_2$
Gibbsite	$Al(OH)_3$
Calcite	$CaCO_3$
Laumontite	$CaAl_2Si_4O_{12} \; 4H_2O$

forsterite, 0.09 mmol/l albite and 0.04 mmol/l anorthite were dissolved, and 0.001 mmol/l illite and 0.07 mmol/l Ca-montmorillonite precipitated (Figure 2).

3.2 Flow path in a hyaloclastitic aquifer

The sample from the spring G117 (Figure 1, 3) representing water from the hyaloclastite cluster was inversely modeled with in input water containing 83% rainwater, and 17% rainwater (5x concentrated).

The selected sample is a typical water from a hyaloclastitic aquifer. It has a low bicarbonate content of 15 mg/l. This is also a typical bicarbonate content of basaltic aquifers, but the pH in the hyaloclastitic aquifer is much higher (pH 8.9). The partial pressure of CO_2 for this sample was calculated to $10^{-4.8}$ bar, which indicates that the hyaloclastitic aquifer system is a closed system. The saturation index for calcite is close to saturation equilibrium. The observed relatively low bicarbonate content is assumed to be due to the precipitation of carbonate. In this model 0.02 mmol/l forsterite, 0.05 mmol/l albite and 0.68 mmol/l anorthite dissolve and 0.02 mmol/l illite, 0.48 mmol/l gibbsite, 0.38 mmol/l Ca-montmorillonite, 0.56 mmol/l calcite precipitate (Figure 3).

4 CONCLUSIONS

Geostatistical methods like factor and cluster analysis were able to differentiate the main factors which determine the water composition of the waters from Snæfellsnes Peninsula (W-Iceland) and were able to separate clusters of samples. Two clusters could be identified as samples from basaltic and from hyaloclastitic aquifers respectively. Hydrochemical inverse modeling allowed a quantification of the mass transfers within the aquifer. The mass transfer in the hyaloclastitic aquifer is higher than in a crystalline basaltic aquifer. Dissolution of anorthite is in the hyaloclastite three times higher than in crystalline basalt. The precipitation of Ca-montmorillonite is 6 times higher. It is shown that the hyaloclastitic ground water system is most possibly a closed system, where carbonate is precipitating. Although the mass transfer is high the TDS is low, because of high dissolution and high precipitation of minerals in the hyaloclastites. H^+ is consumed for precipitation of secondary mineral phases. This leads to high pH-values, although the TDS is similar to samples from basaltic aquifers.

REFERENCES

Bistry, T. 1986. Nat rlicher und anthropogener Stoffeintrag in das Grundwasser der vulkanischen Ozeaninsel La Palma. *Ber.-Rep. Geol.Pal.Inst. Kiel* 85:1-172.

Gíslason S.R. 1985. Meteoric water-basalt interactions. A field and laboratory study. Ph.D.Thesis. John Hopkins Univ., 238p.

Gíslason S.R. & S. Arnórsson 1993. Dissolution of primary basaltic minerals in natural waters: saturation state and kinetics. *Chem. Geology* 105:117-135.

Gíslason S.R. & H.P. Eugster 1987a. Meteoric water-basalt-interactions. I: A laboratory study. *Geochim. Cosmochim. Acta* 51:2827-2840.

Gíslason S.R. & H.P. Eugster 1987b. Meteoric waterbasalt interactions II: A field study in N.E. Iceland. *Geochim. Cosmochim. Acta* 51:2841-2855.

Gíslason, S.R., D.R. Veblen & K.J.T. Livi 1993. Experimental meteoric water-basalt interactions: Characterization and interpretation of alteration products. *Geochim. Cosmochim. Acta* 57:14591471.

Jakobsson, S.P. 1972. Chemistry and distribution pattern of recent basaltic rocks in Iceland. *Lithos* 5:365-386.

Nesbitt, H.W. & G.M. Young 1984. Prediction of some weathering trends of plutonic and volcanic rocks based on thermodynamic and kinetic considerations. *Geochim. Cosmochim. Acta* 48: 1523-1534.

Parkhurst, D.L. 1995. User's guide to PHREEQC - a computer program for speciation, reaction-path, advective-transport, and inverse geochemical calculations. *Water-Resources Invest. Rep.* 95-4227.

Petersen, M., unpubl. Regionalstatistische Untersuchungen der Grundwasserbeschaffenheit auf der westlichen Halbinsel Snæfellsnes (W-Island). Dipl.-Arb. Univ. Kiel 1998.

Sigurðsson, F. 1990. Iceland. In United Nations (eds.), Groundwater in Eastern and Northern Europe. *Nat. Resourc. / Wat. Ser.* 24:123-137.

Wollschläger, U., unpubl. Thermodynamische Modellierung regionaler Einflüsse auf die Grundwasserbeschaffenheit des Snæfellsjökull-Gebietes (W-Island). Dipl.-Arb. Univ. Kiel 1998.

Geochemistry of the Earth's Surface, Ármannsson (ed.) © 1999 Balkema, Rotterdam, ISBN 90 5809 073 6

Multi-isotope geochemistry of the Palmottu hydrosystem, Finland

J. Casanova & Ph. Négrel
BRGM, Department of Geochemistry and Hydrogeology, Orléans, France

S. K. Frape
Department of Earth Sciences, University of Waterloo, Ont., Canada

J. Kaija & R. Blomqvist
GTK, Nuclear Waste Disposal Research, Espoo, Finland

ABSTRACT: The combination of individual isotopic tools in groundwater studies helps to restrain hypotheses on the nature of water-rock interaction and the end-members participating in mixing processes. A toolbox containing strontium, boron, oxygen and chlorine isotopes has been used in deep groundwaters at the Palmottu (Finland) site. This paper presents the results obtained from this complex hydrosystem in order to elucidate the origin of the different types of water and the most likely processes involved in time and space.

1 INTRODUCTION

The Palmottu U-ore deposit is located in a granitic host rock in southern Finland (Figure 1). It provides an excellent framework for analogue studies to assess radionuclide transport from the U-ore deposit along well-defined pathways in the fractured crystalline rock (Blomqvist et al. 1995).

In this context, multi-isotope systematics, including strontium, boron, oxygen and chlorine isotopes, are used to trace the degree of water-rock interaction (WRI) and mixing processes in groundwaters. This study should contribute to understanding the recent hydrogeological and hydrogeochemical conditions of the site, including the design of a conceptual groundwater flow model.

Multi-isotope analyses have been performed on groundwater from the Palmottu hydrosystem and, in particular, from the deep geochemical boreholes R385 and R386 (Blomqvist et al. 1998). The 47 water samples analyzed include surface waters, springs, groundwaters in the overburden, waters from different former boreholes (packer and tube sampling) and the waters sampled during drilling of the R385 and R386 boreholes. The chemical analysis was performed by inductively coupled plasma mass spectrometry for elemental concentrations and by mass spectrometry for isotopic determinations.

Different groundwater types characterize the Palmottu site (Blomqvist et al. 1998). The groundwater from shallow overburden, the Upper Flow System and the Dynamic Deep Flow System are dilute Ca-Na-HCO$_3$-Cl water types. In the intermediate Stagnant Flow System the groundwaters are Na-SO$_4$-Cl types and in the deeper Stagnant Flow System they are of the Na-Cl type.

2 RESULTS AND DISCUSSION

2.1 Strontium isotopic constraints

The Sr isotopes show no detectable fractionation by any natural process. Given the relatively short time scale of the processes studied, the measured differences in the ^{87}Sr/^{86}Sr ratios are due to the contribution of Sr derived from various sources with different isotopic compositions. The ^{87}Sr/^{86}Sr ratio variations within an hydrosystem can provide information about those sources, the mixing proportions of groundwater components and the degree of water rock interaction (McNutt et al. 1990).

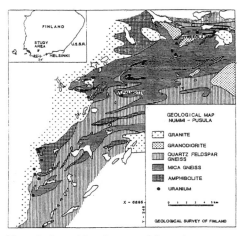

Figure 1. General location map of the study area.

Rb and Sr concentrations in the Palmottu surface waters are in the range of 1.1-2.1 ppb and 16.4-58.1 ppb respectively, whereas in the groundwaters, the ranges are 1.1-11.3 ppb and 14.5-1080 ppb. No direct relationship is observed between the Rb and Sr contents and the samples are scattered between several fields. The $^{87}Sr/^{86}Sr$ ratios range between 0.716910 and 0.735606 in the surface waters and between 0.719991 and 0.750787 in the groundwaters, but are between 0.720 and 0.735 in most of the samples.

The hydrochemical functioning of the Palmottu hydrosystem is very complex and the results show a lack of correlation between the water chemistries determining the classification into different water types (Na-Cl, Na-SO$_4$, etc.) and the results of the strontium contents and Sr isotopic ratios. From a WRI standpoint, this implies that the Sr behavior is independent of the water chemistry; the occurrence of large $^{87}Sr/^{86}Sr$ variations is site specific and mainly dependent on the lithology.

Figure 2 shows the plot of the $^{87}Sr/^{86}Sr$ vs. Ca/Na ratios. The scattering of the data can be explained by the presence of four end-members: a brine component with low $^{87}Sr/^{86}Sr$ and Ca/Na ratios (less than 0.03), a deep granitic component with high $^{87}Sr/^{86}Sr$ ratios and low Ca/Na ratios (close to 0.2-0.6), a subsurface component, from 0 to 50 m depth, with intermediate $^{87}Sr/^{86}Sr$ ratios (around 0.730) associated with high Ca/Na ratios (up to 3.5), and a surface end-member (snow and river drainage) with low $^{87}Sr/^{86}Sr$ and low Ca/Na ratios (close to 1). These extreme end-members define a series of WRI mixing line within a rather complex hydrosystem.

2.2 Boron isotopic constraints

Due to the large relative mass difference between ^{10}B and ^{11}B and the high chemical reactivity of boron, significant isotope fractionation produces large variations in the $^{11}B/^{10}B$ ratios in natural samples from different geological environments. This results in high isotopic contrasts of potential mixing sources, and also in process-specific changes of the isotope signature.

The B concentrations are between 251 and 657 ppb with $\delta^{11}B$ values ranging from 33.6 to 43.5‰. Plotting the samples from different water types on a $\delta^{11}B$ vs. the reverse of the boron content diagram (Figure 3) shows a scatter along a negative trend from high $\delta^{11}B$ values and B contents to low $\delta^{11}B$ values and B contents. Four samples exhibit a $\delta^{11}B$ higher than that of present mean seawater (MSW, Barth, 1993) and of the Baltic Sea.

From the snow to the shallow groundwaters in the overburden, the $\delta^{11}B$ remains constant whereas the boron content increases. This trend suggests a ternary mixing involving shallow groundwaters with low $\delta^{11}B$, brines with $\delta^{11}B$ higher than MSW, and a marine component with $\delta^{11}B$ close to MSW and the Baltic Sea.

This mixing model can also be observed when comparing $\delta^{11}B$ to $^{87}Sr/^{86}Sr$. None of the samples have a $^{87}Sr/^{86}Sr$ ratio close to the Baltic Sea (Andersson et

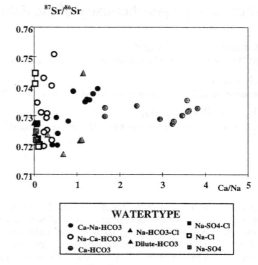

Figure 2. Plot of $^{87}Sr/^{86}Sr$ vs. Ca/Na ratios. The water types are also indicated.

al. 1994), but one sample exhibits $^{87}Sr/^{86}Sr$ close to that of brackish water from Olikuoto. The scattering of the points is in agreement with a family of mixing hyperbola, which characterizes a set of two-component mixing processes. The extreme end-member (Na-SO$_4$-Cl water type), showing a ^{11}B enrichment with respect to MSW, may correspond to freezing processes that occurred during the glacial history of the site (Blomqvist et al. 1998).

2.3 Oxygen isotopic constraints

The groundwaters have meteoric signatures, as evidenced by the relation ^2H vs. ^{18}O (Blomqvist et al. 1998). The Upper Flow System (surface waters, springs, groundwaters in the overburden) has a $\delta^{18}O$ around -11‰ whereas the intermediate and deep saline waters show extreme $\delta^{18}O$ from -16 to -17‰.

Most of the stable isotope data (^2H, ^{18}O) cluster around the groundwaters from the overburden, indicating recent recharge conditions. The depleted $\delta^{18}O$ values (down to -18‰ SMOW) represent older groundwaters of the stagnant Na-SO$_4$ and Na-Cl types. These waters reflect a colder climate recharge which most likely took place during the retreat of the Weichselian ice sheet.

When plotted against $\delta^{11}B$, the $\delta^{18}O$ values show a positive trend (Figure 4). The deep brines characterized by the most negative $\delta^{18}O$ values also show the highest $\delta^{11}B$ values measured in the system. At the other extreme, the lowest $\delta^{11}B$ values observed in the shallow groundwaters correspond to the least ^{18}O-depleted waters.

When the $\delta^{18}O$ values are plotted against the $^{87}Sr/^{86}Sr$ ratios, two opposing trends can be observed. The first, with a roughly constant $^{87}Sr/^{86}Sr$ ratio around

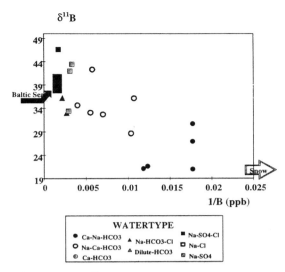

Figure 3. Plot of $\delta^{11}B$ vs. the reverse of B content

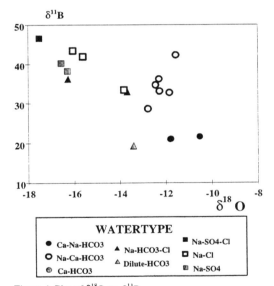

Figure 4. Plot of $\delta^{18}O$ vs. $\delta^{11}B$

0.720, shows large $\delta^{18}O$ variations from around -8 to -18‰. The second, with a slightly variant $\delta^{18}O$, shows a large increase of the $^{87}Sr/^{86}Sr$ ratio (up to 0.75).

2.4 Chlorine isotopic constraints

The stable isotopes of chlorine ($^{37}Cl/^{35}Cl$) are most often used to define different chloride sources. In the Palmottu system the range of signatures have an approximately 1.1 ‰. In plotting the variation of

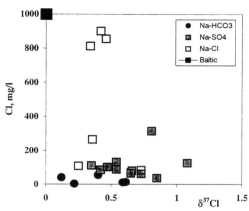

Figure 5. Plot of $\delta^{37}Cl$ vs. the chlorine content

isotopic signature vs. chloride concentration for the major water types (Figure 5), it is seen that the deepest most concentrated Na-Cl waters have a significantly more positive signature (+0.4‰) than Baltic Sea water (0.0‰).

The most positive signatures (+1.1‰) are associated with chloride found in a thin lens of Na-SO$_4$ water. The origin of the Na-SO$_4$ fluid is still in debate, although the chlorine isotopic signature is similar to rock-derived chlorine from other sites in Fennoscandia.

A third end-member is found in the shallow bicarbonate waters and some of the sulphate waters. In some cases the Cl isotopic values in these waters approach the Baltic Sea signature, possibly indicating a dilute component of Baltic water from marine aerosols. The intermediate and other samples appear to be mixtures of the three main chlorine-bearing waters.

3 CONCLUSION

As a working hypothesis, we can consider the local hydrological evolution as follows:

(1) the oldest and deepest groundwater in the Palmottu area was of Na-Cl type and originated from a glacial meltwater.

(2) after this deep recharge, the WRI processes between the granites and the meteoric input generated as many groundwaters as the different types of crystalline rock in the Scandinavian Shield. These shallow granitic groundwaters defined a single mixing line between the snow and the more radiogenic granite.

(3) the exact depth of the boundary between the glacial meltwater and the groundwaters already influenced by WRI processes is unknown, but it was above 220 m during Late Glacial times. At the very beginning of the melting of the local Palmottu glacier, a certain amount of melt water was hydrologically forced into the deep bedrock by means of fractures

(intersected in R385 at a depth of 220 m).

(4) the younger and shallower groundwaters in the Palmottu area above 200 m result from different mixing lines between the glacial meltwater and the shallow granitic groundwaters, including the surface waters.

REFERENCES

Andersson , P.S., G.J. Wasserburg & J. Ingri 1994. The sources and transport of Sr and Nd isotopes in the Baltic Sea. *Earth and Planetary Sci. Letters* 113: 459-472.

Barth, S 1993. Boron isotope variations in nature: a synthesis. Geol. Rundsch 82: 640-641.

Blomqvist, R. & the project team. 1995. The Palmottu natural analogue project. Summary report 1992-1994. *Geological Survey of Finland, report YST-88: 82p.*

Blomqvist, R. & the project team. 1998. The Palmottu natural analogue project. Phase I: Hydrogeological evaluation of the site. *European Commission, EUR18202: 96p.*

McNutt, R.H., S.K. Frape, P. Fritz, M.G. Jones & I.M. MacDonald 1990. The $^{87}Sr/^{86}Sr$ values of the Canadian Shielsd brines and fractures minerals with application to groundwater mixing, fracture history and geochronology. *Geochim. Cosmochim. Acta* 54: 205-215.

Geochemistry of the Earth's Surface, Ármannsson (ed.) © 1999 Balkema, Rotterdam, ISBN 90 5809 073 6

Geochemistry of waters of the Paratunka geothermal area, Kamchatka

O. V. Chudaev – *Far East Geological Inst., Far Eastern Branch of Russian Acad. of Sciences, Vladivostok, Russia*

V. A. Chudaeva – *Pacific Inst. of Geography, Far Eastern Branch of Russ. Acad. of Sciences, Vladivostok, Russia*

W. M. Edmunds & P. Shand – *British Geological Survey, Wallingford, UK*

V. G. Okhapkin – *Hydrogeological Expedition, Elizovo, Kamchatka, Russia*

ABSTRACT: The chemical composition of mineral waters was studied in the Paratunka basin, one of the largest reservoirs of thermal waters in Russia. The main factor controlling chemical composition of thermal waters is water-rock interaction. It seems, Cl, Br, S, I, were added to groundwater from exhalations, which are typical for the volcanic area of Kamchatka. CO_2 as well as He are most likely to be of mantle origin.

1 INTRODUCTION

The Paratunka thermal basin (about 690 km^2) is located in the SE part of Kamchatka peninsula. The basin (depression) is a graben with NE orientation consisting of three tectonic blocks dipping to the north. The Paratunka River flows in the same direction. In the lower part of the Paratunka River the northern block is less deep as compared with the others in the middle and southern (upper) parts of the Paratunka River.

The graben is filled mainly by deposits of Neogene age, represented by basalt, andesite, tuff, sandstone, and gravels. These deposits were intruded by Miocene diorite dykes and have a thickness of about 2000 m. The Quaternary deposits consist of glacio-aqueous materials and products of modern volcanic activity. Neogene deposits (mainly Paratunka Unit) include thermal water. In the central and southern parts of the basin the Neogene deposits are covered by clays which are an insulator for thermal waters from overlying Quaternary deposits.

The hydrothermal system is divided into three main areas of thermal anomaly which are located in the lower (LP), middle (MP), and upper parts (UP) of the Paratunka River basin; these waters are referred to as Lower, Middle, and Upper. The highest temperatures of waters were observed close to the main faults crossing the valley. The maximum temperatures (about 100°C) were found in boreholes to depths of 1500 m depth. The thermal waters, circulating via fractures are moving from the Upper to the Lower group (Manukhin & Vorozeika 1976). The cold groundwaters circulating in Quaternary deposits are characterized by an average temperature of about 5°C.

2 TECHNIQUES

Water samples were collected from thermal and cold springs and boreholes. Using portable field apparatus specific electrical conductance (SEC), pH, dissolved oxygen (DO), and temperature (t°C) were measured; HCO_3 was also measured in the field. Major and trace elements were determined in the BGS laboratory by ICP-OES and/or ICP-MS, and anions by automated colorimetry. Stable isotopes δ^2H, $\delta^{18}O$ and $\delta^{13}C$ were determined by mass spectrometry.

3 RESULTS AND DISCUSSION

The highest temperature (98.4°C) was found in borehole 13, located in LP area. The maximum t°C for MP was 86,6°C and for UP-67°C. Geophysical logging discovered water with > 90 °C at 200 m depth in LP area; water at 80-85 °C at 300 m depth in MP; and water at 80 °C at > 500 m in the UP area (Manukhin & Vorozeikina 1976, Trukhin & Petrova 1976). Geochemical data for boreholes with the highest temperature in each group were collected in Table 1 to compare with the cold underground waters of LP and Poperechnaya River water - background for the Paratunka catchment. The thermal waters of the Paratunka basin are found to be alkaline with pH up to 9.23. The main cations are represented by Na + K > Ca. SEC ranges from 2350 (borehole #502, LP) to 960 µs/cm in borehole # 39 (MP), cold spring (4) and river have much lower SEC (Table 1).

Table 1. Concentrations of some elements in the Paratunka waters, Kamchatka

T°C	pH	SEC μS/cm	Na mg/l	K mg/l	Ca mg/l	Cl mg/l	SO₄ mg/l	Si mg/l	B :g/l	Sr :g/l	Li :g/l
1 67,7	8,69	1680	224,9	3,78	192,3	59,9	823,4	23,66	2830	1250	310
2 86,6	8,5	1620	263,7	7,62	154,4	73,1	814,1	27,21	2830	1370	260
3 98,4	7,39	2270	390	16,50	199,9	356	804,2	40,15	1670	2460	250
4 6,5	7,32	160	9,30	2,4	19,1	7,0	31,5	13,3	110	255	4,76
5 6,3	7,5	56,5	2,1	0,2	9,25	1,5	4,3	4,14	11,04	33,88	2,02

1-Upper Paratunka borehole; 2-Middle Paratunka borehole; 3-Low Paratunka borehole #13;
4-Low Paratunka spring; 5-Poperechnaya River.

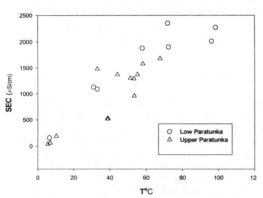
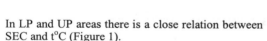

Figure 1. Plot of temperature vs. SEC for all thermal waters in the Paratunka Basin.

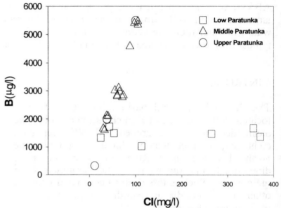

Figure 2. Plot of B vs Cl for the Paratunka thermal waters

In LP and UP areas there is a close relation between SEC and t°C (Figure 1).

SO₄ and Cl are the dominant anions. In general the ionic concentrations depend on the temperature. Thus Cl, Na, and Ca concentrations in LP and UP areas increase with temperature. A weak positive correlation between pH, Na, and Si, on the one hand and T°C on the other was found only for UP. For all groups of thermal waters a strong correlation between Ca and Sr was observed as well as some positive correlation between Cl and B (Figure 2). Na shows a strong positive correlation with Cl in the LP and UP thermal waters. Li and B show a weak correlation in the UP and LP areas. In UP waters a positive correlation exists between Cl and SO₄.

The SO₄/Cl molar ratio is highest in the thermal waters (about 6-7), and decreases from the upper to the lower parts of the basin; in the cold springs it is about 3 and for the Poperechnaya River is close to 2.

Concentrations of Cu, Zn, Co, Zr, Ni, Mo in thermal waters ranged from a few to 10 μg/l and showed. no big differences between groups -16, Isotope data for thermal waters with the highest

temperature show values δ²H =-111°/₀₀ and δ¹⁸O =-16.2 °/₀₀. For cold underground waters and rivers slightly heavier values are found: δ²H= -100 and -99°/₀₀ with δ¹⁸O = -14.0 and -13.6 °/₀₀. The variation in warm waters is most likely to be due to mixing of high-temperature water and shallow ground water with heavier values.

The gases accompanying the thermal waters are composed of N₂ (up to 96%) and CO₂ - a few percent (Manukhin & Vorozeikina 1976). In all boreholes with warm waters dissolved oxygen was absent; in the warm springs it reached 2.7% ppm in underground waters up to 8.7 ppm. The measurements of δ¹³C in CO₂ gas from different places in Kamchatka, including the warm spring of the Lower Paratunka show values about -7°/₀₀ (Chudaev et al. 1998). It is too light for carbon to be from marine carbonates and not light enough to be biogenic and more consistent with a mantle origin.

Thus, the Paratunka basin is an integrated groundwater system where chemical features of underground waters are controlled by a few factors including geological setting and tectonic control, as well as water-rock interaction, residence time which

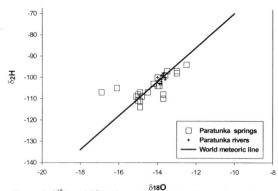

Figure 3. $\delta^{18}O$ and δ^2H plot of waters in the Paratunka valley

is shown especially by the build-up in concentrations of certain trace elements

The main source of water for the Paratunka Basin is meteoric water penetrating through the fractures and dislocations into rocks up to 2-4 km. Interaction of water with host rocks led to an increasing content of a wide range of elements. Volcanic exhalations also input some specific components (including Cl, Br, S, F,). There is a strong relationship between piezometric level of thermal waters and seasonal level of surface waters (Manukhin & Vorozeikina 1976). In Figure 3, the deuterium and oxygen-18 ratios for all waters of the Paratunka Basin are compared with the Global Meteoric Water Line (Craig 1961).

The main issue is the extent to which the chemical components in the Paratunka thermal basin are from primary magmatic sources and what has resulted from water-rock interaction. Manukhin & Vorozeikina (1976) had estimated it as 50% of each. However taking, the Poperechnaya river as initial water and altered greenschist facies basalt as host rock, (Trukhin & Petrova 1976) we can estimate what elements were added to the water by water-rock interaction. According to the experimental data of Bischoff & Dixon (1975), and Kholodkevich (1981) during basalt and sea or fresh water interaction, most chemical elements except Mg are leached out. In general, the increase of the experimental temperature leads to an increase of the chemical element concentrations solution. As was shown above (Figure 1) for UP and LP there is a positive correlation between SEC and t°C. According to our data and calculations (which chemical components a fresh basalt lost to alter itself in a greenschist rock) we can show, that the most chemical elements were added to water during water-rock interaction.

The isotopic ratios of sulfur in the thermal waters of Kamchatka (Vinogradov 1970) indicate that for the Paratunka thermal waters sulfur has two sources: a deep origin and marine deposits. Our data on ^{13}C

allow proposing of a mantle source for carbon (Chudaev et al. 1998). It supports the mantle origin suggested by $^3He/^4He$ ratios in the gases of thermal waters of the Paratunka area (Polyak et al. 1979; Poreda & Craig 1989) for Eastern Kamchatka. It seems that rather high concentrations of Cl, Br, I, F, and a part of S are transported from volcanic centers as a gas and are added to water, changing its chemical characteristics. A good correlation between halogen concentrations suggests a common source. Three groups of thermal waters of the Paratunka basin are connected with each other and the flow of thermal waters is directed from UP to LP. Nevertheless, the waters of MP area have some geochemical differences from UP and LP, that can be due to some differences in depths of faults controlling the elevation of volcanic gases.

CONCLUSION

The Paratunka basin is an integrated hydrogeological system where geochemical features are controlled by a few factors including geological setting and tectonic control, as well as water-rock interaction, and the influence of deep sources of certain constituents

The Paratunka thermal waters have high SEC values ranging from 2350 to 960 µS/cm with Na + K > Ca and SO_4 as the dominant anion. The ion concentrations depend on temperature. Concentrations of Cu, Zn, Co, Zr, Ni, Mo in thermal waters ranged from a few to 10 µg/l and there are no big differences among groups All waters of the Paratunka thermal basin are of meteoric origin. The main factor controlling chemical composition of thermal waters is water-rock interaction. Cl, Br, S, F, B, and I, are likely, to come in mainly as volatiles from the volcanic centres, that are typical for the volcanic area of Kamchatka. Carbon dioxide and helium in the gases dissolved in thermal waters are more likely to be of mantle origin.

ACKNOWLEDGEMENT

We acknowledge financial support from INTAS, contract 94-1592 in support this work.

REFERENCES

Chudaev O.,V.A.Chudaeva, P.Shand. M.Edmunds 1998. New data on the chemical composition of waters in the Paratunka hydrothermal system, Kamchatka. In Water-Rock Interaction. G.E.Arehart and J.R.Hulston (eds) 621-624. Rotterdam: Balkema.
Bischoff J. & F. Dixon. 1975. Seawater-basalt interactions at 200°C and 500 bars. *Earth and Planetary Science Letters*. 25: 385-397.

Craig H. 1961. Isotopic variation in meteoric waters. *Science*.133:1702-1703.

Kholodkevich I. 1981. *Secondary alteration of basalt (experimental and natural data)*. Ph.D. thesis. Vladivostok, FEGI RAS (In Russian).

Manukhin Yi. F. & L. A. Vorozeikina 1976. Hydrogeology of the Paratunka hydro- thermal system and conditions of its origin. In: V.M.Sugrobov (ed), *Hydrothermal system and thermal areas of Kamchatka*: 143-178. Vladivostok. (In Russian).

Polyak B.G., I.N.Tolstikhin & V.P.Yakuzeni 1979. Helium isotopes and heat flow and geophysical aspects of tectogenesis. *Geotectonics* 5: 3- 23. (In Russian).

Poreda R. & H. Craig 1989. Helium isotope ratios in circum-Pacific volcanic arcs. *Nature* 338: 473-478.

Sereznikov A.I. & V.M.Zimin 1976. Geological setting of the Paratunka geothermal area, the influence of the some geological factors on modern hydrothermal activity. In: V.M.Sugrobov (ed), *Hydrothermal systems and thermal areas of Kamchatka:* 115-142.Vladivostok. (In Russian).

Trukhin Yi.& V. Petrova 1976. *Some features of modern hydrothermal process*. Moscow: Nauka. (In Russian).

Vinogradov V.I. 1970. Isotopic composition of sulfur in thermal waters of Kamchatka and Kuril islands and its genetic interpretation. In: *Essay of geochemistry of mercury, molybdenum, and sulfur in the hydrothermal process*: 258-271. Moscow: Nauka. (In Russian).

Geochemistry of the Earth's Surface, Ármannsson (ed.)© 1999 Balkema, Rotterdam, ISBN 90 5809 073 6

Chemical composition of dissolved gases in groundwaters from Mt. Etna, Eastern Sicily

W. D'Alessandro
Istituto di Geochimica dei Fluidi, CNR, Palermo, Italy

S. Inguaggiato
Istituto Nazionale di Geofisica, Roma & Poseidon System, Catania, Italy

C. Federico & F. Parello
Dipartimento CFTA, University of Palermo, Italy

ABSTRACT: Chemistry of dissolved gases of Mt. Etna represents mixing between three end-members: atmospheric air, magmatic gases (CO_2 and He) and crustal gases (CH_4). The geographic distribution of deep gases is strongly affected by the tectonic setting of Mt. Etna. The SW (Paternò) and the E (Zafferana) sectors, recognised as the most active fault zones, display highest CO_2 contents. Methane-rich waters are related to the rising of gases from a hydrocarbon reservoir in the sedimentary sequence, underlying the volcanic pile. These waters mostly issue near Bronte, Adrano, Paternò and Acireale.

1 INTRODUCTION

Mount Etna is an highly active alkali-basaltic strato-volcano which has built upon tensional faults located in eastern Sicily, at the collision boundary between the African and European plates (Barberi et al. 1974). Its activity and growth over the last 0.2 Ma has been controlled by the intersection of two main fault systems, striking respectively NNW-SSE and NNE-SSW, and a shallower E-W one, cutting a 18-20 km thick continental crust whose upper part is made of carbonaceous-terrigenous mesozoic-pleistocenic deposits (Lo Giudice et al. 1982). Intense seismicity occurs along these regional fault systems.

Magmatic gas release is known to happen along such weakness zones, together with an intense degassing from summit craters. The frequency of eruptions and the amount of gases released during inter-eruptive periods make Mt. Etna one of the most active volcanoes in the world. It has been shown that summit crater plume emissions at Etna produce about 2×10^3 tons/day of SO_2 and 35×10^3 tons/day of CO_2, which corresponds to 10% of the total global budget of carbon dioxide of volcanic origin (Allard et al. 1991, Gerlach 1991).

Moreover, an intense soil degassing of magma-derived CO_2 and He occurs along the flanks of the volcano (Allard et al. 1991, D'Alessandro et al. 1992, Giammanco et al. 1998). During their ascent towards the surface, these gases interact with the groundwaters, modifying their composition. Since the 80s, the chemical composition of Etnean

groundwaters and their dissolved gases have been studied more extensively (Anzà et al., 1989, Brusca et al. in press). It has been recognised that the water chemistry reflects the dissolution of magma-derived CO_2 buffered by variable low temperature weathering of basaltic rocks. Some geochemical anomalies (temperature, TDS and pCO_2), observed before the 1991-93 eruption in some sampling sites, testify a sharp increase of magmatic gas inflow into the Etnean aquifers (Bonfanti et al. 1996a, b).

In this context, chemical composition of dissolved gases represent an important tool to evaluate the temporal variation of the magmatic input (mainly CO_2) into the aquifers.

The present paper focuses on the interpretation of the geographic distribution of dissolved gases in groundwaters, providing an outline of the distribution of the magmatic emission along active tectonic lineaments on Mt Etna.

Dissolved gases have been determined following a procedure proposed by Capasso & Inguaggiato (1998). The method is based on the equilibrium partitioning of gases between water and gas (a known volume of Ar is introduced as host gas). Initial concentration of gases in the liquid is derived from their concentration in the gas phase, using the partition coefficient for each gas.

2 HYDROGEOCHEMISTRY

Mt Etna has no real hydrographic network, due to a cover of highly fractured lavas (run-off coefficient is only 5%; Ferrara 1975), through which infiltrating rainwater feeds the underground circulation. The importance of the effective infiltration is highlighted by high outflows at the springs along the perimeter of the volcano, at the contact with the sedimentary rocks of the basement, and especially along the coast-line, where a considerable amount of water discharges into the sea (Ogniben 1966, Ferrara, 1975).

A huge volume of water accumulates every year into the aquifers (~ 0.7 km^3; Ogniben, 1966). This is due to both the high precipitation of rain and snow (about 0.86 km^3; Ogniben, 1966) and the high permeability of volcanic rocks.

The presence of an impermeable sedimentary basement beneath Etna's volcanics and the limited thickness of the latter prevent Etna's groundwaters from reaching considerable depths and thus limits their heating. Temperatures measured in Etna groundwaters are always lower than 25 °C. Only the mud volcanoes known as "Salinelle di Paternò", on the lower SW flank, present some typical hydrothermal features. The waters emitted in this area are characterised by abundant free gas phase and show typical features of waters linked to hydrocarbon reservoirs. Geothermometric estimates carried out on both the liquid and the gas phases emitted at the "Salinelle" indicate temperatures in the range 100 - 150 °C for their last equilibration (Chiodini et al. 1996).

Previous hydrogeological studies (Anzà et al. 1989, Allard et al. 1997, Brusca et al. in press) identified a genetic and spatial relationship between the compositional range of most Etna groudwaters and the fault-controlled input of CO_2-rich magmatic gas. Etna groudwaters with anomalous concentrations of dissolved gases emerge in the most fractured and seismically active zones of the volcano (i.e., Paternò - Belpasso to the SW and Zafferana - S.Venerina to the E). Along their pathway to the surface, such waters strongly interact at low temperature with the host rocks and leach their cations. Such a process is highlighted by the close relationship between the geographic distribution of calculated pCO_2 values and K, Rb, Mg, Ca and Sr, pointing to a greater chemical aggression of water on the host basalt rocks in response to CO_2 inflow into the aquifer.

3 RESULTS AND DISCUSSION

The chemical composition of dissolved He, H_2, O_2, N_2, CO, CH_4 and CO_2 has been determined in 181 water samples, collected in the time span from April 1997 to December 1998.

Air represents the principal component of the dissolved gases, except for high-CO_2 waters emerging in the South-western sector of Mt Etna, where pCO_2 displays values up to 2.6 atm.

Only twelve samples (all but one from the SW sector of Mt. Etna) showed pHe greater than detection limit (~ $1x10^{-5}$ atm): He contents range from $1.3x10^{-5}$ to $29.8x10^{-5}$ atm. H_2 has been detected only in 16 samples (ranging from $6.5x10^{-6}$ to $56.4x10^{-6}$ atm), while CO has been measured in 105 samples (from $0.4x10^{-6}$ to $42.6x10^{-6}$ atm). CH_4 contents, ranging from $2.5x10^{-7}$ to 0.44 atm, have been measured in 120 samples; only 12 samples display pCH_4 values higher than 10^{-4} atm. All samples display pCO_2 values higher than the atmospheric ($3.5x10^{-4}$ atm), ranging from $4.6x10^{-4}$ to 2.6 atm. In the Paternò area, 7 sparkling waters have pCO_2 greater than 1 atm.

N_2, O_2 and CO_2 triangular diagram (Fig. 1) shows that most samples plot along a mixing line between air-saturated water and a CO_2-rich end member.

The composition of free gases emerging at Paternò (Salinelle and Acqua Grassa), Bronte and Fondachello (near Giarre village) is also plotted (Data from D'Alessandro et al. 1997).

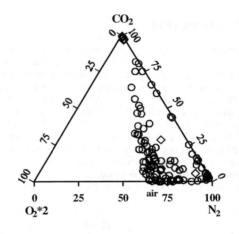

Figure 1. CO_2-O_2-N_2 triangular plot. Circles and diamonds represent groundwaters and free gases respectively. Data are expressed as partial pressures.

In a general way, a slight relative enrichment in N_2 is observed due to the consumption of O_2 during oxidation in sedimentary strata. In some waters this process is more effective, as confirmed by the high CH_4 content (Fig. 3) and lower redox potential values. Some of these waters present a distinct smell of H_2S.

CO_2 is the main component in free gases, except CH_4-rich gases in both Bronte and Fondachello areas.

As shown in Figure 2, CO_2-rich waters mostly emerge in the SW sector of Mt. Etna, in the Paternò area.

High CO_2 contents have also been detected in the Western sector, in the Zafferana area. As suggested by Anzà et al. (1989), Allard et al. (1997), Brusca et al. (in press), these waters occur in the most fractured and seismically active zones of Mt. Etna. pCO_2 values in the described areas indicate higher CO_2 content in the SW sector with respect to the Eastern one. Brusca et al. (in press) observed that the groundwaters of the SW sector also display higher temperatures and saline content. According to these Authors, we consider that these differences are due to the peculiar hydrogeological conditions of the E basin (stronger precipitation, greater steepness of the volcano slopes and lower altitude of the recharge zone) that allow a considerably higher water flow. Assuming that both basins are affected by a roughly similar gas influx from depth, the longer residence time of groundwaters is responsible for the higher deep-gas content.

As shown in the CH_4-CO_2-N_2 plot (Figur 3), most samples plot along the CO_2-N_2 line except CH_4-rich waters, emerging near Bronte, Paternò, Adrano and Acireale villages (Figure 4).

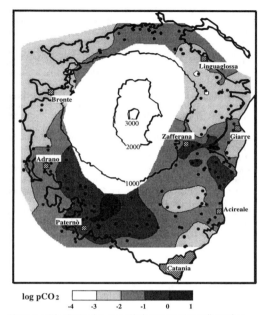

Figure 2. pCO_2 distribution map. Circles represent sampling points.

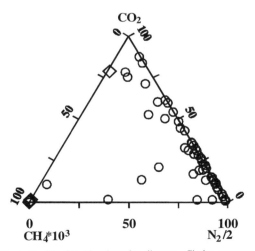

Figure 3. CO_2-CH_4-N_2 triangular diagram. Circles represent groundwaters, while diamonds indicate free gases. Data are expressed as partial pressures.

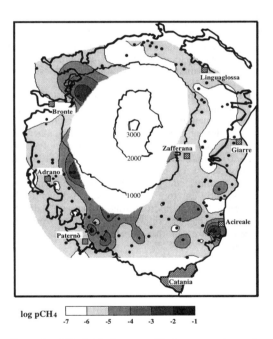

Figure 4. pCH_4 distribution map. Circles represent sampling points.

The genesis of methane from the thermal decomposition of organic carbon, due to the high geothermal gradient, may justify CH_4 contents recorded in the described areas, as suggested by preliminary isotopic data (δ^{13}C from -44 to -64 ‰; δD from -140 to -150 ‰; Bacheca pers. comm.).

In these areas the rise of hydrocarbon gases through faults probably promote the rise of more saline waters, affected by a small contamination of brine from the sedimentary basement. The rising of "Salinelle" - type waters is generally masked by the high flow rate of the meteoric groundwater through the shallower aquifers, except for the Paternò area, where the "Salinelle" brine springs out undiluted. The existence of a brine-component was already observed east of Bronte, north-east of Adrano and south of Acireale (Brusca et al. in press).

4 CONCLUSIONS

Data on dissolved gases indicate that a deep CO_2 magmatic component interacts with groundwaters, mostly in the SW part and in the Eastern part of Mt. Etna. These magmatic emissions are related to the principal tectonic alignments, already recognised on Mt. Etna. High soil CO_2 fluxes (Giammanco et al. 1998) and positive gravimetric anomalies (Loddo et al. 1989) have been detected. CO_2-rich waters also display the highest He contents with high R/Ra values (Allard et al. 1997), confirming a common deep origin for both He and CO_2. The magmatic origin of CO_2 is confirmed by C-isotope data (D'Alessandro et al. 1997, Allard et al. 1997).

The distribution of CH_4-rich groundwaters doesn't match the CO_2 map, indicating a different origin for methane, likely related to the thermal decomposition of organic carbon in a sedimentary setting. Hydrocarbon-rich waters emerge near Paternò, Adrano, Bronte and Acireale villages, where high dissolved salt contents have also been detected. These findings suggest a common origin for these waters, deriving from a deep brine and driven to the surface by rising hydrothermal gases. At Salinelle (Paternò) brine waters emerge undiluted, producing typical features of the so-called "sedimentary volcanism".

REFERENCES

Allard, P., Carbonelle, J., Dajlevic, D., Le Bronec, J., Morel, P., Robe, M. C., Maurenas, J. M., Faivre-Pierret, R., Martin, D., Sabroux, J.C. & Zettwoog, P. 1991. Eruptive and diffuse emissions of CO_2 from Mount Etna. *Nature* 351, 387-391.

Allard, P., Jean-Baptiste, P., D'Alessandro, W., Parello, F., Parisi, B. & Flehoc, C. 1997. Mantle-derived helium and carbon in groundwaters and gases of Mount Etna, Italy. *Earth Planet. Sci. Lett.* 148, 501-516.

Anzà, S., Dongarrà, G., Giammanco, S., Gottini, V., Hauser, S. & Valenza, M. 1989. Geochimica dei fluidi dell'Etna: Le acque sotterranee. *Mineral. Petrogr. Acta* 32, 231-251.

Barberi, F., Civetta, L., Gasparini, P., Innocenti, F., Scandone, R. & Villari, L. 1974. Evolution of a section of the Africa-

Europe plate boundary: paleomagnetic and volcanological evidence from Sicily. *Earth Planet. Sci. Lett.* 22: 123-132.

Bonfanti, P., D'Alessandro, W., Dongarrà, G., Parello, F. & Valenza, M. 1996a. Medium-term anomalies in groundwater temperature before 1991-1993 Mt. Etna Eruption, *J. Volcanol. Geoth. Res.* 73, 303-308.

Bonfanti, P., D'Alessandro, W., Dongarrà, G., Parello, F. & Valenza, M. 1996b. Mt. Etna eruption 1991-93: geochemical anomalies in groundwaters, *Acta Vulcanol,* 8 (1), 107-109.

Brusca, L., Aiuppa, A., D'Alessandro, W., Parello, F., Allard, P. & Michel, A. (in press). Geochemical mapping of magmatic gas-water-rock interactions in the aquifer of Mount Etna volcano. *J. Volcanol. Geoth. Res.*

Capasso, G. & Inguaggiato, S. 1998. A simple method for the determination of dissolved gases in natural waters. An application to thermal waters from Vulcano Island. *Appl. Geochem.* 13, 5, 631-642.

Chiodini, G., D'Alessandro, W. & Parello, F. 1996. Geochemistry of the gases and of the waters discharged by the mud volcanoes of Paternò, Mt. Etna (Italy). *Bull. Volcanol.* 58, 51-58.

D'Alessandro, W., De Domenico, R., Parello, F. & Valenza, M. 1992. Soil degassing in tectonically active areas of Mt. Etna. *Acta Vulcanol.* 2, 175-183.

D'Alessandro, W., De Gregorio, S., Dongarrà, G., Gurrieri, S., Parello, F. & Parisi, B. 1997. Chemical and isotopic characterization of the gases of Mount Etna (Italy). *J. Volcanol. Geotherm. Res,* 78, 65-76.

Ferrara, V. 1975. Idrogeologia del versante orientale dell'Etna, *in Proceedings of the Third International Conference on Groundwaters,* pp. 91-144, Palermo.

Gerlach T. 1991. Etna's greenhouse pump. *Nature* 351: 352-353.

Giammanco, S., Gurrieri, S. & Valenza, M. 1998. Anomalous soil CO_2 degassing in relation to faults and eruptive fissures on Mount Etna (Sicily, Italy), *Bull. Volcanol.* 60(4), 252-259.

Loddo, M., Patella, D., Quarto, R., Ruina, G., Tramacere, A. & Zito, G. 1989. Application of gravity and deep dipole geoelectrics in the volcanic area of Mt. Etna (Sicily), *J. Volcanol. Geotherm. Res.* 39, 17-39.

Lo Giudice, E., Patanè, G., Rasà, R. & Romano, R. 1982. The structural framework of Mount Etna, *Mem. Soc. Geol. It.* 23, 125-158.

Ogniben, L.,1966. Lineamenti idrogeologici dell'Etna, *Rivista Mineraria Siciliana* 100-102, 1-24.

Geochemistry of the Earth's Surface, Ármannsson (ed.) © 1999 Balkema, Rotterdam, ISBN 90 5809 073 6

Chemical and isotopic features of dissolved gases from thermal springs of Sicily, Italy

R. Favara
Istituto di Geochimica dei Fluidi, CNR, Palermo, Italy

F. Grassa
Dipartimento CFTA, University of Palermo, Italy

S. Inguaggiato
Istituto Nazionale di Geofisica, Roma & Poseidon System, Catania, Italy

ABSTRACT: The study of chemical and isotopic composition of dissolved gases is a useful tool for geochemical monitoring both in volcanic and seismically active areas. A preliminary geochemical study of Sicilian thermal springs made it possible to distinguish a group of springs interacting with deep CO_2 and another group receiving CO_2 from dissolution of calcite and organic processes. Knowledge of the geochemistry of the springs connected to deep structures represents the starting point for more detailed investigations for geochemical surveillance of seismic and volcanic activity.

1 INTRODUCTION

Thermal springs are quite widespread both in Sicily and in the minor islands (Pantelleria, Aeolian Arc) and are essentially linked to the geological setting of the island. These springs consist of almost 30 emergence points with temperatures in the range 18-56°C distinguishable in two types: the first is located along the main deep seismogenetic structures and hosted in carbonate rocks or metamorphic rocks. The second is linked to the principal volcanic systems. In Fig. 1 the location of sampling sites is shown.

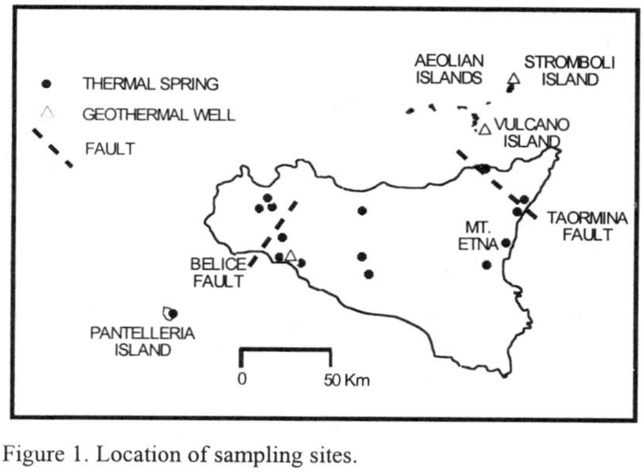

Figure 1. Location of sampling sites.

These hydrothermal waters have already been studied and classified by many authors (Alaimo et al. 1978; Dongarrà & Hauser 1982; Sano et al. 1989; Favara et al. 1998). The aim of this paper is to char-

acterize the thermal springs of Sicily by means of the chemical and isotopic composition of the dissolved gaseous species and to evaluate the interaction processes occurring between deep fluids and thermal reservoirs. The study of the chemical composition of dissolved and free gases plays an important role in the investigation of natural springs. In fact, the high mobility and the different solubility in water of gaseous species makes it possible to use them for geochemical monitoring both in volcanic and in seismically active areas.

2 GEOLOGICAL OUTLINE

Sicily represents the southernmost part of the alpidic orogenesis in the Mediterranean. It comprises a non-deformed carbonatic area (Iblean Plateau); two sedimentary basins (Caltanissetta basin and Castelvetrano basin) made up of evaporite and post-orogenic sediments; the chain that develops along a E-W direction, and is constituted by carbonate rocks. Northeastern Sicily is characterized by the Calabrian-Peloritan arc formed by metamorphic rocks that tectonically overthrusted the chain. In central-eastern Sicily rises Mount Etna, the largest active volcano in Europe that has been built up in an area geodinamically linked to an extensional tectonic regime at the boundary between the Eurasian plate and the African plate. The most recent eruption occurred in 1991-1993 and Strombolian explosions localized at its summit craters characterize its actual activity. Pantelleria Island an intraplate active volcano connected to extensional tectonics within the African plate is located in the Sicily channel. Pantelleria's last eruption in 1891was a submarine one. At present, volcanic activity on the island is represented by hot springs (temperature up to 90°C) and low flux gas manifestations with temperatures below 100°C. Vulcano and Stromboli, two of the seven islands of the Aeolian archipelago, are located in the southern Tyrrhenian sea, near the North Sicilian coast. These islands can be considered a part of an island arc structure, characterized by calc-alcaline and shoshonitic volcanic products. The last eruption in Vulcano Island took place during the period 1888-1890. The present activity is characterized by fumarolic activity located in the crater area and on the Vulcano Porto beach and by several thermal water emergences. Stromboli volcano is characterized by typical explosive activity (Strombolian type) cyclically and persistently mild throughout the years.

3 ANALYTICAL PROCEDURES

In order to determine the chemical composition of the dissolved gases and the isotopic composition of total dissolved carbon (TDC) in thermal waters we utilized the procedure respectively proposed by Capasso & Inguaggiato (1998) and Capasso et al., (1999). These methods involve collecting the sample in a glass bottle, totally filling it and immediately sealing by using a gas-tight rubber/teflon plug. The first technique is based on the partitioning equilibrium of gaseous species between the liquid and the gas phase. The second is based on the chemical and physical stripping of CO_2. The stripped gas is purified by means of a standard technique. The analytical determinations of chemical gas were performed using a Perkin-Elmer 8500 gas chromatograph with argon as carrier and equipped with a double detector (TCD-FID). The isotopic values were measured using a Finnigan delta plus mass spectrometer and the results are reported in δ ‰ vs. V-PDB standard. The standard deviation of $^{13}C/^{12}C$ ratio is ± 0.2 ‰.

4 GAS GEOCHEMISTRY

We have analyzed the chemical composition of the dissolved gases collected from 16 thermal springs and 3 wells in Sicily (Table1). The concentrations of the dissolved gases show that many samples have pCO2 and pHe values higher than those at equilibrium with the atmosphere (ASW). This evidence suggests an input of CO_2 and He rich fluids ascending from greater depth. The physico-chemical conditions of groundwater (pH, temperature) affect the dissolution of CO_2 and He into the thermal reservoirs. The contents of the dissolved gaseous components relative to total samples are plotted together with the ASW values on the $O_2-N_2-CO_2$ (Fig.2) and He-CO_2-N_2 (Fig.3) diagrams.

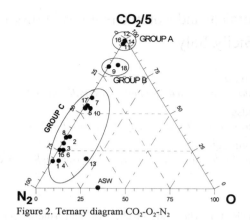

Figure 2. Ternary diagram CO_2-O_2-N_2

In the first diagram we observe that all the samples plot on a line connecting CO_2 and N_2 and are characterized by an O_2/N_2 ratio lower than ASW. It is possible to distinguish at least three degrees of interaction between thermal and deep fluids, corresponding to three different groups of samples:
A) The thermal springs relative to Vigliatore, Cassibile Marino and Grassa samples show a CO_2 content higher than 1000 cm^3/liter STP;
B) Gadir and Kammordino samples that have a relatively high CO_2 content (193 cm^3/liter and 263 cm^3/liter respectively);
C) All other samples that show low CO_2 concentrations (<40 cm^3/liter).

The highest CO_2 values suggest that group A underwent a strong contribution of deep gases rich in CO_2. In fact they are three bubbling springs located along the Taormina fault (Marino, Vigliatore and Cassibile) and in an area of Mount Etna characterized by anomalous CO_2 discharge (Grassa). Kammordino and Gadir belong to group B samples: the first is a water well located along the Belice fault, one of the most important seismogenetic structures of the Western Sicily. The second is a bubbling spring at Pantelleria Island with lower CO_2 values because of its high pH values due to seawater contamination. Low CO_2 contents have been recorded in the samples collected from springs located along secondary tectonic structures, in the Aeolian archipelago (Zurro and Camping Sicilia) and in the Eastern part of Etna (Ponteferro) (Group C). Zurro and Ponteferro are located in volcanic areas with low CO_2 influxes. Moreover Ponteferro is fed by a large reservoir, thus causing dilution of CO_2. The low CO_2 values of the thermal springs hosted in carbonatic reservoirs seem likely to be due to calcite dissolution thus excluding a deep supply. The lower CO_2 content at Camping Sicilia could be linked to the actual, relatively quiet, volcanic activity of Vulcano Island. This well underwent marked variations of physico-chemical parameters (temperature, pH, and conduc-

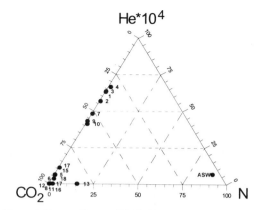

He*10^4

CO$_2$ ... N

Figure 3. Ternary diagram CO$_2$-He-N$_2$

tivity) caused by repeated inputs of fluid of magmatic origin (Capasso et al. 1991).

The diagram in Figure 3 makes it possible to divide the samples in two main groups:

a) samples from thermal springs hosted in carbonate reservoirs and linked to the tectonic structures lie preferentially along the CO$_2$-He line and show the highest He contents;

b) the other samples are located near the CO$_2$ vertex.

The high He contents shown by group A could be related to the presence of abundant quantities of radiogenic helium produced within the continental crust. Because of its high mobility, helium migrates through the fractures and reaches the reservoirs thus increasing its concentration in these thermal waters. This hypothesis seems to be confirmed by previous studies on the isotopic composition of this gas (Sano et al. 1989). The ^{3}He/^{4}He ratio in the Gorga group (0.72 Ra) and the Montevago (1.4 Ra) thermal waters revealed typical crust values. High He contents and low CO$_2$ concentrations suggest that all the thermal springs hosted in carbonatic-evaporitic reservoirs linked to tectonic structures are conductively heated by contact with the host-rocks and apparently without any appreciable contribution from deep fluids.

5 GEOTHERMOMETRY AND ISOTOPE GEOCHEMISTRY

The δ^{13}C of dissolved Carbon species is very important because it represents the result of interaction processes that involve water, carbonatic rocks and CO$_2$ coming from different sources. Each of the main CO$_2$-sources has got a different isotopic marker even if partly overlapping with the others. For these reasons many authors used the δ^{13}C$_{TDC}$ of thermal waters to understand the physico-chemical processes that occur between thermal reservoirs and the deep

CO$_2$-rich fluids. The isotopic composition of δ^{13}C$_{TDC}$ in the studied thermal waters is reported in Table 1 together with sampling and calculated temperatures. The ^{13}C/^{12}C ratios show extremely different values ranging between -6.63 (no.15) and 1.28 (no.13) δ‰ vs.V-PDB. In order to characterize the isotopic composition of the CO$_2$-source for each spring, the deep temperature of the reservoir has been estimated by means of different solute geotermometers. Applying the Ca/Mg (Marini et al. 1986) and the SO$_4$/F (Chiodini et al. 1995) geothermometers the equilibrium temperature of the thermal waters hosted in carbonate rocks has been estimated. The evaluation of deep temperatures in CO$_2$-rich waters is complicated because the uprising of fluids causes changes in the chemical composition of the waters (Fouillac 1983). So, for these samples we have used the Na/Li geothermometer (Fouillac & Michard 1981) and the silica geothermometer (Fournier 1981). For the other springs we used the Na-K-Ca geothermometer (Fournier & Truesdell 1973) and we have compared the estimated temperatures with those from the silica geothermometers. The isotopic composition of CO$_2$ gas was calculated for each thermal reservoir by means of the following equation (Capasso et al. 1999):

$$\delta\,^{13}C_{CO2} = \delta\,^{13}C_{TDC} - \varepsilon_a * X_{HCO3} - \varepsilon_\beta * X_{CO2} \quad (1)$$

where: ε_a and ε_β are the enrichment factors δ^{13}C$_{HCO3}$-$\delta\,^{13}$C$_{CO2gas}$ (Mook et al., 1974) and δ^{13}C$_{CO2aq}$-$\delta\,^{13}$C$_{CO2gas}$ (Deines et al., 1974), respectively; X_{HCO3} and X_{CO2} are the HCO$_3^-$ and CO$_{2aq}$ molar fractions. The diagram in Figure 4 shows the relationship between the isotopic composition of carbon of CO$_2$ gas and the CO$_2$ concentration values.

Table 1. Analytical and calculated data for the thermal springs. pH is expressed in pH units, Temperatures in °C and $\delta\,^{13}$C in δ ‰ vs. V-PDB.

No.	Location	pH	Temp.		δ 13C	
			meas.	estim.	TDC	CO$_2$
1	Gorga 1	7.07	48.3	85	-4.4	-6.5
2	Gorga 2	7.15	49.6	84	-4.9	-6.8
3	Segestane	6.81	44.2	90	-4.6	-6.3
4	Montevago	6.89	39.2	77	-4.0	-6.5
5	Selinuntine	6.42	54.0	80	-2.2	-5.0
6	Fitusa 1	7.22	25.2	62	-2.0	-6.2
7	Fitusa 2	7.49	20.7	92	-1.5	-3.3
8	Sclafani	7.01	32.7	68	-1.8	-5.1
9	Kammordino	6.91	56.4	72	0.4	-0.7
10	Molinelli	6.43	32.1	71	-0.5	-2.3
11	Vigliatore	6.72	31.2	130	-3.8	-2.2
12	Marino	6.14	23.8	77	-1.1	-0.8
13	Ponteferro	7.40	19.6	85	1.2	-0.3
14	Grassa	6.13	18.0	92	-0.7	-0.2
15	Zurro	6.66	36.1	299	-6.6	0.0
16	Cassibile	6.44	40.5	105	0.1	-1.5
17	Camping Sicilia	6.17	55.4	110	0.1	-0.4
18	Gadir	6.86	53.1	160	-2.1	-0.4

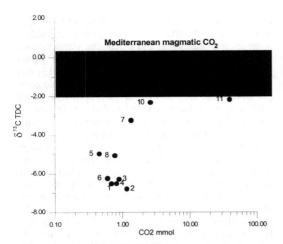

Figure 4. Relationship between carbon isotopic composition and CO_2 gas content. The highlighted area corresponds to Mediterranean magmatic CO_2.

Some samples show a positive correlation between isotopic composition and CO_2 contents. This could be linked to different contributions coming from sources marked by distinct isotopic compositions. The most plausible CO_2 sources are an organic contribution, characterized by negative isotopic value ($\delta^{13}C$ <-10‰) and CO_2 deriving from dissolution of calcite ($\delta^{13}C$ ~0‰). Apart from their CO_2 contents, Cassibile, Vigliatore, Marino and Kammordino and all the samples located in volcanic areas, lie inside the highlighted zone corresponding to the isotopic composition of Mediterranean magmatic CO_2 ($\delta^{13}C_{CO2}$ -2 ÷0.5‰ vs. PDB). These values affirm that a supply of deep fluids seems to reach the reservoirs of these thermal springs with the exception of Kammordino. In fact, both its geological context and its He enrichment seem to exclude it. A cleaner pattern of interaction processes involving these thermal waters may be obtained from prospecting isotopic composition of dissolved He in the study areas.

6 CONCLUSIONS

The geochemical characterization of Sicilian thermal springs by means of the chemical and the isotopic composition of the dissolved gases made it possible to highlight the most important interaction processes occurring between gas phase and thermal waters. Two main groups of thermal springs have been distinguished: a group that is hosted in carbonate reservoirs and characterized by CO_2 coming both from dissolution of calcite and from organic processes. The second group is related to the volcanic systems and tectonically active structures that host the sample waters and show high CO_2 values. The calculated

isotopic composition of the inferred pristine CO_2 highlights an input of deep fluids. Knowledge of the geochemistry of the thermal springs connected to deep structures represents the starting point for more detailed investigations aimed at the geochemical surveillance of seismic and volcanic activity.

REFERENCES

Alaimo R., M.Carapezza, G.Dongarrà & S.Hauser 1978. Geochimica delle sorgenti termali siciliane. Rend. Soc. Mineral. Ital. 34. (2): 577-590.
Capasso G. & S.Inguaggiato 1998. A simple method for the determination of dissolved gases in natural waters. An application to thermal waters from Vulcano Island. Appl. Geochem. 13 (5): 631-642
Capasso G., G.Dongarrà, S.Hauser, R.Favara, & M.Valenza 1991. Chemical changes in waters from Vulcano island: an update. Acta Vulcanol. 1: 199-209.
Capasso G., R.Favara, F.Grassa., S.Inguaggiato & G.Pecoraino 1998. A new method to determine the $\delta^{13}C$ of TDC in natural waters an application to Pantelleria Island (Italy). IAVCEI. International Volcanological Congress. Cape Town. July 1998.
Chiodini G., F.Frondini & L.Marini 1995. Theoretical geothermometers and pCO$_2$ indicators for aqueous solution coming from hydrothermal systems of medium-low temperature hosted in carbonate-evaporite rocks. Application to the thermal springs of the Etruscan Swell, Italy. Appl. Geochem. 10: 337-346
Deines P., D.Langmuir & S.Russell 1974. Stable carbon isotope ratios and the existence of a gas phase in the evolution of carbonate groudwaters. Geochim. Cosmochim. Acta 38: 1147-1164
Dongarrà G., & S.Hauser 1982. Isotopic composition of dissolved sulphate and Hydrogen sulphide from some thermal springs of Sicily. Geothermics 11 (3): 193-200
Favara R., F.Grassa, S.Inguaggiato & F.D'Amore 1998. Geochemical and hydrogeological characterization of thermal springs in Western Sicily. Journ. Volcan. Geotherm. Res. 84: 125-141
Fouillac C. 1983. Chemical geothermometer in CO$_2$-rich thermal waters. Example of the Frenc Massif Central. Geotermics 12: 149-160
Fouillac C. & G.Michard 1981. Sodium/lithium ratio in water applied to geothermometry of geothermal reservoir. Geotermics 10: 55-70
Fournier R.O. 1981. Application of water geochemistry to geothermal exploration and reservoir engineering. In Rybach L. & L.J.P.Muffler (eds), Geothermal systems: Principles and case Histories. New York: John Wiley: 109-143
Fournier R.O. & A.H.Truesdell 1973. An empirical Na-K-Ca geothermometers for natural waters. Geochim. Cosmochim. Acta 37: 1255-1279
Marini L., G.Chiodini & R.Cioni 1986. New geothermometers for carbonate-evaporite geothermal reservoirs. Geothermics, 15, 77-86.
Mook W.G., J.C.Bommerson & W.H.Staverman 1974. Carbon isotope fractionation between dissolved bicarbonate and gaseous carbon dioxide. Earth Planet. Sci. Lett. 22 (2): 169-176
Sano Y., H.Wakita, F.Italiano & P.M.Nuccio 1989. Helium isotopes and tectonics in Southern Italy. Geoph. Res. Lett. 16 (6): 511-514.

Geochemistry of the Earth's Surface, Ármannsson (ed.)© 1999 Balkema, Rotterdam, ISBN 90 5809 073 6

Silica sinters in Snæfellsnes and Hnappadalur, West Iceland

G.Ó. Fridleifsson & H. Ármannsson
Orkustofnun, Geosciences Division, Reykjavík, Iceland

ABSTRACT: Since 1947 hot spring surface deposits in Snæfellsnes have been considered to be travertines, directly related to carbonate rich hot springs in the region. In recent years the presence of these travertines has been seriously questioned, the hot spring deposits being interpreted as early Holocene fossil siliceous sinters witnessing much hotter surface waters than characterize the low-temperature region today. Some of these deposits were sampled in 1993 but the first major element chemical analysis was done last year. Here the first major-, minor- and trace element chemical analyses of 6 test samples are published, showing the hot spring deposits in three areas in Snæfellsnes to be 96-99% silica sinters with only minor $CaCO_3$ content. The silica sinters are unlikely to have precipitated from the present day 30-70°C hot water, equilibrium calculations suggesting boiling of hot water down to about 100°C. The sinters are comparable to the sinters in the Geysir area.

1 INTRODUCTION

Since Kjartansson (1947) described travertines in Iceland, the presence of travertines at Snæfellsnes, West Iceland, has been taken for granted. Kjartanson based his result on a hydrochloric acid test, not questioned by later workers. As a result of a hot- and cold water exploration survey for fish farming a decade ago, the presence of travertines in Snæfellsnes was questioned (Tulinius et al. 1991, Fridleifsson 1992) and it was concluded that the "travertines" looked more like silica sinters comparable to the sinters in the Great Geysir area and other boiling (or near boiling) hot spring areas in Iceland. This was supported by a field examination at Snæfellsnes, showing the sinters to be of early Holocene age, and by XRD analysis of a few sinter samples which showed nothing but a flat opaline silica charater on the XRD-diagrams, and only a small if any calcite peak. Thorough sampling was carried out a year later in 1993 by the first author, especially of the Lýsuhóll sinters, which earlier were interpretated as travertines by Kjartansson (1947) and later of old eroded sinters at Syðri-Rauðamelur and Landbrot in Hnappadalur, 40 – 50 km to the east of Lýsuhóll (Figure 1). The sampling was assisted by local farmers who provided an excavator, and two colleagues at Orkustofnun. Two cross sections through the sinter in Lýsuhóll were sampled (Figure 2) , one in an excavator ditch next to the highest sinter. The excavator could not penetrate the hard

sinter, which from field relations could be 4-5 m thick at the site, to sample but a 1.5 m thick section which consists of 19 layers at least, all of which were sampled (Figure 2). One of the layers consists of peat and a few others of a mixture of peat and a siliceous clayish-like deposit, while most of the layers consists of dark or light coloured sinter deposits, with or without a clay-like mixture, or layering. Four of these samples were analysed for major-, minor- and trace elements. The other section is 1.3 m long, from a river bank, 10-15 m east of drillhole No. 1 and an old swimming pool (Figure 2). The bottom 25 - 30 cm, overlying a

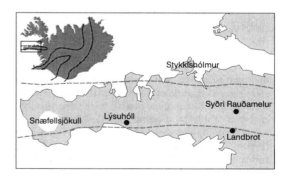

Figure 1. Sinter sampling locations. Broken lines show outlines of Neo-volcanic system

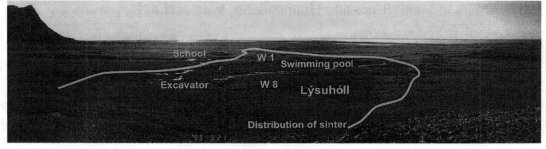

Figure 2. The Lýsuhóll area, showing wells 1 and 8 (W1 and W8) and the distribution of the sinter

hardened early Holocene river gravel, consists of peat. The remaining 1 m consists of 6-8 sinter layers with evidence of episodic deposition, where thin and flaky sinter layers are interpreted to represent the cooling or waning stages in the sinter layering, whereas the massive or porous gray, dark or light coloured sinters represent more vigorous episodes. Both the peat and some of the sinter layers were sampled for ^{14}C dating and chemical analyses.

The erupting activity in the Geysir area, South Iceland where silica sinters have been observed, is often more vigorous in the wake of major earthquakes (M 5-7), which have occurred about once every century in the South Iceland Seismic Zone in late Holocene. The outflowing hot water slowly diminishes as a result of reduced activity, and sinter formation. Therefore, the resulting sinter deposit layering could be used to construct a seismic history combining a detailed study of the sinter layers, ^{14}C analysis of peat layers, and a tephrocronological study. Samples were collected from a 2.5 m long cross section up through the Great Geysir sinter and from a less detailed 2.5 m cross section through a Holocene fossil sinter at Laugar, Hreppar. The cross sections were similar to those at Snæfellsnes but the samples await analysis.

2   S INTER ANALYSIS

In order to prove that at least some of the hot spring deposits at Snæfellsnes are siliceous sinters, 6 samples from the batch were chemically analysed (Table 1). Four of these are from section 1 at Lýsuhóll, samples L-1, L-11, L-12 and L-19 (Figure 3), representing a gray-, dark-, yellowish- and a lightbrown dense sinter layer respectively. Samples H-2 and H-6 in Table 1 are from Syðri Rauðamelur and Landbrot (Figure 1). The 6

chemical analyses shown in Table 1 all reveal 96-99% SiO_2, or almost pure silica sinter. Part of the volatiles (LOI) is due to CO_2, showing the presence of a minor amount of $CaCO_3$ in the samples, which evidently is responsible for the positive response to the HCl-test metioned above. Sample L-19 (the surface sample) is somewhat contaminated by eolian soil as seen by the relatively high Al, Fe, alkaline earths and other element contents. A note is made of the Au, Sb, As, and Hg analyses in Table 1, showing about 1 ppm Au and a clear signal for the related elements, Sb, As and Hg which are comparable to results of analysis for the same elements from the Geysir sinter, carried out by the Málmís Ltd. Au-exploration company. Detailed whole rock analysis of the Geysir sinter was not available to the present authors.

3   FLUID CHEMISTRY

The chemical composition of water from wells at Lýsuhóll and Syðri Rauðamelur and a warm pool at Landbrot is shown in Table 2 along with geothermometer temperatures. The waters from Lýsuhóll and Syðri Rauðamelur are high carbon dioxide waters, but the Syðri Rauðamelur water is characterized by a high magnesium concentration. The well was cased to 200 m depth. All the samples show chalcedony temperatures well in excess of 100°C but the NaK temperature is lower than 100°C at Landbrot. Higher temperatures than 100°C may be expected in the systems and are very likely to have been higher than now in earlier times. In support of this laumontite has been found in the borehole at Syðri Rauðamelur. Deep drilling in areas of carbon dioxide rich fluids has usually resulted in fluids with temperatures as predicted by

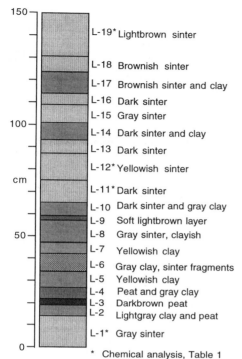

150 —

L-19* Lightbrown sinter

L-18 Brownish sinter

L-17 Brownish sinter and clay

L-16 Dark sinter

L-15 Gray sinter

100 —

L-14 Dark sinter and clay

L-13 Dark sinter

L-12* Yellowish sinter

cm

L-11* Dark sinter

L-10 Dark sinter and gray clay

L-9 Soft lightbrown layer

L-8 Gray sinter, clayish

50 —

L-7 Yellowish clay

L-6 Gray clay, sinter fragments

L-5 Yellowish clay

L-4 Peat and gray clay

L-3 Darkbrown peat

L-2 Lightgray clay and peat

L-1* Gray sinter

0 —

* Chemical analysis, Table 1

Figure 3. Cross section 1 at Lýsuhóll

Table 1. Sinter analysis from Snæfellsnes

Sample	L-1	L-11	L-12	L-19	H-2	H-6	dtl (ppm)
element	%	%	%	%	%	%	o/oo
SiO_2	91,9	93,1	83,69	88,6	96,42	96,2	60
TiO_2	0,08	0,07	0,078	0,31	0,015	0,01	35
Al_2O_3	0,37	0,21	0,24	2,98	<d/l	<d/l	120
Fe_2O_3	0,83	0,75	0,88	2,2	0,15	0,18	30
MnO	0,01	0,23	0,013	0,05	0,003	0	30
MgO	0,16	0,07	0,11	0,42	0,03	0,02	95
CaO	0,34	0,6	0,23	1,13	0,08	0,1	15
Na_2O	0,17	0,02	0,02	0,82	0,03	<d/l	75
K_2O	0,04	0,03	0,03	0,31	0,01	0,01	25
P_2O_5	0,02	0,03	0,014	0,14	0,009	0,01	35
LOI	5,99	4,79	14,62	3,23	3,27	3,34	100
Total	99,9	99,9	99,93	100	99,98	99,8	
CO_2	1,89	4,66	1,17	0,38	2,33	2,51	0,02
BaO	<d/l	<d/l	<d/l	88	<d/l	<d/l	17
	ppm	ppm	ppm	ppm	ppm	ppm	ppm
Cu	34	10	16	28	14	9	2
V	15	<d/l	<d/l	19	<d/l	<d/l	10
Zn	<d/l	<d/l	<d/l	19	<d/l	<d/l	2
Ga	2	1,9	1,7	5,9	1,8	2,5	1
Nb	5,5	5,8	5,8	15	5,9	5,8	1
Pb	<d/l	<d/l	<d/l	<d/l	<d/l	<d/l	1
Rb	2,4	2,1	2,7	9,6	1	1,3	1
Sr	5,4	8,6	5	64,9	3,9	5,3	1
Th	<d/l	<d/l	<d/l	<d/l	<d/l	<d/l	1
U	<d/l	<d/l	<d/l	<d/l	<d/l	<d/l	1
Y	4,9	7,2	6,5	16,4	4,9	4,8	1
Zr	<d/l	1	1,6	94,8	<d/l	<d/l	1
Au	1	1	1	<d/l	<d/l	1	1
Sb	6,9	7,2	7,1	5	6	5,9	1
As	23,1	35,9	23,6	8,2	7,1	4,1	1
Hg	3	9,5	<d/l	11,2	9,8	10,9	1

silica geothermometers (see Ármannsson 1981, Georgsson 1976, Yildirim et al. 1997)

The chemical speciation of the water was studied using WATCH (Arnórsson et al. 1982, Bjarnason1994) and SOLVEQ (Reed & Spycher 1984, Spycher & Reed 1989a), and the effect of heating this water to selected temperatures and then cooling them conductively and adiabatically from these temperatures to 30°C using CHILLER (Spycher & Reed 1989b). The results for heating, conductive cooling from 80°C and boiling and cooling from higher temperatures are presented in Table 2. In all cases the waters are supersaturated with respect to carbonates after heating and would deposit calcite and/or dolomite before being cooled again. The Lýsuhóll and Landbrot waters would deposit amorphous silica at 30-60°C, at higher temperatures after boiling than conductive cooling, and at Lýsuhóll more deposit would be generated at 150°C than 250°C while the reverse is true for Landbrot. Amorphous silica would not be deposited from the Syðri Rauðamelur well water. Calculations suggested that traces of microcline might be formed from it at low temperatures after boiling, and siderite,

pyrophyllite, kaolinite and magnesite as well upon conductive cooling. It is however possible that an original geothermal component from depth is present and that the composition may be partly due to the water encountering sediment layers containing magnesium-rich shell remains thus forming the magnesium-rich water which prevents silica from being deposited. It is possible that geothermal water, relatively low in magnesium, may have reached the surface without encountering such layers and the silica concentration is high enough for silica deposition to take place from a relatively magnesium-free water as is the case at Lýsuhóll and Syðri Rauðamelur (Table 2).

Table 2. Chemical composition and geothermometer temperatures for waters from Lýsuhóll, Landbrot and Syðri Rauðamelur (ppm)

Location	t °C	pH/°C	CO₂	SiO₂	Na	K	Mg	Ca	F	Cl	SO₄	Fe	t_ch [1]	t_NaK
Lýsuhóll	57	6.72/19	1495	219	434	33.9	17.8	93.4	4.60	80.0	43.0	12.4	129	164
Landbrot	54	7.88/18	154	146	162	2.5	0.89	21.3	0.50	122	61.0	0.50	117	60
S. Rauða-melur	43	6.40/21	1939	170	256	7.0	131	163	0.17	38.9	50.9	3.73	151	106

[1] ch: chalcedony

Table 3. Deposition from water from Lýsuhóll, Landbrot and Syðri Rauðamelur after heating and subsequent cooling to 30°C

Location	Heat to °C	Major mineral [1]	Mass g/l	am.s. [1] deposit at °C	Mass am. s. , g/l
Lýsuhóll	80	do/ca	0.27	40	0.07
	150	do/ca	0.32	60	0.10
	250	ca	0.14	55	0.06
Landbrot	80	ca	0.02	30	0.02
	135	ca	0.05	40	0.04
	260	ca	0.06	55	0.06
Syðri Rauða-melur	80	do	0.71	n.d.	
	150	do	0.90	n.d.	
	245	do	0.92	n.d.	

[1] ca: calcite, do: dolomite, am.s.: amorphous silica, n.d.: no deposit

4 CONCLUSIONS AND DISCUSSION

Deposits found at Lýsuhóll in Snæfellsnes and Landbrot and Syðri Rauðamelur in Hnappadalur are silica sinters containing minor amounts of calcite, but not travertines as previously considered. The chemical composition of the water suggests that silica deposition is probably taking place and has been for a long time. The water is supersaturated with carbonates and deposition of calcite and/or dolomite is possible. Such deposits could be formed underground. Carbonate surface deposits are soluble and might have been eroded away to some extent by dissolution in surface water and/or shallow ground water.

ACKNOWLEDGEMENTS

Thanks are due to colleagues at Orkustofnun, especially Hjalti Franzson, Ásgrímur Guðmundsson, Magnús Ólafsson and Bjarni Richter.

REFERENCES

Ármannsson, H. 1981. *Leirá, Borgarfjörður. Chemical composition and deposition potential.* Orkustofnun report, OS-81028/JHD-16. Reykjavík: Orkustofnun (in Icelandic).

Arnórsson, S., S. Sigurðsson & H. Svavarsson 1982. The chemistry of geothermal waters in Iceland I. Calculation of aqueous speciation from 0°C to 370°C. *Geochim. Cosmochim. Acta* 46: 1513-1532.

Bjarnason, J.Ö. 1994. *The speciation program WATCH version 2.1.* Reykjavík: Orkustofnun

Friðleifsson, G. Ó. 1992. *Thoughts on geothermal energy in Snæfellsnes.* Orkustofnun, report, OS-GÓF-92/01. Reykjavík: Orkustofnun (in Icelandic).

Georgsson, L. 1976. *Hot water acquisition for Hallkelshólar and Borg, Grímsnes.* Orkustofnun, OS-JHD-7652. Reykjavík: Orkustofnun (in Icelandic).

Karlsdóttir, R., H. Jóhannesson & J.Benjamínsson, 1981. *Exploration for geothermal energy at Lýsuhóll, Snæfellsnes.* Orkustofnun report OS81004/JHD01. Reykjavík: Orkustofnun (in Icelandic).

Kjartansson, G. 1947. Travertines in Iceland. *Náttúrufræðingurinn* 17: 88-92 (in Icelandic).

Reed, M.H. & N.F. Spycher 1984. Calculation of pH and mineral equilibria in hydrothermal water with application to geothermometry and studies of boiling and dilution. *Geochim. Cosmochim. Acta* 48: 1479-1490.

Spycher, N.F & M.H. Reed 1989a. *SOLVEQ: A computer program for computing aqueous-mineral-gas equilibria.* Eugene, Oregon: University of Oregon.

Spycher, N.F & M.H. Reed 1989b. *CHILLER: A program for computing water-rock reactions, boiling, mixing and other raction processes in aqueous-mineral-gas systems.* Eugene, Oregon: University of Oregon.

Tulinius, H., Á. Hjartarson, G. Ó. Friðleifsson & Guðrún Sverrisdóttir 1991. *Hnappadalur – Cold water and geothermal energy. A specialised fish farming project 1989 -1990.* Orkustofnun report OS-91039/JHD-05). Reykjavík: Orkustofnun (in Icelandic).

Yildirim, N. Z. Demirel, & A.U. Dogan, 1997. Geochemical characteristics and re-injection of the Kizildere-Tekke Hamam geothermal fluids. *Geoenv '97.* Istanbul, Turkey: 48-60.

Geochemistry of the Earth's Surface, Ármannsson (ed.)© 1999 Balkema, Rotterdam, ISBN 90 5809 073 6

Mineral-fluid equilibria in the Krafla and Námafjall geothermal systems, Iceland

B.Th.Gudmundsson
University of Iceland, Reykjavík, Iceland

S.Arnórsson
Science Institute, University of Iceland, Reykjavík, Iceland

ABSTRACT: The WATCH chemical speciation program was used to calculate reservoir fluid composition and aqueous species distribution for 20 producing wells in the Krafla and Námafjall high temperature geothermal areas NE-Iceland. It is concluded that fluid-secondary mineral equilibria are closely approached for all major components incorporated into these minerals.

1 INTRODUCTION

The Krafla geothermal system is located within the caldera of the Krafla central volcano, in the neovolcanic zone of axial rifting in NE-Iceland. Today 32 wells have been drilled in the area providing steam for the 60 MW Krafla power plant. Three wellfields are being exploited at Krafla, Leirbotnar, Sudurhlidar and Hvíthólar. The reservoirs are two phase or sub-boiling. Maximum temperature is ~350°C and well discharge enthalpies are 800- 2700 kJ/kg. The Námafjall high-temperature area, located 10 km south of Krafla, lies astride the main fissure swarm that is associated with the Krafla central volcano. Twelve wells have been drilled in the area. Petrologically and volcanologically it is closely related to Krafla. The reservoir fluid in both areas is dilute, 1000-1500 ppm dissolved solids.

2 GEOLOGY

The Krafla caldera is about 8 to 10 km across, associated with the volcanologically and tectonically active Krafla fissure swarm. Geothermal activity is confined to the caldera. The volcanics in the area are mainly basaltic but icelandite and dacite rocks also occur. The most prominent topographic features in the area are sub-glacially formed hyaloclastite mountains and ridges (Ármannsson et al. 1987).

3 ALTERATION MINERALOGY

The hydrothermal minerals in the Krafla and Námafjall reservoirs display depth-zonal distribution. The main zones are the following with increasing depth:

Smectite-zeolite zone, mixed layer mineral zone, chlorite zone, chlorite-epidote zone and epidote-amphibole zone. The most important hydrothermal minerals identified include calcite, quartz, epidote, smectite, chlorite, albite and pyrite. Anhydrite, adularia, prehnite, actinolite, wollastonite, garnet, pyrrhotite and various zeolites have also been identified (Kristmannsdóttir 1984, Arnórsson 1995).

The present study is based on analysis of 20 water and steam samples collected in 1997 and 1998 from producing wells at Krafla and Námafjall. The WATCH chemical speciation program (Arnórsson et al. 1982) was used to calculate reservoir fluid composition and aqueous species distribution in the fluid. The reference temperature at which the speciation calculations were carried out for each well was selected after careful examination of downhole temperature and aquifer inflows. Saturation indices were calculated for all important hydrothermal minerals in both areas. The thermodynamic properties of these minerals at 25°C and 1 bar are summarised in Table 1, as well as equations describing the temperature dependence of their apparent standard Gibbs energy, valid in the range 0-350°C at P_{sat}.

4 RESULTS

The aquifer water at Krafla and Námafjall is close to equilibrium with quartz, calcite, low albite and microcline (Figure 1). The waters are also close to saturation with wollastonite and anhydrite, or slightly undersaturated. The saturation indices plotted for prehnite show more scatter than similar plots for the minerals discussed above, yet the points are distributed around the equilibrium curve. The solubility of prehnite is pH-dependent. Various simplifying

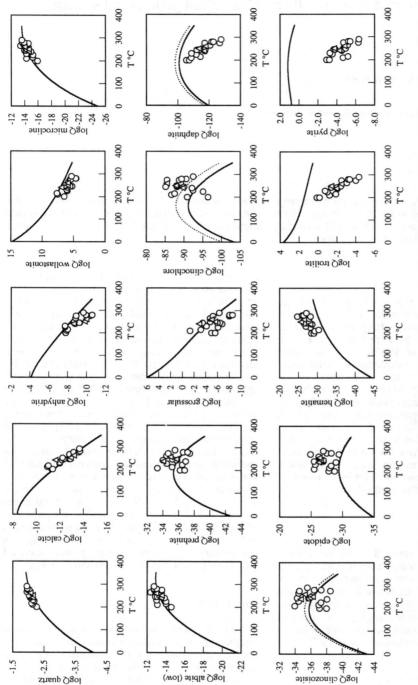

Figure 1. The saturation state of secondary minerals in the Krafla and Námafjall geothermal reservoirs (circles and triangles: solubility product for samples from Krafla and Námafjall, respectively. The solid curves represent solubilities of endmember minerals but broken curves solubilities corresponding to actual mineral composition at Krafla. Clinochlore: 0.46, daphnite: 0.54, clinozoisite: 0.26, epidote: 0.76.

504

Table 1

Thermodynamic properties of secondary minerals at Krafla and Námafjall at 25°C and 1 bar. Apparent standard Gibbs energy-temperature equations are also given which are valid from 0 to 350°C at 1 bar below 100°C and Psat at higher temperatures.

| Mineral | Formulae for the endmembers of the phases used in the internally consitent data set. | ΔG°_f kJ/mol | ΔH°_f kJ/mol | S° J/mol/K | V° J/bar | ΔG°_f $k*_1 xT + k*_2/T + k*_3 + k*_4 xT^2 + k*_5 xT^3 + k*_6 xT^{0.5} + k*_7 xT x logT$ | | | | | |
						$k*_1 \times 10^{-2}$	$k*_2 \times 10^{-4}$	$k*_3 \times 10^{-4}$	$k*_5 \times 10^{6}$	$k*_6 \times 10^{-3}$	$k*_7 \times 10^{-2}$
Albite	$NaAlSi_3O_8$	-3713.038 (a)	-3936.185 (a)	208.20 (a)	10.043 (a)	23.056 (a)	584.950 (a)	-379.65 (a)	64.607 (a)		-8.475 (a)
Anorthite	$CaAl_2Si_2O_8$	-4002.095 (a)	-422.783 (a)	199.30 (a)	10.079 (a)	24.364 (a)	625.160 (a)	-409.597 (a)	69.106 (a)		-8.883 (a)
Clinochlore	$Mg_5Al_2Si_3O_{10}(OH)_8$	-8262.863 (e)	-8929.234	410.50 (b)	21.090 (b)	60.904	1636.055	-852.608	129.765		-21.836
Clinozoisite	$Ca_2Al_3Si_3O_{12}(OH)$	-6496.506 (f)	-6893.743	291.00 (g)	13.659 (h)	38.471	944.125	-664.898	86.995		-13.941
Daphnite	$Fe_5Al_2Si_3O_{10}(OH)_8$	-6533.793 (i)	-7152.012	545.00 (b)	21.34 (b)	59.987	1547.118	-675.350	124.936		-22.013
Enstatite	$MgSiO_3$	-1458.181 (j)	-1545.552 (j)	66.17 (j)	3.131 (j)	9.252 (v)	237.400 (v)		21.227 (v)		-3.336 (v)
Epidote	$Ca_2Al_2FeSi_3O_{12}(OH)$	-6072.543 (k)	-6458.444	328.00 (k)	13.910 (m)	41.096	1069.922	-623.442	104.210		-14.938
Forsterite	Mg_2SiO_4	-2055.023 (j)	-2174.420 (j)	94.01 (j)	4.366 (j)	13.175 (v)	332.550 (v)	-210.915 (v)	32.142 (v)		-4.753 (v)
Grossular	$Ca_3Al_2Si_3O_{12}$	-6274.524 (n)	-6636.476	254.72 (d)	12.535 (d)	209.049	-1707.885	-538.120	256.791	-165.556	-57.160
Hematite	Fe_2O_3	-743.681 (j)	-825.627 (j)	87.44 (j)	3.027 (j)	11.044 (v)	264.650 (v)	-78.611 (v)	24.067 (v)		-4.017 (v)
Microcline	$KAlSi_3O_8$	-3749.230 (a)	-3974.339 (a)	214.20 (a)	10.869 (a)	23.158 (a)	597.255 (a)	-383.221 (a)	69.418 (a)		-8.531 (a)
Prehnite	$Ca_2Al_2Si_3O_{10}(OH)_2$	-5819.455 (p)	-6196.630	292.75 (q)	14.033 (r)	38.018	1003.725	-597.281	95.693		-13.776
Pyrite	FeS_2	-160.229 (c)	-171.544	52.93 (c)	2.394 (c)	9.475		-18.723	42.567		-3.479
Quartz	SiO_2	-856.281 (s)	-910.700	41.44 (s)	2.269 (s)	2.633 (u)		-85.763 (u)	-3.327 (u)		-1.044 (u)
Troilite	FeS	-101.333 (c)	-100.960	60.33 (c)	1.820 (c)	3.282		-10.041	5.648		-1.341
Wollastonite	$CaSiO_3$	-1548.569 (t)	-1633.971	81.69 (d)	3.993 (d)	9.454	227.949	-158.298	26.823		-3.468

Selected and retrieved data on the standard Gibbs energy of formation from the elements for minerals used to calclulate the standard Gibbs energies for cln, daph, czo and pre.

Almandine	$Fe_3Al_2Si_3O_{12}$	-4939.058 (w)
Annite	$KFe_3AlSi_3O_{10}(OH)_2$	-4796.698 (x)
Kyanite	Al_2SiO_5	-2445.120 (y)
Phlogopite	$KMg_3AlSi_3O_{10}(OH)_2$	-5830.900 (z)
Pyrope	$Mg_3Al_2Si_3O_{12}$	-5932.412 (a)
Spinel	$MgAl_2O_4$	-2174.860 (c)
Zoisite	$Ca_2Al_3Si_3O_{12}(OH)$	-6494.743 (f)

(a) Arnórsson & Stefánsson (1999). (b) As in Holland & Powell (1998). (c) As in Robie et al. (1979). (d) As in Zhu et al. (1994). (e) Calculated from phase equilibrium experiments of Jenkins & Chernosky (1986). (f) Calculated from the experiments of Newton (1966). (g) Calculated from the data on zoizite using the entropy difference for zoisite and clinozoisite reported by Jenkins et al. (1985). (h) Chatterjee et al. (1984). (i) Calculated from the experiment of Dickenson & Hewitt (1986). (j) As in Berman (1988). (k) Calculated from the experimental data of Liou (1973). (m) As in Helgeson (1978). (n) Calculated from the experimental results referred to in Zhu et al. (1994). (p) Calculated from the experimental data of Connolly & Kerric (1985). (q) Perkins et al. (1980). (r) As in Hemingway et al. (1982). (s) Richet et al. (1982). (t) Calculated from calorimetric measurements referred to in Zhu et al. (1994). (u) Gunnarsson & Arnórsson (1999). (v) Stefánsson (1999). (w) Calculated from the experiments of Harlov & Newton (1992). (x) Calculated from the experimental data of Ferry & Spear (1978). (y) Robie & Hemingway (1984). (z) Circone & Navrotsky (1992). Some of the references in this table are not given in the reference list below. They can be found in Holland & Powell (1998) or Berman (1988).

Table 2

logK-temperature equations for secondary mineral-solution rections. They are valid from 0° to 350°C at 1 bar below 100°C and at Psat at higher temperatures. The equations are consistent with the data reported in table 1 on the minerals and data on the aqueous species as given by Arnórsson and Stefánsson (1999), Arnórsson and Andrésdóttir (1999), Stefánsson (1999) and Arnórsson et al. (1996).

$$logK = k_1 + k_2/T^2 + k_3/T + k_4*T + k_5*T^2 + k_6/T^{0.5} + K_7*logT$$

Minerals	Reaction	k_1	$k_2 \times 10^{-3}$	$k_3 \times 10^{-2}$	$k_5 \times 10^5$	$k_6 \times 10^{-2}$	k_7
Pyrite	$pyr + 2H^+ + H_{2,aq} = 2H_2S + Fe^{+2}$	-78.294		25.647	-1.750		29.144
Troilite	$tro + 2H^+ = H_2S + Fe^{+2}$	-73.125		43.090	-1.067		25.373
Albite	$alb(l) + 8H_2O = Na^+ + Al(OH)_4^- + 3H_4SiO_4$	-96.267	305.542	-39.847	-2.859		35.790
Microcline	$mic + 8H_2O = K^+ + Al(OH)_4^- + 3H_4SiO_4$	-78.552	311.970	-60.805	-2.778		30.308
Quartz	$qz + 2H_2O = H_4SiO_4$	-34.188		1.975	-0.585		12.245
Clinochlore	$cln + 10H_2O = 5Mg^{+2} + 2Al(OH)_4^- + 3H_4SiO_4 + 8OH^-$	-212.369	854.576	-172.956	-16.355		71.041
Clinozoisite	$czo + 12H_2O = 2Ca^{+2} + 3Al(OH)_4^- + 3H_4SiO_4 + OH^-$	-151.881	493.154	-66.456	-7.671		54.254
Daphnite	$daph + 10H_2O = 5Fe^{+2} + 2Al(OH)_4^- + 3H_4SiO_4 + 8OH^-$	-191.072	808.121	-227.418	-16.554		64.199
Epidote	$ep + 12H_2O = 2Ca^{+2} + Fe(OH)_4^- + 2Al(OH)_4^- + 3H_4SiO_4 + OH^-$	-139.483	558.862	-45.156	-5.445		48.327
Prehnite	$pre + 10H_2O = 2Ca^{+2} + 2Al(OH)_4^- + 2OH^- + 3H_4SiO_4$	-123.243	524.286	-81.340	-7.576		44.731
Wollastonite	$wo + 2H^+ + H_2O = Ca^{+2} + H_4SiO_4$	-41.064	119.067	50.238	-1.061		15.025
Grossular	$gro + 4H^+ + 8H_2O = 3Ca^{+2} + 2Al(OH)_4^- + 3H_4SiO_4$	743.104	-892.096	605.718	-4.864	-86.477	-172.415
Hematite	$hem + 5H_2O = 2Fe(OH)_4^- + 2H^+$	-15.151	138.237	-91.462	-0.705		1.200
Anhydrite (a)	$anh = SO_4^{-2} + Ca^{+2}$	78.414		-32.472	-0.903		-28.723
Calcite (a)							

(a) the logK temperature equation was calculated from the experiment of Yeatts and Marshall (1969) taking into consideration Na-SO4 ion pairing.

(b) Arnórsson et al. (1982).

assumptions are made in the WATCH program, when calculating the aquifer pH that relates to phase segregation and degassing in the aquifer, causing considerable uncertainty in the calculated pH values. This uncertainty is considered to be the principal cause of the scatter of the data points for prehnite. The results are taken to indicate that the reservoir waters in Krafla and Námafjall closely approach equilibrium with this mineral. The picture for the clinozoisite-epidote solid solution and grossular is similar to that of prehnite. In the case of epidote the scatter is much larger than anticipated from known variation in its composition in the Krafla field, ($X_{Fe} =$ 0.59-0.76, Sveinbjörnsdóttir 1992). The samples are saturated to supersaturated with the pure epidote component. This is considered to be an artifact.

The WATCH program assumes sulphide/sulphate redox equilibrium and the redox potential calculated on the basis of that assumption is used to compute ferrous/ferric aqueous ratios. Redox equilibrium between sulphide and sulphate may not be closely approached in the Krafla and Námafjall reservoirs. It appears that the activities of the ferric iron species are overestimated leading to too high values for the solubility products involving such species and in the case of epidote an apparent supersaturation. The same also applies to hematite (Figure 1). Overestimation of the ferric iron species activities leads to simultaneous underestimation of the ferrous iron species concentrations, .causing the large apparent troilite and pyrite undersaturation. Had the ferric iron species been selected when writing the reaction for these sulphides troilite and pyrite under-saturation would still be indicated. Redox disequlibrium between sulphate, sulphide and pyrite is considered to be the reason for the apparent pyrite undersaturation. Equilibrium is indicated for the Mg component in chlorite. The results are uncertain for daphnite due to inadequate evaluation of the Fe-species distribution by the WATCH program.

5 DISCUSSION AND CONCLUSION

The Krafla and Námafjall reservoir fluids seem to have closely approached equilibrium with hydrated Al-silicates of relatively complex composition found as alteration minerals in the rock as well as with minerals of simpler composition, as has been verified by earlier studies (Arnórsson et al. 1983).

According to Kristmannsdóttir (1984) wollastonite and garnet occur in basalts which have undergone contact metamorphism. Thus, equilibrium with these minerals may only be attained under special conditions. Troilite (pyrrhotite) does not seem to be stable although found in the geothermal systems. The formation of this mineral may be linked to boiling and cooling in upflow zones. These processes cause the troilite solubility product to increase, leading to its

precipitation although troilite is not stable in the reservoir fluid below the zone of extensive boiling. It is concluded that fluid-secondary mineral equilibria are attained, or closely approached, for all major components incorporated into these minerals in the Krafla and Námafjall geothermal systems.

REFERENCES

Ármannsson, H., Á. Gudmundsson & B. S. Steingrímsson 1987. Exploration and development of the Krafla geothermal area. *Jökull* 37: 13-30.

Arnórsson, S. 1995: Geothermal systems in Iceland. Structure and conceptual models-1. High temperature areas. *Geothermics* 24: 561-601.

Arnórsson, S., & A. Stefánsson 1999. Assessment of feldspar solubility constants in water in the range 0° to 350°C vapor saturation pressure. *Am. J. Sci.* in press.

Arnórsson, S., & A. Andrésdóttir 1999. The dissociation constants of Al-hydroxy complexes at 0-350°C. *This volume.*

Arnórsson, S., K. Geirsson, A. Andrésdóttir & S. Sigurdsson 1996. *Compilation and evaluation of thermodynamic data on aqueous species and dissociational equilibria in aqueous solutions.* Report of the Science Institute, University of Iceland, RH-17-96: 20 p.

Arnórsson, S., E. Gunnlaugsson & H. Svavarsson 1983. The chemistry of geothermal waters in Iceland. II. Mineral equilibria and independent variables controlling water compositions. *Geochim. Cosmochim. Acta* 47: 547-566.

Arnórsson, S., S. Sigurdsson & H. Svavarsson 1982. The chemistry of geothermal waters in Iceland. I. Calculation of aqueous speciation from 0° to 350°C. *Geochim. Cosmochim. Acta* 46: 1513-1532.

Berman, R. G. 1988. Internally consistent thermodynamic data for minerals in the system $Na_2O-K_2O-CaO-MgO-FeO-Fe_2O_3-Al_2O_3-SiO_2-TiO_2-H_2O-CO_2$. *J. Petrology* 29: 455-522.

Gunnarsson, I. & S. Arnórsson 1999. Amorphous silica solubility, and thermodynamic properties of H_4SiO_4 in the range 0-350°C at P_{sat}. *Geochim. Cosmochim. Acta*, submitted.

Hemingway, B. S., J. L. Haas & G. R. Robinson 1982. Thermodynamical properties of selected minerals in the system $Al_2O_3-CaO-SiO_2-H_2O$ at 298.15 K and 1 bar (10^5 pascal) pressure and at higher temperatures. *U.S. Geol. Surv. Bull.* 1544, 70 p.

Holland, T. J. B. & R. Powell 1998. An internally consistent thermodynamic data set for phases of petrological interest. *J. metamorphic Geol.* 16: 309-343.

Kristmannsdóttir, H. 1984. Chemical evidence from Icelandic geothermal systems as compared to submarine geothermal systems. In *Hydrothermal processes at seafloor spreading centers* (Edited by Rona, Bostrom, Lauber and Smith): 291-319. New York: Plenum Publ. Corp.

Stefánsson, A. 1999. Dissolution of primary minerals of basalt in natural waters I. Mineral solubilities from 0 to 350°C. *Chem. Geol.*, in press.

Sveinbjörnsdóttir, Á. E. 1992. Composition of geothermal minerals from saline and dilute fluids - Krafla and Reykjanes, Iceland. *Lithos* 27: 301-315.

Yeatts, L.B. & W.L. Marshall 1969. Apparent invariance of activity coefficients of calcium sulfate at constant ionic strength and temperature in the system $CaSO_4-Na_2SO_4$ $NaNO_3-H_2O$ to the critical temperature of water. Association equilibria. *J. Phys. Chem.* 73: 81-92.

Geochemistry of the Earth's Surface, Ármannsson (ed.)© 1999 Balkema, Rotterdam, ISBN 90 5809 073 6

The Hekla 1947 eruption, rise and fall of the volcanic plume

Ármann Höskuldsson
South Iceland Institute of Natural History, Westmann Islands, Iceland

ABSTRACT: This paper presents the modelled versus observed evolution of the Plinian plume above Hekla volcano during the Plinian eruption 1947. This eruption is the most powerful eruption of Hekla during this century. The volcano had been dormant since 1845 or for about 102 years. It has been shown that eruptions of Hekla volcano tend to be more explosive if dormancy periods are long than short. First observations of the eruption at Hekla 1947 were made at 06:47 hrs on the morning of March 29th. The plume rose rapidly up to about 30 km height in about 21 minutes. The ash in the plume was th en carried out towards the south, across the British isles and across to Scandinavia.

Field measurements of the deposits on the south coast of Iceland were carried out to obtain distribution patterns of lithics around the volcano. Correlation between observed and calculated parameters for the eruptions is good and suggests that past eruption at Hekla can be modelled with good accuracy in the future.

1 INTRODUCTION

The Hekla eruption in 1947 began early morning on the 29th of March. Little precursor activity occurred prior to the eruption, as is one of the characteristics of Hekla eruptions. First observations of the plume above Hekla were made at 06:47 hrs. In about 21 minutes it rose rapidly up to about 30 km height. Ash was blown towards the south and then eastward all the way to northern British islands and Scandinavia (Figure 1).

In this paper the distribution of fall out deposits from the 1947 Hekla eruption shall be presented. Further, observations during the eruption shall be compared to data obtained from the tephra deposits.

2 PLUME EVOLUTION

Evolution of the volcanic plume during the first hours is quite well documented from photographs (Thorarinsson 1957). The plume had risen up to its maximum height of 27-30 km in only about 21 minutes, it then stabilised at 10 km altitude after 1:13 hours (Figure 2). Several pyroclastic-flows were observed descending from the plume during the first phase of its evolution. Generation of pyroclastic-flows from the volcanic plume can be related to its overload in volcanic glass particles.

3 FIELD OBSERVATIONS

In the summer 1996 the tephra layer from the 1947 eruption was sampled and grain size distribution of lithics and volcanic glass particles measured. Field data were then plotted as a function of distance from Hekla volcano (Figure 1). The results of these measurements were then used to estimate the height of the volcanic plume according to methods given by Sparks et al. (1997).

4 OBSERVED VERSUS CACULATED PLUME

A 20 m/s wind speed was recorded on the 29th of Mars 1947 in the upper layers of the atmosphere. By using the maximum distribution of Md=3 cm size lithics along axis (about 13 km) and recorded wind speed, the calculated maximum height of the volcanic plume is about 26 km (Figure 3). However, applying cross axis distribution of the lithics (about 6 km), the maximum height of the plume is somewhat lower, or around 24 km (Figure 3). This can partly be explained by column collapse and generation of pyroclastic-flows that should effect the tephra distribution close to the volcano. Measurements of effusion rate during the eruption gave a maximum of 13,500 m^3/s (DRE) during the

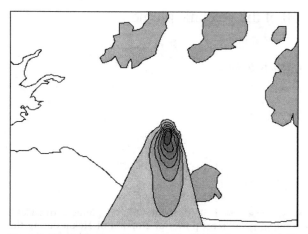

Figure 1. The distribution of lithics from the Hekla volcano. Isopach lines mark 0 to 7 cm size lithics.

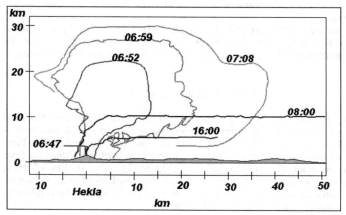

Figure 2 Evolution of the volcanic plume during the first day of the eruption, after Thorarinsson 1957.

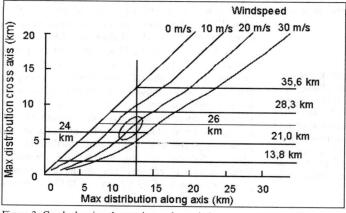

Figure 3. Graph showing the maximum downwind range versus crosswind range for lithic clasts in the size range of 3.2 cm and density of about 2500 kg m[-3].

508

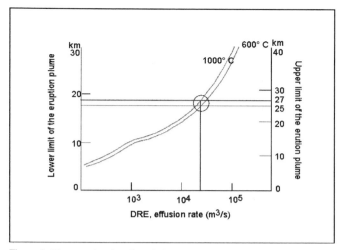

Figure 4. Diagram showing the relationship between upper and lower limits of the eruption column and the effusion rate. This indicates that the eruption column did reach about 27 km height

first hours (Einarsson 1949). The average eruption rate was calculated from the amount of tephra ejected during the first hours. According to the effusion rate, the maximum height of the eruptive column should have reached 27 km with a lower limit at around 18 km (Figure 4).

5 CONCLUSIONS

Results of measurments of the deposits are in good agreement with the observed evolution of the volcanic plume above Hekla early in the morning of March 29. Observations show that the upper limit was between 27 and 30 km and the lower limit at about 18 to 19 km.

Results of measments of the maximum size of lithics within the tephra sector from the eruption
indicates that if the eruption column were to be judged from it, an estimate of some 24-26 km maximum height would be obtained. Using effusion rate, the lower limit of the column is calculated to be about 18 km height. That is close to what was observed during the eruption in 1947. Effusion rate does also correlate well with maximum height observations.

The results show that studies on ancient volcanic eruptions can be used to predict future eruption conditions and evolution of eruption clouds to ensure the safety of people living close to volcanoes and to make future aviation safer.

REFERENCES

Einarsson, T. 1949. The rate of production of material during the eruption. In: T. Einarsson, G. Kjartansson & S. Thorarinsson (eds). *The eruption of Hekla 1947-48 IV,2.* Reykjavík: Vísindafélag Íslendinga.
Sparks, R.S.J., M.I. Bursik, S.N. Carey, J.S. Gilbert, L.S. Glaze, H. Sigurðsson & A.W. Woods. 1997. *Volcanic Plumes.* Wiley: New York.
Thorarinsson, S. 1957. The tephrafall from Hekla on March 29th 1947. In: T. Einarsson, G. Kjartansson & S. Thorarinsson (eds). *The eruption of Hekla 1947-48 II,3.* Reykjavík: Vísindafélag Íslendinga,

Geochemistry of the Earth's Surface, Ármannsson (ed.) © 1999 Balkema, Rotterdam, ISBN 90 5809 073 6

Redistribution of REE in aquifers in the Netherlands

D.J. Huisman, B.J.H. van Os & G.Th. Klaver
NITG-TNO, Geological Survey of the Netherlands, Haarlem, Netherlands

J.J. van Loef
Interfacultair Reactor Instituut, Delft, Netherlands

ABSTRACT: REE-patterns in shallow aquifers in the Netherlands are probably derived from the weatheirng of clay minerals, but due to differing compexing behaviour, the LREE concentrations are relatively low. Deep groundwater has low REE-concentrations, and they appear to be derived mainly from feldspar weathering. Sinks for REE, such as organic-rich layers and Fe-oxides show REE patterns comparable to the shallow groundwater. However, locally short-distance translocation of REE is determined.

1 INTRODUCTION

The rare earth elements (REE) are used increasingly to trace groundwater, and to study mixing of waters from different sources (e.g. Smedley 1991, Johannesson et al. 1997). The main reasons for this are that REE often differ between waters from different rock types, and that they are little influenced by fractionation due to variations in pH, Eh and salinity. This makes it possible to link groundwater to its source, even after it has been influenced by diagenetic processes. REE may also be used for studies on paleo-groundwater composition: Fe-oxides and subsurface organic-rich layers can function as sinks for a whole range of elements that are present in groundwater, including REE (Huisman 1998). These sinks may therefore retain a REE-pattern similar to paleo groundwater composition, after the groundwater-REE has changed.

In this paper we present preliminary results of a study on REE behaviour in the subsurface of the Nehterlands. To do so, we compare and discuss groundwater data from a well-field, and solid-phase data from sections with organic-rich layers and from Fe-oxide concretions (including "rattle-stones"; Van Loeff submitted).

2 REE IN GROUNDWATER

In samples from various depths (30-170 m.) in the well-field at Nuland, three different types of REE-patterns can be discerned (Figure 1).

Shallow groundwater samples show a pattern with low values for the light REE (LREE), and a strong increase in NASC-normalised concentrations from La to Eu. In the range from Eu to Lu, this increase is

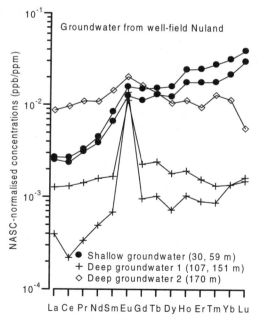

Figure 1. Examples of REE-pattern from groundwater from various depths in the Nuland well-field.

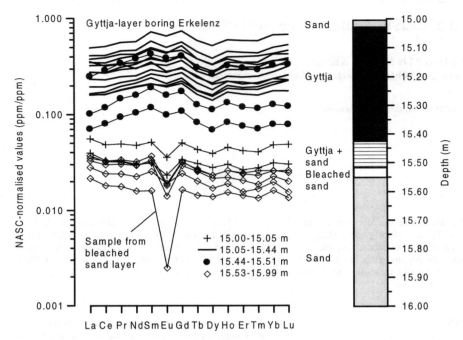

Figure 2. REE-patterns from an organic-rich layer and the immediately surrounding sand; boring Erkelenz (KB 4903/10; courtesy A. Prüfert, Geological Survey of Nordrhein-Westphalen, Krefeld, D).

much less. This pattern is comparable to groundwater from literature (McLennan 1989). The relatively low LREE-values can be attributed to differences in the stability of carbonate complexes. The REE are probably derived from the weathering of clay minerals. This weathering may occur in the aquifer itself, but it is also possible that the REE are inherited from soil weathering processes.

Deep groundwater samples (> 100 m), are very different: they usually have very low concentrations, and show a more or less horizontal pattern, with an extremely high positive Eu-anomaly. This Eu-anomaly indicates that weathering of feldspars is the major source for the dissolved metals. The low REE-contents, compared to shallower groundwater, suggests that in deep groundwater aquifers hardly any weatherable mineral is available.

Figure 2 shows an example of a possible source for the REE in high-Eu groundwater; a (calcite-rich) Pleistocene gyttja-layer which is sandwiched between two reduced, water-bearing sandlayers, shows horizontal, flat REE patterns. The surrounding sand, however, has REE-patterns with small to large negative Eu-anomalies. The most extreme negative Eu-anomaly is from a bleached layer directly underneath the gyttja, which probably has been subject to more severe leaching than the rest of the sand. Such continuous leaching of solid

phases by groundwater would have produced the type of high-Eu groundwater encountered in Nuland (deep samples).

One deep groundwater sample from the Nuland well field (Figure 1, "deep groundwater 2") was found to contain higher contents of REE (more in the range of shallow groundwater), and a more or less flat pattern. The flat pattern indicates that weathering of clays was the most important source for the dissoveld elements. This, and the higher REE-contents, suggest that in the deep groundwater aquifers weathering of clays still occurs locally.

3 REE IN THE SOLID PHASE; SINKS

In several diagenetically influenced settings, REE-patterns are comparable to the shallow groundwater composition. This is the case for browncoal (Figure 3a), Fe-oxide encrusted sand (Figure 3b) and for the mantles of so-called rattle-stones (hollow Fe-oxide concretions with a kernel of clay; Van Loeff, submitted; Figure 3c). Only in the rattle-stone mantles has further fractionation apparently occurred ; redox-processes are probably responsible for the Ce-anomalies. The groundwater-like REE-patterns indicate that in these cases, the trace elements originate from the weathering of clays, and are

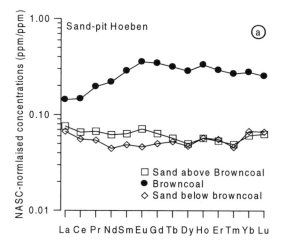

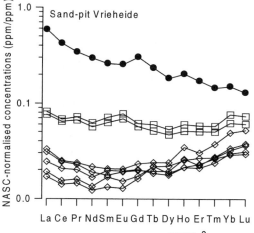

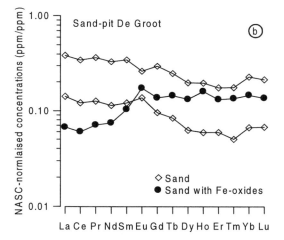

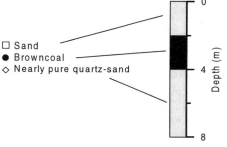

Figure 4 REE-patterns in material from sand-pit Vrieheide.

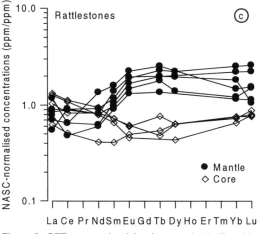

Figure 3. REE-patterns in sinks; browncoal (a), Fe-oxide-encrusted sand (b) and rattle-stones (c).

supplied by carbonate-bearing shallow groundwater.

In the Vrieheide sand pit (Figure 4), the REE have probably been transported only a short distance: Nearly pure quartz sand is depleted in light REE. The overlying browncoal is enriched in REE, and shows a pattern of high light REE that decrease steadily towards th heavy REE. The overlying sand shows horizontal REE patterns. Apparently, the lower sand layer underwent steady leaching during long periods of time. Some of the metals leached were immobilized in the browncoal.

4 CONCLUSIONS

REE in shallow groundwater are mainly derived from the weathering of clays, but fractionation due to carbonate complexation results in lower contents of LREE. Local sinks of REE like Fe-oxides and organic-rich layers usually show similar fractionation patterns.

In the deep aquifers, the groundwater REE are mainly derived from feldspars. This suggests that other weatherable phases (esp. clay minerals) are either already completely weathered, or unavailable for weathering through physical conditions. Feldspars form the most important weatherable phases in the deeper aquifers; the large amounts of deep groundwater compensate for the overall low REE-contents. The occurrence (at least once) of higher amounts of shale-derived REE is probably only of local importance.

Organic-rich layers and Fe-oxides form major sinks for REE and other trace elements, probably due to their absorption onto this matter although they may also be present as phosphates. In these sinks, the REE patterns are indicative for the groundwater composition at the time of the REE-accumulation. In some cases REE-patterns in sinks cannot be linked to a known groundwaterpatterns. They are probably related to short-distance redistribution.

REFERENCES

Huisman, D.J. 1998. *Geochemical characterization of subsurface sediments in the Netherlands.* Ph.D.-thesis Wageningen; Agricultural University

Johannesson, K.H., K.J. Stetzenbach, V.F. Hodge, D.K. Kreamer & X. Zhou 1997. Delineation of groundwater flow systems in the Southern Great Basin using aqueous Rare Earth Element distributions, *Ground Water* 35: 807-819

McLennan, S.M. 1989, Rare earth elements in sedimentary rocks: Influence of provenance and sedimenary processes. In B.R. Lipin & G.A. McKay (eds.) *Geochemistry and Mineralogy of Rare Earth Elements:* 169-200 Reviews in Mineralogy 21, Mineralogical Society of America

Van Loeff, J.J. submitted, *The composition and formation of rattlestones from Dutch soils studied by Mössbauer spectroscopy, INAA and XRD*

Geochemistry of the Earth's Surface, Ármannsson (ed.)© 1999 Balkema, Rotterdam, ISBN 90 5809 073 6

Crustal fluids: CO_2 of mantle and crustal origins in the San Andreas fault system, California

Yousif K. Kharaka, James J. Thordsen & William C. Evans
US Geological Survey, Menlo Park, Calif., USA

ABSTRACT: Chemical and isotopic analyses of water and gases were determined from 44 thermal and saline springs and wells located near the San Andreas fault system to investigate the role of fluids in the dynamics of this major plate-bounding fault. Results indicate that the waters are mainly of meteoric origin, with shallow to moderate circulation depths (down to 6 km). The chemical compositions of water and gases are highly variable, controlled mainly by the enclosing rock types. However, compositions and isotope abundances of noble gases and $\delta^{13}C$ values of HCO_3 indicate a significant (up to 50%) mantle component for volatiles. Because upper mantle gases are dominated by CO_2 ($C/^3He \sim 10^{10}$), model calculations yield high CO_2 flux values of 0.001-1 kg $CO_2/(km^2\text{-s})$. The highest CO_2 fluxes are from fluids in the Franciscan assemblage, where C and He isotopes indicate a dominantly crustal source for CO_2. Presently we are measuring the surficial and dissolved CO_2 fluxes and C isotopes over two large drainage areas where data from springs are available. Preliminary results yield total (uncorrected for soil CO_2) values of ~0.001-1 kg $CO_2/(km^2\text{-s})$. Numerical simulations indicate that the CO_2 flux values, extrapolated to seismogenic depths, are sufficient to generate lithostatic fluid pressures, and thus explain the mechanical weakness of this fault. Furthermore, the model times required to increase fluid pressures to lithostatic values are comparable to those of earthquake cycles.

1 INTRODUCTION

Faults and shear zones generally provide conduits focussing the flow of water and gases in the upper crust. Fault hosted gold-quartz and other metalliferous mineral veins together with associated hydrothermal rock alteration, for example, provide good evidence for large fluxes of deeply sourced fluids along fault systems and shear zones (Kerrich et al. 1984). Detailed site investigations have indicated that fluid flow in these faults, which have experienced repeated fault failure, is episodic and driven by transient lithostatic fluid pressure, consistent with Sibson's (1992) 'fault valve' model during the seismic cycle (Byerlee 1993)

These fluids, on the other hand, can play a critical role in a variety of faulting processes, including earthquakes (Irwin & Barnes 1980). The physical role of fluid pressure in controlling the effective stress and the strength of faults is well recognised (Rice 1992), and the chemical effects of fluids on fault zone rheology are probably as important as their physical role (Kerrich et al. 1984, Hickman et al. 1995). Specific water-rock interactions, including pressure solution, crack growth and healing, and clay formation through retrograde reactions, have been identified that could reduce or increase fault strengths principally by modifying the cohesion strength and coefficient of friction of its gouge (Wintsch et al. 1995, Moore et al. 1996).

We have carried out detailed chemical and isotopic analyses of water and gases from 44 thermal and saline springs and wells located near the San Andreas and associated faults between San Francisco and San Bernardino, California (Kennedy et al. 1997, Kharaka et al. 1998). Currently we are measuring the surficial and dissolved CO_2 fluxes and carbon isotopes over relatively large areas at two locations where data on water and gas compositions from springs are available. Our principal goal is to improve our understanding of the role of fluids on the physics and geochemistry of faulting at seismogenic depths in this system. In this report, we emphasize results of the surficial CO_2 fluxes and the general role that CO_2 plays in the dynamics of the San Andreas fault system.

2 REGIONAL GEOLOGIC SETTING

The San Andreas fault is a major transform plate boundary that has been the locus of tectonic activity in California for more than 30 Ma. This fault system, as described by Wallace (1990), refers to the network of faults with predominantly right-lateral strike slip that collectively has accommodated most of the relative motion (300-350 km) between the North American and Pacific plates. Our study area encompasses that section of the San Andreas fault system that extends from San Francisco to Arrowhead Hot Springs east of San Bernardino.

The principal tectonic units in our study area are: (1) The Salinian block granitoids of Cretaceous age; (2) The Franciscan assemblage, consisting of dis-membered sequences of graywacke, shale, and lesser amounts of mafic volcanic rock, chert, and rare limestone, that are early Jurassic to Cretaceous in age; (3) The Great Valley sequence which comprises an enormous thickness (>15 km) of monotonous sections of interbedded marine mudstone, sandstone, and conglomerate of Jurassic to Cretaceous age. The Coast Range ophiolites are present at the base of the Great Valley sequence, and clastic Cenozoic sedimentary rocks of variable thickness overlie much of these basement units (Irwin 1990).

Tertiary volcanics occur in many localities in the Coast Ranges, notably the andesitic Quien Sabe volcanics in the center of the Diablo Range. The age of these volcanic rocks generally decreases from south to north, being 25-22 Ma for Neenach, 12-7 Ma for the Quien Sabe and 2.2-0.04 Ma for the Clear Lake volcanics (McLaughlin et al. 1996).

3 GEOCHEMISTRY OF WATER AND GASES

Results of chemical and isotopic analyses of water and gases from 44 thermal and saline springs and wells located near the San Andreas fault system are summarized in Kennedy et al. (1997) and Kharaka et al. (1998). The important results are: (1) The δD and $\delta^{18}O$ values establish that waters are mainly of meteoric origin, with no significant mantle or metamorphic components; (2) The chemical compositions of water and gases are highly variable, controlled mainly by the enclosing rock types; (3) Chemical geothermometry gives reservoir temperatures of 80-150°C, indicating shallow to moderate circulation depths for the water of up to 6 km; and (4) The compositions and isotopes of noble gases, especially the $^3He/^4He$ and $\delta^{13}C$ values of HCO_3 indicate a significant (up to 50%) mantle component for CO_2 and other volatiles.

3.1 Chemical and Isotopic Composition of Gases

Gas compositions are variable and controlled mainly by geology and rock types. Samples from the Salinian block granitoids are composed mainly of N_2 of atmospheric origin, and minor CH_4 and CO_2. Samples from the Great Valley sequence are dominated by CH_4 of thermogenic origin. Gas samples from Franciscan rocks are CO_2-rich but also have relatively high concentrations of CH_4 and N_2. The ^{14}C values of bicarbonate (Figure 1) indicate relatively old ages, but • ^{13}C values range from -20 to +23‰, spanning the main known sources of carbon, including organic, marine carbonate and mantle. Samples obtained from CO_2-rich Franciscan rocks have $\delta^{13}C$ range of -6 to -11‰, which are closer to that expected from a mantle ($\delta^{13}C = -4$ to -10‰) source (Trull et al. 1993).

The fluid samples are enriched in total He relative to air saturated water, and have Ra (Ra is the $^3He/^4He$ normalized to air) values from 0.06 to 4.0. Almost all the values are higher than crustal (<0.1 Ra), indicating enrichment in ^3He of mantle (~8 Ra in MORB) origin. Our main conclusion is that the high $^3He/^4He$ ratios indicate a pervasive and currently active mantle degassing throughout the San Andreas fault system that does not correlate with rock provenance, with water types, with the age or distance of nearby volcanic rocks (Kennedy et al. 1997).

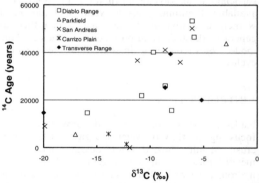

Figure 1. Carbon isotopes of bicarbonate in water from springs and wells.

4 CO_2 FLUX IN THE SAN ANDREAS FAULT

Our model estimation gives the flux of ^3He in the Phillips-Varian well at Parkfield to be ~2×10^{-15} mol/cm^2-year, and extrapolating this rate to the entire San Andreas fault system, as defined by Wallace (1990), yields a modest ~6 mol/year (Kennedy et al. 1997). More importantly, the mantle ^3He is coupled to other more abundant volatiles, especially CO_2. Mid-oceanic ridge basalts originating in the upper mantle, for instance, have a relatively constant $CO_2/^3He$ ratio of ~10^{10} (Trull et al. 1993). Using this ratio and the ^3He flux, yields a very high CO_2 flux of ~10^{11} mol/year for the entire San Andreas system. Because the density, specific heat and viscosity of CO_2 are comparable to those of water at seismogenic depths, CO_2 flux into the fault zone will generate high pore pressures. Bredehoeft & Ingebritsen (1990) used a computer model to show that global CO_2 fluxes, focussed along seismically active areas, could generate high pressures in the crust in time scales of 10^4-10^6 years.

We used the same simulation model, but with temperature, pressure and CO_2 flux data obtained from our study, to model the generation of high fluid pressure by CO_2 flux into the San Andreas fault. Results indicate that for reasonable geologic and hydrologic parameters (Figure 2) operating at a 10 km depth, fluid pressures could increase from hydrostatic (~74 MPa) to lithostatic (~250 MPa) values in a short period of only ~ 200 years using the highest CO_2 flux; a CO_2 flux, that is lower by a factor of 10 from the maximum, yields a period of ~2000 years. Because the stress relieved by an earthquake is probably of the order of ~10 MPa (Lachenbruch & Sass 1992), time periods required to increase fluid pressure to values that would cause fault rupture and trigger an earthquake are probably less than 1/10 of the values in Figure 2. These model results indicate that deep CO_2 flux alone could account for the generation of high fluid pressures that may be responsible for triggering earthquakes in parts of the San Andreas fault; the times required for increased fluid pressures to lithostatic values are comparable to those of earthquake recurrence.

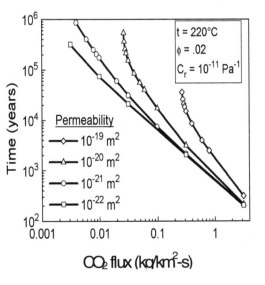

Figure 2. Computed times required to generate lithostatic fluid pressures from hydrostatic values by CO_2 flux at a 10 km depth.

4.1 CO_2 flux at Alum Rock Park

We are currently measuring the surficial and dissolved CO_2 fluxes and C isotopes in the drainage area of Alum Rock Park, located near San Jose, where data on the chemical composition of water and gases from several mineral and thermal springs are available. The CO_2 flux at ground level was measured by the methodology described in Norman et al. (1992) using a LI-COR infrared gas analyser. Measurement sites were selected to cover most of the park, but were concentrated in an area that is ~500 m by 50 m located on both sides of the Penitencia Creek, where the bulk of the mineralized springs are located. Water and gas samples for chemical and isotope analyses were collected from several locations in the Penitencia Creek and other streams and creeks in the drainage area. Water discharge values were determined at the time of sampling to compute the budgets for C and other solutes in the different sections of the drainage.

Preliminary results yield total CO_2 flux values of ~0.05-1 kg CO_2/(km²-s). The higher values are generally, but not exclusively, obtained from the area of the mineralized springs. The calculated load of dissolved CO_2 from a 13.5 km² drainage area was ~340,000 kg/yr, resulting in a dissolved CO_2 flux of ~0.001 kg CO_2/(km²-s). A higher dissolved CO_2 flux of ~0.2 kg CO_2/(km²-s) was calculated from the ~500 m by 50 m area of the mineralized springs.

The surficial and dissolved CO_2 flux values measured at Alum Rock Park, especially those from the mineralized area, are comparable to those calculated from thermal springs and wells (Fig. 2). However, these CO_2 flux values have to be extrapolated to seismogenic depths (5-15 km), which would require decreasing the measured values to account for contributions from soil CO_2 or CO_2 obtained from the shallow zones in the section. The measured values, on the other hand, could be increased substantially at seismogenic depths, because fluid flow would be focussed in a narrower fault gouge zone.

We are currently conducting chemical and isotope analyses of gases obtained from selected sites at Alum Rock Park to investigate the contributions of soil and near-surface CO_2 to the measued flux. It is important to mention that measurements of CO_2 flux on the San Andreas fault, carried out by one of us (Evans) near Hollister and reported by Lewicki & Brantley (1998) from the Parkfield segment, obtained comparable values to our results from Alum Rock Park; the ^{14}C and ^{13}C values, however, indicated the gas to be predominantly soil CO_2. Our present study was carried out at the Alum Rock Park, because our previous results indicated high (2-5 L/s) mineral and thermal water discharges from 20-30 springs and seeps in the area. The gas from springs is dominantly CO_2 (Table 1), with minimal contribution from soil gas as indicated from ^{14}C ages of ~50,000 years and $\delta^{13}C$ values of ~ -6 ‰. A deep source, including a mantle component, for the CO_2 in the mineralized area is also indicated by He isotopes that yield an Ra value of 0.34.

Table 1. Chemical composition (in volume percent) of gases obtained from two Alum Rock springs.

Gas	Spring #4 94-SAF-21a	Spring #10 94-SAF-22
He	0.0001	0.0001
H_2	0.0005	0.0006
Ar	0.0306	0.0163
O_2	0.1470	<0.0005
N_2	3.765	3.687
CH_4	3.446	12.148
CO_2	93.14	83.41
C_2H_6	0.0139	0.0229
H_2S	<0.0005	0.0975
Sum	100.5	99.4

5 CONCLUSIONS

To investigate the role of fluids in the dynamics of the San Andreas fault system, we carried out chemical and isotopic analysis of water and gases from (44) thermal and saline springs and wells located between San Francisco and Los Angeles. Results establish that waters are of meteoric origin, with shallow to moderate circulation depths of up to 6 km. The compositions and isotope abundances of noble gases, especially the $^3He/^4He$, and $\delta^{13}C$ values of HCO_3, however, indicate a high (up to 50%) mantle component for CO_2 and other volatiles. Presently we are measuring the surficial and dissolved CO_2 fluxes and C isotopes from the Alum Rock Park drainage areas, where data from several mineralized springs are available. Results yield total (uncorrected for soil CO_2) flux values of ~0.001-1 kg $CO_2/(km^2\text{-}s)$, which are comparable to those calculated from thermal springs and wells. Numerical simulations indicate that the CO_2 flux values, extrapolated to seismogenic depths, are sufficient to generate lithostatic fluid pressures, and thus explain the mechanical weakness of this fault. Furthermore, the model times required to increase fluid pressures to lithostatic values are comparable to those of earthquake cycles on this fault.

REFERENCES

Bredehoeft, J.D. & S.E. Ingebritsen 1990. Degassing of carbon dioxide as a possible source of high pore pressure in the crust: In *The Role of Fluids in Crustal Processes*: 158-164. Washington D.C.: National Academy Press.

Byerlee, J.D. 1993. A model for episodic flow of high pressure water in fault zones before earthquakes. *Geology* 21:303-306.

Hickman, S.H., R. Sibson. & R. Bruhn 1995. Introduction to special section: Mechanical involvement of fluids in faulting. *J. Geophys. Res.* 100B:13,113-13,132.

Irwin, W.P. & I. Barnes 1980. Tectonic relations of carbon dioxide discharges and earthquakes. *J. Geophys. Res.* 85:3115-3121.

Irwin, W.P. 1990. Geology and plate-tectonic development: In R.E. Wallace (ed), *The San Andreas Fault System, California*, U.S. Geol. Surv. Prof. Paper 1515:61-80.

Kennedy, B.M., Y.K. Kharaka, W.C. Evans, A. Ellwood, D.J. DePaolo, J. Thordsen, G. Ambats & R.H. Mariner 1997. Mantle fluids in the San Andreas fault system, California. *Science* 278:1278-1281.

Kerrich, R., T.E. La Tour & L. Willmore 1984, Fluid participation in deep fault zones: Evidence from geological, geochemical, and $^{18}O/^{16}O$ relations. *J. Geophys. Res.* 89:4331-4343.

Kharaka, Y.K., J.J. Thordsen, W.C. Evans & B.M. Kennedy 1998. Fluids and faults: The chemistry, origin and interactions of fluids associated with the San Andreas fault system, California, USA. In G.B. Arehart & J. Hulston (eds) *Proc. Water-Rock Interaction*: 781-784. Rotterdam: Balkeema.

Lachenbruch, A.H. & J.H. Sass 1992. Heat flow from Cajon Pass, fault strength, and tectonic implications. *J.Geophys. Res.* 97:4995-5015.

Lewicki, J.L. & S.L. Brantley 1998. Measurement of CO_2 degassing and its origin along the San Andreas fault, Parkfield, CA. *Eos, Transactions*, 79(45):F928.

Mclaughlin, R.J., W.V. Sliter, D.H. Sorg & others 1996. Large-scale right-slip displacement on the East San Francisco Bay Region fault system, California: implications for location of late Miocene to Pliocene Pacific plate boundary. *Tectonics* 15:1-18.

Moore, D.E., D.A. Lockner, R. Summers & others 1996. Strength of chrysotile-serpentinite gouge under hydrothermal conditions: Can it explain a weak San Andreas fault? *Geology* 24:1041-1044.

Norman, J.M., R. Garcia & S.B. Verma 1992. Soil surface CO_2 fluxes and the carbon budget of a grassland. *J. Geophys. Res.* 97D:18,845-18,853.

Rice, J.R. 1992. Fault stress states, pore pressure distributions, and the weakness of the San Andreas fault: In B. Evans & T.-F. Wong (eds) *Fault Mechanics and Transport Properties of Rocks*: 475-503. San Diego: Academic Press.

Sibson, R.H. 1992. Implications of fault-valve behavior for rupture nucleation and recurrence. *Tectonophysics* 211:283-293.

Trull, T., S. Nadeau, F. Pineau, M. Polve & M. Javoy 1993. C-He systematics in hotspot xenoliths: implications for mantle carbon contents and C recycling. *Earth Planet. Sci. Lett.* 118:43-64.

Wallace, R.E. 1990. General features: In R.E. Wallace (ed), *The San Andreas Fault System, California*, U.S. Geol. Surv. Prof. Paper 1515:3-12.

Wintsch, R.P., R. Christofferson & A.K. Kronenberg 1995. Fluid-rock reaction weakening of fault zones. *J. Geophys. Res.* 100B:13,021-13,032.

Geochemistry of the Earth's Surface, Ármannsson (ed.)© 1999 Balkema, Rotterdam, ISBN 90 5809 073 6

Crystal-chemistry of Mg-Si and Al-Si scales in geothermal waters, Iceland

H. Kristmannsdóttir – *Orkustofnun, Reykjavík, Iceland*

Ph. Ildefonse – *Laboratoire de Minéralogie-Cristallographie, Université Paris, UMR 7590 et IPGP, France*

J. Bertaux – *Laboratoire des Formations Superficielles, IRD, Bondy, France*

A. M. Flank – *LURE, CNRS / CEA / MEN, Orsay, France*

ABSTRACT: Poorly crystalline silicate scales are frequently encountered in Icelandic geothermal utilizations. Magnesium silicate scales form in two kinds of water used in Icelandic district heating systems: heated fresh waters and mixtures of geothermal and fresh waters. Alumino-silicate scales form in some geothermal waters. The scales have fixed compositions, but are poorly crystalline. They have been studied by XRD, FTIR, ^{27}Al MAS NMR and XANES at Mg- and Si-K edge. Three kinds of mineral precipitates were recognized: serpentine-like structure, talc- or saponite-like structures, and Al-bearing opal-like structures. Local structure of these precipitates is controled by water chemistry, those with serpentine-like structures being associated with waters having Mg/Si ratios ranging from 0.4 to 1.25. In contrast, waters with low Mg/Si ratios (0.05) gives talc- or smectite-like structure components. The alumino-silicates appear to precipitate from medium and high-temperature waters at somewhat higher temperature than opaline saturation and precipitation of amorphous silica occurs.

1 INTRODUCTION

Where high temperature geothermal fluids are used for space heating, precipitation of mineral scales frequently occurs and can be a limiting factor for the utilization (Manceau et al. 1995). In Iceland, heating of fresh waters or mixing of fresh water with geothermal water gives rise to Mg-Si scales (Kristmannsdóttir et al. 1980, 1989, Sverrisdóttir et al. 1992) and the direct use of high- to medium-temperature water for district heating may give rise to Al-Si scales (Thorhallsson et al. 1975, Kristmannsdóttir 1989). The scales consist mainly of poorly ordered mineraloids and an accurate knowledge of their intimate structure can only be resolved by using spectroscopic methods. New data on mineral scales have been obtained by a combination of complementary methods including spectroscopic methods sensitive to local order and local molecular environment of elements. The analysed samples are correlated to variations in water chemistry.

2 MATERIALS AND METHODS

Scale samples studied originate from six *hitaveitas*. a) Hitaveita Sudurnesja. Mg-scale #9119 is related

Table 1 Chemical composition of geothermal water in mg/l

Loc-ation	Sudur nes	Reykj ahlid	Laug. nes	Nesja vellir	Laugar ás	Hvera gerði	Seltj. nes
Sample number	94020 7	88016 0	82070	91756	88142	91131	95077
Temp °C	105	96	130	83	98	185	107
pH/°C	8.9/24	7.5/18	9.3/23	8.8/25	9.7/20	7.07*	8.5/24
CO2 #	6	48	20	36	18.5	126	8.0
H2S	0.08	1.3	0.22	0.23	0.63	26	0.09
B	-	-	-	-	-	0.6	0.22
SiO2	13	28	146	21	112	270	98
TOD	153	89	331	-	349	707	3598
(Na	34	9.1	62	9.3	81	146	633
K	1.5	1.3	2.9	1.1	2.1	12	16.5
Mg	6.9	5.5	0.007	4.9	0.108	0.06	0.22
Ca	8.1	9.4	3.1	8.8	4.8	2.4	549
F	0.07	0.12	1.13	0.12	1.88	1.8	0.62
Cl	75	3.4	46	8.4	48	126	1691
SO4	11	5.8	29	8.2	61	40	292
Al	0.003	0.084	-	-	-	0.48	0.021
Fe	0.006	-	-	-	<0.005	0.005	0.008
Mn	0.001	-	-	-	-	-	0.007

* at ambient temperature
\# Total carbonate

to an example of heated fresh water (Table 1) with a high Mg/Si ratio (1.2). b) Hitaveita Reykjahlidar uses heated fresh water (Table 1) with a moderate Mg/Si ratio (0.4). Mg-scale #9108 was collected there. c) Hitaveita Laugarass was used as an example of low temperature geothermal water with a low Mg/Si ratio (4×10^{-5}) and "contaminated" (Table 1) by a natural inflow of fresh water, causing precipitation of Mg-scale #8710. c) Several samples were collected from Hitaveita Reykjavíkur (Laugarnes and Nesjavellir in Table 1), which for a few months in 1991 used mixed medium-temperature geothermal waters and heated freshwater. e) Hitaveita Hveragerðis now also uses heated cold water in part of the system (Mg/Si=0.3) and mixed steam and water in other parts. Before that high-temperature water (Table 1) was used directly for a short time. Mg-scale #8204 and two Al-Si scales, #8111 and #9414 were collected at that site. f) One sample # 8910 was collected from a snow melting system in hitaveita Seltjarnarnes (Table 1), which utilizes slightly saline geothermal water. Chemical data on scales were obtained by using SEM-EDS analyses. Mineralogical data were performed both by powder X-ray diffraction (XRD) and FTIR on bulk samples. The ^{27}Al NMR spectra was collected at Ho=11.7T (130.3 MHz ^{27}Al Larmor frequency) under magic angle-spinning (MAS) conditions at spinning frequencies of 11 kHz. The ^{27}Al chemical-shifts values at the maximum intensity, δ_p, are expressed relative to an external standard of 1M $Al(H_2O)Cl_3$ in H_2O. XAS measurements were performed at the LURE-SUPERACO radiation facility (Orsay, France). Mg-K and Si-K spectra were collected on the SA32 station using beryl and SbIn monochromators respectively. Experimental details are presented by Ildefonse et al. (1995) and Cabaret et al. (1998).

3 RESULTS AND DISCUSSION

3.1 Chemical analyses

SEM chemical analyses are reported in Table 2 for some scales studied (Sverrisdóttir et al. 1992, Kristmannsdóttir 1989). Sample #8710, from the "contaminated" geothermal water has a low Mg/Si ratio of 0.68 close to that of ideal talc (0.75) or saponite-stevensite (0.8-0.67). On the contrary, scales associated with heated fresh waters Mg/Si=1.2 to 0.4) have higher Mg/Si ratios of 1.22-1.14 close to that known as serpentine like minerals (1.5).

Table 2. Chemical composition of precipitates

Samples	8710	9119	9108	7401	8910
SiO$_2$	64.85	50.07	56.21	61.50	61.0
Al$_2$O$_3$	3.24	2.31	0.29	11.70	6.8
FeO	0.71	0.66	0.16	3.62	-
MgO	29.68	41.20	41.81	-	-
CaO	1.50	0.27	0.49	4.11	-
Total	99.98	94.51	99.96	-	-
Mg/Si	0.68	1.22	1.14	-	-
Al/Si				0.22	0.13

Also shown in Table 2 are semiquantitative XRF analyses of two alumino-silicates. Sample #7401 is

from Hitaveita Hveragerðis during the time of direct use of high-temperature geothermal water. Sample #8910 is from a snow melting system in Hitaveita Seltjarnarness, sample #7401 is typical for the scales formed in high-temperature waters with Al/Si about 0.2, whereas the ones precipitated in medium-temperature waters, like #8910, have lower Al/Si ratio of about 0.1. The relative Al/Si ratios in those two types of waters are about 0.2 and 0.004.

3.2 Powder X-ray diffraction data

XRD patterns of the various scale samples are showed in Figure 1. Three kinds of XRD traces may be distinguished among the samples studied.

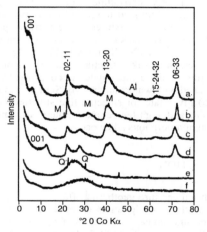

Figure 1: XRD patterns of scales studied. a: #8204, b: #8710, c: #9108, d: #9119, e: #9414, f: #8111. Q: quartz, M: magnetite, Al sample holder

The XRD trace of #8710 and #8204 closely resemble those of trioctahedral smectites (Decarreau et al. 1987). They display a broad 001 basal reflection between 5,1 and 6,5 ° 2θ and two-dimensional hk diffraction bands at 4.56Å, 2.61Å, 1.7Å and 1.536 Å. These features are characteristic of 02-11, 13-20, 15-24-31 and 06-33 diffraction bands of a turbostratic layer silicate (Brindley & Brown 1980, Decarreau et al. 1987). The b dimension deduced from d (06-33) equals 9.216 Å. These XRD patterns also display a broad diffuse band near 25-27° 2θ (3.6-3.3Å), which was tentatively attributed to diffusion by adsorbed water by Decarreau et al. (1987), but which could also correspond to the 004 reflection. Number #8710 differs from #8204 by narrower diffraction bands related to a better stacking order. Numbers #9119 and #9107 show very similar XRD patterns to those of previous samples but (001) diffraction bands occur at higher 2θ angle (7.9-7.57 Å). (06-33) bands at 1.536 and 1.542 Å are related to trioctahedral mineral. These sets of diffraction bands are close to the XRD patterns of serpentine like minerals. XRD trace for Al-Si scales, #8111 and #9414, yield an intense scattering band centered at at 30.2 and 26.2 ° 2θ respectively, characteristic of A-opal. Quartz is present at trace level in #9414, and diffuse bands at the same 2θ positions as those of #8204 and #8710 but with a very weak amplitude can be recognized in #8111.

3.3 FTIR data

IR spectra for scale samples are presented in Figure 2 along with those of chrysotile and talc. Spectra for Mg-Si silicates with low Mg/Si ratios (#8204 and #8710) resemble that of talc or trioctahedral smectites (saponite-stevensite). Spectra are dominated by Si-O-Si and Si-O modes at 1016 and 467 cm⁻¹. Bending Mg$_3$OH occur at 654-664 cm⁻¹ close to that of talc at 670 cm⁻¹. FTIR features of #8204 are larger than those observed for #8710 in agreement with the more disordered character evidenced by XRD. FTIR spectra of #9108 and #9119 are close to that of chrysotile but with larger bands indicating the disordered character of these scales. Si-O-Si and Si-O modes dominate the spectra at 996-1006 cm⁻¹ and 457 cm⁻¹ respectively.

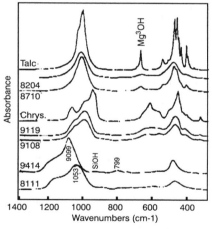

Figure 2: FTIR spectra of scales, chrysotile (chrys) and talc.

OH bending modes at 632, 602 and 560 cm⁻¹ closely resemble the observed features in chrysotile spectrum. On all scale spectra, a small Si-OH contribution around 900 cm⁻¹ may also be considered. Spectra for Al-Si scales resemble that of opal (Webb & Finlayson 1987) displaying specifically an antisymmetric Si-O-Si stretching band at 1099 (#9414) and 1053 cm⁻¹ (#8111) and a

symmetric Si-O-Si band at 799 and 794 cm⁻¹ (Moenke 1974). In tectosilicates, the antisymmetric band has been recognized to shift downfield with increasing lattice substitution of tetrahedral Al (Milkey 1960, Webb & Finlayson 1987). In opals, this peak varies from 1100 cm⁻¹ (no Al) to 1090 cm⁻¹ (Al-opal). Thus the value found for the scale #8111 (1053 cm⁻¹) suggests the presence of 4-fold coordinated Al in the opal framework. Weak bands at 938 and 900 cm⁻¹ are related to Si-OH groups.

3.4 ²⁷Al-MAS NMR spectroscopy

The ²⁷Al NMR spectra of AlSi-scales, #9414 and #8111, yield a main resonance with an apparent maximum at δ = 53.7 ppm and 55 ppm respectively (Figure 3). These values are characteristic of tetrahedrally coordinated Al (Müller et al. 1981). The resonance near -6 ppm, observed in #9414, is related to octahedral Al. The resonance near -50 ppm, and quoted "ssb" is a spinning side band. The coordination number of Al derived from NMR is in agreement with IR data presented above for the same sample.

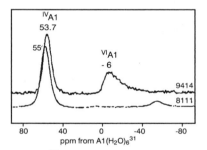

Figure 3: ²⁷Al MAS NMR spectra of AlSi scales

3.5 Mg-K and Si-K XAS data

Mg-Xanes spectra for scale samples are presented in Figure 4 along with those of saponite and talc. #8710 and #8204 spectra present three main edge structures at 1311, 1314 and 1318 eV and a third one at higher energy at 1331 eV. A fourth feature with a low amplitude is also present at 1325 eV. These spectra are close to those of talc and saponite and indicate that the molecular environment of Mg in these two scales is similar to that of Mg in 2:1 trioctahedral phyllosilicates. Spectra of the two others scales slightly differ mainly with respect to the respective amplitude of the three main structures at low energy, by the absence of the low amplitude feature and finally by a small shift at lower energy of the high energy feature.

Si-K Xanes spectra of scales studied together with those of model compounds (saponite,

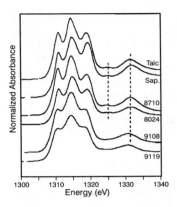

Figure 4: Mg-K Xanes of Mg-scales, talc and saponite (sap.)

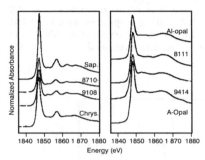

Figure 5: Si-K edge spectra of Mg- and AlSi-scales, chrysotile, saponite (sap.), A-opal and Al-opal (S2 in Manceau et al.1995)

chrysotile, A-opal, and Al-opal from Salton sea geothermal field) are presented in Figure 5. Two sets of spectra may be considered with reference to the chemistry of samples considered.

Mg-scales yield close spectra characterized by an intense white line at 1847 eV, and low amplitude multiple scattering features related to a sheet structures as evidenced by the comparison with spectra of saponite and chrysotile. Only slight differences may be observed between TO and TOT lattices. Si-K Xanes of Si-scales differ strongly from those of Mg-scales. They present a sharp white line at 1847 eV, and very week features at higher energy related to a disordered silica polymorphs (A-opal and Al-opal).

4 CONCLUSION

Crystal chemistry of the poorly crystalline scales found in Iceland differ according to kind of waters used in hitaveita. Local structure of such precipitates is controlled by water chemistry, those with serpentine-like structures being associated with

waters having Mg/Si ratios ranging from 0.47 to 1.25. In contrast, waters with low Mg/Si ratios (0.05) give talc- or smectite-like structure components. The alumino-silicates seem to precipitate from medium and high-temperature waters at temperatures above opaline saturation and precipitation of amorphous silica occurs.

ACKNOWLEDGMENTS

We thank J.B d'Espinose (ESPCI) for the ^{27}Al NMR data. This is an IPGP contribution number xxx.

REFERENCES

Brindley, G. W. & G. Brown 1980. *Crystal structures of clay minerals and their X-ray identification.* London :Mineralogical Society.
Cabaret, D., Ph. Ildefonse, Ph. Sainctavit & A.M. Flank, 1998. Full multiple scattering calculations on pyroxenes at Mg-K edge. *American Miner.* 83: 300-304.
Decarreau, A., D. Bonnin, D. Badaut-Trauth, R. Couty & P. Kaiser 1987. Synthesis and crystallogenesis of ferric smectite by evolution of Si-Fe coprecipitates in oxidizing conditions. *Clay Miner.* 22: 207-223.
Ildefonse, Ph., G. Calas, A.M. Flank, & P. Lagarde 1995. Low Z elements (Mg, Al, and Si) K-edge X-ray absorption spectroscopy in minerals and disordered systems. *Nuclear Instr. Method*s in *Phys. Res.* B97: 172-175.
Kristmannsdóttir, H. 1980. Magnesium silicate scaling in Icelandic district heating systems. *Proc. 3rd. Int. Symp. Water-Rock Interaction*: 110-111.
Kristmannsdóttir, H. 1989. Types of scaling occuring by geothermal utilization in Iceland. *Geothermics*, 18: 183-190.
Kristmannsdóttir, H., M. Ólafsson & S. Thorhallsson, 1989. Magnesium silicate scaling in district heating sysstems in Iceland. *Geothermics* 18: 191-198.
Manceau, A., Ph. Ildefonse, D. Gallup & A.M. Flank, 1995. Crystal Chemistry of Hydrous Iron Silicate Scales Deposits at the Salton Sea Geothermal Field. *Clays and Clay Minerals*: 43, 304-317.
Milkey, R. G. 1960. Infrared spectra of some tectosilicates. *Amer. Mineral.* 45: 990-1007.
Moenke, H. H. W. 1974. Silica, the three-dimensional silicates, borosilicates and beryllium silicates. In V.C. Farmer, Ed., The infra-red spectra of minerals. *Mineralogical Society Monograph*, 4: 365-382.
Müller, D., W. Gessner, H.J. Behrens, & G. Scheler 1981. Determination of the aluminium coordination in aluminium-oxygen compounds by solid-state high resolution ^{27}Al NMR. *Chem. Phys. Letters*, 79: 59-62.
Sverrisdóttir, G., H. Kristmannsdóttir & M. Ólafsson, 1992. Magnesium silicate scales in geothermal utilization. *Proc. 7th. Int. Symp. Water-Rock Interaction*: 1431-1434.
Thorhallsson, S., K. Ragnars, S. Arnorsson & H. Kristmannsdóttir 1975. Rapid scaling in two district heating systems. *Proc. Second UN Symposium on the development and use of geothermal resources*, San Francisco, U.S.A.: 1445-1449.
Webb, J. A. & B. L. Finlayson 1987. Incorporation of Al, Mg, and water in opal-A: Evidence from speleothems. *Amer. Min.*, 72: 1204-1210.

Geochemistry of the Earth's Surface, Ármannsson (ed.)© 1999 Balkema, Rotterdam, ISBN 90 5809 073 6

Thermal springs around the Quaternary volcano Ankaratra, Madagascar

A. Minissale – *CNR, Study Center for Minerogenesis and Applied Geochemistry, Florence, Italy*

O. Vaselli & F. Tassi – *Department of Earth Sciences, University of Florence, Italy*

G. Magro – *CNR, Institute of Geochronology and Isotopic Geochemistry, Pisa, Italy*

F. Pezzotta – *Museum of Natural History, Milan, Italy*

ABSTRACT: Madagascar has a large number of discharging thermal springs. Those emerging at the foot of the Quaternary volcanic Massif of the Ankaratra Mts. are characterized by Na-HCO$_3$ composition, high silica content and associated CO$_2$-rich gas phase marked by ^{3}He/^{4}He ratio as high as 3.9 R/Ra. All these characteristics suggest the presence of active hydrothermal systems at relatively shallow depth and active degassing from mantle-derived magmas. On the contrary, far from the Quaternary volcanic area, thermal springs are characterized by much lower salinity, Na-Cl composition, a N$_2$-rich gas phase with near crustal ^{3}He/^{4}He ratio. These springs do not relate with the occurrence of active hydrothermal systems but are typical of areas with active tectonics.

1 INTRODUCTION

Madagascar is a continental island located in an intraplate geological context between the Mozambique Channel and the Indian Ocean. The western part of the island hosts three contiguous sedimentary basins formed after the break-up of Godwana, whereas the central-eastern regions consist of cratonic folded, metamorphosed, migmatized and reworked Precambrian rocks (Besairie 1964), some of which are more than 3,000 Ma in age (Cahen et al. 1984).

In spite of its old crust, Madagascar is also a site of substantial seismic activity (M>5.0), with earthquakes prevalently concentrated in the axial central and northern parts of the island (Bertil & Regnoult 1998). Besides seismicity, such N-S axial areas are also characterized by the presence of both Quaternary alkali-basaltic volcanism and active fault systems, as well as several thermal springs. All these tectonic and thermal features have been triggered by extensive Pliocene tectonics, possibly related to the presence of a bulge of thermal origin located at the centre of the island (Bertil & Regnoult 1998).

This report deels with the chemical composition of some of the thermal springs located in the central part of the island and their associated gas phases. Relations with active tectonics of the island and the presence of active hydrothermal systems at shallow depth in the central volcanic area are investigated.

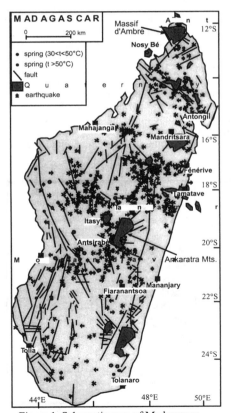

Figure 1. Schematic map of Madagascar

Table 1 Chemical composition (in mg/kg) of thermal springs from central Madagascar

		t	pH	Na	K	Ca	Mg	HCO$_3$	Cl	SO$_4$	SiO$_2$	Li	F	NH$_4$
1	Spring in the lake	36	6.7	880	137	100	48.0	3099	9	3.4	119	3.3	0.03	9.6
2	Manandona	24	7.8	7	1	3	0.7	31	1	0.1	34	0.02	0.05	7.3
3	Ranomafana II	48	9.3	64	1	3	0.1	82	4	74.0	60	0.02	1.50	7.5
4	Ranomafana I	45	9.3	67	1	3	0.1	75	3	81.0	39	0.03	1.80	7.3
5	Ranomafana cold	18	7.2	3	1	1	0.3	6	2	0.6	9	0.02	0.03	1.7
6	Old Spa	52	6.9	1328	152	42	51.0	3447	485	190.0	131	3.70	0.02	4.4
7	Visy Gasy	28	6.5	801	91	116	50.0	2348	360	79.0	98	1.90	0.75	4.5
8	Visy Gasy b	29	6.5	849	94	121	50.0	2318	295	77.5	113	2.00	0.20	-
9	Tsivatrinikamo	22	5.5	5	2	32	9.6	134	18	0.1	35	0.02	0.05	-
10	Hospital spring	43	6.7	1526	154	129	53.0	3507	570	190.0	138	3.80	0.01	19.3
11	Betafo	55	7.8	133	4	17	1.4	170	6	185.0	64	0.10	1.75	19.3
12	Betafo well	22	7.1	19	2	20	6.9	125	4	1.2	32	0.03	0.05	-
13	Miarinavaratra	50	9.3	48	2	2	0.1	101	4	17.5	36	0.04	1.60	19.2
14	Miarinavaratra cold	18	6.0	2	2	1	0.2	10	2	0.2	11	0.03	0.03	-
15	Ranomafana coast	50	9.2	99	2	2	0.1	110	33	75.0	67	0.08	7.00	-
16	Ranomafana coast	50	9.4	93	2	2	0.1	114	26	72.0	68	0.09	8.50	-
17	Brickaville	20	6.0	7	1	4	2.8	31	8	2.55	17	0.04	0.07	-

2 THERMAL SPRINGS OF MADAGASCAR

A map showing location of thermal springs, earthquakes, active faults and Quaternary volcanics in Madagascar is reported in Figure 1. Although thermal springs (>30 °C) are scattered all over the island suggesting a widespread regional active tectonics, most of them (all springs with t > 60°C; Gunnlaugsson et al. 1981) are concentrated around the Quaternary volcanic Massif d'Ambre (1,475 m) and the Nosy Bé-Antongil lineament (2876 m) in the northern part of the island, as well as around the Ankaratra and Itasy Massifs (2642 m in elevation), north of Antsirabé. Location of earthquakes follows the distribution of thermal springs and are both concentrated in the northern parts of the island and radially distributed around the Ankaratra Massif, suggesting a clear relation among them.

The presence of thermal springs around the city of

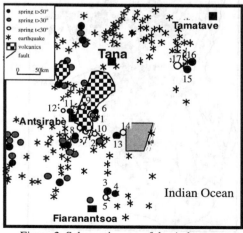

Figure 2. Schematic map of the Ankaratra volcanic Massif.

Antsirabé is well known since the 19th century (Bocquillon-Limousin 1859) and have been mapped by Gunnlaugsson et al. (1981) and studied, from a chemical point of view by Sarazin et al. (1986). In a sampling campaign in central Madagascar in 1998, 9 thermal springs (5 near Antsirabé, 2 east of Fiaranantsoa and 2 south of Tamatave), 6 gas samples (4 associated to thermal waters), 8 cold waters (2 with gas) have been collected. Sampling locations are reported in Figure 2 and water and gas analyses are reported in Tables 1 and 2, respectively.

3 CHEMISTRY

By looking at the Langelier-Ludwig diagram in Figure 3 is evident that the thermal springs located near Antsirabé (# 1, 6 and 10) and some cold springs nearby (# 7 and 8) have a clear Na-HCO$_3$ composition. They also have high salinity (>6000 mg/kg), relatively low pH (<7.0) and elevated silica contents (98-130 mg/kg). On the contrary, thermal springs east of Fiaranantsoa (# 3 and 4) and south of Tamatave (# 15 and 16) have very low salinity (<250 mg/kg), Na-SO$_4$-HCO$_3$ composition, lower silica concentration (<70 mg/kg) and basic pH (>9.0). Cold springs in the latter areas (# 5 and 17) lie on a mixing line between the usual Ca-HCO$_3$ composition of groundwaters and the above mentioned thermal waters. The remaining thermal spring sampled north of Antsirabé (# 11), although located near the volcanic area, has a chemistry very similar to the springs far from Antsirabé.

By considering the composition of the gas phase associated with the springs, at Ansirabè it is >98% CO$_2$ with total He<70 ppm (# 1, 7, 9 and 10) whereas at Fiaranantsoa (# 4) and south of Tamatave (# 15) it is >97% N$_2$ with He as high as 6700 ppm. Hydrogen, H$_2$S, CO and CH$_4$ are averywhere either low or below their instrumental detection limit. A crustal enrichment in [4]He by moving away from the Ankaratra Massif (# 4, 9 and 15) or, alternatively, a

Table 2. Chemical composition of gases (in % by vol.) from central Madagascar

		He	Ar	O₂	N₂	CO₂	CO	H₂	H₂S	CH₄	3/4He
1	Spring in the lake	0.00001	0.0170	0.061	0.65	98.74	<0.00001	<0.00001	<0.005	0.0009	
4	Ranomafana I	0.57060	1.3320	0.571	97.8	0.01	<0.0001	<0.00001	<0.005	0.0496	0.21
7	Visy Gasy	0.00037	0.0051	0.034	0.22	99.89	<0.00001	<0.00001	<0.005	0.0002	1.21
9	Tsivatrinikamo	0.00070	0.0027	0.042	0.19	99.20	<0.00001	0.00001	<0.005	0.0043	
10	Hospital spring	0.00001	0.0022	0.048	0.15	99.68	<0.00001	<0.00001	<0.005	0.0001	3.89
15	Ranomafana coast	0.67750	1.4390	1.731	96.8	<0.01	<0.0001	<0.00001	<0.005	0.1382	

dilution of "crustal" gas with low-He rising gas phase rich in CO_2 around Antsirabé is evident in the triangular N_2-He-Ar diagram of Figure 4 (Giggenbach et al. 1983). Apart from the CO_2-rich thermal springs that precipitate travertine, the rising of CO_2 in the volcanic area near Antsirabé is evidenced also by the presence of cold dry emissions of CO_2 and cold shallow springs rich in CO_2 (# 9). This high CO_2 flow causes, around the Ankaratra Mts., greater water-rock interaction in both deep and shallow aquifers with consequent acquisition by the solutions of a marked Na-HCO3 composition.

4 GEOTHERMOMETRY

Considering the poor industrial potential of the island a geothermal resource would be a good opportunity for the development of Madagascar. Based on the composition of waters and gases, as briefly described in the previous paragraph, the most

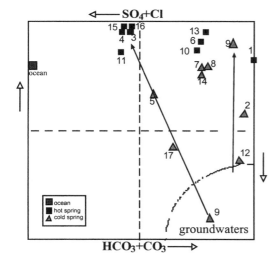

Figure 3. Langelier-Ludwig diagram

promising area from a geothermal point of view is the thermal area around Antsirabé.

By applying the quartz geothermometer (Fournier and Potter 1982) to the less diluted springs of

Antsirabé (# 6 and 10) a minimum deep temperature of 150 °C can be assessed for their feeding aquifers, likely located south of the Ankaratra Massif. Similar temperature estimates have already been proposed in this area by Sarazin et al. (1986). Although at

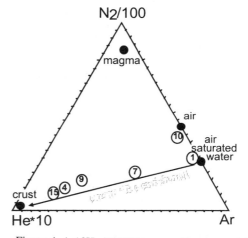

Figure 4. Ar10He-N2/100 ternary diagram rre44

present a potential reservoir cannot be proposed, the presence of shallow localized CO_2-producing hydrothermal reservoir seems to be very likely. Such a speculation is based on the fact that the thermal spring at Betafo (# 11), not far from Antsirabé, does not discharge CO_2.

Lower temperature, around 100 °C, has been assessed for the remaining areas far from the volcanic massif of Ankaratra.

5 HELIUM-3

The ³He/⁴He ratio has been determined in three selected samples, two from the area of Antsirabé (# 7 and 10) and one from near Fiaranantsoa (# 4). The high R/Ra (R= ³He/⁴He$_{(sample)}$; Ra= ³He/⁴He$_{(air)}$) value found in sample # 10 (3.9) near Antsirabé suggests that the source magma that erupted the Quaternary alkali-basalt of the Ankaratra Massif is still degassing primordial ³He to the surface. By assuming a value of 8.0 for the deep mantle R/Ra ratio, a rising of 50% mantle gas can be assessed. The same magma source for the anomalous [3]He can be indirectly invoked to justify the previous described flow of CO_2 in the area around Antsirabé. If mantle-derived magmas are still cooling inside the crust below the Ankaratra Massif and, as suggested by Bertil & Regnoult (1998), they are also the main cause of earthquakes in central Madagascar, it is very likely to hypothesize the presence of shallow hydrothermal systems at depth below the Ankaratra Massif activated by such mantle magmas not completely cooled.

In addition, an indirect evidence of the hypothesized hydrothermal systems can be seen in the other gas sample near Antsirabé. Although not as high as sample # 10, sample # 7 has a R/Ra value of 1.21. This spring is a cold emergence in the same area, and it is possible to hypothesize a secondary cold reservoir for CO_2 in shallower aquifers, much less affected by mantle derived fluids. If this is the case, hydrothermal systems producing CO_2 at more shallow levels with respect to the mantle should be present at the southern boundary of the Ankaratra Mts.

6 CONCLUSIONS

Differences in the chemical composition of springs and gases from near Antsirabé and the two areas south of Tamatave and east of Fiaranantsoa, respectively, have suggested that active CO_2-degassing hydrothermal systems are probably present in association with the Quaternary volcanics of the Ankaratra Massif. Actually, several not sampled thermal springs are present north of the Ankaratra Massif and around Itasy (Figure 2). Although such areas were not accessible during our visit, we know from the literature that some of these springs actively precipitate travertine (Gunnlaugsson et al. 1981). Such precipiation indirectly suggests that emission of CO_2 affects also the thermal springs discharging at the northern boundary of the Ankaratra Massif, from a system that is probably much larger than what we can suppose at present after having analyzed the springs around Antsirabé only.

The presence of thermal springs far from the Quaternary volcanics, such as those N_2-rich near Fiaranantsoa and Tamatave, suggests the rising of deep seated, old (high He concentration) waters along active faults. Such emissions together with those located around the Ankaratra Massif are in line with the seismotectonic model of Madagascar recently proposed by Bertil & Regnault (1998) that have proposed the presence of a thermal bulge in central Madagascar. Based on the anomalous [4]He/He ratio found at Antsirabé, the presence of such a mantle thermal bulge below the Ankaratra

Massif, speculated by the latter authors, seems a very likely possibility.

REFERENCES

Bertil D. & Regnoult J.M. 1998. Seismotectonics of Madagascar. *Tectonophysics* 294: 57-74.
Bésairie H. 1964. Carte géologique de Madagascar (1:1000000). *Service Geologique de Madagascar.*
Bocquillon-Limousin M.H. 1859. Analyses d'eaux minérales de Madagascar. *Soc. Hydrologie et Climatologie Méd. Paris Annales* 6: 320-326.
Cahen L., Snelling N.J., Delhal J., Vail J.R., Bonhomme M. & Ledent D. 1984. Madagascar. In: *The Geochronology and Evolution of Africa*: 100-112. Oxford: Clarendon Press,.
Fournier R.O & Potter R.W. 1982. A revised and expanded silica (quartz) geothermometer. *Geothermal Resources Council Trans. 11*: 3-12.
Giggenbach W.F., Gonfiantini R., Jangi B.L. & Truesdell A.H. 1983. Isotopic and chemical composition of Parbaty Valley geothermal discharges, NW Himalaya, India. *Geothermics* 12: 199-222.
Gunnlaugsson E., Arnorsson S. & Matthiasson M. 1981. *Madagascar: reconnaissance survey for geothermal resources.* U.N. Report: Virkir Consulting Group, Reykjavik, Iceland.
Sarazin G., Michard G., Rakotonindrainy & Pastor L. 1986. Geochemical study of the geothermal field of Antsirabé (Madagascar). *Geochem. J.* 20: 41-50.

Geochemistry of the Earth's Surface, Ármannsson (ed.)© 1999 Balkema, Rotterdam, ISBN 90 5809 073 6

Rare earth elements and yttrium in mineral and geothermal waters from crystalline rocks

P. Möller & P. Dulski
GeoForschungsZentrum Potsdam, Germany

ABSTRACT: Abundances of rare earth elements REE and Y (combined to REY) in mineral and geothermal waters are compared with those in the aquifer rocks. All REY/Ca patterns of waters from felsic rocks are more similar than those from mafic rocks and this is attributed to different modes of crystallisation. Anomalous Eu, Y and Ce contents yield information on the mode of fluid-rock interaction. Eu is sensitive to temperature, whereas Y is not. Anomalous Eu content is inherited and may be enhanced at temperatures above 250°C. Ce is sensitive to oxygen fugacity and pH. Y seems to be sensitive to pH and to ligands dominating REY complexation in solution and on surfaces. In general, Y is released more easily from the rocks and is less retained by sorption onto mineral surfaces than REEs.

1 INTRODUCTION

The main alteration reactions controlling the chemical composition of mineral and geothermal water are the decomposition of plagioclase to kaolinite and smectite (Garrels 1967). Although this might explain the molar ratios of Na^+/Ca^{2+} and HCO_3^-/H_4SiO_4 of most waters, these reactions are not relevant to trace elements like the group of the rare earth elements (REE) and yttrium (Y), henceforth termed REY. They originate largely from dissolution of accessory minerals.

It is often difficult to sample (i) the water in its in-situ PT conditions and uncontaminated by surface water or exsolution of CO_2 if present, and (ii) the representative rock that controls the REY abundances in the water. If CO_2 escapes, pH values increase and carbonates precipitate. Since REY are co-precipitated with calcite and aragonite, the changes in PTX conditions during ascent of natural waters will change the original composition, if CO_2 is exsolved. Another aspect is that most aquifers are petrologically inhomogeneous and the residence time of water is locally variable. REY contents of the water seldom represent equilibria with the host rock sampled because of the formation of metastable components, surface coatings on minerals, and kinetics of ion exchange. At best, steady state conditions are reached.

2 REY/Ca PATTERNS OF WATERS

It is appropriate to normalise REY in waters to REY of their individual host rocks (Banks et al. 1999). Because most REYs originate from soluble Ca minerals, it may be also helpful to relate REY to the major element Ca. Additional presence of complexing ions such as phosphate may control the REY distribution.

The source-rock-normalised REY/Ca patterns of some felsic waters from Moldanubian and Variscan granites from Bohemia, Czech Republic, and Black Forest, Germany, (Figure 1) represent a condensed field with a somewhat wider spread for the light REEs (LREEs) than for the heavy REEs (HREEs). The very narrow spread of HREEs and Y indicates that the distribution of these elements between soluble Ca- and REY-bearing minerals is comparable in the sampled granites. The gneiss-related waters from Peterstal deviate from the granitic trend. The mica-schist-related REY/Ca pattern of the geothermal water from Kizildere, Menderes graben, Turkey, is more horizontal than the gneiss-related ones. The marble –related REY/Ca ratios from Kizildere yield the highest REY level of all samples. In general, waters derived from granites show significant similarities, whereas those from marble from metamorphic rocks show higher REY/Ca ratios.

Table 1: Compilation of studied mineral and geothermal waters and corresponding aquifer rocks.[1] Möller (unpubl.); [2] Möller et al. (1998); [3] Möller et. al (1997); [4] Möller et al. (1994); [5] Klinkhammer et al. (1994); [6] Bau [unpubl.]; [7] Michard et al. (1993); [8] Hemond et al. (1994); [9] Bau et al. (1998).

Sampling locality of water	Abbrev.	Bottom hole/sampling temperature °C	Type of water	CO_2+HCO_3 mg/kg	pH below 50°C	Host rock	Abbrev.
Kizildere, Turkey [1]	Kiz	220/100	Na-HCO₃	13 bars	6.8	mica schist [1]	Igd
Kizildere, Turkey [1]	Kiz	220/100	Na-HCO₃	13 bars	6.8	marble [1]	Saz
Kyselka, Czech Republic [2]	Kys-BJ	17	Na-Ca-Cl	2000	5.8	granite [2]	Gr
Kyselka, Czech Republic [2]	Kys-V5	14	Na-Ca-Cl	1800	5.7	granite [2]	Gr
Jachymov, Czech Republic [2]	Jach	34	Ca-Na-HCO₃	14	7.5	granite [2]	Gr
Säckingen, Germany [3]	Säck	25	Na-Ca-Cl	450	6.8	granite [3]	Gr
Bad Peterstal, Germany [3]	Pet	17	Na-Ca-Cl	1610	6.2	gneiss [3]	Gn
Hermersberg, Germany [3]	Herm	17	Na-Ca-Cl	<5	8.2	gneiss [3]	Gn
Bad Wildbad, Germany [3]	Wild	37	Ca-Na-HCO₃	19	7	granite [3]	Gr
Kyselka, Czech Republic [2]	Kys-HJ	11	Na-Ca-Cl	2300	6.2	alkalibasalt [2]	Bas
Continental Deep Drilling Project, Germany [4]	KTB	119/30	Ca-Na-Cl	<100	5.8-8.3	amphibolite [4]	Amph
Vent, East Pacific Rise, 11°, 13°, 21°N [5]	Vent	400	Na-Cl	Low	4	MOR basalt, 21°S [6]	MORB
Dispersed Flow, Teahitia, Society Islands [7]	Disp	30	Na-Cl	Low	5-6	ocean island basalt [8]	OIB
Hokkaido, Japan [9]	Hokk	8	Ca-SO₄	Low	4	basaltic andesite [9]	And

REY/Ca patterns of waters from basic rocks show an extremely wide range of REY levels (Figures 2). The few examples available at the moment can be devided into two groups: (i) waters from basalts and andesite with high levels of REY, and (ii) waters from amphibolite and ocean island basalts (OIB) with low REY contents. They differ by 3 orders of magnitude. With the exception of the high-temperature vent fluids and the low-temperature mineral water from Kyselka, all other patterns are relatively flat.

3. ANOMALOUS BEHAVIOUR OF Ce, Eu, AND Y

Among the REY, Ce, Eu, and Y often behave in an anomalous manner relative to the REE series. Ce and Eu anomalies are defined by the ratio of actual to expected values interpolated between the neighbouring REE. According to its ionic size, Y plots at the place of Ho, the ratio of Y/Ho is preferred to any type of interpolation method.

In the forthcoming discussion anomalies inherited from the host rocks and those acquired by intensive interaction of water with minerals will be distinguished. Ions with different valencies (e.g. Eu^{2+} and Eu^{3+}) or ions with different electron configurations (such as Y and Ho) may be fractionated by differences in sorption kinetics, thus acquiring an anomaly.

3.1. *Europium*

At high temperatures Eu^{2+} is less sorbed onto surfaces and moves faster through pores than the trivalent species (Möller & Holzbecher 1998). Thus, a positive Eu anomaly is acquired with time in

migrating fluids. At temperatures below about 200°C Eu anomalies can only be inherited from host rocks (Bau and Möller 1992) because Eu^{2+} is virtually absent.

Since rocks rarely host large-ion-dominated minerals, into which Eu^{2+} would fit, Eu is enriched in the residual fluids or is sorbed onto mineral surfaces. This fraction of "loosely-bound" Eu on mineral surfaces, often referred to as excess Eu, is more

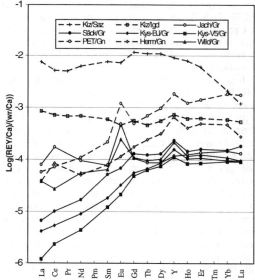

Figure 1: Host-rock-normalised REY/Ca patterns of mineral and thermal waters from felsic rocks. Solid and broken lines represent waters from magmatic and metamorphic rocks, respectively.

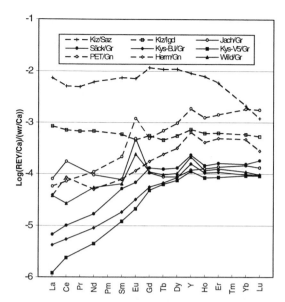

Figure 2: Host-rock-normalised REY/Ca patterns of mineral and thermal waters from mafic rocks. Solid and broken lines represent waters from magmatic and metamorphic rocks, respectively.

easily leachable from rocks by percolating water than the structurally incorporated REE, although at low temperatures all released Eu is trivalent. This excess Eu, is part of any inherited positive anomaly as is the contribution by altered magmatic feldspars, particularly plagioclase, with their positive Eu anomalies. For instance, the high temperature alteration of plagioclase (>250°C) is considered as the source of the strong positive Eu anomalies in black smoker fluids (Figure 2) at mid-oceanic ridges (Klinkhammer et al. 1994). The inherited anomaly is which is further enlarged due to little sorption of Eu^{2+} at high temperatures. Separated biotite from felsic rocks, which host numerous tiny solid inclusions of accessory minerals, shows strongly negative Eu anomalies, although pure biotite has low REY abundances and negligible Eu anomalies (Bea et al. 1994). When biotite in felsic rocks is chloritized (low-temperature reaction), the fluid inherits the high abundance levels of REYs and the negative Eu anomaly from the tiny inclusions (Möller and Giese 1997). Because some Eu in high-temperature rocks is present as excess Eu in the intergranular space, this quantity is leached faster than REYs from dissolving minerals. Thus, Eu in source-rock-normalised REY patterns of waters start with an initially positive Eu anomaly that vanishes with time. This may be the reason, why the thermal waters from Kyselka, Bad Säckingen, and Kizildere

only show insignificant Eu anomalies (Figure 1), although all host granites are characterised by negative Eu anomalies.

3.2 *Yttrium*

Due to the close physical similarity in size and charge, Y behaves similarly to Ho in magmatic systems, even though Y is not a 4*f* element. In fresh to poorly altered igneous rocks, Y and Ho behave in the same way. In alteration processes, however, the small dissimilarity of sorption onto mineral surfaces, controlled by surface complexation (Bau et al. 1996) is multiplied in migrating fluids, and Y-Ho fractionation occurs. Unlike anomalous Eu, the Y/Ho ratio is not principally dependent on temperature. In some groundwaters and thermal waters Y is enriched (see Y and Ho in Fig. 1 and 2). This anomaly is acquired, whilst the rock is not in a transient equilibrium with the migrating water. If such a process lasts long enough, a steady state equilibrium might be reached and Y anomalies may vanish. As shown in Figures 1 and 2, many waters exhibit acquired anomalies of Y.

Where determined, the patterns show enhanced Y abundances in water indicating that the corresponding aquifer rocks are either in disequilibrium with the percolating water, or fresh rocks are constantly involved in the water-rock interaction due to processes of weathering or tectonic events.

3.3 *Cerium*

Strongly negative Ce anomalies are observed in the thermal water of Pamukkale, Turkey (unpubl.) and groundwater from the Calcareous Alps (unpubl.). It is common to all these bicarbonate-rich waters that their pH values are below 7 and their Eh values are positive. It is assumed that the precipitation of FeOOH along the migration paths of the infiltrating meteoric waters induce oxidation to Ce^{4+} which is preferentially sorbed onto freshly precipitated oxyhydroxides. Because most igneous and metamorphic rocks yield waters with low Eh values, negative Ce anomalies are absent from waters derived from such rocks. Among the waters from felsic rocks studied, positive Ce anomalies have been found in the thermal water from Jachymov and the mineral water from Hermersberg. This might be explained by the type of alteration in these areas. For instance, the granite of Jachymov has been altered along with the post-Variscan uranium mineralisation. Since U is only mobilised by oxidising fluids, it is suggested that Ce was also partly oxidised and fixed at mineral surfaces of the altered granite. The present-day water is chemically reducing. If this type of water passes the Jachymov granite, some of the surfacially fixed Ce^{4+} is leached as Ce^{3+} and added to the Ce fraction from dissolved minerals of the leached rocks.

In general, Ce anomalies are not typical features of primary rocks but the result of alteration reactions.

4· CONCLUSIONS

Waters from magmatic and metamorphic rocks exhibit widely varying source-rock-normalised REY/Ca patterns. Waters from felsic rocks are similar, whereas those from mafic rocks show a wide spread of REY/Ca levels. This may be attributed to different modes of crystallisation of the aquifer rocks. The felsic rocks studied are all coarse grained, whereas the basaltic and andesitic rocks are fine grained with variable amounts of glass.

Anomalies of Eu, Y and Ce reveal the history of the fluid-rock interaction. Eu is sensitive to temperature, whereas Y is not. Anomalous Eu is inherited and may be further enhanced at temperatures above 250°C. Ce is sensitive to oxygen fugacity and pH. Y seems to be sensitive to pH and to ligands dominating REY complexation in solution and on surfaces. Some waters acquired enhanced Y/Ho ratios irrespective of their source rocks. Using the anomalies of REY patterns of waters, temperature-dependent reactions can be discussed. Metamorphic waters are geologically old and could have interacted intensively with the aquifer rocks, leading to total removal of excess Eu. Such waters would not show Eu anomalies in source-rock-normalised patterns. Ce anomalies may still occur depending on the development of Eh with time. Although groundwater is geologically very young, it might be in a steady state equilibrium with the aquifer rocks. In such conditions, an acquired positive Y and inherited Eu anomaly indicate temperatures of water-rock interaction below 200°C.

In summary, REYs represent a unique tool to study the behaviour of trace elements in water-rock interaction with time.

REFERENCES

Banks, D., G. Hall, C. Reimann,. & U. Siewers 1999. Distribution of rare earth elements in crystalline bedrock groundwaters: Oslo and Bergen regions, Norway. *Appl. Geochem.* 14:27-39.

Bau, M. & P. Möller 1992. Rare earth element frac-tionation in metamorphogenic hydrothermal calcite, magnesite and siderite. *Min. Petrol.* 45:231-246

Bau, M., A. Koschinsky, P. Dulski & J.R. Hein 1996. Comparison of the partitioning behavoiurs of yttrium, rare earth elements, and titanium between hydrogenetic marine ferromanganese crusts and seawater. *Geochim. Cosmochim. Acta* 60:1709-1725.

Bau, M., A. Usui, B. Pracejus, N. Mita, Y.Kanai, W. Irber & P. Dulski 1998. Geochemistry of low-temperature water-rock interaction: Evidence from natural waters, andesites and Fe-oxyhydroxide precipitates at Nishiki-numa iron-spring, Kokkaido, Japan. *Chem. Geol.* 15:293-307.

Bea, F., M.D. Pereira, L.G. Corretge, & G.B. Fershitater 1994. Differentiation of strongly peraluminous, perphosporous granites: The Pedrobernardo pluton, Central Spain. *Geochim. Cosmochim. Acta* 58:2609-2627.

Bilal, B.A. 1991. Thermodynamic study of Eu^{3+}/Eu^{2+} redox reaction in aqueous solutions at elevated temperatures and pressures by means of cyclic voltammetry. *Z. Naturforsch.*, 46a: 1108-1116.

Garrels R. M. 1967. Genesis of some ground waters from igneous rocks. In P. H. Abelson (ed.) *Researches in Geochemistry* 2:405-420.

Hemond, C., C.W. Devey, & C. Chauvel 1994. Source compositions and melting processes in the Society and Austral plumes (South Pacific Ocean): Element and isotope (Sr, Nd, Pb, Th) geochemistry. *Chem. Geol.* 115:7-45.

Klinkhammer, G.P., H. Elderfield, J.M. Edmond & A. Mitra 1994. Geochemical implications of rare earth element patterns in hydrothermal fluids from mid-ocean ridges. *Geochim. Cosmochim. Acta* 58: 5105-5113.

Michard, A., G. Michard, D. Stüben, P. Stoffers, J.-L. Cheminee & N. Binard 1993. Submarine thermal springs associated with young volcanoes: The Teahitia vents, Society Island, Pacific Ocean. *Geochim. Cosmochim. Acta* 57:4977-4986.

Möller, P., P. Dulski & U. Giese 1994. Rare earth elements in KTB-VB fluids. *Sci. Drill.* 4:113-122.

Möller, P. & U. Giese 1997. Determination of easily accessible metal fractions in rocks by batch leaching with acid cation-exchange resin. *Chem. Geol.* 137:41-55.

Möller P. & E. Holzbecher 1998. Eu anomalies in hydrothermal fluids and minerals: A combined thermochemical and dynamic phenomenon. *Freib. Forsch.-H.* C475:73-84.

Möller, P., G. Morteani, A. Fuganti, P. Dulski. & H. Gerstenberger 1998. Rare earth elements, yttrium and H, O, C, Sr, Nd, and Pb isotope studies in mineral waters and corresponding rocks from NW Bohemia, Czech Republic, *Appl. Geochem.* 13, 975-994.

Möller P., I. Stober & P. Dulski 1997. Seltenerdelement-, Yttrium-Gehalte und Bleiisotope in Thermal- und Mineralwässern des Schwarzwaldes. *Grundwasser* 2:118-132.

Geochemistry of the Earth's Surface, Ármannsson (ed.)© 1999 Balkema, Rotterdam, ISBN 90 5809 073 6

Isotope geochemistry of mineral spring waters in the Massif Central, France

Ph. Négrel, J. Casanova & M. Azaroual
BRGM, Department of Geochemistry and Hydrogeology, Orléans, France

C. Guerrot & A. Cocherie
BRGM, Department of Isotope Chemistry, Orléans, France

Ch. Fouillac
BRGM, Research Division, Orléans, France

ABSTRACT: A multi-isotope (O, D, C, Sr, Nd) study of mineral spring waters in the Massif Central was made in order to characterize the sources and fluid paths of the mineralized spring waters, and the mixing processes within the crystalline bedrocks. The O and D isotopes indicate waters of local meteoric origin. The Sr isotopes reveal three end-members, two of which are mineralized and define a mixing trend. The Nd isotopes generally show signatures similar to those of the parent rocks.

1 INTRODUCTION

Many diffusive mineral springs are scattered throughout the geothermal area of the French Massif Central (Grande Limagne, Cézallier and Margeride regions, Figure 1), where granitoids, gneiss and volcanic deposits constitute the main bedrock lithologies. The Cézallier region, in the centre, is primarily made up of a metamorphic basement (ortho- and paragneiss and leptyno-amphibolite) partially covered by alkaline basalt series. The Grande Limagne (or Limagne) region to the north consists of calc-alkaline monzogranite, and the Margeride massif to the south consists of light and dark granite facies.

The studied mineral springs are characterized by neutral pH (6-7), high CO_2 concentrations (PCO_2 = 0.1-1.0 bar), high dissolved Na, Cl and HCO_3, and high Fe and Al concentrations.

The springs emerging from granitic bedrock in the Limagne are of the Na-Cl-HCO_3 type (Négrel et al. 1997a). The springs emerging from the intensively fractured basement in the Cézallier are CO_2-rich Na-HCO_3 cold springs with a low discharge and are generally accompanied by travertine deposits (Criaud & Fouillac 1986a, b) based on their chemical composition, they have been classified into four groups (Criaud & Fouillac 1986a, b): the north group (7 springs), the centre group (25 springs), the south group (13 springs) and the southwest group (3 springs). The few springs studied in the Margeride are of the Na-HCO_3 type and emerge with a low discharge near a major fault in the west of the massif.

Our study of these mineral springs is based on the use of isotope systematics (O, D, Sr and Nd) in order to characterize the sources and fluid paths of the waters, and the mixing processes within the crystalline bedrocks. Our previous studies in the Massif Central showed the potential of Sr isotopes for deconvoluting

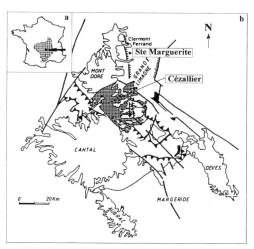

Figure 1. General location map of the study area of the French Massif Central

mixing in small areas (Négrel et al. 1997a 1997b), and it has been our intent here to expand this approach with a more detailed study of a larger area of the Massif Central (Casanova et al. in press, Négrel et al. submitted,).

2 ISOTOPIC CHARACTERIZATION OF THE MINERAL SPRING WATERS

2.1 *Using stable isotope systematics ($\delta^{18}O$, δD, $\delta^{13}C$)*

Deuterium and ^{18}O isotope measurements of the mineral spring waters in the Cézallier indicate that they are of a local meteoric origin. A few waters plot close to the local meteoric water line (LMWL, $\delta D = 8\delta^{18}O + 11.4$, Vuataz et al. 1987), but most plot significantly (up to +2.3‰) to the left of this line (Figure 2). This ^{18}O depletion is known to be linked to a low-temperature equilibration of ^{18}O between dissolved CO_2 and the water (Fouillac et al. 1975). The waters plot within a very narrow δD range implying a common recharge origin, whereas the variable ^{18}O depletion reflects a variable CO_2-H_2O exchange rate. Given that individual springs exhibited a δD fluctuation of up to 7‰ during the period 1985-1997, the δD signature of the mineral springs in 1997, which is common to the whole Cézallier region, suggests that the deep circulation of these fluids may preserve the climatic signature of the recharge waters.

The waters collected at the outlet of the springs and free of atmospheric CO_2 were analysed for the ^{13}C content of their Total Dissolved Inorganic Carbon (TDIC). The measured $\delta^{13}C_{TDIC}$ ranges from +1.9 to +4.2% vs. PDB. When compared to the $\delta^{13}C_{TDIC}$ of the parent solution, calculated for each spring from the presently forming carbonates, the measured $\delta^{13}C_{TDIC}$ data set shows a ^{13}C enrichment of the order of +8‰. This is accounted for by the kinetic fractionation that is linked to the strong CO_2 degassing, which probably occurred upon opening the sampling bottle for analysis. The calculated data set shows a strong covariance between the stable carbon and oxygen isotope contents that reflects both partial equilibrium with atmospheric CO_2 and a temperature effect. The initial $\delta^{13}C_{TDIC}$ signature of the springs, which is probably more negative than the most depleted calculated water (e.g. about -6‰), is difficult to estimate considering that the $\delta^{13}C_{CO_2}$ values for the free CO_2 of the regional hydrothermal system range from -12.4 to -7.3‰.

2.2 Using Sr isotopic systematics

The strontium isotopic composition of the mineral waters in the Massif Central range from near 0.713 in the Limagne and Margeride to 0.718 in the Cézallier (Négrel et al. submitted). The Sr isotopic ratios appear to fall into different groups and the lowest $^{87}Sr/^{86}Sr$ ratios are observed in mineral spring waters that circulated within granitic areas (Limagne and Margeride) whereas highest ratios are observed in the gneissic Cézallier. However, four waters from the Margeride have higher $^{87}Sr/^{86}Sr$ ratios than the other springs emerging from the same granitic bedrock.

The $^{87}Sr/^{86}Sr$ ratios of the mineral spring waters from the Limagne (Négrel et al. 1997a), and the Margeride are plotted vs. 1/Sr in Figure 3.

Three fields corresponding to the Limagne, the Margeride (Mazel spring) and the Margeride (F.B. spring) can be discriminated and reveal two main types of water in the granitic area.

The radiogenic and less-mineralized water from the Margeride may correspond to a shallow groundwater end-member that originates by rapid and sub-superficial water circulation in the granitic bedrock. In

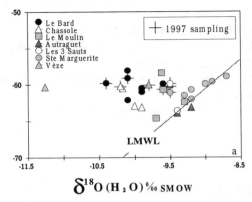

Figure. 2. Isotope composition of the Cézallier and Ste Marguerite springs in a plot of δD (H_2O) vs. $\delta^{18}O$ (H_2O), LMWL: local meteoric water line.

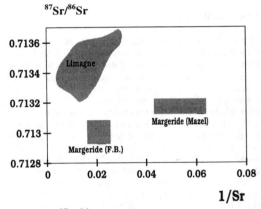

Figure 3. A $^{87}Sr/^{86}Sr$ vs. 1/Sr discrimination diagram for the mineralized waters from Margeride and Limagne.

this reservoir, the water interacts with the most altered part of the granite, which includes the most radiogenic mineral phases. Conversely, concentrated waters with lower $^{87}Sr/^{86}Sr$ ratios, observed in some parts of the Margeride and Limagne, may reflect deeper circulation with a water-rock interaction constrained by equilibrium with Sr-bearing phases with low $^{87}Sr/^{86}Sr$ ratios.

The $^{87}Sr/^{86}Sr$ ratios of the mineral springs from the Cézallier are plotted vs. 1/Sr in Figure 4 and the scatter of points can be explained by three end-member compositions. The most dilute end-member (denominated EM1) agrees with poorly mineralized waters of surficial origin described by Négrel et al. (1997b). The two mineralized end-members (denominated EM2 and EM3) have very different isotopic compositions and Sr concentrations. EM3, the most mineralized, is characterized by a more radiogenic signature and a lower Sr concentration than EM2.

Numerous samples from the north group plot close to EM2 along the trend between EM2 and EM3. Some points from this group show a shift towards lower Sr concentrations associated with a slight decrease in $^{87}Sr/^{86}Sr$; this shift may be influenced by mixing with diluted EM1 waters. Notwithstanding the effect of the poorly mineralized waters, it is obvious that at least two end-members constrain the Sr concentrations and $^{87}Sr/^{86}Sr$ ratios in this area. Selected mineralized spring data can be used to illustrate this binary mixing and samples plot along a straight line between EM2 and EM3, with EM2 contributing 60% of Sr to the north group, 40% to the centre group, and 30% to the south group.

2.3 Using Nd isotopic systematics

The concentrations of individual dissolved REEs and total dissolved REEs in the mineral waters vary over several orders of magnitude and fall in the range 60 to 1700 ng/L (Négrel et al. submitted). The behaviour of dissolved REEs is not sensitive to physico-chemical parameters (pH, total organic carbon, TDS or Na or Cl concentrations). When dissolved REE concentrations are presented as upper continental crust normalized patterns, most water samples are characterized by HREE enrichment [$(La/Yb)_N$ < 0.4, large positive $(Er/Nd)_N$ values] as predicted thermodynamically (Négrel et al., submitted).

In contrast to strontium isotopes, Nd isotopes have not been extensively employed in hydrogeological studies. The $^{143}Nd/^{144}Nd$ ratios are expressed as $\varepsilon_{Nd}(0)$ and Figure 5 illustrates the range in $\varepsilon_{Nd}(0)$ for the studied mineralized waters (Négrel et al., submitted) along with the range for the parent rocks (granites and gneisses). The waters fall in a $\varepsilon_{Nd}(0)$ range of -12 to +4, with 75% in the range of -6 to -10, which is in complete agreement with the range for the parent rocks. The five springs with $\varepsilon_{Nd}(0)$ above -4 are outside the range of the parent rocks, which may reflect a contribution from volcanic rocks.

The lack of general relationship when the $^{87}Sr/^{86}Sr$ ratios are plotted vs. the $\varepsilon_{Nd}(0)$ reflects the discrepancy between strontium and neodymium in crustal material exposed to weathering and the lack of simple mixing of waters with different $^{87}Sr/^{86}Sr$, $\varepsilon_{Nd}(0)$ and Nd/Sr ratios. Although a relatively linear trend appears for some samples, this relationship shows constant $^{87}Sr/^{86}Sr$ ratios and a wide variation in the $\varepsilon_{Nd}(0)$.

The only known rocks in the surface geology capable of inducing a shift in $\varepsilon_{Nd}(0)$ are the basalts, although a surprising feature is the lack of strontium isotopic shift, because basalts have $^{87}Sr/^{86}Sr$ ratios of around 0.703-0.704. An explanation can be found in the concentration fluctuations. Weathering of volcanic basement yields a larger proportion of Nd in solution than Sr and therefore induces an increase in $\varepsilon_{Nd}(0)$ without drastically changing the $^{87}Sr/^{86}Sr$ ratio.

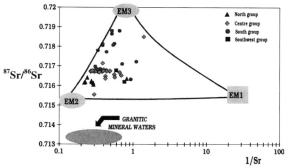

Figure 4. Relationship between $^{87}Sr/^{86}Sr$ and 1/Sr for the mineralized springs from the Cézallier. Mineralized waters from the Margeride and Limagne, i.e. 'granitic waters', are also represented in the shaded field. The 1/Sr scale is expressed in logarithmic units for easier viewing

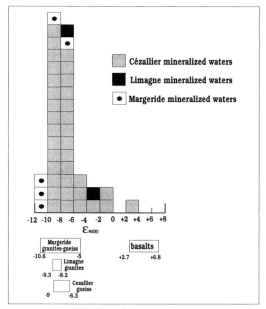

Figure 5. Histogram of $\varepsilon_{Nd}(0)$ in the mineralized waters. The range of parent rocks is also represented (see text).

3 CONCLUSION

The concluding remarks emerging from the multi-isotope investigations in the mineralized waters from the Massif Central are :

1- the D and ^{18}O isotope measurements indicate that the waters are of a local meteoric origin but ^{18}O depletion indicates CO_2-H_2O exchange,

2- the Sr isotopes reveal three end-members, one of which is poorly mineralized. The two, mineralized end-members define a good mixing trend,

3- the Nd isotopes show $\varepsilon_{Nd}(0)$ signatures similar to those of the parent rocks, except when the water circulation encompasses volcanic rocks, which impart a positive shift in the $\varepsilon_{Nd}(0)$.

REFERENCES

Casanova, J., F. Bodénan, Ph. Négrel & M. Azaroual. 1999. Microbial control on the precipitation of ferrihydrite and carbonate modern deposits from the Cézallier hydrothermal springs (Massif Central, France). *Journal of Sedimentary Geology, in press.*

Criaud, A. & C. Fouillac. 1986a. Etude des eaux thermominérales carbogazeuses du Massif Central Français. I. Potentiel d'oxydo-réduction et comportement du fer. *Geochim. Cosmochim. Acta* 50: 525-533.

Criaud, A. & C. Fouillac. (1986b) Etude des eaux thermominérales carbogazeuses du Massif Central Français. II. Comportement de quelques métaux en trace, de l'arsenic, de l'antimoine et du germanium. *Geochim. Cosmochim. Acta* 50: 1573-1582.

Fouillac, C., C. Cailleaux, G. Michard & L. Merlivat. 1975. Premières études de sources thermales du Massif Central français au point de vue géothermique. *2nd Symposium on Development and Utilization of Geothermal Resources, San Francisco* 721-729.

Négrel, Ph., C. Fouillac & M. Brach. 1997a. Variations spatio-temporelles de la composition chimique et des rapports $^{87}Sr/^{86}Sr$ des eaux minérales de la Limagne d'Allier. *C.R. Acad. Sci.*, Paris, série II 324: 119-124.

Négrel, Ph, C. Fouillac & M. Brach. 1997b. A strontium isotopic study of mineral and surface waters from the Cezallier (Massif Central, France) : implications for the mixing processes in areas of disseminate emergences of mineral waters. *Chem. Geol.* 135: 89-101.

Négrel, Ph., C. Guerrot, A. Cocherie, M. Azaroual, M. Brach & C. Fouillac. Rare Earth Elements, neodymium and strontium isotopic systematics in mineral waters : evidence from the Massif Central, France. *Applied Geochemistry,* submitted.

Vuataz, F.D., A.M. Fouillac, C. Fouillac, Michard, G. & M. Brach. 1987. Etude isotopique et suivi géochimique des eaux des sondages de Chassole et de quelques sources minérales du Cézallier. *Géologie de la France* 4: 121-131.

Geochemistry of the Earth's Surface, Ármannsson (ed.) © 1999 Balkema, Rotterdam, ISBN 90 5809 073 6

Helium isotopic composition of fluid inclusions from two Ethiopian geothermal fields

G. Ruggieri, E. Droghieri, G. Gianelli & C. Panichi
International Institute for Geothermal Research, CNR-IIRG, Pisa, Italy

G. Magro
Institute for Geochronology and Isotopic Geochemistry, CNR-IGGI, Pisa, Italy

ABSTRACT: He isotopic analyses have been carried out on the gas extracted from fluid inclusions which trapped geothermal fluids of the Aluto-Langano and Tendaho geothermal fields (Ethiopia). The results suggest an important He contribution to the past geothermal fluids from the high ^{3}He Ethiopian plume that also determined the high ^{3}He/^{4}He ratios of the present-day volcanic and geothermal gas and of the Ethiopian Rift and Afar basalts. The ^{3}He/^{4}He ratios of both present-day and past geothermal fluids were also influenced by crustal and/or upper mantle sources. The He/Ne ratios indicate a moderate or very low input of atmospheric He from meteoric waters. The exceptions are represented by a fluid discharged from a shallow-depth geothermal well and by the gas extracted from fluid inclusions of a shallow-depth sample from Tendaho, which show the lowest He/Ne ratios.

1 INTRODUCTION

The Horn of Africa is characterized by the presence of three active rift systems: the Gulf of Aden, the Red Sea and the Ethiopian rifts. These systems join up in the Southern Afar of Ethiopia, where they constitute a complex triple junction. In this area, tensional tectonics and uplift started in the Upper Eocene and was accompanied by widespread flood basalt eruptions. The main volcanic phase took place during the Oligocene; younger volcanism, mainly basaltic, occurred during the Miocene and Quaternary along the Ethiopian Rift and in the Afar area. Volcanological, petrological and geochemical studies have been carried out to investigated the origin and evolution of this magmatic activity and the possible presence of a mantle plume below the Afar triple junction (see Marty et al. 1996, Scarsi and Craig 1996, and reference therein). Recent analyses on He isotopes extracted from olivines and pyroxenes in basalts of the Ethiopian Rift and the Afar depression showed a wide range of R/Ra (Figure 1; Marty et al. 1996, Scarsi and Craig 1996). These data were interpreted to be the result of the contributions of different sources and in particular they indicated the occurrence of a lower mantle plume which influenced a large area of the Ethiopian volcanic province.

Several geothermal areas, related to the magmatic activity, occur in the Horn of Africa. We have

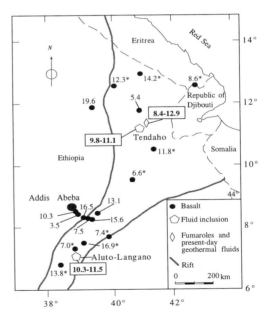

Figure 1. Location of the studied geothermal fields and ranges of R/Ra ratios of the gas extracted from fluid inclusions. The R/Ra ratios of present-day geothermal fluids and of selected basalts are also shown. The data on present-day geothermal fluid and on basalts are from Marty et al. (1996) except those with * that are from Scarsi et al. (1996).

measured the He isotopic compositions and the He/Ne ratios of the fluids trapped in hydrothermal minerals found in drill cores of two Ethiopian geothermal fields, Aluto-Langano and Tendaho. The results have been compared with the existing data on rocks, present-day fluids discharged from geothermal wells, and fumaroles, and have been used to estimate some information on the different contribution to the He of past geothermal fluids.

2 DESCRIPTION OF THE GEOTHERMAL SYSTEMS

2.1 *The Tendaho geothermal system*

This geothermal system, described by Gianelli et al. (1998), is located about 400 km NE of Addis Ababa, in the Afar region (Figure 1), within the NW-SE trending Tendaho Rift, which is a southward extension of the Red Sea Rift system.

Drilling operation are in progress and data of four wells (T1, T2, T3 and T4), with a maximum depth 2200 m below ground level (b.g.l.), are available so far. The subsurface geology consists of an upper unit (Pleistocene to Present) of fluvio-lacustrine sediments interlayered with basalts, and a lower unit (Pliocene to early Pleistocene) of basalt flows. The highest measured in-hole temperature (270°C) was recorded in well T1. Well T3 is characterized by relatively low temperature, with a maximum in-hole temperature of 195°C.

T2 and T4 are productive wells, with an average flow rate of 13 kg/s and 70 kg/s, respectively; well T1 has a very low flow rate, and well T3 is unproductive. The discharged waters from wells T1, T2 and T4 are low salinity (1620-2070 ppm TDS), sodium-chloride geothermal waters with reservoir temperature ranging between 220 and 270°C.

The hydrothermal mineralogy at Tendaho includes calcite and chlorite, which are ubiquitous throughout most levels of the wells, and quartz, albite, adularia, epidote, garnet, pyroxene, laumontite, heulandite, stilbite, wairakite and prehnite (Gianelli et al. 1998). In addition, corrensite, vermiculite and smectite usually occur at shallow depths, whereas in well T3 these clay minerals and low-temperature zeolites were also found in the deep part of the well where they precipitated after higher temperature hydrothermal minerals.

The U/Th absolute age (12 ± 5 ka) of hydrothermal quartz-calcite veins outcropping in the Tendaho area, reported by Gianelli et al. (1998), suggests that geothermal activity is relatively recent in this area.

Fluid inclusion data, compared with present-day measured temperatures indicate either heating (well T1) or cooling (well T2 at 804 m b.g.l.) processes.

Primary and secondary inclusions clearly documented the cooling process that affected well T3.

2.2 *The Aluto-Langano geothermal system*

The Aluto-Langano geothermal field is located about 200 km south of Addis Ababa (Figure 1), close to the eastern margin of the Ethiopian Rift and well inside the presently active axis of the Rift. At Aluto-Langano, eight deep exploratory wells (with a maximum depth of 2500 m b.g.l.) were drilled, five of which are productive (L3, L4, L6, L7, L8). Subsurface geology consists of (from top to bottom): a sequence of silicic volcanic products (Quaternary) of the Aluto volcano, lake sediments (Quaternary), a basaltic unit (Pliocene to Pleistocene) and an ignimbrite unit (Miocene to Pliocene).

The Aluto-Langano geothermal system is characterized by a high-temperature (maximum temperature around 335°C) upflow zone (wells L3 and L6) and by a zone of lateral outflow (wells L4, L5, L7, L8) with maximum temperatures in the range of 180-270°C and local temperature inversions. The geothermal wells produce a two-phase fluid (water + vapour) with a total flow rate of 10.4-25.6 kg/s and partial pressure of CO_2 between 0.6 and 5.8 MPa (Teklemariam et al. 1996). The discharged fluids are alkali-bicarbonate-chloride waters (TDS: 2430-3990 ppm) with reservoir temperatures ranging between 225 and 335°C.

At Aluto-Langano is present a propylitic alteration consisting of quartz, calcite, epidote, garnet, prehnite and actinolite. This alteration is replaced by an argillic alteration (smectite, illite-smectite mixed layers, chloritic intergrades, chlorite-smectite mixed layers, kaolinite and vermiculite) and calcite in the marginal areas of the field, where there are relatively low temperatures and high CO_2 concentrations (Teklemariam et al. 1996).

Comparison of fluid inclusion data with present-day field temperatures indicated that a slight heating process occurred in the upflow zone, whereas the fluid inclusions of the lateral outflow zone were usually trapped at temperatures higher than present-day (Teklemariam et al. 1996).

3 SAMPLES

A total of nine samples (six from the Aluto-Langano and three from the Tendaho geothermal fields) have been selected for gas extraction. The selected samples are quartz and calcite contained in drill cores from variable depths (787 to 2500 m b.g.l.). The gas was extracted from fluid inclusions, of mainly primary and pseudosecondary origin, trapped in hydrothermal quartz and calcite and from secondary fluid inclusions, filled with geothermal

Table 1. Sample description, fluid inclusion characteristics, and results.

Sample/depth b.g.l. (m)	Mineral	Fluid inclusion types	Average Th (°C)	R/Ra	He/Ne
L4/1001	hydrothermal calcite	primary and pseudosecondary	223	11.2	6.3
L4/1001	hydrothermal quartz	primary and pseudosecondary	238	10.9	7.1
L7/798	magmatic quartz	secondary	250	11.4	4.0
L7/2040	magmatic quartz	secondary	286	11.0	4.7
L8/787	magmatic quartz	secondary	227	10.3	5.2
L8/2500	hydrothermal calcite	primary and pseudosecondary	309	11.5	175
T1/1500	hydrothermal quartz	primary	215	11.1	2.2
T1/1500	hydrothermal calcite	primary and pseudosecondary	-	9.8	10.2
T2/804	hydrothermal calcite	primary and pseudosecondary	230	9.9	9.9

b.g.l. = below the ground level, Th = homogenization temperature

fluids, trapped in magmatic quartz.

4 METHODS AND RESULTS

The gas trapped in the fluid inclusions was extracted by under-vacuum crushing of 1-2 g of quartz and calcite chips separated from the matrix. The $^3He/^4He$ and the He/Ne ratios were measured by a rare-gas mass spectrometer (MAP 251-50). Several blank analyses, repeated throughout this work, excluded the possibility of air contamination during gas extraction and analyses.

The $^3He/^4He$ (R) isotopic ratios, normalised to the atmospheric $^3He/^4He$ (Ra = 1.4×10^{-6}), the He/Ne ratios and average homogenization temperature of fluid inclusions are reported in Table 1. In Figure 1 are summarized the ranges of R/Ra. In Figure 2 the R/Ra and He/Ne ratios of the gas extracted from fluid inclusions are plotted. Figure 2 also shows the R/Ra and He/Ne for the geothermal fluids of two wells (T2 and T4) and two fumaroles (DB3 and DB4) of Tendaho (Marty et al.1996). No R/Ra data are available for the present-day fluids discharged at Aluto-Langano.

5 DISCUSSION

Previous studies documented an extremely variable R/Ra ratio, from 0.035 to 19.6, for the gas extracted from olivines and pyroxenes of the basalts of the Ethiopian volcanic province (Marty et al. 1996, Scarsi and Craig 1996). Large variations of R/Ra (from 2 to 15) were also found for the volcanic and geothermal gases of the Ethiopian Rift Valley and Afar depression (Scarsi and Craig 1996). These values were interpreted to be the result of different contributions from the lower mantle (R/Ra ≥ 32), upper mantle (R/Ra = 8) and continental crust (R/Ra = 0.01-0.2). In particular, the relatively high R/Ra ratios, more than 8 ± 1 (typical of N-MORB), indicated the presence of a lower mantle component

and supports the hypothesis of the presence of a large plume (Ethiopian plume), characterized by a maximum R/Ra around 20, that originates in the lower mantle under the Afar triple junction (Marty et al. 1996). Lower R/Ra values were explained by mixing between the Ethiopian plume end-member and upper mantle and/or crustal sources.

The R/Ra values of the gas extracted from fluid inclusions and the present-day geothermal fluids and fumaroles of the Tendaho area are also relatively high (Table 1, Figure 2), which suggests that the high R/Ra mantle plume influenced the He isotopic compositions of both the past and present-day geothermal fluids.

In general, the R/Ra values of the geothermal fluids could be also affected by the presence of atmospheric He carried by meteoric waters (air saturated water: ASW). The contribution of He from meteoric waters in geothermal fluids can be evaluated from the He/Ne ratio, which is a marker of atmospheric He. In Figure 2, some mixing lines have been drawn with the purpose of evaluating the different contributions to the He of the geothermal fluids. These lines represent the mixing between the atmospheric end-member (ASW) and sources characterized by different degrees of mixing between the Ethiopian plume (R/Ra = 19.6) and the crustal source (R/Ra = 0.02) end-members. However, the possibility cannot be excluded that the non-atmospheric source is also influenced by upper mantle contribution, as with the basalts. Both the past and present-day geothermal gas compositions fall between the curves representing 50 and 70% of the Ethiopian plume component, with most of the values around the 60% curve. This suggests a mixing, in relatively constant proportions, of the He from the plume and from the crust, both at Aluto-Langano and at Tendaho. In particular, the relatively similar distribution of values at Tendaho of past and modern gases, shown in Figure 2, indicates that the proportions of the mixed deep and crustal He did not change with time in this geothermal field.

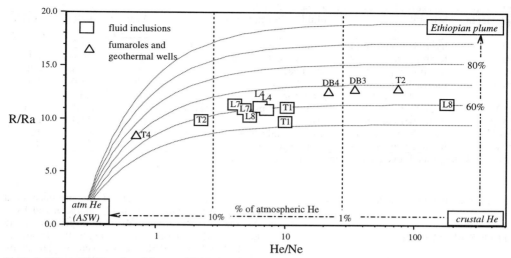

Figure 2. R/Ra vs. He/Ne ratios of the gas of fluid inclusions, and of geothermal wells (T2 and T4) and fumaroles (DB3 and DB4) of Tendaho from Marty et al. (1996). Mixing lines between atmospheric He (from air saturated water: ASW) and different proportions of Ethiopian plume and crustal sources are also shown.

The He/Ne ratio indicates that the atmospheric He content of the gas extracted from fluid inclusions is, in most samples, modest (1-10 %) and even lower in fumaroles and in present-day geothermal fluid from well T2 (Figure 2). This amount of atmospheric He could enter the ascending geothermal fluids by means of surface waters that infiltrated the geothermal systems. It is worth noting that the deepest sample (L8 from 2500 m b.g.l.) exhibits the highest He/Ne content (i.e. the lowest atmospheric He concentration) and the highest average homogenization temperature (309°C) of the fluid inclusions (Table 1), which suggests that the inclusions of this sample probably trapped the deep geothermal fluid not affected by contamination of atmospheric He. On the other hand, at Tendaho the T2 sample from 804 m b.g.l., with moderate fluid inclusion average homogenization temperature (230°C) and the present-day fluid discharged from shallow well T4 (bottomhole is at 470 m b.g.l.) are characterized by a low He/Ne ratio. This suggests that a significant amount of atmospheric He was introduced at shallow levels in this geothermal field.

6 CONCLUSIONS

Geothermal fluids trapped in fluid inclusions from the Aluto-Langano and Tendaho geothermal fields exhibit relatively high R/Ra (9.8-11.5) ratios, close to the values of present-day gas at Tendaho discharged from fumaroles and geothermal wells.

These high values are probably related to the supply of He from the high ^{3}He Ethiopian plume, which can have mixed with He of crustal (and/or upper mantle) origin.

Atmospheric He is in general present in relatively small amounts (< 10 %), with the exception of the geothermal fluids from well T4 and the fluid inclusions of T2 well, which indicate the introduction of atmospheric He, probably from infiltrating surface water at relatively shallow depths (up to 800 m b.g.l.).

REFERENCES

Gianelli, G., Mekuria, N., Battaglia, S., Chersicla, A., Garofalo, P., Ruggieri, G., Manganelli, M. & Z. Gebregziabher 1998. Water-rock interaction and hydrothermal mineral equilibria in the Tendaho geothermal system. *J. Volcanol. Geotherm. Res.* 86: 253-276.

Marty, B., Pik, R. & Y. Gezahegn 1996. Helium isotopic variations in Ethiopian plume lavas: nature of magmatic sources and limit of lower mantle contribution. *Earth Planet. Sci. Lett.* 144: 223-237.

Scarsi, P. & H. Craig 1996. Helium isotope in Ethiopian Rift basalts. *Earth Planet. Sci. Lett.* 144: 505-516.

Teklemariam, M., Battaglia, S., Gianelli, G. & G. Ruggieri 1996. Hydrothermal alteration in the Aluto-Langano geothermal field, Ethiopia. *Geothermics* 25: 679-702.

Geochemistry of the Earth's Surface, Ármannsson (ed.) © 1999 Balkema, Rotterdam, ISBN 90 5809 073 6

Flood from Grímsvötn subglacial caldera 1996: Composition of suspended matter

M. B. Stefánsdóttir
University of Iceland, Reykjavík, Iceland

S. R. Gíslason & S. Arnórsson
Science Institute, University of Iceland, Reykjavík, Iceland

ABSTRACT: Since the settlement of Iceland (870 AD) at least 80 subglacial volcanic eruptions have been reported, many of them causing tremendous outburst floods or jökulhlaups. The 1996 Gjálp subglacial eruption under the Vatnajökull ice cap triggered a catastrophic jökulhlaup. During the eruption 3 km^3 of ice were melted. The meltwater flowed along the glacier bed into the Grímsvötn caldera lake for five weeks, before draining in a sudden jökulhlaup. The fate of the volcanics in the meltwater and the composition of the suspended jökulhlaup material were the focus of this study. The 1996 volcanics were not carried to any significant extent to the center of Lake Grímsvötn. Most of the 180 million tons of suspended jökulhlaup material was altered basalt (palagonite) eroded from the subglacial bedrock during the 42 hours of the jökulhlaup, suggesting greater change in subglacial landscape than previously anticipated. The composition of the suspended material matches that of the Grímsvötn system and the Sídufjöll mountains.

1 INTRODUCTION

The Vatnajökull ice cap in southeastern Iceland, the largest ice cap in Europe is temperate and covers an area of about 8300 km^2. Its western part lies astride the volcanic fissure systems of the Mid-Atlantic ridge plate boundaries and the Icelandic mantle plume. Volcanic activity in the area has been extensive throughout historic time (Figure 1).

After a 60 year recess in the history of cata-strophic jökulhlaups a tremendous jökulhlaup occurred on November 5 1996 on the Skeidarár-sandur flood plain, following a subglacial eruption in the Gjálp eruption fissure north of the Grímsvötn caldera (Figure 1). A mixture of meltwater and quenched ash fragments, of temperature 15°-20°C, flowed along a narrow channel in the glacier bed for five weeks during and after the eruption, and accumulated in the Grímsvötn caldera, which is dammed by the surrounding ice (Gudmundsson et al. 1997).

At least 180 million tons of suspended material flowed across the coastal plain in 42 hours (Jónsson et al. 1997). This translates to 103×10^6 t/day, compared with the Ganges/Brahmaputra which transports 4.58×10^6 t/day and the Amazon 2.47×10^6 t/day (Milliman & Meade 1983). The sediment transport affected the land and land formation, and the ecosystem of the sea (Figure 1). In the last 200 years, 15 catastrophic floods have emerged on the Skeidarársandur plain, usually in connection with volcanic eruptions under the Vatnajökull ice cap.

The jökulhlaup transformed the Skeidará River into one of the largest rivers in the world for a few hours. The 3.6 km^3 of water flowed across the coastal plain in 42 hours (Snorrason et al. 1997, Björnsson 1997), or 24,000 m^3/sec on average. At peak discharge the flow rate was 45,000 m^3/sec (Björnsson 1997). The average discharge of the Yangtse-kiang is 35,000 m^3/sec, of the Congo 42,000m^3/sec, and the Amazon carries 190,000 m^3/sec (Baumgartner & Reichel 1975).

This paper describes the petrologic, chemical, spatial, and temporal variations in the suspended material from the eruption fissure, Lake Grímsvötn, and the rivers on the Skeidarársandur flood plain. The overall goal is to unravel the origin and fate of the suspended particles before, during, and after the 1996 jökulhlaup.

2 METHODS

Samples were collected from the Skeidará, Gígja and Súla Rivers from the beginning of the eruption, October 1 - December 20. During the jökulhlaup, samples were collected at the base of the glacier. Crushed ice floating in the water was sieved through a high-density polyethylene funnel with a coarse polyethylene sieve (5 mm) into a high-density polyethylene container. Water was pumped by peristaltic pump from the container through 0.2 μm pore membranes (cellulose acetate) and a polypropylene filter holder into the sampling bottles used for analysis of dissolved constituents. The rest of the water, along with the suspended material, was poured into polyethylene bottles. The sampling bucket, container, sieve, funnel, filter holder, and bottles were washed thoroughly with the sample water between samplings. The flood water temperature was measured with a Thermistor thermometer with an accuracy of 0.1°C (Gíslason et al. 1997).

During the summers of 1997 and 1998, members of the Icelandic Glaciological Society drilled through the ice shelf on the Grímsvötn caldera lake, using a hot water drill (Taylor 1984), producing a hole 30 mm in diameter and 270 m in depth. Samples of water and bottom sediment were obtained through the borehole. Water samples were obtained with a teflon bailer. The water was filtered on site through 0.2 μm pore membranes (cellulose nitrate) with a PSF (polysulfone) filter holder into sampling bottles. Samples of the bottom sediment were obtained with a gravity corer. The thickness of the ice shelf and the water temperature were measured.

The polyethylene bottles containing flood water and suspended material were stored in the laboratory until the suspended material had settled to the bottom. The water was then sucked from the bottle, and the suspended material was poured into a plastic beaker and freeze dried.

Sixteen samples of a total of 43 collected were selected for microscopic examination. Transparent minerals were examined with a transmitted light microscope and opaque minerals with a reflected light microscope. A few samples were examined and photographed by scanning electron microscopy (SEM). Microprobe analysis of fresh volcanic glass fragments in the suspended material was used to determine their origin by comparing them with the known chemical composition of several volcanic systems in the vicinity of the Vatnajökull ice cap.

3 RESULTS AND INTERPRETATION

Microscopic examination revealed that before the jökulhlaup the major part of the suspended material in the Skeidará River was fresh volcanic glass (Figure 2). On the other hand, only 5% of the suspended material during the jökulhlaup was fresh volcanic material. The major part of the suspended material during and after the jökulhlaup consisted of palagonite. The SEM microscopic photographs also showed that before the jökulhlaup the suspended material consisted mostly of fresh volcanic glass fragments with little or no alteration minerals in the cavities, and most of the suspended material during the jökulhlaup was altered basaltic grains, their cavities loaded with alteration minerals. That was also the case during the first weeks after the flood. Microprobe analysis was used to determine the origin of fresh volcanic glass fragments in the suspended material from the rivers on Skeidarársandur and from Lake Grímsvötn. Many

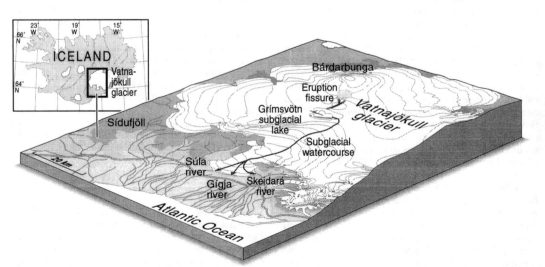

Figure 1. Map of the western part of the Vatnajökull ice cap, Iceland, the eruption site, the subglacial caldera lake Grímsvötn, and the path of the flood at the base of the glacier.

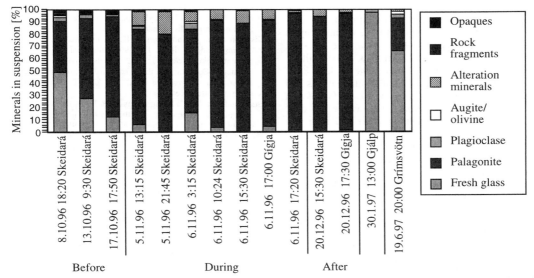

Figure 2. Histogram showing the petrologic composition of volcanic glass in the Gjálp eruption fissure, bottom sediment from the center of Grímsvötn caldera lake, and suspended material in the rivers on the Skeidarársandur flood plain before, during, and after the jökulhlaup in November 1996.

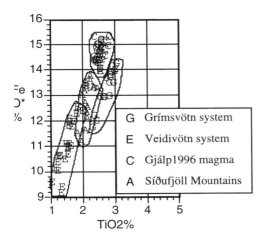

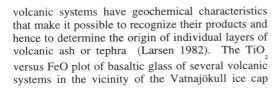

Figure 3. The TiO₂/FeO weight percent composition of several volcanic systems in the vicinity of Vatnajökull (Steinthórsson 1977, Jakobsson 1979, Bergh 1985, Larsen 1982, Larsen pers. comm., Grönvold pers. comm.).

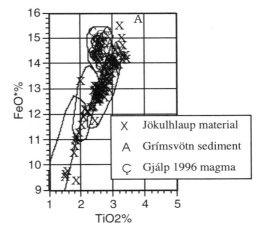

Figure 4. The TiO₂/FeO weight percent composition of suspended material in the November 1996 jökulhlaup and the composition of sediment from the bottom of the Grímsvötn caldera. The composition of volcanic glass from the 1996 Gjálp eruption is plotted for comparison.

volcanic systems have geochemical characteristics that make it possible to recognize their products and hence to determine the origin of individual layers of volcanic ash or tephra (Larsen 1982). The TiO_2 versus FeO plot of basaltic glass of several volcanic systems in the vicinity of the Vatnajökull ice cap shows that each volcanic system plots in a separate area (Figure 3). By comparing the chemical composition of the fresh volcanic glass in the jökulhlaup (Figure 4) with Figure 3, the origin of the fresh glass in the jökulhlaup was traced mostly to the Grímsvötn system as well as to the Síðufjöll mountain ridges, southwest of Vatnajökull.

However, the composition of none of the samples was close to the Gjálp 1996 magma. The ice shelf on Lake Grímsvötn was 270 m thick at the drilling location, and in June 1997 and 1998 the lake itself was 40-50 m deep. Before the jökulhlaup it was 215 m deep. The extent and volume of Lake Grímsvötn was still decreasing in June 1998, as the ice shelf has gradually been thickening from 150 m in the 1950´s (Björnsson 1997).

The impacts of jökulhlaups on life on the continental shelf south of the Skeiðarársandur flood plain are not yet well known. A large amount of dissolved constituents such as nutrients and heavy metals was carried into the sea with the jökulhlaup water, and possibly also elements adsorbed onto suspended material (Gíslason et al. 1998). The Vatnajökull area is very isolated and remote, and constitutes an ideal site to investigate the origin of nutrients from eruptions, and their flux through an abiotic watershed into the sea.

Studies of the desorption and dissolution kinetics of the suspended jökulhlaup matter in seawater, along with the Lake Grímsvötn water chemistry, will be presented later.

ACKNOWLEDGEMENTS

The Icelandic Public Road Administration, the Icelandic Glaciological Society, and the Science Institute of the University of Iceland have supported the work reported in this paper. We wish to express our appreciation to the Marine Research Institute of Iceland for assistance with gravity corer construction. We are grateful to Karl Grönvold for his assistance with the microprobe analysis, and we would like to thank him and Gudrún Larsen for the use of unpublished data on the composition of volcanic ash in the Gjálp 1996 eruption, and for fruitful discussion.

REFERENCES

Baumgartner, A. & E. Reichel, 1975. *The world water balance*. Amsterdam: Elsevier.

Bergh, S. G. 1985. *Structure, depositional environment and mode of emplacement of basaltic hyaloclastites and related lavas and sedimentary rocks: The Plio-Pleistocene of the eastern volcanic rift zone, southern Iceland*. Reykjavík: Nordic Volcanological Institute 8502, Reykjavík.

Björnsson, H. 1997. Grímsvötn Lake past and present. In H. Haraldsson (ed.), *Vatnajökull: Gos og hlaup 1996*: 61-77. Reykjavík: The Icelandic Public Road Administration (In Icelandic).

Gíslason, S. R., H. Kristmannsdóttir, S. Hauksdóttir & I. Gunnarsson, 1997. Studies of the chemical composition of the river water on Skeiðarársandur after the eruption under the Vatnajökull. In H. Haraldsson (ed.), *Vatnajökull: Gos og hlaup 1996*: 139-171. Reykjavík: The Icelandic Public Road Admini-stration (In Icelandic).

Gíslason, S. R., Á. Snorrason, H. K. Kristmannsdóttir & Á. E. Sveinbjörnsdóttir, 1998. The 1996 subglacial eruption from the Vatnajökull glacier, Iceland: effects of volcanoes on the transient CO_2 storage in the ocean. In J. Schott et al. (eds.). *Mineralogical magazine* 62A: 523-524. Toulouse, Goldschmidt conference.

Gudmundsson, M. T., F. Sigmundsson & H. Björnsson 1997. Ice-volcano interaction of the 1996 Gjálp subglacial eruption, Vatnajökull, Iceland. *Nature* 389: 954-957.

Jónsson, P., Á. Snorrason & S. Pálsson 1997. The discharge graph and the suspended sediment on Skeiðarársandur in the autumn of 1996. In Á. Sveinbjörnsdóttir, G. Larsen & M. T. Gudmundsson (eds.). *Eldgos í Vatnajökli 1996*: 27. Reykjavík: The Icelandic geological society. (In Icelandic).

Jakobsson, S. P. 1979. Petrology of recent basalts of the Eastern Volcanic Zone, Iceland. *Acta Naturalia Islandica* 26: 1-103.

Larsen, G. 1982. Ash layer chronology of Jökuldalur valley and its vicinity. In H. Thórarinsdóttir, Ó. H. Óskarsson, S. Steinthórsson & Th. Einarsson, (eds.). *Eldur er í nordri*: 52-65. Reykjavík: The Historical Society (In Icelandic).

Milliman, J. D. & R. H. Meade 1983: Worldwide delivery of river sediment to the oceans. *J. Geol.* 91: 1-21.

Snorrason, Á., P. Jónsson, S. Pálsson, S. Árnason, O. Sigurdsson, S. Víkingsson, Á. Sigurdsson, & S. Zophoníasson 1997: The jökulhlaup on Skeiðarársandur in the autumn of 1996: Distribution, discharge and sediment . In H. Haraldsson (ed.), *Vatnajökull: Gos og hlaup 1996*: 79-36. Reykjavík: The Icelandic Public Road Administration (In Icelandic).

Steinthórsson, S. 1977. Tephra layers in a drill core from the Vatnajökull ice cap. *Jökull* 27: 2-27.

Taylor, P. L. 1984: A hot water drill for temperate ice. In G. Holdsworth et al. (eds.). *Ice drilling Technology, Cold Reg. Res. Eng.Lab., Spec. Rep.* 84-34: 105-117. US Army Corps of Engineers.

Geochemistry of the Earth's Surface, Ármannsson (ed.)© 1999 Balkema, Rotterdam, ISBN 90 5809 073 6

Feldspar equilibria in geothermal waters

A. Stefánsson
Institut für Mineralogie und Petrographie, ETH Zentrum, Zürich, Switzerland

S. Arnórsson
Science Institute, University of Iceland, Reykjavík, Iceland

ABSTRACT: Reactions between feldspars and geothermal waters were investigated. Geothermal waters when above 200 °C, are saturated with microcline and low-albite and equilibrium with these minerals is considered to control the aqueous Na^+/K^+ activity ratios. Supersaturation is indicated for waters below 200 °C. However, the Na^+/K^+ activity ratio at temperature as low as 50 °C closely approaches that predicted from simultaneous equilibrium of the waters with low-albite and microcline suggesting that Na and K concentrations in geothermal waters are also controlled by equilibrium with thsee two feldspars. The waters considered in the present study are undersaturated with primary disordered plagioclases and alkali-feldspars of composition typically found in basaltic to silicic volcanic rocks, which may therefore may be dissolving simultaneously with precipitation of secondary albite and microcline. Ordered or partly ordered feldspars, as found in deep-seated plutonic and metamorphic rocks may, on the other hand, be stable.

1 INTRODUCTION

Albite and K-feldspar are frequently identified as secondary minerals in hydrothermal systems when above 200 °C (Browne 1978), and it has been considered that Na and K concentrations of geothermal waters are controlled by equilibrium with low albite and microcline (Giggenbach, 1981, Arnórsson et al. 1983). Yet, it has not been convincingly demonstrated that geothermal waters are in equilibrium with these minerals. Such a demonstration requires the study of the state of saturation between each of these feldspars and waters of different compositions. The primary reason for this relates to uncertainties in the calculated solubilities of feldspars as a function of composition and ordering, and of uncertainties in aqueous alumium speciation calculations. However, the thermodynamics of the feldspars and the aqueous species formed upon their dissolution has received considerable attention during the last decade, resulting in better understanding of the complex thermodynamics of the feldspar minerals and more accurate thermodynamic data for aqueous silica and aluminum species under hydrothermal conditions.

The aim of the present contribution is to assess the saturation state of feldspars in geothermal waters as a function of cation ordering and composition. For this purpose the solubilities of feldspars, both end-members and solid solutions, were compared with calculated reaction quotients of 115 geothermal waters of varying temperature, pH and salinity from Iceland,

China, Japan, New Zealand, Bolivia, Nicaragua, Guatemala and Alaska.

2 DATA BASE AND DATA HANDLING

Of the 115 geothermal samples considerd, 57 are from Iceland and 58 from other countries. Table 1 summarizes the types of waters used.

The waters selected come from a range of geological environments. Most of them are dilute, the TDS being less than 1000 ppm. This is a characteristic of waters in the basaltic environment of Iceland. However, waters associated with andesitic and silicic volcanics from regions of converging crustal plate boundaries have also been included. There are typically moderately saline with Cl concentrations ranging from 1,000 ppm to 10,000 ppm.

The reaction of feldspars of any composition with water may be written as

$$Na_xK_yCa_zAl_{(1+z)}Si_{(3-z)}O_8 + 8H_2O = xNa^+ + yK^+ \\ + zCa^{2+} + (1+z)Al(OH)_4^- + (3-z)H_4SiO_4^o \quad (1)$$

where x+y+z=1. For reaction (1) the reaction quotient, is

$$Q_{feld} = a_{Na^+}^x \cdot a_K^y \cdot a_{Ca^{2+}}^z \cdot a_{Al(OH)_4^-}^{(1+z)} \cdot a_{H_4SiO_4^o}^{(3-z)} \quad (2)$$

where a_i is the activity of i-th aqueous species.

The WATCH program of Arnórsson et al. (1982),

Table 1. Chemical compostion of reservoir waters calculated by the WATCH program (conentrations are in ppm)*

Site	Fushime	Bolivia	Zhangs.	Wairakei	G.Bight	Zunil	Monot.	Selfoss	Reykjav.	Reykjan.	Krafla
Country	Japan	Bolivia	China	NZ	Alaska	Guatem.	Nicarag.	Iceland	Iceland	Iceland	Iceland
well/spr.no	SKG-7	AP1	16	24	H6	5	38	13	11	8	20
t °C **	272 [qz]	244 [qz]	120 [mme]	237 [qz]	146 [cd]	255 [qz]	250 [qz]	75 [m]	129 [m]	244 [m]	283 [a]
pH	4.59	6.48	6.28	6.40	6.79	5.97	5.75	7.63	8.11	5.39	6.29
SiO_2	560	443	52.7	408	171	427	468	68.7	148	582	692
B	43.8	102	-	19.2	46.6	48.2	25.3	0.13	0.05	8.04	1.20
Na	9709	2597	6797	921	372	937	1524	159	57.1	10276	151
K	2749	453	167	147	17.4	164	232	3.66	2.60	1585	31.7
Ca	1214	133	2208	19.6	30.2	21.9	32.7	31.1	2.56	1571	0.45
Mg	5.25	0.054	363	0.015	<0.16	0.236	0.063	0.079	0.024	1.33	0.009
Al	0.18	0.07	0.42	0.23	0.09	1.47	0.23	0.12	0.18	0.06	0.93
ΣCO_2	418.5	94.1	284.7	412.0	61.3	1158	1522	21.0	17.5	466.7	8594
SO_4	23.6	120.5	1159	25.0	136.1	24.4	29.8	59.1	19.4	-	17.1
H_2S	118	0.39	-	4.30	0.75	19.5	29.2	0.02	0.41	15.7	329
Cl	19942	4663	13667	1600	508	1434	2691	235.2	36.4	21045	102.2
F	3.35	1.00	-	5.06	1.28	4.89	-	0.24	0.82	0.19	0.92

* For ref. see Stefánsson et al. (1998) and Stefánsson & Arnórsson (1999); ** qz=quartz temp., mme=multi mineral equilibrium temp., cd=chalcedony temp., m=measured temp., a=averge NaK, NaKCa & qz temp.

version 2.1A (Bjarnason, 1994), was used to calculate the aqueous speciation for the waters selected for the present study. In the new version the thermodyn-amic data base for Al-hydroxy complex dissociation and gas solubility constants has been reviewed. Also, aqueous $KAl(OH)_4^\circ$ and $NaAl(OH)_4^\circ$ (Arnórsson & Andrésdóttir 1998) have been incorporated into the alumium speciation calculations in addition to the aluminum hydroxy, fluoride and sulphide species in the WATCH program.

The feldspar solubility constants used in the present study are those given by Arnórsson & Stefánsson (1999) and are valid from 0° to 350 °C at water vapour saturation pressure.

3 RESULTS

3.1 *End-member feldspars*

High-temperature thermal waters (>200 °C) are close to low-albite and microcline saturation (Figure 1) with almost equal numbers of super- and undersaturated data points; the average deviation being 0.16 and 0.07 log units for low-albite and microcline, respectively. This suggests that equilibrium with these minerals controls the Na^+/K^+ activity of the waters.

Considerably more scatter is observed for data points in the temperature range 50° to 200 °C. This may be due to some extent to errors associated with empirical correction of CO_2 loss from springs, mixing and/or that the waters have not come very close to equilib-rium with these minerals. Despite the scatter of these data, supersaturation is generally observed for waters below 200 °C with respect to low-albite and microcline, but saturation and undersaturation with high-albite and sanidine.

The saturation state of anorthite is somewhat differ-ent compared to albite and the K-feldspars. Under-

saturation is observed with only few exceptions at high temperatures, indicating that anorthite is unstable. However, the fact that $\sqrt{Ca^{2+}}/H^+$ activity ratios of the waters are a function of temperature alone (see Arnórsson et al. 1983) suggests that the aqueous con-centrations of calcium are controlled by a mineral equi-librium not involving anorthite.

The main uncertainty in constructing the picture for feldspar saturation is associated with aqueous aluminum speciation. To avoid this problem, simul-taneous equilibrium with low-albite and microcline may be considered:

$$low\text{-}albite + K^+ = microcline + Na^+ \tag{3}$$

As seen in Figure 2, the Na^+/K^+ activity ratio of the waters closely approaches that predicted by equilibrium according to reaction (3). It is therefore clear, that Na^+/K^+ ratios are controlled by equilibrium with low-albite and microcline above 150 °C and down to tempera-tures as low as 50 °C in many cases.

3.2 *Feldspar solid solution*

The saturation state of 5 water samples with respect to alkali feldspars and plagioclases of different composi-tion is shown in Figure 3 including both fully ordered and fully disordered series

Similar trend is observed for all the waters with re-spect to the alkali-feldspars despite variation both in their salinity and temperature. Yet, slight differences are seen between low- and high-temperature waters as logK varies less with composition at high temperatures than at low ones because the difference in the solubil-ity of the end-members decreases with rising tempera-ture. Intermediate compositions are the least stable.

More variability is observed between samples from the plagioclase series. For some samples the value of the saturation index is about the same for all composit-

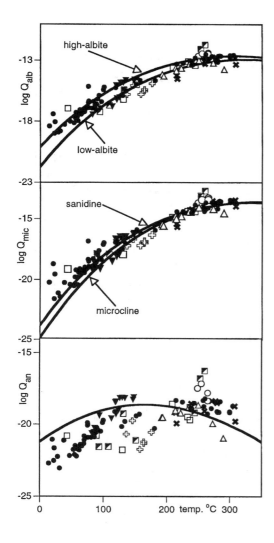

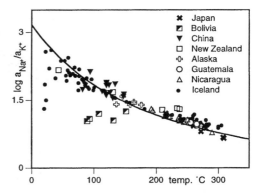

Figure 2. The Na/K activity ratio vs. temperature. The solid line represents the ratio assuming simultaneous equilibrium with low-albite and microcline.

Figure 1. The saturation state of end-member feldspar as a function of temperature. The solid lines represent feldspar solubility (Arnórsson and Stefánsson 1999). Symbols have the same notation as in Figure 2.

ions but for others the stability of anorthite risch compositions is less than of albite rich compositions. This pattern can be related to temperature, at least to some extent, because waters below 200 °C tend to be anorthite undersaturated whereas waters above 200 °C are closer to saturation (Figure 1).

Plagioclases in volcanic rocks range from andesine to bytownite compositions. Due to their high formation temperatures they can be assumed to be highly disordered. The waters considered in this study are usually undersaturated with such plagioclases (Figure 3), the degree of undersaturation increasing with decreasing temperature. Less disordered or ordered feldspars, found in deep-seated plutonic and metamorphic rocks, may, however, been stable.

4 DISCUSSION

Albite and K-feldspar are frequently found as alteration minerals in active geothermal systems above 200 °C, as has been demonstrated for the Ohaaki-Broadlands geothermal systems, New Zealand in zones of good permeability (Browne 1978). These feldspars have also been identified in systems of lower temperatures and the Na^+/K^+ activity ratios in thermal waters suggest that these waters may at times attain equilibrium with low-albite and microcline at temperature as low as 50 °C.

Experimental work has shown that the aluminate ion complexes with aqueous silica in alkaline solutions (Pokrovski et al. 1998). For the water samples used in the present study the Si-Al complex would account for 60-80% of the total dissolved alumium. However, incorporation of the complex into the speciation calculations was found to result in systematic undersaturation of low-albite and microcline above 200 °C, which is in disagreement with field data. The systematic undersaturation is larger than the anticipated error of the calculated feldspar solubilities from thermodynamic data.

Hydrothermal Na-feldspars range in composition from pure albite to mid oligoclase (Larsson, pers. comm., Browne, pers. comm.). It is therefore possible that the Na concentrations of geothermal waters are controlled by a plagioclase of this compositional range. Such slight variations in the albite composition are not considered to change the conclusions of the present study, as the saturation state for ordered plagioclases is little affected by their composition at temperatures above 200 °C.

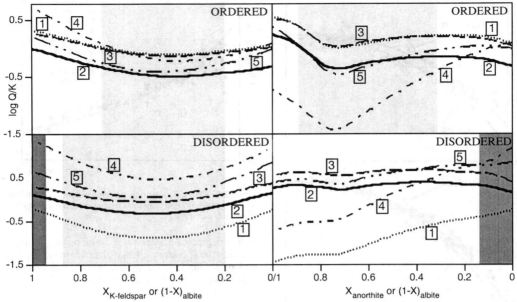

Figure 3. Saturation state of feldspar vs composition. [1] Reykjan. [2] Wairakei [3] Krafla [4] Selfoss [5] Zhangzhou. Light-gray and dark-gray shadings represent the compositional range of primary and hydrothermal feldspars (Deer et al., 1996)

CONCLUSIONS

1. Geothermal waters from variable geological settings when above 200 °C, are in equilibrium with microcline and low-albite, the two feldspars controlling the aqueous Na^+/K^+ activity ratios in these waters.

2. Na^+/K^+ activity ratios at temperatures as low as 50 °C closely approach those predicted from simultaneous equilibrium of the waters with low-albite and microcline suggesting that these minerals control the Na and K concentrations in such geothermal waters.

3. The waters considered in this study are understaurated with primary disordered plagioclases and alkali-feldspars of composition typically found in volcanic rocks ranging in composition from basaltic to silicic. These primary feldspars may be dissolving simultaneously with precipitation of low-albite and microcline. Partly ordered and fully ordered feldspars as found in deep-seated plutonic and metamorphic rocks may, on the other hand, be stable.

REFERENCES

Arnórsson, S. & A. Andrésdóttir 1998. Review and assessment of the thermodynamic properties of aqueous alumium hydroxy complexes at 0-350 ˚C and Psat. *Sci. Inst. Univ. Iceland Rep. RH-15-98*: 25 p.

Arnórsson, S., E. Gunnlaugsson & H. Svavarsson 1983. The chemistry of geothermal waters in Iceland. III. Chemical geothermometry in geothermal investigation. *Geochim. Cosmochim. Acta* 47: 567-577.

Arnórsson S., S. Sigurdsson & H. Svavarsson 1982. The chemistry of geothermal waters in Iceland. I. calculation of aqueous speciation from 0˙ to 370 °C. *Geochim. Cosmochim. Acta* 46: 1513-1532.

Arnórsson, S. & A. Stefánsson 1999. Assessment of feldspar solubility in water in the range 0-350 °C at P_{sat}. *Am. J. Sci.*, in press.

Bjarnason, J.O. 1994. *The speciation program WATCH vesrion 2.1*. Reykjavik. The National Energy Authority.

Browne, P.R.L. 1978. Hydrothermal alteration in active geothermal fields. *An. Rev. Earth Plan. Sci.* 6: 209-250.

Deer, W.A., R.A. Howie & J. Zussman 1996. *An introduction to the rock-forming minerals*. Longman Scientific & Technical 696 p.

Giggenbach, W.F. 1981. Geothermal mineral equilibria. *Geochim. Cosmochim. Acta* 45: 393-410.

Pokrovski, G.S., J. Schott, S. Salvi, R. Gout & J.D. Kubicki 1998. Structure and stability of aluminum-silica complexes in neutral to basic solutions. Experimental study and molecular orbital calculations. *Min. Mag.* 62A: 1194-1195.

Stefánsson, A. & S. Arnórsson 1999. Saturation state of feldspars in natural waters. submitted.

Stefánsson, A., S.R. Gíslason & S. Arnórsson 1999. Dissolution of primary minerals of basalt in natural waters. II. Mineral saturation state. submitted.

Geochemistry of the Earth's Surface, Ármannsson (ed.) © 1999 Balkema, Rotterdam, ISBN 90 5809 073 6

Chemical monitoring during reinjection in the Laugaland geothermal system, N-Iceland

Gudrún Sverrisdóttir
Orkustofnun & Nordic Volcanological Institute, Reykjavík, Iceland

Steinunn Hauksdóttir & Gudni Axelsson
Orkustofnun, Reykjavík, Iceland

Árni Árnason
Hita- og Vatnsveita Akureyrar, Akureyri, Iceland

ABSTRACT: During a long term reinjection experiment which is now underway to extract thermal energy from 90-100°C hot rock in the Laugaland geothermal system, North-Iceland, extensive chemical monitoring has taken place. Since the startup of reinjection in September 1997 until late February 1999, 590,000 m^3 or 12.5 l/s on average, of 6-21°C warm return water from the district heating system of Akureyri has been pumped down into the geothermal system. No chemical changes have been observed in the geothermal water pumped from the production wells in the area during this 17 months period of reinjection. This indicates that the reinjection of this water does not induce significant cooling and subsequent precipitation of secondary minerals in the geothermal system.

1 INTRODUCTION

The Laugaland geothermal system in N-Iceland is the largest of 5 separate geothermal fields utilized for space heating for the town of Akureyri for 20 years (Flóvenz, et al. 1995) (Figure 1). The geothermal systems are embedded in fractured, low-grade hydrothermally altered basaltic rocks. The Laugaland geothermal field comprises three production wells, LJ-05, LJ-07 and LN-12 and two reinjection wells LJ-08 and LN-10 (Figure 2). At Ytri Tjarnir field one production well (TN-04) is currently in use.

1.1 The reinjection project

The productivity of the Laugaland system is limited by insufficient recharge and continuously increasing pressure draw-down. A reinjection project, aimed at extracting some of the thermal energy stored in the 95-100 °C hot rocks, is underway and will continue through the year 1999. The work is a cooperative project of companies and institutions in Iceland, Sweden and Denmark supported by the European Commission. Preliminary results regarding the thermal extraction have been published in a Mid-Term report to the E.C. (Hita- og Vatnsveita Akureyrar et al. 1998). The focus of this work will be on the chemical monitoring during the reinjection project.

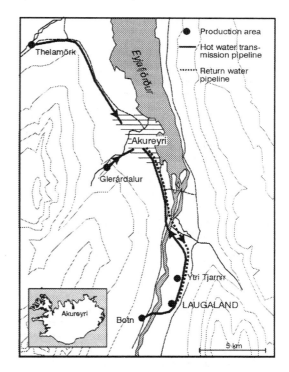

Figure 1. Location of the Laugaland geothermal field.

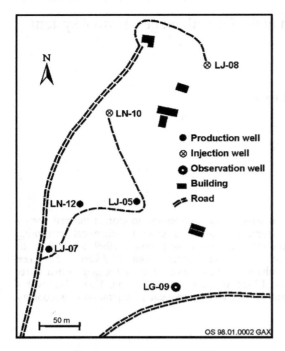

Figure 2. Wells in the Laugaland geothermal field.

1.2 *Tracer tests*

Important parts of the reinjection project are three tracer tests, two of which have already been carried out. The first one started two weeks after the beginning of the reinjection as 10 kg of sodium fluorescein dye were injected into well LJ-08. In the second tracer test 45.3 kg of potassium iodide were injected into well LN-10 simultaneously with the start of injection into that well. The water from LN-12 contained the highest concentration of the fluorescein dye of the wells sampled. Within a week from the tracer test the concentration reached a maximum of 6 μg/l. This means that extensive dilution is taking place in the reservoir; the total recovery by February 1999 was 30 %. On the contrary, the concentration of iodide reached a maximum of 60 μg/l in water from LJ-05, as late as 5 weeks after injection. Total recovery of potassium iodide in February 1999 was 45 %. As the dilution of the tracers by the pressured pumping into the system is so effective, no lasting effects on the chemistry of the geothermal water are expected.

The preliminary conclusions about the two tracer tests indicate that the injection- and production wells of the Laugaland area are not directly connected through the fracture-zone, which supplies the major feed-zones of the latter (Hita- og Vatnsveita Akureyrar et al. 1998). These tracer tests also indicate a connection between the Laugaland and Ytri-Tjarnir geothermal fields.

1.3 *This work*

Chemical monitoring has included regular sampling and analysis of selected elements in the geothermal water pumped from production wells in the Laugaland area. Samples from nearby areas were analysed as well. The main aim of the chemical monitoring is to detect whether if some precipitation of secondary minerals or cooling in the geothermal system is immediately induced by the reinjection.

2 CHEMISTRY OF INJECTED WATER

On first consideration it would seem possible to use local groundwater for the injection. This idea was soon rejected, because severe problems of magnesium-silicate precipitation have been experienced elsewhere by mixing of geothermal water and the relatively Mg-rich Icelandic groundwater (Kristmannsdóttir et al. 1989, Sverrisdóttir et al. 1992). Such deposition might cause the injection wells and its feed zones to clog up and probably cause serious problems for the production of the geothermal system. This was later confirmed by observations and model calculations for the geothermal water in the area done by Bi Erping (1998).

Table 1. Chemical composition of the return water (mg/l).

Date	03.04.1997 A	03.04.1997 B	18.02.1998 Mixed
Temp. (°C)	26,5	25,0	19,9
pH/°C	9,83/20,5	9,83/20,5	9,82/21,9
CO_2	21,2	22,0	19,4
H_2S	<0,03	<0,03	0,09
SiO_2	88,6	94,4	95,3
Na	53,0	53,1	55,3
K	0,96	1,00	0,99
Ca	3,15	2,82	2,96
Mg	<0,001	<0,001	0,002
SO_4	39,7	35,7	37,5
F	0,44	0,49	0,45
Cl	13,5	12,7	12,9
B	0,16	0,17	0,18
O_2	0	0	0,01

Using return water from the district space heating system appeared to be the best choice because its chemical composition is almost identical to the Laugaland geothermal water. Although originally produced from 5 separate geothermal systems, the difference in water chemistry is very small. The chemical composition of three samples from the return water is shown in Table 1. Two samples are from two separate parts of the domestic heating system respectively (A and B); they are mixed before injection. These samples were taken before the reinjection program started but the third sample is taken a year after the project started. This sample is from the mixed return water after it has been piped 13 km from the town of Akureyri to the Laugaland area. The earlier samples of return water were analysed for major elements as well as for various organic solvents, heavy metals and other elements which the water could plausibly assimilate from the heating system. No such chemicals were found in significant amount.

Table 2. Chemical composition of the geothermal fluid from LN-12 (mg/l).

Date	08.09.1997	18.02.1998
Temp. (°C)	95,8	94,9
pH/°C	9,76/21,9	9,79/21,7
CO_2	18,2	19,0
H_2S	0,08	0,10
SiO_2	99,2	97,3
Na	50,8	54,0
K	1,11	1,16
Ca	2,91	3,00
Mg	0,004	0,001
SO_4	37,9	39,2
F	0,37	0,30
Cl	11,6	11,6
B	0,16	0,16
O_2	0	0

The major element composition of Laugaland geothermal water prior to injection was established by sampling and analyseis of water from LN-12, the production well at that time. The results of the analysis of this sample and one collected from the well 6 months later are presented in Table 2. No significant changes in the water chemistry are detected.

3 CHEMICAL MONITORING

From the start of injection, several samples of water from production wells in Laugaland and nearby geothermal fields have been collected and analysed for Si, Cl, Ca and K. Even small changes in the concentration of these elements would give an indication of changes in the geothermal system. At the beginning of the reinjection project these samples were collected daily from the Laugaland area but less frequently in other areas. Few months after the project started sampling frequency was decreased until after the start of the second tracer test in February 1998 when it was increased again for a while.

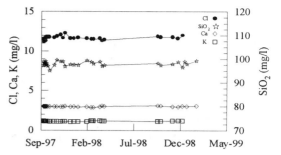

Figure 3. SiO_2, Ca, K, and Cl concentrations of geothermal water from LN-12.

Figure 3 shows the concentration of SiO_2, Ca, K, and Cl in the geothermal fluid of LN-12 as a function of time. The variation observed is less than expected for most of the elements in relation to the production of the well. A slight increase in potassium concentration in February 1998 can be attributed to the injection of the potassium iodide tracer. This increase amounted to up to 10% of the potassium concentration and was observed for about 4 weeks after the injection. After that the potassium concentration has been diluted to what is normal for the water. Production from LN-12 was discontinued during the summer 1998 but monitoring of other wells in the area confirmed that chemical changes during the period had not occurred.

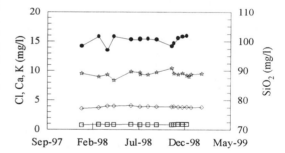

Figure 4. SiO_2, Ca, K, and Cl concentrations of geothermal water from TN-4. Legend as in Figure 3.

As the preliminary results from the first tracer test were acknowledged, a decision was made to sample the water from TN-4 for analysis of selected elements. This started simultaneously with the second tracer test. The chemical composition of the water from TN-4 has not changed during the time of the reinjection test (Figure 4). Variations observed do not exceed those observed during geochemical monitoring in recent years (Axelsson et al. 1998).

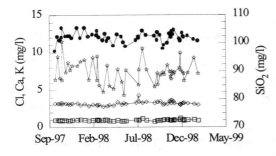

Figure 5. SiO₂, Ca, K, and Cl concentrations of return water. Legend as in Figure 3.

Figure 5 shows the chemical variations of the selected constituents in the injected return water. The variations are greater than observed for the fluid from production wells, LN-12 and TN-4. This is a result of mixing of water from the 5 production fields in various ratios within the district-heating system.

4 CONCLUSIONS

No lasting chemical changes are observed in geothermal fluids of the Laugaland system or nearby geothermal fields during the first 17 months of an ongoing reinjection experiment in the Laugaland geothermal reservoir. Neither downpumping of a considerable amount of return water from the district heating system, or injection of two different chemical tracers seem to affect the chemical properties of the thermal water. Consequently, no deposition is expected to have occurred in the reservoir during the reinjection.

Although this work presents only preliminary results of the chemical part of the project, they support the contention that return water from the space heating system is the most appropriate fluid for reinjection, at least in the Laugaland system.

ACKNOWLEDGEMENTS

The Nordic Volcanological Institute is thanked for supporting the presentation of this work.

REFERENCES

Axelsson, G., G. Sverrisdóttir, & Ó.G. Flóvenz 1998. The Akureyri District Heating Service. Geothermal monitoring in 1997 (in icelandic). *Report OS-98032.* Reykjavík: National Energy Authority, Orkustofnun.

Bi Erping 1998. Chemical aspects of fluid injection into the geothermal fields in Eyjafjörður, N-Iceland. In L. Georgsson (ed), *Geothermal Training in Iceland 1997*: 81-112. Reykjavík: The United Nations University, Orkustofnun.

Flóvenz, Ó.G., F. Árnason, M. Finnsson & G. Axelsson 1995. Direct utilization of geothermal water for space-heating in Akureyri, N-Iceland. In E. Barbier, G. Frye, E. Iglesias & G. Pálmason (eds), *Proceedings of the World Geothermal Congress 1995*: 2233-2238. Florence: International Geothermal Association.

Hita- og Vatnsveita Akureyrar HVA, Orkustofnun National Energy Authority, Uppsala University, Hoechst Danmark A/S and Rarik Iceland State Electricity 1998. Demonstration of improved energy extraction from a fractured geothermal reservoir. *Mid-Term Report for Thermie Project GE-0060/96, Report OS-98050.* Reykjavík: National Energy Authority, Orkustofnun.

Kristmannsdóttir, H., M. Ólafsson & S. Thórhallsson 1989. Magnesium silicate scaling in district heating systems in Iceland. *Geothermics* 18 (1/2): 191-198.

Sverrisdóttir, G., H. Kristmannsdóttir & M. Ólafsson 1992. Magnesium silicale scales in geothermal utilization. In Y.K. Kharaka & A.S. Maest (eds), *Water-Rock Interaction 2*: 1431-1434. Rotterdam: Balkema.

Geochemistry of the Earth's Surface, Ármannsson (ed.) © 1999 Balkema, Rotterdam, ISBN 90 5809 073 6

Fluid geochemistry at Rincon de la Vieja volcano, Costa Rica

F. Tassi, O. Vaselli & C. Giolito – *Department of Earth Sciences, Florence, Italy*

G. Magro – *CNR, Institute of Geochronology and Isotopic Geochemistry, Pisa, Italy*

E. Duarte-Gonzales & E. Fernandez – *Seismological and Volcanological Observatory, Heredia, Costa Rica*

B. Capaccioni – *Institute of Volcanology and Geochemistry, Urbino, Italy*

A.A. Minissale – *CNR, Study Center for Minerogenesis and Applied Geochemistry, Florence, Italy*

ABSTRACT: Costa Rica is characterized by the presence of several active and quiescent volcanoes which belong to the so-called "ring of fire". Rincon de la Vieja volcano, located in the northwestern part of the country, is one of the most active, as testified by the numerous phreatic eruptions, vigorous jets of hot water, ash and water vapor, occurring since 1991. This paper presents a preliminar geochemical study of fluid discharges emerging at Rincon de la Vieja volcano. Thermal waters and associated gases have been collected for the analysis of major and minor constituents and isotopic measurements (^{3}He/^{4}He in gases). The chemical composition of the thermal waters is clearly affected by the influence of deep magmatic fluids, such as CO_2, HCl, H_2S and partly SO_2, in agreement with the composition of the associated gas phases.

1 INTRODUCTION

Rincon de la Vieja, one of the six active volcanoes of Costa Rica, is located in the northwestern corner of the country (10.2 N, 85.5 W). The edifice of the volcano consists of 9 coalescent eruptive centers that form a northwest trending chain with elevations of individual cones from 1670 to 1920 m.

Phreatic and phreatomagmatic activity has been reported since 1851 (Tristan 1921, Barquero & Segura 1983). From 1991 to 1998, 3 phreatic eruptions from the acidic lake at the top of the active crater have occurred. Presently, the volcano is daily affected by ash and vapor emissions which can be seen from several kilometers.

Thermal springs, emerging in the surrounding area of the volcanic apparatus, follow the northwest oriented tectonic lineations parallel to the volcano ridge. During February 1998, in collaboration with the Seismological and Vulcanological Observatory of Heredia (Costa Rica), thermal water discharges and associated gas phases from this area, have been sampled.

In this work results obtained from chemical and isotopic analysis of water and gas samples are presented and discussed.

2 VOLCANOLOGICAL SETTING

Central America can be subdivided, on the basis of the crustal composition, into a northern area (Guatemala, Honduras, El Salvador, Belize and northern Nicaragua), considered part of the continental block of North America, and a southern area (southern Nicaragua, Costa Rica and Panama) which is constituted by an uplifted lower and upper Cretaceous oceanic crust (McBirney & Williams 1965, Dengo 1973). The subduction of the Cocos plate beneath those of Central America and Mexico, marked by the Middle America Trench, has caused the Tertiary and Quaternary volcanic activity, that resulted in the formation of the Central American volcanic arc which covers both the continental and oceanic crust (Bourgais et al. 1984). In Costa Rica 6 geological provinces can be distinguished from west to east: 1) Cretaceous-Middle Tertiary Ophiolitic Suite; 2) Tertiary basin; 3) Tertiary volcanic range: Cordillera de Talamanca and Montes del Aguacate to the northwest, Cordillera de Talamanca to the southeast; 4) Active Quaternary volcanic range: Cordillera de Guanacaste to the northwest, Cordillera Central to the southeast; 5) Intra-arc basin; 6) Caribbean coastal plain (Mora 1983). The northwestern part of Costa Rica is characterized by a caldera-related volcanism that, during Pleistocene, formed a wide ignimbrite plateau from the Pacific to the Caribbean coast (Dengo 1962, Tournon 1984). The Tertiary volcanic basement is covered by volcanics ejected from Quaternary volcanoes. Rincon de la Vieja, pertaining to the Cordillera de Guanacaste, is a composite stratovolcano. Its stratigraphy is dominated by Plinian and sub-Plinian tephra deposits with interbedded lava flows (Kempter et al. 1996). The last eruption involving juvenile magma occurred at about 3500 yr BP

(Alvarado et al. 1992). Since then, volcanic activity has produced only minor amounts of tephra and ash material. Historical eruptions consist of hot water jets mixed with ash giving rise to hot mudflows and lahars to the north of the active crater (Kempter et al. 1996).

3 SAMPLING AND ANALYTICAL METODS

Water and gas sampling points are shown in Figure 1. Las Pailas thermal springs (1, 2 and 3) emerge along a deep valley in the southwest area of the volcano; sample 8 is collected from the lake inside the active crater, while samples 4, 5, 6, 7 (Aguas Tibias) and 9 and 10 (Volcancito) are situated in the northern sector.

Emergence temperatures and pH were measured in the field. Gas sampling has been performed following the procedure described by Giggenbach (1975). Chemical components of the sampled waters were determined in the laboratory by using a Perkin-Elmer Analyst 100 and a Dionex DX 100 ion chromatograph. Major and trace gases were analyzed with Shimadzu 15A and Shimadzu 14A gaschromatographs, equipped with thermal conductivity and flame ionization detectors, respectively. Light hydrocarbon contents were determined by the gaschromatographic method of Mangani et al. (1991). $^3He/^4He$ ratios were determined using a Map 215-50 magnetic mass spectrometer.

4 WATER CHEMISTRY

Chemical data for the thermal water discharges are reported in Table 1. Emergence temperatures range between 22.6 (3) and 94^{0}C (2). pH values vary from 1 (8) to 7.81 (10). Sample 8 from the crater lake shows a particularly high salinity, almost 500 times greater than the lower salinity values (2 and 3). The Cl-SO$_4$-HCO$_3$ triangular diagram is reported in Figure 2. Chemical compositions are Ca(Mg)-SO$_4$, for samples 1 to 3 and Ca(Mg)-Cl for the remaining samples. Due to the acidity of the solutions HCO$_3$ is absent or very low. Such acidity, due to the probable presence of rising H$_2$S and SO$_2$ together with boron and ammonia, enriched in sample 8 and partly in samples 1, 2 and 9, suggests the occurrence of steam and gas inputs from a magmatic source. In fact, both boron and ammonia easily mobilize in the vapor phase at high temperature and they are correlated with the emergence temperatures (r = 0.73 and 0.51, respectively).

5 GAS CHEMISTRY

The analytical data for major and minor gases are reported in Table 2. All gas samples have high CO$_2$ contents (>89%), while H$_2$S is detectable only in samples 1 and 2 (9820 and 11810 ppm, respectively). Atmospheric components (Ar, Ne and O$_2$) are high in samples 9 and 10, which are also enriched in CH$_4$, probably derived from surficial alteration of organic material. N$_2$/Ar ratio in these samples ranges between air (83) and air saturated water (38) values, suggesting an atmospheric origin

Figure 1. Map of Costa Rica and location of the sampling sites at Rincon de la Vieja volcano.

Table 1. Chemical composition of thermal waters at Rincon de la Vieja volcano (values for chemical species are in ppm).

	1	2	3	4	5	6	7	8	9	10
pH	2.35	4.8	4.64	4.1	3.7	3.51	3.7	1	5.65	7.81
T°C	92.6	94	22.6	29	32	35	36	47	62	28.2
Ca	52	10.5	10	130	175	200	193	1567	459	46
Mg	4.5	5.5	2.8	40	61	72	79	2037	343	30
Na	1.8	6.1	4.8	51	66	79	92	609	389	36
K	1.5	2.4	2	9.4	9	12	14	292	132	12
HCO$_3$	n.d.	15.7	n.d.	n.d.	n.d.	n.d.	n.d.	642	101	n.d.
SO$_4$	1087	65	52.5	330	400	437	425	12500	1000	100
Cl	6	2.5	5.5	1125	300	400	437	18800	1150	135
Li	n.d.	n.d.	n.d.	0.04	0.1	0.06	0.1	0.78	0.24	0.02
NH$_4$	9.2	2.3	0.65	1.26	1.8	0.81	2.4	254	3.2	0.28
Fe	18	2.6	0.06	0.32	0.1	n.d.	0.1	3720	n.a.	n.d.
Mn	0.08	0.08	0.03	3	3.8	4.1	4.9	70	n.a.	0.03
F	0.26	0.4	0.35	8	11	12.3	11	1400	0.4	0.27
Br	n.d.	0.1	0.02	0.6	1	1.5	0.8	80	2.5	0.3
NO$_3$	0.02	0.2	0.19	0.9	1.8	1.5	0.5	10	0.2	0.35
B	2	3	0.01	0.27	0.3	0.39	0.4	22	1.7	0.18
SiO$_2$	n.a.	66	n.a.	58	87	53	68	n.a.	87	35

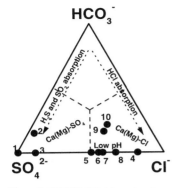

Figure 2. Cl/HCO₃/SO₄ triangular plot.

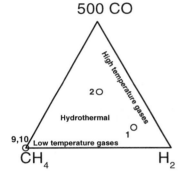

Figure 3. H₂/500CO/CH₄ triangular plot.

Table 2. Gas chemical compositions at Rincon de la Vieja volcano (values are in ppm).

	1	2	9	10
CO_2	974310	972720	890660	950653
H_2S	9820	11810	n.d.	n.d.
N_2	13430	14690	104380	40850
CH_4	55.89	43.78	1405	643
O_2	n.d.	n.d.	1620	910
Ar	300	270	1930	1050
Ne	0.24	0.26	1.35	0.65
H_2	2146	473	n.d.	n.d.
He	2.7	100	16.2	6.6
CO	1.04	1.44	n.d.	n.d.
Σ Alkenes	9.846	4.472	n.d.	n.d.
Σ Alkanes	4.395	1.526	0.016	0.006
Σ Aromatics	4.215	3.055	n.d.	n.d.
R/Ra	5.31	n.a.	4.56	n.a.

for N_2. Relatively high amounts of H_2 and CO have been measured in samples 1 and 2 (2146 and 473 and 1.0 and 1.4, respectively). In the CH_4-CO-H_2 ternary diagram (Figure 3) samples 9 and 10 plot close to the CH_4 corner (low temperature gases), while samples 1 and 2 fall in the field pertaining to the high temperature volcanic gases. Light hydrocarbon contents, reported as Σ alkane, Σ alkene and Σ aromatic groups, give rise to similar conclusions. In fact alkene compounds (propene, i-butene, 1-butene, t-2-butene, c-2-butene), which are considered typical products of high temperature reactions (Capaccioni et al., 1993, 1995), are enriched in samples 1 and 2. Gas geothermometric calculations, based on H_2/Ar ratios (Giggenbach

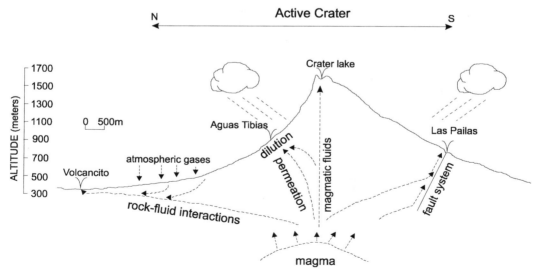

Figure 4. Scheme of fluid circulation at Rincon de la Vieja volcano.

1991), indicate minimum equilibrium temperatures of 200-250^{0}C for deep fluids at sites 1 and 2. ^{3}He/^{4}He ratios, expressed as R/Ra (where R is ^{3}He/^{4}He in the sample and Ra is ^{3}He/^{4}He in the air = 1.34×10^{-6}) are 5.31 and 4.56 for samples 1 and 9 respectively, indicating a strong mantle component in the gas phase, the typical crustal value being 0.02 (O'Nions & Oxburgh 1988).

6 CONCLUSIONS

The chemical and isotopic data of water and gas samples briefly presented here, have provided some constraints on the origin and the hydrogeochemistry of the thermal fluids of the Rincon de la Vieja volcano hydrothermal system. In the proximity of the active volcanic crater, thermal water discharges are strongly influenced by acidic magmatic components such as CO_2, HCl, H_2S and SO_2, especially those from both the crater lake (8) and Las Pailas (1, 2 and 3), the latter emerging along a deep fault system which allows the magmatic fluids to reach the surface easily. The high emergence temperatures, high chlorine, boron, ammonia and sulphate contents
in the waters, the presence of typical volcanic compounds in the gases (H_2, CO and alkenes) and the ^{3}He/^{4}He ratios, support this hypothesis. The deep-originated fluids permeate the volcanic cone and mix with shallow aquifers in the north area of the volcano as is apparent from samples 4, 5, 6 and 7. The springs that seep out further away from the active crater (Volcancito and rio Volcancito, 9 and 10 respectively), show differences in water and gas compositions (higher pH and HCO_3 contents in the waters, lower H_2, CO, and alkene concentrations in the gases), probably due to lower temperature rock-fluid interactions and to the absorption of atmospheric gas compounds that partially dilute the magmatic component. On the basis of such considerations, a schematic view of fluids circulation at Rincon de la Vieja volcanic system has been drafted in Figure 4.

REFERENCES

Alvarado, G.E., S. Kussmaul, S. Chiesa, P.Y. Gillot, H. Appel, G. Worner & C. Rundle 1992. Resumen cronoestratigraphico de las rocas igneas de Costa Rica basado en dataciones radiometricas. *J. S. Am. Earth Sci.* 6: 151-168.

Barquero, J. & J. Segura 1983. La actividad del volcan Rincon de la Vieja. *Bol. Volcanol.* 13: 5-10.

Bourgois, J., J. Azema, P.O. Baumgartner, J. Tourton, A. Desmet & J. Aubouin 1984. The geologic history of the Caribbean-Cocos plate. Boundary with special reference to the Nicoya ophiolite complex (Costa Rica) and

D.S.D.P. results (legs 67 and 84 off Guatemala): a synthesis. *Tectonophysics* 108: 1-32.

Capaccioni. B., M. Martini, F. Mangani, L. Giannini, G. Nappi & F. Prati 1993. Light hydrocarbons in gas emissions from volcanic areas and geothermal fields. *Geochem. J.* 27: 7-17.

Capaccioni, B., M. Martini & F. Mangani 1995. Light hydrocarbons in hydrothermal and magmatic fumaroles: hints of catalytic and thermal reactions. *Bull. Volcanol.* 56: 593-600.

Dengo, G. 1962. *Estudio geologico de la region de Guanacaste, Costa Rica.* Costa Rica: Inst. Geografico Nacional, San José.

Dengo, G. 1973. *Estructura geologica, historia tectonica y morfologica de America Central.* Guatemala: Inst. Centroamericano de Investigacion y Technologia Industrial.

Giggenbach W.F. 1975. A simple method for the collection and analysis of volcanic gas sample. *Bull. Volcanol.* 39: 132-145.

Giggenbach, W.F. 1991. Chemical tecniques in geothermal exploration. In D'Amore (coordinator), *Application of Geochemical Techniques in Geothermal Research and Development*: 119-144. Rome: UNITAR,.

Kempter, K.A., S.G. Benner & S.N. Williams 1996. Rincon de la Vieja volcano, Guanacaste province, Costa Rica: geology of the southwestern flank and hazard implications. *J. Volcanol. Geotherm. Res.* 71: 109-127.

Mangani, F., A. Cappiello, B. Capaccioni & M. Martini 1991. Sampling and analysis of light hydrocarbons in volcanic gases. *Chromatographia* 32: 441-444.

McBirney, A.R. & H. Williams 1965. Volcanic history of Nicaragua. *Univ. Calif. Publ. Geol. Sci.* 55: 1-65.

Mora, S. 1983. Una revision y actualizacion de la clasificacion morfotectonica de Costa Rica, segun la teoria de la tectonica de placas. *Bol. Volcanol.* Costa Rica 13: 18-36.

O'Nions, R.K. & E.R. Oxburgh 1988. Helium, volatile fluxes and development of continental crust. *Earth Planet Sci. Lett.* 90: 331-347.

Tournon, J. 1984. *Magmatismes du Mesozoique a l'actuel en Amerique Centrale: l'exemple de Costa Rica, des ophiolites aux andesites, Thesis.* Paris: Univ. Pierre et Marie Curie.

Tristan, J.K. 1921. Apuntes sobre el volcan Rincon de la Vieja. *Rev. Costa Rica* II(6): 161-168.

Geochemistry of the Earth's Surface, Ármannsson (ed.)© 1999 Balkema, Rotterdam, ISBN 90 5809 073 6

Running and ground water geochemistry from Chiavenna Valley, Northern Italy

O.Vaselli, A.Casiglia, F.Tassi & L.Gallorini
Department of Earth Sciences, Florence, Italy

A.A.Minissale
CNR, Study Center for Minerogenesis and Applied Geochemistry, Florence, Italy

ABSTRACT: Running and ground waters from Chiavenna Valley (Northern Italy) have been collected in order to evaluate the geochemical background of major and trace elements in the Alpine environment. The Chiavenna Valley waters can be classified as Ca(Mg)-HCO$_3$ although some of them can be regarded as Ca(Mg)-SO$_4$. The latter are located in the north-eastern sector of the valley where the highest conductivity values have been measured, as well. Heavy metal contents are generally low, suggesting that anthropogenic pollution can be considered negligble. Slight enrichments in As, Sb, and, to a minor extent, U have been found and attributed to the leaching of the host rocks.

1 INTRODUCTION

Environmental geochemistry plays a key-role in the study of air, water and soil pollution. However, levels of toxicity for essential, trace and/or toxic elements are not well defined, yet. This is also due to the fact that the background for major and trace compounds is somehow difficult to establish, expecially when dealing with highly industrialised countries. Nevertheless, natural pollution (excess or paucity of a certain element) can be very important locally, as well.

This work presents a preliminary report on the geochemistry of running and ground waters from Chiavenna Valley (Northern Italy), this valley being located in the central-western sector of the Alpine chain and relatively free from both highly industrialised and agricultural areas. Chiavenna Valley is a NNW-SSE elongated basin and is a relatively open valley since the northern watershed is intersected by the Spluga Valley while water flows into the Mezzola Lake and then, to Como Lake to the south (Figure 1). Furthermore, the Chiavenna Valley has a surface of _700km^2 and, geologically, consists of schistose to crystalline rocks, sedimentary formations and a relatively well developed Quaternary cover. These features also make this valley suitable for reasonable comparisons with other Alpine basins.

In the Chiavenna Valley, extensive running and ground water sampling has been carried out. Major, minor and trace compound analyses have been performed in order to assess: i) the

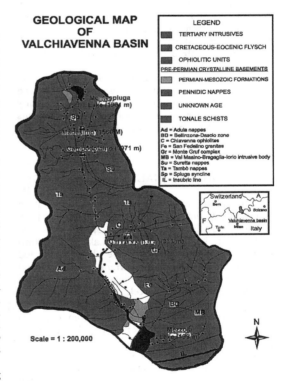

Figure 1. Schematic geological map of Chiavenna Valley with sampling location sites (simplified after Sciesa 1991).

geochemical characterization of these waters; ii) major and trace element background, including toxic elements such as Cd, Pb, As, Sb, Hg, etc.

2 GEOLOGICAL BACKGROUND

Chiavenna Valley is located in the central-western part of the Alpine belt (Figure 1), north to the Insubric Line which tectonically divides the south-Alpine basement from the Alpine nappes s.s., and constitutes terraines belonging to three distinct Pennide Nappes: Adula, Tambò and Surretta. They are composed of the pre-Alpine crystalline basement and separated by Permian to Mesozoic meta-sediments (Sciesa 1991). The crystalline basement consists mainly of para-gneiss and meta-granitoid rocks. Relatively extensive allochthonous Thetys-related ophiolitic rocks (Schmutz 1976) and Alpine orogenesis-related outcrops of Tertiary intrusive bodies (Giger & Hurford 1989) occur in the Chiavenna Valley, e.g. Chiavenna ophiolites and Val Masino-Bregaglia-Iorio and San Fedelino granites, respectively. A smaller acidic intrusive body crops out between Novate and Samolaco. Most of the Chiavenna Valley rocks have secondary permeability and, owing to the steep slope meteoric waters do not tend to infiltrate the sub-stratum and short-term ground water circulation is highly likely. The strong differences in composition of the outcropping rocks make this area particularly interesting to evaluate the occurrence of naturally derived toxic elements. Ultramafic rocks and granites and their differentiated products generally have, indeed high concentrations of toxic elements such as Cr, Ni, Co and As,U, Hg, As, Sb, Se, etc., respectively.

3 SAMPLING AND ANALYTICAL METHODS

Five sampling sessions have been carried out in Chiavenna Valley between July and October 1998 (relatively dry season) from the Spluga Pass (ca. 2200 m in height) to Novate Lake (ca. 200m) (Figure 1). Water samples have been collected along the main valley axes (Mera and Liro rivers) and at the confluences of the main tributaries. Main dextral and sinistral creek waters have been sampled, too. Wells are mainly present in the southern part of the main valley (ca. 300m in height) and supply drinkable water for the population who is prevalically concentrated, here.

About 150 running and ground water sampless have been collected and analysed for major, minor and trace compounds. All the samples have been filtered (0.45µm) and acidified (0.1% HNO_3) in the field. In-situ analyses have been carried out for pH, $T(°C)$, HCO_3, dissolved O_2, NO_2, SiO_2 and NH_4.

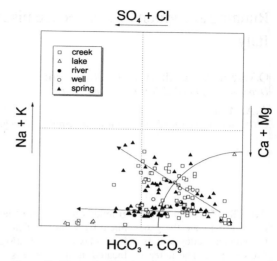

Figure 2. Water classification diagram for the Chiavenna Valley waters.

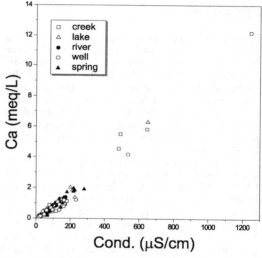

Figure 3. Ca (meq/L) vs. conductivity (µS/cm) diagram.

The major and some minor constituents have been analysed by routine methods such as atomic flame absorption spectrometry (Na, K, Ca and Mg), or ion chromatography (F, Cl, NO_3, Br, PO_4, SO_4). Forty-nine trace elements were determined by ICP-MS at the XRAL Laboratories (Canada). Sampling sites are schematically shown in Figure 1.

4 GEOCHEMICAL DATA

Representative mean values for selected parameters are reported in Table 1. Samples have been divided into 5 groups: i) main rivers (Mera and Liro) (closed circle); ii) main tributaries (open square); iii) springs and fountains (closed triangle); iv) lakes and dams (open triangle) and v) wells (open circle).

Chemically, the Chiavenna Valley waters have been classified using the diagram of Langelier & Ludwig (1942) (Figure 2) and can be regarded as Ca(Mg)-HCO$_3$ waters, though a few samples have a Ca(Mg)-SO$_4$ composition with a clear trend which relates them to the Ca(Mg)-HCO$_3$ waters. pH ranges between 7.1 and 8.0 while conductivity values are generally <200 μS/cm, even if a group of samples, located close to Madesimo (Figure 1), have values up to 1250 μS/cm and these waters are those which show a Ca(Mg)-SO$_4$ composition. Some of them precipitate travertine at the emergence. This can be related to the leaching of gypsum-bearing units belonging to the Permo-Mesozoic formations occurring in this area (Sciesa, 1991). The strong enrichments in Ca and Sr and Ba (up to 240 mg/L, and 4110 and 13 μg/L, respectively) in these waters are also attributed to the presence of Middle to Upper Triassic marble and dolomitic units

On the whole, the main chemical features indicate that all the waters considered have been subjected to short-term circulation only as indicated by the low content of Na (<3.2mg/L); the main process is likely to be due to a simple dissolution of carbonatic (and subordinately, sulphatic) formations which occur widely in the valley along with granitic (hardly insoluble at low temperatures) and, to a very minor extent, ophiolitic complexes.

Calcium, SO$_4$ and HCO$_3$ are the main species, their sum generally being >80% of the whole cation and anion content. This is supported by the Ca (meq/L) vs. conductivity (μS/cm) diagram (Figure 3) where a highly significant correlation is obtained, the high Ca samples being related to the depositing-travertine waters. The sulphatic character of the above mentioned waters shows up better in the Ca (meq/L) vs HCO$_3$ (meq/L) diagram (Figure 4) where the results for the samples trend away from the main pattern.

As far as the trace compounds and elements are concerned, their contents are generally below either the instrumental detection limits or toxic thresholds. In Table 1, mean values for selected elements, including NO$_3$ and NH$_4$, are reported. Nitrogenated species are generally below 3.5 (NO$_3$) and 0.5 mg/L (NH$_4$).

Significant positive correlations between geochemically similarly-behaved element pairs have been observed for Ca-Sr, Ca-Ba, K-Rb, etc. thus, suggesting their common source with no external anthropogenic-derived supply.

Table 1. Mean values of selected parameters for the Chiavenna Valley waters. 1. Main rivers; 2. Creeks; 3. Springs and fountains; 4. Lakes and dams; 5. Wells.

	1 (n=8)	2 (n=66)	3 (n=61)	4 (n=7)	5 (n=7)
T°C	15.7	10.7	11.6	16.1	11.7
pH	7.68	7.57	7.46	7.90	7.51
Cond. μS/cm	102	114	94	226	145
Namg/L	1.4	1.4	1.5	3.2	2.6
K	1.4	1.3	1.5	3.2	2.9
Ca	16.6	18.3	13.4	39	17
Mg	2.3	3.2	2.3	4.9	4.4
HCO$_3$	38	36	36	60	53
SO$_4$	22.3	34.7	17.5	80	26
Cl	1.1	0.9	0.9	1	1.8
F	0.09	0.1	0.09	0.06	0.15
NO$_3$	1	1.5	1.4	0.8	3.6
NH$_4$	0.25	0.4	0.3	0.4	0.4
Sc μg/L	1.3	2.1	2.3	1.8	4.1
V	0.4	0.5	0.4	0.4	0.9
Cr	0.3	0.3	0.3	1.9	0.8
Mn	4.9	6.9	0.8	4.7	9.7
Ni	1.1	1.6	1.3	2.3	1.2
Cu	0.8	1.1	1.3	1.6	1.9
Zn	4.5	7.2	8.3	4.1	86
As	3.5	3.4	3.6	3.8	11.6
Sr	153	223	106	665	178
Zr	0.4	0.8	0.6	0.6	0.5
Cd	0.4	0.8	0.5	0.5	0.5
Ba	5.5	7.8	6.2	10	12
Ce	0.04	0.05	0.03	0.03	0.12
Hf	0.06	0.3	0.47	0.12	0.4
Hg	<0.2	0.8	0.9	0.6	0.9
Pb	0.25	0.4	0.5	0.3	8.4

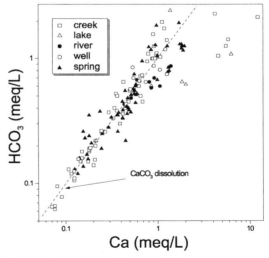

Figure 4. Ca (meq/L) vs. HCO3 (meq/L) binary diagram.

Among the heavy metals (all characterised by extremely low values), binary significant correlations are observed although, generally speaking, they present a more scattered character with respect to those observed for the main

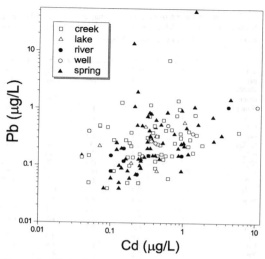

Figure 5. Cd (μg/L) vs. Pb (μg/L) log diagram for the Chiavenna Valley running and groundwaters.

components as shown in the Pb vs. Cs diagram (Figure 5).

It is noteworthy to point out the relatively high contents of trace elements and heavy metals such as Sc, V, Mn, Cu, Zn, As, Ba, Pb and U in the 7 wells sampled with respect to the other groups of water. Some wells are utilised for drinking water supply. One well sample, at -140m, not yet plumbed to the main water system supply, has As content of 53 μg/L which is above the W.H.O. recommended value (50 μg/L).

However, the general homogeneous and low heavy metal contents throughout the whole Chiavenna Valley waters suggest that anthropogenic pollution can possibly be ruled out and their source is mostly probably associated with interaction with host or leached rocks.

5 CONCLUDING REMARKS

The present analytical data for the Chiavenna Valley running and ground waters indicate that this basin can be considered a relatively pollution-free area, though small, presumably natural, enrichments in toxic elements have been detected.

On the whole the geochemical data gathered do
• represent an useful data-base from which water evolution and pollution sources might be monitored.

A better characterization of the water reservoir bodies will be performed when the whole set of $\delta^{18}O$ and δ^2H data, presently being processed, will be available.

ACKNOWLEDGEMENTS

This work has financially been supported by the Italian National Council of Research (C.N.R.). OV wishes to thank N. Coradossi (University of Florence) for her help and encouragement in this work.

REFERENCES

Giger, M. & A.J. Hurford 1989. Tertiary intrusives of the Central Alps: their Tertiary uplift, erosion, redeposition and burial in the south-Alpin foreland. *Eclogae Geol. Helv.* 82: 857-866.

Langelier, W.F. & H.F. Ludwig 1942. Graphical methods for indicating the mineral character of natural waters. *J. Am. W. W. Assoc.* 34: 335-352.

Schmutz, H.U. 1976. Petrography and structural evolution of ophiolitic remnants in the Bellinzona Zone, Southern Steep Belt, Central Alps (CH/I). *Schweiz. Mineral. Petrogr. Mitt.* 69: 393-405.

Sciesa, E. 1991. Geologia della Alpi Centrali lungo la traversa Colico-Passo dello Spluga (Province di Sondrio e Como). *Atti Mus. Civ. Stor. Nat. Morbegno* 2: 3-34.

Geochemistry of the Earth's Surface, Ármannsson (ed.)© 1999 Balkema, Rotterdam, ISBN 90 5809 073 6

Mineralogy of hydrothermally altered tuffs at Phlegrean Fields, Italy

B. Yven & M. Zamora
Laboratoire des Géomatériaux, Université Paris, ESA CNRS 7046 et IPGP, France

Ph. Ildefonse
Laboratoire de Minéralogie-Cristallographie, Université Paris, UMR 7590 et IPGP, France

ABSTRACT: Mineralogical, physical and chemical measurements were carried out on 5 samples of altered tuffs coming from a 3046 m depth well (Phlegrean Fields, Italy). Different secondary mineralogical assemblages have been distinguished as a function of depth and temperature. At depth (2860 m, 390 °C), iron-rich biotites are associated with epidote. At 2130 m (300 °C), iron-rich chlorites (chamosite) predominate with minor amounts of biotite and epidote. At 1415 m (260 °C), chamosite prevails with minor amounts of analcime. A shallow sample (805 m, 130 °C) is composed of a mixture of vermiculite and vermiculite/smectite mixed layers which is associated with analcime. At the surface, phillipsite and chabazite were identified. These drastic changes in clay mineralogy control the physical and chemical properties of bulk tuffs. Two sets of electrical surface conductivity, σ_s, were observed: $\sigma_s < 0.01$ Sm^{-1} in biotite- and chlorite-bearing tuffs and $\sigma_s > 0.02$ Sm^{-1} in tuffs containing vermiculite and vermiculite/smectite and zeolites.

1 INTRODUCTION

The exploration and exploitation of the geothermal possibilities of any area need geophysical measurements. Electrical properties are widely used to con-

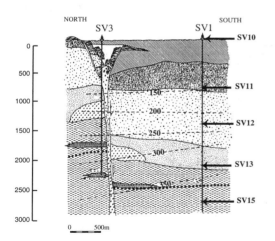

Figure 1. Vertical cross-section of San Vito area showing the isotherms and the location of the samples studied.

strain *in situ* fluid saturation. In altered rocks, the contribution of electrical surface conductivity, at the interface of clays and fluid, may be important (Revil & Glover 1997). This contribution depends of the nature and content of clay minerals. In this paper we present mineralogical and chemical data obtained on altered tuffs coming from San Vito 1 borehole (Figure 1), located in Phlegrean Fields (Italy). The effect of clay mineralogy on electrical properties have been studied in these samples. This geothermal area is well known for its recent volcanic (35,000 BP to present day) and seismic activity (Barberi et al. 1984).

The highest thermal gradient (to 420°C at the bottom) is recorded in Well San Vito 1, drilled by AGIP (Italian Oil Company) from the surface to 3046 m, penetrating the main lithologies of Phlegrean Fields. Since temperature, pressure and lithology are well constrained, this well offers a unique possibility to correlate physical properties and clay mineralogy.

2 MATERIAL AND METHOD

Five samples of altered tuff have been selected along borehole San Vito 1 (Figure 1). These samples are located between 2865 m depth (T= 390°C) and the surface. Bulk rock apparent mass density, porosity and specific surface area (BET) have been measured in samples. The electrical conductivity has been measured with a two-electrode technique in saturated samples, with increasing salinity of the fluid (ranging from 0 to 3.18 mole/l NaCl). At low salinity, the surface conductivity is predominant and controlled by the interaction between fluid and clay minerals. Cation exchange capacities (obtained by the cobalt hexamine chloride method) and exchangeable cation concentrations were measured in < 2 µm fractions isolated by sedimentation after grinding and dispersion by sonification.

Mineralogical characterization of the samples has been made by a petrographic study of thin sections and by in-situ microprobe analysis (Camebax). X-ray diffraction (XRD) data (Philips PW1710, CoKα) were recorded both for bulk samples and <2 µm fractions. The classical procedure for clay analysis was used (Moore and Reynolds, 1997). Fourier transform infrared (FTIR) data (Nicolet Magna IR560) were recorded by transmission in < 2 µm fractions (KBr pellets).

3 RESULTS AND DISCUSSION

3.1 Bulk properties

Variations of temperature (T), bulk mass density (ρ), porosity (ϕ), specific surface area (BET), cation exchange capacity (CEC), exchangeable cation concentrations and surface conductivity (σ_s) are presented as a function of depth in Table 1.

The rock mass density increases with depth from 1.16 g cm^{-3} to 2.31 g cm^{-3} and is inversely correlated to the porosity (decreasing from 46% to 15%). Specific surface area values show a similar trend as porosity.

Table 1. Bulk properties of samples studied

Sample	Depth (m)	T °C	ρ gcm^{-3}	ϕ %	BET m^{2}g^{-1}	CEC 10^{-2} mmole/ g	Ca	Na	σ_s Sm^{-1}
SV10	0	nd	1.16	46	17.2	18.3	2.9	10.6	0.028
SV11	805	130	1.68	33	7.4	50.5	48.0	0.8	0.024
SV12	1415	260	1.98	23	5.4	12.1	9.3	0.7	0.009
SV13	2130	300	2.20	22	2.5	5.5	3.1	0.3	0.004
SV15	2860	390	2.31	15	0.3	11.1	3.4	0.5	0.002

nd not determined

The highest value for CEC (50.5 10^{-2} mmole/g) is obtained for SV11 at 805 m depth and the lowest one (5.5 10^{-2} mmole/g) for SV13 at 2130 m depth. Important variations of exchangeable cation concentrations are observed.

Na is the prevailing cation in surface sample whereas Ca dominates other ions in deeper samples with a maximum value of 48.0 10^{-2} mmole/g for SV11. Surface conductivity values range from 0.002 S m^{-1} at depth to 0.028 S m^{-1} at the surface. This parameter is controlled both by porosity, CEC and the amount of clay minerals and zeolites.

3.2 XRD data

XRD powder measurements, carried out on bulk samples, show that SV10, at the surface, only contains two zeolites: phillipsite, which prevails, and chabazite. Analcime was identified in SV11 and SV12 with a maximum concentration in SV11. K-feldspar and plagioclase are present in all samples but SV10, and their concentrations increase with depth. Drastic differences between samples concern clay mineralogy. Figure 2 presents X-ray dif-

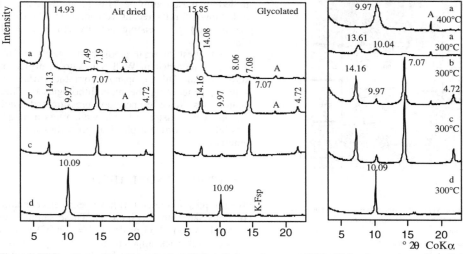

Figure 2. XRD patterns of the <2 µm fraction of all studied samples except SV10 which only contains zeolites. Only SV11 has been heated and analysed at 400°C. a) SV11, b) SV12, c) SV13 and d) SV15.

fraction patterns of fine fractions between 3 and 30° 2θ CoKα. SV11 consists of a mixture of two clay minerals. Type 1 is an expandable phase which is characterized by (001) and (002) reflections at 15.85 Å and 8.06 Å respectively from a glycolated sample. Type 2 does not expand and yields a 14.08 Å reflection on glycolated sample. By heating the sample to 300°C, two diffraction peaks are observed at 13.61 Å and 10.04 Å. Heated sample at 400 °C gives a unique reflection at 9.97 Å. This behavior may be related to a mixture of vermiculite and a random mixed-layer vermiculite/smectite (Reynolds 1980).

Figure 3 presents X-ray diffraction patterns of fine fractions between 69 and 75° 2θ CoKα.

XRD patterns of < 2 μm fraction of SV11 exhibit two (060) reflections, at 1.54 Å and 1.50 Å, indicating a mixture of trioctahedral and dioctahedral layers. XRD patterns of SV12 and SV13 are similar and contain a prevailing trioctahedral Fe-chlorite, characterized by three main reflections in the 2θ domain considered, 7.07, 14.13 and 4.72 Å, reflections whose amplitude decreased in this order. These reflections are neither changed by glycolation nor heating to 300 °C. The (060) and (062) peaks are indexed at 1.552 and 1.516 Å respectively, which gives a 9.312 Å b parameter. An additional weak amplitude reflection at 9.97 Å may be related to a mica phase. The XRD pattern of SV15 yields a unique peak at 10.09 Å, which does not move by glycolation or heating. The (060) and (061) reflections are located at 1.546 and 1.528 Å respectively. This behavior characterizes a trioctahedral mica.

3.3 FTIR data

FTIR spectra of < 2 μm fractions (SV11 to SV15) are diplayed in Figure 4. SV11, SV12 and SV13 yield closed vibration modes but in 2800-3800cm⁻¹ region. In SV11, three OH stretching modes are evidenced, at 3625, 3542 and 3427 cm⁻¹. The 3625 cm⁻¹ mode is related to analcime (also identified by XRD) whereas the two other modes may be attributed to vermiculite and vermiculite/smectite mixed layers. SV12 and SV13 display vibrational features close to that reported for chamosite.

The 3546 and 3402 cm⁻¹ modes characterize OH-silicate and OH-brucite layers respectively. These energies are indicative of the ferriferous character of the chloritic layers (Farmer, 1974).

The 640 and 672 cm⁻¹ vibration modes are related to OH-bending in chlorites. SV15 displays different features. The 3630 cm⁻¹ OH-stretching mode characterizes an iron-rich biotite. In all spectra, additional features at 1132 and 589 cm⁻¹ are related to Si-O-Si and Si-O vibration modes of alkali-feldspars which are detected as impurities in XRD powder patterns (not shown).

3.4 Chemical data

Secondary minerals were analysed in situ by microprobe in thin sections. They occur mainly as alteration products of the glassy matrix. Figure 5 displays chemical data represented in the MR3-2R3-3R2 system (Velde, 1985).

Phillipsite (K>Na>>Ca) and chabazite (Ca>>Na>K) observed in SV10 are located near the MR3 pole. In SV11, vermiculite and vermiculite/smectite (total iron expressed as Fe^{3+}) are located near the 2R3 pole with a range of interlayer

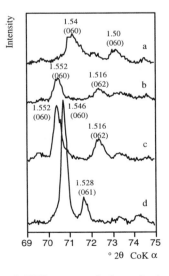

Figure 3. XRD patterns of <2 μm fractions a) SV11, b) SV12, c) SV13 and d) SV15.

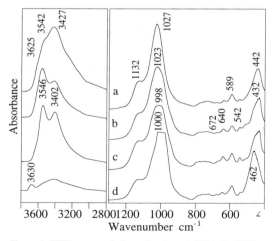

Figure 4. FTIR spectra of <2 μm fractions a) SV11, b) SV12, c) SV13 and d) SV15.

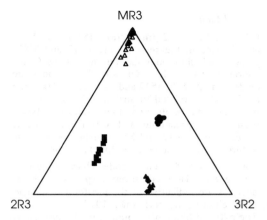

MR3

2R3   3R2

Figure 5. Chemical data obtained on secondary clay minerals on MR3-2R3-3R2 system. Filled triangle: SV12, Filled diamond: SV13, Filled squares: SV11, Filled circles: SV15, open triangles: SV10

charge in agreement with smectite-vermiculite. In SV12 and SV13, clay minerals are characterized by a low charge, an iron-rich character (Mg/Mg+Fe= 0.36-0.41) and occur in the chemical field of chlorite (total iron expressed as Fe^{2+}). In SV15, chemical data obtained on mica flakes characterize iron-rich biotites (Mg/Mg+Fe = 0.66) with a high interlayer charge. Epidote was found in association with chlorite and biotite in SV13 (Al/Fe = 2.7) and SV15 (Al/Fe = 3.0).

4 CONCLUSIONS

Different secondary mineralogical assemblages have been distinguished as a function of depth and temperature. At depth (2860 m, 390 °C), iron-rich biotites are associated with epidote. In SV13 (2130 m, 300 °C), iron-rich chlorites (chamosite) prevail with minor amounts of biotite and epidote. In SV12 (1415 m, 260 °C), chamosite prevails with minor amounts of analcime. In SV11 (805 m, 130 °C), a mixture of vermiculite and vermiculite/smectite mixed layers is associated with analcime. At the surface, in SV10, phillipsite and chabazite were identified. These drastic changes in clay mineralogy control the physical and chemical properties of bulk tuffs. Two sets of electrical surface conductivity (σ_s) values may be considered: $\sigma_s < 0.1$ Sm^{-1} in biotite- and chlorite-bearing tuffs (CEC = 0.05-0.12 mmole/g), and $\sigma_s > 0.02$ S m^{-1} in tuffs containing vermiculite and vermiculite/smectite, and zeolites (CEC = 0.18-0.5 mmole/g).

ACKNOWLEDGEMENTS

This work was supported by the PNRN/CNRS program (Programme National pour les Risques Naturels) and by project TEKVOLC (European Union). We thank W. Chelini (AGIP) for providing samples. This is an IPGP contribution and an INSU contribution.

REFERENCES

Barberi, F., Hill, D.P., Innocenti, F., Luongo, G & Treuil, M. 1984. The 1982-1984 Bradyseismic Crisis at Phlegrean Fields (Italy). *Bull. Volcanol.* (special issue) 47(2), 173-411.

Farmer, V.C. 1974. The I.R. spectra of minerals. Miner. Soc. London, 539 p.

Moore, D.M., and Reynolds, R.C. Jr. 1997. X-ray diffraction and identification and analyses of clay minerals. Oxford University press, 378 p.

Revil, A. & Glover, P.W.J. 1997. Theory of ionic surface electrical conduction in porous media. *Phys. Rev. B*, 55, 1757-1773.

Reynolds, R.C., Jr. 1980. Interstratified clay minerals: in G.W. Brindley and G. Brown eds, Crystal structures of clay minerals and their X-ray identification, monograph No. 5, Mineral. Soc., Lodon, 249-303.

Velde, B. 1985. Clay minerals. A physico-chemical explanation of their occurence. Dev. in Sediment. 40, 427 p.

9 Supplement

Geochemistry of the Earth's Surface, Ármannsson (ed.)© 1999 Balkema, Rotterdam, ISBN 90 5809 073 6

Calcite precipitation and interaction with phosphorus cycle in Lake Bourget, France

A. Groleau, B. Vinçon-Leite & B. Tassin
CEREVE, Ecole Nationale des Ponts et Chaussées, Ecole Nationale du Génie Rural des Eaux et Forêts, Université Paris Val de Marne, Champs-sur-Marne, France

G. Sarazin & C. Quiblier-Lloberas
Laboratoire de Géochimie des Eaux, Université Denis Diderot-Paris 7 & Institut de Physique du Globe de Paris, France

ABSTRACT: A two years survey of chemical composition of epilimnitic waters and settling particles of Lake Bourget shows that calcite precipitation is a major geochemical process occuring each year during Spring and early Summer. Specific conductance is used as a calcite precipitation tracer. The results allow to calculate a steady $CaCO_3$ precipitation of 14,000 tons year^{-1}. A photosynthetic-respiration budget points out that phosphorus behaves quite differently from other nutrients. Sedimentation fluxes data show that calcite precipitation and phosphorus cycle are closely associated. Understanding of mechanisms of calcite and phosphorus interactions still require further studies but they certainly act toward an improvement of the trophic status of the lake by enhancing the internal phosphorus sink.

1 INTRODUCTION

Lake Bourget is a deep (max. depth: 145 m), large (surface area : 42 km²), subalpine lake (Figure 1). The watershed encloses two cities (Chambery and Aix-les-Bains) which industrial and domestic sewage have led to eutrophication of the water body during the 70's. Since 1980 an important restoration plan has reduced drastically phosphorus input. Results of biogeochemical research obtained between 1988 and 1991 show that long term phosphorus decrease cannot be explained only by input reduction and sedimentation of algal biomass (Vinçon-Leite et al. 1995). One important feature of alpine lakes is the occurrence of authigenic calcite precipitation during summer. This major process is initiated and sustained both by temperature elevation of the epilimnion and developpement of the phytoplankton (Gaillard 1995). Calcite precipitation is known to interact with the internal phosphorus cycle (Dittrich et al. 1997, Hartley et al. 1997, Kleiner 1988). The aims of this paper are to :
- Quantify and depict calcite precipitation process in Lake Bourget.
- Check if chemical interactions between calcite and phosphorus occur in the lake and if they represent a significant sink for phosphorus.

2 MATERIAL AND METHODS

In situ measurements (pH, temperature, conductivity and dissolved oxygen) and water sampling have been processed monthly or biweekly during two years (April 97 to December 98) at point B in the epilimnion (Figure 1). Four points (0, 2, 15 and 30 m) describe the upper layer of the lake. Major cations were analysed by FAAS, major anions by ionic chromatography, SRP analysis was performed by the standard colorimetric method, and alkalinity by a spectroscopic method described by Sarazin et al 1998. Sediment traps contents have been recovered biweekly during the same period at three different depths (10, 30 and 80 m) at point T (Figure 1).

3 RESULTS AND DISCUSSION

3.1 *Evidence for calcite precipitation during Spring and early Summer*

3.1.1 *Origin of the epilimnion calcium depletion*
Calcium concentration profiles (Figure 2) obviously show a decrease in the epilimnion between May and June. This trend is going on until August ; afterwards calcium depletion stops. If calcite precipitation occurs during this period, the decrease in calcium concentration must be related to a corresponding decrease of alkalinity.

In this geochemical system, the expression of alkalinity can be reduced to :

$$Alk = [HCO_3^-] + 2[CO_3^{2-}] + [OH^-] - [H^+] \qquad (1)$$

Calcite precipitation corresponds to the reaction :

$$Ca^{2+} + HCO_3^- \Leftrightarrow CaCO_{3\,(s)} + H^+ \qquad (2)$$

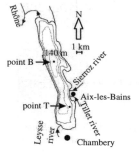

Figure 1. Situation map and sampling points.

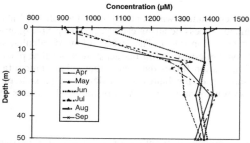

Figure 2. Calcium concentration profiles in the epilimnion from April to September 1997.

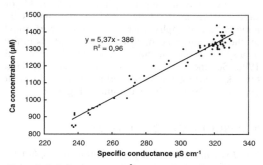

Figure 3. Relation between Ca^{2+} and specific conductance.

According to Equations 1 and 2, a decrease of calcium (ΔCa) leads to a decrease of alkalinity (ΔAlk) following Equation 3 :

$$\Delta[Ca^{2+}] = 2 \, \Delta Alk \qquad (3)$$

Spring and Summer data of calcium and alkalinity show over a 2-year period a linear relationship (Equation 4):

$$[Ca^{2+}] = (1.9 \pm 0.2)Alk \quad \text{with } \rho^2 = 0.96 \text{ and } n = 75 \, (4)$$

Since the confidence interval lies within the experimental uncertainties we can therefore infer that the calcium decrease in the water column of the epilimnion is mainly related to calcite precipitation.

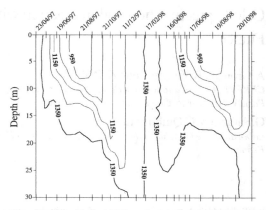

Figure 4. Time-depth evolution of calcium (μM) in the epilimnion.

3.1.2 Assessement of the spatio-temporal pattern of calcite precipitation using specific conductance

In order to improve the assessment of spatio-temporal conditions of calcite precipitation, we used the specific conductance profiles.

They are obtained at a biweekly time step with a spatial resolution of 15 cm from the lake surface to the bottom. In Lake Bourget the water composition is roughly equivalent to a $Ca(HCO_3)_2$ solution (> 83 % of the total dissolved species), therefore the observed specific conductance variations are related to calcite precipitation. This is confirmed by a fairly good correlation between calcium concentration and measured specific conductance (Figure 3). These results allow us to compute calcium profiles from the specific condutance data according to the linear relation (Figure 3 and Equation 5.) :

$$[Ca^{2+}] = 5.37 \, S - 386 \qquad (5)$$

where $[Ca^{2+}]$ is the calculated calcium concentration in μM and S is the measured specific conductance in μS cm^{-1}.

Calculated Ca profiles are plotted in a time-depth evolution diagram in the epilimnion from April 97 to December 98 (Figure 4).

Some conclusions can be drawn from this diagram :

- a fair similarity is observed between the two years concerning the spatial pattern of the precipitation process.

- probably due to a different meteorological forcing from one year to another, calcite precipitation starts one month earlier in 1998.

- integration of calcium profiles infers that the amount of produced authigenic calcite is the same in 1997 and in 1998 within experimental and calculation uncertainties ($\pm$ 10 %).

Extrapolating to the whole surface of the lake gives an evaluation of this production to 14,000 $\pm$ 1400 tons year^{-1}.

3.2 Interaction of calcite precipitation with biological processes and phosphorus cycle

3.2.1 Stoichiometric modelling of photosynthesis and respiration

Photosynthesis and respiration are two opposite reactions involving the same species :

$$xCO_2 + y\ NO_3^- + w\ H_3PO_4 + y\ H^+ + (x+y)\ H_2O$$

$$\underset{1}{\overset{2}{\rightleftharpoons}} (x+2y)\ O_2 + (CH_2O)_x(NH_3)_y\ (H_3PO_4)_w \quad (6)$$

Hence the dissolved inorganic carbon (ΣCO_2 calculated from pH and alkalinity measurements ; Stumm & Morgan 1981) budget in the water column takes into account both photosynthesis and respiration processes (Equation 6) and the inorganic carbon associated with calcite precipitation (Equation 2). Then, the difference between photosynthesis and respiration (P-R), in terms of ΣCO_2 can be written as :

$$(P\text{-}R) = - \{[\Sigma CO_{2\ (t\,;\,Z)} - \Sigma CO_{2\ (t0\,;\,Z0)}] - [Ca_{(t\,;\,Z)} -$$

$$Ca_{(t0\,;\,Z0)}]\} \quad\quad\quad\quad (7)$$

where Ca concentration difference represents the ΣCO_2 depletion linked to calcite precipitation.

Assuming a time and space reference where these two reactions are globally balanced and when no calcite precipitation occurs (April 98 as t_0 and 30 m as Z_0), we can calculate at any time (t) and depth (Z) the prevalence of photosynthesis (P-R>0) or respiration (P-R<0).

The stoechiometry of the Equation 6 shows that linear relations must be observed between the (P-R) values and nutrients variations if photosynthesis, respiration and calcite precipitation are the main biogeochemical processes occurring in the water column.

On one hand, Figure 5a shows that this linear relation is observed for nitrate at any time and depth, for the whole '98 year. The fairly good correlation coefficient ($\rho^2 = 0.94$) validates the stoiechiometric model for nitrate. The slope of the line (Figure 5a) leads to a C:N ratio of 9.4 which belong to the range of usual values for lake phytoplankton (Hecky et al. 1993). On the other hand, no linearity is observed for soluble reactive phosphorus (SRP). In particular, significant SRP uptake can occur in the water column without noticeable photosynthesis. Two hypothesis can be stated in order to explain this depletion :

- SRP can be uptaken by heterotrophic bacterial activity (Capblancq 1990, Currie 1990).
- SRP can be scavenged by authigenic calcite particles settling down in the water column.

3.2.2 Sedimentation fluxes in the lake

Sediment trap data during calcite precipitation period can supply additional insights into the interac-

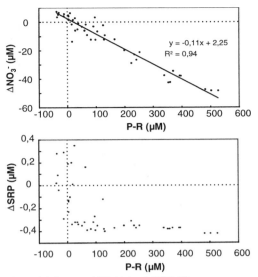

Figure 5. P-R versus ΔNO_3^- (a) and ΔSRP (b)

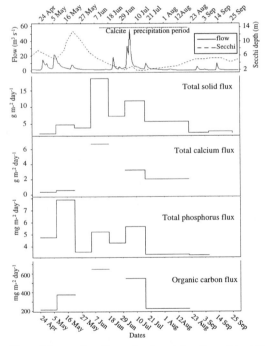

Figure 6. Leysse river flow and sedimentation fluxes data at 30 m depth in Lake Bourget from April 97 to September 97.

tions between phosphorus and calcite. Figure 6 shows the sedimentation patterns at 30 meters depth during spring and summer 1997.

567

During this period, two significant flood events occured in Leysse river at the beginning of May and at the beginning of July. During these floods, sedimentation fluxes are increased by riverine particulate matter, transported by the river plume to the sediment traps station (Vinçon-Leite et al. 1998). Total sedimentation flux reaches a high level dominated by calcitic minerals and organic particles coming from the watershed.

The highest sedimentation flux of total solid is observed between the 4th and the 18th of June although Leysse river inflow is low and should not influence the sedimentation flux measurements. This period corresponds exactly with the beginning of calcite precipitation. Therefore highest fluxes are mainly autochtonous. Calculation of calcite flux, obtained from the calcium flux data, shows that calcite weights more than 80% of the total solid. Simultaneously, phosphorus flux exhibits a significant increase. This trend is similar for organic carbon flux, indicating that organic particulate matter is associated with the authigenic calcite flux. Based on 1997 sedimentation fluxes, autochtonous calcite formation and sedimentation are closely associated with phosphorus flux. Discrimination between association process like P-adsorption onto calcite reactive surfaces, or biologic process like incorporation in biomass, is not yet achieved.

4 CONCLUSION

1. Calcite precipitation in Lake Bourget happened each year with a reproducible intensity over a two year period. The order of magnitude obtained by extrapolation to the whole lake surface is about 14,000 tons. This process is restricted to a quite short period from May-June to August-September.

2. Sedimentation of authigenic calcite corresponds to the main particulate matter flux and occurs simultaneously with significant organic matter and phosphorus fluxes. Therefore it probably contributes to the recovery of a mesotrophic status of the lake.

REFERENCES

Capblancq, J. 1990. Nutrient dynamics and pelagic food web interactions in oligotrophic and eutrophic environments : an overview. In D.J. Bonin and H.L. Golterman, *Fluxes between trophic levels and through the water sediment interface* : 1-14, Kluwer Academic Publishers.

Currie, D.J. 1990. Large scale variability and interactions among phytoplankton, bacterioplankton and phosphorus. *Limnol. Oceanogr.* 35 :1437-1455.

Dittrich, M., Th. Dittrich, I. Seiber & R. Koschel 1997. A balance analysis of elimination of phosphorus by artificial calcite precipitation in stratified hardwater lake. *Wat. Res.* 31 :237-248.

Gaillard, J.F. 1995. Limnologie chimique : principes et processus. In Pourriot R. & M. Meybeck eds. *Limnologie Générale*, p. 115-156. Paris : Masson.

Hartley, A.M., W.A. House, M. Callow & B.S.C. Leadbetter 1997. Coprecipitation of phosphates with calcite in the presence of photosynthesizing green algae. *Wat. Res.* 31 :2261-2268.

Hecky, R.E., P. Campbell & L.L. Hendzel 1993. The stoichiometry of carbon, nitrogen and phosphorus in particulate matter of lakes and oceans. *Limnol. Oceanogr.* 38 :709-724.

Kleiner, J. 1988. Coprecipitation of phosphates with calcite in lake water : a laboratory experiment modelling phosphorus removal with calcite in lake Constance. *Wat. Res.* 22 :1259-1265.

Sarazin, G., G. Michard & F. Prevot 1998. A rapid and accurate spectroscopic method for alkalinity measurements in sea water samples. *Wat. Res.*33 : 290-294.

Stumm, W. & J.J. Morgan 1981. *Aquatic Chemistry.* New York : Wiley Interscience.

Vinçon-Leite, B. , B. Tassin & J.M. Jaquet 1995. Contribution of mathematical modelling to lake ecosystem understanding : Lake Bourget (Savoie, France). *Hydrobiologia.* 300-301 : 433-442.

Vinçon-Leite, B., P.E. Bournet, X. Gayte, D. Fontvieille & B. Tassin 1998. Impact of a flood event on the biogeochemical behavior of a mesotrophic alpine lake : Lake Bourget (Savoy). *Hydrobiologia.* 373-374 : 361-377.

Geochemistry of the Earth's Surface, Ármannsson (ed.)© 1999 Balkema, Rotterdam, ISBN 90 5809 073 6

Boron continental input to the ocean: Global flux and isotopic assessment

D. Lemarchand, J. Gaillardet & C. J. Allègre
Laboratoire de Géochimie et Cosmochimie, URA CNRS 1758, IPGP, Université Paris 7, France

ABSTRACT: We have improved the sensitivity for boron isotopic measurement using the classical Positive Thermal Ionization Mass Spectrometry with $Cs_2BO_2^+$ method and developed a new chemical procedure to separate boron from organic rich media. We have analysed both boron content and boron isotopic composition of ten of the world largest rivers. We estimate a discharge weighted isotopic composition for the river input to the ocean to about 9 ‰ and a total boron flux of 30.10^{10} g/yr, which is 10 times higher than previously proposed.

1 INTRODUCTION

Natural boron is composed of two stable isotopes ^{10}B and ^{11}B. Based on the large relative mass difference between these two isotopes and the high geochemical reactivity of boron, significant isotope fractionation can occur in natural samples, producing large variations in $^{11}B/^{10}B$ ratios.

The geochemical cycle of boron and its secular evolution in the ocean is still poorly constrained. Although no consensus has been reached, the river boron flux seems to be the major input to the ocean (You et al. 1993, Rose et al. 1998). Due to analytical difficulties, the systematics of boron concentration and isotopic composition in rivers has not been addressed so far.

For a better understanding of the boron input to the ocean, we have focused on the largest river basins. This offers the possibility of getting a good estimate of the boron riverine input to the ocean. Additionally, large river basins will be integrated over great continental surfaces and should therefore be more representative samples.

2 ANALYTICAL METHODS

The samples were collected in acid washed polypropylene containers to prevent contamination. After collection, the samples were filtered on site through 0.2 μm acetate cellulose filters with a pressurised Sartorius® Teflon filtration unit in order to separate the dissolved and suspended loads. Blanks for the whole procedure have been shown to be negligible.

We have improved the sensitivity for boron isotope measurement using the classical Positive Thermal Ionization Mass Spectrometry technique using $Cs_2BO_2^+$ method with graphite and mannitol. This improvement allows to load 250 ng of boron on an outgassed tungsten filament. We have developed a new chemical procedure to separate boron from organic rich media which was the major difficulty for analysing boron isotopic composition in natural waters. Prior to measurement of isotopic ratio, the boron content was determined by isotope dilution using standard NIST SRM 952 enriched in ^{10}B as a spike. The total procedure blank ranges between 1.5 and 4 ng.

The boron isotopic ratio was determined with a thermal ionization mass spectrometer (Thomson CSF Type THN 206), and is expressed as a permil deviation ($\delta^{11}B$) relative to NIST SRM 951 boric acid. The oxygen contribution on the mass 309 is corrected as follows :

$$^{11}B/^{10}B = \text{measured ratio} - 0.00078$$

$$\delta^{11}B \,(‰) = [(^{11}B/^{10}B)_{sample} / (^{11}B/^{10}B)_{SRM951} - 1]^* 1000.$$

Eighteen separate analyses of NBS SRM 951 carried out during this study gave a mean $^{11}B/^{10}B$ ratio of 4.0529 ± 0.0003 ($2\sigma_{mean}$).

Four separate analyses of ocean water (CASS2) was used as a control for the procedure for analysing natural waters (Table 1).

Table 1. Isotopic composition and reproducibility of CASS2

Sample	$\delta^{11}B$ (‰) ($2\sigma_{mean}$)
Run 1	39.6 ± 0.1
Run 2	39.6 ± 0.3
Run 3	39.6 ± 0.2
Run 4	39.6 ± 0.2
Mean value	39.6 ± 0.1

Table 2. Boron isotopic composition of some of the world largest rivers

Samples	$\delta^{11}B$ (‰)		
	Run 1	Run 2	Run 3
Amazon	11.4 ± 0.3		
Ganges	6.0 ± 0.2		
Huanghe	3.1 ± 0.1	3.3 ± 0.1	3.3 ± 0.1
Lena	8.6 ± 0.2		
Mekong	2.3 ± 0.2		
Niger	35.2 ± 0.2	35.2 ± 0.2	
Pearl River	18.3 ± 0.1		
Slave River	16.6 ± 0.1	16.7 ± 0.2	16.5 ± 0.2
St Lawrence	8.2 ± 0.1	8.3 ± 0.1	
Changjiang	4.0 ± 0.2		

3 RESULTS AND DISCUSSION

3.1 Boron isotopic composition

Measured boron isotopic compositions of some of the world largest rivers are listed in Table 2. The external reproducibility is better than 0.2 ‰ (2σ). $\delta^{11}B$ values show large variations illustrating the

Table 4. Oceanic boron budget

	Flux (10^{10} g/yr)	$\delta^{11}B$ (‰)
Outputs from ocean		
Oceanic crust alteration (1)	14	3.7
Adsorption on sediment (2)	10	15 ± 1
Coprecipitation in carbonate (3)	6	20 ± 5
	30	
Input to ocean		
Weathering (4)	30.5	9
Hydrothermalism (5)	0.9	6.5 ± 8
Fluid expulsion (6)	3	10 ± 10
	34.4	

Sources are : 1, Smith et al. (1995); 2, Spivack (1986); 3, Vengosh et al. (1991); 4, this work; 5, Spivack et al. (1987); 6, You et al. (1993)

high geochemical reactivity of boron. The isotopic compositions range from 2.3 ‰ (Mekong) to 35.2 ‰ (Niger).

This data, combined with the boron content and water discharge of associated rivers (Table 3) allow to calculate a weighted isotopic composition for the river input to the ocean of about 9 ‰.

3.2 Oceanic boron budget

The total water discharge of the ten analysed samples is 10836 km³/yr, which represents around 30% of the world water discharge. When compared to the weighted world average for the major dissolved load (Table 3), we consider that the set of analysed samples here is a reasonable estimate for the weighted world average. Boron content ranges from 3.1 ppb

Table 3. Boron isotopic composition of some of the largest world rivers

	Water discharge (km³/yr)	Suspended load (Mt/yr)	[B] ppb	$\delta^{11}B$ (‰)	Dissolved load (μmol/l)*		
					Na	Mg	Ca
Amazon	6590	1200	5.8	11.4	195	65	208
Changjiang	928	480	11.0	4.0	205	208	719
Lena	525	17.6	4.6	8.6	315	190	464
Ganges	493	520	5.5	6.0	57	105	371
Mekong	467	150	14.5	2.3	245	193	646
Pearl River	363		5.4	18.3	68	116	720
St Lawrence	337	4	25.0	8.3	502	357	905
Niger	154	40	3.1	35.2	78	78	138
Slave River	101		16.0	16.7	227	265	694
Huanghe	41	1100	201.0	3.2	2826	905	1225
samples total	10836			weighted samples average :	210	305	106
world total	37400			weighted world average:	240	359	144

* concentration measured by Ion Chromatography.

(Niger) to 201 ppb (Huanghe) with an analytical uncertainty of about 1%.

Thus we estimate the global dissolved boron discharge to the ocean to about 30.10^{10} g/yr, which is 10 times higher than the flux given by You et al. (1993) and consistent with the estimate of Rose et al. (1998). We calculate the boron residence time in the ocean at about 20 Ma.

Preliminary results concerning adsorbed boron on sediment give a rough content of 1 ppm. We estimate the flux of adsorbed boron on the suspended load to the ocean to about $0.5.10^{10}$ g/yr which seems to be negligible compared to the dissolved flux. This last result is consistent with the highly soluble behaviour of boron.

4 CONCLUSION

The river boron flux is the major input to the ocean, around 90 % of the total input. Both boron content and boron isotopic composition for the dissolved load are highly variable. We estimate the discharge weighted isotopic composition for the river input to the ocean at about 9 ‰. For the purpose of having a better understanding of the geochemical cycle of boron, we have to constrain mechanisms which control this flux. In particular, the sediment behaviour in estuaries. On the other hand, the understanding of the boron isotopic composition of large rivers should give tools for constraining the secular evolution of the ocean.

REFERENCES

Rose E., M. Chaussidon & C. France-Lanord 1998 B analysis in natural waters by ion microprobe : A new technique for the study of B transfer from the continent to the ocean. *Goldschmidt Conference Abstract*

Smith H.J., A.J. Spivack H., Staudigel & S.R. Stanley 1995 The boron isotopic composition of altered oceanic crust . *Chemical Geology* . 126: 119-135.

Spivack A.J. 1986 Boron Isotopic geochemistry *Ph.D. Dissertation, Massachussetts Institute of Technology, Woods Hole Oceanographic Institute, Woods Hole, Mass.*

Spivack A.J., M.R. Palmer & J.M. Edmond 1987 The sedimentary cycle of boron isotopes. *Geochim. Cosmochim. Acta.* 51: 1939-1949.

Vengosh A., Y. Kolodny, A. Starinsky, A. Chivas & M. McCulloch 1991 Coprecipitation and isotopic fractionation of boron in modern biogenic carbonate. *Geochim. Cosmochim. Acta* . 55: 2901-2910.

You C.F., A.J. Spivack, H.J Smith & J.M. Gieske 1993 Mobilisation of boron in convergent margins : implications for the boron geochemical cycle. *J. Geol..* 21: 207-210.

Author index